STUDENT'S SOLUTIONS MANUAL

INTERMEDIATE ALGEBRA
CONCEPTS AND APPLICATIONS
TENTH EDITION

Marvin L. Bittinger
Indiana University Purdue University Indianapolis

David J. Ellenbogen
Community College of Vermont

Barbara L. Johnson
Ivy Tech Community College of Indiana

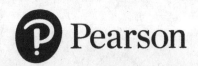

Copyright © 2018 Pearson Education, Inc.
Publishing as Pearson, 501 Boylston Street, Boston, MA 02116.

ISBN-13: 978-0-13-449753-2
ISBN-10: 0-13-449753-8

1   17

www.pearsonhighered.com

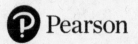

CONTENTS

CONTENTS

Chapter 1

Algebra and Problem Solving

Exercise Set 1.1

1. A letter that can be any one of a set of numbers is called a *variable*.

3. When $x = 10$, the *value* of the expression $4x$ is 40.

5. When all variables in a variable expression are replaced by numbers and a result is calculated, we say that we are *evaluating* the expression.

7. A number that can be written in the form a/b, where a and b are integers (with $b \neq 0$) is said to be a *rational* number.

9. Division can be used to show that $\dfrac{7}{40}$ can be written as a *terminating* decimal.

11. Five less than some number
Let n represent the number; $n - 5$

13. Twice a number
Let x represent the number; $2x$

15. Twenty-nine percent of some number
Let x represent the number; $0.29x$, or $\dfrac{29}{100}x$

17. Six less than half of a number
Let y represent the number; $\frac{1}{2}y - 6$

19. Seven more than ten percent of some number
Let s represent the number; $0.1s + 7$, or $\dfrac{10}{100}s + 7$

21. One less than the product of two numbers
Let m and n represent the numbers; $mn - 1$

23. Ninety miles per every four gallons of gas
We have
$$90 \div 4, \text{ or } \frac{90}{4}.$$

25. The area of a square is given by $A = s^2$. We substitute $s = 6$ and solve.
$$A = s^2 = 6^2 = 36 \text{ ft}^2$$

27. The area of a square is given by $A = s^2$. We substitute $s = 0.5$ and solve.
$$A = s^2 = 0.5^2 = 0.25 \text{ m}^2$$

29. The area of a triangle is given by $A = \frac{1}{2}bh$. We substitute $b = 5$, $h = 7$ and solve.
$$A = \frac{1}{2}bh = \frac{1}{2}(5)(7) = 17.5 \text{ ft}^2$$

31. The area of a triangle is given by $A = \frac{1}{2}bh$. We substitute $b = 7$, $h = 3.2$ and solve.
$$A = \frac{1}{2}bh = \frac{1}{2}(7)(3.2) = 11.2 \text{ ft}^2$$

33. Substitute and carry out the operations indicated.
$$\begin{aligned}
3(x - 7) + 2 &= 3(10 - 7) + 2 \\
&= 3(3) + 2 \\
&= 9 + 2 \\
&= 11
\end{aligned}$$

35. Substitute and carry out the operations indicated.
$$\begin{aligned}
12 + 3(n + 2)^2 &= 12 + 3(1 + 2)^2 \\
&= 12 + 3(3)^2 \\
&= 12 + 3(9) \\
&= 12 + 27 \\
&= 39
\end{aligned}$$

37. Substitute and carry out the operations indicated.
$$\begin{aligned}
4x + y &= 4 \cdot 2 + 3 \\
&= 8 + 3 \\
&= 11
\end{aligned}$$

39. Substitute and carry out the operations indicated.
$$\begin{aligned}
20 + r^2 - s &= 20 + (5)^2 - 10 \\
&= 20 + 25 - 10 \\
&= 35
\end{aligned}$$

41. Substitute and carry out the operations indicated.
$$\begin{aligned}
2c \div 3b &= 2 \cdot 6 \div 3 \cdot 2 \\
&= 12 \div 3 \cdot 2 \\
&= 4 \cdot 2 \\
&= 8
\end{aligned}$$

43. $3n^2 p - 3pn^2 = 3 \cdot 5^2 \cdot 9 - 3 \cdot 9 \cdot 5^2$

Observe that $3 \cdot 5^2 \cdot 9$ and $3 \cdot 9 \cdot 5^2$ represent the same number, so their difference is 0.

45. Substitute and carry out the operations indicated.
$$\begin{aligned}
5x \div (2 + x - y) &= 5 \cdot 6 \div (2 + 6 - 2) \\
&= 5 \cdot 6 \div (8 - 2) \\
&= 5 \cdot 6 \div 6 \\
&= 30 \div 6 \\
&= 5
\end{aligned}$$

47. Substitute and carry out the operations indicated.
$$\begin{aligned}
[10 - (a - b)]^2 &= [10 - (7 - 2)]^2 \\
&= [10 - 5]^2 \\
&= 5^2 \\
&= 25
\end{aligned}$$

49. Substitute and carry out the operations indicated.

$$\begin{aligned}
[5(r+s)]^2 &= [5(1+2)]^2 \\
&= [5(3)]^2 \\
&= 15^2 \\
&= 225
\end{aligned}$$

51. Substitute and carry out the operations indicated.

$$\begin{aligned}
x^2 - [3(x-y)]^2 &= 6^2 - [3(6-4)]^2 \\
&= 6^2 - [3(2)]^2 \\
&= 6^2 - 6^2 \\
&= 0
\end{aligned}$$

53. Substitute and carry out the operations indicated.

$$\begin{aligned}
(m-2n)^2 - 2(m+n) &= (8-2\cdot 1)^2 - 2(8+1) \\
&= (8-2)^2 - 2(9) \\
&= 6^2 - 2(9) \\
&= 36 - 2(9) \\
&= 36 - 18 \\
&= 18
\end{aligned}$$

55. List the letters in the set: {a, l, g, e, b, r}

57. List the numbers in the set: {1, 3, 5, 7, ... }

59. List the numbers in the set: {10, 20, 30, 40, ... }

61. Specify the conditions under which a number is in the set: $\{x | x$ is an even number between 9 and 99$\}$

63. Specify the conditions under which a number is in the set: $\{x | x$ is whole number less than 5$\}$

65. Specify the conditions under which a number is in the set: $\{x | x$ is an odd number between 10 and 20$\}$

67.
a. 0 and 6 are whole numbers.
b. –3, 0, and 6 are integers.
c. –8.7, –3, 0, $\frac{2}{3}$, and 6 are rational numbers.
d. $\sqrt{7}$ is an irrational number.
e. All of the given numbers are real numbers.

69.
a. 0 and 8 are whole numbers.
b. –17, 0, and 8 are integers.
c. –17, –0.01, 0, $\frac{5}{4}$, and 8 are rational numbers.
d. $\sqrt{77}$ is an irrational number.
e. All of the given numbers are real numbers.

71. Since 196 is a natural number, the statement is true.

73. Since every whole number is an integer, the statement is true.

75. Since $\frac{2}{3}$ is not an integer, the statement is false.

77. Since $\sqrt{10}$ is an irrational number, and every member of the set of irrational numbers is a member of the set of real numbers, the statement is true.

79. Since the set of integers includes some numbers that are not natural numbers, the statement is true.

81. Since the set of rational numbers includes some numbers that are not integers, the statement is false.

83. *Writing Exercise.*

85. *Writing Exercise.*

87. The quotient of the sum of two numbers and their difference
Let a and b represent the numbers. Then we have
$\dfrac{a+b}{a-b}$.

89. Half of the difference of the squares of two numbers
Let r and s represent the numbers. Then we have
$\dfrac{1}{2}\left(r^2 - s^2\right)$, or $\dfrac{r^2 - s^2}{2}$.

91. The only whole number that is not also a natural number is 0. Using roster notation to name the set, we have {0}.

93. List the numbers in the set: {5, 10, 15, 20, ...}

95. List the numbers in the set: {1, 3, 5, 7, ...}

97. Recall from geometry that when a right triangle has legs of length 2 and 3, the length of the hypotenuse is

$$\sqrt{2^2 + 3^2} = \sqrt{4+9} = \sqrt{13}.$$ We draw such a triangle:

Exercise Set 1.2

1. It is true that the sum of the two negative number is always negative.

3. It is true that the product of a negative number and a positive number is always negative.

5. The statement is false. Consider –5+2 = –3, for example.

7. The statement is false. Let $a = 2$ and $b = 6$. Then $2 < 6$ and $|2| < |6|$.

9. It is true that the associative law of multiplication states that for all real numbers a, b, and c, $(ab)c$ is equivalent to $a(bc)$.

11. $|-10| = 10$ –10 is 10 units from 0

13. $|7| = 7$ 7 is 7 units from 0

15. $|-46.8| = 46.8$ –46.8 is 46.8 units from 0

17. $|0| = 0$ 0 is 0 units from itself

19. $\left|1\frac{7}{8}\right| = 1\frac{7}{8}$ $1\frac{7}{8}$ is $1\frac{7}{8}$ units from 0

21. $|-4.21| = 4.21$ –4.21 is 4.21 units from 0

23. $-5 \le -4$
-5 is less than or equal to -4, a true statement since -5 is to the left of -4.

25. $-9 > 1$
-9 is greater than 1, a false statement since -9 is to the left of 1.

27. $0 \ge -5$
0 is greater than or equal to -5, a true statement since -5 is to the left of 0.

29. $-8 < -3$
-8 is less than -3, a true statement since -8 is to the left of -3.

31. $-4 \ge -4$
-4 is greater than or equal to -4. Since $-4 = -4$ is true, $-4 \ge -4$ is true.

33. $-5 < -5$
-5 is less than -5, a false statement since -5 does not lie to the left of itself.

35. $4 + 8$
Two positive numbers: Add the numbers getting 12. The answer is positive 12.

37. $(-3) + (-9)$
Two negative numbers: Add the absolute values, getting 12. The answer is negative, -12.

39. $-5.3 + 2.8$
A negative number and a positive number: The absolute values are 5.3 and 2.8. Subtract 2.8 from 5.3 to get 2.5. The negative number is farther from 0, so the answer is negative, -2.5.

41. $\frac{2}{7} + \left(-\frac{3}{5}\right) = \frac{10}{35} + \left(-\frac{21}{35}\right)$
A positive and negative number. The absolute values are $\frac{10}{35}$ and $\frac{21}{35}$. Subtract $\frac{10}{35}$ from $\frac{21}{35}$ to get $\frac{11}{35}$. The negative number is farther from 0, so the answer is negative, $-\frac{11}{35}$.

43. $-3.26 + (-5.8)$
Two negative numbers: Add the absolute values, getting 9.06. The answer is negative, -9.06.

45. $-\frac{1}{9} + \frac{2}{3} = -\frac{1}{9} + \frac{6}{9}$
A negative and positive number. The absolute values are $\frac{1}{9}$ and $\frac{6}{9}$. Subtract $\frac{1}{9}$ from $\frac{6}{9}$ to get $\frac{5}{9}$. The positive number is farther from 0, so the answer is positive, $\frac{5}{9}$.

47. $-6.25 + 0$
One number is zero: The sum is the other number, -6.25.

49. $4.19 + (-4.19)$
A negative and a positive number: The numbers have the same absolute value, 4.19, so the answer is 0.

51. $-18.3 + 22.1$
A negative and a positive number; The absolute values are 18.3 and 22.1. Subtract 18.3 from 22.1 to get 3.8. The positive number is farther from 0, so the answer is positive, 3.8.

53. The opposite of 2.37 is -2.37, because $2.37 + (-2.37) = 0$.

55. The opposite of -56 is 56, because $-56 + 56 = 0$.

57. The opposite of 0 is 0, because $0 + 0 = 0$.

59. If $x = 8$, then $-x = -8$. (The opposite of 8 is -8.)

61. If $x = -\frac{1}{10}$, then $-x = -\left(-\frac{1}{10}\right) = \frac{1}{10}$. (The opposite of $-\frac{1}{10}$ is $\frac{1}{10}$.)

63. If $x = -4.67$, then $-x = -(-4.67) = 4.67$. (The opposite of -4.67 is 4.67.)

65. If $x = 0$, then $-x = 0$. (The opposite of 0 is 0.)

67. $10 - 4 = 10 + (-4)$ Change the sign and add.
$= 6$

69. $4 - 10 = 4 + (-10)$ Change the sign and add.
$= -6$

71. $-5 - (-12) = -5 + 12$ Change the sign and add.
$= 7$

73. $-5 - 14 = -5 + (-14) = -19$

75. $2.7 - 5.8 = 2.7 + (-5.8) = -3.1$

77. $-\frac{3}{5} - \frac{1}{2} = -\frac{3}{5} + \left(-\frac{1}{2}\right)$
$= -\frac{6}{10} + \left(-\frac{5}{10}\right)$ Finding a common denominator
$= -\frac{11}{10}$

79. $-31 - (-31) = 0$
A negative and a positive number: The numbers have the same absolute value, 31, so the answer is 0.

81. $0 - (-5.37) = 0 + 5.37$ Change the sign and add.
$= 5.37$

83. $(-3)8$
Two numbers with unlike signs: Multiply their absolute values, getting 24. The answer is negative, -24.

85. $(-2)(-11)$
Two numbers with the same sign: Multiply their absolute values, getting 22. The answer is positive, 22.

87. $(4.2)(-5)$
Two numbers with unlike signs: Multiply their absolute values, getting 21. The answer is negative, -21.

89. $\frac{3}{7}(-1)$
Two numbers with unlike signs: Multiply their absolute values, getting $\frac{3}{7}$. The answer is negative, $-\frac{3}{7}$.

91. $(-17.45) \cdot 0 = 0$

93. $-\frac{2}{3}\left(\frac{3}{4}\right)$

Two numbers with unlike signs: Multiply their absolute values, getting $\frac{1}{2}$. The answer is negative, $-\frac{1}{2}$.

95. $\frac{-28}{-7}$

Two numbers with same sign: Divide their absolute values, getting 4. The answer is positive, 4.

97. $\frac{-100}{25}$

Two numbers with unlike signs: Divide their absolute values getting 4. The answer is negative, −4.

99. $\frac{73}{-1}$

Two numbers with unlike signs: Divide their absolute values, getting 73. The answer is negative, −73.

101. $\frac{0}{-7} = 0$

103. The reciprocal of 8 is $\frac{1}{8}$, because $8 \cdot \frac{1}{8} = 1$.

105. The reciprocal of $-\frac{5}{7}$ is $-\frac{7}{5}$, because $-\frac{5}{7} \cdot \left(-\frac{7}{5}\right) = 1$.

107. Does not exist

109. $\frac{3}{5} \div \frac{6}{7}$

$= \frac{3}{5} \cdot \frac{7}{6}$ Multiplying by the reciprocal of $\frac{6}{7}$

$= \frac{21}{30}$, or $\frac{7}{10}$

111. $\left(-\frac{3}{5}\right) \div \frac{1}{2}$

$= -\frac{3}{5} \cdot \frac{2}{1}$ Multiplying by the reciprocal of $\frac{1}{2}$

$= -\frac{6}{5}$

113. $\left(-\frac{2}{9}\right) \div (-8)$

$= -\frac{2}{9} \cdot \left(-\frac{1}{8}\right)$ Multiplying by the reciprocal of -8

$= \frac{2}{72}$, or $\frac{1}{36}$

115. $-\frac{12}{7} \div \left(-\frac{12}{7}\right)$

This is a number divided by itself so the quotient is 1. We would also do this exercise as follows.

$-\frac{12}{7} \div \left(-\frac{12}{7}\right) = -\frac{12}{7} \cdot \left(-\frac{7}{12}\right)$ Multiplying by the reciprocal of $-\frac{12}{7}$

$= 1$

117. $-4^2 = -(4 \cdot 4) = -16$ Squaring 4 and then taking the opposite

119. $-(-3)^2 - (-3)(-3) = -9$ Squaring (-3) and then taking the opposite

121. $(2-5)^2 = (-3)^2$ Working within the
$= 9$ parentheses first

123. $9 - (8 - 3 \cdot 2^3) = 9 - (8 - 3 \cdot 8)$ Working within the
$= 9 - (8 - 24)$ parentheses first
$= 9 - (-16)$
$= 9 + 16$
$= 25$

125. $\frac{5 \cdot 2 - 4^2}{27 - 2^4} = \frac{5 \cdot 2 - 16}{27 - 16} = \frac{10 - 16}{11} = \frac{-6}{11}$, or $-\frac{6}{11}$

127. $\frac{3^4 - (5-3)^4}{8 - 2^3} = \frac{3^4 - 2^4}{8 - 8} = \frac{81 - 16}{0}$

Since division by 0 is undefined, this quotient is undefined.

129. $\frac{(2-3)^3 - 5|2-4|}{7 - 2 \cdot 5^2} = \frac{(-1)^3 - 5|-2|}{7 - 2 \cdot 25} = \frac{-1 - 5(2)}{7 - 50}$

$= \frac{-1 - 10}{-43} = \frac{-11}{-43} = \frac{11}{43}$

131. $\left|2^2 - 7\right|^3 + 4 = |4 - 7|^3 + 4 = |-3|^3 + 4 = 3^3 + 4$
$= 27 + 4 = 31$

133. $32 - (-5)^2 + 15 \div (-3) \cdot 2$
$= 32 - 25 + 15 \div (-3) \cdot 2$ Evaluating the
exponential expression
$= 32 - 25 - 5 \cdot 2$ Dividing
$= 32 - 25 - 10$ Multiplying
$= -3$ Subtracting

135. Using the commutative law of addition, we have
$6 + xy = xy + 6$.
Using the commutative law of multiplication, we have
$6 + xy = 6 + yx$.

137. Using the commutative law of multiplication, we have
$-9(ab) = (ab)(-9)$
or $-9(ab) = -9(ba)$

139. $(3x)y$
$= 3(xy)$ Associative law of multiplication

141. $(3y + 4) + 10$
$= 3y + (4 + 10)$ Associative law of addition

143. $7(x+1) = 7 \cdot x + 7 \cdot 1$ Using the distributive law
$= 7x + 7$

145. $5(m - n) = 5 \cdot m - 5 \cdot n$ Using the distributive law
$= 5m - 5n$

147. $-5(2a + 3b)$
$= -5 \cdot 2a + (-5) \cdot 3b$
$= -10a - 15b$

149. $9a(b - c + d)$
$= 9a \cdot b - 9a \cdot c + 9a \cdot d$
$= 9ab - 9ac + 9ad$

151. $5x + 50 = 5 \cdot x + 5 \cdot 10 = 5(x + 10)$

153. $9p - 3 = 3 \cdot 3p - 3 \cdot 1 = 3(3p - 1)$

155. $7x - 21y + 14z = 7 \cdot x - 7 \cdot 3y + 7 \cdot 2z = 7(x - 3y + 2z)$

157. $255 - 34b = 17 \cdot 15 - 17 \cdot 2b = 17(15 - 2b)$

159. $xy + x = x \cdot y + x \cdot 1 = x(y + 1)$

161. *Writing Exercise.*

163. *Writing Exercise.*

165. $(8 - 5)^3 + 9 = 36$

167. $5 \cdot 2^3 \div (3 - 4)^4 = 40$

169. $17 - \sqrt{11 - (3 + 4)} \div [-5 - (-6)]^2$
$= 17 - \sqrt{11 - 7} \div [-5 + 6]^2$
$= 17 - \sqrt{4} \div [1]^2$
$= 17 - 2 \div 1$
$= 17 - 2$
$= 15$

171. Any value of a such that $a \leq -6.2$ satisfies the given conditions. The largest of these values is -6.2.

173. *Writing Exercise.*

175. a. Let t represent the temperature at midnight, in °F; $-16 - 5 = t$; $t = -21$. The temperature at midnight was -21°F.

   b. Let x represent the temperature outside Ethan's jet, in °F; $42 - 3.5(20) = x$; $x = -28$. The temperature outside Ethan's jet is -28°F.

Exercise Set 1.3

1. Two equations are *equivalent* if they have the same solutions.

3. A *contradiction* is an equation that is never true.

5. By the distributive law, the expression $2(x + 7)$ is equivalent to the expression $2x + 14$, so they are equivalent expressions.

7. $4x - 9 = 7$ and $4x = 16$ are equations and they have the same solution, 4, so they are equivalent equations.

9. Combining like terms in the expression $8t + 5 - 2t + 1$, we get the expression $6t + 6$, so these are equivalent expressions.

11. $3t = 21$ and $t + 4 = 11$
Each equation has only one solution, the number 7. Thus, the equations are equivalent.

13. $12 - x = 3$ and $2x = 20$
When x is replaced by 9, the first equation is true, but the second equation is false. Thus the equations are not equivalent.

15. $5x = 2x$ and $\dfrac{4}{x} = 3$
When x is replaced by 0, the first equation is true, but the second equation is not defined. Thus the equations are not equivalent.

17. $x - 2.9 = 13.4$
   $x - 2.9 + 2.9 = 13.4 + 2.9$ Adding princple; adding 2.9
   $x + 0 = 13.4 + 2.9$ Law of opposites
   $x = 16.3$
Check: $x - 2.9 = 13.4$
   $\begin{array}{c|c} 16.3 - 2.9 & 13.4 \\ \hline & \overset{?}{} \\ 13.4 = 13.4 & \text{TRUE} \end{array}$
The solution is 16.3.

19. $8t = 72$
   $\dfrac{1}{8} \cdot 8t = \dfrac{1}{8} \cdot 72$ Multiplication principle; multiplying by $\frac{1}{8}$, the reciprocal of 8
   $1t = 9$
   $t = 9$
Check: $8t = 72$
   $\begin{array}{c|c} 8 \cdot 9 & 72 \\ \hline & \overset{?}{} \\ 72 = 72 & \text{TRUE} \end{array}$
The solution is 9.

21. $\dfrac{2}{3}x = 30$
   $\dfrac{3}{2} \cdot \dfrac{2}{3}x = \dfrac{3}{2} \cdot 30$
   $1x = 45$
   $x = 45$
Check: $\dfrac{2}{3}x = 30$
   $\begin{array}{c|c} \frac{2}{3} \cdot 45 & 30 \\ \hline & \overset{?}{} \\ 30 = 30 & \text{TRUE} \end{array}$
The solution is 45.

23. $4a + 25 = 9$
   $4a + 25 - 25 = 9 - 25$
   $4a = -16$
   $\dfrac{1}{4} \cdot 4a = \dfrac{1}{4}(-16)$
   $1a = -4$
   $a = -4$
Check: $4a + 25 = 9$
   $\begin{array}{c|c} 4(-4) + 25 & 9 \\ -16 + 25 & \\ \hline & \overset{?}{} \\ 9 = 9 & \text{TRUE} \end{array}$
The solution is -4.

25.
$$2y - 8 = 9$$
$$2y - 8 + 8 = 9 + 8$$
$$2y = 17$$
$$\frac{1}{2} \cdot 2y = \frac{1}{2} \cdot 17$$
$$1y = \frac{17}{2}$$
$$y = \frac{17}{2}$$

Check:

$$
\begin{array}{c|c}
2y - 8 = 9 \\
\hline
2\left(\frac{17}{2}\right) - 8 & 9 \\
17 - 8 & 9 \\
& \overset{?}{9 = 9} \quad \text{TRUE}
\end{array}
$$

The solution is $\frac{17}{2}$.

27. $9t^2 + t^2 = (9+1)t^2 = 10t^2$

29. $16a - a = (16-1)a = 15a$

31. $n - 8n = (1-8)n = -7n$

33. $5x - 3x + 8x = (5 - 3 + 8)x = 10x$

35. $18p - 12 + 3p + 8$
$= 18p + 3p - 12 + 8$ Commutative law of addition
$= (18+3)p + (-12+8)$
$= 21p - 4$

37. $-7t^2 + 3t + 5t^3 - t^3 + 2t^2 - t$
$= (-7+2)t^2 + (3-1)t + (5-1)t^3$
$= -5t^2 + 2t + 4t^3$

39. $2x + 3(5x - 7)$
$= 2x + 15x - 21$
$= 17x - 21$

41. $7a - (2a + 5)$
$= 7a - 2a - 5$
$= 5a - 5$

43. $m - (6m - 2)$
$= m - 6m + 2$
$= -5m + 2$

45. $3d - 7 - (5 - 2d)$
$= 3d - 7 - 5 + 2d$
$= 5d - 12$

47. $2(x - 3) + 4(7 - x)$
$= 2x - 6 + 28 - 4x$
$= -2x + 22$

49. $3p - 4 - 2(p + 6)$
$= 3p - 4 - 2p - 12$
$= p - 16$

51. $-2(a - 5) - [7 - 3(2a - 5)]$
$= -2a + 10 - [7 - 6a + 15]$
$= -2a + 10 - [22 - 6a]$
$= -2a + 10 - 22 + 6a$
$= 4a - 12$

53. $5\{-2x + 3[2 - 4(5x + 1)]\}$
$= 5\{-2x + 3[2 - 20x - 4]\}$
$= 5\{-2x + 3[-20x - 2]\}$
$= 5\{-2x - 60x - 6\}$
$= 5\{-62x - 6\}$
$= -310x - 30$

55. $8y - \{6[2(3y - 4) - (7y + 1)] + 12\}$
$= 8y - \{6[6y - 8 - 7y - 1] + 12\}$
$= 8y - \{6[-y - 9] + 12\}$
$= 8y - \{-6y - 54 + 12\}$
$= 8y - \{-6y - 42\}$
$= 8y + 6y + 42$
$= 14y + 42$

57.
$$4x + 5x = 63$$
$$9x = 63$$
$$\frac{1}{9} \cdot 9x = \frac{1}{9} \cdot 63$$
$$x = 7$$

Check:

$$
\begin{array}{c|c}
4x + 5x = 63 \\
\hline
4 \cdot 7 + 5 \cdot 7 & 63 \\
28 + 35 & \\
& \overset{?}{63 = 63} \quad \text{TRUE}
\end{array}
$$

The solution is 7.

59.
$$\frac{1}{4}y - \frac{2}{3}y = 5$$
$$\frac{3}{12}y - \frac{8}{12}y = 5$$
$$-\frac{5}{12}y = 5$$
$$\left(-\frac{12}{5}\right)\left(-\frac{5}{12}y\right) = \left(-\frac{12}{5}\right) \cdot 5$$
$$1y = -12$$
$$y = -12$$

Check:

$$
\begin{array}{c|c}
\frac{1}{4}y - \frac{2}{3}y = 5 \\
\hline
\frac{1}{4}(-12) - \frac{2}{3}(-12) & 5 \\
-3 + 8 & 5 \\
& \overset{?}{5 = 5} \quad \text{TRUE}
\end{array}
$$

The solution is -12.

61.
$$4(t - 3) - t = 6$$
$$4t - 12 - t = 6$$
$$3t - 12 = 6$$
$$3t - 12 + 12 = 6 + 12$$
$$3t = 18$$
$$\frac{1}{3} \cdot 3t = \frac{1}{3} \cdot 18$$
$$t = 6$$

Check:

$$
\begin{array}{c|c}
4(t - 3) - t = 6 \\
\hline
4(6 - 3) - 6 & 6 \\
4(3) - 6 & \\
12 - 6 & \\
& \overset{?}{6 = 6} \quad \text{TRUE}
\end{array}
$$

The solution is 6.

63.
$$3(x+4) = 7x$$
$$3x+12 = 7x$$
$$3x+12-3x = 7x-3x$$
$$12 = 4x$$
$$\frac{1}{4}\cdot 12 = \frac{1}{4}\cdot 4x$$
$$3 = x$$

Check:
$$\begin{array}{c|c} 3(x+4) = 7x \\ \hline 3(3+4) & 7\cdot 3 \\ 3\cdot 7 & \\ & \overset{?}{21=21} \quad \text{TRUE} \end{array}$$

The solution is 3.

65.
$$70 = 10(3t-2)$$
$$70 = 30t-20$$
$$70+20 = 30t-20+20$$
$$90 = 30t$$
$$\frac{1}{30}\cdot 90 = \frac{1}{30}\cdot 30t$$
$$3 = t$$

Check:
$$\begin{array}{c|c} 70 = 10(3t-2) \\ \hline 70 & 10(3\cdot 3-2) \\ & 10(9-2) \\ & 10\cdot 7 \\ & \overset{?}{70=70} \quad \text{TRUE} \end{array}$$

The solution is 3.

67.
$$1.8(2-n) = 9$$
$$3.6-1.8n = 9$$
$$3.6-1.8n-3.6 = 9-3.6$$
$$-1.8n = 5.4$$
$$\left(-\frac{1}{1.8}\right)(-1.8n) = \left(-\frac{1}{1.8}\right)\cdot 5.4$$
$$n = -3$$

Check:
$$\begin{array}{c|c} 1.8(2-n) = 9 \\ \hline 1.8(2-(-3)) & 9 \\ 1.8(2+3) & \\ 1.8(5) & \\ & \overset{?}{9=9} \quad \text{TRUE} \end{array}$$

The solution is –3.

69.
$$5y-(2y-10) = 25$$
$$5y-2y+10 = 25$$
$$3y+10 = 25$$
$$3y+10-10 = 25-10$$
$$3y = 15$$
$$\frac{1}{3}\cdot 3y = \frac{1}{3}\cdot 15$$
$$y = 5$$

Check:
$$\begin{array}{c|c} 5y-(2y-10) = 25 \\ \hline 5\cdot 5-(2\cdot 5-10) & 25 \\ 25-(10-10) & \\ 25-0 & \\ & \overset{?}{25=25} \quad \text{TRUE} \end{array}$$

The solution is 5.

71.
$$\frac{9}{10}y-\frac{7}{10} = \frac{21}{5}$$
$$\frac{9}{10}y-\frac{7}{10}+\frac{7}{10} = \frac{21}{5}+\frac{7}{10}$$
$$\frac{9}{10}y = \frac{42}{10}+\frac{7}{10}$$
$$\frac{9}{10}y = \frac{49}{10}$$
$$\frac{10}{9}\cdot\frac{9}{10}y = \frac{10}{9}\cdot\frac{49}{10}$$
$$y = \frac{49}{9}$$

Check:
$$\begin{array}{c|c} \frac{9}{10}y-\frac{7}{10} = \frac{21}{5} \\ \hline \frac{9}{10}\cdot\frac{49}{9}-\frac{7}{10} & \frac{21}{5} \\ \frac{49}{10}-\frac{7}{10} & \\ \frac{42}{10} & \\ & \overset{?}{\frac{21}{5}=\frac{21}{5}} \quad \text{TRUE} \end{array}$$

The solution is $\frac{49}{9}$.

73.
$$7r-2+5r = 6r+6-4r$$
$$12r-2 = 2r+6$$
$$12r-2-2r = 2r+6-2r$$
$$10r-2 = 6$$
$$10r-2+2 = 6+2$$
$$10r = 8$$
$$\frac{1}{10}\cdot 10r = \frac{1}{10}\cdot 8$$
$$r = \frac{8}{10}$$
$$r = \frac{4}{5}$$

Check:
$$\begin{array}{c|c} 7r-2+5r = 6r+6-4r \\ \hline 7\cdot\frac{4}{5}-2+5\cdot\frac{4}{5} & 6\cdot\frac{4}{5}+6-4\cdot\frac{4}{5} \\ \frac{28}{5}-2+4 & \frac{24}{5}+6-\frac{16}{5} \\ & \overset{?}{\frac{38}{5}=\frac{38}{5}} \quad \text{TRUE} \end{array}$$

The solution is $\frac{4}{5}$.

75.
$$\frac{2}{3}(x-2)-1 = \frac{1}{4}(x-3)$$
$$\frac{2}{3}x-\frac{4}{3}-1 = \frac{1}{4}x-\frac{3}{4}$$
$$\frac{2}{3}x-\frac{7}{3} = \frac{1}{4}x-\frac{3}{4}$$
$$\frac{2}{3}x-\frac{7}{3}-\frac{1}{4}x = \frac{1}{4}x-\frac{3}{4}-\frac{1}{4}x$$
$$\frac{5}{12}x-\frac{7}{3} = -\frac{3}{4}$$
$$\frac{5}{12}x-\frac{7}{3}+\frac{7}{3} = -\frac{3}{4}+\frac{7}{3}$$
$$\frac{5}{12}x = \frac{19}{12}$$
$$\frac{12}{5}\cdot\frac{5}{12}x = \frac{12}{5}\cdot\frac{19}{12}$$
$$x = \frac{19}{5}$$

The check is left to the student. The solution is $\frac{19}{5}$.

77. $2(t-5)-3(2t-7)=12-5(3t+1)$

$2t-10-6t+21=12-15t-5$

$-4t+11=-15t+7$

$-4t+11+15t=-15t+7+15t$

$11t+11=7$

$11t+11-11=7-11$

$11t=-4$

$\frac{1}{11}\cdot 11t=\frac{1}{11}(-4)$

$t=-\frac{4}{11}$

Check:

$$2(t-5)-3(2t-7)=12-5(3t+1)$$

$2\left(-\frac{4}{11}-5\right)-3\left(2\left(-\frac{4}{11}\right)-7\right)$	$12-5\left(3\left(-\frac{4}{11}\right)+1\right)$
$2\left(-\frac{59}{11}\right)-3\left(-\frac{8}{11}-7\right)$	$12-5\left(-\frac{12}{11}+1\right)$
$-\frac{118}{11}-3\left(-\frac{85}{11}\right)$	$12-5\left(-\frac{1}{11}\right)$
$-\frac{118}{11}+\frac{255}{11}$	$12+\frac{5}{11}$

$$\frac{137}{11}\overset{?}{=}\frac{137}{11}\quad\text{TRUE}$$

The solution is $-\frac{4}{11}$.

79. $3[2-4(x-1)]=3-4(x+2)$

$3[2-4x+4]=3-4x-8$

$3[6-4x]=-4x-5$

$18-12x=-4x-5$

$18-12x+12x=-4x-5+12x$

$18=8x-5$

$18+5=8x-5+5$

$23=8x$

$\frac{1}{8}\cdot 23=\frac{1}{8}(8x)$

$\frac{23}{8}=x$

Check: $3[2-4(x-1)]=3-4(x+2)$

$3\left[2-4\left(\frac{23}{8}-1\right)\right]$	$3-4\left(\frac{23}{8}+2\right)$
$3\left[2-4\left(\frac{15}{8}\right)\right]$	$3-4\left(\frac{39}{8}\right)$
$3\left[2-\frac{15}{2}\right]$	$3-\frac{39}{2}$
$3\left(-\frac{11}{2}\right)$	$\frac{6}{2}-\frac{39}{2}$

$$-\frac{33}{2}\overset{?}{=}-\frac{33}{2}\quad\text{TRUE}$$

The solution is $\frac{23}{8}$.

81. $2x+2=2(x+1)$

$2x+2=2x+2$

$2x+2-2x=2x+2-2x$

$2=2$

All real numbers are solutions. The solution set is the set of all real numbers. The equation is an identity.

83. $7x-2-3x=4x$

$4x-2=4x$

$4x-2-4x=4x-4x$

$-2=0$

Since the original equation is equivalent to the false equation $-2=0$, there is no solution. The solution set is $\varnothing$. The equation is a contradiction.

85.

$2+9x=3(4x+1)-1$

$2+9x=12x+3-1$

$2+9x=12x+2$

$2+9x-2=12x+2-2$

$9x=12x$

$9x-9x=12x-9x$

$0=3x$

$\frac{1}{3}\cdot 0=\frac{1}{3}\cdot 3x$

$0=x$

The solution set is $\{0\}$. The equation is a conditional equation.

87. $3x-(8-x)=6x-2(x+4)$

$3x-8+x=6x-2x-8$

$4x-8=4x-8$

$4x-8-4x=4x-8-4x$

$-8=-8$

All real numbers are solutions. The solution set is the set of all real numbers. The equation is an identity.

89. $-9t+2=-9t-7(6\div 2(49)+8)$

Observe that $-7(6\div 2(49)+8)$ is a negative number. Then on the left side we have $-9t$ plus a positive number and on the right side we have $-9t$ plus a negative number. This is a contradiction, so the solution set is $\varnothing$.

91. $2\{9-3[-2x-4]\}=12x+42$

$2\{9+6x+12\}=12x+42$

$2\{21+6x\}=12x+42$

$42+12x=12x+42$

$42+12x-12x=12x+42-12x$

$42=42$

The original equation is equivalent to the equation $42=42$, which is true for all real numbers. Thus the solution set is the set of all real numbers. The equation is an identity.

93. *Writing Exercise.*

95. *Writing Exercise.*

97. $-0.00458y+1.7787=13.002y-1.005$

$-13.00658y+1.7787=-1.005$

$-13.00658y=-2.7837$

$y=\frac{-2.7837}{-13.00658}$

$0.2140224409\approx y$

The check is left to the student. The solution is approximately 0.2140224409.

99. $6x - \{5x - [7x - (4x - (3x+1))]\} = 6x + 5$

$6x - \{5x - [7x - (4x - 3x - 1)]\} = 6x + 5$

$6x - \{5x - [7x - (x - 1)]\} = 6x + 5$

$6x - \{5x - [7x - x + 1]\} = 6x + 5$

$6x - \{5x - [6x + 1]\} = 6x + 5$

$6x - \{5x - 6x - 1\} = 6x + 5$

$6x - \{-x - 1\} = 6x + 5$

$6x + x + 1 = 6x + 5$

$7x + 1 = 6x + 5$

$x = 4$

101. $23 - 2\{4 + 3[x - 1]\} + 5\{x - 2(x + 3)\}$
$\qquad = 7\{x - 2[5 - (2x + 3)]\}$

$23 - 2\{4 + 3x - 3\} + 5\{x - 2x - 6\}$
$\qquad = 7\{x - 2[5 - 2x - 3]\}$

$23 - 2\{3x + 1\} + 5\{-x - 6\} = 7\{x - 2[-2x + 2]\}$

$23 - 6x - 2 - 5x - 30 = 7\{x + 4x - 4\}$

$-11x - 9 = 7\{5x - 4\}$

$-11x - 9 = 35x - 28$

$-9 = 46x - 28$

$19 = 46x$

$\dfrac{19}{46} = x$

The check is left to the student. The solution is $\dfrac{19}{46}$.

103. *Writing Exercise.*

Mid-Chapter Review

1. $3x - 2(x - 1) = 3x - 2x + 2$
$\qquad = x + 2$

2. $3x - 2(x - 1) = 6x$

$3x - 2x + 2 = 6x$

$x + 2 = 6x$

$2 = 5x$

$\dfrac{2}{5} = x$

3. Five less than three times a number
Let n represent the number; $3n - 5$

4. $2a \div 3x - a + x = 2(3) \div 3(5) - 3 + 5$
$\qquad = 6 \div 3(5) - 3 + 5$
$\qquad = 2(5) - 3 + 5$
$\qquad = 10 - 3 + 5$
$\qquad = 12$

5. The area of a triangle is given by $A = \frac{1}{2}bh$. We

substitute $b = \frac{1}{2}$, $h = 3$ and solve.

$A = \frac{1}{2}bh = \frac{1}{2}\left(\frac{1}{2}\right)(3) = \frac{3}{4}$ ft^2

6. $\frac{1}{2} - \left(-\frac{1}{3}\right) = \frac{5}{6}$

7. $-32 \div (-0.8) = \dfrac{-32}{-0.8} = 40$

Two numbers with same sign: Divide their absolute
values, getting 40. The answer is positive, 40.

8. 2.52

9. $\left(\dfrac{3}{10}\right)\left(-\dfrac{2}{5}\right) = -\dfrac{6}{50}$, or $-\dfrac{3}{25}$

Two numbers with unlike signs: Multiply their absolute

values, getting $\frac{3}{25}$. The answer is negative, $-\frac{3}{25}$.

10. $8 - 2^3 \div 4 \cdot (-2) + 1 - 2 = 8 - 8 \div 4 \cdot (-2) + 1 - 2$
$\qquad = 8 - 2 \cdot (-2) + 1 - 2$
$\qquad = 8 + 4 + 1 - 2$
$\qquad = 11$

11. $(x + 3) + y$
$\qquad = x + (3 + y)$ $\qquad$ Associative law of addition

12. $3x - 5 - x + 12 = 2x + 7$

13. $4t - (3t - 1) = 4t - 3t + 1$
$\qquad = t + 1$

14. $8x + 2[x - (x - 1)] = 8x + 2[x - x + 1]$
$\qquad = 8x + 2[1]$
$\qquad = 8x + 2$

15. $-(p - 4) - [3 - (9 - 2p)] + p$
$\qquad = -p + 4 - [3 - 9 + 2p] + p$
$\qquad = -p + 4 - [-6 + 2p] + p$
$\qquad = -p + 4 + 6 - 2p + p$
$\qquad = -2p + 10$

16. $2x - 6 = 3x + 5$

$2x - 6 - 2x = 3x + 5 - 2x$

$-6 = x + 5$

$-6 - 5 = x + 5 - 5$

$-11 = x$

17. $5 - (t - 2) = 6$

$5 - t + 2 = 6$

$7 - t = 6$

$-t = -1$

$t = 1$

18. $6(y - 1) - 2(y + 1) = 4(y - 2)$

$6y - 6 - 2y - 2 = 4y - 8$

$4y - 8 = 4y - 8$

$-8 = -8$

All real numbers are solutions. The solution set is the
set of all real numbers. The equation is an identity.

19. $3(x - 1) - 2(2x + 1) = 5(x - 1)$

$3x - 3 - 4x - 2 = 5x - 5$

$-x - 5 = 5x - 5$

$-x - 5 + x = 5x - 5 + x$

$-5 = 6x - 5$

$-5 + 5 = 6x - 5 + 5$

$0 = 6x$

$\dfrac{1}{6} \cdot 0 = \dfrac{1}{6} \cdot 6x$

$0 = x$

20. $\frac{1}{3}t - 2 = \frac{1}{6} + t$

$\qquad -2 = \frac{1}{6} + \frac{4}{3}t$

$\qquad -\frac{13}{6} = \frac{4}{3}t$

$\qquad -\frac{13}{4} = t$

Exercise Set 1.4

1. The five steps of the problem solving process:
1. Familiarize
2. Translate
3. Carry out
4. Check
5. State

3. *State* Give the answer clearly.

5. *Familiarize* Read the problem carefully.

7. *Familiarize*. We want to find two numbers. We can let x represent the first number and note that the second number is 9 more than the first. Also the sum of the numbers is 91.
Translate. The second number can be named $x + 9$. We translate to an equation:

First number	plus	second number	is	91.
↓	↓	↓	↓	↓
x	$+$	$(x+9)$	$=$	91

9. *Familiarize*. Let t = the time, in hours, it will take Noah to paddle 8 mi. We will use the formula Distance = Speed × Time. Noah's speed paddling against the current is $(4.6 - 2.1)$ mph.
Translate. We substitute in the formula
$\qquad 8 = (4.6 - 2.1)t$.

11. *Familiarize*. There are three angle measures involved, and we want to find all three. We can let x represent the smallest angle measure and note that the second is one more than x and the third is one more than the second, or two more than x. We also note that the sum of the three angle measures must be 180°.
Translate. The three angle measures are x, $x + 1$, and $x + 2$. We translate to an equation:

First	plus	second	plus	third	is	180°.
↓	↓	↓	↓	↓	↓	↓
x	$+$	$(x+1)$	$+$	$(x+2)$	$=$	180

13. *Familiarize*. Since the escalator's speed is 105 ft/min and Dominik's walking speed is 100 ft/min, Dominik will move at a speed of $100 + 105$ ft/min on the escalator. Let t = the time, in minutes, it takes to reach the top.
Translate. We will use the formula Distance = Speed × Time.

Distance	=	Speed	×	Time
↓	↓	↓	↓	↓
205	$=$	$(100+105)$	$×$	t

15. *Familiarize*. Let w represent the wholesale price, in dollars. Then the wholesale price raised by 50% is $w + 0.5w$.
Translate.

Wholesale price raised 50%	plus	$1.50	is	selling price.
↓	↓	↓	↓	↓
$w + 0.5w$	$+$	1.50	$=$	22.50

17. *Familiarize*. Let t = the time, in minutes, required for the plane to reach 29,000 ft. Since Distance = Speed × Time, the plane will travel $3500 \times t$ ft in t min. Note that the plane starts at an altitude of 8000 ft.
Translate.

Current altitude	plus	distance climbed	is	29,000 ft.
↓	↓	↓	↓	↓
8000	$+$	$3500t$	$=$	$29,000$

19. *Familiarize*. Note that each odd integer is 2 more than the one preceding it. If we let n represent the first odd integer then $n + 2$ represents the next odd integer and $(n + 2) + 2$, or $n + 4$ is the third odd integer.
Translate.

First	plus	twice the second	plus	three times the third	is	70.
↓	↓	↓	↓	↓	↓	↓
n	$+$	$2(n+2)$	$+$	$3(n+4)$	$=$	70

21. *Familiarize*. The perimeter of an equilateral triangle is 3 times the length of a side. Let s = the length of a side of the smaller triangle. Then $2s$ = the length of a side of the larger triangle. The sum of the two perimeters is 90 cm.
Translate.

Perimeter of smaller triangle	plus	perimeter of larger triangle	is	90 cm.
↓	↓	↓	↓	↓
$3s$	$+$	$3 \cdot 2s$	$=$	90

23. *Familiarize*. Let c represent the number of calls Cody will need on his next shift if he is to average 3 calls per shift. We find the average by adding the number of calls on each of the 5 shifts and then dividing by the number of addends.
Translate.

Average number of calls per shift	is	3.
↓	↓	↓
$\dfrac{5+2+1+3+c}{5}$	$=$	3

25. *Familiarize*. Let p represent the price Tony paid for his graphing calculator.
Translate.

Price Tess paid	is	price Tony paid	less	$13.
↓	↓	↓	↓	↓
124	$=$	p	$-$	13

Carry out. We solve the equation.
$$124 = p - 13$$
$$137 = p \qquad \text{Adding 13 to both sides}$$

Check. The price Tess paid, \$124, is \$13 less than \$137, so the answer checks.

State. Tony paid \$137 for his graphing calculator.

27. **Familiarize**. Let r represent the average monthly rent of an apartment in Charlotte, in dollars.
Translate.

Rent in Greenville	is	$\frac{4}{5}$	of	rent in Charlotte
↓	↓	↓	↓	↓
1100	=	$\frac{4}{5}$	·	r

Carry out. We solve the equation.
$$1100 = \frac{4}{5}r$$
$$\frac{5}{4} \cdot 1100 = \frac{5}{4} \cdot \frac{4}{5}r$$
$$1375 = r$$

Check. Since $\frac{4}{5}$ of \$1375 is \$1100, the answer checks.

State. The average monthly apartment rent in Charlotte is \$1375.

29. **Familiarize**. Let x represent the number of flu shots given by Mike. Then $x + 11$ represents the number of flu shots given by Vance.
Translate.

Total flu shots	is	53.
↓	↓	↓
$x + (x + 11)$	=	53

Carry out. We solve the equation.
$$x + (x + 11) = 53$$
$$2x + 11 = 53$$
$$2x = 42 \qquad \text{Subtracting 11 from both sides}$$
$$x = 21$$
If $x = 21$, then $x + 11 = 21 + 11$, or 32.

Check. 32 is 11 more than 21. Also, $21 + 32 = 53$ flu shots. The answer checks.

State. Vance gave 32 flu shots.

31. **Familiarize**. Let w represent the width of the mirror, in cm. Then $3w$ represents the length. Recall that the formula for the perimeter P of a rectangle with length l and width w is $P = 2l + 2w$.
Translate.

Perimeter	is	120 cm,
↓	↓	↓
$2 \cdot 3w + 2 \cdot w$	=	120

Carry out. We solve the equation.
$$2 \cdot 3w + 2 \cdot w = 120$$
$$6w + 2w = 120$$
$$8w = 120$$
$$w = 15$$
When $w = 15$, then $3w = 3 \cdot 15 = 45$.

Check. If the length is 45 cm and the width is 15 cm, then the length is three times the width. Also
$P = 2 \cdot 45 + 2 \cdot 15 = 90 + 30 = 120$ cm. The answer checks.

State. The length of the mirror is 45 cm, and the width is 15 cm.

33. **Familiarize**. Let l represent the length of the greenhouse, in meters. Then $\frac{1}{4}l$ represents the width. Recall that the formula for the perimeter P, of a rectangle with length l and width w is $P = 2l + 2w$.
Translate.

Perimeter	is	130 m.
↓	↓	↓
$2l + 2 \cdot \frac{1}{4}l$	=	130

Carry out. We solve the equation.
$$2l + 2 \cdot \frac{1}{4}l = 130$$
$$2l + \frac{1}{2}l = 130$$
$$\frac{5}{2}l = 130$$
$$\frac{2}{5} \cdot \frac{5}{2}l = \frac{2}{5} \cdot 130$$
$$l = 52$$
When $l = 52$, then $\frac{1}{4}l = 13$.

Check. If the length is 52 m and the width is 13 m, then the width is $\frac{1}{4}$ of the length. Also,
$P = 2 \cdot 52 + 2 \cdot 13 = 104 + 26 = 130$ m. The answer checks.

State. The length of the greenhouse is 52 m, and the width is 13 m.

35. The Familiarize and Translate steps were done in Exercise 9.
$8 = (4.6 - 2.1)t$.
Carry out. We solve the equation.
$$8 = (4.6 - 2.1)t$$
$$8 = 2.5t$$
$$\frac{8}{2.5} = \frac{2.5}{2.5}t$$
$$3.2 = t$$

Check. At a speed of 2.5 mph, in 3.2 hr, Noah paddles 2.5(3.2), or 8 mi. Our answer checks.

State. Noah takes 3.2 hr to paddle 8 mi.

37. The Familiarize and Translate steps were done in Exercise 11.
Carry out. We solve the equation.
$$x + (x + 1) + (x + 2) = 180$$
$$3x + 3 = 180$$
$$3x = 180 - 3$$
$$3x = 177$$
$$\frac{1}{3} \cdot 3x = \frac{1}{3} \cdot 177$$
$$x = 59$$
When $x = 59$, then $x + 1 = 59 + 1 = 60$, and
$x + 2 = 59 + 2 = 61$.

Check. The angles, 59°, 60°, and 61° are consecutive integers. Also $59° + 60° + 61° = 180°$. The answer checks.

State. The measures of the angles are 59°, 60°, and 61°.

39. The Familiarize and Translate steps were done in Exercise 16.

Carry out. We solve the equation.

$$c - 0.05c = 142.50$$
$$0.95c = 142.50$$
$$\frac{0.95c}{0.95} = \frac{142.50}{0.95}$$
$$c = 150$$

Check. 5% of $150 is $7.50 and $150 − $7.50 = $142.50.

The answer checks.

State. The cost would have been $150 if the bill had not been paid promptly.

41. The Familiarize and Translate steps were done in Exercise 15.

Carry out. We solve the equation.

$$1.5w + 1.50 = 22.50$$
$$1.5w = 21$$
$$w = \frac{1}{1.5} \cdot 21$$
$$w = 14$$

Check. If a wholesale price of $14 is raised by 50%, we have $14 + 0.5($14) = $14 + $7 = $21. When $1.50 is added to this figure, we have $21 + $1.50 = $22.50. The answer checks.

State. The wholesale price is $14.

43. *Writing Exercise.*

45. *Writing Exercise.*

47. **Familiarize.** The average score on the first four tests is $\frac{83 + 91 + 78 + 81}{4}$, or 83.25. Let x = the number of points above this average that Tico scores on the next test. Then the score on the fifth test is $83.25 + x$.

Translate.

| Average score on 5 tests | is | 2 | more than | average score on 4 tests. |

$$\frac{83 + 91 + 78 + 81 + (83.25 + x)}{5} = 2 + 83.25$$

Carry out. Carry out some algebraic manipulation.

$$\frac{83 + 91 + 78 + 81 + (83.25 + x)}{5} = 2 + 83.25$$
$$\frac{416.25 + x}{5} = 85.25$$
$$416.25 + x = 426.25$$
$$x = 10$$

Check. If Tico scores 10 points more than the average of the first four tests on the fifth test, his score will be $83.25 + 10$, or 93.25. Then the five-test average will be $\frac{83 + 91 + 78 + 81 + 93.25}{5}$, or 85.25. This is 2 points above the four-test average, so the answer checks.

State. Tico must score 10 points above the four-test average in order to raise the average 2 points.

49. **Familiarize.** Let a = the number of animals adopted in 2011. From 2011 to 2012 the number of adoptions decreased 1.1%, so $a - 0.011a$, or $0.989a$. From 2012 to 2013 the number of adoptions decreased 3.6%, so $0.989a - 0.036(0.989a)$, or $0.964(0.989a)$. From 2013 to 2014 the numbers decreased 8.5%, so $0.964(0.989a) - 0.085(0.964)(0.989a)$, or $0.915(0.964)(0.989a)$. From 2014 to 2015, the numbers increased 11.0%, so the number of adoptions became $0.915(0.964)(0.989a) + 0.11(0.915)(0.964)(0.989a)$, or $1.11(0.915)(0.964)(0.989a)$.

Translate.

| Adoptions in 2015 | were | 2879. |

$$1.11(0.915)(0.964)(0.989a) = 2879$$

Carry out. We solve the equation.

$$1.11(0.915)(0.964)(0.989a) = 2879$$
$$a = \frac{2879}{1.11(0.915)(0.964)(0.989)}$$
$$a \approx 2974$$

Check. If the number of adoptions in 2011 was 2974, then in 2012 there were 0.989(2974), or 2941. In 2013 there were 0.964(2941), or 2835. In 2014 there were 0.915(2835), or 2594, and in 2015 there were 1.11(2594), or 2879. Our answer checks.

State. There were 2974 animal adoptions in 2011.

Connecting the Concepts

1. $2x - 3 = 7$
$$2x = 10$$
$$x = 5$$

2. $2x - c = h$
$$2x = c + h$$
$$x = \frac{c + h}{2}$$

3. $8 - 3(y - 7) = 2y$
$$8 - 3y + 21 = 2y$$
$$-3y + 29 = 2y$$
$$29 = 5y$$
$$\frac{29}{5} = y$$

4. $8 - n(y - c) = ay$
$$8 - ny + nc = ay$$
$$8 + nc = ay + ny$$
$$8 + nc = y(a + n)$$
$$\frac{8 + nc}{a + n} = y$$

5. $\frac{a}{3} - \frac{1}{2} = 4a$
$$6\left(\frac{a}{3} - \frac{1}{2}\right) = 6(4a)$$
$$2a - 3 = 24a$$
$$-3 = 22a$$
$$-\frac{3}{22} = a$$

6. $\dfrac{a}{3} - \dfrac{n}{2} = xa$

$6\left(\dfrac{a}{3} - \dfrac{n}{2}\right) = 6(xa)$

$2a - 3n = 6xa$

$-3n = 6xa - 2a$

$-3n = a(6x - 2)$

$\dfrac{-3n}{6x - 2} = a$ or $a = \dfrac{3n}{2 - 6x}$

Exercise Set 1.5

1. A formula is an *equation* that uses letters to represent a relationship between two or more quantities.

3. The formula $C = \pi d$ is used to calculate the *circumference* of a circle.

5. The formula $\underline{A = bh}$ is used to calculate the area of a parallelogram of height h and base length b.

7. In the formula for the area of a trapezoid,

$A = \dfrac{h}{2}(b_1 + b_2)$ the numbers 1 and 2 are referred to as

subscripts.

9. $E = wA$

$\dfrac{1}{w} \cdot E = \dfrac{1}{w} \cdot wA$ Multiplying both sides by $\dfrac{1}{w}$

$\dfrac{E}{w} = A$ Simplifying

11. $d = rt$

$\dfrac{1}{t} \cdot d = \dfrac{1}{t} \cdot rt$ Multiplying both sides by $\dfrac{1}{t}$

$\dfrac{d}{t} = r$ Simplifying

13. $V = lwh$

$\dfrac{1}{lw} \cdot V = \dfrac{1}{lw} \cdot lwh$ Multiplying both sides by $\dfrac{1}{lw}$

$\dfrac{V}{lw} = h$ Simplifying

15. $L = \dfrac{k}{d^2}$

$d^2 \cdot L = d^2 \cdot \dfrac{k}{d^2}$ Multiplying both sides by d^2

$d^2 L = k$ Simplifying

17. $G = w + 150n$

$G - w = 150n$ Subtracting w from both sides

$\dfrac{1}{150}(G - w) = \dfrac{1}{150} \cdot 150n$ Multiplying both sides by $\dfrac{1}{150}$

$\dfrac{G - w}{150} = n$ Simplifying

19. $2w + 2h + l = p$

$l = p - 2w - 2h$ Adding $-2w - 2h$ to both sides

21. $2x + 3y = 4$

$3y = 4 - 2x$ Subtracting $2x$ from both sides

$\dfrac{1}{3} \cdot 3y = \dfrac{1}{3}(4 - 2x)$ Multiplying both sides by $\dfrac{1}{3}$

$y = \dfrac{4 - 2x}{3}$ Simplifying

or $y = -\dfrac{2}{3}x + \dfrac{4}{3}$

23. $Ax + By = C$

$By = C - Ax$ Subtracting Ax from both sides

$\dfrac{1}{B} \cdot By = \dfrac{1}{B}(C - Ax)$ Multiplying both sides by $\dfrac{1}{B}$

$y = \dfrac{C - Ax}{B}$ Simplifying

25. $C = \dfrac{5}{9}(F - 32)$

$\dfrac{9}{5} \cdot C = \dfrac{9}{5} \cdot \dfrac{5}{9}(F - 32)$ Multiplying both sides by $\dfrac{9}{5}$

$\dfrac{9}{5}C = F - 32$ Simplifying

$\dfrac{9}{5}C + 32 = F$

27. $V = \dfrac{4}{3}\pi r^3$

$\dfrac{3}{4\pi} \cdot V = \dfrac{3}{4\pi} \cdot \dfrac{4}{3}\pi r^3$ Multiplying both sides by $\dfrac{3}{4\pi}$

$\dfrac{3V}{4\pi} = r^3$ Simplifying

29. $np + nm = t$

$n(p + m) = t$ Factoring

$n = \dfrac{t}{p + m}$ Dividing both sides by $p + m$

31. $uv + wv = x$

$v(u + w) = x$ Factoring

$v = \dfrac{x}{u + w}$ Dividing both sides by $u + w$

33. $A = \dfrac{q_1 + q_2 + q_3}{n}$

$n \cdot A = n \cdot \dfrac{q_1 + q_2 + q_3}{n}$ Clearing the fraction

$nA = q_1 + q_2 + q_3$

$nA \cdot \dfrac{1}{A} = (q_1 + q_2 + q_3) \cdot \dfrac{1}{A}$ Multiplying both sides by $\dfrac{1}{A}$

$n = \dfrac{q_1 + q_2 + q_3}{A}$

35. $v = \dfrac{d_2 - d_1}{t}$

$t \cdot v = t \cdot \dfrac{d_2 - d_1}{t}$ Clearing the fraction

$tv = d_2 - d_1$

$tv \cdot \dfrac{1}{v} = (d_2 - d_1) \cdot \dfrac{1}{v}$ Multiplying both sides by $\dfrac{1}{v}$

$t = \dfrac{d_2 - d_1}{v}$

37.
$$v = \frac{d_2 - d_1}{t}$$
$$t \cdot v = t \cdot \frac{d_2 - d_1}{t} \quad \text{Clearing the fraction}$$
$$tv = d_2 - d_1$$
$$tv - d_2 = -d_1 \quad \text{Subtracting } d_2 \text{ from both sides}$$
$$-1 \cdot (tv - d_2) = -1(-d_1) \quad \text{Multiplying both sides by } -1$$
$$-tv + d_2 = d_1,$$
$$\text{or } d_2 - tv = d_1$$

39.
$$bd = c + ba$$
$$bd - ba = c \quad \text{Adding } -ba \text{ to both sides}$$
$$b(d - a) = c \quad \text{Factoring}$$
$$b = \frac{c}{d - a} \quad \text{Dividing both sides by } d - a$$

41.
$$v - w = uvw$$
$$v = uvw + w \quad \text{Adding } w \text{ to both sides}$$
$$v = w(uv + 1) \quad \text{Factoring}$$
$$\frac{v}{uv + 1} = w \quad \text{Dividing both sides by } uv + 1$$

43.
$$n - mk = mt^2$$
$$n = mt^2 + mk \quad \text{Adding } mk \text{ to both sides}$$
$$n = m(t^2 + k) \quad \text{Factoring}$$
$$\frac{n}{t^2 + k} = m \quad \text{Dividing both sides by } t^2 + k$$

45. *Familiarize*. In Example 2, we find the formula for simple interest, $I = Prt$, when I is the interest, P is the principal, r is the interest rate, and t is the time, in years.

***Translate*.** We want to find the interest rate, so we solve the formula for r.
$$I = Prt$$
$$\frac{1}{Pt} \cdot I = \frac{1}{Pt} \cdot Prt$$
$$\frac{I}{Pt} = r$$

***Carry out*.** The model $r = \frac{I}{Pt}$ can be used to find the rate of interest at which an amount (the principal) must be invested in order to earn a given amount. We substitute \$2600 for P, $\frac{1}{2}$ for t (6 months = $\frac{1}{2}$ yr), and \$104 for I.

$$\frac{I}{Pt} = r$$
$$\frac{\$104}{\$2600\left(\frac{1}{2}\right)} = r$$
$$\frac{104}{1300} = r$$
$$0.08 = r$$
$$8\% = r$$

***Check*.** Since $\$2600(0.08)\left(\frac{1}{2}\right) = \$104$, the answer checks.

***State*.** The interest rate must be 8%.

47. *Familiarize*. In the text, we find the formula for the area of a parallelogram, $A = bh$, where b is the base and h is the height.

***Translate*.** We solve the formula for h.
$$A = bh$$
$$\frac{1}{b} \cdot A = \frac{1}{b} \cdot bh$$
$$\frac{A}{b} = h$$

***Carry out*.** The model $h = \frac{A}{b}$ can be used to find the height of any parallelogram for which the area and base are known. We substitute 96 for A and 6 for b.

$$h = \frac{A}{b}$$
$$h = \frac{96}{6}$$
$$h = 16$$

***Check*.** We repeat the calculation. The answer checks.

***State*.** The height is 16 cm.

49. *Familiarize and Translate*. In Example 4, the formula for body mass index is solved for W:
$$W = \frac{IH^2}{704.5}$$

***Carry out*.** We substitute 30.8 for I and 74 for H (6 ft 2 in. is 74 in.) and calculate W.
$$W = \frac{30.8(74)^2}{704.5} \approx 239$$

***Check*.** We could repeat the calculations or substitute in the original formula solving for W. The answer checks.

***State*.** Arnold Schwarzenegger weighs about 239 lb.

51. *Familiarize and Translate*. We will use the model developed in Example 5, $m = \pi r^2 h D$ to find the weight of the salt. Then we will add 28 g, the weight of the empty canister, to find the weight of the filled canister.

***Carry out*.** We substitute 4 for r, 13.6 for h, and 2.16 for D and calculate m.
$$m = \pi r^2 h D$$
$$= \pi(4)^2(13.6)(2.16)$$
$$\approx 1476.6$$

Add the weight of the empty canister:
$$1476.6 + 28 = 1504.6$$

***Check*.** We repeat the calculations. The answer checks.

***State*.** The filled canister weighs about 1504.6 g.

53. *Familiarize*. The formula for the area of a trapezoid is $A = \frac{1}{2}h(b_1 + b_2)$, where A is the area, h is the height, and b_1 and b_2 are the bases.

***Translate*.** The unknown dimension is the height, so we solve the formula for h. We have $h = \frac{2A}{b_1 + b_2}$.

***Carry out*.** We substitute.
$$h = \frac{2A}{b_1 + b_2}$$
$$h = \frac{2 \cdot 90}{8 + 12} = \frac{180}{20}$$
$$h = 9$$

***Check*.** We repeat the calculation. The answer checks.

***State*.** The unknown dimension, the height of the trapezoid, is 9 ft.

55. Observe that 4% of $1000 is $40, so $40 is the amount of simple interest that would be earned in 1 yr. Thus, it will take 1 yr for the investment to be worth $1040.

57. *Familiarize*. We use the formula given in the text,

$$f = \frac{2r+c}{2L}.$$

Translate. The unknown quantity is the pipe length. We solve the formula for L.

$$f = \frac{2r+c}{2L}.$$

$$L = \frac{2r+c}{2f}$$

Carry out. We substitute 40 for b and 7.8 for r and calculate c.

$$L = \frac{2(1)+13,500}{2(27.5)} \approx 246$$

Check. We repeat the calculations. The answer checks.
State. The pipe should be 246 inches long.

59. *Familiarize*. We will use the formula given in the text,

$$R = r + \frac{400(W-L)}{N}.$$

Translate. We solve the formula for r.

$$R = r + \frac{400(W-L)}{N}$$

$$R - \frac{400(W-L)}{N} = r$$

Carry out. We substitute 1305 for R, 5 for w, 3 for L, and $5+3$, or 8, for N and calculate r.

$$1305 - \frac{400(5-3)}{8} = r$$

$$1205 = r$$

Check. We can repeat the calculation or substitute in the original formula and then solve for r. The answer checks.
State. The average rating of Ulana's opponents was 1205.

61. *Familiarize*. We will use the formula given in the text,
$K = 917 + 6(w+h-a)$.
Translate. We solve the formula for h.

$$K = 917 + 6(w+h-a)$$

$$K = 917 + 6w + 6h - 6a$$

$$K - 917 - 6w + 6a = 6h$$

$$\frac{K - 917 - 6w + 6a}{6} = h$$

Carry out. We substitute 1901 for K, 120 for w, and 23 for a and calculate h.

$$\frac{1901 - 917 - 6\cdot120 + 6\cdot23}{6} = h$$

$$67 = h$$

Check. We can repeat the calculation or substitute in the original formula and then solve for h. The answer checks.
State. Julie is 67 in., or 5 ft 7 in., tall.

63. *Familiarize*. We use the formula given in the text.
$g = 0.0778n + 4.55s - 2.2029$.
Translate. We solve the formula for n.

$$g = 0.0778n + 4.55s - 2.2029$$

$$n = \frac{g - 4.55s + 2.2029}{0.0778}$$

Carry out. We substitute 3.0 for g and 1.02 for s and calculate n.

$$n = \frac{3.0 - 4.55(1.02) + 2.2029}{0.0778}$$

$$n \approx 7.22$$

Check. We can repeat the calculations or substitute in the original formula and solve for n. The answer checks.
State. There should be 7.22 words per sentence.

65. *Familiarize*. We use the formula given in the text.

$$S = \frac{HR(W_i - W_n)}{1000}.$$

Translate. We solve the formula for R.

$$S = \frac{HR(W_i - W_n)}{1000}$$

$$R = \frac{1000S}{H(W_i - W_n)}$$

Carry out. We substitute 100 for W_i, 15 for W_n, 2000 for H, and $20.40 for S, and calculate R.

$$R = \frac{1000(20.40)}{2000(100-15)} = 0.12$$

Check. We can repeat the calculations or substitute in the original formula and solve for R. The answer checks.
State. The cost is $0.12 per kWh.

67. *Familiarize*. We use the formula given in the text.

$$r = \frac{tmap}{hs}$$

Translate. We solve the formula for t.

$$r = \frac{tmap}{hs}$$

$$t = \frac{rhs}{map}$$

Carry out. We substitute 30 for s, 4 for h, 0.05 for m, 100 for a, 0.15 for p, 3.2 for r and solve for t.

$$t = \frac{3.2\cdot4\cdot30}{0.05\cdot100\cdot0.15}$$

$$t = 512$$

Check. We repeat the calculation or substitute in the original formula and solve for t. The answer checks.
State. The average daily blog traffic is 512 visits/day.

69. *Familiarize*. We will use Goiten's model,
$I = 1.08(T/N)$. Note that 8 hr $= 8 \times 1$ hr $= 8 \times 60$ min $= 480$ min.
Translate. We solve the formula for N.

$$I = 1.08\left(\frac{T}{N}\right)$$

$$N\cdot I = N(1.08)\left(\frac{T}{N}\right)$$

$$NI = 1.08T$$

$$N = \frac{1.08T}{I}$$

Carry out. We substitute 480 for T and 15 for I.

$$N = \frac{1.08T}{I}$$

$$N = \frac{1.08(480)}{15}$$

$$N = 34.56$$

$$N \approx 34 \quad \text{Rounding down}$$

Check. We repeat the calculations. The answer checks.
State. Dr. Cruz should schedule 34 appointments in one day.

71. Familiarize. We will use Thurnau's model,
$P = 9.337da - 299$.
Translate. Since we want to find the diameter of the fetus' head, we solve for d.
$$P = 9.337da - 299$$
$$P + 299 = 9.337da$$
$$\frac{P + 299}{9.337a} = d$$
Carry out. Substitute 1614 for P and 24.1 for a in the formula and calculate:
$$\frac{1614 + 299}{9.337(24.1)} = d$$
$$8.5 \approx d$$
Check. We repeat the calculation. The answer checks.
State. The diameter of the fetus' head at 29 weeks is about 8.5 cm.

73. Writing Exercise.

75. Writing Exercise.

77. Familiarize. First we find the volume of the ring. Note that the inner diameter is 2 cm, so the inner radius is 2/2 or 1 cm. Then the volume of the ring is the volume of a right circular cylinder with height 0.5 cm and radius 1 + 0.15, or 1.15 cm, less the volume of a right circular cylinder with height 0.5 cm and radius 1 cm. Recall that the formula for the volume of a right circular cylinder is $V = \pi r^2 h$. Then the volume of the ring is
$$\pi(1.15)^2(0.5) - \pi(1)^2(0.5) = 0.16125\pi \text{ cm}^3.$$
Translate. To find the weight of the ring we will use the formula $D = \frac{m}{V}$. Solving for m, we get
$$D = \frac{m}{V}$$
$$V \cdot D = V \cdot \frac{m}{V}$$
$$V \cdot D = m$$
Carry out. We substitute in the formula $m = V \cdot D$
$$m = 0.16125\pi(21.5)$$
$$m \approx 10.9$$
Check. We repeat the calculations. The answer checks.
State The ring will weigh about 10.9 g.

79. Writing Exercise.

81.
$$s = v_i t + \frac{1}{2}at^2$$
$$s - v_i t = \frac{1}{2}at^2$$
$$2(s - v_i t) = at^2$$
$$\frac{2(s - v_i t)}{t^2} = a, \text{ or}$$
$$\frac{2s - 2v_i t}{t^2} = a$$

83.
$$b = \frac{h + w + p}{a + w + p + f}$$
$$b(a + w + p + f) = h + w + p \quad \text{Multiplying both sides by } a+w+p+f$$
$$ab + bw + bp + bf = h + w + p$$
$$bw - w = h + p - ab - bp - bf$$
$$w(b-1) = h + p - b(a + p + f)$$
$$w = \frac{h + p - b(a + p + f)}{b - 1} \quad \text{Multiplying both sides by } \frac{1}{b-1}$$

85.
$$\frac{b}{a - b} = c$$
$$b = c(a - b)$$
$$b + bc = ac$$
$$b(1 + c) = ac$$
$$b = \frac{ac}{1 + c}$$

87. $s + \dfrac{s+t}{s-t} = \dfrac{1}{t} + \dfrac{s+t}{s-t}$

Observe that if we subtract $\dfrac{s+t}{s-t}$ from both sides we are left with an equivalent equation, $s = \dfrac{1}{t}$. We solve this equation for t.
$$s = \frac{1}{t}$$
$$st = 1 \quad \text{Multiplying both sides by } t$$
$$t = \frac{1}{s} \quad \text{Multiplying both sides by } \frac{1}{s}$$

Exercise Set 1.6

1. The power rule

3. Raising a product to a power

5. The product rule

7. Raising a quotient to a power

9. The quotient rule

11. $6^4 \cdot 6^7 = 6^{4+7} = 6^{11}$

13. $m^0 \cdot m^8 = m^{0+8} = m^8$

15. $5x^4 \cdot 4x^3 = 5 \cdot 4 \cdot x^4 \cdot x^3 = 20x^{4+3} = 20x^7$

17. $(-3a^2)(-8a^6) = (-3)(-8)a^2 \cdot a^6 = 24a^{2+6} = 24a^8$

19. $(m^5n^2)(m^3np^0) = (m^5m^3)(n^2n)(p^0) = m^{5+3}n^{2+1} \cdot 1$
$$= m^8n^3$$

21. $\dfrac{t^8}{t^3} = t^{8-3} = t^5$

23. $\dfrac{15a^7}{3a^2} = \dfrac{15}{3}a^{7-2} = 5a^5$

25. $\dfrac{m^7 n^9}{m^2 n^5} = m^{7-2} \cdot n^{9-5} = m^5 n^4$

27. $\dfrac{32 x^8 y^5}{8 x^2 y} = \dfrac{32}{8} \cdot x^{8-2} \cdot y^{5-1} = 4 x^6 y^4$

29. $\dfrac{28 x^{10} y^9 z^8}{-7 x^2 y^3 z^2} = \dfrac{28}{-7} \cdot x^{10-2} \cdot y^{9-3} \cdot z^{8-2} = -4 x^8 y^6 z^6$

31. $-x^0 = -(-2)^0 = -(1) = -1$

33. $(4x)^0 = (4(-2))^0 = (-8)^0 = 1$

35. $t^{-9} = \dfrac{1}{t^9}$

37. $6^{-2} = \dfrac{1}{6^2} = \dfrac{1}{36}$

39. $(-3)^{-2} = \dfrac{1}{(-3)^2} = \dfrac{1}{9}$

41. $-3^{-2} = -\dfrac{1}{3^2} = -\dfrac{1}{9}$

43. $-1^{-10} = -\dfrac{1}{1^{10}} = -1$

45. $\dfrac{1}{10^{-3}} = 10^3 = 1000$

47. $6 x^{-1} = 6 \cdot \dfrac{1}{x} = \dfrac{6}{x}$

49. $3 a^8 b^{-6} = 3 a^8 \cdot \dfrac{1}{b^6} = \dfrac{3 a^8}{b^6}$

51. $\dfrac{2 z^{-3}}{x^5} = \dfrac{2}{x^5} \cdot \dfrac{1}{z^3} = \dfrac{2}{x^5 z^3}$

53. $\dfrac{3 y^2}{z^{-4}} = 3 y^2 \cdot z^4 = 3 y^2 z^4$

55. $\dfrac{a b^{-1}}{c^{-1}} = a \cdot \dfrac{1}{b} \cdot c = \dfrac{ac}{b}$

57. $\dfrac{p q^{-2} r^{-3}}{2 u^5 v^{-4}} = \dfrac{p}{2 u^5} \cdot \dfrac{1}{q^2} \cdot \dfrac{1}{r^3} \cdot v^4 = \dfrac{p v^4}{2 q^2 r^3 u^5}$

59. $\dfrac{1}{x^3} = x^{-3}$

61. $\dfrac{1}{(-10)^3} = (-10)^{-3}$

63. $8^{10} = \dfrac{1}{8^{-10}}$

65. $4 x^2 = 4 \cdot \dfrac{1}{x^{-2}} = \dfrac{4}{x^{-2}}$

67. $\dfrac{1}{(5y)^3} = (5y)^{-3}$

69. $\dfrac{1}{3 y^4} = \dfrac{1}{3} \cdot \dfrac{1}{y^4} = \dfrac{1}{3} \cdot y^{-4} = \dfrac{y^{-4}}{3}$

71. $6^{-3} \cdot 6^{-5} = 6^{-3+(-5)} = 6^{-8}$, or $\dfrac{1}{6^8}$

73. $a \cdot a^{-8} = a^{1+(-8)} = a^{-7}$, or $\dfrac{1}{a^7}$

75. $x^{-7} \cdot x^2 \cdot x^5 = x^{-7+2+5} = x^0 = 1$

77. $(4 m n^3)(-2 m^3 n^2) = 4(-2) \cdot m \cdot m^3 \cdot n^3 \cdot n^2$
$= -8 m^{1+3} n^{3+2}$
$= -8 m^4 n^5$

79. $(-7 x^4 y^{-5})(-5 x^{-6} y^8) = (-7)(-5) \cdot x^4 \cdot x^{-6} \cdot y^{-5} \cdot y^8$
$= 35 x^{4+(-6)} y^{-5+8}$
$= 35 x^{-2} y^3$, or $\dfrac{35 y^3}{x^2}$

81. $(5 a^{-2} b^{-3})(2 a^{-4} b) = 5 \cdot 2 \cdot a^{-2} \cdot a^{-4} \cdot b^{-3} \cdot b$
$= 10 a^{-2+(-4)} b^{-3+1}$
$= 10 a^{-6} b^{-2}$, or $\dfrac{10}{a^6 b^2}$

83. $\dfrac{10^{-3}}{10^6} = 10^{-3-6} = 10^{-9}$, or $\dfrac{1}{10^9}$

85. $\dfrac{2^{-7}}{2^{-5}} = 2^{-7-(-5)} = 2^{-7+5} = 2^{-2}$, or $\dfrac{1}{2^2}$, or $\dfrac{1}{4}$

87. $\dfrac{y^4}{y^{-5}} = y^{4-(-5)} = y^{4+5} = y^9$

89. $\dfrac{24 a^5 b^3}{-8 a^4 b} = \dfrac{24}{-8} a^{5-4} b^{3-1} = -3 a b^2$

91. $\dfrac{15 m^5 n^3}{10 m^{10} n^{-4}} = \dfrac{15}{10} m^{5-10} n^{3-(-4)} = \dfrac{3}{2} m^{-5} n^7$, or $\dfrac{3 n^7}{2 m^5}$

93. $\dfrac{-6 x^{-2} y^4 z^8}{-24 x^{-5} y^6 z^{-3}} = \dfrac{-6}{-24} x^{-2-(-5)} y^{4-6} z^{8-(-3)}$
$= \dfrac{1}{4} x^3 y^{-2} z^{11}$, or $\dfrac{x^3 z^{11}}{4 y^2}$

95. $(x^4)^3 = x^{4 \cdot 3} = x^{12}$

97. $(9^3)^{-4} = 9^{3(-4)} = 9^{-12}$, or $\dfrac{1}{9^{12}}$

99. $(t^{-8})^{-5} = t^{-8(-5)} = t^{40}$

101. $(-5 x y)^2 = (-5)^2 x^2 y^2 = 25 x^2 y^2$

103. $\left(-2a^{-2}b\right)^{-3} = \left(-2\right)^{-3} \cdot \left(a^{-2}\right)^{-3} \cdot b^{-3} = \left(-2\right)^{-3} a^{6}b^{-3}$

$\qquad = \dfrac{1}{\left(-2\right)^{3}} \cdot a^{6} \cdot \dfrac{1}{b^{3}} = -\dfrac{a^{6}}{8b^{3}}$

105. $\left(\dfrac{m^{2}n^{-1}}{4}\right)^{3} = \dfrac{m^{6}n^{-3}}{4^{3}} = \dfrac{m^{6}n^{-3}}{64}$, or $\dfrac{m^{6}}{64n^{3}}$

107. $\dfrac{\left(2a^{3}\right)^{3}4a^{-3}}{\left(a^{2}\right)^{5}} = \dfrac{2^{3}a^{3\cdot3}4a^{-3}}{a^{2\cdot5}} = \dfrac{8a^{9}4a^{-3}}{a^{10}}$

$\qquad = 8 \cdot 4a^{9+(-3)-10} = 32a^{-4}$, or $\dfrac{32}{a^{4}}$

109. $\left(8x^{-3}y^{2}\right)^{-4}\left(8x^{-3}y^{2}\right)^{4} = \left(8x^{-3}y^{2}\right)^{-4+4} = \left(8x^{-3}y^{2}\right)^{0} = 1$

111. $\dfrac{\left(5a^{3}b\right)^{2}}{10a^{2}b} = \dfrac{5^{2}\left(a^{3}\right)^{2}b^{2}}{10a^{2}b} = \dfrac{25a^{6}b^{2}}{10a^{2}b}$

$\qquad = \dfrac{25}{10}a^{6-2}b^{2-1} = \dfrac{5}{2}a^{4}b$, or $\dfrac{5a^{4}b}{2}$

113. $\left(\dfrac{2x^{3}y^{-2}}{3y^{-3}}\right)^{3} = \left(\dfrac{2}{3}x^{3}y^{-2+3}\right)^{3} = \left(\dfrac{2}{3}x^{3}y\right)^{3} = \dfrac{2^{3}}{3^{3}}\left(x^{3}\right)^{3}y^{3}$

$\qquad = \dfrac{8x^{9}y^{3}}{27}$

115. $\left(\dfrac{21x^{5}y^{-7}}{14x^{-2}y^{-6}}\right)^{0} = 1$

(Any nonzero real number raised to the zero power is 1.)

117. $\left(\dfrac{5x^{0}y^{-7}}{2x^{-2}y^{4}}\right)^{-2} = \left(\dfrac{5}{2}x^{0+2}y^{-7-4}\right)^{-2} = \left(\dfrac{5x^{2}y^{-11}}{2}\right)^{-2}$

$\qquad = \left(\dfrac{2}{5x^{2}y^{-11}}\right)^{2} = \dfrac{2^{2}}{5^{2}\left(x^{2}\right)^{2}\left(y^{-11}\right)^{2}}$

$\qquad = \dfrac{4}{25x^{4}y^{-22}} = \dfrac{4}{25}x^{-4}y^{22}$, or $\dfrac{4y^{22}}{25x^{4}}$

119. *Writing Exercise.*

121. *Writing Exercise.*

123. $\dfrac{8a^{x-2}}{2a^{2x+2}} = \dfrac{8}{2} \cdot a^{x-2-(2x+2)} = 4a^{x-2-2x-2} = 4a^{-x-4}$

125. $\left\{\left[\left(8^{-a}\right)^{-2}\right]^{b}\right\}^{-c} \cdot \left[\left(8^{0}\right)^{a}\right]^{c} = 8^{-2abc} \cdot 8^{0} = 8^{-2abc}$

127. $\dfrac{-28x^{b+5}y^{4+c}}{7x^{b-5}y^{c-4}} = -4x^{b+5-(b-5)}y^{4+c-(c-4)}$

$\qquad = -4x^{b+5-b+5}y^{4+c-c+4} = -4x^{10}y^{8}$

129. $\dfrac{3^{q+3} - 3^{2}\left(3^{q}\right)}{3\left(3^{q+4}\right)} = \dfrac{3^{q+3} - 3^{q+2}}{3^{q+5}} = \dfrac{3^{q+2}(3-1)}{3^{q+2}\left(3^{3}\right)} = \dfrac{2}{3^{3}} = \dfrac{2}{27}$

131. $\left[\left(\dfrac{a^{-2c}}{b^{7c}}\right)^{-3}\left(\dfrac{a^{4c}}{b^{-3c}}\right)^{2}\right]^{-a} = \left(\dfrac{a^{6c}}{b^{-21c}} \cdot \dfrac{a^{8c}}{b^{-6c}}\right)^{-a} = \left(\dfrac{a^{14c}}{b^{-27c}}\right)^{-a}$

$\qquad = \dfrac{a^{-14ac}}{b^{27ac}}$

Exercise Set 1.7

1. The number 27×10^{16} *is not* written in scientific notation.

3. The number 4.587×10^{5} has *four* significant digits.

5. The length of an Olympic marathon, in centimeters, is a large number so its representation in scientific notation would include a positive power of 10.

7. The mass of a hydrogen atom, in grams, is a small number so its representation in scientific notation would include a negative power of 10.

9. The time between leap years, in seconds, is a large number so its representation in scientific notation would include a positive power of 10.

11. $64,000,000,000 = \dfrac{64,000,000,000}{10^{10}} \cdot 10^{10}$

$\qquad$ Multiplying by 1 $\left(10^{10}/10^{10} = 1\right)$

$\qquad = 6.4 \times 10^{10}$ This is scientific notation.

13. $0.0000013 = \dfrac{0.0000013}{10^{6}} \cdot 10^{6}$ $\quad$ Multiplying by 1: $\quad 10^{6}/10^{6} = 1$

$\qquad = \dfrac{1.3}{10^{6}}$

$\qquad = 1.3 \times 10^{-6}$ $\quad$ This is scientific notation.

15. $0.00009 = \dfrac{0.0009}{10^{5}} \cdot 10^{5}$ $\quad$ Multiplying by 1: $\quad 10^{5}/10^{5} = 1$

$\qquad = \dfrac{9}{10^{5}}$

$\qquad = 9 \times 10^{-5}$ $\quad$ This is scientific notation.

17. $803,000,000,000 = \dfrac{803,000,000,000}{10^{11}} \cdot 10^{11}$

$\qquad = 8.03 \times 10^{11}$

19. $0.000000904 = \dfrac{0.000000904}{10^{7}} \cdot 10^{7}$

$\qquad = \dfrac{9.04}{10^{7}}$

$\qquad = 9.04 \times 10^{-7}$

21. $431,700,000,000 = \dfrac{431,700,000,000}{10^{11}} \cdot 10^{11}$

$\qquad = 4.317 \times 10^{11}$

23. $4 \times 10^{5} = 400,000$ $\quad$ Moving the decimal point 5 places to the right.

25. $1.2 \times 10^{-4} = 0.00012$ $\quad$ Moving the decimal point 4 places to the left.

27. $3.76 \times 10^{-9} = 0.00000000376$ Moving the decimal point 9 places to the left.

29. $8.056 \times 10^{12} = 8,056,000,000,000$ Moving the decimal point 12 places to the right.

31. $7.001 \times 10^{-5} = 0.00007001$ Moving the decimal point 5 places to the left.

33. $(3.4 \times 10^{-8})(2.6 \times 10^{15})$
$= (3.4 \times 2.6)(10^{-8} \times 10^{15})$
$= 8.84 \times 10^{7}$
$= 8.8 \times 10^{7}$ Rounding to 2 significant digits

35. $(2.36 \times 10^{6})(1.4 \times 10^{-11})$
$= (2.36 \times 1.4)(10^{6} \times 10^{-11})$
$= 3.304 \times 10^{-5}$
$= 3.3 \times 10^{-5}$ Rounding to 2 significant digits

37. $(5.2 \times 10^{6})(2.6 \times 10^{4}) = (5.2 \times 2.6)(10^{6} \times 10^{4})$
$= 13.52 \times 10^{10}$
$= (1.352 \times 10) \times 10^{10}$
$= 1.352 \times (10 \times 10^{10})$
$= 1.352 \times 10^{11}$
$= 1.4 \times 10^{11}$ (2 significant digits)

39. $(7.01 \times 10^{-5})(6.5 \times 10^{-7})$
$= (7.01 \times 6.5)(10^{-5} \times 10^{-7})$
$= 45.565 \times 10^{-12}$
$= (4.5565 \times 10) \times 10^{-12}$
$= 4.5565 \times 10^{-11}$
$= 4.6 \times 10^{-11}$ (2 significant digits)

41. $(2.0 \times 10^{6})(3.02 \times 10^{-6})$

Observe that $10^{6} \times 10^{-6} = 1$, so the product is $2.0(3.02)$, or 6.04, or 6.0 rounded to two significant digits.

43. $\dfrac{6.5 \times 10^{15}}{2.6 \times 10^{4}} = \dfrac{6.5}{2.6} \times \dfrac{10^{15}}{10^{4}}$
$= 2.5 \times 10^{11}$ (2 significant digits)

45. $\dfrac{9.4 \times 10^{-9}}{4.7 \times 10^{-2}} = \dfrac{9.4}{4.7} \times \dfrac{10^{-9}}{10^{-2}}$
$= 2.0 \times 10^{-7}$ (2 significant digits)

47. $\dfrac{3.2 \times 10^{-7}}{8.0 \times 10^{8}} = \dfrac{3.2}{8.0} \times \dfrac{10^{-7}}{10^{8}}$
$= 0.40 \times 10^{-15}$ (2 significant digits)
$= (4.0 \times 10^{-1}) \times 10^{-15}$
$= 4.0 \times 10^{-16}$

49. $\dfrac{9.36 \times 10^{-11}}{3.12 \times 10^{11}} = \dfrac{9.36}{3.12} \times \dfrac{10^{-11}}{10^{11}}$
$= 3.00 \times 10^{-22}$ (3 significant digits)

51. $\dfrac{6.12 \times 10^{19}}{3.06 \times 10^{-7}} = \dfrac{6.12}{3.06} \times \dfrac{10^{19}}{10^{-7}}$
$= 2.00 \times 10^{26}$ (3 significant digits)

53. *Familiarize*. Let s = the number of stars. We have 8 trillion cubic light-years.
Translate. We multiply.
$$s = \dfrac{0.025 \text{ star}}{\text{light-year}^{3}} \cdot 8,000,000,000,000 \text{ light-years}^{3}$$
Carry out.
$$s = \dfrac{0.025 \text{ star}}{\text{light-year}^{3}} \cdot 8,000,000,000,000 \text{ light-years}^{3}$$
$= (2.5 \times 10^{-2})(8 \times 10^{12}) \text{ stars}$
$= 2 \times 10^{11} \text{ stars}$
Check. Recheck the translation and calculations. The answer checks.
State. There are 2×10^{11} stars in the Milky Way.

55. *Familiarize*. We have a cylinder with diameter 4.0×10^{-10} in. and length 100 yd. We will use the formula for the volume of a cylinder $V = \pi r^{2} h$. The radius is $\dfrac{4.0 \times 10^{-10}}{2}$, or 2.0×10^{-10} in. We convert 100 yd to inches:
$100 \text{ yd} = 100 \times 1 \text{ yd} = 100 \times 36 \text{ in.} = 3600 \text{ in.,}$ or 3.6×10^{3} in.
Translate. We substitute in the formula.
$V = \pi r^{2} h$
$V = \pi (2.0 \times 10^{-10})^{2}(3.6 \times 10^{3})$
Carry out. We do the calculation.
$V = \pi (2.0 \times 10^{-10})^{2}(3.6 \times 10^{3})$
$= \pi \times 4.0 \times 10^{-20} \times 3.6 \times 10^{3}$
$= (\pi \times 4.0 \times 3.6) \times (10^{-20} \times 10^{3})$
$\approx 45.2 \times 10^{-17}$
$\approx (4.52 \times 10) \times 10^{-17}$
$\approx 4.52 \times 10^{-16}$
$\approx 4.5 \times 10^{-16}$ (2 significant digits)
Check. Recheck the translation and the calculations. The answer checks.
State. The volume of a 100-yd carbon nanotube is about 4.5×10^{-16} in^{3}.

57. *Familiarize*. Let p = the number of pages of information per person. Recall 1 exabyte is 10^{15} kilobytes. We convert from exabytes to kilobytes:
2.5 exabytes $= 2.5 \times 1$ exabytes
$= 2.5 \times 10^{15}$ kilobytes
$= 2.5 \times 10^{15}$ kilobytes
Translate. We divide.
$$p = \dfrac{2.5 \times 10^{15} \text{ kilobytes}}{7.1 \text{ billion people}} \cdot \dfrac{1 \text{ page}}{2 \text{ kilobytes}}$$
Carry out. We perform the calculations and write scientific notation for the answer.

$$p = \frac{2.5 \times 10^{15} \text{ kilobytes}}{7.1 \text{ billion people}} \cdot \frac{1 \text{ page}}{2 \text{ kilobytes}}$$

$$= \frac{2.5}{7.1} \times \frac{10^{15}}{10^9} \cdot \frac{1}{2} \frac{\text{pages}}{\text{person}}$$

$$\approx 1.76 \times 10^5 \text{ pages/person}$$

$$= 1.8 \times 10^5 \quad \text{(2 significant digits)}$$

Check. Recheck the translation and calculation. The answer checks.

State. Each person generates an average of 1.8×10^5 pages of information.

59. Familiarize. Let w = the weight of each sheet. 1 ream = 500 sheets.
Translate. We divide.

$$w = \frac{2.25}{500}$$

Carry out.

$$w = \frac{2.25}{5 \times 10^2}$$

$$= 0.450 \times 10^{-2}$$

$$= 4.50 \times 10^{-3}$$

Check. Recheck the translation and calculations. The answer checks.
State. Each sheet of copier paper weighs 4.50×10^{-3} kg, or 4.50 g.

61. Familiarize. We know that 1 light year = 5.88×10^{12} mi. Let y = the number of light years from one and of the Milky Way galaxy to the other.
Translate. The distance is $\left(5.88 \times 10^{12}\right) y$ mi. It is also given by 5.88×10^{17} mi. We write the equation:

$$\left(5.88 \times 10^{12}\right) y = 5.88 \times 10^{17}$$

Carry out. We solve the equation.

$$\left(5.88 \times 10^{12}\right) y = 5.88 \times 10^{17}$$

$$y = \frac{5.88 \times 10^{17}}{5.88 \times 10^{12}}$$

$$y = \frac{5.88}{5.88} \times \frac{10^{17}}{10^{12}}$$

$$y = 1.00 \times 10^5$$

Check. Since light travels 5.88×10^{12} mi in one light year, in 1.00×10^5 yr it will travel $\left(1.00 \times 10^5\right) \times \left(5.88 \times 10^{12}\right) = 5.88 \times 10^{17}$ mi. The answer checks.
State. The distance from one end of the galaxy to the other is 1.00×10^5 light years.

63. Familiarize. We are told that 1 Angstrom $= 1 \times 10^{-10}$ m, 1 parsec ≈ 3.26 light years, and 1 light year $= 9.46 \times 10^{15}$ m. Let a represent the number of Angstroms in one parsec.
Translate The length of one parsec is $a \times 10^{-10}$ m. It can also be expressed as 3.26 light years, or $3.26 \times 9.46 \times 10^{15}$ m. Since these quantities represent

the same number, we can write the equation.

$$a \times 10^{-10} = 3.26 \times 9.46 \times 10^{15}.$$

Carry out. Solve the equation:

$$a \times 10^{-10} = 3.26 \times 9.46 \times 10^{15}$$

$$a \times 10^{-10} \times \frac{1}{10^{-10}} = 3.26 \times 9.46 \times 10^{15} \times \frac{1}{10^{-10}}$$

$$a = \frac{3.26 \times 9.46 \times 10^{15}}{10^{-10}}$$

$$= (3.26 \times 9.46) \times \frac{10^{15}}{10^{-10}}$$

$$= 30.8396 \times 10^{25}$$

$$= (3.08396 \times 10) \times 10^{25}$$

$$= 3.08396 \times \left(10 \times 10^{25}\right)$$

$$= 3.08 \times 10^{26} \quad \text{(Rounding to 3 significant digits)}$$

Check. We recheck the translation and calculation.
State. There are about 3.08×10^{26} Angstroms in one parsec.

65. Familiarize. We have a very long cylinder. Its length is the average distance from Earth to the sun, 1.5×10^{11} m, and the diameter of its base is 3 Å. We will use the formula for the volume of a cylinder, $V = \pi r^2 h$ (See Example 8.)
Translate. We will express all distances in Angstroms.

Height (length): 1.5×10^{11} m $= \dfrac{1.5 \times 10^{11}}{10^{-10}}$ Å, or

$$1.5 \times 10^{21} \text{ Å}$$

Diameter: 3 Å
The radius is half the diameter:
Radius: $\dfrac{1}{2} \times 3$ Å $= 1.5$ Å

Now substitute into the formula (using 3.14 for π):

$$V = \pi r^2 h$$

$$V = 3.14 \times 1.5^2 \times 1.5 \times 10^{21}$$

Carry out. Do the calculations.

$$V = 3.14 \times 1.5^2 \times 1.5 \times 10^{21}$$

$$= 10.5975 \times 10^{21}$$

$$= 1.05975 \times 10^{22}$$

$$= 1 \times 10^{22} \quad \text{Rounding to 1 significant digit}$$

We can convert this result to cubic meters, if desired.

1 Å $= 10^{-10}$ m, So 1 cu Å $= \left(10^{-10}\right)^3$ m³ $= 10^{-30}$ m³.

Then 1×10^{22} cu Å $= 1 \times 10^{22}$ cu Å $\times \dfrac{10^{-30} \text{ m}^3}{1 \text{ cu Å}}$

$$= 1 \times 10^{22} \times 10^{-30} \times \frac{\text{cu Å}}{\text{cu Å}} \times \text{m}^3 = 1 \times 10^{-8} \text{ m}^3.$$

Check. We recheck the translation and the calculations.
State. The volume of the sunbeam is about 1×10^{22} cu Å, or 1×10^{-8} m³.

67. Familiarize. First we will find d, the number of drops in a pound. Then we will find b, the number of bacteria in a drop of U.S. mud.
Translate. To find d we convert 1 pound to drops:

$$d = 1 \text{ lb} \cdot \frac{16 \text{ oz}}{1 \text{ lb}} \cdot \frac{6.0 \text{ tsp}}{1 \text{ oz}} \cdot \frac{60.0 \text{ drops}}{1 \text{ tsp}}.$$

Then we divide to find b:

$$b = \frac{4.55 \times 10^{11}}{d}.$$

Carry out. We do the calculations.

$$d = 1 \text{ lb} \cdot \frac{16 \text{ oz}}{1 \text{ lb}} \cdot \frac{6.0 \text{ tsp}}{1 \text{ oz}} \cdot \frac{60.0 \text{ drops}}{1 \text{ tsp}}$$
$$= 5760 \text{ drops}$$

Now we find b.

$$b = \frac{4.55 \times 10^{11}}{d} = \frac{4.55 \times 10^{11}}{5.760 \times 10^{3}} \approx 0.790 \times 10^{8}$$
$$\approx (7.90 \times 10^{-1}) \times 10^{8} = 7.9 \times 10^{7}$$

(Our answer must have 2 significant digits.)

Check. If there are about 7.9×10^{7} bacteria in a drop of U.S. mud, then in a pound there are about

$$\frac{7.9 \times 10^{7}}{1 \text{ drop}} \cdot \frac{60.0 \text{ drops}}{1 \text{ tsp}} \cdot \frac{6.0 \text{ tsp}}{1 \text{ oz}} \cdot \frac{16 \text{ oz}}{1 \text{ lb}}$$
$$= \frac{45,504 \times 10^{7}}{1 \text{ lb}} \approx 4.55 \times 10^{11} \text{ bacteria per pound.}$$

The answer checks.

State. About 7.9×10^{7} bacteria live in a drop of U.S. mud.

69. Familiarize. First we will find the distance C around Jupiter at the equator, in km. Then we will use the formula Speed $\times$ Time = Distance to find the speed s at which Jupiter's equator is spinning.

Translate. We will use the formula for the circumference of a circle to find the distance around Jupiter at the equator:

$$C = \pi d = \pi (1.43 \times 10^{5}).$$

Then we find the speed s at which Jupiter's equator is spinning:

Speed	$\times$	Time	$=$	Distance
$\downarrow$	$\downarrow$	$\downarrow$	$\downarrow$	$\downarrow$
s	$\times$	10	$=$	C

Carry out. First we find C.

$$C = \pi(1.43 \times 10^{5}) \approx 4.49 \times 10^{5}$$

Then we find s.

$$s \times 10 = C$$
$$s \times 10 = 4.49 \times 10^{5}$$
$$s = \frac{4.49 \times 10^{5}}{10}$$
$$s = 4.49 \times 10^{4}$$

Check. At 4.49×10^{4} km/h. in 10 hr, Jupiter's equator travels $4.49 \times 10^{4} \times 10$, or 4.49×10^{5} km. A circle with circumference 4.49×10^{5} km has a diameter of

$$\frac{4.49 \times 10^{4}}{\pi} \approx 1.43 \times 10^{5} \text{ km.}$$ The answer checks.

State. Jupiter's equator spins at a speed of about 4.49×10^{4} km/h.

71. Writing Exercise.

73. Writing Exercise.

75. Familiarize. Let d = the average density, v = the volume and m = the mass.
Translate. We divide.

$$d = \frac{m}{v}$$

To convert the volume from km^{3} to cm^{3},

$$v = 1.08 \times 10^{12} \text{ km}^{3} \cdot \left(\frac{1000 \text{ m}}{1 \text{ km}}\right)^{3} \cdot \left(\frac{100 \text{ cm}}{1 \text{ m}}\right)^{3}$$

To convert the mass from kg to g

$$m = 5.976 \times 10^{24} \text{ kg} \cdot \frac{1000 \text{ g}}{1 \text{ kg}}$$

Carry out. We do the calculations.

$$v = 1.08 \times 10^{12} \cdot (1000)^{3} \cdot (100)^{3}$$
$$= 1.08 \times 10^{27}$$
$$m = 5.976 \times 10^{24} \cdot 1000$$
$$= 5.976 \times 10^{27}$$
$$d = \frac{m}{v} = \frac{5.976 \times 10^{27}}{1.08 \times 10^{27}}$$
$$= \frac{5.976}{1.08} \times \frac{10^{27}}{10^{27}}$$
$$= 5.53 \quad \text{(3 significant digits)}$$

Check. We recheck the translation and calculations.
State. The average density of Earth is approximately 5.53 g/cm^{3}.

77. Familiarize. Let r = the radius, v = the speed, and t = the time.
Translate. Convert the time 1 year to hours.

$$1 \text{ year} = 365 \text{ days} = 365 \times 24 \text{ hr} = 8.76 \times 10^{3} \text{ hr}$$

Recall that 1 meter $\approx 6.21 \times 10^{-4}$ miles. Convert the radius from meters to miles.

$$1.5 \times 10^{11} \text{ m} = 1.5 \times 10^{11} \times 6.21 \times 10^{-4} \text{ mi}$$
$$= 9.315 \times 10^{7} \text{ mi}$$

To find the orbital speed, divide circumference by time.

$$v = \frac{C}{t} = \frac{2\pi r}{t}$$

Carry out. We do the calculations.

$$v = \frac{2\pi(9.315 \times 10^{7})}{8.76 \times 10^{3}}$$
$$v \approx 6.7 \times 10^{4} \quad \text{(2 significant digits)}$$

Check. We recheck the translation and calculations.
State. Earth's orbital speed around the sun is approximately 6.7×10^{4} mph.

79. The larger number is the one in which the power of ten has the larger exponent. Since −90 is larger than −91, 8×10^{-90} is larger than 9×10^{-91}.

$$8 \times 10^{-90} - 9 \times 10^{-91} = 10^{-90}(8 - 9 \times 10^{-1})$$
$$= 10^{-90}(8 - 0.9)$$
$$= 7.1 \times 10^{-90}$$

Thus, 8×10^{-90} is larger by 7.1×10^{-90}.

81. $(4096)^{0.05}(4096)^{0.2} = 4096^{0.25}$

$$= \left(2^{12}\right)^{0.25}$$
$$= 2^3$$
$$= 8$$

83. **_Familiarize._** Observe that there are 2^{n-1} grains of sand on the nth square of the chessboard. Let g represent this quantity. Recall that a chessboard has 64 squares. Note also that $2^{10} \approx 10^3$.

Translate. We write the equation

$$g = 2^{n-1}.$$

To find the number of grains of sand on the last (or 64th) square, substitute 64 for n: $g = 2^{64-1}$.

Carry out. Recheck the translation and the calculations.

$$g = 2^{64-1} = 2^{63} = 2^3\left(2^{10}\right)^6$$
$$\approx 2^3\left(10^3\right)^6 \approx 8 \times 10^{18}$$

State. Approximately 8×10^{18} grains of sand are required for the last square.

85. Answers will vary.

Chapter 1 Review

1. The correct choice is (e), since the equation $2x - 1 = 9$ is equivalent to $2x = 10$.

2. The correct choice is (g), since the expression $2x - 1$ is equivalent to $5x - 1 - 3x$.

3. The correct choice is (j), since the equation $\frac{3}{4}x = 5$ is equivalent to $\frac{4}{3} \cdot \frac{3}{4}x = \frac{4}{3} \cdot 5$.

4. The correct choice is (a) since the expression $\frac{3}{4}x - 5$ is equivalent to $2 + \frac{3}{4}x - 7$.

5. The correct choice is (i), since the expression $2(x + 7)$ is equivalent to $2x + 14$.

6. The correct choice is (b), since the equation $2(x + 7) = 6$ is equivalent to $2x + 14 = 6$.

7. The correct choice is (f), since the equation $4x - 3 + 2x = 5$ is equivalent to $6x - 3 = 5$.

8. The correct choice is (c), since the expression $4x - 3 + 2x$ is equivalent to $6x - 3$.

9. The correct choice is (d), since the expression $6 + 2x$ is equivalent to $2(3 + x)$.

10. The correct choice is (h), since the equation $6 = 2x$ is equivalent to $3 = x$.

11. Eight less than the quotient of two numbers
Let x and y represent the numbers.

Then we have $\frac{x}{y} - 8$.

12. Substitute and carry out the operations.
$$7x^2 - 5y \div zx = 7(-2)^2 - 5(3) \div (-5)(-2)$$
$$= 7 \cdot 4 - 15 \div (-5) \cdot (-2)$$
$$= 28 + 3 \cdot (-2)$$
$$= 28 - 6$$
$$= 22$$

13. Name the set consisting of the first five odd natural numbers
$\{1, 3, 5, 7, 9\}$
or $\{x \mid x \text{ is an odd natural number less than } 10\}$

14. $A = \frac{1}{2}(50)(70) = 1750 \text{ cm}^2$

15. $|-19| = 19$ -19 is 19 units from 0

16. $|0| = 0$ 0 is 0 units from itself

17. $|6.08| = 6.08$ 6.08 is 6.08 units from 0

18. $-2.3 + (-8.7)$
Two negative numbers: Add the absolute values, getting 11. The answer is negative, -11.

19. $-\frac{3}{4} - \left(-\frac{4}{5}\right) = -\frac{3}{4} + \frac{4}{5} = -\frac{15}{20} + \frac{16}{20} = \frac{1}{20}$

20. $10 + (-5.6) = 4.4$

21. $12.3 - 16.1 = 12.3 + (-16.1) = -3.8$

22. $(-12)(-8)$
Two numbers with the same sign: Multiply their absolute values, getting 96. The answer is positive, 96.

23. $\left(-\frac{2}{3}\right)\left(\frac{5}{8}\right)$
Two numbers with unlike signs: Multiply their absolute values to get $\frac{10}{24}$, or $\frac{5}{12}$. The answer is negative, $-\frac{5}{12}$.

24. $\frac{72.8}{-8}$
Two numbers with unlike signs. Divide their absolute values, getting 9.1. The answer is negative, -9.1.

25. $-7 \div \frac{4}{3} = -7 \cdot \left(\frac{3}{4}\right)$ Multiplying by the reciprocal of $\frac{4}{3}$
$$= -\frac{21}{4}$$

26. If $a = -6.28$, then $-a = -(-6.28) = 6.28$.

27. $12 + x = x + 12$ Using the commutative law of addition

28. $5x + y = x \cdot 5 + y$ Using the commutative law of multiplication
or $= y + 5x$ Using the commutative law of addition

29. $(4 + a) + b = 4 + (a + b)$ Using the associative law of addition

30. $x(yz) = (xy)z$ Using the associative
law of multiplication

31. $12m + 4n - 2 = 2 \cdot 6m + 2 \cdot 2n - 2 \cdot 1$
$= 2(6m + 2n - 1)$

32. $3x^3 - 6x^2 + x^3 + 5$
$= (3+1)x^3 - 6x^2 + 5$
$= 4x^3 - 6x^2 + 5$

33. $7x - 4[2x + 3(5 - 4x)]$
$= 7x - 4[2x + 15 - 12x]$
$= 7x - 4[-10x + 15]$
$= 7x + 40x - 60$
$= 47x - 60$

34. $3(t+1) - t = 4$
$3t + 3 - t = 4$
$2t + 3 = 4$
$2t + 3 - 3 = 4 - 3$
$2t = 1$
$\frac{1}{2} \cdot 2t = \frac{1}{2} \cdot 1$
$t = \frac{1}{2}$

The solution is $\frac{1}{2}$.

35. $\frac{2}{3}n - \frac{5}{6} = \frac{8}{3}$
$\frac{2}{3}n - \frac{5}{6} + \frac{5}{6} = \frac{8}{3} + \frac{5}{6}$
$\frac{2}{3}n = \frac{16}{6} + \frac{5}{6}$
$\frac{2}{3}n = \frac{21}{6}$
$\frac{3}{2} \cdot \frac{2}{3}n = \frac{3}{2} \cdot \frac{21}{6}$
$n = \frac{21}{4}$

The solution is $\frac{21}{4}$.

36. $-9x + 4(2x - 3) = 5(2x - 3) + 7$
$-9x + 8x - 12 = 10x - 15 + 7$
$-x - 12 = 10x - 8$
$-x - 12 + x = 10x - 8 + x$
$-12 = 11x - 8$
$-12 + 8 = 11x - 8 + 8$
$-4 = 11x$
$\frac{1}{11}(-4) = \frac{1}{11} \cdot 11x$
$-\frac{4}{11} = x$

The solution is $-\frac{4}{11}$.

37. $3(x - 4) + 2 = x + 2(x - 5)$
$3x - 12 + 2 = x + 2x - 10$
$3x - 10 = 3x - 10$
$-10 = -10$

All real numbers are solutions. The solution set is the
set of all real numbers. The equation is an identity.

38. $5t - (7 - t) = 4t + 2(9 + t)$
$5t - 7 + t = 4t + 18 + 2t$
$6t - 7 = 6t + 18$
$-7 = 18$ False equation

The solution set is $\varnothing$. The equation is a contradiction.

39. *Familiarize*. Let x represent the number.
Translate.

Twice the number	plus	15	is	21.
$\downarrow$	$\downarrow$	$\downarrow$	$\downarrow$	$\downarrow$
$2x$	$+$	15	$=$	21

40. *Familiarize*. Let $x =$ one number, then $x - 19 =$ the
other number.
Translate.

First number	plus	second number	is	115.
$\downarrow$	$\downarrow$	$\downarrow$	$\downarrow$	$\downarrow$
x	$+$	$(x - 19)$	$=$	115

Carry out. We solve the equation.
$x + x - 19 = 115$
$2x - 19 = 115$
$2x - 19 + 19 = 115 + 19$
$2x = 134$
$x = 67$

When $x = 67$, $x - 19 = 67 - 19 = 48$
Check. 48 is 19 less than 67. Also $48 + 67 = 115$. The
answer checks.
State. The smaller number is 48.

41. *Familiarize*. Let x represent the measure of the second
angle. Then the first angle is three times x, and the third
angle is two times x. The sum of the three angles is
$180°$.
Translate. The first angle is $3x$, the second angle is x,
and the third angle is $2x$. Translate to an equation:

First	plus	second	plus	third	is	180°.
$\downarrow$	$\downarrow$	$\downarrow$	$\downarrow$	$\downarrow$	$\downarrow$	$\downarrow$
$3x$	$+$	x	$+$	$2x$	$=$	180

Carry out. We solve the equation.
$3x + x + 2x = 180$
$6x = 180$
$x = 30$

When $x = 30$, then $2x = 2 \cdot 30 = 60$, and
$3x = 3 \cdot 30 = 90$.
Check. One angle, $90°$, is 3 times the $30°$ angle, and the
third angle, $60°$, is two times the $30°$ angle. Also $60° +
30° + 90° = 180°$. The answer checks.
State. The measures of the angles are $90°$, $30°$, and $60°$.

42. $x = \frac{bc}{t}$
$tx = bc$ Multiplying both sides by t
$\frac{tx}{b} = c$ Multiplying both sides by $\frac{1}{b}$

43. $c = mx - rx$
$c = x(m - r)$
$\frac{c}{m - r} = x$

44. *Familiarize*. Recall the formula for the volume of a right circular cylinder is $V = \pi r^2 h$.

Translate. We solve the formula for h.

$$V = \pi r^2 h$$
$$h = \frac{V}{\pi r^2}$$

Carry out. We substitute 3.5 for r, 538.51 for V, and 3.14 for π and calculate.

$$h = \frac{538.51}{3.14(3.5)^2}$$
$$h = 14$$

Check. We can repeat the calculation or substitute into the original formula and then solve for h. The answer checks.

State. The height of the candle is 14 cm.

45. $(-4mn^8)(7m^3n^2)$
$= (-4)(7)m \cdot m^3 \cdot n^8 \cdot n^2$
$= -28m^{1+3}n^{8+2}$
$= -28m^4n^{10}$

46. $\dfrac{12x^3y^8}{3x^2y^2} = \dfrac{12}{3}x^{3-2}y^{8-2} = 4xy^6$

47. $a^0 = (-8)^0 = 1$
$a^2 = (-8)^2 = 64$
$-a^2 = -(-8)^2 = -64$

48. $3^{-5} \cdot 3^7 = 3^{-5+7} = 3^2$, or 9

49. $(2t^4)^3 = 2^3 \cdot (t^4)^3 = 8t^{12}$

50. $(-5a^{-3}b^2)^{-3} = (-5)^{-3}(a^{-3})^{-3}(b^2)^{-3}$
$= -\dfrac{1}{5^3} \cdot a^9 \cdot b^{-6}$
$= -\dfrac{a^9}{125b^6}$

51. $\left(\dfrac{x^2y^3}{z^4}\right)^{-2} = \dfrac{(x^2)^{-2}(y^3)^{-2}}{(z^4)^{-2}} = \dfrac{x^{-4}y^{-6}}{z^{-8}} = \dfrac{z^8}{x^4y^6}$

52. $\left(\dfrac{3m^{-5}n}{9m^2n^{-2}}\right)^4 = \left(\dfrac{1}{3}m^{-5-2}n^{1-(-2)}\right)^4$
$= \left(\dfrac{1}{3}m^{-7}n^3\right)^4$
$= \dfrac{m^{-28}n^{12}}{81}$
$= \dfrac{n^{12}}{81m^{28}}$

53. $\dfrac{4(9-2\cdot3)-3^2}{4^2-3^2} = \dfrac{4(9-6)-9}{16-9} = \dfrac{4(3)-9}{7} = \dfrac{12-9}{7} = \dfrac{3}{7}$

54. $1-(2-5)^2 + 5 \div 10 \cdot 4^2$
$= 1-(-3)^2 + 5 \div 10 \cdot 16$
$= 1-9 + \dfrac{1}{2} \cdot 16$
$= 1-9+8$
$= 0$

55. $0.000307 = \dfrac{0.000307 \times 10^4}{10^4} = \dfrac{3.07}{10^4} = 3.07 \times 10^{-4}$

56. $30{,}860{,}000{,}000{,}000$
$= \dfrac{30{,}860{,}000{,}000{,}000}{10^{13}} \times 10^{13}$
$= 3.086 \times 10^{13}$

57. $(8.7 \times 10^{-9}) \times (4.3 \times 10^{15})$
$= (8.7 \times 4.3)(10^{-9} \times 10^{15})$
$= 37.41 \times 10^6$
$= (3.741 \times 10) \times 10^6$
$= 3.7 \times 10^7$ (2 significant digits)

58. $\dfrac{1.2 \times 10^{-12}}{6.1 \times 10^{-7}} = \dfrac{1.2}{6.1} \times \dfrac{10^{-12}}{10^{-7}}$
$= 0.1967 \times 10^{-5}$
$= (1.967 \times 10^{-1}) \times 10^{-5}$
$= 1.967 \times 10^{-6}$
$\approx 2.0 \times 10^{-6}$ (2 significant digits)

59. *Familiarize*. Recall the formula for the volume is $V = lwh$. We convert 1.2 m to mm.

$$1.2 \text{ m} = 1.2 \text{ m} \times \frac{1000 \text{ mm}}{1 \text{ m}} = 1200 \text{ mm}$$

We convert 79 m to mm.

$$79 \text{ m} = 79 \text{ m} \times \frac{1000 \text{ mm}}{1 \text{ m}} = 79{,}000 \text{ mm}$$

Translate. We substitute into the formula.
$$V = lwh$$
$$V = (1200) \times (79{,}000) \times (0.00015)$$

Carry out. We do the calculations.
$$V = (1200) \times (79{,}000) \times (0.00015)$$
$$= 14{,}220 \text{ mm}^3,$$
or 1.4×10^4 mm^3 (2 significant digits)
or 1.4×10^{-5} m^3

Check. We recheck the translation and the calculations.

State. The volume of the sheet is about 1.4×10^4 mm^3, or or 1.4×10^{-5} m^3.

60. *Writing Exercise*. To write an equation that has no solution, begin with a simple equation that is false for any value of x, such as $x = x + 1$. Then add or multiply by the same quantities on both sides of the equation to construct a more complicated equation with no solution.

61. *Writing Exercise.*

a. $-(-x)$ is positive when x is positive; the opposite of the opposite of a number is the number itself;

b. $-x^2$ is never positive; x^2 is always nonnegative, so the opposite of x^2 is always nonpositive;

c. $-x^3$ is positive when x is negative; x^3 is negative when x is negative, and the opposite of a negative number is positive;

d. $(-x)^2$ is positive when $x \neq 0$; the square of any nonzero number is positive;

e. x^{-2} is positive when $x \neq 0$; $x^{-2} = \frac{1}{x^2}$ and x^2 is positive when x is nonzero.

62. First we express the quotient, 3 parts per billion, as the fraction and then express as a percent.

$$\frac{3}{1,000,000,000}$$
$$= \frac{3}{1,000,000,000} \cdot \frac{100}{100}$$
$$= \frac{300}{1,000,000,000}\%$$
$$= 0.0000003\%$$

63. $a + b(c - a^2)^0 + (abc)^{-1}$

$$= 3 + (-2)(-4 - (3)^2)^0 + (3(-2)(-4))^{-1}$$
$$= 3 + (-2)(1) + (24)^{-1}$$
$$= 3 - 2 + \frac{1}{24}$$
$$= 1 + \frac{1}{24}$$
$$= 1\frac{1}{24}, \text{ or } \frac{25}{24}$$

64. *Familiarize.* First we find the area of each pizza. Then we find the price per square inch of each pizza. Recall, the formula for the area of a circle is $A = \pi r^2$ and the radius is $\frac{d}{2}$.

Translate. For each pizza, the price per square inch is the cost of each pizza divided by the area of each pizza.

$$P = \frac{C}{A}$$

Carry out. We substitute in the formula $A = \pi r^2$. For the first pizza, which costs \$12, $r = \frac{13}{2}$, or 6.5. We use 3.14 for π.

$$A = 3.14(6.5)^2 = 132.665$$

So the price per square inch is

$$P = \frac{12}{132.665} \approx 0.09$$

For the second pizza, which costs \$15, $r = \frac{17}{2}$, or 8.5.

We use 3.14 for π.

$$A = 3.14(8.5)^2 = 226.865$$

So the price per square inch is

$$P = \frac{15}{226.865} \approx 0.07$$

Check. Repeat the calculations. The answer checks.
State. The 17-inch pizza is about 9¢ per square inch, while the 13-inch pizza is about 7¢ per square inch. The 17-inch pizza is a better deal.

65. *Familiarize and Translate.* The surface area of a cube is the sum of the areas of the six sides, each a square, or

$$S_a = 6s^2.$$

We can solve for the side s, by substituting.

$$486 = 6s^2$$
$$81 = s^2$$
$$9 = s$$

Recall the formula for the volume of a cube is $V = s^3$.
Carry out. We substitute.

$$V = 9^3 = 729$$

Check. We repeat the calculations. The answer checks.

State. The volume of the cube is 729 cm^3.

66.
$$m = \frac{x}{y - z}$$
$$m(y - z) = x$$
$$my - mz = x$$
$$-mz = -my + x$$
$$z = \frac{-my + x}{-m}, \text{ or } \frac{my - x}{m}$$

67. $\dfrac{\left(3^{-2}\right)^a \cdot \left(3^b\right)^{-2a}}{\left(3^{-2}\right)^b \cdot \left(9^{-b}\right)^{-3a}} = \dfrac{\left(3^{-2}\right)^a \cdot \left(3^b\right)^{-2a}}{\left(3^{-2}\right)^b \cdot \left[\left(3^2\right)^{-b}\right]^{-3a}} = \dfrac{3^{-2a} \cdot 3^{-2ab}}{3^{-2b} \cdot 3^{6ab}}$

$$= 3^{-2a - 2ab + 2b - 6ab}$$
$$= 3^{-2a + 2b - 8ab}$$

68. $5x - 7(x + 3) - 4 = 2(7 - x) + \underline{\quad}$
$\quad 5x - 7x - 21 - 4 = 14 - 2x + \underline{\quad}$
$\quad\quad\quad -2x - 25 = 14 - 2x + \underline{\quad}$
$\quad\quad\quad\quad\quad -25 = 14 + \underline{\quad}$
$\quad\quad\quad\quad\quad -39 = \underline{\quad}$

69. $20 - 7[3(2x + 4) - 10] = 9 - 2(x - 5) + \underline{\quad}$
$\quad 20 - 7[6x + 12 - 10] = 9 - 2x + 10 + \underline{\quad}$
$\quad\quad 20 - 7[6x + 2] = 19 - 2x + \underline{\quad}$
$\quad\quad 20 - 42x - 14 = 19 - 2x + \underline{\quad}$
$\quad\quad\quad 6 - 42x = 19 - 2x + \underline{\quad}$
$\quad\quad\quad 6 - 40x = 19 + \underline{\quad}$
$\quad\quad\quad\quad -40x = \underline{\quad}$

70. $a \cdot 2 + cb + cd + ad$
$= ad + a \cdot 2 + cb + cd$
$= a(d + 2) + c(b + d)$

71. Answers will vary; $\sqrt{5}/4$ is one example.

Chapter 1 Test

1. Let m and n represent the numbers; $mn - 4$

2. $a^3 - 5b + b \div ac$

$= (-2)^3 - 5(6) + 6 \div (-2) \cdot (3)$

$= -8 - 5(6) + 6 \div (-2) \cdot 3$

$= -8 - 30 + (-3) \cdot 3$

$= -8 - 30 - 9$

$= -47$

3. We substitute 7.8 for b and 46.5 for h and multiply.

$A = \frac{1}{2} \cdot b \cdot h = \frac{1}{2}(7.8)(46.5) = 181.35$ sq m

4. $-15 + (-16)$

Two negative numbers: Add the absolute values, getting 31. The answer is negative, -31.

5. $-7.5 + 3.8$

A negative and positive number: The absolute values are 7.5 and 3.8. Subtract 3.8 from 7.5 to get 3.7. The negative number is farther from 0, so the answer is negative, -3.7.

6. $29.5 - 43.7 = 29.5 + (-43.7) = -14.2$

7. $-6.4(5.3)$

Two numbers with unlike signs: Multiply their absolute values getting 33.92. The answer is negative, -33.92.

8. $-\frac{7}{6} - \left(-\frac{5}{4}\right) = -\frac{7}{6} + \frac{5}{4}$

$= -\frac{14}{12} + \frac{15}{12}$

$= \frac{1}{12}$

9. $-\frac{2}{7}\left(-\frac{5}{14}\right)$

Two numbers with the same sign: Multiply their absolute values getting $\frac{10}{98}$, or $\frac{5}{49}$. The answer is positive, $\frac{5}{49}$.

10. $\frac{-42.6}{-7.1}$

Two numbers with the same sign: Divide their absolute values, getting 6. The answer is positive, 6.

11. $\frac{2}{5} \div \left(-\frac{3}{10}\right) = \frac{2}{5} \cdot \left(-\frac{10}{3}\right)$

Two numbers with unlike signs: Divide their absolute values getting $\frac{4}{3}$. The answer is negative, $-\frac{4}{3}$.

12. $7 + (1 - 3)^2 - 9 \div 2^2 \cdot 6$

$= 7 + (-2)^2 - 9 \div 4 \cdot 6$

$= 7 + 4 - \frac{9}{4} \cdot 6$

$= 7 + 4 - \frac{27}{2}$

$= 11 - \frac{27}{2}$

$= -\frac{5}{2}$

13. Using the commutative law of addition, we have $3 + x = x + 3$.

14. $4y - 10 - 7y - 19 = (4 - 7)y - 10 - 19 = -3y - 29$

15. $10x - 7 = 38x + 49$

$-7 = 28x + 49$

$-56 = 28x$

$-2 = x$

The solution is -2.

16. $13t - (5 - 2t) = 5(3t - 1)$

$13t - 5 + 2t = 15t - 5$

$15t - 5 = 15t - 5$

$-5 = -5$ True

All real numbers are solutions. The equation is an identity.

17. $2p = sp + t$

$2p - sp = t$

$p(2 - s) = t$

$p = \dfrac{t}{2 - s}$

18. *Familiarize*. Let x = the number of points on the sixth test.

Translate

$$\underbrace{\text{Average score on 6 tests}}_{\dfrac{84 + 80 + 76 + 96 + 80 + x}{6}} \quad \underbrace{\text{is}}_{=} \quad \underbrace{85.}_{85}$$

Carry out. We solve the equation.

$\dfrac{416 + x}{6} = 85$

$416 + x = 510$

$x = 94$

Check. $\dfrac{84 + 80 + 76 + 96 + 80 + 94}{6}$, or 85. The answer checks.

State. Linda must score 94 so her average will be 85.

19. *Familiarize*. Let x represent the first odd integer. Then the second odd integer is $x + 2$ and the third is $x + 2 + 2$, or $x + 4$.

Translate.

$$\underbrace{\text{Four times the first}}_{4x} \text{ plus } \underbrace{\text{Three times the second}}_{3(x+2)} \text{ plus } \underbrace{\text{Two times the third}}_{2(x+4)} \text{ is } \underbrace{167.}_{167}$$

Carry out. We solve the equation.

$4x + 3(x + 2) + 2(x + 4) = 167$

$4x + 3x + 6 + 2x + 8 = 167$

$9x + 14 = 167$

$9x = 153$

$x = 17$

$x + 2 = 19$

$x + 4 = 21$

Check. $4 \cdot 17 + 3 \cdot 19 + 2 \cdot 21$, or $68 + 57 + 42$, or 167. The answer checks.

State. The consecutive odd integers are 17, 19, and 21.

20. $3x - 7 - (4 - 5x) = 3x - 7 - 4 + 5x = 8x - 11$

21. $6b - [7 - 2(9b - 1)]$
$= 6b - [7 - 18b + 2]$
$= 6b - [9 - 18b]$
$= 6b - 9 + 18b$
$= 24b - 9$

22. $(7x^{-4}y^{-7})(-6x^{-6}y)$
$= (7)(-6)(x^{-4}x^{-6})(y^{-7}y)$
$= -42x^{-4+(-6)}y^{-7+1}$
$= -42x^{-10}y^{-6}$, or $-\dfrac{42}{x^{10}y^{6}}$

23. $-6^{-2} = -\dfrac{1}{6^{2}}$, or $-\dfrac{1}{36}$

24. $(-5x^{-1}y^{3})^{3} = (-5)^{3}(x^{-1})^{3}(y^{3})^{3}$
$= -125x^{-3}y^{9}$
$= -\dfrac{125y^{9}}{x^{3}}$

25. $\left(\dfrac{2x^{3}y^{-6}}{-4y^{-2}}\right)^{-2} = \left(-\dfrac{2}{4}x^{3}y^{-6-(-2)}\right)^{-2} = \left(-\dfrac{1}{2}x^{3}y^{-4}\right)^{-2}$
$= \left(-\dfrac{1}{2}\right)^{-2}(x^{3})^{-2}(y^{-4})^{-2}$
$= (-2)^{2}x^{-6}y^{8} = \dfrac{4y^{8}}{x^{6}}$

26. $(7x^{3}y)^{0} = 1$

27. $(9.05 \times 10^{-3})(2.22 \times 10^{-5})$
$= (9.05 \times 2.22) \times (10^{-3} \times 10^{-5})$
$= 20.091 \times 10^{-8}$
$= (2.0091 \times 10^{1}) \times 10^{-8}$
$= 2.0091 \times 10^{-7}$
$\approx 2.01 \times 10^{-7}$ (3 significant digits)

28. $\dfrac{1.8 \times 10^{-4}}{4.8 \times 10^{-7}} = \dfrac{1.8}{4.8} \times \dfrac{10^{-4}}{10^{-7}}$
$= 0.375 \times 10^{3}$
$= (3.75 \times 10^{-1}) \times 10^{3}$
$= 3.75 \times 10^{2}$
$\approx 3.8 \times 10^{2}$ (2 significant digits)

29. ***Familiarize***. Let n = the smallest number of neutrinos that could have the same mass as an alpha particle of mass 3.62×10^{-27} kg.
Translate. We divide.
$$n = \dfrac{3.62 \times 10^{-27} \text{ kg}}{1.8 \times 10^{-36} \text{ kg}}$$
Carry out. We perform the calculation and write scientific notation for the answer.

$$n = \dfrac{3.62 \times 10^{-27} \text{ kg}}{1.8 \times 10^{-36} \text{ kg}}$$
$$= \dfrac{3.62}{1.8} \times \dfrac{10^{-27}}{10^{-36}}$$
$$\approx 2.0 \times 10^{9} \quad \text{(2 significant digits)}$$

Check. We multiply the answer by the maximum mass of a neutrino.
$$(2.0 \times 10^{9})(1.8 \times 10^{-36}) = 3.6 \times 10^{-27} \approx 3.62 \times 10^{-27}$$
The answer checks.

State. The smallest number of neutrinos that could have the same mass as an alpha particle of mass 3.62×10^{-27} kg is about 2.0×10^{9} neutrinos.

30. $(2x^{3a}y^{b+1})^{3c} = 2^{3c} \cdot (x^{3a})^{3c} \cdot (y^{b+1})^{3c} = 8^{c}x^{9ac}y^{3bc+3c}$

31. $\dfrac{-27a^{x+1}}{3a^{x-2}} = \dfrac{-27}{3} \cdot a^{(x+1)-(x-2)} = -9a^{3}$

32. $-\dfrac{5x+2}{x+10} = 1$
$-(5x+2) = 1(x+10)$
$-5x - 2 = x + 10$
$-2 = 6x + 10$
$-12 = 6x$
$-2 = x$

Chapter 2

Graphs, Functions, and Linear Equations

Exercise Set 2.1

1. The two perpendicular number lines that are used for graphing are called <u>axes</u>.

3. In the <u>third</u> quadrant, both coordinates of a point are negative.

5. To graph an equation means to make a drawing that represents all <u>solutions</u> of the equation.

7. *A* is 5 units right of the origin and 3 units up, so its coordinates are (5, 3).
   B is 4 units left of the origin and 3 units up, so its coordinates are (–4, 3).
   C is 0 units right or left of the origin and 2 units up, so its coordinates are (0, 2).
   D is 2 units left of the origin and 3 units down, so its coordinates are (–2, –3).
   E is 4 units right of the origin and 2 units down, so its coordinates are (4, –2).
   F is 5 units left of the origin and 0 units up or down, so its coordinates are (–5, 0).

9.

   $A(3, 0)$ is 3 units right and 0 units up or down.

   $B(4, 2)$ is 4 units right and 2 units up.

   $C(5, 4)$ is 5 units right and 4 units up.

   $D(6, 6)$ is 6 units right and 6 units up.

   $E(3, -4)$ is 3 units right and 4 units down.

   $F(3, -3)$ is 3 units right and 3 units down.

   $G(3, -2)$ is 3 units right and 2 units down.

   $H(3, -1)$ is 3 units right and 1 unit down.

11.

   A triangle is formed. The area of a triangle is found by using the formula $A = \frac{1}{2}bh$. In this triangle the base and height are 7 units and 6 units, respectively.

   $A = \frac{1}{2}bh = \frac{1}{2} \cdot 7 \cdot 6 = \frac{42}{2} = 21$ square units

13. The smallest first coordinate is –75 and the largest is 9. The smallest second coordinate is –4 and the largest is 5. The points are in quadrants II, III, and IV.

15. The smallest first coordinate is –100 and the largest is 800. The smallest second coordinate is –5 and the largest is 37. The points are in quadrants I and III.

17. The first coordinate is positive and the second negative, so the point (7, –2) is in quadrant IV.

19. Both coordinates are negative, so the point (–4, –3) is in quadrant III.

21. The first coordinate is zero, so the point (0, –3) is on the *y*-axis.

23. The first coordinate is negative and the second positive, so the point (–4.9, 8.3) is in quadrant II.

25. The second coordinate is zero, so the point $\left(-\frac{5}{2}, 0\right)$ is on the *x*-axis.

27. Both coordinates are positive, so the point (160, 2) is in quadrant I.

29. $y = 3x - 7$

   $$\begin{array}{c|c} \hline -1 & 3 \cdot 2 - 7 \\ & 6 - 7 \\ & ? \\ \hline -1 = -1 \end{array}$$

   Substituting 2 for *x* and –1 for *y* (alphabetical order of variables)

   Since $-1 = -1$ is true, (2, –1) is a solution of $y = 3x - 7$.

31. $2x - y = 5$

   $$\begin{array}{c|c} \hline 2 \cdot 3 - 2 & 5 \\ 6 - 2 & \\ ? & \\ \hline 4 = 5 \end{array}$$

   Substituting 3 for *x* and 2 for *y* (alphabetical order of variables)

   Since $4 = 5$ is false, (3, 2) is not a solution of $2x - y = 5$.

33. $a - 5b = 8$

   $$\begin{array}{c|c} \hline 3 - 5(-1) & 8 \\ 3 + 5 & \\ ? & \\ \hline 8 = 8 \end{array}$$

   Substituting 3 for *a* and –1 for *b*

   Since $8 = 8$ is true, (3, –1) is a solution of $a - 5b = 8$.

35.
$$6x + 8y = 4$$

$$\begin{array}{c|c} 6 \cdot \frac{2}{3} + 8 \cdot 0 & 4 \\ 4 + 0 & \\ & ? \\ 4 & = 4 \end{array}$$ Substituting $\frac{2}{3}$ for x and 0 for y

Since $4 = 4$ is true, $\left(\frac{2}{3}, 0\right)$ is a solution of $6x + 8y = 4$.

37.
$$r - s = 4$$

$$\begin{array}{c|c} 6 - (-2) & 4 \\ 6 + 2 & \\ & ? \\ 8 & = 4 \end{array}$$ Substituting 6 for r and -2 for s

Since $8 = 4$ is false, $(6, -2)$ is not a solution of $r - s = 4$.

39. $y = 2x^2$

$$\begin{array}{c|c} 1 & 2(2)^2 \\ ? \\ 1 & = 8 \end{array}$$ Substituting 2 for x and 1 for y

Since $1 = 8$ is false, $(2, 1)$ is not a solution of $y = 2x^2$.

41.
$$x^3 + y = 1$$

$$\begin{array}{c|c} (-2)^3 + 9 & 1 \\ -8 + 9 & \\ & ? \\ 1 & = 1 \end{array}$$ Substituting -2 for x and 9 for y

Since $1 = 1$ is true, $(-2, 9)$ is a solution of $x^3 + y = 1$.

43. $y = 3x$

To find an ordered pair, we choose any number for x and then determine y by substitution.

When $x = 0$, $y = 3 \cdot 0 = 0$.

When $x = 1$, $y = 3 \cdot 1 = 3$.

When $x = -2$, $y = 3 \cdot (-2) = -6$.

x	y	(x, y)
0	0	(0, 0)
1	3	(1, 3)
-2	-6	$(-2, -6)$

Plot these points, draw the line they determine, and label the graph $y = 3x$.

45. $y = x + 4$

To find an ordered pair, we choose any number for x and then determine y by substitution.

When $x = 0$, $y = 0 + 4 = 4$.

When $x = 1$, $y = 1 + 4 = 5$.

When $x = -2$, $f = -2 + 4 = 2$.

x	y	(x, y)
0	4	(0, 4)
1	5	(1, 5)
-2	2	$(-2, 2)$

Plot these points, draw the line they determine, and

label the graph $y = x + 4$.

47. $y = x - 4$

To find an ordered pair, we choose any number for x and then determine y by substitution. For example, if we choose 1 for x, then $y = 1 - 4 = -3$. We find several ordered pairs, plot them and draw the line.

x	y	(x, y)
1	-3	$(1, -3)$
0	-4	$(0, -4)$
-2	-6	$(-2, -6)$

49. $y = -2x + 3$

To find an ordered pair, we choose any number for x and then determine y. For example, if $x = 1$, then $y = -2 \cdot 1 + 3 = -2 + 3 = 1$. We find several ordered pairs, plot them, and draw the line.

x	y
1	1
3	-3
-1	5
0	3

51. $y + 2x = 3$

$$y = -2x + 3 \quad \text{Solving for } y$$

Observe that this is the equation that was graphed in Exercise 49. The graph is shown above.

53. $y = -\frac{3}{2}x$

To find an ordered pair, we choose any number for x and then determine y by substitution. For example, if $x = 2$, then $y = -\frac{3}{2} \cdot 2 = -3$. We find several ordered pairs, plot them and draw the line.

x	y
2	-3
0	0
-4	6

55. $y = \frac{3}{4}x - 1$

To find an ordered pair, we choose any number for x and then determine y by substitution. For example, if $x = 4$, then $y = \frac{3}{4} \cdot 4 - 1 = 3 - 1 = 2$. We find several ordered pairs, plot them and draw the line.

x	y
4	2
0	-1
-8	-7

57. $y = |x| + 2$

We select x-values and find the corresponding y-values. The table lists some ordered pairs. We plot these points.

x	y
3	5
1	3
0	2
−1	3
−3	5

Note that the graph is V-shaped, centered at (0, 2).

59. $y = |x| - 2$

We select x-values and find the corresponding y-values. The table lists some ordered pairs. We plot these points.

x	y
3	1
1	−1
0	−2
−1	−1
−4	2

61. $y = x^2 + 2$

To find an ordered pair, we choose any number for x and then determine y. For example, if $x = 2$, then

$y = 2^2 + 2 = 6$. We find several ordered pairs, plot them, and connect them with a smooth curve.

x	y
2	6
1	3
0	2
−1	3
−2	6

63. $y = x^2 - 2$

To find an ordered pair, we choose any number for x and then determine y. For example, if $x = 2$, then

$y = 2^2 - 2 = 4 - 2 = 2$. We find several ordered pairs, plot them, and connect them with a smooth curve.

x	y
2	2
1	−1
0	−2
−1	−1
−2	2

65. *Writing Exercise.*

67. $3 - 2(1 - 4)^2 \div 6 \cdot 2 = 3 - 2(-3)^2 \div 6 \cdot 2$
$$= 3 - 2(9) \div 6 \cdot 2$$
$$= 3 - 18 \div 6 \cdot 2$$
$$= 3 - 3 \cdot 2$$
$$= 3 - 6$$
$$= -3$$

69. $\left(2x^6 y\right)^2 = 4x^{12} y^2$

71. *Writing Exercise.*

73. a. Graph III seems most appropriate for this situation. It reflects less than 40 hours of work per week until September, 40 hours per week from September until December, and more than 40 hours per week in December.

b. Graph II seems most appropriate for this situation. It reflects 40 hours of work per week until September, 20 hours per week from September until December, and 40 hours per week again in December.

c. Graph I seems most appropriate for this situation. It reflects more than 40 hours of work per week until September, 40 hours per week from September until December, and more than 40 hours per week again in December.

d. Graph IV seems most appropriate for this situation. It reflects less than 40 hours of work per week until September, approximately 20 hours per week from September until December, and about 40 hours per week in December.

75. a. Graph III seems most appropriate for this situation. It reflects a constant speed for a lakeshore loop.

b. Graph II seems most appropriate for this situation. In climbing a monster hill, the speed gradually decreases, the speed bottoms out at the top of the hill, and increases for the downhill ride.

c. Graph IV seems most appropriate for this situation. For interval training, the speed is constant for an interval, then increases to a new constant speed for an interval, then returns to the first speed. The sequence repeats.

d. Graph I seems most appropriate for this situation. For a variety of intervals, the speed varies with no particular pattern of increase or decrease in speed.

77. Plot (−10,−2), (−3, 4), and (6, 4), and sketch a parallelogram.

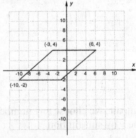

Since (6, 4) is 9 units directly to the right of (−3, 4), a fourth vertex could lie 9 units directly to the right of (−10,−2). Then its coordinates are (−10 + 9,−2), or (−1,−2).

If we connect the points in a different order, we get a second parallelogram.

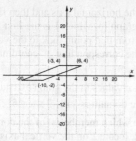

Since (−3, 4) is 9 units directly to the left of (6, 4), a fourth vertex could lie 9 units directly to the left of (−10,−2). Then its coordinates are (−10 − 9,−2), or (−19,−2).

If we connect the points in yet a different order, we get a third parallelogram.

Since (6, 4) lies 16 units directly to the right of and 6 units above (−10, −2), a fourth vertex could lie 16 units to the right of and 6 units above (−3, 4). Its coordinates are (−3 + 16, 4 + 6), or (13, 10).

79. $y = \dfrac{1}{x^2}$

Choose x-values from −4 to 4 and use a calculator to find the corresponding y-values. (Note that we cannot choose 0 as a first coordinate since $1/0^2$, or 1/0, is not defined.) Plot the points and draw the graph. Note that it has two branches, one on each side of the y-axis.

x	y
−4	0.0625
−3	0.1̄
−2	0.25
−1	1
−0.5	4
0.5	4
1	1
2	0.25
3	0.1̄
4	0.0625

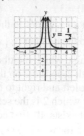

81. $y = \dfrac{1}{x} + 3$

Choose x-values from −4 to 4 and use a calculator to find the corresponding y-values. (Note that we cannot choose 0 as a first coordinate since 1/0 is not defined.) Plot the points and draw the graph. Note that it has two branches, one in the first quadrant and the other in the

second and third quadrants.

x	y
−4	2.75
−3	2.67
−2	2.5
−1	2
−0.25	−1
0.25	7
1	4
2	3.5
3	3.33
4	3.25

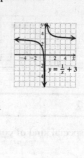

83. $y = \sqrt{x} + 1$

Choose x-values from 0 to 10 and use a calculator to find the corresponding y-values. Plot the points and draw the graph.

x	y
0	1
1	2
2	2.414
3	2.732
4	3
5	3.236
6	3.449
7	3.646
8	3.828
9	4
10	4.162

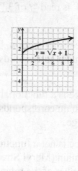

85. $y = x^3$

Choose x-values from −2 to 2 and use a calculator to find the corresponding y-values. Plot the points and draw the graph.

x	y
−2	−8
−1.5	−3.375
−1	−1
−0.5	−0.125
−0.25	−0.016
0	0
0.5	0.125
1	1
1.5	3.375
2	8

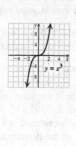

87. a.

$y = 0.375x^3$

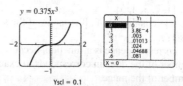

Yscl = 0.1

b.

$y = -3.5x^2 + 6x - 8$

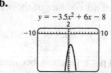

c.

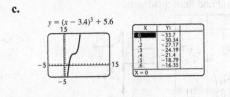

Exercise Set 2.2

1. A function is a special kind of <u>correspondence</u> between two sets.

3. For any function, the set of all inputs, or first values, is called the <u>domain</u>.

5. When a function is graphed, members of the domain are located on the <u>horizontal</u> axis.

7. The notation $f(3)$ is read <u>"f of 3," "f at 3," or "the value of f at 3."</u>

9. The correspondence is a function, because each member of the domain corresponds to exactly one member of the range.

11. The correspondence is a function, because each member of the domain corresponds to exactly one member of the range.

13. The correspondence is not a function because a member of the domain (Meryl Streep) corresponds to more than one member of the range.

15. The correspondence is a function, because each member of the domain corresponds to exactly one member of the range.

17. The correspondence is a function, because each flash drive has only one storage capacity.

19. This correspondence is a function, because each player has only one uniform number.

21. **a.** The domain is the set of all x-values of the set. It is $\{-3, -2, 0, 4\}$.
    b. The range is the set of all y-values of the set. It is $\{-10, 3, 5, 9\}$.
    c. The correspondence is a function, because each member of the domain corresponds to exactly one member of the range.

23. **a.** The domain is the set of all x-values of the set. It is $\{1, 2, 3, 4, 5\}$.
    b. The range is the set of all y-values of the set. It is $\{1\}$.
    c. The correspondence is a function, because each member of the domain corresponds to exactly one member of the range.

25. **a.** The domain is the set of all x-values of the set. It is $\{-2, 3, 4\}$.
    b. The range is the set of all y-values of the set. It is $\{-8, -2, 4, 5\}$.
    c. The correspondence is not a function, because a member of the domain (4) corresponds to more than one member of the range.

27. **a.** Locate 1 on the horizontal axis and then find the point on the graph for which 1 is the first coordinate. From that point, look to the vertical axis to find the corresponding y-coordinate, –2. Thus, $f(1) = -2$.
    b. The set of all x-values in the graph extends from –2 to 5, so the domain is $\{x \mid -2 \le x \le 5\}$.
    c. To determine which member(s) of the domain are paired with 2, locate 2 on the vertical axis. From there look left and right to the graph to find any points for which 2 is the second coordinate. One such point exists. Its first coordinate is 4. Thus, the x-value for which $f(x) = 2$ is 4.
    d. The set of all y-values in the graph extends from –3 to 4, so the range is $\{y \mid -3 \le y \le 4\}$.

29. **a.** Locate 1 on the horizontal axis and then find the point on the graph for which 1 is the first coordinate. From that point, look to the vertical axis to find the corresponding y-coordinate, –2. Thus, $f(1) = -2$.
    b. The set of all x-values in the graph extends from –4 to 2, so the domain is $\{x \mid -4 \le x \le 2\}$.
    c. To determine which member(s) of the domain are paired with 2, locate 2 on the vertical axis. From there look left and right to the graph to find any points for which 2 is the second coordinate. One such point exists. Its first coordinate is –2. Thus, the x-value for which $f(x) = 2$ is –2.
    d. The set of all y-values in the graph extends from –3 to 3, so the range is $\{y \mid -3 \le y \le 3\}$.

31. **a.** Locate 1 on the horizontal axis and then find the point on the graph for which 1 is the first coordinate. From that point, look to the vertical axis to find the corresponding y-coordinate, 3. Thus, $f(1) = 3$.
    b. The set of all x-values in the graph extends from –4 to 3, so the domain is $\{x \mid -4 \le x \le 3\}$.
    c. To determine which member(s) of the domain are paired with 2, locate 2 on the vertical axis. From there look left and right to the graph to find any points for which 2 is the second coordinate. One such point exists. Its first coordinate is –3. Thus, the x-value for which $f(x) = 2$ is –3.
    d. The set of all y-values in the graph extends from –2 to 5, so the range is $\{y \mid -2 \le y \le 5\}$.

33. **a.** Locate 1 on the horizontal axis and then find the point on the graph for which 1 is the first coordinate. From that point, look to the vertical axis to find the corresponding y-coordinate, 3. Thus, $f(1) = 3$.
    b. The domain is the set of all x-values in the graph. It is $\{-4, -3, -2, -1, 0, 1, 2\}$.
    c. To determine which member(s) of the domain are paired with 2, locate 2 on the vertical axis. From there look left and right to the graph to find any

points for which 2 is the second coordinate. There are two such points, (–2, 2) and (0, 2). Thus, the *x*-values for which $f(x) = 2$ are –2 and 0.

d. The range is the set of all *y*-values in the graph. It is {1, 2, 3, 4}.

35. a. Locate 1 on the horizontal axis and then find the point on the graph for which 1 is the first coordinate. From that point, look to the vertical axis to find the corresponding *y*-coordinate, 4. Thus, $f(1) = 4$.

b. The set of all *x*-values in the graph extends from –3 to 4, so the domain is $\{x \mid -3 \le x \le 4\}$.

c. To determine which member(s) of the domain are paired with 2, locate 2 on the vertical axis. From there look left and right to the graph to find any points for which 2 is the second coordinate. There are two such points, (–1, 2) and (3, 2). Thus, the *x*-values for which $f(x) = 2$ are –1 and 3.

d. The set of all *y*-values in the graph extends from –4 to 5, so the range is $\{y \mid -4 \le y \le 5\}$.

37. a. Locate 1 on the horizontal axis and then find the point on the graph for which 1 is the first coordinate. From that point, look to the vertical axis to find the corresponding *y*-coordinate, 2. Thus, $f(1) = 2$.

b. The set of all *x*-values in the graph extends from –4 to 4, so the domain is $\{x \mid -4 \le x \le 4\}$.

c. To determine which member(s) of the domain are paired with 2, locate 2 on the vertical axis. From there look left and right to the graph to find any points for which 2 is the second coordinate. All points in the set $\{x \mid 0 < x \le 2\}$ satisfy this condition. These are the *x*-values for which $f(x) = 2$.

d. The range is the set of all *y*-values in the graph. It is {1, 2, 3, 4}.

39. The domain of *f* is the set of all *x*-values that are used in the points on the curve.

The domain is $\{x \mid x \text{ is a real number}\}$ or $\mathbb{R}$.

The range of *f* is the set of all *y*-values that are used in the points on the curve.

The range is $\{y \mid y \text{ is a real number}\}$ or $\mathbb{R}$.

41. The domain of *f* is the set of all *x*-values that are used in the points on the curve.

The domain is $\{x \mid x \text{ is a real number}\}$ or $\mathbb{R}$.

The range of *f* is the set of all *y*-values that are used in the points on the curve.

The range is {4}.

43. The domain of *f* is the set of all *x*-values that are used in the points on the curve.

The domain is $\{x \mid x \text{ is a real number}\}$ or $\mathbb{R}$.

The range of *f* is the set of all *y*-values that are used in

the points on the curve.

The range is $\{y \mid y \ge 1\}$.

45. The domain of *f* is the set of all *x*-values that are used in the points on the curve. The domain is $\{x \mid x \ge 0\}$. The range of *f* is the set of all *y*-values that are used in the points on the curve. The range is $\{y \mid y \ge 0\}$.

47. We can use the vertical-line test:

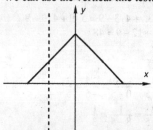

Visualize moving this vertical line across the graph. No vertical line will intersect the graph more than once. Thus, the graph is a graph of a function.

49. We can use the vertical-line test:

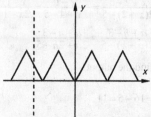

Visualize moving this vertical line across the graph. No vertical line will intersect the graph more than once. Thus, the graph is a graph of a function.

51. We can use the vertical line test.

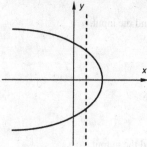

It is possible for a vertical line to intersect the graph more than once. Thus this is not the graph of a function.

53. $g(x) = 2x + 5$

a. $g(0) = 2(0) + 5 = 0 + 5 = 5$

b. $g(-4) = 2(-4) + 5 = -8 + 5 = -3$

c. $g(-7) = 2(-7) + 5 = -14 + 5 = -9$

d. $g(8) = 2(8) + 5 = 16 + 5 = 21$

e. $g(a + 2) = 2(a + 2) + 5 = 2a + 4 + 5 = 2a + 9$

f. $g(a) + 2 = 2(a) + 5 + 2 = 2a + 7$

55. $f(n) = 5n^2 + 4n$

   a. $f(0) = 5 \cdot 0^2 + 4 \cdot 0 = 0 + 0 = 0$

   b. $f(-1) = 5(-1)^2 + 4(-1) = 5 - 4 = 1$

   c. $f(3) = 5 \cdot 3^2 + 4 \cdot 3 = 45 + 12 = 57$

   d. $f(t) = 5t^2 + 4t$

   e. $f(2a) = 5(2a)^2 + 4 \cdot 2a = 5 \cdot 4a^2 + 8a = 20a^2 + 8a$

   f. $f(3) - 9 = 5 \cdot 3^2 + 4 \cdot 3 - 9 = 5 \cdot 9 + 4 \cdot 3 - 9$
$$= 45 + 12 - 9 = 48$$

57. $f(x) = \dfrac{x-3}{2x-5}$

   a. $f(0) = \dfrac{0-3}{2 \cdot 0 - 5} = \dfrac{-3}{0-5} = \dfrac{-3}{-5} = \dfrac{3}{5}$

   b. $f(4) = \dfrac{4-3}{2 \cdot 4 - 5} = \dfrac{1}{8-5} = \dfrac{1}{3}$

   c. $f(-1) = \dfrac{-1-3}{2(-1)-5} = \dfrac{-4}{-2-5} = \dfrac{-4}{-7} = \dfrac{4}{7}$

   d. $f(3) = \dfrac{3-3}{2 \cdot 3 - 5} = \dfrac{0}{6-5} = \dfrac{0}{1} = 0$

   e. $f(x+2) = \dfrac{x+2-3}{2(x+2)-5} = \dfrac{x-1}{2x+4-5} = \dfrac{x-1}{2x-1}$

   f. $f(a+h) = \dfrac{a+h-3}{2(a+h)-5} = \dfrac{a+h-3}{2a+2h-5}$

59. Given the input, find the output.
$$f(x) = 2x - 5$$
$$f(8) = 2(8) - 5 = 16 - 5 = 11$$

61. Given the output, find the input.
$$f(x) = 2x - 5$$
$$-5 = 2x - 5$$
$$0 = 2x$$
$$0 = x$$

63. Given the output, find the input.
$$f(x) = \tfrac{1}{3}x + 4$$
$$\tfrac{1}{2} = \tfrac{1}{3}x + 4$$
$$-\tfrac{7}{2} = \tfrac{1}{3}x$$
$$-\tfrac{21}{2} = x$$

65. Given the input, find the output.
$$f(x) = \tfrac{1}{3}x + 4$$
$$f\!\left(\tfrac{1}{2}\right) = \tfrac{1}{3}\!\left(\tfrac{1}{2}\right) + 4 = \tfrac{1}{6} + 4 = \tfrac{25}{6}$$

67. Given the output 7, find the input.
$$f(x) = 4 - x$$
$$7 = 4 - x$$
$$3 = -x$$
$$-3 = x$$

69. Given the output -3, find the input.
$$f(x) = 0.1x - 0.5$$
$$-3 = 0.1x - 0.5$$
$$-2.5 = 0.1x$$
$$-25 = x$$

71. $A(s) = s^2 \dfrac{\sqrt{3}}{4}$
$$A(4) = 4^2 \dfrac{\sqrt{3}}{4} = 4\sqrt{3} \approx 6.93$$

The area is $4\sqrt{3}$ cm$^2 \approx 6.93$ cm^2.

73. $V(r) = 4\pi r^2$
$$V(3) = 4\pi(3)^2 = 36\pi$$

The area is 36π in$^2 \approx 113.10$ in^2.

75. $H(x) = 2.75x + 71.48$
$$H(34) = 2.75(34) + 71.48 = 164.98$$
The predicted height is 164.98 cm.

77. Locate the point that is directly above 225. Then estimate its second coordinate by moving horizontally from the point to the vertical axis. The rate is about 75 heart attacks per 10,000 men.

79. Locate the point that is directly to the right of 100. Then estimate its first coordinate by moving vertically from the point to the horizontal axis. The blood cholesterol is about 250 milligrams per deciliter.

81. $f(x) = \dfrac{5}{x-3}$

Since $\dfrac{5}{x-3}$ cannot be computed when the denominator is 0, we find the x-value that causes $x - 3$ to be 0:
$$x - 3 = 0$$
$$x = 3 \quad \text{Adding 3 to both sides}$$

Thus, 3 is not in the domain of f, while all other real numbers are. The domain of f is
$\{x \mid x \text{ is a real number and } x \neq 3\}$.

83. $g(x) = 2x + 1$

Since we can compute $2x + 1$ for any real number x, the domain is the set of all real numbers.

85. $h(x) = |6 - 7x|$

Since we can compute $|6 - 7x|$ for any real number x, the domain is the set of all real numbers.

87. $f(x) = \dfrac{3}{8 - 5x}$

Solve: $8 - 5x = 0$
$$x = \tfrac{8}{5}$$

The domain is $\left\{x \middle| x \text{ is a real number and } x \neq \tfrac{8}{5}\right\}$.

89. $h(x) = \dfrac{x}{x+1}$

Solve: $x + 1 = 0$

$x = -1$

The domain is $\{x | x$ is a real number and $x \neq -1\}$.

91. $f(x) = \dfrac{3x+1}{2}$

Since we can compute $\dfrac{3x+1}{2}$ for any real number x, the domain is the set of all real numbers.

93. $g(x) = \dfrac{1}{2x}$

Solve: $2x = 0$

$x = 0$

The domain is $\{x | x$ is a real number and $x \neq 0\}$.

95. $f(x) = \begin{cases} x, & \text{if } x < 0, \\ 2x+1, & \text{if } x \geq 0 \end{cases}$

a. Since $-5 < 0$, we use $f(x) = x$. Thus,

$f(-5) = -5$

b. Since $0 \geq 0$, we use $f(x) = 2x + 1$. Thus,

$f(0) = 2(0) + 1 = 1$

c. Since $10 \geq 0$, we use $f(x) = 2x + 1$. Thus,

$f(10) = 2(10) + 1 = 20 + 1 = 21$

97. $G(x) = \begin{cases} x - 5, & \text{if } x \leq -1, \\ x, & \text{if } x > -1 \end{cases}$

a. Since $-10 \leq -1$, we use $G(x) = x - 5$. Thus,

$G(-10) = -10 - 5 = -15$

b. Since $0 > -1$, we use $G(x) = x$. Thus,

$G(0) = 0$

c. Since $-1 \leq -1$, we use $G(x) = x - 5$. Thus,

$G(-1) = -1 - 5 = -6$

99. $f(x) = \begin{cases} x^2 - 10, & \text{if } x < -10, \\ x^2, & \text{if } -10 \leq x \leq 10 \\ x^2 + 10, & \text{if } x > 10 \end{cases}$

a. Since $-10 \leq -10$, we use $f(x) = x^2$. Thus,

$f(-10) = (-10)^2 = 100$

b. Since $10 \leq 10$, we use $f(x) = x^2$. Thus,

$f(10) = (10)^2 = 100$

c. Since $11 > 10$, we use $f(x) = x^2 + 10$. Thus,

$f(11) = (11)^2 + 10 = 131$

101. *Writing Exercise.* For the function $f(x) = \dfrac{x+3}{2}$, the denominator is a constant, so the domain is $\mathbb{R}$. For the function $g(x) = \dfrac{2}{x+3}$, the denominator is $x + 3$, so the domain cannot include -3, and therefore is not $\mathbb{R}$.

103. Let n represent the number; $n - 7$

105. $7 + t = t + 7$

107. *Writing Exercise.*

109. To find $f(g(-4))$, we first find $g(-4)$:

$g(-4) = 2(-4) + 5 = -8 + 5 = -3$.

Then

$f(g(-4)) = f(-3) = 3(-3)^2 - 1 = 3 \cdot 9 - 1 = 27 - 1 = 26$.

To find $g(f(-4))$, we first find $f(-4)$:

$f(-4) = 3(-4)^2 - 1 = 3 \cdot 16 - 1 = 48 - 1 = 47$.

Then $g(f(-4)) = g(47) = 2 \cdot 47 + 5 = 94 + 5 = 99$.

111. $f(\text{tiger}) = \text{dog}$

$f(\text{dog}) = f(f(\text{tiger})) = \text{cat}$

$f(\text{cat}) = f(f(f(\text{tiger}))) = \text{fish}$

$f(\text{fish}) = f(f(f(f(\text{tiger})))) = \text{worm}$

113. Domain: $\{x | x$ is a real number and $x \neq 5\}$

Range: $\{y | y$ is a real number and $y \neq 2\}$

115. Locate the highest point on the graph. Then move vertically to the horizontal axis and read the corresponding time. It is about 2 min, 50 sec.

117. The two largest contractions occurred at about 2 minutes, 50 seconds and 5 minutes, 40 seconds. The difference in these times, is 2 minutes, 50 seconds, so the frequency is about 1 every 3 minutes.

119. a. Locate 1 on the horizontal axis and then find the point on the graph for which 1 is the first coordinate. From that point, look to the vertical axis to find the corresponding y-coordinate, 2. Thus, $f(1) = 2$.

b. Locate 2 on the horizontal axis and then find the point on the graph for which 2 is the first coordinate. From that point, look to the vertical axis to find the corresponding y-coordinate, 2. Thus, $f(2) = 2$.

c. To determine which member(s) of the domain are paired with 2, locate 2 on the vertical axis. From there look left and right to the graph to find any points for which 2 is the second coordinate. All points in the set $\{x | 0 < x \leq 2\}$ satisfy this condition. These are the x-values for which $f(x) = 2$.

121.

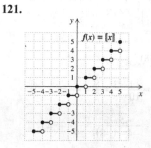

Exercise Set 2.3

1. A y-intercept is of the form (f), $(0, b)$.

3. Rise is (e), the difference in y.

5. Slope-intercept form is (a), $y = mx + b$.

7. Graph: $f(x) = 2x - 1$.

We make a table of values. Then we plot the corresponding points and connect them.

x	$f(x)$
1	1
2	3
0	−1
−1	−3

9. Graph: $g(x) = -\frac{1}{3}x + 2$.

We make a table of values. Then we plot the corresponding points and connect them.

x	$g(x)$
−3	3
0	2
3	1
6	0

11. Graph: $h(x) = \frac{2}{5}x - 4$.

We make a table of values. Then we plot the corresponding points and connect them.

x	$h(x)$
−5	−6
0	−4
5	−2

13. $y = 5x + 3$
The y-intercept is $(0, 3)$.

15. $g(x) = -x - 1$
The y-intercept is $(0, -1)$.

17. $y = -\frac{3}{8}x - 4.5$
The y-intercept is $(0, -4.5)$.

19. $f(x) = 1.3x - \frac{1}{4}$

The y-intercept is $\left(0, -\frac{1}{4}\right)$.

21. $y = 17x + 138$
The y-intercept is $(0, 138)$.

23. Slope $= \dfrac{\text{difference in } y}{\text{difference in } x} = \dfrac{11 - 3}{10 - 8} = \dfrac{8}{2} = 4$

25. Slope $= \dfrac{\text{difference in } y}{\text{difference in } x} = \dfrac{-5 - 3}{-4 - (-8)} = \dfrac{-8}{4} = -2$

27. Slope $= \dfrac{\text{difference in } y}{\text{difference in } x} = \dfrac{4 - (-7)}{13 - (-20)} = \dfrac{11}{33} = \dfrac{1}{3}$

29. Slope $= \dfrac{\text{difference in } y}{\text{difference in } x} = \dfrac{-\frac{2}{3} - \frac{1}{6}}{\frac{1}{2} - \frac{1}{6}} = \dfrac{-\frac{5}{6}}{\frac{2}{6}} = -\dfrac{5}{2}$

31. Slope $= \dfrac{\text{difference in } y}{\text{difference in } x} = \dfrac{43.6 - 43.6}{4.5 - (-9.7)} = \dfrac{0}{14.2} = 0$

33. $y = \frac{2}{3}x + 4$

Since $m = \frac{2}{3}$, the slope is $\frac{2}{3}$. Since $b = 4$, the y-intercept is $(0, 4)$.

35. $2x - y = 3$
$$-y = -2x + 3$$
$$y = 2x - 3$$

Since $m = 2$, the slope is 2. Since $b = -3$, the y-intercept is $(0, -3)$.

37. $y = x - 2$
Since $m = 1$, the slope is 1. Since $b = -2$, the y-intercept is $(0, -2)$.

39. $4x + 5y = 8$
$$5y = -4x + 8$$
$$y = -\frac{4}{5}x + \frac{8}{5}$$

Since $m = -\frac{4}{5}$, the slope is $-\frac{4}{5}$. Since $b = \frac{8}{5}$, the y-intercept is $\left(0, \frac{8}{5}\right)$.

41. Use the slope-intercept equation, $f(x) = mx + b$, with $m = 2$ and $b = 5$.
$$f(x) = mx + b$$
$$f(x) = 2x + 5$$

43. Use the slope-intercept equation, $f(x) = mx + b$, with $m = -\frac{2}{3}$ and $b = -2$.
$$f(x) = mx + b$$
$$f(x) = -\frac{2}{3}x - 2$$

45. Use the slope-intercept equation, $f(x) = mx + b$, with $m = -7$ and $b = \frac{1}{3}$.
$$f(x) = mx + b$$
$$f(x) = -7x + \frac{1}{3}$$

47. $y = \frac{5}{2}x - 3$

Slope is $\frac{5}{2}$; y-intercept is $(0, -3)$.

From the y-intercept, we go *up* 5 units and to the *right* 2 units. This gives us the point $(2, 2)$. We can now draw the graph.

As a check, we can rename the slope and find another point.

$$\frac{5}{2} = \frac{5}{2} \cdot \frac{-1}{-1} = \frac{-5}{-2}$$

From the y-intercept, we go *down* 5 units and to the *left* 2 units. This gives us the point $(-2, -8)$. Since $(-2, -8)$ is on the line, we have a check.

49. $f(x) = -\frac{5}{2}x + 2$

Slope is $-\frac{5}{2}$, or $\frac{-5}{2}$; y-intercept is $(0, 2)$.

From the y-intercept, we go *down* 5 units and to the *right* 2 units. This gives us the point $(2, -3)$. We can now draw the graph.

$f(x) = -\frac{5}{2}x + 2$

As a check, we can rename the slope and find another point.

$$\frac{-5}{2} = \frac{5}{-2}$$

From the y-intercept, we go *up* 5 units and to the *left* 2 units. This gives us the point $(-2, 7)$. Since $(-2, 7)$ is on the line, we have a check.

51. $F(x) = 2x + 1$

Slope is 2, or $\frac{2}{1}$; y-intercept is $(0, 1)$.

From the y-intercept, we go *up* 2 units and to the *right* 1 unit. This gives us the point $(1, 3)$. We can now draw the graph.

$F(x) = 2x + 1$

As a check, we can rename the slope and find another point.

$$2 = \frac{2}{1} \cdot \frac{3}{3} = \frac{6}{3}$$

From the y-intercept, we go *up* 6 units and to the *right* 3 units. This gives us the point $(3, 7)$. Since $(3, 7)$ is on the line, we have a check.

53. Convert to a slope-intercept equation.

$$4x + y = 3$$
$$y = -4x + 3$$

Slope is -4, or $\frac{-4}{1}$; y-intercept is $(0, 3)$.

From the y-intercept, we go *down* 4 units and to the *right* 1 unit. This gives us the point $(1, -1)$. We can now draw the graph.

$4x + y = 3$,
or
$y = -4x + 3$

As a check, we can rename the slope and find another point.

$$\frac{-4}{1} = \frac{-4}{1} \cdot \frac{-1}{-1} = \frac{4}{-1}$$

From the y-intercept, we go *up* 4 units and to the *left* 1 unit. This gives us the point $(-1, 7)$. Since $(-1, 7)$ is on the line, we have a check.

55. Convert to a slope-intercept equation.

$$6y + x = 6$$
$$6y = -x + 6$$
$$y = -\frac{1}{6}x + 1$$

Slope is $-\frac{1}{6}$, or $\frac{-1}{6}$; y-intercept is $(0, 1)$.

From the y-intercept, we go *down* 1 unit and to the *right* 6 units. This gives us the point $(6, 0)$. We can now draw the graph.

$6y + x = 6$,
or
$y = -\frac{1}{6}x + 1$

As a check, we choose some other value for x, say -6, and determine y:

$$y = -\frac{1}{6}(-6) + 1 = 1 + 1 = 2$$

We plot the point $(-6, 2)$ and see that it *is* on the line.

57. $g(x) = -0.25x$

Slope is -0.25, or $\frac{-1}{4}$; y-intercept is $(0, 0)$.

From the y-intercept, we go *down* 1 unit and to the *right* 4 units. This gives us the point $(4, -1)$. We can now draw the graph.

$g(x) = -0.25x$

As a check, we can rename the slope and find another point. $\frac{-1}{4} = \frac{-1}{4} \cdot \frac{-1}{-1} = \frac{1}{-4}$

From the y-intercept, we go *up* 1 unit and to the *left* 4 units. This gives us the point $(-4, 1)$. Since $(-4, 1)$ is on the line, we have a check.

59. Convert to a slope-intercept equation.

$$4x - 5y = 10$$
$$-5y = -4x + 10$$
$$y = \frac{4}{5}x - 2$$

Slope is $\frac{4}{5}$; y-intercept is $(0, -2)$.

From the y-intercept, we go *up* 4 units and to the *right* 5 units. This gives us the point $(5, 2)$. We can now draw the graph.

$4x - 5y = 10, \text{ or } y = \frac{4}{5}x - 2$

As a check, we choose some other value for x, say -5, and determine y:

$$y = \frac{4}{5}(-5) - 2 = -4 - 2 = -6$$

We plot the point $(-5, -6)$ and see that it *is* on the line.

61. Convert to a slope-intercept equation.

$$2x + 3y = 6$$
$$3y = -2x + 6$$
$$y = -\frac{2}{3}x + 2$$

Slope is $-\frac{2}{3}$; y-intercept is $(0, 2)$.

From the y-intercept, we go *down* 2 units and to the *right* 3 units. This gives us the point $(3, 0)$. We can now draw the graph.

$2x + 3y = 6,$
or
$y = -\frac{2}{3}x + 2$

As a check, we choose some other value for x, say -3, and determine y:

$$y = -\frac{2}{3}(-3) + 2 = 2 + 2 = 4$$

We plot the point $(-3, 4)$ and see that it *is* on the line.

63. Convert to a slope-intercept equation.

$$5 - y = 3x$$
$$-y = 3x - 5$$
$$y = -3x + 5$$

Slope is -3, or $\frac{-3}{1}$; y-intercept is $(0, 5)$.

From the y-intercept, we go *down* 3 units and to the *right* 1 unit. This gives us the point $(1, 2)$. We can now draw the graph.

$5 - y = 3x,$
or
$y = -3x + 5$

As a check, we choose some other value for x, say -1, and determine y:

$$y = -3(-1) + 5 = 3 + 5 = 8$$

We plot the point $(-1, 8)$ and see that it *is* on the line.

65. $g(x) = 4.5 = 0x + 4.5$

Slope is 0; y-intercept is $(0, 4.5)$.

From the y-intercept, we go up or down 0 units and any number of nonzero units to the left or right. Any point on the graph will lie on a horizontal line 4.5 units above the x-axis. We draw the graph.

67. We can use the coordinates of any two points on the line. Let's use $(0, 5)$ and $(4, 6)$.

$$\text{Rate of change} = \frac{\text{change in } y}{\text{change in } x} = \frac{6 - 5}{4 - 0} = \frac{1}{4}$$

The distance from home is increasing at a rate of $\frac{1}{4}$ km per minute.

69. We can use the coordinates of any two points on the line. We'll use $(0, 100)$ and $(9, 40)$.

$$\text{Rate of change} = \frac{\text{change in } y}{\text{change in } x} = \frac{40 - 100}{9 - 0} = \frac{-60}{9} = -\frac{20}{3},$$

or $-6\frac{2}{3}$

The distance from the finish line is decreasing at a rate of $6\frac{2}{3}$ m per second.

71. We can use the coordinates of any two points on the line. We'll use $(3, 2.5)$ and $(6, 4.5)$.

$$\text{Rate of change} = \frac{\text{change in } y}{\text{change in } x} = \frac{2.5 - 4.5}{3 - 6} = \frac{-2}{-3} = \frac{2}{3}$$

The number of bookcases stained is increasing at a rate of $\frac{2}{3}$ bookcase per quart of stain used.

73. We can use the coordinates of any two points on the line. We'll use $(30, 480)$ and $(50, 500)$.

$$\text{Rate of change} = \frac{\text{change in } y}{\text{change in } x} = \frac{480 - 500}{30 - 50} = \frac{-20}{-20}, \text{ or } 1$$

The average SAT math score is increasing at a rate of 1 point per thousand dollars of family income.

75. a. Graph II indicated that 200 ml of fluid was dripped in the first 3 hr, a rate of $\frac{200}{3}$ ml/hr. It also indicates that 400 ml of fluid was dripped in the next 3 hr, a rate of $\frac{400}{3}$ ml/hr, and that this rate continues until the end of the time period shown. Since the rate of $\frac{400}{3}$ ml/hr is double the rate of $\frac{200}{3}$ ml/hr, this graph is appropriate for the given situation.

b. Graph IV indicates that 300 ml of fluid was dripped in the first 2 hr, a rate of 300/2, or 150 ml/hr. In the next 2 hr, 200 ml was dripped. This is a rate of 200/2, or 100 ml/hr. Then 100 ml was dripped in the next 3 hr, a rate of 100/3, or $33\frac{1}{3}$ ml/hr. Finally, in the remaining 2 hr, 0 ml of fluid was dripped, a rate of 0/2, or 0 ml/hr. Since the rate at which the fluid was given decreased as time progressed and eventually became 0, this graph is appropriate for the given situation.

c. Graph I is the only graph that shows a constant rate for 5 hours, in this case from 3 PM to 8 PM. Thus, it is appropriate for the given situation.

d. Graph III indicates that 100 ml of fluid was dripped in the first 4 hr, a rate of 100/4, or 25 ml/hr. In the next 3 hr, 200 ml was dripped. This is a rate of 200/3, or $66\frac{2}{3}$ ml/hr. Then 100 ml was dripped in the next hour, a rate of 100 ml/hr. In the last hour 200 ml was dripped, a rate of 200 ml/hr. Since the rate at which the fluid was given gradually increased, this graph is appropriate for the given situation.

77. The average rate of change is given by

$$\frac{\text{change in percentage}}{\text{change in time}}$$

$$\frac{\text{change in percentage}}{\text{change in time}} = \frac{394.8 - 142.8}{2011 - 2001} = \frac{252}{10} = 25.2$$

The average rate of change is increasing at a rate of 25.2 messages per year.

79. The average rate of descent is given by

$$\frac{\text{change in altitude}}{\text{change in time}}.$$ We will express time in minutes:

$$1\frac{1}{2} \text{ hr} = \frac{3}{2} \text{ hr} \cdot \frac{60 \text{ min}}{1 \text{ hr}} = 90 \text{ min}$$

$$2 \text{ hr}, 10 \text{ min} = 2 \text{ hr} + 10 \text{ min}$$

$$= 2 \text{ hr} \cdot \frac{60 \text{ min}}{1 \text{ hr}} + 10 \text{ min} = 120 \text{ min} + 10 \text{ min} = 130 \text{ min}$$

Then

$$\frac{\text{change in altitude}}{\text{change in time}} = \frac{0 - 12,000}{130 - 90} = \frac{-12,000}{40} = -300.$$

The average rate of descent is 300 ft/min.

81. The rate at which the number of hits is increasing is given by $\dfrac{\text{change in number of hits}}{\text{change in time}}$.

$$\frac{\text{change in number of hits}}{\text{change in time}} = \frac{430,000 - 80,000}{2017 - 2015}$$

$$= \frac{350,000}{2} = 175,000$$

The number of hits is increasing at a rate of 175,000 hits/yr.

83. $C(d) = 0.75d + 30$

0.75 signifies the cost per mile is \$0.75; 30 signifies that the minimum cost to rent a truck is \$30.

85. $L(t) = \frac{1}{2}t + 5$

$\frac{1}{2}$ signifies that Lauren's hair grows $\frac{1}{2}$ in. per month; 5 signifies that her hair is 5 in. long immediately after she gets it cut.

87. $A(t) = \frac{1}{7}t + 75.5$

$\frac{1}{7}$ signifies that the life expectancy of American women increases $\frac{1}{7}$ yr per year, for years after 1970;

75.5 signifies that the life expectancy in 1970 was 75.5 yr.

89. $P(t) = 0.67t + 23.21$

0.67 signifies that the average price of a ticket increases by \$0.67 per year, for years after 2006; 23.21 signifies that the cost of a ticket is \$23.21 in 2006.

91. $c(t) = 8.5t + 550$

8.5 signifies that the amount of carbon emissions from building operations increases 8.5 metric tons per year, for years after 1984; 550 signifies that 550 metric tons of carbon was emitted from buildings in 1984.

93. $F(t) = -5000t + 90,000$

a. −5000 signifies that the truck's value depreciates \$5000 per year; 90,000 signifies that the original value of the truck was \$90,000.

b. We find the value of t for which $F(t) = 0$.

$$0 = -5000t + 90,000$$
$$5000t = 90,000$$
$$t = 18$$

It will take 18 yr for the truck to depreciate completely.

c. The truck's value goes from \$90,000 when $t = 0$ to \$0 when $t = 18$, so the domain of F is $\{x \mid 0 \leq t \leq 18\}$.

95. $v(n) = -200n + 1800$

a. −200 signifies that the depreciation is \$200 per year; 1800 signifies that the original value of the bike was \$1800.

b. We find the value of n for which $v(n) = 600$.

$$600 = -200n + 1800$$
$$-1200 = -200n$$
$$6 = n$$

The trade-in value is \$600 after 6 yrs of use.

c. First we find the value of n for which $v(n) = 0$.

$$0 = -200n + 1800$$
$$-1800 = -200n$$
$$9 = n$$

The value of the bike goes from \$1800 when $n = 0$, to \$0 when $n = 9$, so the domain of v is $\{n \mid 0 \leq n \leq 9\}$.

97. *Writing Exercise.*

99. $-\frac{2}{3} - \frac{1}{2} = -\frac{4}{6} - \frac{3}{6} = -\frac{7}{6}$

101. $-12.9 \div (-3) = 4.3$

103. *Writing Exercise.*

105. a. Graph III indicates that the first 2 mi and the last 3 mi were traveled in approximately the same length of time and at a fairly rapid rate. The mile following the first two miles was traveled at a much slower rate. This could indicate that the first two miles were driven, the next mile was swum and the last three miles were driven, so this graph is most appropriate for the given situation.

b. The slope in Graph IV decreases at 2 mi and again at 3 mi. This could indicate that the first two miles were traveled by bicycle, the next mile was run, and the last 3 miles were walked, so this graph is most appropriate for the given situation.

c. The slope in Graph I decreases at 2 mi and then increases at 3 mi. This could indicate that the first two miles were traveled by bicycle, the next mile was hiked, and the last three miles were traveled by bus, so this graph is most appropriate for the given situation.

d. The slope in Graph II increases at 2 mi and again at 3 mi. This could indicate that the first two miles were hiked, the next mile was run, and the last three miles were traveled by bus, so this graph is most appropriate for the given situation.

107. The longest uphill climb is the widest rising line. It is the trip from Sienna to Castellina in Chianti.

109. Reading from the graph the trip from Castellina in Chianti to Ponte sul Pesa is downhill, then to Panzano is uphill and then to Creve in Chianti is downhill. All sections are about the same grade. So the trip began at Castellina in Chianti.

111. $rx + py = s - ry$

$ry + py = -rx + s$

$y(r + p) = -rx + s$

$y = -\dfrac{r}{r+p}x + \dfrac{s}{r+p}$

The slope is $-\dfrac{r}{r+p}$, and the y-intercept is $\left(0, \dfrac{s}{r+p}\right)$.

113. Since (x_1, y_1) and (x_2, y_2) are two points on the graph of $y = mx + b$, then $y_1 = mx_1 + b$ and $y_2 = mx_2 + b$. Using the definition of slope, we have

$$\text{Slope} = \frac{y_2 - y_1}{x_2 - x_1}$$

$$= \frac{(mx_2 + b) - (mx_1 + b)}{x_2 - x_1}$$

$$= \frac{m(x_2 - x_1)}{x_2 - x_1}$$

$$= m.$$

115. Let $c = 1$ and $d = 2$. Then

$f(c + d) = f(1 + 2) = f(3) = 3m + b$, but

$f(c) + f(d) = (m + b) + (2m + b) = 3m + 2b$.

The given statement is false.

117. Let $k = 2$. Then $f(kx) = f(2x) = 2mx + b$, but $kf(x) = 2(mx + b) = 2mx + 2b$. The given statement is false.

119. a. $\dfrac{-c - (-6c)}{b - 5b} = \dfrac{5c}{-4b} = -\dfrac{5c}{4b}$

b. $\dfrac{(d + e) - d}{b - b} = \dfrac{e}{0}$

Since we cannot divide by 0, the slope is undefined.

c. $\dfrac{(-a - d) - (a + d)}{(c - f) - (c + f)} = \dfrac{-a - d - a - d}{c - f - c - f}$

$$= \frac{-2a - 2d}{-2f}$$

$$= \frac{-2(a + d)}{-2f}$$

$$= \frac{a + d}{f}$$

121.
$y_1 = 1.4x + 2, \quad y_2 = 0.6x + 2,$
$y_3 = 1.4x + 5, \quad y_4 = 0.6x + 5$

123. *Writing Exercise.*

Exercise Set 2.4

1. Every *horizontal* line has a slope of 0.

3. The slope of a vertical line is *undefined*.

5. To find the *x*-intercept, we let $y = 0$ and solve the original equation for x.

7. To solve $3x - 5 = 7$, we can graph $f(x) = 3x - 5$ and $g(x) = 7$ and find the *x*-value at the point of *intersection*.

9. Only *linear* equations have graphs that are straight lines.

11. $y - 2 = 6$

$y = 8$

The graph of $y = 8$ is a horizontal line. Since $y - 2 = 6$ is equivalent to $y = 8$, the slope of the line $y - 2 = 6$ is 0.

13. $8x = 6$

$x = \dfrac{3}{4}$

The graph of $x = \dfrac{3}{4}$ is a vertical line. Since $8x = 6$ is equivalent to $x = \dfrac{3}{4}$, the slope of the line $8x = 6$ is undefined.

15. $3y = 28$

$y = \dfrac{28}{3}$

The graph of $y = \dfrac{28}{3}$ is a horizontal line. Since $3y = 28$ is equivalent to $y = \dfrac{28}{3}$, the slope of the line $3y = 28$ is 0.

17. $5 - x = 12$

$x = -7$

The graph of $x = -7$ is a vertical line. Since $5 - x = 12$ is equivalent to $x = -7$, the slope of the line $5 - x = 12$ is undefined.

19. $2x - 4 = 3$
$$2x = 7$$
$$x = \frac{7}{2}$$

The graph of $x = \frac{7}{2}$ is a vertical line. Since $2x - 4 = 3$ is equivalent to $x = \frac{7}{2}$, the slope of the line $2x - 4 = 3$ is undefined.

21. $5y - 4 = 35$
$$5y = 39$$
$$y = \frac{39}{5}$$

The graph of $y = \frac{39}{5}$ is a horizontal line. Since $5y - 4 = 35$ is equivalent to $y = \frac{39}{5}$, the slope of the line $5y - 4 = 35$ is 0.

23. $4y - 3x = 9 - 3x$
$$y = \frac{9}{4}$$

The graph of $y = \frac{9}{4}$ is a horizontal line. Since $4y - 3x = 9 - 3x$ is equivalent to $y = \frac{9}{4}$, the slope of the line $4y - 3x = 9 - 3x$ is 0.

25. $5x - 2 = 2x - 7$
$$5x = 2x - 5$$
$$3x = -5$$
$$x = -\frac{5}{3}$$

The graph of $x = -\frac{5}{3}$ is a vertical line. Since $5x - 2 = 2x - 7$ is equivalent to $x = -\frac{5}{3}$, the slope of the line $5x - 2 = 2x - 7$ is undefined.

27. $y = -\frac{2}{3}x + 5$

The equation is written in slope-intercept form. We see that the slope is $-\frac{2}{3}$.

29. Graph $y = 4$.

This is a horizontal line that crosses the y-axis at $(0,\ 4)$. If we find some ordered pairs, note that, for any x-value chosen, y must be 4.

x	y
-2	4
0	4
3	4

31. Graph $x = 3$.

This is a vertical line that crosses the x-axis at $(3, 0)$. If we find some ordered pairs, note that, for any y-value

chosen, x must be 3.

x	y
3	-1
3	0
3	2

33. Graph $f(x) = -2$.

This is a horizontal line that crosses the y-axis at $(0, -2)$.

35. Graph $3x = -15$.

Since y does not appear, we solve for x.
$$3x = -15$$
$$x = -5$$

This is a vertical line that crosses the x-axis at $(-5, 0)$.

$$3x = -15$$

37. Graph $3 \cdot g(x) = 15$.

First solve for $g(x)$.
$$3 \cdot g(x) = 15$$
$$g(x) = 5$$

This is a horizontal line that crosses the vertical axis at $(0,\ 5)$.

$$3 \cdot g(x) = 15$$

39. We first solve for y and determine the slope of each line.
$$x + 2 = y$$
$$y = x + 2 \qquad \text{Reversing the order}$$

The slope of $y = x + 2$ is 1.
$$y - x = -2$$
$$y = x - 2$$

The slope of $y = x - 2$ is 1.

The slopes are the same; the lines are parallel.

41. We first solve for y and determine the slope of each line.
$$y + 9 = 3x$$
$$y = 3x - 9$$

The slope of $y = 3x - 9$ is 3.
$$3x - y = -2$$
$$3x + 2 = y$$
$$y = 3x + 2 \qquad \text{Reversing the order}$$

The slope of $y = 3x + 2$ is 3.

The slopes are the same; the lines are parallel.

43. We determine the slope of each line.

The slope of $f(x) = 3x + 9$ is 3.

$$2y = 8x - 2$$
$$y = 4x - 1$$

The slope of $y = 4x - 1$ is 4.

The slopes are not the same; the lines are not parallel.

45. We determine the slope of each line.

$$x - 2y = 3$$
$$-2y = -x + 3$$
$$y = \frac{1}{2}x - \frac{3}{2}$$

The slope of $x - 2y = 3$ is $\frac{1}{2}$.

$$4x + 2y = 1$$
$$2y = -4x + 1$$
$$y = -2x + \frac{1}{2}$$

The slope of $4x + 2y = 1$ is -2.

The product of their slopes is $\left(\frac{1}{2}\right)(-2)$, or -1; the lines are perpendicular.

47. We determine the slope of each line.

The slope of $f(x) = 3x + 1$ is 3.

$$6x + 2y = 5$$
$$2y = -6x + 5$$
$$y = -3x + \frac{5}{2}$$

The slope of $6x + 2y = 5$ is -3.

The product of their slopes is $3(-3)$, or $-9 \neq -1$, so the lines are not perpendicular.

49. Graph $x + y = 4$.

To find the y-intercept, let $x = 0$ and solve for y.

$$0 + y = 4$$
$$y = 4$$

The y-intercept is (0, 4).

To find the x-intercept, let $y = 0$ and solve for x.

$$x + 0 = 4$$
$$x = 4$$

The x-intercept is (4, 0).

Plot these points and draw the line. A third point could be used as a check.

51. Graph $f(x) = 2x - 6$.

Because the function is in slope-intercept form, we know that the y-intercept is (0, –6). To find the x-intercept, let $f(x) = 0$ and solve for x.

$$0 = 2x - 6$$
$$6 = 2x$$
$$3 = x$$

The x-intercept is (3, 0).

Plot these points and draw the line. A third point could

be used as a check.

53. Graph $3x + 5y = -15$.

To find the y-intercept, let $x = 0$ and solve for y.

$$3 \cdot 0 + 5y = -15$$
$$5y = -15$$
$$y = -3$$

The y-intercept is $(0, -3)$.

To find the x-intercept, let $y = 0$ and solve for x.

$$3x + 5 \cdot 0 = -15$$
$$3x = -15$$
$$x = -5$$

The x-intercept is $(-5, 0)$.

Plot these points and draw the line. A third point could be used as a check.

55. Graph $2x - 3y = 18$.

To find the y-intercept, let $x = 0$ and solve for y.

$$2 \cdot 0 - 3y = 18$$
$$-3y = 18$$
$$y = -6$$

The y-intercept is $(0, -6)$.

To find the x-intercept, let $y = 0$ and solve for x.

$$2x - 3 \cdot 0 = 18$$
$$2x = 18$$
$$x = 9$$

The x-intercept is (9, 0).

Plot these points and draw the line. A third point could be used as a check.

57. Graph $3y = -12x$.

To find the y-intercept, let $x = 0$ and solve for y.

$$3y = -12 \cdot 0$$
$$3y = 0$$
$$y = 0$$

The y-intercept is (0, 0). This is also the x-intercept.

We find another point on the line. Let $x = 1$ and find the corresponding value of y.

$$3y = -12 \cdot 1$$
$$3y = -12$$
$$y = -4$$

The point (1, –4) is on the graph.

Plot these points and draw the line. A third point could

be used as a check.

59. Graph $f(x) = 3x - 7$.

Because the function is in slope-intercept form, we know that the y-intercept is $(0, -7)$. To find the x-intercept, let $f(x) = 0$ and solve for x.

$$0 = 3x - 7$$
$$7 = 3x$$
$$\frac{7}{3} = x$$

The x-intercept is $\left(\frac{7}{3}, 0\right)$.

Plot these points and draw the line. A third point could be used as a check.

61. Graph $5y - x = 5$.

To find the y-intercept, let $x = 0$ and solve for y.

$$5y - 0 = 5$$
$$5y = 5$$
$$y = 1$$

The y-intercept is $(0, 1)$.

To find the x-intercept, let $y = 0$ and solve for x.

$$5 \cdot 0 - x = 5$$
$$-x = 5$$
$$x = -5$$

The x-intercept is $(-5, 0)$.

Plot these points and draw the line. A third point could be used as a check.

63. $0.2y - 1.1x = 6.6$
  $2y - 11x = 66$ Multiplying by 10

Graph $2y - 11x = 66$.

To find the y-intercept, let $x = 0$.

$$2y - 11x = 66$$
$$2y - 11 \cdot 0 = 66$$
$$2y = 66$$
$$y = 33$$

The y-intercept is $(0, 33)$.

To find the x-intercept, let $y = 0$.

$$2y - 11x = 66$$
$$2 \cdot 0 - 11x = 66$$
$$-11x = 66$$
$$x = -6$$

The x-intercept is $(-6, 0)$.

Plot these points and draw the line. A third point could

be used as a check.

65. $x + 2 = 3$

Graph $f(x) = x + 2$ and $g(x) = 3$ on the same grid.

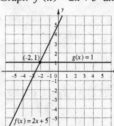

The lines appear to intersect at $(1, 3)$, so the solution is apparently 1.

Check: $\dfrac{x + 2 = 3}{1 + 2 \ \vert \ 3}$
$$\overset{?}{3 = 3} \quad \text{TRUE}$$

The solution is 1.

67. $2x + 5 = 1$

Graph $f(x) = 2x + 5$ and $g(x) = 1$ on the same grid.

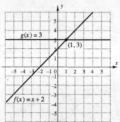

The lines appear to intersect at $(-2, 1)$, so the solution is apparently -2.

Check: $\dfrac{2x + 5 = 1}{2(-2) + 5 \ \vert \ 1}$
$$-4 + 5 \ \vert$$
$$\overset{?}{1 = 1} \quad \text{TRUE}$$

The solution is -2.

69. $\frac{1}{2}x + 3 = 5$

Graph $f(x) = \frac{1}{2}x + 3$ and $g(x) = 5$ on the same grid.

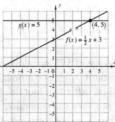

The lines appear to intersect at $(4, 5)$, so the solution is apparently 4.

Check:

$$\frac{1}{2}x + 3 = 5$$

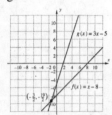

5 = 5   TRUE

The solution is 4.

71. $x - 8 = 3x - 5$

Graph $f(x) = x - 8$ and $g(x) = 3x - 5$ on the same grid.

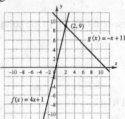

The lines appear to intersect at $\left(-\frac{3}{2}, -\frac{19}{2}\right)$, so the

solution is apparently $-\frac{3}{2}$.

Check:

$$
\begin{array}{c|c}
\multicolumn{2}{c}{x - 8 = 3x - 5} \\
\hline
-\frac{3}{2} - 8 & 3\left(-\frac{3}{2}\right) - 5 \\
-\frac{19}{2} & -\frac{9}{2} - 5 \\
& -\frac{19}{2} \stackrel{?}{=} -\frac{19}{2} \quad \text{TRUE}
\end{array}
$$

The solution is $-\frac{3}{2}$.

73. $4x + 1 = -x + 11$

Graph $f(x) = 4x + 1$ and $g(x) = -x + 11$ on the same grid.

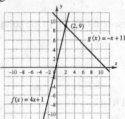

The lines appear to intersect at (2, 9), so the solution is apparently 2.

Check: $4x + 1 = -x + 11$

$$
\begin{array}{c|c}
4 \cdot 2 + 1 & -2 + 11 \\
8 + 1 & \\
& 9 \stackrel{?}{=} 9 \quad \text{TRUE}
\end{array}
$$

The solution is 2.

75. *Familiarize*. After an initial fee of $75, an additional fee of $35 is charge each month. After one month, the total cost is $75 + $35 = $110. After two months, the total cost is $75 + 2 \cdot 35 = \$145$. We can generalize this with a model, letting $C(t)$ represent the total cost, in dollars, for t months of membership.

Translate. We reword the problem and translate.

Total Cost	is	initial cost	plus	$35 per month.
↓	↓	↓	↓	↓
$C(t)$	=	75	+	$35t$

Carry out. First we write the model in slope-intercept form. $C(t) = 35t + 75$. The vertical intercept is (0, 75), and the slope, or rate, is $35 per month. Plot (0, 75) and from there go up $35 and to the right 1 month. This takes us to (1, 110). Draw a line passing through both points.

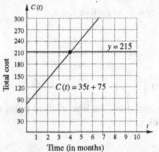

To estimate the time required for the total cost to reach $215, we are estimating the solution of $35t + 75 = 215$. We do this by graphing $y = 215$ and finding the point of intersection of the graphs. This point appears to be (4, 215). Thus, we estimate that it takes 4 months for the total cost to reach $215.

Check. We evaluate.

$$
\begin{aligned}
C(4) &= 35 \cdot 4 + 75 \\
&= 140 + 75 \\
&= 215
\end{aligned}
$$

The estimate is precise.

State. It takes 4 months for the total cost to reach $215.

77. *Familiarize*. After paying the $150 for paint, Gavin charges 70¢ per square foot. Let x = the number of square feet.

Translate. We reword the problem and translate.
For a $750 bill, the Gavin charges $150 + 0.70x = 750$.

Total charge	is	$150	plus	0.70	times	the number of square feet.
↓	↓	↓	↓	↓	↓	↓
$C(x)$	=	150	+	0.70	×	x

Carry out. First we rewrite the model in slope-intercept form.

$$
\begin{aligned}
C(x) &= 150 + 0.7x \\
&= 0.7x + 150
\end{aligned}
$$

The vertical intercept is (0, 150) and the slope is 0.7. We plot (0, 150) and, from there, to up 7 units and right 10 units to (10, 157). Draw a line through both points.

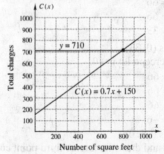

First we will find the value of x for which $C(x) = 710$. We find the solution of $0.7x + 150 = 710$. We graph $y = 710$ and find the point of intersection of the graphs. This point appears to be $(800, 710)$. Thus, Gavin painted 800 square feet.

Check. We evaluate.

$C(800) = 0.7(800) + 150 = 560 + 150 = 710$

The estimate is precise.

State. Gavin painted 800 square feet.

79. **Familiarize**. After paying the first \$3000, the patient must pay $\frac{1}{4}$ of all additional charges. For a \$8500 bill, the patient pays $\$3000 + \frac{1}{4}(\$8500 - \$3000)$, or \$4375.

We can generalize this with a model, letting $C(b)$ represent the total cost to the patient, in dollars, for a hospital bill of b dollars.

Translate. We reword the problem and translate.

Total Cost to patient	is	\$3000	plus	$\frac{1}{4}$	of	Cost in excess of \$3000.
↓	↓	↓	↓	↓	↓	↓
$C(b)$	=	3000	+	$\frac{1}{4}$	×	$(b - 3000)$

Carry out. First we rewrite the model in slope-intercept form.

$$C(b) = 3000 + \frac{1}{4}(b - 3000)$$
$$= 3000 + \frac{1}{4}b - 750$$
$$= \frac{1}{4}b + 2250$$

The vertical intercept is $(0, 2250)$ and the slope is $\frac{1}{4}$.

We plot $(0, 2250)$. If we evaluate the function at 7000,

$C(7000) = \frac{1}{4}(7000) + 2250 = 1750 + 2250 = 4000$, so

we get the point $(7000, 4000)$. Draw a line through both points.

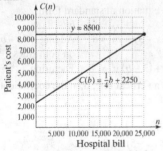

First we will find the value of b for which $C(b) = 8500$. Then we will subtract 3000 to find by how much Nancy's bill exceeded \$3000. We find the solution of $\frac{1}{4}b + 2250 = 8500$. We graph $y = 8500$ and find the point of intersection of the graphs. This point appears to be $(25,000, 8500)$ Thus, Nancy's total hospital bill was \$25,000 and it exceeded \$3000 by \$25,000 − \$3000, or \$22,000.

Check. We evaluate.

$C(25,000) = \frac{1}{4}(25,000) + 2250 = 6250 + 2250 = 8500$

The estimate is precise.

State. Nancy's hospital bill exceeded \$3000 by \$22,000.

81. **Familiarize**. After an initial \$5.00 parking fee, an additional 50¢ fee is changed for each 15-min unit of time. After one 15-min unit of time, the cost is \$5.00 + \$0.50, or \$5.50. After two 15-min units, or 30 min, the cost is \$5.00 + 2(\$0.50), or \$6.00. We can generalize this with a model if we let $C(t)$ represent the total cost, in dollars, for t 15-min units of time.

Translate. We reword the problem and translate.

Total Cost	is	initial cost	plus	\$0.50 per 15-min time unit.
↓	↓	↓	↓	↓
$C(t)$	=	5.00	+	$0.50t$

Carry out. First write the model in slope-intercept form: $C(t) = 0.50t + 5$. The vertical intercept is $(0, 5)$ and the slope, or rate, is 0.50, or $\frac{1}{2}$. Plot $(0, 5)$ and from there go up \$1, and to the right 2 15-min units of time. This takes us to $(2, 6)$. Draw a line passing through both points.

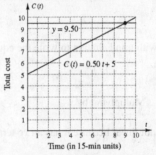

To estimate how long someone can park for \$9.50, we are estimating the solution of $0.50t + 5 = 9.50$. We do this by graphing $y = 9.50$ and finding the point of intersection of the graphs. The point appears to be $(9, 9.50)$. Thus, we estimate that someone can park for nine 15-min units of time, or 2 hr, 15 min, for \$9.50.

Check. We evaluate.

$$C(9) = 0.50(9) + 5$$
$$= 4.50 + 5$$
$$= 9.50$$

The estimate is precise.

State. Someone can park for 2 hr 15 min for \$9.50.

83. $5x - 3y = 15$

This equation is in the standard form for a linear equation, $Ax + By = C$, with $A = 5$, $B = -3$, and $C = 15$. Thus, it is a linear equation.

Solve for y to find the slope.

$$5x - 3y = 15$$
$$-3y = -5x + 15$$
$$y = \frac{5}{3}x - 5$$

The slope is $\frac{5}{3}$.

85. $8x + 40 = 0$
$$8x = -40$$
$$x = -5$$
The equation is linear. Its graph is a vertical line.

87. $4g(x) = 6x^2$

Replace $g(x)$ with y and attempt to write the equation in standard form.
$$4y = 6x^2$$
$$-6x^2 + 4y = 0$$

The equation is not linear, because it has an x^2-term.

89. $3y = 7(2x - 4)$
$$3y = 14x - 28$$
$$-14x + 3y = -28$$

The equation can be written in the standard form for a linear equation, $Ax + By = C$, with $A = -14$, $B = 3$, and $C = -28$. Thus, it is a linear equation. Solve for y to find the slope.
$$-14x + 3y = -28$$
$$3y = 14x - 28$$
$$y = \frac{14}{3}x - \frac{28}{3}$$

The slope is $\frac{14}{3}$.

91. $f(x) - \frac{5}{x} = 0$

Replace $f(x)$ with y and attempt to write the equation in standard form.
$$y - \frac{5}{x} = 0$$
$$xy - 5 = 0 \qquad \text{Multiplying by } x$$
$$xy = 5$$

The equation is not linear because it has an xy-term.

93. $\frac{y}{3} = x$
$$y = 3x$$
$$3x - y = 0$$

The equation can be written in standard form for the linear equation $Ax + By = C$, with $A = 3$, $B = -1$, and $C = 0$. Thus, it is a linear equation. Solve for y to find the slope.
$$y = 3x$$

The slope is 3.

95. $xy = 10$

The equation is not linear, because it has an xy-term.

97. *Writing Exercise.*

99. $2(x - 7) = 3 - x$
$$2x - 14 = 3 - x$$
$$3x = 17$$
$$x = \frac{17}{3}$$

101. $n - (8 - n) = 2(n - 4)$
$$n - 8 + n = 2n - 8$$
$$2n - 8 = 2n - 8$$

This is an identity. The solution set is $\mathbb{R}$.

103. $4(t - 6) = 7t - 15 - 3t$
$$4t - 24 = 4t - 25$$
$$-24 = -25 \quad \text{FALSE}$$

This is a contradiction. The solution set is $\varnothing$.

105. *Writing Exercise.*

107. The line contains the points $(5, 0)$ and $(0, -4)$. We use the points to find the slope.
$$\text{Slope} = \frac{-4 - 0}{0 - 5} = \frac{-4}{-5} = \frac{4}{5}$$

Then the slope-intercept equation is $y = \frac{4}{5}x - 4$. We rewrite this equation in standard form.
$$y = \frac{4}{5}x - 4$$
$$5y = 4x - 20 \quad \text{Multiplying by 5 on both sides}$$
$$-4x + 5y = -20 \qquad \text{Standard form}$$

This equation can also be written as $4x - 5y = 20$.

109. $rx + 3y = p^2 - s$

The equation is in standard form with $A = r$, $B = 3$, and $C = p^2 - s$. It is linear.

111. Try to put the equation in standard form.
$$r^2 x = py + 5$$
$$r^2 x - py = 5$$

The equation is in standard form with $A = r^2$, $B = -p$, and $C = 5$. It is linear.

113. Let equation A have intercepts $(a, 0)$ and $(0, b)$. Then equation B has intercepts $(2a, 0)$ and $(0, b)$.
$$\text{Slope of } A = \frac{b - 0}{0 - a} = -\frac{b}{a}$$
$$\text{Slope of } B = \frac{b - 0}{0 - 2a} = -\frac{b}{2a} = \frac{1}{2}\left(-\frac{b}{a}\right)$$

The slope of equation B is $\frac{1}{2}$ the slope of equation A.

115. First write the equation in standard form.
$$ax + 3y = 5x - by + 8$$
$$ax - 5x + 3y + by = 8 \qquad \text{Adding } -5x + by$$
$$\text{on both sides}$$
$$(a - 5)x + (3 + b)y = 8 \qquad \text{Factoring}$$

If the graph is a vertical line, then the coefficient of y is 0.
$$3 + b = 0$$
$$b = -3$$

Then we have $(a - 5)x = 8$.

If the line passes through $(4, 0)$, we have:
$$(a - 5)4 = 8 \qquad \text{Substituting 4 for } x$$
$$a - 5 = 2$$
$$a = 7$$

117. We graph $C(t) = 0.50t + 5$, where t represents the number of 15-min units of time, as a series of steps. The cost is constant within each 15-min unit of time. Thus,

for $0 < t \leq 1$, $C(t) = 0.5(1) + 5 = \$5.50$;
for $1 < t \leq 2$, $C(t) = 0.5(2) + 5 = \$6.00$;
for $2 < t \leq 3$, $C(t) = 0.5(3) + 5 = \$6.50$;
and so on. We draw the graph. An open circle at a point indicates that the point is not on the graph.

119. Graph $y_1 = 4x - 1$ and $y_2 = 3 - 2x$ in the same window and use the Intersect feature to find the first coordinate of the point of intersection, 0.66666667. We check by solving the equation algebraically.

$$4x - 1 = 3 - 2x$$
$$6x - 1 = 3$$
$$6x = 4$$
$$x = \frac{2}{3} = 0.\overline{6} \approx 0.66666667$$

121. Graph $y_1 = 8 - 7x$ and $y_2 = -2x - 5$ in the same window and use the Intersect feature to find the first coordinate of the point of intersection, 2.6. We check by solving the equation algebraically.

$$8 - 7x = -2x - 5$$
$$8 - 5x = -5$$
$$-5x = -13$$
$$x = \frac{13}{5} = 2.6$$

123. Graph $y = 38 + 4.25x$. Use the Intersect feature to find the x-coordinate that corresponds to the y-coordinate 671.25. We find that 149 shirts were printed.

Mid-Chapter Review

1. y-intercept: $y - 3 \cdot 0 = 6$
$$y = 6$$
The y-intercept is (0, 6).
x-intercept: $0 - 3x = 6$
$$-3x = 6$$
$$x = -2$$
The x-intercept is (−2, 0).

2. $m = \dfrac{y_2 - y_1}{x_2 - x_1} = \dfrac{-1 - 5}{3 - 1}$
$$= \dfrac{-6}{2}$$
$$= -3$$

3. $\left(-16, \dfrac{1}{2}\right)$ is in Quadrant II.

4. $f(6) = 6 - 6^2 = -30$

5. $g(x) = \dfrac{2x}{x - 7}$
Solve: $x - 7 = 0$
$$x = 7$$
The domain is $\{x | x$ is a real number and $x \neq 7\}$.

6. $m = \dfrac{8 - (-2)}{1 - (-5)} = \dfrac{10}{6} = \dfrac{5}{3}$

7. $m = \dfrac{-2 - 0}{0 - 0} = \dfrac{-2}{0}$ undefined

8. 0

9. The line $x = -7$ is a vertical line. The slope is undefined.

10. $x - 3y = 1$
$$-3y = -x + 1$$
$$y = \frac{1}{3}x - \frac{1}{3}$$
Slope is $\dfrac{1}{3}$; y-intercept is $\left(0, -\dfrac{1}{3}\right)$.

11. $f(x) = -3x + 7$

12. We determine the slope of each line.
The slope of $f(x) = \frac{1}{4}x - 3$ is $\frac{1}{4}$.
$$4x + y = 8$$
$$y = -4x + 8$$
The slope of $4x + y = 8$ is −4.
The product of their slopes is $\left(\dfrac{1}{4}\right)(-4)$, or −1; the lines are perpendicular.

13. $\frac{1}{3}x - 2 = 2x + 3$

Graph $f(x) = \frac{1}{3}x - 2$ and $g(x) = 2x + 3$ on the same grid.

The lines appear to intersect at (−3,−3), so the solution is apparently −3.
A check verifies that the solution is −3.

14. Yes

15. Graph $y = 2x - 1$.

16. Graph $3x + y = 6$.

17. Graph $y = |x| - 4$.

18. Graph $f(x) = 4$.

19. Graph $f(x) = -\frac{3}{4}x + 5$.

20. Graph $3x = 12$.

Connecting the Concepts

1. $2x + 5y = 8$ is in standard form.

2. $y = \frac{2}{3}x - \frac{11}{3}$ is in slope-intercept form.

3. $x - 13 = 5y$ is none of these forms.

4. $y - 2 = \frac{1}{3}(x - 6)$ is in point-slope form.

5. $x - y = 1$ is in standard form.

6. $y = -18x + 3.6$ is in slope-intercept form.

7.
$$y = \frac{2}{5}x + 1$$
$$5y = 2x + 5$$
$$-2x + 5y = 5 \quad \text{or} \quad 2x - 5y = -5$$

8.
$$y - 1 = -2(x - 6)$$
$$y - 1 = -2x + 12$$
$$2x + y = 13$$

9.
$$3x - 5y = 10$$
$$-5y = -3x + 10$$
$$y = \frac{3}{5}x - 2$$

10.
$$y + 2 = \frac{1}{2}(x - 3)$$
$$y + 2 = \frac{1}{2}x - \frac{3}{2}$$
$$y = \frac{1}{2}x - \frac{7}{2}$$

Exercise Set 2.5

1. False

3. False; given just one point, there is an infinite number of lines that can be drawn through it.

5. True

7.
$$y - 3 = \frac{1}{4}(x - 5)$$
$$y - y_1 = m(x - x_1)$$

$m = \frac{1}{4}$, $x_1 = 5$, $y_1 = 3$, so the slope is $\frac{1}{4}$ and a point (x_1, y_1) on the graph is (5, 3).

9.
$$y + 1 = -7(x - 2)$$
$$y - (-1) = -7(x - 2)$$
$$y - y_1 = m(x - x_1)$$

$m = -7$, $x_1 = 2$, $y_1 = -1$, so the slope is -7 and a point (x_1, y_1) on the graph is (2, –1).

11.
$$y - 6 = -\frac{10}{3}(x + 4)$$
$$y - 6 = -\frac{10}{3}(x - (-4))$$
$$y - y_1 = m(x - x_1)$$

$m = -\frac{10}{3}$, $x_1 = -4$, $y_1 = 6$, so the slope is $-\frac{10}{3}$ and a point (x_1, y_1) on the graph is (–4, 6).

13.
$$y = 5x$$
$$y - 0 = 5(x - 0)$$
$$y - y_1 = m(x - x_1)$$

$m = 5$, $x_1 = 0$, $y_1 = 0$, so the slope is 5 and a point (x_1, y_1) on the graph is (0, 0).

15.
$$y - y_1 = m(x - x_1) \quad \text{Point-slope equation}$$
$$y - 2 = 3(x - 5) \quad \text{Substituting 3 for } m,$$
$$\text{5 for } x_1, \text{ and 2 for } y_1$$

To graph the equation, we count off a slope of $\frac{3}{1}$, starting at (5, 2), and draw the line.

17. $y - y_1 = m(x - x_1)$ Point-slope equation

$y - 2 = -4(x - 1)$ Substituting -4 for m,
1 for x_1, and 2 for y_1

To graph the equation, we count off a slope of $-\dfrac{4}{1}$,

starting at $(1, 2)$, and draw the line.

$y - 2 = -4(x - 1)$

19. $y - y_1 = m(x - x_1)$ Point-slope equation

$y - (-4) = \dfrac{1}{2}[x - (-2)]$ Substituting $\dfrac{1}{2}$ for m,
-2 for x_1, and -4 for y_1

To graph the equation, we count off a slope of $\dfrac{1}{2}$,

starting at $(-2, -4)$, and draw the line.

$y - (-4) = \frac{1}{2}(x - (-2))$,
or $y + 4 = \frac{1}{2}(x + 2)$

21. $y - y_1 = m(x - x_1)$ Point-slope equation

$y - 0 = -1(x - 8)$ Substituting -1 for m,
8 for x_1, and 0 for y_1

To graph the equation, we count off a slope of $\dfrac{-1}{1}$,

starting at $(8, 0)$, and draw the line.

$y - 0 = -1(x - 8)$, or
$y = -(x - 8)$

23. From $y = 3x - 7$, the parallel slope is 3.
The y-intercept is $(0, 4)$.

$y = mx + b$
$y = 3x + 4$

25. From $y = -\dfrac{3}{4}x + 1$, the perpendicular slope is $\dfrac{4}{3}$.

The y-intercept is $(0, -12)$.

$y = mx + b$
$y = \dfrac{4}{3}x - 12$

27. $2x - 3y = 4$

$-3y = -2x + 4$

$y = \dfrac{2}{3}x - \dfrac{4}{3}$

The parallel slope is $\dfrac{2}{3}$. The y-intercept is $\left(0, \dfrac{1}{2}\right)$.

$y = mx + b$
$y = \dfrac{2}{3}x + \dfrac{1}{2}$

29. $x + y = 18$

$y = -x + 18$

The perpendicular slope is 1. The y-intercept is $(0, -32)$.

$y = mx + b$
$y = x - 32$

31. $m = 6,\ (7, 1)$

$y - y_1 = m(x - x_1)$ Point-slope equation
$y - 1 = 6(x - 7)$

33. $m = -5,\ (3, 4)$

$y - y_1 = m(x - x_1)$ Point-slope equation
$y - 4 = -5(x - 3)$

35. $m = \dfrac{1}{2},\ (-2, -5)$

$y - y_1 = m(x - x_1)$ Point-slope equation
$y - (-5) = \dfrac{1}{2}(x - (-2))$

37. $m = -1,\ (9, 0)$

$y - y_1 = m(x - x_1)$ Point-slope equation
$y - 0 = -1(x - 9)$

39. $y - y_1 = m(x - x_1)$ Point-slope equation

$y - (-4) = 2(x - 1)$ Substituting 2 for m,
1 for x_1, and -4 for y_1

$y + 4 = 2x - 2$ Simplifying
$y = 2x - 6$ Subtracting 4 from both sides
$f(x) = 2x - 6$ Using function notation

To graph the equation, we count off a slope of $\dfrac{2}{1}$,

starting at $(1, -4)$ or $(0, -6)$, and draw the line.

$f(x) = 2x - 6$

41. $y - y_1 = m(x - x_1)$ Point-slope equation

$y - 8 = -\dfrac{3}{5}[x - (-4)]$ Substituting $-\dfrac{3}{5}$ for m,
-4 for x_1, and 8 for y_1

$y - 8 = -\dfrac{3}{5}(x + 4)$

$y - 8 = -\dfrac{3}{5}x - \dfrac{12}{5}$ Simplifying

$y = -\dfrac{3}{5}x + \dfrac{28}{5}$ Adding 8 to both sides

$f(x) = -\dfrac{3}{5}x + \dfrac{28}{5}$ Using function notation

To graph the equation, we count off a slope of $\dfrac{-3}{5}$,

starting at $(-4, 8)$, and draw the line.

$f(x) = -\frac{3}{5}x + \frac{28}{5}$

43.
$$y - y_1 = m(x - x_1) \quad \text{Point-slope equation}$$
$$y - (-4) = -0.6[x - (-3)] \quad \text{Substituting } -0.6 \text{ for } m,$$
$$\qquad\qquad\qquad\qquad -3 \text{ for } x_1, \text{ and } -4 \text{ for } y_1$$
$$y + 4 = -0.6(x + 3)$$
$$y + 4 = -0.6x - 1.8$$
$$y = -0.6x - 5.8$$
$$f(x) = -0.6x - 5.8 \quad \text{Using function notation}$$

To graph the equation, we count off a slope of –0.6, or $\frac{-3}{5}$, starting at (–3, –4), and draw the line.

$$f(x) = -0.6x - 5.8$$

45. $m = \frac{2}{7}$; $(0, -6)$

Observe that the slope is $\frac{2}{7}$ and the y-intercept is $(0, -6)$. Thus, we have $f(x) = \frac{2}{7}x - 6$.

To graph the equation, we count off a slope of $\frac{2}{7}$, starting at $(0, -6)$, and draw the line.

$$f(x) = \tfrac{2}{7}x - 6$$

47.
$$y - y_1 = m(x - x_1) \quad \text{Point-slope equation}$$
$$y - 6 = \frac{3}{5}[x - (-4)] \quad \text{Substituting } \frac{3}{5} \text{ for } m,$$
$$\qquad\qquad\qquad\qquad -4 \text{ for } x_1, \text{ and } 6 \text{ for } y_1$$
$$y - 6 = \frac{3}{5}(x + 4)$$
$$y - 6 = \frac{3}{5}x + \frac{12}{5}$$
$$y = \frac{3}{5}x + \frac{42}{5}$$
$$f(x) = \frac{3}{5}x + \frac{42}{5} \quad \text{Using function notation}$$

To graph the equation, we count off a slope of $\frac{3}{5}$, starting at (–4, 6), and draw the line.

$$f(x) = \tfrac{3}{5}x + \tfrac{42}{5}$$

49. First solve the equation for y and determine the slope of the given line.
$$x - 2y = 3 \quad \text{Given line}$$
$$-2y = -x + 3$$
$$y = \frac{1}{2}x - \frac{3}{2}$$

The slope of the given line is $\frac{1}{2}$.

The slope of every line parallel to the given line must also be $\frac{1}{2}$. We find the equation of the line with slope $\frac{1}{2}$ and containing the point (2, 5).
$$y - y_1 = m(x - x_1) \quad \text{Point-slope equation}$$
$$y - 5 = \frac{1}{2}(x - 2) \quad \text{Substituting}$$
$$y - 5 = \frac{1}{2}x - 1$$
$$y = \frac{1}{2}x + 4$$

51. First solve the equation for y and determine the slope of the given line.
$$x + y = 7 \quad \text{Given line}$$
$$y = -x + 7$$

The slope of the given line is –1.
The slope of every line parallel to the given line must also be –1. We find the equation of the line with slope –1 and containing the point (–3, 2).
$$y - y_1 = m(x - x_1) \quad \text{Point-slope equation}$$
$$y - 2 = -1(x - (-3)) \quad \text{Substituting}$$
$$y - 2 = -1(x + 3)$$
$$y - 2 = -x - 3$$
$$y = -x - 1$$

53. First solve the equation for y and determine the slope of the given line.
$$2x + 3y = -7 \quad \text{Given line}$$
$$3y = -2x - 7$$
$$y = -\frac{2}{3}x - \frac{7}{3}$$

The slope of the given line is $-\frac{2}{3}$.

The slope of every line parallel to the given line must also be $-\frac{2}{3}$. We find the equation of the line with slope $-\frac{2}{3}$ and containing the point $(-2, -3)$.
$$y - y_1 = m(x - x_1) \quad \text{Point-slope equation}$$
$$y - (-3) = -\frac{2}{3}[x - (-2)] \quad \text{Substituting}$$
$$y + 3 = -\frac{2}{3}(x + 2)$$
$$y + 3 = -\frac{2}{3}x - \frac{4}{3}$$
$$y = -\frac{2}{3}x - \frac{13}{3}$$

55. $x = 2$ is a vertical line. A line parallel to it that passes through $(5, -4)$ is the vertical line 5 units to the right of the y-axis, or $x = 5$.

57. First solve the equation for y and determine the slope of the given line.
$$2x - 3y = 4 \quad \text{Given line}$$
$$-3y = -2x + 4$$
$$y = \frac{2}{3}x - \frac{4}{3}$$

The slope of the given line is $\frac{2}{3}$.

The slope of perpendicular line is given by the opposite

of the reciprocal of $\frac{2}{3}$, $-\frac{3}{2}$. We find the equation of

the line with slope $-\frac{3}{2}$ and containing the point (3, 1).

$$y - y_1 = m(x - x_1) \qquad \text{Point-slope equation}$$
$$y - 1 = -\frac{3}{2}(x - 3) \qquad \text{Substituting}$$
$$y - 1 = -\frac{3}{2}x + \frac{9}{2}$$
$$y = -\frac{3}{2}x + \frac{11}{2}$$

59. First solve the equation for y and determine the slope of the given line.

$$x + y = 6 \qquad \text{Given line}$$
$$y = -x + 6$$

The slope of the given line is -1.
The slope of perpendicular line is given by the opposite of the reciprocal of -1, 1. We find the equation of the line with slope 1 and containing the point $(-4, 2)$.

$$y - y_1 = m(x - x_1) \qquad \text{Point-slope equation}$$
$$y - 2 = 1(x - (-4)) \qquad \text{Substituting}$$
$$y - 2 = x + 4$$
$$y = x + 6$$

61. First solve the equation for y and determine the slope of the given line.

$$3x - y = 2 \qquad \text{Given line}$$
$$-y = -3x + 2$$
$$y = 3x - 2$$

The slope of the given line is 3.
The slope of perpendicular line is given by the opposite of the reciprocal of 3, $-\frac{1}{3}$. We find the equation of the line with slope $-\frac{1}{3}$ and containing the point $(1, -3)$.

$$y - y_1 = m(x - x_1) \qquad \text{Point-slope equation}$$
$$y - (-3) = -\frac{1}{3}(x - 1) \qquad \text{Substituting}$$
$$y + 3 = -\frac{1}{3}x + \frac{1}{3}$$
$$y = -\frac{1}{3}x - \frac{8}{3}$$

63. First solve the equation for y and find the slope of the given line.

$$3x - 5y = 6$$
$$-5y = -3x + 6$$
$$y = \frac{3}{5}x - \frac{6}{5}$$

The slope of the given line is $\frac{3}{5}$. The slope of a perpendicular line is given by the opposite of the reciprocal of $\frac{3}{5}$, $-\frac{5}{3}$.

We find the equation of the line with slope $-\frac{5}{3}$ and

containing the point $(-4, -7)$.

$$y - y_1 = m(x - x_1) \qquad \text{Point-slope equation}$$
$$y - (-7) = -\frac{5}{3}[x - (-4)] \qquad \text{Substituting}$$
$$y + 7 = -\frac{5}{3}(x + 4)$$
$$y + 7 = -\frac{5}{3}x - \frac{20}{3}$$
$$y = -\frac{5}{3}x - \frac{41}{3}$$

65. $y = 5$ is a horizontal line, so a line perpendicular to it must be vertical. The equation of the vertical line containing $(-3, 7)$ is $x = -3$.

67. First find the slope of the line:

$$m = \frac{7 - 3}{3 - 2} = \frac{4}{1} = 4$$

Use the point-slope equation with $m = 4$ and $(2, 3) = (x_1, y_1)$. (We could let $(3, 7) = (x_1, y_1)$ instead to obtain an equivalent equation.)

$$y - 3 = 4(x - 2)$$
$$y - 3 = 4x - 8$$
$$y = 4x - 5$$
$$f(x) = 4x - 5 \qquad \text{Using function notation}$$

69. First find the slope of the line:

$$m = \frac{5 - (-4)}{3.2 - 1.2} = \frac{5 + 4}{2} = \frac{9}{2} = 4.5$$

Use the point-slope equation with $m = 4.5$ and $(1.2, -4) = (x_1, y_1)$.

$$y - (-4) = 4.5(x - 1.2)$$
$$y + 4 = 4.5x - 5.4$$
$$y = 4.5x - 9.4$$
$$f(x) = 4.5x - 9.4 \qquad \text{Using function notation}$$

71. First find the slope of the line:

$$m = \frac{-1 - (-5)}{0 - 2} = \frac{-1 + 5}{0 - 2} = \frac{4}{-2} = -2$$

Use the slope-intercept equation with $m = -2$ and $b = -1$.

$$y = -2x - 1$$
$$f(x) = -2x - 1 \qquad \text{Using function notation}$$

73. First find the slope of the line:

$$m = \frac{-5 - (-10)}{-3 - (-6)} = \frac{-5 + 10}{-3 + 6} = \frac{5}{3}$$

Use the point-slope equation with $m = \frac{5}{3}$ and $(-3, -5) = (x_1, y_1)$.

$$y - (-5) = \frac{5}{3}(x - (-3))$$
$$y + 5 = \frac{5}{3}(x + 3)$$
$$y + 5 = \frac{5}{3}x + 5$$
$$y = \frac{5}{3}x$$
$$f(x) = \frac{5}{3}x \qquad \text{Using function notation}$$

75. $y = -6$

77. $x = -10$

79. Plot the given information and connect the three dots. We let the horizontal axis represent the Incandescent wattage and the vertical axis represent LED wattage. Extend a dotted line.

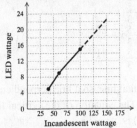

To estimate the wattage of a LED bulb equivalent to a 75-watt incandescent bulb: we use interpolation. Locate the point directly above 75. Then estimate its second coordinate by moving horizontally from that point to the y-axis. We see that 75-watt incandescent bulb is equivalent to a LED with 11 watts.

To estimate the wattage of a LED bulb equivalent to a 120-watt incandescent bulb: we use extrapolation. Locate the point directly above 120 on the dotted line. Then estimate its second coordinate by moving horizontally from that point to the y-axis. We see that 120-watt incandescent bulb is equivalent to a LED with 18 watts.

81. Plot the given information and connect the three dots. We let the horizontal axis represent the Body weight (in pounds) and the vertical axis represent Number of drinks. Extend a dotted line.

To estimate the number of drinks for intoxication of 140-lb person: we use interpolation. Locate the point directly above 140. Then estimate its second coordinate by moving horizontally from that point to the y-axis. We see that a 140-lb person can consume 3.5 drinks in order to be considered intoxicated.

To estimate the number of drinks for intoxication of 230-lb person: we use extrapolation. Locate the point directly above 230. Then estimate its second coordinate by moving horizontally from that point to the y-axis. We see that a 230-lb person can consume 6 drinks in order to be considered intoxicated.

83. Plot the given information and connect the two dots. We let the horizontal axis represent the Year and the vertical axis represent Total sales. Extend a dotted line.

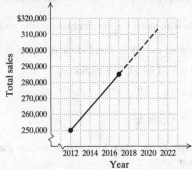

To estimate the sales from 2012: we use interpolation. Locate the point directly above 2012. Then estimate its second coordinate by moving horizontally from that point to the y-axis. We see that sales from 2012 is estimated at $257,000.

To estimate the sales from 2020: we use extrapolation. Locate the point directly above 2020 on the dotted line. Then estimate its second coordinate by moving horizontally from that point to the y-axis. We see that sales from 2020 is estimated at $306,000.

85. a. Let t represent the number of years after 2010, and form the pairs $(0, 85)$ and $(3, 87.1)$. First we find the slope of the function that fits the data:

$$m = \frac{87.1 - 85}{3 - 0} = \frac{2.1}{3} = 0.7.$$

We know the y-intercept is $(0, 85)$, so we write a function in slope-intercept form.

$$N(t) = 0.7t + 85$$

b. In 2020, $t = 2020 - 2010 = 10$

$$N(10) = 0.7(10) + 85 = 92 \text{ million tons}$$

In 2020, the amount recycled will be approximately 92 million tons.

87. a. Let t represent the number of years since 2000, and form the pairs $(0, 79.7)$ and $(10, 81.1)$. First we find the slope of the function that fits the data:

$$m = \frac{81.1 - 79.7}{10 - 0} = \frac{1.4}{10} = 0.14.$$

Use the point-slope equation with $m = 0.14$ and $(0, 79.7) = (t_1, E_1)$.

$$E - 79.7 = 0.14(t - 0)$$
$$E - 79.7 = 0.14t$$
$$E = 0.14t + 79.7$$
$$E(t) = 0.14t + 79.7$$

b. In 2020, $t = 2020 - 2000 = 20$

$$E(20) = 0.14(20) + 79.7 = 82.5$$

The life expectancy of females in 2020 is 82.5 yrs.

89. a. Let t represent the number of years since 1650, and form the pairs $(10, 671)$ and $(35, 1137)$. First we find the slope of the function that fits the data:

$$m = \frac{1137 - 671}{35 - 10} = \frac{466}{25}.$$

Use the point-slope equation with $m = \frac{466}{25}$ and $(10, 671) = (t_1, S_1)$.

$$S - 671 = \frac{466}{25}(t - 10)$$
$$S - 671 = \frac{466}{25}t - \frac{932}{5}$$
$$S = \frac{466}{25}t + \frac{2423}{5}$$
$$S(t) = \frac{466}{25}t + \frac{2423}{5}$$

b. In 1670, $t = 1670 - 1650 = 20$

$$S(20) = \frac{466}{25}(20) + \frac{2423}{5} = 857.4$$

The average displacement of a ship in 1670 was 857.4 tons.

91. a. $\quad m = \dfrac{7 - 4}{200 - 100} = 0.03$
$$P - 4 = 0.03(d - 100)$$
$$P(d) = 0.03d + 1$$

b. $P(690) = 0.03(690) + 1 = 21.7$ atm

93. a. Let t represent the number of years since 2008, and form the pairs (2, 10.05) and (6, 5.47). First we find the slope of the function that fits the data:
$$m = \frac{5.47 - 10.05}{6 - 2} = \frac{-4.58}{4} = -1.145.$$
Use the point-slope equation with $m = -1.145$ and $(2, 10.05) = (t_1, R_1)$.
$$R - 10.05 = -1.145(t - 2)$$
$$R - 10.05 = -1.145t + 2.29$$
$$R = -1.145t + 12.34$$
$$R(t) = -1.145t + 12.34$$

b. In 2016, $t = 2016 - 2008 = 8$
$$R(8) = -1.145(8) + 12.34 = 3.18$$
In 2016, the revenue from sales of physical video and computer games will be \$3.18 billion.

c. We substitute 0 for $R(t)$ and solve for t.
$$0 = -1.145t + 12.34$$
$$1.145t = 12.34$$
$$t \approx 10.8$$
There will be no sales from physical video and computer games approximately 11 years after 2008, or 2019.

95. *Writing Exercise.*

97. $-6x - x - 2x = -9x$

99. $x = \dfrac{mp}{c}$
$$xc = mp$$
$$\frac{xc}{p} = m$$

101. *Writing Exercise.*

103. Familiarize. Celsius temperature C corresponding to a Fahrenheit temperature F can be modeled by a line that contains the points (32, 0) and (212, 100).
Translate. We find an equation relating C and F.
$$m = \frac{100 - 0}{212 - 32} = \frac{100}{180} = \frac{5}{9}$$

$$C - 0 = \frac{5}{9}(F - 32)$$
$$C = \frac{5}{9}(F - 32)$$

Carry out. Using function notation we have
$C(F) = \frac{5}{9}(F - 32)$. Now we find $C(70)$:
$$C(70) = \frac{5}{9}(70 - 32) = \frac{5}{9}(38) \approx 21.1.$$

Check. We can repeat the calculations. We could also graph the function and determine that (70, 21.1) is on the graph.
State. A temperature of about 21.1°C corresponds to a temperature of 70°F.

105. Familiarize. The total cost C of the phone in dollars, after t months, can be modeled by a line that contains the points (5, 410) and (9, 690).
Translate. We find an equation relating C and t.
$$m = \frac{690 - 410}{9 - 5} = \frac{280}{4} = 70$$
$$C - 410 = 70(t - 5)$$
$$C - 410 = 70t - 350$$
$$C = 70t + 60$$
Carry out. Using function notation, we have
$C(t) = 70t + 60$. To find the costs already incurred when the service began, we find $C(0)$:
$$C(0) = 70 \cdot 0 + 60 = 60.$$
Check. We can repeat the calculations. We could also graph the function and determine that (0, 60) is on the graph.
State. Tam had already incurred \$60 in costs when her service just began.

107. We find the value of p for which $A(p) = -\frac{1}{6}p + \frac{17}{2}$

and $A(p) = \frac{2}{3}p - 3$ are the same.

$$-\frac{1}{6}p + \frac{17}{2} = \frac{2}{3}p - 3$$
$$-p + 51 = 4p - 18 \quad \text{Multiplying by 6}$$
$$-5p = -69$$
$$p = 13.8$$
Supply will equal demand at a price of \$13.80 per pound.

109. The price must be a positive number, so we have $p > 0$. Furthermore, the amount of coffee supplied must be a positive number. We have:

$$\frac{2}{3}p - 3 > 0$$
$$\frac{2}{3}p > 3$$
$$p > 4.5$$
Thus, the domain is $\{p \mid p > 4.5\}$.

111. Find the slope of $5y - kx = 7$:
$$5y - kx = 7$$
$$5y = kx + 7$$
$$y = \frac{k}{5}x + \frac{7}{5}$$

The slope is $\frac{k}{5}$.

Find the slope of the line containing $(7, -3)$ and $(-2, 5)$:

$$m = \frac{5 - (-3)}{-2 - 7} = \frac{5 + 3}{-9} = -\frac{8}{9}$$

If the lines are parallel, their slopes must be equal:

$$\frac{k}{5} = -\frac{8}{9}$$

$$k = -\frac{40}{9}$$

113. *Graphing Calculator and Writing Exercise.*

117. Find the slope of the Walking function.

$$m = \frac{300 - 210}{3.75 - 2.5} = \frac{90}{1.25} = 72$$

Using $(2.5, 210)$, we find the Walking function.

$$y - 210 = 72(x - 2.5)$$
$$y = 72x + 30$$
$$f(x) = 72x + 30$$

Using the function, we find the calories per hour for walking at 4.5 mph.

$$f(4.5) = 72(4.5) + 30 = 354 \text{ calories per hour}$$

For 2 hours, that is $2(354) = 708$ calories.

Find the slope of the Bicycling function.

$$m = \frac{660 - 210}{13 - 5.5} = \frac{450}{7.5} = 60$$

Using $(13, 660)$, we find the Bicycling function.

$$y - 660 = 60(x - 13)$$
$$y = 60x - 120$$
$$g(x) = 60x - 120$$

Using the function, we find the calories per hour for bicycling at 14 mph.

$$g(14) = 60(14) - 120 = 720 \text{ calories per hour}$$

Therefore, bicycling at 14 mph burns more calories (720) than walking 4.5 mph for 2 hours (708).

Exercise Set 2.6

1. If f and g are functions then $(f + g)(x)$ is the *sum* of the functions.

3. One way to compute $(f - g)(2)$ is to simplify $f(x) - g(x)$ and then *evaluate* the result for $x = 2$.

5. The domain of f / g is the set of all values common to the domains of f and g, *excluding* any values for which $g(x)$ is 0.

7. Since $f(3) = -2 \cdot 3 + 3 = -3$ and $g(3) = 3^2 - 5 = 4$, we have $f(3) + g(3) = -3 + 4 = 1$.

9. Since $f(1) = -2 \cdot 1 + 3 = 1$ and $g(1) = 1^2 - 5 = -4$, we have $f(1) - g(1) = 1 - (-4) = 5$.

11. Since $f(-2) = -2 \cdot (-2) + 3 = 7$ and $g(-2) = (-2)^2 - 5 = -1$ we have $f(-2) \cdot g(-2) = 7(-1) = -7$.

13. Since $f(-4) = -2 \cdot (-4) + 3 = 11$ and $g(-4) = (-4)^2 - 5 = 11$, we have $\frac{f(-4)}{g(-4)} = \frac{11}{11} = 1$.

15. Since $g(1) = 1^2 - 5 = -4$ and $f(1) = -2 \cdot 1 + 3 = 1$, we have $g(1) - f(1) = -4 - 1 = -5$.

17. $(f + g)(x) = f(x) + g(x) = (-2x + 3) + (x^2 - 5)$
$$= x^2 - 2x - 2$$

19. $(g - f)(x) = g(x) - f(x) = (x^2 - 5) - (-2x + 3)$
$$= x^2 + 2x - 8$$

21. $(F + G)(x) = F(x) + G(x)$
$$= x^2 - 2 + 5 - x$$
$$= x^2 - x + 3$$

23. $(F - G)(x) = F(x) - G(x)$
$$= x^2 - 2 - (5 - x)$$
$$= x^2 - 2 - 5 + x$$
$$= x^2 + x - 7$$

Then we have

$(F - G)(3) = 3^2 + 3 - 7$
$$= 9 + 3 - 7$$
$$= 5.$$

25. $(F \cdot G)(x) = F(x) \cdot G(x)$
$$= (x^2 - 2)(5 - x)$$
$$= 5x^2 - x^3 - 10 + 2x$$

Then we have

$(F \cdot G)(-3) = 5(-3)^2 - (-3)^3 - 10 + 2(-3)$
$$= 5 \cdot 9 - (-27) - 10 - 6$$
$$= 45 + 27 - 10 - 6$$
$$= 56.$$

27. $(F / G)(x) = F(x) / G(x)$
$$= \frac{x^2 - 2}{5 - x}, \; x \neq 5$$

29. $(G / F)(-2) = \frac{5 - (-2)}{(-2)^2 - 2} = \frac{5 + 2}{4 - 2} = \frac{7}{2}$.

31. $(F + F)(x) = F(x) + F(x)$
$$= (x^2 - 2) + (x^2 - 2)$$
$$= 2x^2 - 4$$

$(F + F)(1) = 2(1)^2 - 4$
$$= -2$$

33. $N(2015) = (C + B)(2015) = C(2015) + B(2015)$
$$\approx 1.3 + 2.7 = 4.0 \text{ million}$$

We estimate the number of births in 2015 to be 4.0 million.

35. $(B - C)(2015) = B(2015) - C(2015)$
$\approx 2.7 - 1.3 = 1.4$ million
This represents how many more non-Caesarian section births than Caesarian section births there were in 2015.

37. $(p + r)('09) = p('09) + r('09)$
$\approx 25 + 60 = 85$ million
This represents the number of tons of municipal solid waste that was composted or recycled in 2009.

39. $F('00) \approx 240$ million
This represents the number of tons of municipal solid waste in 2000.

41. $(F - p)('08) = F('08) - p('08)$
$\approx 240 - 20 = 220$ million
This represents the number of tons of municipal solid waste that was not composted in 2008.

43. The domain of f and of g is all real numbers. Thus,
Domain of $f + g$ = Domain of $f - g$ = Domain of $f \cdot g$
$= \{x \mid x \text{ is a real number}\}$.

45. Because division by 0 is undefined, we have
Domain of $f = \{x \mid x \text{ is a real number } and \ x \neq -5\}$,
and Domain of $g = \{x \mid x \text{ is a real number}\}$.
Thus,
Domain of $f + g$ = Domain of $f - g$ = Domain of $f \cdot g$
$= \{x \mid x \text{ is a real number } and \ x \neq -5\}$.

47. Because division by 0 is undefined, we have
Domain of $f = \{x \mid x \text{ is a real number } and \ x \neq 0\}$,
and Domain of $g = \{x \mid x \text{ is a real number}\}$.
Thus, Domain of
$f + g$ = Domain of $f - g$ = Domain of $f \cdot g$
$= \{x \mid x \text{ is a real number } and \ x \neq 0\}$.

49. Because division by 0 is undefined, we have
Domain of $f = \{x \mid x \text{ is a real number } and \ x \neq 1\}$,
and Domain of $g = \{x \mid x \text{ is a real number}\}$.
Thus,
Domain of $f + g$ = Domain of $f - g$ = Domain of $f \cdot g$
$= \{x \mid x \text{ is a real number } and \ x \neq 1\}$.

51. Because division by 0 is undefined, we have
Domain of $f = \left\{x \middle| x \text{ is a real number } and \ x \neq -\frac{9}{2}\right\}$,
and Domain of $g = \{x \mid x \text{ is a real number } and \ x \neq 1\}$.
Thus, Domain of
$f + g$ = Domain of $f - g$ = Domain of $f \cdot g$
$= \left\{x \middle| x \text{ is a real number } and \ x \neq -\frac{9}{2} \text{ and } x \neq 1\right\}$.

53. Domain of
f = Domain of $g = \{x \mid x \text{ is a real number}\}$.
Since $g(x) = 0$ when $x - 3 = 0$, we have $g(x) = 0$

when $x = 3$. We conclude that
Domain of $f / g = \{x \mid x \text{ is a real number } and \ x \neq 3\}$.

55. Domain of
f = Domain of $g = \{x \mid x \text{ is a real number}\}$.
Since $g(x) = 0$ when $2x + 8 = 0$, we have $g(x) = 0$
when $x = -4$. We conclude that
Domain of $f / g = \{x \mid x \text{ is a real number } and \ x \neq -4\}$.

57. Domain of $f = \{x \mid x \text{ is a real number and } x \neq 4\}$.
Domain of $g = \{x \mid x \text{ is a real number}\}$.
Since $g(x) = 0$ when $5 - x = 0$, we have $g(x) = 0$
when $x = 5$. We conclude that Domain of
$f / g = \{x \mid x \text{ is a real number and } x \neq 4 \text{ and } x \neq 5\}$.

59. Domain of $f = \{x \mid x \text{ is a real number and } x \neq -1\}$.
Domain of $g = \{x \mid x \text{ is a real number}\}$.
Since $g(x) = 0$ when $2x + 5 = 0$, we have $g(x) = 0$
when $x = -\frac{5}{2}$. We conclude that Domain of f / g
$= \left\{x \middle| x \text{ is a real number and } x \neq -1 \text{ and } x \neq -\frac{5}{2}\right\}$.

61. $(F + G)(5) = F(5) + G(5) = 1 + 3 = 4$
$(F + G)(7) = F(7) + G(7) = -1 + 4 = 3$

63. $(G - F)(7) = G(7) - F(7) = 4 - (-1) = 4 + 1 = 5$
$(G - F)(3) = G(3) - F(3) = 1 - 2 = -1$

65. From the graph we see that Domain of
$F = \{x \mid 0 \leq x \leq 9\}$ and Domain of $G = \{x \mid 3 \leq x \leq 10\}$.
Then
Domain of $F + G = \{x \mid 3 \leq x \leq 9\}$. Since $G(x)$ is
never 0,
Domain of $F / G = \{x \mid 3 \leq x \leq 9\}$.

67. We use $(F + G)(x) = F(x) + G(x)$.

69. *Writing Exercise.*

71. *Familiarize*. Let x = the second angle of a triangle. Then the first angle is $2x$. The third angle is $3x$.
Translate.

First angle	plus	second angle	plus	third angle	is	180.
↓	↓	↓	↓	↓	↓	↓
$2x$	+	x	+	$3x$	=	180

Carry out. Solve the equation.

$$2x + x + 3x = 180$$
$$6x = 180$$
$$x = 30$$

Then $2x = 2(30°) = 60°$
and $3x = 3(30°) = 90°$

Check. If the angles are 60°, 30° and 90°, the sum of the three angles is 60° + 30° + 90° = 180°. The answer checks.

State. The angles of the triangle are 60°, 30°, and 90°.

73. To find the weight of each molecule, divide the weight of a mole of water by the number of molecules.

$$\frac{18.015 \text{ g}}{6.022 \times 10^{23}} = 2.992 \times 10^{-23} \text{ g}.$$

Therefore, each molecule weighs 2.992×10^{-23} g.

75. *Writing Exercise.*

77. Domain of $F = \{x \mid x \text{ is a real number and } x \neq 4\}$.

Domain of $G = \{x \mid x \text{ is a real number and } x \neq 3\}$.

$G(x) = 0$ when $x^2 - 4 = 0$, or when $x = 2$ or $x = -2$. Then Domain of $F / G = \{x \mid x \text{ is a real number and } x \neq 4 \text{ and } x \neq 3 \text{ and } x \neq 2 \text{ and } x \neq -2\}$.

79. Answers may vary.

81. The problem states that Domain of $m = \{x \mid -1 < x < 5\}$. Since $n(x) = 0$ when $2x - 3 = 0$, we have $n(x) = 0$ when $x = \frac{3}{2}$. We conclude that Domain of m / n

$= \left\{ x \mid x \text{ is a real number and } -1 < x < 5 \text{ and } x \neq \frac{3}{2} \right\}$.

83. Answers may vary. $f(x) = \frac{1}{x+2}$, $g(x) = \frac{1}{x-5}$

85. *Graphing Calculator Exercise*

Chapter 2 Review

1. True

2. False

3. True

4. False

5. False

6. True

7. True

8. True

9. False

10. True

11.

$$\begin{array}{c|c} x = 2y + 12 \\ \hline -2 & 2 \cdot 8 + 12 \\ & \overset{?}{} \\ -2 = 28 \end{array}$$

No

12.

$$\begin{array}{c|c} 3a - 4b = 2 \\ \hline 3(0) - 4\left(-\frac{1}{2}\right) & 2 \\ 0 + 2 & \\ & \overset{?}{} \\ 2 = 2 \end{array}$$

Yes

13. The first coordinate is negative and the second is positive, so the point $(-3, 5)$ is in quadrant II.

14. $y = -x^2 + 1$

To find an ordered pair, we choose any number for x and then determine y. We find several ordered pairs, plot them, and connect them with a smooth curve.

x	y
2	−3
1	0
0	1
−1	0
−2	−3

$y = -x^2 + 1$

15. We can use the coordinates of any two points on the line. Let's use (2, 75) and (8, 120).

$$\text{Rate of change} = \frac{\text{change in } y}{\text{change in } x} = \frac{75 - 120}{2 - 8} = \frac{-45}{-6} = 7.5$$

The value of the apartment is increasing at a rate of $7500 per year.

16. The rate of new homes sold is given by

$$\frac{\text{change in number of homes}}{\text{change in time}} = \frac{352,000 - 134,000}{4}$$
$$= \frac{218,000}{4} = 54,500$$

The number of homes sold is increasing by 54,500 homes per month.

17. $\text{Slope} = \dfrac{\text{difference in } y}{\text{difference in } x} = \dfrac{5 - 1}{4 - (-3)} = \dfrac{4}{7}$

18. $\text{Slope} = \dfrac{3.5 - 2.8}{-16.4 - (-16.4)} = \dfrac{0.7}{0}$

The slope is undefined.

19. $\text{Slope} = \dfrac{-1 - (-2)}{-5 - (-1)} = \dfrac{1}{-4} = -\dfrac{1}{4}$

20. $\text{Slope} = \dfrac{\frac{1}{2} - \frac{1}{2}}{\frac{1}{3} - \frac{1}{6}} = \dfrac{0}{\frac{1}{6}} = 0$

21. $g(x) = -5x - 11$

Slope is −5; y-intercept is $(0, -11)$

22. Convert to slope-intercept equation.
$$-6y + 5x = 10$$
$$-6y = -5x + 10$$
$$y = \frac{5}{6}x - \frac{5}{3}$$

Slope is $\frac{5}{6}$; y-intercept is $\left(0, -\frac{5}{3}\right)$.

23. $s(t) = \frac{4}{7}t + 2$

$\frac{4}{7}$ signifies the number of students taking at least one online course increases by $\frac{4}{7}$ million students per year, for years after 2003.
2 signifies that 2 million students took at least one online course in 2003.

24. $y + 3 = 7$
$$y = 4$$

The graph of $y = 4$ is a horizontal line. Since $y + 3 = 7$ is equivalent to $y = 4$, the slope is 0.

25. $-2x = 9$
$$x = -\frac{9}{2}$$

The graph of $x = -\frac{9}{2}$ is a vertical line. Since $-2x = 9$ is equivalent to $x = -\frac{9}{2}$, the slope is undefined.

26. $3x - 2y = 8$
To find the y-intercept, let $x = 0$ and solve for y.
$$3 \cdot 0 - 2y = 8$$
$$-2y = 8$$
$$y = -4$$

The y-intercept is $(0, -4)$.
To find the x-intercept, let $y = 0$ and solve for x.
$$3x - 2 \cdot 0 = 8$$
$$3x = 8$$
$$x = \frac{8}{3}$$

The x-intercept is $\left(\frac{8}{3}, 0\right)$.

27. Graph $f(x) = -3x + 2$.

Slope is -3 or $\frac{-3}{1}$; y-intercept is $(0, 2)$.

From the y-intercept, we go down 3 units and to the right 1 unit. This gives us $(1, -1)$. We can now draw the graph.

$y = -3x + 2$

28. Graph $-2x + 4y = 8$.
To find the y-intercept, let $x = 0$ and solve for y.
$$-2 \cdot 0 + 4y = 8$$
$$y = 2$$

The y-intercept is $(0, 2)$.
To find the x-intercept, let $y = 0$ and solve for x.
$$-2x + 4 \cdot 0 = 8$$
$$-2x = 8$$
$$x = -4$$
The x-intercept is $(-4, 0)$.
Plot these points and draw the line.

$-2x + 4y = 8$

29. Graph $y = 6$.
This is a horizontal line that crosses the y-axis at $(0, 6)$. If we find some ordered pairs, note that, for any x-value chosen, y must be 6.

x	y
3	6
0	6
6	6

$y = 6$

30. $y + 1 = \frac{3}{4}(x - 5)$
$$y - (-1) = \frac{3}{4}(x - 5)$$

To graph the equation, we count off a slope of $\frac{3}{4}$, starting at $(5, -1)$, and draw the line.

$y + 1 = \frac{3}{4}(x - 5)$

31. Graph $8x + 32 = 0$.
Since y does not appear, we solve for x.
$$8x + 32 = 0$$
$$x = -4.$$
This is a vertical line that crosses the x-axis at $(-4, 0)$.

$8x + 32 = 0$

32. Graph $g(x) = 15 - x$ or $g(x) = -x + 15$.

Slope is $-1 = \frac{-1}{1}$; y-intercept is $(0, 15)$.

From the y-intercept, we go down 1 unit and right 1 unit. This gives us the point $(1, 14)$. We can now draw the graph.

$g(x) = 15 - x$

33. Graph $f(x) = \frac{1}{2}x - 3$.

Slope is $\frac{1}{2}$; y-intercept is $(0, -3)$.

From the y-intercept, we go *up* 1 unit and *right* 2 units. This gives us the point $(2, -2)$. We can now draw the graph.

$f(x) = \frac{1}{2}x - 3$

34. Graph $f(x) = 0$.

This is a horizontal line that crosses the vertical axis at $(0, 0)$.

$f(x) = 0$

35. $2 - x = 5 + 2x$

Graph $f(x) = 2 - x$ and $g(x) = 5 + 2x$ on the same grid.

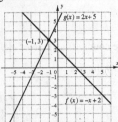

$g(x) = 2x + 5$

$(-1, 3)$

$f(x) = -x + 2$

The lines appear to intersect at $(-1, 3)$ so the solution is apparently -1.

Check:
$$\begin{array}{c|c} 2 - x = 5 + 2x \\ \hline 2 - (-1) & 5 + 2(-1) \\ 2 + 1 & 5 - 2 \\ \end{array}$$
$$3 \overset{?}{=} 3 \quad \text{TRUE}$$

The solution is -1.

36. Let $t =$ the number of tee shirts printed and $C(t)$ is the total cost. Graph $C(t) = 8t + 80$ and $y = 200$.

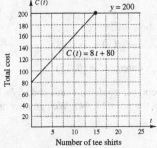

$y = 200$

$C(t) = 8t + 80$

Total cost

Number of tee shirts

The graph appears to intersect at about $(15, 200)$. The number 15 checks. So $15 - 5$, or 10 additional tee shirts were printed.

37. First solve for y and determine the slope of each line.
$$y + 5 = -x$$
$$y = -x - 5$$
The slope of $y + 5 = -x$ is -1.
$$x - y = 2$$
$$y = x - 2$$
The slope of $x - y = 2$ is 1.
The product of their slopes is $(-1)(1)$, or -1; the lines are perpendicular.

38. First solve for y and determine the slope of each line.
$$3x - 5 = 7y$$
$$y = \frac{3}{7}x - \frac{5}{7}$$
The slope of $3x - 5 = 7y$ is $\frac{3}{7}$.
$$7y - 3x = 7$$
$$y = \frac{3}{7}x + 1$$
The slope of $7y - 3x = 7$ is $\frac{3}{7}$.

The slopes are the same, so the lines are parallel.

39. Use the slope-intercept equation $f(x) = mx + b$, with $m = \frac{2}{9}$ and $b = -4$.
$$f(x) = mx + b$$
$$f(x) = \frac{2}{9}x - 4$$

40. $y - y_1 = m(x - x_1)$ Point-slope equation
$y - 10 = -5(x - 1)$ Substituting -5 for m,
                    1 for x_1, and 10 for y_1

41. First find the slope of the line:
$$m = \frac{6 - 5}{-2 - 2} = -\frac{1}{4}$$
Use the point-slope equation with $m = -\frac{1}{4}$ and $(2, 5) = (x_1, y_1)$.
$$y - y_1 = m(x - x_1)$$
$$y - 5 = -\frac{1}{4}(x - 2)$$
$$y - 5 = -\frac{1}{4}x + \frac{1}{2}$$
$$y = -\frac{1}{4}x + \frac{11}{2}$$
$$f(x) = -\frac{1}{4}x + \frac{11}{2} \quad \text{Using function notation}$$

42. $3x - 5y = 9$
$$y = \frac{3}{5}x - \frac{9}{5} \qquad m = \frac{3}{5}$$
$$y - (-5) = \frac{3}{5}(x - 2)$$
$$y + 5 = \frac{3}{5}x - \frac{6}{5}$$
$$y = \frac{3}{5}x - \frac{31}{5}$$

43. First solve the equation for y and determine the slope of the given line.

$$3x - 5y = 9$$
$$y = \frac{3}{5}x - \frac{9}{5}$$

The slope of the given line is $\frac{3}{5}$.

The slope of a perpendicular line is given by the opposite of the reciprocal of $\frac{3}{5}$, $-\frac{5}{3}$.

We find the equation of the line with slope $-\frac{5}{3}$ containing the point $(2, -5)$.

$$y - (-5) = -\frac{5}{3}(x - 2)$$
$$y + 5 = -\frac{5}{3}x + \frac{10}{3}$$
$$y = -\frac{5}{3}x - \frac{5}{3}$$

44. Plot and connect the points, using the years as the first coordinate and the corresponding outstanding student loan debt, in billions of dollars, as the second coordinate.

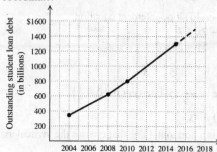

To estimate the debt in 2007, first locate the point that is directly above 2007. Then move horizontally from the point to the vertical axis and read the approximate function value there. We estimate the outstanding student loan debt to be $550 billion in 2007.

45. To estimate the outstanding student loan debt in 2017, extend the graph and extrapolate. It appears that in 2017, the outstanding student loan debt is about $1500 billion or $1.5 trillion.

46. a. Letting t represent the number of years since 1980, we form the pairs $(3, 19.75)$ and $(31, 19.19)$. First we find the slope of the function that fits the data.

$$m = \frac{19.19 - 19.75}{31 - 3} = \frac{-0.56}{28} = -0.02$$

Using $m = -0.02$ and $t_1 = 3$ and $R_1 = 19.75$ we substitute into point-slope equation.

$$R - R_1 = m(t - t_1)$$
$$R - 19.75 = -0.02(t - 3)$$
$$R = -0.02t + 19.81$$
$$R(t) = -0.02t + 19.81$$

b. In 2015, $t = 2015 - 1980 = 35$

$$R(35) = -0.02(35) + 19.81 = 19.11 \text{ sec}$$

In 2020, $t = 2020 - 1980 = 40$

$$R(40) = -0.02(40) + 19.81 = 19.01 \text{ sec}$$

47. $2x - 7 = 0$

$$x = \frac{7}{2}$$

The equation is linear. The graph is a vertical line.

48. $3x - \frac{y}{8} = 7$

$$24x - y = 56$$

This equation is in standard form for a linear equation, $Ax + By = C$. Thus, it is a linear equation.

49. $2x^3 - 7y = 5$

The equation is not linear, because it has an x^3-term.

50. $\frac{2}{x} = y$

$$2 = xy \qquad \text{Multiplying by } x$$

The equation is not linear, because it has an xy-term.

51. a. Locate 2 on the horizontal axis and find the point on the graph for which 2 is the first coordinate. From this point, look to the vertical axis to find the corresponding y-coordinate, 3. Thus $f(2) = 3$.

b. The set of all x-values in the graph extends from -2 to 4, so the domain is $\{x | -2 \le x \le 4\}$.

c. To determine which member(s) of the domain are with 2, locate 2 on the vertical axis. From there, look left and right to the graph to find any points for which 2 is the second coordinate. One such point exists. Its first coordinate is -1. Thus $f(-1) = 2$.

d. The set of all y-values in the graph extends from 1 to 5, so the range is $\{y | 1 \le y \le 5\}$..

52. a. Domain of g is $\{x | x \text{ is a real number}\}$.

b. Range of g is $\{y | y \ge 0\}$.

53. We can use the vertical-line test. Visualize moving a vertical line across the graph. No vertical line will intersect the graph more than once. Thus, the graph is a graph of a function.

54. We use the vertical-line test. It is possible for a vertical line to intersect the graph more than once. Thus, this is not the graph of a function.

55. $g(0) = 3 \cdot 0 - 6 = 0 - 6 = -6$

56. $h(-5) = (-5)^2 + 1 = 25 + 1 = 26$

57. $g(a + 5) = 3(a + 5) - 6 = 3a + 15 - 6 = 3a + 9$

58. $(g \cdot h)(4) = g(4) \cdot h(4) = [3(4) - 6][(4)^2 + 1]$
$$= [12 - 6][16 + 1]$$
$$= 6 \cdot 17 = 102$$

59. $\left(\frac{g}{h}\right)(-1) = \frac{g(-1)}{h(-1)} = \frac{3(-1) - 6}{(-1)^2 + 1} = \frac{-3 - 6}{1 + 1} = -\frac{9}{2}$

60. $(g + h)(x) = g(x) + h(x) = (3x - 6) + (x^2 + 1)$
$$= x^2 + 3x - 5$$

61. $g(x) = 3x - 6$

Since we can compute $3x - 6$ for any real number x, the domain is the set of all real numbers.

62. The domain of g and h is all real numbers. Thus, Domain of $g + h = \{x | x \text{ is a real number}\}$.

63. Domain of g = Domain of $h = \{x | x \text{ is a real number}\}$.

Since $g(x) = 0$ when $3x - 6 = 0$, we have $g(x) = 0$ when $x = 2$. We conclude that

Domain of $h / g = \{x | x \text{ is a real number and } x \neq 2\}$.

64. *Writing Exercise.* To find $f(a) + h$, we find the output that corresponds to a and add h. To find $f(a + h)$, we find the output that corresponds to $a + h$.

65. *Writing Exercise.* The slope of a line is the rise between two points on the line divided by the run between those points. For a vertical line, there is no run between any two points, and division by 0 is undefined; therefore, the slope is undefined. For a horizontal line, there is no rise between any two points, so the slope is 0/run, or 0.

66. To find the y-intercept, choose $x = 0$.
$$f(0) + 3 = 0.17(0)^2 + (5 - 2(0))^0 - 7$$
$$f(0) + 3 = 0 + 1 - 7$$
$$f(0) = -9$$

67. Write both equations in slope-intercept form.
$$y = \frac{3}{4}x - 3 \quad m = \frac{3}{4}$$
$$y = -\frac{a}{6}x - \frac{3}{2} \quad m = -\frac{a}{6}$$

For parallel lines, the slopes are the same, or
$$\frac{3}{4} = -\frac{a}{6}$$
$$a = -\frac{9}{2}$$

68. Let x represent the number of packages.
Total cost = package charge + shipping charges
$$f(x) = 7.99x + (2.95x + 20)$$
$$f(x) = 10.94x + 20$$

69. a. Graph III indicates the beginning and end, walking at the same rate, with a rapid rate in between.

b. Graph IV indicates fast bike riding, followed by not as fast running, finishing with slower walking.

c. Graph I indicates fast rate of motorboat followed by constant rate of 0 for fishing, finishing with fast rate of motorboat.

d Graph II indicates constant rate of 0 while waiting, followed by fast rate of train, and finishing with not as fast run.

Chapter 2 Test

1.
$$\frac{x + 4y = -20}{12 + 4(-3) \,\big|\, -20}$$
$$12 - 12 \,\big|$$
$$\overset{?}{0 = -20}$$
Since $0 = -20$ is false, $(12, -3)$ is not a solution.

2. $y = x^2 + 3$
To find an ordered pair, we choose any number for x

and then determine y. We find several ordered pairs, plot them, and connect them with a smooth curve.

x	y
2	7
1	4
0	3
−1	4
−2	7

$f(x) = x^2 + 3$

3. Rate of change = $\dfrac{\text{change in } y}{\text{change in } x} = \dfrac{400 - 250}{8 - 4} = \dfrac{150}{4} = 37.5$

The total cost of the monitored security system is increasing at a rate of $37.50 per month.

4. Slope = $\dfrac{\text{change in } y}{\text{change in } x} = \dfrac{3 - (-2)}{6 - (-2)} = \dfrac{3 + 2}{6 + 2} = \dfrac{5}{8}$

5. Slope = $\dfrac{\text{change in } y}{\text{change in } x} = \dfrac{5.2 - 5.2}{-4.4 - (-3.1)} = \dfrac{0}{-4.4 + 3.1} = 0$

6. $f(x) = -\dfrac{3}{5}x + 12$

Slope is $-\dfrac{3}{5}$; y-intercept is (0, 12).

7. Convert to slope-intercept equation.
$$-5y - 2x = 7$$
$$-5y = 2x + 7$$
$$y = -\frac{2}{5}x - \frac{7}{5}$$
Slope is $-\dfrac{2}{5}$; y-intercept is $\left(0, -\dfrac{7}{5}\right)$.

8. $f(x) = -3$
This is a horizontal line that crosses the vertical axis at $(0, -3)$. The slope is 0.

9. $x - 5 = 11$
$$x = 16$$
The graph of $x = 16$ is a vertical line. Since $x - 5 = 11$ is equivalent to $x = 16$, the slope of $x - 5 = 11$ is undefined.

10. $5x - y = 15$
To find the y-intercept, let $x = 0$ and solve for y.
$$5 \cdot 0 - y = 15$$
$$y = -15$$
The y-intercept is $(0, -15)$.
To find the x-intercept, let $y = 0$ and solve for x.
$$5x - 0 = 15$$
$$x = 3$$
The x-intercept is $(3, 0)$.

11. Graph $f(x) = -3x + 4$.

Slope is -3 or $\dfrac{-3}{1}$; y-intercept is (0, 4).

From the y-intercept, we go *down* 3 units and to the *right* 1 unit. This gives us (1, 1). We can now draw the graph.

$y = -3x + 4$

12. Convert to slope-intercept form.

$$y - 1 = -\frac{1}{2}(x + 4)$$

$$y = -\frac{1}{2}x - 1$$

Slope is $-\frac{1}{2}$; y-intercept is $(0, -1)$.

From the y-intercept, we go *down* 1 unit and *right* 2 units. This gives us the point $(2, -2)$. We can now draw the graph.

$y - 1 = -\frac{1}{2}(x + 4)$

13. Convert to slope-intercept form.

$$-2x + 5y = 20$$

$$y = \frac{2}{5}x + 4$$

Slope is $\frac{2}{5}$; y-intercept is $(0, 4)$.

From the y-intercept, we go *down* 2 units and *left* 5 units. This gives us the point $(-5, 2)$. We can now draw the graph.

$-2x + 5y = 20$

14. Graph $3 - x = 9$.

Since y does not appear, we solve for x.

$$x = -6.$$

This is a vertical line that crosses the x-axis at $(-6, 0)$.

15. $x + 3 = 2x$

Graph $f(x) = x + 3$ and $g(x) = 2x$ on the same grid.

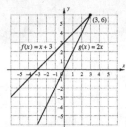

The lines appear to intersect at $(3, 6)$ so the solution is apparently 3.

Check: $x + 3 = 2x$

$$\frac{3 + 3 \mid 2 \cdot 3}{}$$

$$\overset{?}{6 = 6} \quad \text{TRUE}$$

The solution is 3.

16. Plot and connect the points using the SAT math score as the first coordinate and the family income, in thousands of dollars, as the second coordinate.

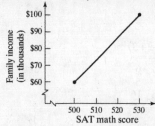

To estimate the SAT math score for a family income of $75,000, locate the point directly to the right of $75. Then estimate the second coordinate by moving vertically from the point to the horizontal axis. The SAT math score appears to be approximately 510.

17. a. $8x - 7 = 0$

$$x = \frac{7}{8}$$

The equation is linear. (Its graph is a vertical line.)

b. $4x - 9y^2 = 12$

The equation is not linear, because it has a y^2-term.

c. $2x - 5y = 3$

The equation is linear.

18. Write both equations in slope-intercept form.

$$4y + 2 = 3x \qquad -3x + 4y = -12$$

$$y = \frac{3}{4}x - \frac{1}{2} \qquad y = \frac{3}{4}x - 3$$

$$m = \frac{3}{4} \qquad\qquad m = \frac{3}{4}$$

The slopes are the same, so the lines are parallel.

19. Write both equations in slope-intercept form.

$$y = -2x + 5 \qquad 2y - x = 6$$

$$m = -2 \qquad\qquad y = \frac{1}{2}x + 3$$

$$m = \frac{1}{2}$$

The product of their slopes is $(-2)\left(\frac{1}{2}\right)$, or -1; the lines are perpendicular.

20. Use the slope-intercept equation, $f(x) = mx + b$, with $m = -5$ and $b = -1$.

$$f(x) = -5x - 1$$

21. $y - (-4) = 4[x - (-2)]$ or $y + 4 = 4(x + 2)$

22. $m = \dfrac{-2 - (-1)}{4 - 3} = \dfrac{-1}{1} = -1$

$$y - (-2) = -1(x - 4)$$

$$y + 2 = -x + 4$$

$$y = -x + 2$$

$$f(x) = -x + 2$$

23. $2x - 5y = 8$

$$y = \frac{2}{5}x - \frac{8}{5}$$

The slope is $\frac{2}{5}$.

$$y - 2 = \frac{2}{5}(x + 3)$$

$$y - 2 = \frac{2}{5}x + \frac{6}{5}$$

$$y = \frac{2}{5}x + \frac{16}{5}$$

24. $2x - 5y = 8$

$$y = \frac{2}{5}x - \frac{8}{5}$$

The slope is $\frac{2}{5}$.

The slope of a perpendicular line is given by the opposite of the reciprocal of $\frac{2}{5}$, $-\frac{5}{2}$.

$$y - 2 = -\frac{5}{2}[x - (-3)]$$

$$y = -\frac{5}{2}x - \frac{11}{2}$$

25. a. Let m = the number of miles, $C(m)$ represents the cost. We form the pairs (250, 100) and (300, 115).

$$\text{Slope} = \frac{115 - 100}{300 - 250} = \frac{15}{50} = \frac{3}{10} = 0.3$$

$$y - y_1 = m(x - x_1)$$

$$C - 100 = 0.3(m - 250)$$

$$C = 0.3m + 25$$

$$C(m) = 0.3m + 25$$

b. $C(500) = 0.3(500) + 25 = \$175$

26. a. $f(-2) = 1$

b. Domain is $\{x | -3 \le x \le 4\}$.

c. If $f(x) = \frac{1}{2}$, then $x = 3$.

d. Range is $\{y | -1 \le y \le 2\}$.

27. $h(-5) = 2(-5) + 1 = -9$

28. $(g + h)(x) = g(x) + h(x) = \frac{1}{x} + 2x + 1$

29. Domain of g: $\{x | x \text{ is a real number and } x \ne 0\}$

30. Domain of h: $\{x | x \text{ is a real number}\}$

Domain of $g + h$: $\{x | x \text{ is a real number and } x \ne 0\}$

31. $h(x) = 2x + 1 = 0$, if $x = -\frac{1}{2}$

Domain of g / h:

$$\left\{ x \middle| x \text{ is a real number and } x \ne 0 \text{ and } x \ne -\frac{1}{2} \right\}$$

32. a. 1 hr and 40 min is equal to $1\frac{40}{60} = 1\frac{2}{3} = \frac{5}{3}$ hr.

We must find $f\left(\frac{5}{3}\right)$.

$$f\left(\frac{5}{3}\right) = 5 + 15 \cdot \frac{5}{3} = 5 + 25 = 30$$

The cyclist will be 30 mi from the starting point 1 hr and 40 min after passing the 5-mi marker.

b. From the equation $f(t) = 5 + 15t$, we see that the cyclist is advancing 15 mi for every hour he travels. So the rate is 15 mph.

33. If the graph is to be parallel to the line $3x - 2y = 7$, then we must determine the slope:

$$3x - 2y = 7$$

$$-2y = -3x + 7$$

$$y = \frac{3}{2}x - \frac{7}{2} \qquad m = \frac{3}{2}$$

The line must contain the two points $(r, 3)$ and $(7, s)$.

The slope of this line must be $\frac{3}{2}$.

$$\frac{3}{2} = \frac{s - 3}{7 - r}$$

$$3(7 - r) = 2(s - 3)$$

$$21 - 3r = 2s - 6$$

$$21 - 3r + 6 = 2s$$

$$27 - 3r = 2s$$

$$2s = -3r + 27$$

$$s = -\frac{3}{2}r + \frac{27}{2} \text{ or } s = \frac{27 - 3r}{2}$$

34. Answers may vary. In order to have the restriction on the domain of $f / g / h$ that $x \ne \frac{3}{4}$ and $x \ne \frac{2}{7}$, we need to find some function $h(x)$ such that $h(x) = 0$ when $x = \frac{2}{7}$.

$$x = \frac{2}{7}$$

$$7x = 2$$

$$7x - 2 = 0$$

One possible answer is $h(x) = 7x - 2$.

Chapter 3

Systems of Linear Equations and Problem Solving

Exercise Set 3.1

1. False; see Example 4(b).

3. True

5. True; see Example 4(b).

7. False; see page 154 in the text.

9. We use alphabetical order for the variables. We replace x by 2 and y by 3.

$$\begin{array}{c|c} 2x - y = 1 \\ \hline 2 \cdot 2 - 3 & 1 \\ 4 - 3 & \\ & \overset{?}{1 = 1} \quad \text{TRUE} \end{array} \qquad \begin{array}{c|c} 5x - 3y = 1 \\ \hline 5 \cdot 2 - 3 \cdot 3 & 1 \\ 10 - 9 & \\ & \overset{?}{1 = 1} \quad \text{TRUE} \end{array}$$

The pair (2, 3) makes both equations true, so is it a solution of the system.

11. We use alphabetical order for the variables. We replace x by −5 and y by 1.

$$\begin{array}{c|c} x + 5y = 0 \\ \hline -5 + 5 \cdot 1 & 0 \\ -5 + 5 & \\ & \overset{?}{0 = 0} \quad \text{TRUE} \end{array} \qquad \begin{array}{c|c} y = 2x + 9 \\ \hline 1 & 2(-5) + 9 \\ & -10 + 9 \\ & \overset{?}{1 = -1} \quad \text{FALSE} \end{array}$$

The pair (−5, 1) is not a solution of $y = 2x + 9$. Therefore, it is not a solution of the system of equations.

13. We replace x by 0 and y by −5.

$$\begin{array}{c|c} x - y = 5 \\ \hline 0 - (-5) & 5 \\ 0 + 5 & \\ & \overset{?}{5 = 5} \quad \text{TRUE} \end{array} \qquad \begin{array}{c|c} y = 3x - 5 \\ \hline -5 & 3 \cdot 0 - 5 \\ & 0 - 5 \\ & \overset{?}{-5 = -5} \quad \text{TRUE} \end{array}$$

The pair (0, −5) makes both equations true, so is it a solution of the system.

15. Observe that if we multiply both sides of the first equation by 2, we get the second equation. Thus, if we find that the given points makes the one equation true, we will also know that it makes the other equation true. We replace x by 3 and y by −1 in the first equation.

$$\begin{array}{c|c} 3x - 4y = 13 \\ \hline 3 \cdot 3 - 4(-1) & 13 \\ 9 + 4 & \\ & \overset{?}{13 = 13} \quad \text{TRUE} \end{array}$$

The pair (3, −1) makes both equations true, so is it a solution of the system.

17. Graph both equations.

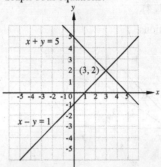

The solution (point of intersection) is apparently (3, 2).
Check:

$$\begin{array}{c|c} x - y = 1 \\ \hline 3 - 2 & 1 \\ & \overset{?}{1 = 1} \quad \text{TRUE} \end{array} \qquad \begin{array}{c|c} x + y = 5 \\ \hline 3 + 2 & 5 \\ & \overset{?}{5 = 5} \quad \text{TRUE} \end{array}$$

The solution is (3, 2).

19. Graph the equations.

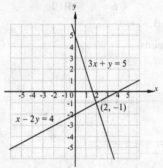

The solution (point of intersection) is apparently (2, −1).
Check:

$$\begin{array}{c|c} 3x + y = 5 \\ \hline 3 \cdot 2 + (-1) & 5 \\ 6 - 1 & \\ & \overset{?}{5 = 5} \quad \text{TRUE} \end{array} \qquad \begin{array}{c|c} x - 2y = 4 \\ \hline 2 - 2(-1) & 4 \\ 2 + 2 & \\ & \overset{?}{4 = 4} \quad \text{TRUE} \end{array}$$

The solution is (2, −1).

21. Graph both equations.

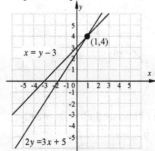

The solution (point of intersection) is apparently (1, 4).
Check:

$$\frac{2y = 3x + 5}{2 \cdot 4 \mid 3 \cdot 1 + 5}$$
$$8 \mid 3 + 5$$
$$\overset{?}{8 = 8} \quad \text{TRUE}$$

$$\frac{x = y - 3}{1 \mid 4 - 3}$$
$$\overset{?}{1 = 1} \quad \text{TRUE}$$

The solution is (1, 4).

23. Graph both equations.

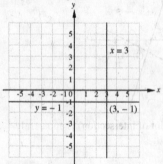

The solution (point of intersection) is apparently
(−3, −2).
Check:

$$\frac{x = y - 1}{-3 \mid -2 - 1}$$
$$\overset{?}{-3 = -3} \quad \text{TRUE}$$

$$\frac{2x = 3y}{2(-3) \mid 3(-2)}$$
$$\overset{?}{-6 = -6} \quad \text{TRUE}$$

The solution is (−3, −2).

25. Graph both equations.

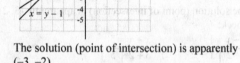

The ordered pairs (3, −1) checks in both equations. It is
the solution.

27. Graph both equations.

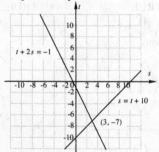

The solution (point of intersection) is apparently
(3, −7).
Check:

$$\frac{t + 2s = -1}{-7 + 2 \cdot 3 \mid -1}$$
$$-7 + 6 \mid$$
$$\overset{?}{-1 = -1} \quad \text{TRUE}$$

$$\frac{s = t + 10}{3 \mid -7 + 10}$$
$$\overset{?}{3 = 3} \quad \text{TRUE}$$

The solution is (3, −7).

29. Graph both equations.

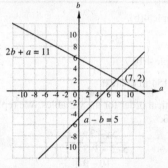

The solution (point of intersection) is apparently (7, 2).
Check:

$$\frac{2b + a = 11}{2 \cdot 2 + 7 \mid 11}$$
$$4 + 7 \mid$$
$$\overset{?}{11 = 11} \quad \text{TRUE}$$

$$\frac{a - b = 5}{7 - 2 \mid 5}$$
$$\overset{?}{5 = 5} \quad \text{TRUE}$$

The solution is (7, 2).

31. Graph both equations.

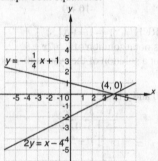

The solution (point of intersection) is apparently (4, 0).
Check:

$$\frac{y = -\frac{1}{4}x + 1}{0 \mid -\frac{1}{4} \cdot 4 + 1}$$
$$-1 + 1 \mid$$
$$\overset{?}{0 = 0} \quad \text{TRUE}$$

$$\frac{2y = x - 4}{2 \cdot 0 \mid 4 - 4}$$
$$\overset{?}{0 = 0} \quad \text{TRUE}$$

The solution is (4, 0).

33. Graph both equations.

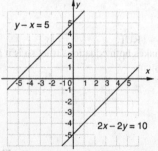

The lines are parallel. The system has no solution.

35. Graph both equations.

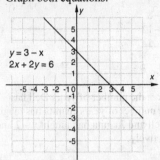

$y = 3 - x$
$2x + 2y = 6$

The graphs are the same. Any solution of one equation is a solution of the other. Each equation has infinitely many solutions. The solution set is the set of all pairs (x, y) for which $y = 3 - x$, or $\{(x, y) \mid y = 3 - x\}$. (In place of $y = 3 - x$ we could have used $2x + 2y = 6$ since the two equations are equivalent.)

37. A system of equations is consistent if it has at least one solution. Of the systems under consideration, only the one in Exercise 33 has no solution. Therefore, all except the system in Exercise 33 are consistent.

39. A system of two equations in two variables is dependent if it has infinitely many solutions. Only the system in Exercise 35 is dependent.

41. *Familiarize.* Let $x =$ the first number and $y =$ the second number.
Translate.

The sum of the numbers is 10.
$$x + y = 10$$

The first number is $\frac{2}{3}$ of the second number.
$$x = \frac{2}{3} \times y$$

We have a system of equations:
$$x + y = 10,$$
$$x = \frac{2}{3}y.$$

43. *Familiarize.* Let $p =$ the number of endangered plant species and $a =$ the number of endangered animal species.
Translate.

plants and animals is 1223.
$$p + a = 1223$$

plants are 243 more than animals.
$$p = 243 + a$$

We have a system of equations:
$$p + a = 1223,$$
$$p = a + 243.$$

45. *Familiarize.* Let $x =$ the measure of one angle and $y =$ the measure of the other angle.
Translate.
Two angles are supplementary.
Rewording: The sum of the measures is 180°.
$$x + y = 180$$

One angle is 3 less than twice the other.
Rewording:

One angle is twice the other angle minus 3°.
$$x = 2y - 3$$

We have a system of equations:
$$x + y = 180,$$
$$x = 2y - 3$$

47. Familiarize. Let $x =$ the number of two-point shots and $y =$ the number of foul shots made.

Translate. We organize the information in a table.

Kind of shot	2-point shot	Foul shot	Total
Number scored	x	y	64
Points per score	2	1	
Points scored	$2x$	x	100

From the "Number scored" row of the table we get one equation:
$$x + y = 64$$
The "Points scored" row gives us another equation:
$$2x + y = 100$$
We have a system of equations:
$$x + y = 64,$$
$$2x + y = 100$$

49. *Familiarize.* Let $w =$ the number of wrapped strings and $u =$ the number of unwrapped strings.
Translate. We organize the information in a table.

Strings	Wrapped	Unwrapped	Total
Number	w	u	32
Price	$4.49	$2.99	
Amount paid	$4.49w$	$2.99u$	107.68

The "Number" row of the table gives us one equation:
$$w + u = 32.$$
The "Amount paid" row gives us a second equation:
$$4.49w + 2.99u = 107.68.$$

51. *Familiarize.* Let $h =$ the number of hats knitted and $s =$ the number of scarves knitted.
Translate. We organize the information in a table.

Items knitted	Hats	Scarves	Total
Number	h	s	110
Time	8	12	
Time spent	$8h$	$12s$	1072

The "Number" row of the table gives us one equation:
$$h + s = 110.$$
The "Time spent" row gives us a second equation:
$$8h + 12s = 1072.$$

53. *Familiarize*. The lacrosse field is a rectangular with perimeter 340 yd. Let l = the length, in yards, and w = the width, in yards. Recall that for a rectangle with length l and width w, the perimeter P is given by $P = 2l + 2w$.

Translate. The formula for perimeter gives us one equation:
$$2l + 2w = 340.$$
The statement relating length and width gives us another equation:
$$l = w + 50.$$
We have a system of equations:
$$2l + 2w = 340,$$
$$l = w + 50.$$

55. *Writing Exercise*.

57. $-\dfrac{1}{2} - \dfrac{3}{10} = -\dfrac{5}{10} - \dfrac{3}{10} = -\dfrac{8}{10} = -\dfrac{4}{5}$

59. $-10^{-2} = -\dfrac{1}{10^2} = -\dfrac{1}{100}$

61. $(-3)^2 - 2 - 4 \cdot 6 \div 2 \cdot 3$
$= 9 - 2 - 4 \cdot 6 \div 2 \cdot 3$
$= 9 - 2 - 24 \div 2 \cdot 3$
$= 9 - 2 - 12 \cdot 3$
$= 9 - 2 - 36$
$= -29$

63. *Writing Exercise*.

65. a. There are many correct answers. One can be found by expressing the sum and difference of the two numbers:
$$x + y = 6,$$
$$x - y = 4.$$

b. There are many correct answers. For example, write an equation in two variables. Then write a second equation by multiplying the left side of the first equation by one nonzero constant and multiplying the right side by another nonzero constant.
$$x + y = 1,$$
$$2x + 2y = 3.$$

c. There are many correct answers. One can be found by writing an equation in two variables and then writing a nonzero constant multiple of that equation:
$$x + y = 1,$$
$$2x + 2y = 2.$$

67. Substitute 4 for x and -5 for y in the first equation:
$$A(4) - 6(-5) = 13$$
$$4A + 30 = 13$$
$$4A = -17$$
$$A = -\dfrac{17}{4}$$
Substitute 4 for x and -5 for y in the second equation:

$$4 - B(-5) = -8$$
$$4 + 5B = -8$$
$$5B = -12$$
$$B = -\dfrac{12}{5}$$
We have $A = -\dfrac{17}{4}$, $B = -\dfrac{12}{5}$.

69. *Familiarize*. Let x = the number of years Dell has taught and y = the number of years Juanita has taught. Two years ago, Dell and Juanita had taught $x - 2$ and $y - 2$ years, respectively.

Translate.

Together, the number of years of service	is	46.
↓	↓	↓
$x + y$	=	46

Two years ago Dell had taught 2.5 times as many years as Juanita.
$$x - 2 = 2.5(y - 2)$$
We have a system of equations:
$$x + y = 46,$$
$$x - 2 = 2.5(y - 2)$$

71. *Familiarize*. Let s = the number of ounces of baking soda and v = the number of ounces of vinegar to be used. The amount of baking soda in the mixture will be four times the amount of vinegar.

Translate.

The amount of baking soda	is	four times the amount of vinegar.
↓	↓	↓
s	=	$4v$

The total amount	is	16 oz.
↓	↓	↓
$s + v$	=	16

We have a system of equations.
$$s = 4v,$$
$$s + v = 16$$

73. From Exercise 44, graph both equations:
$$v + m = 16$$
$$m = 2v + 4$$

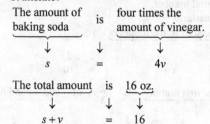

The lines intersect at $v = 4$, $m = 12$. The solution is 4 oz of vinegar and 12 oz of mineral oil.

75. Graph both equations.

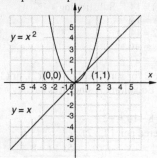

The solutions are apparently (0, 0) and (1, 1). Both pairs check.

77. (0.07, −7.95)

79. (0.00, 1.25)

Connecting the Concepts

1. $x = y$,   (1)
   $x + y = 2$ (2)

Substitute y for x in Equation (2) and solve for y.
$$x + y = 2 \quad (2)$$
$$y + y = 2 \quad \text{Substituting}$$
$$2y = 2$$
$$y = 1$$

Substitute 1 for y in Equation (1) and solve for x.
$$x = y \quad (1)$$
$$x = 1$$

We obtain $(1, 1)$ as the solution.

2. $x + y = 10$ (1)
   $\dfrac{x - y = 8}{2x = 18}$ (2)   Adding
      $x = 9$

Substitute 9 for x in Equation (1) and solve for y.
$$x + y = 10 \quad (1)$$
$$9 + y = 10 \quad \text{Substituting}$$
$$y = 1$$

We obtain (9, 1) as the solution.

3. $y = \dfrac{1}{2}x + 1$,   (1)
   $y = 2x - 5$ (2)

Substitute $2x - 5$ for y in Equation (1) and solve for x.
$$y = \frac{1}{2}x + 1 \quad (1)$$
$$2x - 5 = \frac{1}{2}x + 1 \quad \text{Substituting}$$
$$4x - 10 = x + 2 \quad \text{Clearing fractions}$$
$$3x = 12$$
$$x = 4$$

Substitute 4 for x in Equation (2) and solve for y.
$$y = 2x - 5 \quad (2)$$
$$y = 2(4) - 5 = 3$$

We obtain (4, 3) as the solution.

4. $y = 2x - 3$,   (1)
   $x + y = 12$ (2)

Substitute $2x - 3$ for y in Equation (2) and solve for x.

$$x + y = 12 \quad (2)$$
$$x + (2x - 3) = 12 \quad \text{Substituting}$$
$$x + 2x - 3 = 12$$
$$3x - 3 = 12$$
$$3x = 15$$
$$x = 5$$

Substitute 5 for x in Equation (2) and solve for y.
$$y = 2x - 3 \quad (1)$$
$$y = 2(5) - 3 = 7$$

We obtain (5, 7) as the solution.

5. $12x - 19y = 13$ (1)
   $\dfrac{8x + 19y = 7}{20x = 20}$ (2)   Adding
      $x = 1$

Substitute 1 for x in Equation (2) and solve for y.
$$8x + 19y = 7 \quad (2)$$
$$8(1) + 19y = 7 \quad \text{Substituting}$$
$$19y = -1$$
$$y = -\frac{1}{19}$$

We obtain $\left(1, -\dfrac{1}{19}\right)$ as the solution.

6. $2x - 5y = 1$,   (1)
   $3x + 2y = 11$ (2)

We multiply Equation (1) by 2 and Equation (2) by 5.
$$4x - 10y = 2 \quad \text{Multiplying (1) by 2}$$
$$\frac{15x + 10y = 55}{19x = 57} \quad \text{Multiplying (2) by 5}$$
$$x = 3$$

Substitute 3 for x in Equation (2) and solve for y.
$$3x + 2y = 11 \quad (2)$$
$$3(3) + 2y = 11$$
$$9 + 2y = 11$$
$$2y = 2$$
$$y = 1$$

We obtain (3, 1) as the solution.

7. $y = \dfrac{5}{3}x + 7$,   (1)
   $y = \dfrac{5}{3}x - 8$ (2)

We see that both lines have the same slope and are parallel. There is no solution.

8. $x = 2 - y$,   (1)
   $3x + 3y = 6$ (2)

Substitute $2 - y$ for x in Equation (2) and solve for y.
$$3x + 3y = 6 \quad (2)$$
$$3(2 - y) + 3y = 6 \quad \text{Substituting}$$
$$6 - 3y + 3y = 6$$
$$6 = 6$$

There are many solutions. The solution set is
$$\{(x, y) \mid x = 2 - y\} \text{ or } \{(x, y) \mid 3x + 3y = 6\}.$$

Exercise Set 3.2

1. To use the *substitution* method, a variable must be isolated.

3. To eliminate a variable by adding, two terms must be *opposites*.

5. Adding the equations, we get $8x = 7$, so choice (d) is correct.

7. Multiplying the first equation by -5 gives us the system of equations in (a), so choice (a) is correct.

9. Substituting $4x - 7$ for y in the second equation gives us $6x + 3(4x - 7) = 19$, so choice (c) is correct.

11. $y = 3 - 2x,$ (1)
  $3x + y = 5$ (2)

We substitute $3 - 2x$ for y into the second equation and solve for x.
$$3x + y = 5 \quad (2)$$
$$3x + (3 - 2x) = 5 \quad \text{Substituting}$$
$$x + 3 = 5$$
$$x = 2$$

Next substitute 2 for x in either equation of the original system and solve for y.
$$y = 3 - 2x \quad (1)$$
$$y = 3 - 2 \cdot 2 \quad \text{Substitute}$$
$$y = 3 - 4$$
$$y = -1$$

We check the ordered pair $(2, -1)$.

$y = 3 - 2x$	$3x + y = 5$
$-1 \mid 3 - 2 \cdot 2$	$3 \cdot 2 + (-1) \mid 5$
$3 - 4$	$6 - 1$
$-1 = -1$ TRUE	$5 = 5$ TRUE

Since $(2, -1)$ checks, it is the solution.

13. $3x + 5y = 3,$ (1)
  $x = 8 - 4y$ (2)

We substitute $8 - 4y$ for x in the first equation and solve for y.
$$3x + 5y = 3 \quad (1)$$
$$3(8 - 4y) + 5y = 3 \quad \text{Substituting}$$
$$24 - 12y + 5y = 3$$
$$24 - 7y = 3$$
$$-7y = -21$$
$$y = 3$$

Next we substitute 3 for y in either equation of the original system and solve for x.
$$x = 8 - 4y \quad (2)$$
$$x = 8 - 4 \cdot 3 = 8 - 12 = -4$$

We check the ordered pair $(-4, 3)$.

$x = 8 - 4y$	$3x + 5y = 3$
$-4 \mid 8 - 4 \cdot 3$	$3(-4) + 5 \cdot 3 \mid 3$
$8 - 12$	$-12 + 15$
$-4 = -4$ TRUE	$3 = 3$ TRUE

Since $(-4, 3)$ checks, it is the solution.

15. $3s - 4t = 14,$ (1)
  $5s + t = 8$ (2)

We solve the second equation for t.
$$5s + t = 8 \quad (2)$$
$$t = 8 - 5s \quad (3)$$

We substitute $8 - 5s$ for t in the first equation and solve for s.

$3s - 4t = 14$ (1)
$3s - 4(8 - 5s) = 14$ Substituting
$3s - 32 + 20s = 14$
$23s - 32 = 14$
$23s = 46$
$s = 2$

Next we substitute 2 for s in Equation (1), (2), or (3). It is easiest to use Equation (3) since it is already solved for t.
$$t = 8 - 5 \cdot 2 = 8 - 10 = -2$$

We check the ordered pair $(2, -2)$.

$3s - 4t = 14$	$5s + t = 8$
$3 \cdot 2 - 4(-2) \mid 14$	$5 \cdot 2 + (-2) \mid 8$
$6 + 8$	$10 - 2$
$14 = 14$ TRUE	$8 = 8$ TRUE

Since $(2, -2)$ checks, it is the solution.

17. $4x - 2y = 6,$ (1)
  $2x - 3 = y$ (2)

We substitute $2x - 3$ for y in the first equation and solve for x.
$$4x - 2y = 6 \quad (1)$$
$$4x - 2(2x - 3) = 6$$
$$4x - 4x + 6 = 6$$
$$6 = 6$$

We have an identity, or an equation that is always true. The equations are dependent and the solution set is infinite: $\{(x, y) \mid 2x - 3 = y\}$.

19. $-5s + t = 11,$ (1)
  $4s + 12t = 4$ (2)

We solve the first equation for t.
$$-5s + t = 11 \quad (1)$$
$$t = 5s + 11 \quad (3)$$

We substitute $5s + 11$ for t in the second equation and solve for s.
$$4s + 12t = 4 \quad (2)$$
$$4s + 12(5s + 11) = 4$$
$$4s + 60s + 132 = 4$$
$$64s + 132 = 4$$
$$64s = -128$$
$$s = -2$$

Next we substitute -2 for s in Equation (3).
$$t = 5s + 11 = 5(-2) + 11 = -10 + 11 = 1$$

We check the ordered pair $(-2, 1)$.

$-5s + t = 11$	$4s + 12t = 4$
$-5(-2) + 1 \mid 11$	$4(-2) + 12 \cdot 1 \mid 4$
$10 + 1$	$-8 + 12$
$11 = 11$ TRUE	$4 = 4$ TRUE

Since $(-2, 1)$ checks, it is the solution.

21. $2x + 2y = 2,$ (1)
  $3x - y = 1$ (2)

We solve the second equation for y.
$$3x - y = 1 \quad (2)$$
$$-y = -3x + 1$$
$$y = 3x - 1 \quad (3)$$

We substitute $3x - 1$ for y in the first equation and solve for x.

$$2x + 2y = 2 \quad (1)$$
$$2x + 2(3x - 1) = 2$$
$$2x + 6x - 2 = 2$$
$$8x - 2 = 2$$
$$8x = 4$$
$$x = \tfrac{1}{2}$$

Next we substitute $\tfrac{1}{2}$ for x in Equation (3).

$$y = 3x - 1 = 3 \cdot \tfrac{1}{2} - 1 = \tfrac{3}{2} - 1 = \tfrac{1}{2}$$

The ordered pair $\left(\tfrac{1}{2}, \tfrac{1}{2}\right)$ checks in both equations. It is the solution.

23. $2a + 6b = 4, \quad (1)$
$3a - b = 6 \quad (2)$

We solve the second equation for b.

$$3a - b = 6 \quad (2)$$
$$-b = -3a + 6$$
$$b = 3a - 6 \quad (3)$$

We substitute $3a - 6$ for b in the first equation and solve for a.

$$2a + 6b = 4 \quad (1)$$
$$2a + 6(3a - 6) = 4$$
$$2a + 18a - 36 = 4$$
$$20a - 36 = 4$$
$$20a = 40$$
$$a = 2$$

We substitute 2 for a in Equation 3 and solve for y.

$$b = 3a - 6 \quad (3)$$
$$b = 3 \cdot 2 - 6$$
$$b = 6 - 6$$
$$b = 0$$

The ordered pair $(2, 0)$ checks in both equations. It is the solution.

25. $2x - 3 = y \quad (1)$
$y - 2x = 1, \quad (2)$

We substitute $2x - 3$ for y in the second equation and solve for x.

$$y - 2x = 1 \quad (2)$$
$$2x - 3 - 2x = 1 \quad \text{Substituting}$$
$$-3 = 1 \quad \text{Collecting like terms}$$

We have a contradiction, or an equation that is always false. Therefore, there is no solution.

27. $x + 3y = 7 \quad (1)$
$\underline{-x + 4y = 7} \quad (2)$
$0 + 7y = 14 \quad \text{Adding}$
$7y = 14$
$y = 2$

Substitute 2 for y in one of the original equations and solve for x.

$$x + 3y = 7 \quad (1)$$
$$x + 3 \cdot 2 = 7 \quad \text{Substituting}$$
$$x + 6 = 7$$
$$x = 1$$

Check:

$$\begin{array}{c|c}
x + 3y = 7 & -x + 4y = 7 \\
\hline
1 + 3 \cdot 2 \mid 7 & -1 + 4 \cdot 2 \mid 7 \\
1 + 6 & -1 + 8 \\
 \overset{?}{7 = 7} \quad \text{TRUE} & \overset{?}{7 = 7} \quad \text{TRUE}
\end{array}$$

Since $(1, 2)$ checks, it is the solution.

29. $x - 2y = 11 \quad (1)$
$\underline{3x + 2y = 17} \quad (2)$
$4x = 28 \quad \text{Adding}$
$x = 7$

Substitute 7 for x in Equation (1) and solve for y.

$$x - 2y = 11 \quad (1)$$
$$7 - 2y = 11 \quad \text{Substituting}$$
$$-2y = 4$$
$$y = -2$$

We obtain $(7, -2)$. This checks, so it is the solution.

31. $9x + 3y = -3 \quad (1)$
$\underline{2x - 3y = -8} \quad (2)$
$11x + 0 = -11 \quad \text{Adding}$
$11x = -11$
$x = -1$

Substitute -1 for x in Equation (1) and solve for y.

$$9x + 3y = -3$$
$$9(-1) + 3y = -3 \quad \text{Substituting}$$
$$-9 + 3y = -3$$
$$3y = 6$$
$$y = 2$$

We obtain $(-1, 2)$. This checks, so it is the solution.

33. $5x + 3y = 19, \quad (1)$
$x - 6y = 11 \quad (2)$

We multiply Equation (1) by 2 to make two terms become opposites.

$$10x + 6y = 38 \quad \text{Multiplying (1) by 2}$$
$$\underline{x - 6y = 11}$$
$$11x = 49$$
$$x = \frac{49}{11}$$

Substitute $\frac{49}{11}$ for x in Equation (1) and solve for y.

$$5x + 3y = 19$$
$$5 \cdot \frac{49}{11} + 3y = 19$$
$$3y = -\frac{36}{11}$$
$$y = -\frac{12}{11}$$

We obtain $\left(\frac{49}{11}, -\frac{12}{11}\right)$. This checks, so it is the solution.

35. $5r - 3s = 24, \quad (1)$
$3r + 5s = 28 \quad (2)$

We multiply twice to make two terms become additive inverses.

From (1): $25r - 15s = 120 \quad \text{Multiplying by 5}$
From (2): $\underline{9r + 15s = 84} \quad \text{Multiplying by 3}$
$34r + 0 = 204 \quad \text{Adding}$
$r = 6$

Substitute 6 for r in Equation (2) and solve for s.

$$3r + 5s = 28$$
$$3 \cdot 6 + 5s = 28 \quad \text{Substituting}$$
$$18 + 5s = 28$$
$$5s = 10$$
$$s = 2$$

We obtain $(6, 2)$. This checks, so it is the solution.

37. $6s + 9t = 12$,   (1)
$4s + 6t = 5$ (2)

We multiply twice to make two terms become opposites.

From (1): $12s + 18t = 24$ Multiplying by 2
From (2): $\underline{-12s - 18t = -15}$ Multiplying by -3
                $0 = 9$

We get a contradiction, or an equation that is always false. The system has no solution.

39. $\dfrac{1}{2}x - \dfrac{1}{6}y = 10$,   (1)

$\dfrac{2}{5}x + \dfrac{1}{2}y = 8$ (2)

We first multiply each equation by the LCM of the denominators to clear fractions.

$3x - y = 60$,   (3)   Multiplying (1) by 6
$4x + 5y = 80$ (4)   Multiplying (2) by 10

We multiply Equation (3) by 5 and then add.

$15x - 5y = 300$ Multiplying (3) by 5
$\underline{4x + 5y = 80}$ (4)
$19x \qquad = 380$
$\qquad x = 20$

Substitute 20 for x in one of the equations in which the fractions were cleared and solve for y.

$$3x - y = 60 \quad (3)$$
$$3 \cdot 20 - y = 60$$
$$60 - y = 60$$
$$-y = 0$$
$$y = 0$$

We obtain $(20,\ 0)$. This checks, so it is the solution.

41. $\dfrac{x}{2} + \dfrac{y}{3} = \dfrac{7}{6}$,   (1)

$\dfrac{2x}{3} + \dfrac{3y}{4} = \dfrac{5}{4}$ (2)

We first multiply each equation by the LCM of the denominators to clear fractions.

$3x + 2y = 7$ (3)   Multiplying (1) by 6
$8x + 9y = 15$ (4)   Multiplying (2) by 12

We multiply twice to make two terms become opposites.

From (3):
From (4):

$27x + 18y = 63$ Multiplying by 9
$\underline{-16x - 18y = -30}$ Multiplying by -2
$11x \qquad = 33$ Adding
$\qquad x = 3$

Substitute 3 for x in one of the equations in which the fractions were cleared and solve for y.

$3x + 2y = 7$ (3)
$3 \cdot 3 + 2y = 7$ Substituting
$9 + 2y = 7$
$2y = -2$
$y = -1$

We obtain $(3, -1)$. This checks, so it is the solution.

43. $12x - 6y = -15$,   (1)
$-4x + 2y = 5$ (2)

Observe that, if we multiply Equation (1) by $-\dfrac{1}{3}$, we obtain Equation (2). Thus, any pair that is a solution of Equation (1) is also a solution of Equation (2). The equations are dependent and the solution set is infinite: $\{(x,\ y)\,|\,-4x + 2y = 5\}$.

45. $0.3x + 0.2y = 0.3$,
$0.5x + 0.4y = 0.4$

We first multiply each equation by 10 to clear decimals.

$3x + 2y = 3$,   (1)
$5x + 4y = 4$ (2)

We multiply Equation (1) by -2.

$-6x - 4y = -6$ Multiplying (1) by -2
$\underline{5x + 4y = 4}$ (2)
$-x \qquad = -2$
$\quad x = 2$

Substitute 2 for x in Equation (1) and solve for y.

$$3x + 2y = 3 \quad (1)$$
$$3 \cdot 2 + 2y = 3$$
$$6 + 2y = 3$$
$$2y = -3$$
$$y = -\frac{3}{2}$$

We obtain $\left(2, -\dfrac{3}{2}\right)$. This checks, so it is the solution.

47. $a - 2b = 16$,   (1)
$b + 3 = 3a$ (2)

We will use the substitution method. First solve Equation (1) for a.

$$a - 2b = 16$$
$$a = 2b + 16 \quad (3)$$

Now substitute $2b + 16$ for a in Equation (2) and solve for b.

$$b + 3 = 3a \qquad (2)$$
$$b + 3 = 3(2b + 16) \quad \text{Substituting}$$
$$b + 3 = 6b + 48$$
$$-45 = 5b$$
$$-9 = b$$

Substitute -9 for b in Equation (3).

$$a = 2(-9) + 16 = -2$$

We obtain $(-2, -9)$. This checks, so it is the solution.

49. $10x + y = 306$,   (1)
$10y + x = 90$ (2)

We will use the substitution method. First solve Equation (1) for y.

$10x + y = 306$

$y = -10x + 306$ (3)

Now substitute $-10x + 306$ for y in Equation (2) and solve for y.

$10y + x = 90$ (2)

$10(-10x + 306) + x = 90$ Substituting

$-100x + 3060 + x = 90$

$-99x + 3060 = 90$

$-99x = -2970$

$x = 30$

Substitute 30 for x in Equation (3).

$y = -10 \cdot 30 + 306 = 6$

We obtain (30, 6). This checks, so it is the solution.

51. $6x - 3y = 3,$ (1)

    $4x - 2y = 2$ (2)

Observe that, if we multiply Equation (1) by $\frac{3}{2}$, we obtain Equation (2). Thus, any pair that is a solution of Equation (1) is also a solution of Equation (2). The equations are dependent and the solution set infinite: $\{(x, y) \mid 4x - 2y = 2\}$.

53. $3s - 7t = 5,$

    $7t - 3s = 8$

First we rewrite the second equation with the variables in a different order. Then we use the elimination method.

$3s - 7t = 5$ (1)

$\dfrac{-3s + 7t = 8}{0 = 13}$ (2)

We get a contradiction, so the system has no solution.

55. $0.05x + 0.25y = 22,$

    $0.15x + 0.05y = 24$

We first multiply each equation by 100 to clear decimals.

$5x + 25y = 2200,$ (1)

$15x + 5y = 2400$ (2)

We multiply by -5 on both sides of the second equation and add.

$5x + 25y = 2200$ (1)

$\dfrac{-75x - 25y = -12,000}{-70x = -9800}$ Multiplying (2) by -5

                  Adding

$x = \dfrac{-9800}{-70}$

$x = 140$

Substitute 140 for x in one of the equations in which the decimals were cleared and solve for y.

$5x + 25y = 2200$ (1)

$5 \cdot 140 + 25y = 2200$ Substituting

$700 + 25y = 2200$

$25y = 1500$

$y = 60$

We obtain (140, 60). This checks, so it is the solution.

57. $13a - 7b = 9,$ (1)

    $2a - 8b = 6$ (2)

We will use the elimination method. First we multiply the equations so that the b-terms can be eliminated.

From (1): $104a - 56b = 72$ Multiplying by 8

From (2): $\dfrac{-14a + 56b = -42}{90a = 30}$ Multiplying by -7

                Adding

$a = \dfrac{1}{3}$

Substitute $\frac{1}{3}$ for a in one of the equations and solve for b.

$2a - 8b = 6$ (2)

$2 \cdot \dfrac{1}{3} - 8b = 6$ Substituting

$\dfrac{2}{3} - 8b = 6$

$-8b = \dfrac{16}{3}$

$b = -\dfrac{2}{3}$

We obtain $\left(\dfrac{1}{3}, -\dfrac{2}{3}\right)$. This checks, so it is the solution.

59. $a - \dfrac{1}{2}c = 6,$

    $c + 2a = 8$

We first multiply first equation by 2 to clear fractions.

$2a - c = 12,$ (1)

$2a + c = 8$ (2)

We will use the elimination method.

$2a - c = 12$ (1)

$\dfrac{2a + c = 8}{4a = 20}$ (2)

            Adding

$a = 5$

Substitute 5 for a in Equation (2) and solve for c.

$c + 2a = 8$ (2)

$c + 2 \cdot 5 = 8$ Substituting

$c + 10 = 8$

$c = -2$

We obtain (5, -2). This checks, so it is the solution.

61. $8x = y - 14,$

    $6(y - x) = 63$

We will use the elimination method. First write each equation in the form $Ax + By = C$.

$8x - y = -14,$ (1)

$-6x + 6y = 63$ (2)

We multiply by 6 on both sides of the first equation and add.

$48x - 6y = -84$ Mutiplying (1) by 6

$\dfrac{-6x + 6y = 63}{42x = -21}$ (2)

             Adding

$x = -\dfrac{1}{2}$

Substitute $-\dfrac{1}{2}$ for x in one of the equations solve for y.

$-6x + 6y = 63$ (2)

$-6\left(-\dfrac{1}{2}\right) + 6y = 63$ Substituting

$3 + 6y = 63$

$6y = 60$

$y = 10$

We obtain $\left(-\dfrac{1}{2}, 10\right)$. This checks, so it is the solution.

63. $2m + 6n = 4,$ (1)
$4m - 2n = 6$ (2)

We will use the elimination method. We multiply Equation (2) by 3.

$2m + 6n = 4$ (1)
$\underline{12m - 6n = 18}$ Multiplying (2) by 3
$14m \qquad = 22$ Adding
$m = \dfrac{11}{7}$

Substitute $\dfrac{11}{7}$ for m in Equation (1) and solve for n.

$2m + 6n = 4$

$2\left(\dfrac{11}{7}\right) + 6n = 4$ Substituting

$\dfrac{22}{7} + 6n = 4$

$6n = \dfrac{6}{7}$

$n = \dfrac{1}{7}$

We obtain $\left(\dfrac{11}{7}, \dfrac{1}{7}\right)$. This checks, so it is the solution.

65. $23x - y = 5,$
$11x - 10 = 2y$

We will use the elimination method. First write each equation in the form $Ax + By = C$.

$23x - y = 5,$ (1)
$11x - 2y = 10$ (2)

We multiply by -2 on both sides of the first equation and add.

$-48x + 2y = -10$ Mutiplying (1) by -2
$\underline{11x - 2y = 10}$ (2)
$-37x \qquad = 0$ Adding
$x = 0$

Substitute 0 for x in one of the equations solve for y.

$11x - 10 = 2y$ (2)
$11(0) - 10 = 2y$ Substituting
$-10 = 2y$
$-5 = y$

We obtain $(0, -5)$. This checks, so it is the solution.

67. *Writing Exercise.*

69. $(4 + m) + n = 4 + (m + n)$

71. $8x - 3[5x + 2(6 - 9x)] = 8x - 3[5x + 12 - 18x]$
$= 8x - 15x - 36 + 54x$
$= 47x - 36$

73. $30{,}050{,}000 = 3.005 \times 10^7$

75. *Writing Exercise.*

77. First write $f(x) = mx + b$ as $y = mx + b$. Then substitute 1 for x and 2 for y to get one equation and also substitute -3 for x and 4 for y to get a second equation:

$2 = m \cdot 1 + b$
$4 = m(-3) + b$

Solve the resulting system of equations.

$2 = m + b$
$4 = -3m + b$

Multiply the second equation by -1 and add.

$2 = m + b$
$\underline{-4 = 3m - b}$
$-2 = 4m$
$-\dfrac{1}{2} = m$

Substitute $-\dfrac{1}{2}$ for m in the first equation and solve for b.

$2 = -\dfrac{1}{2} + b$

$\dfrac{5}{2} = b$

Thus, $m = -\dfrac{1}{2}$ and $b = \dfrac{5}{2}$.

79. Substitute -4 for x and -3 for y in both equations and solve for a and b.

$-4a - 3b = -26,$ (1)
$-4b + 3a = 7$ (2)

$-12a - 9b = -78$ Multiplying (1) by 3
$\underline{12a - 16b = 28}$ Multiplying (2) by 4
$-25b = -50$
$b = 2$

Substitute 2 for b in Equation (2).

$-4 \cdot 2 + 3a = 7$
$3a = 15$
$a = 5$

Thus, $a = 5$ and $b = 2$.

81. $\dfrac{x+y}{2} - \dfrac{x-y}{5} = 1,$

$\dfrac{x-y}{2} + \dfrac{x+y}{6} = -2$

After clearing fractions we have:

$3x + 7y = 10,$ (1)
$4x - 2y = -12$ (2)

$6x + 14y = 20$ Multiplying (1) by 2
$\underline{28x - 14y = -84}$ Multiplying (2) by 7
$34x \qquad = -64$
$x = -\dfrac{32}{17}$

Substitute $-\dfrac{32}{17}$ for x in Equation (1).

$3\left(-\dfrac{32}{17}\right) + 7y = 10$

$7y = \dfrac{266}{17}$

$y = \dfrac{38}{17}$

The solution is $\left(-\dfrac{32}{17}, \dfrac{38}{17}\right)$.

83. $\dfrac{2}{x}+\dfrac{1}{y}=0,$ $2\cdot\dfrac{1}{x}+\dfrac{1}{y}=0,$.

$$\text{or}$$

$\dfrac{5}{x}+\dfrac{2}{y}=-5$ $5\cdot\dfrac{1}{x}+2\cdot\dfrac{1}{y}=-5$

Substitute u for $\dfrac{1}{x}$ and v for $\dfrac{1}{y}$.

$2u+v=0,$ (1)
$5u+2v=-5$ (2)

$\underline{\begin{array}{l}-4u-2v=0 \qquad \text{Multiplying (1) by } -2\\ 5u+2v=-5 \quad (2)\end{array}}$
$u=-5$

Substitute -5 for u in Equation (1).

$2(-5)+v=0$
$-10+v=0$
$v=10$

If $u=-5$, then $\dfrac{1}{x}=-5$. Thus $x=-\dfrac{1}{5}$.

If $v=10$, then $\dfrac{1}{y}=10$. Thus $y=\dfrac{1}{10}$.

The solution is $\left(-\dfrac{1}{5},\ \dfrac{1}{10}\right)$.

85. Familiarize. Let $w=$ the number of kilowatt hours of electricity used each month by the toaster oven. Then $4w=$ the number of kilowatt hours used by the convection oven. The sum of these two numbers is 15 kilowatt hours.
Translate.

$\underbrace{\text{kilowatt hours for toaster oven}}$ plus $\underbrace{\text{kilowatt hours for convection oven}}$ is $\underbrace{15.}$

$\qquad\quad\downarrow\qquad\quad\downarrow\qquad\qquad\downarrow\qquad\quad\downarrow\ \downarrow$
$\qquad\quad w\qquad\quad +\qquad\qquad 4w\qquad\quad =\ \ 15$

Carry out. We solve the equation.

$w+4w=15$
$5w=15$
$w=3$

If $w=3$, then $4w=12$.
Check. We have 3 kWh + 4(3) kWh = 15 kWh. The answer checks.
State. For the month, the toaster oven uses 3 kWh and the convection oven uses 12 kWh.

87. *Writing Exercise.*

Exercise Set 3.3

1. If 10 coffee mugs are sold for $8 each, the *total value* of the mugs is $80.

3. To solve a motion problem, we often use the fact that *distance* divided by rate equals time.

5. The Familiarize and Translate steps were done in Exercise 41 of Exercise Set 3.1.
Carry out. We solve the system of equations.

$x+y=10,$ (1)
$x=\dfrac{2}{3}y$ (2)

where x is the first number and y is the second number.

We use substitution.

$\dfrac{2}{3}y+y=10$

$\dfrac{5}{3}y=10$

$y=6$

Now substitute 6 for y in Equation (2).

$x=\dfrac{2}{3}(6)=4$

Check. The sum of the numbers is $4+6$, or 10 and $\dfrac{2}{3}$ times the second number, 6, is the first number 4. The answer checks.
State. The first number is 4, and the second number is 6.

7. The Familiarize and Translate steps were done in Exercise 43 of Exercise Set 3.1.
Carry out. We solve the system of equations.

$p+a=1223,$ (1)
$p=a+243$ (2)

where p is the number of endangered plant species and a is the number of endangered animal species. We use substitution. Substitute $a+243$ for p in Equation (1) and solve for a.

$(a+243)+a=1223$
$2a+243=1223$
$2a=980$
$a=490$

Now substitute 490 for a in Equation (2).

$p=(490)+243=733$

Check. The sum of the endangered plant and animal species is $733+490$, or 1223. The answer checks.
State. There are 733 endangered plant species and 490 endangered animal species.

9. The Familiarize and Translate steps were done in Exercise 45 of Exercise Set 3.1.
Carry out. We solve the system of equations

$x+y=180,$ (1)
$x=2y-3$ (2)

where $x=$ the measure of one angle and $y=$ the measure of the other angle. We use substitution. Substitute $2y-3$ for x in (1) and solve for y.

$2y-3+y=180$
$3y-3=180$
$3y=183$
$y=61$

Now substitute 61 for y in (2).

$x=2\cdot61-3=122-3=119$

Check. The sum of the angle measures is $119°+61°$, or $180°$, so the angles are supplementary. Also $2\cdot61°-3°=122°-3°=119°$. The answer checks.
State. The measures of the angles are $119°$ and $61°$.

11. The Familiarize and Translate steps were done in Exercise 47 of Exercise Set 3.1.
Carry out. We solve the system of equations

$x+y=64,$ (1)
$2x+y=100$ (2)

where $x=$ the number of two-point shots and $y=$ the

number of foul shots made. We use elimination.

$$-x - y = -64 \quad \text{Multiplying (1) by} -1$$
$$\underline{2x + y = 100}$$
$$x = 36$$

Substitute 36 for x in (1) and solve for y.

$$36 + y = 64$$
$$y = 28$$

Check. The total number of scores was $36 + 28$, or 64. The total number of points was
$2 \cdot 36 + 28 = 72 + 28 = 100$. The answer checks.

State. Chamberlain made 36 two-point shots and 28 foul shots.

13. The Familiarize and Translate steps were done in Exercise 49 of Exercise Set 3.1.
We can multiply both sides of the second equation by 100 to clear the decimals.

$$w + u = 32,$$
$$449w + 299u = 10,768.$$

Carry out. We solve the system of equations.

$$w + u = 32, \quad (1)$$
$$449x + 299y = 10,768 \quad (2)$$

where w is the number of wrapped strings and u is the number of unwrapped strings. We use elimination. Begin by multiplying Equation (1) by -299.

$$-299w - 299u = -9568 \quad \text{Multiplying (1) by} -299$$
$$\underline{449w + 299u = 10,768 \quad (2)}$$
$$150w \qquad = 1200$$
$$w = 8$$

Substitute 8 for w in Equation (1) and solve for u.

$$8 + u = 32$$
$$u = 24$$

Check. The number of wrapped and unwrapped strings is $8 + 24$, or 32. The amount paid was
$\$14.49(8) + \$2.99(24) = \$35.92 + \$71.76 = \$107.68$. The answer checks.

State. 8 wrapped and 24 unwrapped strings were bought.

15. The Familiarize and Translate steps were done in Exercise 51 of Exercise Set 3.1.

$$h + s = 110,$$
$$8h + 12s = 1072.$$

Carry out. We solve the system of equations.

$$h + s = 110, \quad (1)$$
$$8h + 12s = 1072 \quad (2)$$

where h = the number of hats and s = the number of scarves knitted.. We use elimination.

$$-8h - 8s = -880 \quad \text{Multiplying (1) by} -8$$
$$\underline{8h + 12s = 1072 \quad (2)}$$
$$4s = 192$$
$$s = 48$$

Substitute 48 for s in Equation (1) and solve for h.

$$h + 48 = 110$$
$$h = 62$$

Check. A total of $48 + 62$, or 110 items were knitted. The time spent knitting, in hours, was
$8(62) + 12(48) = 496 + 576 = 1072$. The answer

checks.

State. 62 hats and 48 scarves were knitted.

17. The Familiarize and Translate steps were done in Exercise 53 of Exercise Set 3.1.

Carry out. We solve the system of equations.

$$2l + 2w = 340, \quad (1)$$
$$l = w + 50 \quad (2)$$

where l = the length, in yards, and w = the width, in yards of the lacrosse field. We use substitution. We substitute $w + 50$ for l in Equation (1) and solve for w.

$$2(w + 50) + 2w = 340$$
$$2w + 100 + 2w = 340$$
$$4w + 100 = 340$$
$$4w = 240$$
$$w = 60$$

Now substitute 60 for w in Equation (2).

$$l = 60 + 50 = 110$$

Check. The perimeter is
$2 \cdot 110 + 2 \cdot 60 = 220 + 120 = 340$. The length, 110 yards, is 50 yards more than the width, 60 yards. The answer checks.

State. The length of the lacrosse field is 110 yards, and the width is 60 yards.

19. **Familiarize**. Let x = the number of MWH (in thousands) of wind and y = the number of MWH (in thousands) of solar generated.

Translate.

The total MWH is 218.
$$x + y = 218$$

The wind is 2 more than 7 times the solar.
$$x = 2 + 7y$$

We have translated to a system of equations:

$$x + y = 218, \quad (1)$$
$$x = 2 + 7y \quad (2)$$

Carry out. We use the substitution method to solve the system of equations. Substitute $2 + 7y$ for x in Equation (1) and solve for y.

$$2 + 7y + y = 218$$
$$8y + 2 = 218$$
$$8y = 216$$
$$y = 27$$

Substitute 27 for y in Equation (1) and solve for x.

$$x + 27 = 218$$
$$x = 191$$

Check. A total of $27 + 191$, or 218 MWH were generated. The wind was $2 + 7(27) = 191$. The answer checks.

State. 191 thousand MWH of wind and 27 thousand MWH of solar were generated.

21. **Familiarize**. Let x = the number of 3-credit coursess and y = the number of 4-credit courses taken.

Translate. We organize the information in a table.

	3-credit courses	4-credit courses	Total
Number taken	x	y	48
Credits	3	4	
Total credits	$3x$	$4y$	155

We get one equation from the "Number taken" row of the table:

$x + y = 48$

The "Total credits" row yields a second equation:

$3x + 4y = 155$

We have translated to a system of equations:

$x + y = 48 \quad (1)$
$3x + 4y = 155 \quad (2)$

Carry out. We use the elimination method to solve the system of equations.

$\begin{array}{l} -3x - 3y = -144 \quad \text{Multiplying (1) by } -3 \\ \underline{3x + 4y = 155 \quad (2)} \\ y = 11 \end{array}$

Substitute 11 for y in Equation (1) and solve for x.

$x + 11 = 48$
$x = 37$

Check. A total of $37 + 11$, or 48 courses, were taken. The total credits were $3(37) + 4(11) = 111 + 44 = 155$. The answer checks.

State. 37 3-credit courses and 11 4-credit courses were taken.

23. **Familiarize.** Let $x =$ the number of cases of regular paper used and $y =$ the number of cases of recycled paper.

Translate. We organize the information in a table.

	Regular	Recycled paper	Total
Number purchased	x	y	27
Price	$46.99	$61.99	
Total cost	$46.99x$	$61.99y$	1433.73

We get one equation from the "Number purchased" row of the table: $x + y = 27$

The "Total cost" row yields a second equation. All costs are expressed in dollars.

$46.99x + 61.99y = 1433.73.$

We have the problem translated to a system of equations.

$x + y = 27 \quad (1)$
$46.99x + 61.99y = 1433.73 \quad (2)$

Carry out. We use the elimination method to solve the system of equations.

$\begin{array}{l} -4699x - 4699y = -126{,}873 \quad \text{Multiplying (1) by } -4699 \\ \underline{4699x + 6199y = 143{,}373 \quad \text{Multiplying (2) by 100}} \\ 1500y = 16{,}500 \\ y = 11 \end{array}$

Substitute 11 for y in Equation (1) and solve for x.

$x + 11 = 27$
$x = 16$

Check. A total of $16 + 11$, or 27 cases of paper were used. The total cost was $\$46.99(16) + \$61.99(11)$ $= \$751.84 + \$681.89 = \$1433.73$. The answer checks.

State. 16 cases of regular paper and 11 cases of recycled paper were purchased.

25. **Familiarize.** Let $x =$ the number of 8.5-watt bulbs and $y =$ the number of 18-watt bulbs purchased.

Translate. We organize the information in a table.

	8.5-watt bulbs	18-watt bulbs	Total
Number purchased	x	y	200
Price	$3.97	$8.97	
Total cost	$3.97x$	$8.97y$	1494

We get one equation from the "Number purchased" row of the table:

$x + y = 200$

The "Total cost" row yields a second equation:

$3.97x + 8.97y = 1494$

We have translated to a system of equations:

$x + y = 200 \quad (1)$
$3.97x + 8.97y = 1494 \quad (2)$

Carry out. We use the elimination method to solve the system of equations.

$\begin{array}{l} -397x - 397y = -79{,}400 \quad \text{Multiplying (1) by } -397 \\ \underline{397x + 897y = 149{,}400 \quad \text{Mulitplying (2) by 100}} \\ 500y = 70{,}000 \\ y = 140 \end{array}$

Substitute 140 for y in Equation (1) and solve for x.

$x + 140 = 200$
$x = 60$

Check. A total of $60 + 140$, or 200 bulbs, were purchased. The total cost was

$\$3.97(60) + \$8.97(140) = \$238.20 + \$1255.80 = \$1494$

The answer checks.

State. 60 8.5-watt bulbs and 140 18-watt bulbs were purchased.

27. **Familiarize.** Let $s =$ the number of Starter customers and $a =$ the number of Already Composting customers.

Translate.

	Starter	Already Composting	Total
Customers	s	a	215
Price	$160	$105	
Total revenue	$160s$	$105a$	26,975

We get one equation from the "Customers" row of the table:

$s + a = 215$

The "Total revenue" row yields a second equation:

$160s + 105a = 26{,}975$

We have translated to a system of equations:

$s + a = 215 \quad (1)$
$160s + 105a = 26{,}975 \quad (2)$

Carry out. We use the elimination method to solve the system of equations.

$\begin{array}{l} -105s - 105a = -22{,}575 \quad \text{Multiplying (1) by } -105 \\ \underline{160s + 105a = 26{,}975 \quad (2)} \\ 55s = 4400 \\ s = 80 \end{array}$

Substitute 80 for s in Equation (1) and solve for a.

$s + 80 = 215$
$s = 135$

Check. There are a total of $80 + 135$, or 215 customers. The total revenue was $\$160(80) + \$105(135)$

$= \$12,800 + \$14,175 = \$26,975$. The answer checks.

State. There are 135 Starters and 80 Already Composting customers.

29. An immediate solution can be determined from the fact that $19 is the average of equal parts of $18 and $20. So there is 14 lb of Mexican and 14 lb of Peruvian. Or we can solve using the usual method.

Familiarize. Let $m =$ the number of pounds of Mexican coffee and $p =$ the number of pounds of Peruvian coffee to be used in the mixture. The value of the mixture will be $\$19(28)$, or $432.

Translate. We organize the information in a table.

	Mexican	Peruvian	Mixture
Number of pounds	m	p	28
Price per pound	$20	$18	$19
Value of coffee	$20m$	$18p$	532

The "Number of pounds" row of the table gives us one equation:

$$m + p = 28$$

The "Value of coffee" row yields a second equation:

$$20m + 18p = 532$$

We have translated to a system of equations:

$$m + p = 28 \quad (1)$$
$$20m + 18p = 532 \quad (2)$$

Carry out. We use the elimination method to solve the system of equations.

$$\begin{array}{ll} -18m - 18p = -504 & \text{Multiplying (1) by } -18 \\ \underline{20m + 18p = 532} & (2) \\ 2m \qquad\quad = 28 & \\ m = 14 & \end{array}$$

Substitute 14 for m in Equation (1) and solve for p.

$$14 + p = 28$$
$$p = 14$$

Check. The total mixture contains 14 lb + 14 lb, or 28 lb. Its value is $\$20 \cdot 14 + \$18 \cdot 14$

$= \$280 + \$252 = \$532$. The answer checks.

State. 14 lb of Mexican coffee and 14 lb of Peruvian coffee should be used.

31. **Familiarize**. Let $x =$ the number of ounces of custom printed M&Ms and $y =$ the number of ounces of bulk M&Ms. Converting lb to oz: 20 lb = 20(16) = 320 oz. The mixture's value will be $\$0.59(320)$, or $188.80.

Translate. We organize the information in a table.

	Custom printed	Bulk	Mixture
Number of ounces	x	y	320
Price per ounce	$1.04	$0.32	$0.59
Value of M&Ms	$1.04x$	$0.32y$	188.80

The "Number of ounces" row of the table gives us one equation:

$$x + y = 320$$

The "Value of M&Ms" row yields a second equation:

$$1.04x + 0.32y = 188.80$$

After clearing decimals, we have the problem translated to a system of equations:

$$x + y = 320 \quad (1)$$
$$104x + 32y = 18,880 \quad (2)$$

Carry out. We use the elimination method to solve the system of equations.

$$\begin{array}{ll} -32x - 32y = -10,240 & \text{Multiplying (1) by } -32 \\ \underline{104x + 32y = 10,240} & (2) \\ 72x \qquad\quad = 8640 & \\ x = 120 & \end{array}$$

Substitute 120 for x in Equation (1) and solve for y.

$$120 + y = 320$$
$$y = 200$$

Check. The total mixture contains 120 oz + 200 oz, or 320 oz. Its value is

$\$1.04 \cdot 120 + \$0.32 \cdot 200 = \$188.80$. The answer checks.

State. 120 oz of custom-printed M&Ms and 200 oz of bulk M&Ms should be used.

33. **Familiarize**. Let $x =$ the number of mL of 50% acid solution and $y =$ the number of mL of 80% acid solution. The amount of acid in the mixture is 68%(200 mL), or 0.68(200 mL) =136 mL.

Translate. We organize the information in a table.

	50% acid	80% acid	Mixture
Amount of solution	x	y	200
Percent acid	50%	80%	68%
Amount of acid in solution	$0.50x$	$0.80y$	136

The "Amount of solution" row of the table gives us one equation:

$$x + y = 200$$

The last row of the table yields a second equation:

$$0.50x + 0.80y = 136$$

After clearing decimals, we have the problem translated to a system of equations:

$$x + y = 200 \quad (1)$$
$$5x + 8y = 1360 \quad (2)$$

Carry out. We use the elimination method to solve the system of equations.

$$\begin{array}{ll} -5x - 5y = -1000 & \text{Multiplying (1) by } -5 \\ \underline{5x + 8y = 1360} & (2) \\ 3y = 360 & \\ y = 120 & \end{array}$$

Substitute 120 for y in Equation (1) and solve for y.

$$x + 120 = 200$$
$$x = 80$$

Check. The amount of the mixture is 80 mL + 120 mL, or 200 mL. The amount of acid is

$0.50(80) + 0.80(120) = 40 \text{ mL} + 96 \text{ mL} = 136 \text{ mL}$. The answer checks.

State. 80 mL of 50% acid solution and 120 mL of 80% acid solution should be used.

35. *Familiarize.* Let x = the number of pounds 80% bluegrass and y = the number of pounds of 30% bluegrass to be used in the mixture. The blend of the mixture is 60%(50 lb), or $0.60(50 \text{ lb}) = 30 \text{ lb}$.

Translate. We organize the information in a table.

	80%	30%	Mixture
Number of pounds	x	y	50
Percent of blend	80%	30%	60%
Amount of blend	$0.80x$	$0.30y$	30 lb

We get one equation from the "Number of pounds" row of the table:
$$x + y = 50$$

The last row of the table yields a second equation:
$$0.8x + 0.3y = 30$$

After clearing decimals, we have the problem translated to a system of equations:
$$x + y = 50, \quad (1)$$
$$8x + 3y = 300 \quad (2)$$

Carry out. We use the elimination method to solve the system of equations.
$$\begin{array}{ll} -3x - 3y = -150 & \text{Multiplying (1) by } -3 \\ \underline{8x + 3y = 300} & \\ 5x = 150 & \\ x = 30 & \end{array}$$

Substitute 30 for x in (1) and solve for y.
$$30 + y = 50$$
$$y = 20$$

Check. The amount of the mixture is 30 lb + 20 lb, or 50 lb. The amount of bluegrass in the mixture is $0.80(30 \text{ lb}) + 0.30(20 \text{ lb}) = 24 \text{ lb} + 6 \text{ lb} = 30 \text{ lb}$. The answer checks.

State. 30 lb of 80% bluegrass blend and 20 lb of 30% bluegrass blend should be mixed.

37. *Familiarize.* Let x = the amount of the 6.5% loan and y = the amount of the 7.2% loan. Recall that the formula for simple interest is
$$\text{Interest} = \text{Principal} \times \text{Rate} \times \text{Time}$$

Translate. We organize the information in a table.

	6.5% loan	7.2% loan	Total
Principal	x	y	12,000
Interest rate	6.5%	7.2%	
Time	1 yr	1 yr	
Interest	$0.065x$	$0.072y$	811.50

The "Principal" row of the table gives us one equation:
$$x + y = 12,000$$

The last row of the table yields a second equation:
$$0.065x + 0.072y = 811.50$$

After clearing decimals, we have the problem translated to a system of equations:
$$x + y = 12,000 \quad (1)$$
$$65x + 72y = 811,500 \quad (2)$$

Carry out. We use the elimination method to solve the system of equations.

$$\begin{array}{ll} -65x - 65y = -780,000 & \text{Multiplying (1) by } -65 \\ \underline{65x + 72y = 811,500} & (2) \\ 7y = 31,500 & \\ y = 4500 & \end{array}$$

Substitute 4500 for y in Equation (1) and solve for x.
$$x + 4500 = 12,000$$
$$x = 7500$$

Check. The loans total $7500 + $4500, or $12,000. The total interest is $0.065(\$7500) + 0.072(\$4500)$
$$= \$487.50 + \$324 = \$811.50. \text{ The answer checks.}$$

State. The 6.5% loan was for $7500 and the 7.2% loan was for $4500.

39. *Familiarize.* Let x = the number of liters of Steady State and y = the number of liters of Even Flow in the mixture. The amount of alcohol in the mixture is $0.15(20 \text{ L}) = 3 \text{ L}$.

Translate. We organize the information in a table.

	18% solution	10% solution	Mixture
Number of liters	x	y	20
Percent of alcohol	18%	10%	15%
Amount of alcohol	$0.18x$	$0.10y$	3

We get one equation from the "Number of liters" row of the table:
$$x + y = 20$$

The last row of the table yields a second equation:
$$0.18x + 0.1y = 3$$

After clearing decimals we have the problem translated to a system of equations:
$$x + y = 20, \quad (1)$$
$$18x + 10y = 300 \quad (2)$$

Carry out. We use the elimination method to solve the system of equations.
$$\begin{array}{ll} -10x - 10y = -200 & \text{Multiplying (1) by } -10 \\ \underline{18x + 10y = 300} & (2) \\ 8x = 100 & \\ x = 12.5 & \end{array}$$

Substitute 12.5 for x in (1) and solve for y.
$$12.5 + y = 20$$
$$y = 7.5$$

Check. The total amount of the mixture is 12.5 L + 7.5 L or 20 L. The amount of alcohol in the mixture is
$$0.18(12.5 \text{ L}) + 0.1(7.5 \text{ L}) = 2.25 \text{ L} + 0.75 \text{ L} = 3 \text{ L}. \text{ The answer checks.}$$

State. 12.5 L of Steady State and 7.5 L of Even Flow should be used.

41. *Familiarize.* Let x = the number of gallons of 87-octane gas and y = the number of gallons of 95-octane gas in the mixture. The amount of octane in the mixture can be expressed as 93(10), or 930.

Translate. We organize the information in a table.

	87-octane	95-octane	Mixture
Number of gallons	x	y	10
Octane rating	87	95	93
Total octane	$87x$	$95y$	930

We get one equation from the "Number of gallons" row of the table :

$$x + y = 10$$

The last row of the table yields a second equation:

$$87x + 95y = 930$$

We have a system of equations:

$$x + y = 10 \quad (1)$$
$$87x + 95y = 930 \quad (2)$$

Carry out. We use the elimination method to solve the system of equations.

$$\begin{array}{rl} -87x - 87y = -870 & \text{Multiplying (1) by } -87 \\ \underline{87x + 95y = 930} & (2) \\ 8y = 60 \\ y = 7.5 \end{array}$$

Substitute 7.5 for y in Equation (1) and solve for x.

$$x + 7.5 = 10$$
$$x = 2.5$$

Check. The total amount of the mixture is 2.5 gal + 7.5 gal, or 10 gal. The amount of octane can be expressed as $87(2.5) + 95(7.5) = 217.5 + 712.5 = 930$. The answer checks.

State. 2.5 gal of 87-octane gas and 7.5 gal of 95-octane gas should be used.

43. Familiarize. From the bar graph we see that whole milk is 4% milk fat, milk for cream cheese is 8% milk fat, and cream is 30% milk fat. Let x = the number of pounds of whole milk and y = the number of pounds of cream to be used. The mixture contains 8%(200 lb), or 0.08(200 lb) = 16 lb of milk fat.

Translate. We organize the information in a table.

	Whole milk	Cream	Mixture
Number of pounds	x	y	200
Percent of milk fat	4%	30%	8%
Amount of milk fat	$0.04x$	$0.30y$	16 lb

We get one equation from the "Number of pounds" row of the table:

$$x + y = 200$$

The last row of the table yields a second equation:

$$0.04x + 0.3y = 16$$

After clearing decimals, we have the problem translated to a system of equations:

$$x + y = 200, \quad (1)$$
$$4x + 30y = 1600 \quad (2)$$

Carry out. We use the elimination method to solve the system of equations.

$$\begin{array}{rl} -4x - 4y = -800 & \text{Multiplying (1) by } -4 \\ \underline{4x + 30y = 1600} & (2) \\ 26y = 800 \\ y = \frac{400}{13}, \text{ or } 30\frac{10}{13} \end{array}$$

Substitute $\frac{400}{13}$ for y in (1) and solve for x.

$$x + \frac{400}{13} = 200$$
$$x = \frac{2200}{13}, \text{ or } 169\frac{3}{13}$$

Check. The total amount of the mixture is

$\frac{2200}{13}$ lb $+ \frac{400}{13}$ lb $= \frac{2600}{13}$ lb $= 200$ lb . The amount

of milk fat in the mixture is $0.04\left(\frac{2200}{13} \text{ lb}\right) +$

$0.3\left(\frac{400}{13} \text{ lb}\right) = \frac{88}{13}$ lb $+ \frac{120}{13}$ lb $= \frac{208}{13}$ lb $= 16$ lb . The

answer checks.

State. $169\frac{3}{13}$ lb of whole milk and $30\frac{10}{13}$ lb of cream should be mixed.

45. Familiarize. We first make a drawing.

Slow train,
d kilometers,   75 km/h $(t+2)$ hr

Fast train,
d kilometers,   125 km/h t hr

From the drawing we see that the distances are the same. Now complete the chart.

$$d = r \cdot t$$

	Distance	Rate	Time	
Slow train	d	75	$t+2$	$\rightarrow d = 75(t+2)$
Fast train	d	125	t	$\rightarrow d = 125t$

Translate. Using $d = rt$ in each row of the table, we get a system of equations:

$$d = 75(t+2),$$
$$d = 125t$$

Carry out. We solve the system of equations.

$$\begin{array}{rl} 125t = 75(t+2) & \text{Using substitution} \\ 125t = 75t + 150 \\ 50t = 150 \\ t = 3 \end{array}$$

Then $d = 125t = 125 \cdot 3 = 375$

Check. At 125 km/h, in 3 hr the fast train will travel $125 \cdot 3 = 375$ km. At 75 km/h, in $3 + 2$, or 5 hr the slow train will travel $75 \cdot 5 = 375$ km. The numbers check.

State. The trains will meet 375 km from the station.

47. Familiarize. We first make a drawing. Let d = the distance and r = the speed of the canoe in still water. Then when the canoe travels downstream, its speed is $r + 6$, and its speed upstream is $r - 6$. From the drawing we see that the distances are the same.

Downstream, 6 mph current
d mi, $r + 6$, 4 hr

Upstream, 6 mph current
d mi, $r - 6$, 10 hr

Organize the information in a table.

	Distance	Rate	Time
Downstream	d	$r+6$	4
Upstream	d	$r-6$	10

Translate. Using $d = rt$ in each row of the table, we get a system of equations:

$d = 4(r + 6),$ or $d = 4r + 24,$
$d = 10(r - t)$ $d = 10r - 60$

Carry out. Solve the system of equations.

$$4r + 24 = 10r - 60$$
$$24 = 6r - 60$$
$$84 = 6r$$
$$14 = r$$

Check. When $r = 14$, then $r + 6 = 14 + 6 = 20$, and the distance traveled in 4 hours is $4 \cdot 20 = 80$ km. Also $r - 6 = 14 - 6 = 8$, and the distance traveled in 10 hours is $10 \cdot 8 = 80$ km. The answer checks.

State. The speed of the canoe in still water is 14 km/h.

49. **Familiarize**. We make a drawing. Note that the plane's speed traveling toward London is $360 + 50$, or 410 mph, and the speed traveling toward New York City is $360 - 50$, or 310 mph. Also, when the plane is d mi from New York City, it is $3458 - d$ mi from London.

New York City   London
310 mph t hours t hours   410 mph

|———————— 3458 mi ————————|

|——— d ———|——— 3458 mi $- d$ ———|

Organize the information in a table.

	Distance	Rate	Time
Toward NYC	d	310	t
Toward London	$3458 - d$	410	t

Translate. Using $d = rt$ in each row of the table, we get a system of equations:

$$d = 310t, \quad (1)$$
$$3458 - d = 410t \quad (2)$$

Carry out. We solve the system of equations.

$$3458 - 310t = 410t \quad \text{Using substitution}$$
$$3458 = 720t$$
$$4.8028 \approx t$$

Substitute 4.8028 for t in (1).

$$d \approx 310(4.8028) \approx 1489$$

Check. If the plane is 1489 mi from New York City, it can return to New York City, flying at 310 mph, in $1489 / 310 \approx 4.8$ hr. If the plane is $3458 - 1489$, or 1969 mi from London, it can fly to London, traveling at 410 mph, in $1969 / 410 \approx 4.8$ hr. Since the times are the same, the answer checks.

State. The point of no return is about 1489 mi from New York City.

51. **Familiarize**. Let $l = $ the length, in feet, and $w = $ the width, in feet. Recall that the formula for the perimeter P of a rectangle with length l and width w is $P = 2l + 2w$.

Translate.

The perimeter   is   860 ft.
↓   ↓   ↓
$2l + 2w$ $=$ 860

The length   is   100 ft.   more than   the width.
↓   ↓   ↓   ↓   ↓
l $=$ 100 $+$ w

We have translated to a system of equations:

$$2l + 2w = 860, \quad (1)$$
$$l = 100 + w \quad (2)$$

Carry out. We use the substitution method to solve the system of equations.

Substitute $100 + w$ for l in (1) and solve for w.

$$2(100 + w) + 2w = 860$$
$$200 + 2w + 2w = 860$$
$$200 + 4w = 860$$
$$4w = 660$$
$$w = 165$$

Now substitute 165 for w in (2).

$$l = 100 + 165 = 265$$

Check. The perimeter is $2 \cdot 265 \text{ ft} + 2 \cdot 165 \text{ ft} = 530 \text{ ft} + 330 \text{ ft} = 860 \text{ ft}$. The length, 265 ft, is 100 ft more than the width, 165 ft. The answer checks.

State. The length is 265 ft, and the width is 165 ft.

53. **Familiarize**. Let $x = $ the number of minutes to a landline and $y = $ the number of minutes to a wireless.

Translate.

$5.00   and   Landline   and   Wireless   is   Total
              call   call   bill.
↓   ↓   ↓   ↓   ↓   ↓
5.00   +   0.09x   +   0.15y   = $59.90

Total minutes   is   400 minutes.
↓   ↓   ↓
$x + y$ $=$ 400

We have a system of equations:

$$5 + 0.09x + 0.15y = 59.90 \quad (1)$$
$$x + y = 400 \quad (2)$$

Carry out. We use the elimination method to solve the system of equations. Subtract 5 from both sides of (1).

$$9x + 15y = 5490 \quad \text{Multiplying (1) by 100}$$
$$\underline{-9x - 9y = -3600} \quad \text{Multiplying (2) by } -9$$
$$6y = 1890$$
$$y = 315$$

Now substitute 315 for y in Equation (2).

$$x + 315 = 400$$
$$x = 85$$

Check. The total number of minutes is $85 + 315 = 400$. The total bill is $\$5 + \$0.09(85) + \$0.15(315) = \$59.90$. The answer checks.

State. There were 85 minutes to a landline and 315 minutes to a wireless number.

55. *Familiarize*. Let x = the number of Basic $7.99 plans and y = the number of Standard $7.99 + $2 plans.
Translate.

$\underbrace{\text{Total number of plans}}$ is $\underline{280}$,

$$x + y \quad = \quad 280$$

$\underbrace{\text{Total value of plans}}$ is $\underline{2417.20}$,

$$7.99x + 9.99y \quad = \quad 2417.20$$

After clearing decimals, we have a system of equations:

$$x + y = 280 \qquad (1)$$
$$799x + 999y = 241,720 \quad (2)$$

Carry out. We use the elimination method to solve the system of equations.

$$\begin{array}{ll} -799x - 799y = -223,720 & \text{Multiplying (1) by } -799 \\ \underline{799x + 999y = 241,720} & (2) \\ \quad\quad\quad 200y = 18,000 \\ \quad\quad\quad\quad\; y = 90 \end{array}$$

Substitute 90 for y in Equation (1) and solve for x.

$$x + 90 = 280$$
$$x = 190$$

Check. The total number of plans is $190 + 90$, or 280. The total value of the plans is $7.99(190) + 9.99(90)$
$= 1518.10 + 899.10 = 2417.20$. The answer checks.
State. 190 of the $7.99 Basic plans and 90 of the Standard $7.99 +$2 plans were purchased.

57. *Familiarize*. The change from the $9.25 purchase is
$20 - $9.25, or $10.75. Let x = the number of quarters and y = the number of fifty-cent pieces. The total value of the quarters, in dollars, is $0.25x$ and the total value of the fifty-cent pieces, in dollars, is $0.50y$.
Translate.

$\underbrace{\text{The total number of coins}}$ is   30,

$$x + y \quad = \quad 30$$

$\underbrace{\text{The total value of the coins}}$ is $10.75.

$$0.25x + 0.50y \quad = \quad 10.75$$

After clearing decimals we have the following system of equations:

$$x + y = 30, \qquad (1)$$
$$25x + 50y = 1075 \quad (2)$$

Carry out. We use the elimination method to solve the system of equations.

$$\begin{array}{ll} -25x - 25y = -750 & \text{Multiplying (1) by } -25 \\ \underline{25x + 50y = 1075} \\ \quad\quad\; 25y = 325 \\ \quad\quad\quad y = 13 \end{array}$$

Substitute 13 for y in (1) and solve for x.

$$x + 13 = 30$$
$$x = 17$$

Check. The total number of coins is $17 + 13$, or 30. The total value of the coins is $0.25(17) + 0.50(13)$
$= 4.25 + $6.50 = $10.75. The answer checks.
State. There were 17 quarters and 13 fifty-cent pieces.

59. *Writing Exercise*.

61. $h(0) = 0 - 7 = -7$

63. $(h \cdot f)(7) = h(7)f(7) = (7 - 7)(7^2 + 2) = 0 \cdot 51 = 0$

65. From Exercise 64, $h + f = x^2 + x - 5$
Domain: $\mathbb{R}$

67. *Writing Exercise*.

69. *Familiarize*. Let x = the number of reams of 0% post-consumer fiber paper purchased and y = the number of reams of 30% post-consumer fiber paper.
Translate. We organize the information in a table.

	0% post-consumer	30% post-consumer	Total
Reams purchased	x	y	60
Percent of post-consumer fiber	0%	30%	20%
Total post-consumer fiber	$0 \cdot x$, or 0	$0.3y$	$0.2(60)$, or 12

We get one equation from the "Reams purchased" row of the table:

$$x + y = 60$$

The last row of the table yields a second equation:

$$0x + 0.3y = 12, \text{ or } 0.3y = 12$$

After clearing the decimal we have the problem translated to a system of equations.

$$x + y = 60, \quad (1)$$
$$3y = 120 \quad (2)$$

Carry out. First we solve (2) for y.

$$3y = 120$$
$$y = 40$$

Now substitute 40 for y in (1) and solve for x.

$$x + 40 = 60$$
$$x = 20$$

Check. The total purchase is $20 + 40$, or 60 reams. The post-consumer fiber can be expressed as
$0 \cdot 20 + 0.3(40) = 12$. The answer checks.
State. 20 reams of 0% post-consumer fiber paper and 40 reams of 30% post-consumer fiber paper would have to be purchased.

71. *Familiarize*. Let x = the number of ounces of pure silver.
Translate. We organize the information in a table.

	Coin	Pure silver	New Mixture
Number of ounces	32	x	$32 + x$
Percent silver	90%	100%	92.5%
Amount of silver	$0.9(32)$	$1 \cdot x$	$0.925(32 + x)$

The last row gives the equation

$$0.9(32) + x = 0.925(32 + x).$$

Carry out. We solve the equation.

$$28.8 + x = 29.6 + 0.925x$$
$$0.075x = 0.8$$
$$x = \frac{32}{3} = 10\frac{2}{3}$$

Check. 90% of 32, or $28.8 + 100\%$ of $10\frac{2}{3}$, or $10\frac{2}{3}$ is

$39\frac{7}{15}$ and 92.5% of $\left(32 + 10\frac{2}{3}\right)$ is $0.925\left(42\frac{2}{3}\right)$, or

$39\frac{7}{15}$. The answer checks.

State. $10\frac{2}{3}$ ounces of pure silver should be added.

73. Familiarize. Let x = the number of 6-cupcake boxes and y = the number of single cupcakes. Note that $6x$ is the number of cupcakes in the gift box.
Translate.

Total number of cupcakes   is   256,
$$\downarrow \qquad\qquad \downarrow \quad \downarrow$$
$$6x + y \qquad\qquad = \quad 256$$

Total value sales   is   701.67,
$$\downarrow \qquad\quad \downarrow \quad \downarrow$$
$$15.99x + 3y \qquad = \quad 701.67$$

We have a system of equations:
$$6x + y = 256 \qquad (1)$$
$$15.99x + 3y = 701.67 \quad (2)$$

Carry out. We use the elimination method to solve the system of equations.

$$-18x - 3y = -768 \qquad \text{Multiplying (1) by } -3$$
$$\underline{15.99x + 3y = 701.67 \quad (2)}$$
$$-2.01x \qquad\quad = -66.33$$
$$x = 33$$

Although the problem asks for the number of 6-cupcake boxes, we will also find y in order to check. Substitute 33 for x in Equation (1) and solve for y.
$$6(33) + y = 256$$
$$y = 58$$

Check. The total number of cupcakes is $6(33) + 58 = 198 + 58 = 256$. The total sales is $\$15.99(33) + \$3(58) = \$527.67 + \$174 = \$701.67$. The answer checks.
State. 33 6-cupcake boxes were sold.

75. Familiarize. Let x = the number of gallons of pure brown and y = the number of gallons of neutral stain that should be added to the original 0.5 gal. Note that a total of 1 gal of stain needs to be added to bring the amount of stain up to 1.5 gal. The original 0.5 gal of stain contains 20%(0.5 gal), or 0.2(0.5 gal) = 0.1 gal of brown stain. The final solution contains 60%(1.5 gal), or 0.6(1.5 gal) = 0.9 gal of brown stain. This is composed of the original 0.1 gal and the x gal that are added.
Translate.

The amount of stain added   was   1 gal.
$$\downarrow \qquad\qquad \downarrow \quad \downarrow$$
$$x + y \qquad\qquad\quad = \quad 1$$

The amount of brown stain in the final solution   is   0.9 gal.
$$\downarrow \qquad\qquad \downarrow \quad \downarrow$$
$$0.1 + x \qquad\qquad = \quad 0.9$$

We have a system of equations.
$$x + y = 1, \qquad (1)$$
$$0.1 + x = 0.9 \quad (2)$$

Carry out. First we solve (2) for x.
$$0.1 + x = 0.9$$
$$x = 0.8$$

Then substitute 0.8 for x in (1) and solve for y.
$$0.8 + y = 1$$
$$y = 0.2$$

Check. Total amount of stain: $0.5 + 0.8 + 0.2 = 1.5$ gal
Total amount of brown stain: $0.1 + 0.8 = 0.9$ gal
Total amount of neutral stain:
$0.8(0.5) + 0.2 = 0.4 + 0.2 = 0.6$ gal $= 0.4(1.5$ gal$)$
The answer checks.
State. 0.8 gal of pure brown and 0.2 gal of neutral stain should be added.

77. Familiarize. Let x and y represent the number of city miles and highway miles that were driven, respectively.

Then in city driving, $\frac{x}{18}$ gallons of gasoline are used;

in highway driving, $\frac{y}{24}$ gallons are used.

Translate. We organize the information in a table.

Type of driving	City	Highway	Total
Number of miles	x	y	465
Gallons of gasoline used	$\frac{x}{18}$	$\frac{y}{24}$	23

The first row of the table gives us one equation:
$$x + y = 465$$
The second row gives us another equation:

$$\frac{x}{18} + \frac{y}{24} = 23$$

After clearing fractions, we have the following system of equations:
$$x + y = 465, \qquad (1)$$
$$24x + 18y = 9936 \quad (2)$$

Carry out. We solve the system of equations using the elimination method.

$$-18x - 18y = -8370 \quad \text{Multiplying (1) by } -18$$
$$\underline{24x + 18y = 9936 \quad (2)}$$
$$6x \qquad\quad = 1566$$
$$x = 261$$

Now substitute 261 for x in Equation (1) and solve for y.
$$261 + y = 465$$
$$y = 204$$

Check. The total mileage is $261 + 204$, or 465. In 216 city miles, 261/18, or 14.5 gal of gasoline are used; in 204 highway miles, 204/24, or 8.5 gal are used. Then a total of $14.5 + 8.5$, or 23 gal of gasoline are used. The answer checks.
State. 261 miles were driven in the city, and 204 miles were driven on the highway.

79. The 1.5 gal mixture contains $0.1 + x$ gal of pure brown stain. (See Exercise 75.) Thus, the function $P(x) = \dfrac{0.1 + x}{1.5}$ gives the percentage of brown in the mixture as a decimal quantity. Using the Intersect feature, we confirm that when $P(x) = 0.6$, or 60%, then $x = 0.8$.

Exercise Set 3.4

1. The equation is equivalent to one in the form $Ax + By + Cz = D$, so the statement is true.

3. False

5. True; see Example 6.

7. Substitute $(2, -1, -2)$ into the three equations, using alphabetical order.

$$
\begin{array}{c|c}
x + y - 2z = 5 & \\
\hline
2 + (-1) - 2(-2) & 5 \\
2 - 1 + 4 & \\
& \overset{?}{5 = 5} \quad \text{TRUE}
\end{array}
$$

$$
\begin{array}{c|c}
2x - y - z = 7 & \\
\hline
2 \cdot 2 - (-1) - (-2) & 7 \\
4 + 1 + 2 & \\
& \overset{?}{7 = 7} \quad \text{TRUE}
\end{array}
$$

$$
\begin{array}{c|c}
-x - 2y - 3z = 6 & \\
\hline
-2 - 2(-1) - 3(-2) & 6 \\
-2 + 2 + 6 & \\
& \overset{?}{6 = 6} \quad \text{TRUE}
\end{array}
$$

Since the triple $(2, -1, -2)$ is true in all three equations, it is a solution of the system.

9.
$$
\begin{aligned}
x - y - z &= 0, \quad (1) \\
2x - 3y + 2z &= 7, \quad (2) \\
-x + 2y + z &= 1 \quad (3)
\end{aligned}
$$

1., 2. The equations are already in standard form with no fractions or decimals.

3. Use Equations (1) and (2) to eliminate x.

$$
\begin{array}{ll}
-2x + 2y + 2z = 0 & \text{Multiplying (1) by } -2 \\
\underline{2x - 3y + 2z = 7} & (2) \\
-y + 4z = 7 & (4) \text{ Adding}
\end{array}
$$

4. Use a different pair of equations and eliminate x.

$$
\begin{array}{ll}
x - y - z = 0 & (1) \\
\underline{-x + 2y + z = 1} & (3) \\
y \qquad = 1 & \text{Adding}
\end{array}
$$

5. When we used Equation (1) and (3) to eliminate x, we also eliminated z and found $y = 1$. Substitute 1 for y in Equation (4) to find z.

$$
\begin{aligned}
-1 + 4z &= 7 \quad \text{Substituting 1 for } y \text{ in (4)} \\
4z &= 8 \\
z &= 2
\end{aligned}
$$

6. Substitute in one of the original equations to find x.

$$
\begin{aligned}
x - 1 - 2 &= 0 \\
x &= 3
\end{aligned}
$$

We obtain $(3, 1, 2)$. This checks, so it is the solution.

11.
$$
\begin{aligned}
x - y - z &= 1, \quad (1) \\
2x + y + 2z &= 4, \quad (2) \\
x + y + 3z &= 5 \quad (3)
\end{aligned}
$$

1., 2. The equations are already in standard form with no fractions or decimals.

3. Use Equations (1) and (2) to eliminate y.

$$
\begin{array}{ll}
x - y - z = 1 & (1) \\
\underline{2x + y + 2z = 4} & (2) \\
3x \qquad + z = 5 & (4) \text{ Adding}
\end{array}
$$

4. Use a different pair of equations and eliminate y.

$$
\begin{array}{ll}
x - y - z = 1 & (1) \\
\underline{x + y + 3z = 5} & (3) \\
2x \qquad + 2z = 6 & (5) \text{ Adding}
\end{array}
$$

5. Now solve the system of Equations (4) and (5).

$$
\begin{aligned}
3x + z &= 5 \quad (4) \\
2x + 2z &= 6 \quad (5)
\end{aligned}
$$

$$
\begin{array}{ll}
-6x - 2z = -10 & \text{Multiplying (4) by } -2 \\
\underline{2x + 2z = 6} & (5) \\
-4x \qquad = -4 & \text{Adding} \\
x = 1
\end{array}
$$

$$
\begin{aligned}
3 \cdot 1 + z &= 5 \quad \text{Substituting 1 for } x \text{ in (4)} \\
3 + z &= 5 \\
z &= 2
\end{aligned}
$$

6. Substitute in one of the original equations to find y.

$$
\begin{aligned}
1 + y + 3 \cdot 2 &= 5 \quad \text{Substituting 1 for } x \text{ and} \\
& \qquad\qquad 2 \text{ for } z \text{ in (3)} \\
1 + y + 6 &= 5 \\
y + 7 &= 5 \\
y &= -2
\end{aligned}
$$

We obtain $(1, -2, 2)$. This checks, so it is the solution.

13.
$$
\begin{aligned}
3x + 4y - 3z &= 4, \quad (1) \\
5x - y + 2z &= 3, \quad (2) \\
x + 2y - z &= -2 \quad (3)
\end{aligned}
$$

1., 2. The equations are already in standard form with no fractions or decimals.

3., 4. We eliminate y from two different pairs of equations.

$$
\begin{array}{ll}
3x + 4y - 3z = 4 & (1) \\
\underline{20x - 4y + 8z = 12} & \text{Multiplying (2) by 4} \\
23x \qquad + 5z = 16 & (4)
\end{array}
$$

$$
\begin{array}{ll}
10x - 2y + 4z = 6 & \text{Multiplying (2) by 2} \\
\underline{x + 2y - z = -2} & (3) \\
11x \qquad + 3z = 4 & (5)
\end{array}
$$

5. Now solve the system of Equations (4) and (5).

$$
\begin{aligned}
23x + 5z &= 16 \quad (4) \\
11x + 3z &= 4 \quad (5)
\end{aligned}
$$

$$
\begin{array}{ll}
-69x - 15z = -48 & \text{Multiplying (4) by } -3 \\
\underline{55x + 15z = 20} & \text{Multiplying (5) by 5} \\
-14x \qquad = -28 & \text{Adding} \\
x = 2
\end{array}
$$

$$
\begin{aligned}
11 \cdot 2 + 3z &= 4 \quad \text{Substituting 2 for } x \text{ in (5)} \\
3z &= -18 \\
z &= -6
\end{aligned}
$$

6. Substitute in one of the original equations to find y.

$3 \cdot 2 + 4y - 3(-6) = 4$ Substituting 2 for x and
                       -6 for z in (1)

$6 + 4y + 18 = 4$

$4y = -20$

$y = -5$

We obtain $(2, -5, -6)$. This checks, so it is the solution.

15. $x + y + z = 0,$ (1)

     $2x + 3y + 2z = -3,$ (2)

     $-x - 2y - z = 1$ (3)

1., 2. The equations are already in standard form with no fractions or decimals.

3., 4. We eliminate x from two different pairs of equations.

$\begin{array}{l} -2x - 2y - 2z = 0 \qquad \text{Multiplying (1) by } -2 \\ \underline{2x + 3y + 2z = -3} \quad \text{(2)} \\ \qquad\quad y \qquad\quad = -3 \end{array}$

We eliminated not only x but also z and found that $y = -3$.

5., 6. Substitute -3 for y in two of the original equations to produce a system of two equations in two variables. Then solve this system.

$x - 3 + z = 0$ Substituting in (1)

$-x - 2(-3) - z = 1$ Substituting in (3)

Simplifying we have

$\begin{array}{l} x + z = 3 \\ \underline{-x - z = -5} \\ \quad 0 = -2 \end{array}$

We get a false equation, so there is no solution.

17. $2x - 3y - z = -9,$ (1)

     $2x + 5y + z = 1,$ (2)

     $x - y + z = 3$ (3)

1., 2. The equations are already in standard form with no fractions or decimals.

3., 4. We eliminate z from two different pairs of equations.

$\begin{array}{ll} 2x - 3y - z = -9 & \text{(1)} \\ \underline{2x + 5y + z = 1} & \text{(2)} \\ 4x + 2y \qquad = -8 & \text{(4)} \end{array}$

$\begin{array}{ll} 2x - 3y - z = -9 & \text{(1)} \\ \underline{x - y + z = 3} & \text{(3)} \\ 3x - 4y \qquad = -6 & \text{(5)} \end{array}$

5. Now solve the system of Equations (4) and (5).

$4x + 2y = -8$ (4)

$3x - 4y = -6$ (5)

$\begin{array}{ll} 8x + 4y = -16 & \text{Multiplying (4) by 2} \\ \underline{3x - 4y = -6} & \text{(5)} \\ 11x \qquad = -22 & \text{Adding} \\ \quad x = -2 \end{array}$

$4(-2) + 2y = -8$ Substituting -2 for x in (4)

$-8 + 2y = -8$

$y = 0$

6. Substitute in one of the original equations to find z.

$2(-2) + 5 \cdot 0 + z = 1$ Substituting -2 for x and
                      0 for y in (2)

$-4 + z = 1$

$z = 5$

We obtain $(-2, 0, 5)$. This checks, so it is the solution.

19. $a + b + c = 5,$ (1)

     $2a + 3b - c = 2,$ (2)

     $2a + 3b - 2c = 4$ (3)

1., 2. The equations are already in standard form with no fractions or decimals.

3., 4. We eliminate a from two different pairs of equations.

$\begin{array}{ll} -2a - 2b - 2c = -10 & \text{Multiplying (1) by } -2 \\ \underline{2a + 3b \quad - c = 2} & \text{(2)} \\ \qquad b - 3c = -8 & \text{(4)} \end{array}$

$\begin{array}{ll} -2a - 3b + c = -2 & \text{Multiplying (2) by } -1 \\ \underline{2a + 3b - 2c = 4} & \text{(3)} \\ \qquad\quad -c = 2 \\ \qquad\quad\; c = -2 \end{array}$

We eliminate not only a, but also b and found $c = -2$.

5. Substitute -2 for c in Equation (4) to find b.

$b - 3(-2) = -8$ Substituting -2 for c in (4)

$b + 6 = -8$

$b = -14$

6. Substitute in one of the original equations to find a.

$a - 14 - 2 = 5$ Substituting -2 for c and
                  -14 for b in (1)

$a - 16 = 5$

$a = 21$

We obtain $(21, -14, -2)$. This checks, so it is the solution.

21. $-2x + 8y + 2z = 4,$ (1)

     $x + 6y + 3z = 4,$ (2)

     $3x - 2y + z = 0$ (3)

1., 2. The equations are already in standard form with no fractions or decimals.

3., 4. We eliminate z from two different pairs of equations.

$\begin{array}{ll} -2x + 8y + 2z = 4 & \text{(1)} \\ \underline{-6x + 4y - 2z = 0} & \text{Multiplying (3) by } -2 \\ -8x + 12y \quad = 4 & \text{(4)} \end{array}$

$\begin{array}{ll} x + 6y + 3z = 4 & \text{(2)} \\ \underline{-9x + 6y - 3z = 0} & \text{Multiplying (3) by } -3 \\ -8x + 12y \quad = 4 & \text{(5)} \end{array}$

5. Now solve the system of Equations (4) and (5).

$-8x + 12y = 4$ (4)

$-8x + 12y = 4$ (5)

$\begin{array}{ll} -8x + 12y = 4 & \text{(4)} \\ \underline{8x - 12y = -4} & \text{Multiplying (5) by } -1 \\ \qquad 0 = 0 & \text{(6)} \end{array}$

Equation (6) indicates that Equations (1), (2), and (3) are dependent. (Note that if Equation (1) is subtracted from Equation (2), the result is Equation (3).) We could also have concluded that the equations are dependent by observing that Equations (4) and (5) are identical.

23.
$$2u - 4v - w = 8, \quad (1)$$
$$3u + 2v + w = 6, \quad (2)$$
$$5u - 2v + 3w = 2 \quad (3)$$

1., 2. The equations are already in standard form with no fractions or decimals.

3., 4. We eliminate w from two different pairs of equations.

$$\begin{array}{ll} 2u - 4v - w = 8 & (1) \\ 3u + 2v + w = 6 & (2) \\ \hline 5u - 2v \quad\quad = 14 & (4) \end{array}$$

$$\begin{array}{ll} 6u - 12v - 3w = 24 & \text{Multiplying (1) by 3} \\ 5u - 2v + 3w = 2 & (3) \\ \hline 11u - 14v \quad\quad = 26 & (5) \end{array}$$

5. Now solve the system of Equations (4) and (5).

$$5u - 2v = 14 \quad (4)$$
$$11u - 14v = 26 \quad (5)$$

$$\begin{array}{ll} -35u + 14v = -98 & \text{Multiplying (4) by } -7 \\ 11u - 14v = 26 & (5) \\ \hline -24u \quad\quad = -72 \\ u = 3 \end{array}$$

$$5 \cdot 3 - 2v = 14 \quad \text{Substituting 3 for } u \text{ in (4)}$$
$$15 - 2v = 14$$
$$-2v = -1$$
$$v = \frac{1}{2}$$

6. Substitute in one of the original equations to find w.

$$2 \cdot 3 - 4\left(\frac{1}{2}\right) - 2 = 8 \quad \text{Substituting 3 for } u \text{ and } \frac{1}{2}$$
$$\text{for } v \text{ in (1)}$$
$$6 - 2 - w = 8$$
$$w = -4$$

We obtain $\left(3, \frac{1}{2}, -4\right)$. This checks, so it is the solution.

25.
$$r + \frac{3}{2}s + 6t = 2,$$
$$2r - 3s + 3t = 0.5,$$
$$r + s + t = 1$$

1. All equations are already in standard form.
2. Multiply the first equation by 2 to clear the fraction. Also, multiply the second equation by 10 to clear the decimal.

$$2r + 3s + 12t = 4, \quad (1)$$
$$20r - 30s + 30t = 5, \quad (2)$$
$$r + s + t = 1 \quad (3)$$

3., 4. We eliminate s from two different pairs of equations.

$$\begin{array}{ll} 20r + 30s + 120t = 40 & \text{Multiplying (1) by 10} \\ 20r - 30s \quad + 30t = 5 & (2) \\ \hline 40r \quad\quad + 150t = 45 & (4) \text{ Adding} \end{array}$$

$$\begin{array}{ll} 2r + 3s + 12t = 4 & (1) \\ -3r - 3s - 3t = -3 & \text{Multiplying (3) by } -3 \\ \hline -r \quad\quad + 9t = 1 & (5) \text{ Adding} \end{array}$$

5. Solve the system of Equations (4) and (5).

$$40r + 150t = 45, \quad (4)$$
$$-r + 9t = 1, \quad (5)$$

$$\begin{array}{ll} 40r + 150t = 45 & (4) \\ -40r + 360t = 40 & \text{Multiplying (5) by 40} \\ \hline 510t = 85 \\ t = \frac{85}{510} \\ t = \frac{1}{6} \end{array}$$

$$40r + 150\left(\frac{1}{6}\right) = 45 \quad \text{Substituting } \frac{1}{6} \text{ for } t \text{ in (4)}$$
$$40r + 25 = 45$$
$$40r = 20$$
$$r = \frac{1}{2}$$

6. Substitute in one of the original equations to find s.

$$\frac{1}{2} + s + \frac{1}{6} = 1 \quad \text{Substituting } \frac{1}{2} \text{ for } r \text{ and } \frac{1}{6} \text{ for } t \text{ in (3)}$$
$$s + \frac{2}{3} = 1$$
$$s = \frac{1}{3}$$

We obtain $\left(\frac{1}{2}, \frac{1}{3}, \frac{1}{6}\right)$. This checks, so it is the solution.

27.
$$4a + 9b \quad\quad = 8, \quad (1)$$
$$8a \quad + 6c = -1, \quad (2)$$
$$6b + 6c = -1 \quad (3)$$

1., 2. The equations are already in standard form with no fractions or decimals.

3., 4. Note that there is no c in Equation (1). We will use Equations (2) and (3) to obtain another equation with no c-term.

$$\begin{array}{ll} 8a \quad + 6c = -1 & (2) \\ -6b - 6c = 1 & \text{Multiplying (3) by } -1 \\ \hline 8a - 6b \quad\quad = 0 & (4) \text{ Adding} \end{array}$$

5. Now solve the system of Equations (1) and (4).

$$\begin{array}{ll} -8a - 18b = -16 & \text{Multiplying (1) by } -2 \\ 8a - 6b = 0 \\ \hline -24b = -16 \\ b = \frac{2}{3} \end{array}$$

$$8a - 6\left(\frac{2}{3}\right) = 0 \quad \text{Substituting } \frac{2}{3} \text{ for } b \text{ in (4)}$$
$$8a - 4 = 0$$
$$8a = 4$$
$$a = \frac{1}{2}$$

6. Substitute in Equation (2) or (3) to find c.

$$8\left(\frac{1}{2}\right) + 6c = -1 \quad \text{Substituting } \frac{1}{2} \text{ for } a \text{ in (2)}$$
$$4 + 6c = -1$$
$$6c = -5$$
$$c = -\frac{5}{6}$$

We obtain $\left(\frac{1}{2}, \frac{2}{3}, -\frac{5}{6}\right)$. This checks, so it is the solution.

29.
$$x + y + z = 57, \quad (1)$$
$$-2x + y \quad\quad = 3, \quad (2)$$
$$x \quad - z = 6 \quad (3)$$

1., 2. The equations are already in standard form with

no fractions or decimals.

3., 4. Note that there is no z in Equation (2). We will use Equations (1) and (3) to obtain another equation with no z-term.

$$\begin{array}{ll} x+y+z=57 & (1) \\ \underline{x \quad -z=6} & (3) \\ 2x+y \quad =63 & (4) \end{array}$$

5. Now solve the system of Equations (2) and (4).

$$\begin{array}{ll} -2x+y=3 & (2) \\ \underline{2x+y=63} & (4) \\ 2y=66 \\ y=33 \end{array}$$

$$\begin{array}{ll} 2x+33=63 & \text{Substituting 33 for } y \text{ in (4)} \\ 2x=30 \\ x=15 \end{array}$$

6. Substitute in Equation (1) or (3) to find z.

$$\begin{array}{ll} 15-z=6 & \text{Substituting 15 for } x \text{ in (3)} \\ 9=z \end{array}$$

We obtain (15, 33, 9). This checks, so it is the solution.

31. $\quad \begin{array}{ll} a \quad -3c=6, & (1) \\ b+2c=2, & (2) \\ 7a-3b-5c=14 & (3) \end{array}$

1., 2. The equations are already in standard form with no fractions or decimals.

3., 4. Note that there is no b in Equation (1). We will use Equations (2) and (3) to obtain another equation with no b-term.

$$\begin{array}{ll} 3b+6c=6 & \text{Multiplying (2) by 3} \\ \underline{7a-3b-5c=14} & (3) \\ 7a \quad +c=20 & (4) \end{array}$$

5. Now solve the system of Equations (1) and (4).

$$\begin{array}{ll} a-3c=6 & (1) \\ 7a+c=20 & (4) \end{array}$$

$$\begin{array}{ll} a-3c=6 & (1) \\ \underline{21a+3c=60} & \text{Multiplying (4) by 3} \\ 22a \quad =66 \\ a=3 \end{array}$$

$$\begin{array}{ll} 3-3c=6 & \text{Substituting in (1)} \\ -3c=3 \\ c=-1 \end{array}$$

6. Substitute in Equation (2) or (3) to find b.

$$\begin{array}{ll} b+2(-1)=2 & \text{Substituting in (2)} \\ b-2=2 \\ b=4 \end{array}$$

We obtain (3, 4, -1). This checks, so it is the solution.

33. $\quad \begin{array}{ll} x+y+z=83, & (1) \\ y=2x+3, & (2) \\ z=40+x & (3) \end{array}$

Observe, from Equations (2) and (3), that we can substitute $2+3x$ for y and $40+x$ for z in Equation (1) and solve for x.

$$\begin{array}{l} x+y+z=83 \\ x+(2x+3)+(40+x)=83 \\ 4x+43=83 \\ 4x=40 \\ x=10 \end{array}$$

Now substitute 10 for x in Equation (2).

$$y=2x+3=2 \cdot 10+3=20+3=23$$

Finally, substitute 10 for x in Equation (3).

$$z=40+x=40+10=50 \,.$$

We obtain (10, 23, 50). This checks, so it is the solution.

35. $\quad \begin{array}{ll} x \quad +z=0, & (1) \\ x+y+2z=3, & (2) \\ y \quad +z=2 & (3) \end{array}$

1., 2. The equations are already in standard form with no fractions or decimals.

3., 4. Note that there is no y in Equation (1). We use Equations (2) and (3) to obtain another equation with no y-term.

$$\begin{array}{ll} x+y+2z=3 & (2) \\ \underline{-y-z=-2} & \text{Multiplying (3) by } -1 \\ x \quad +z=1 & (4) \text{ Adding} \end{array}$$

5. Now solve the system of Equations (1) and (4).

$$\begin{array}{ll} x+z=0 & (1) \\ \underline{-x-z=-1} & \text{Multiplying (4) by } -1 \\ 0=-1 & \text{Adding} \end{array}$$

We get a false equation, or contradiction. There is no solution.

37. $\quad \begin{array}{ll} x+y+z=1, & (1) \\ -x+2y+z=2, & (2) \\ 2x-y \quad =-1 & (3) \end{array}$

1., 2. The equations are already in standard form with no fractions or decimals.

3. Note that there is no z in Equation (3). We will use Equations (1) and (2) to eliminate z:

$$\begin{array}{ll} x+y+z=1 & (1) \\ \underline{x-2y-z=-2} & \text{Multiplying (2) by } -1 \\ 2x-y \quad =-1 & (4) \end{array}$$

Equations (3) and (4) are identical, so Equations (1), (2), and (3) are dependent. (We have seen that if Equation (2) is multiplied by -1 and added to Equation (1), the result is Equation (3).)

39. *Writing Exercise.*

41. $\quad x-3y=7$

$$-3y=-x+7$$

$$y=\frac{1}{3}x-\frac{7}{3}$$

The slope is $\frac{1}{3}$.

The y-intercept is $\left(0,-\frac{7}{3}\right)$.

43. To find the y-intercept, let $x=0$ and solve for y.

$$0-5y=20$$

$$y=-4$$

The y-intercept is $(0,-4)$.

To find the x-intercept, let $y=0$ and solve for x.

$$2x+0=20$$

$$x=10$$

The x-intercept is $(10, 0)$.

45. $3x - y = 12$

$-y = -3x + 12$

$y = 3x - 12$

The slope is $m = 3$.

$y = 3x + 7$

The slope is $m = 3$.

The lines are parallel.

47. *Writing Exercise.*

49. $\dfrac{x+2}{3} - \dfrac{y+4}{2} + \dfrac{z+1}{6} = 0$,

$\dfrac{x-4}{3} + \dfrac{y+1}{4} - \dfrac{z-2}{2} = -1$,

$\dfrac{x+1}{2} + \dfrac{y}{2} + \dfrac{z-1}{4} = \dfrac{3}{4}$

1., 2. We clear fractions and write each equation in standard form.

To clear fractions, we multiply both sides of each equation by the LCM of its denominators. The LCM's are 6, 12, and 4, respectively.

$6\left(\dfrac{x+2}{3} - \dfrac{y+4}{2} + \dfrac{z+1}{6}\right) = 6 \cdot 0$

$2(x+2) - 3(y+4) + (z+1) = 0$

$2x + 4 - 3y - 12 + z + 1 = 0$

$2x - 3y + z = 7$

$12\left(\dfrac{x-4}{3} + \dfrac{y+1}{4} - \dfrac{z-2}{2}\right) = 12 \cdot (-1)$

$4(x-4) + 3(y+1) - 6(z-2) = -12$

$4x - 16 + 3y + 3 - 6z + 12 = -12$

$4x + 3y - 6z = -11$

$4\left(\dfrac{x+1}{2} + \dfrac{y}{2} + \dfrac{z-1}{4}\right) = 4 \cdot \dfrac{3}{4}$

$2(x+1) + 2(y) + (z-1) = 3$

$2x + 2 + 2y + z - 1 = 3$

$2x + 2y + z = 2$

The resulting system is

$2x - 3y + z = 7$,   (1)

$4x + 3y - 6z = -11$,   (2)

$2x + 2y + z = 2$ (3)

3., 4. We eliminate z from two different pairs of equations.

$12x - 18y + 6z = 42$ Multiplying (1) by 6

$\underline{4x + 3y - 6z = -11}$ (2)

$16x - 15y \quad\quad = 31$ (4) Adding

$2x - 3y + z = 7$ (1)

$\underline{-2x - 2y - z = -2}$ Multiplying (3) by -1

$-5y = 5$ (5) Adding

5. Solve (5) for y: $-5y = 5$

$y = -1$

Substitute -1 for y in (4):

$16x - 15(-1) = 31$

$16x + 15 = 31$

$16x = 16$

$x = 1$

6. Substitute 1 for x and -1 for y in (1):

$2 \cdot 1 - 3(-1) + z = 7$

$5 + z = 7$

$z = 2$

We obtain $(1, -1, 2)$. This checks, so it is the solution.

51. $w + x + y + z = 2$,   (1)

$w + 2x + 2y + 4z = 1$,   (2)

$w - x + y + z = 6$,   (3)

$w - 3x - y + z = 2$ (4)

The equations are already in standard form with no fractions or decimals.

Start by eliminating w from three different pairs of equations.

$w + x + y + z = 2$ (1)

$\underline{-w - 2x - 2y - 4z = -1}$ Multiplying (2) by -1

$-x - y - 3z = 1$ (5) Adding

$w + x + y + z = 2$ (1)

$\underline{-w + x - y - z = -6}$ Multiplying (3) by -1

$2x \quad\quad = -4$ (6) Adding

$w + x + y + z = 2$ (1)

$\underline{-w + 3x + y - z = -2}$ Multiplying (4) by -1

$4x + 2y \quad\quad = 0$ (7) Adding

We can solve (6) for x:

$2x = -4$

$x = -2$

Substitute -2 for x in (7):

$4(-2) + 2y = 0$

$-8 + 2y = 0$

$2y = 8$

$y = 4$

Substitute -2 for x and 4 for y in (5):

$-(-2) - 4 - 3z = 1$

$-2 - 3z = 1$

$-3z = 3$

$z = -1$

Substitute -2 for x, 4 for y, and -1 for z in (1):

$w - 2 + 4 - 1 = 2$

$w + 1 = 2$

$w = 1$

We obtain $(1, -2, 4, -1)$. This checks, so it is the solution.

53. $\dfrac{2}{x} - \dfrac{1}{y} - \dfrac{3}{z} = -1$,

$\dfrac{2}{x} - \dfrac{1}{y} + \dfrac{1}{z} = -9$,

$\dfrac{1}{x} + \dfrac{2}{y} - \dfrac{4}{z} = 17$

Let u represent $\dfrac{1}{x}$, v represent $\dfrac{1}{y}$, and w represent $\dfrac{1}{z}$.

Substituting, we have

$2u - v - 3w = -1$,   (1)

$2u - v + w = -9$,   (2)

$u + 2v - 4w = 17$ (3)

1., 2. The equations in u, v, and w are in standard form with no fractions or decimals.

3., 4. We eliminate v from two different pairs of equations.

$2u - v - 3w = -1$ (1)
$\underline{-2u + v\ \ - w = 9}$ Multiplying (2) by -1
$\qquad -4w = 8$ (4) Adding

$4u - 2v - 6w = -2$ Multiplying (1) by 2
$\underline{u + 2v - 4w = 17}$ (3)
$5u\ \ \ \ -10w = 15$ (5) Adding

5. We can solve (4) for w:
$-4w = 8$
$\quad w = -2$

Substitute -2 for w in (5):
$5u - 10(-2) = 15$
$\quad 5u + 20 = 15$
$\qquad\quad 5u = -5$
$\qquad\quad u = -1$

6. Substitute -1 for u and -2 for w in (1):
$2(-1) - v - 3(-2) = -1$
$\qquad -v + 4 = -1$
$\qquad\quad -v = -5$
$\qquad\quad\ v = 5$

Solve for x, y, and z. We substitute -1 for u, 5 for v and -2 for w.

$u = \dfrac{1}{x}$ $v = \dfrac{1}{y}$ $w = \dfrac{1}{z}$

$-1 = \dfrac{1}{x}$ $5 = \dfrac{1}{y}$ $-2 = \dfrac{1}{z}$

$x = -1$ $y = \dfrac{1}{5}$ $z = -\dfrac{1}{2}$

We obtain $\left(-1, \dfrac{1}{5}, -\dfrac{1}{2}\right)$. This checks, so it is the

solution.

55. $5x - 6y + kz = -5$,   (1)
$\quad x + 3y - 2z = 2$,   (2)
$\quad 2x - y + 4z = -1$ (3)

Eliminate y from two different pairs of equations.
$5x - 6y + kz = -5$ (1)
$\underline{2x + 6y - 4z = 4}$ Multiplying (2) by 2
$7x + (k - 4)z = -1$ (4)

$x + 3y\ \ - 2z = 2$ (2)
$\underline{6x - 3y + 12z = -3}$ Multiplying (3) by 3
$7x\ \ \ \ \ + 10z = -1$ (5)

Solve the system of Equations (4) and (5).
$7x + (k - 4)z = -1$ (4)
$\quad 7x + 10z = -1$ (5)

$-7x - (k - 4)z = 1$ Multiplying (4) by -1
$\underline{\quad 7x\ \ \ \ + 10z = -1}$ (5)
$(-k + 14)z = 0$ (6)

The system is dependent for the value of k that makes Equation (6) true. This occurs when $-k + 14$ is 0. We solve for k:
$-k + 14 = 0$
$\qquad 14 = k$

57. $z = b - mx - ny$

Three solutions are $(1, 1, 2)$, $(3, 2, -6)$, and $\left(\dfrac{3}{2}, 1, 1\right)$.

We substitute for x, y, and z and then solve for b, m, and n.

$2 = b - m - n$,
$-6 = b - 3m - 2n$,
$1 = b - \dfrac{3}{2}m - n$

1., 2. Write the equations in standard form. Also, clear the fraction in the last equation.
$\quad b\ \ - m\ \ - n = 2$,   (1)
$\quad b - 3m - 2n = -6$,   (2)
$2b - 3m - 2n = 2$ (3)

3., 4. Eliminate b from two different pairs of equations.
$\quad b\ \ - m\ \ - n = 2$ (1)
$\underline{-b + 3m + 2n = 6}$ Multiplying (2) by -1
$\qquad 2m\ \ + n = 8$ (4) Adding

$-2b + 2m + 2n = -4$ Multiplying (1) by -2
$\underline{\ 2b - 3m - 2n = 2}$ (3)
$\qquad -m\ \ \ \ \ = -2$ (5) Adding

5. We solve Equation (5) for m:
$-m = -2$
$\ m = 2$

Substitute in Equation (4) and solve for n.
$2 \cdot 2 + n = 8$
$\quad 4 + n = 8$
$\qquad\quad n = 4$

6. Substitute in one of the original equations to find b.
$b - 2 - 4 = 2$ Substituting 2 for m and 4 for n in (1)
$\quad b - 6 = 2$
$\qquad\quad b = 8$

The solution is $(8, 2, 4)$, so the equation is
$z = 8 - 2x - 4y$.

Mid-Chapter Review

1. $2x - 3y = 5$,   (1)
$\quad y = x - 1$ (2)
Substitute $x - 1$ for y in Equation (2) and solve for y.
$2x - 3(x - 1) = 5$ Substituting
$\quad 2x - 3x + 3 = 5$
$\qquad\quad -x + 3 = 5$
$\qquad\qquad -x = 2$
$\qquad\qquad\ x = -2$
Substitute -2 for x in Equation (2) and solve for y.
$y = x - 1$ (2)
$y = -2 - 1$
$y = -3$
The solution is $(-2, -3)$.

2. $2x - 5y = 1$ (1)
$\underline{\ x + 5y = 8}$ (2)
$3x\ \ \ \ = 9$ Adding
$\ x = 3$
Substitute 3 for x in Equation (2) and solve for y.
$\quad x + 5y = 8$ (1)
$\quad 3 + 5y = 8$ Substituting
$\qquad 5y = 5$
$\qquad\ y = 1$
The solution is $(3, 1)$.

3. $x = y$,   (1)
   $x + y = 2$ (2)

Substitute y for x in Equation (2) and solve for y.
$$x + y = 2 \quad (2)$$
$$y + y = 2 \quad \text{Substituting}$$
$$2y = 2$$
$$y = 1$$

Substitute 1 for y in Equation (1) and solve for x.
$$x = y \quad (1)$$
$$x = 1$$
We obtain $(1, 1)$ as the solution.

4. $x + y = 10$ (1)
   $\underline{x - y = 8}$ (2)
   $\quad 2x \quad = 18$ Adding
   $\quad\quad x = 9$

Substitute 9 for x in Equation (1) and solve for y.
$$x + y = 10 \quad (1)$$
$$9 + y = 10 \quad \text{Substituting}$$
$$y = 1$$
We obtain $(9, 1)$ as the solution.

5. $y = \frac{1}{2}x + 1$,   (1)
   $y = 2x - 5$ (2)

Substitute $2x - 5$ for y in Equation (1) and solve for x.
$$y = \frac{1}{2}x + 1 \quad (1)$$
$$2x - 5 = \frac{1}{2}x + 1 \quad \text{Substituting}$$
$$4x - 10 = x + 2 \quad \text{Clearing fractions}$$
$$3x = 12$$
$$x = 4$$

Substitute 4 for x in Equation (2) and solve for y.
$$y = 2x - 5 \quad (2)$$
$$y = 2(4) - 5 = 3$$
We obtain $(4, 3)$ as the solution.

6. $y = 2x - 3$,   (1)
   $x + y = 12$ (2)

Substitute $2x - 3$ for y in Equation (2) and solve for x.
$$x + y = 12 \quad (2)$$
$$x + (2x - 3) = 12 \quad \text{Substituting}$$
$$x + 2x - 3 = 12$$
$$3x - 3 = 12$$
$$3x = 15$$
$$x = 5$$

Substitute 5 for x in Equation (1) and solve for y.
$$y = 2x - 3 \quad (1)$$
$$y = 2(5) - 3 = 7$$
We obtain $(5, 7)$ as the solution.

7. $x = 5$,   (1)
   $y = 10$ (2)

We obtain $(5, 10)$ as the solution.

8. $3x + 5y = 8$ (1)
   $\underline{3x - 5y = 4}$ (2)
   $\quad 6x \quad = 12$ Adding
   $\quad\quad x = 2$

Substitute 2 for x in Equation (1) and solve for y.
$$3x + 5y = 8 \quad (1)$$
$$3(2) + 5y = 8 \quad \text{Substituting}$$
$$6 + 5y = 8$$
$$5y = 2$$
$$y = \frac{2}{5}$$
We obtain $\left(2, \dfrac{2}{5}\right)$ as the solution.

9. $2x - y = 1$,   (1)
   $2y - 4x = 3$ (2)

We multiply Equation (1) by 2.
$$4x - 2y = 2 \quad \text{Multiplying (1) by 2}$$
$$\underline{-4x + 2y = 3} \quad (2)$$
$$0 = 5 \quad \text{FALSE}$$
There is no solution.

10. $x = 2 - y$,   (1)
    $3x + 3y = 6$ (2)

Substitute $2 - y$ for x in Equation (2) and solve for y.
$$3x + 3y = 6 \quad (2)$$
$$3(2 - y) + 3y = 6 \quad \text{Substituting}$$
$$6 - 3y + 3y = 6$$
$$6 = 6$$
There are many solutions.
The solution set is $\{(x, y) | x = 2 - y\}$.

11. $1.1x - 0.3y = 0.8$ (1)
    $\underline{2.3x + 0.3y = 2.6}$ (2)
    $\quad 3.4x \quad\quad = 3.4$ Adding
    $\quad\quad\quad x = 1$

Substitute 1 for x in Equation (1) and solve for y.
$$1.1x - 0.3y = 0.8 \quad (1)$$
$$1.1(1) - 0.3y = 0.8 \quad \text{Substituting}$$
$$-0.3y = -0.3$$
$$y = 1$$
We obtain $(1, 1)$ as the solution.

12. $\frac{1}{4}x = \frac{1}{3}y$,   (1)
    $\frac{1}{2}x - \frac{1}{15}y = 2$ (2)

We solve Equation (1) for x.
$$\frac{1}{4}x = \frac{1}{3}y \quad (1)$$
$$x = \frac{4}{3}y$$

Substitute $\frac{4}{3}y$ for x in Equation (2) and solve for y.

$$\frac{1}{2}x - \frac{1}{15}y = 2 \quad (2)$$
$$\frac{1}{2}\left(\frac{4}{3}y\right) - \frac{1}{15}y = 2$$
$$\frac{2}{3}y - \frac{1}{15}y = 2$$
$$10y - y = 30 \qquad \text{Multiplying by 15}$$
$$9y = 30$$
$$y = \frac{10}{3}$$

Substitute $\frac{10}{3}$ for y in Equation (1) and solve for x.

$$\frac{1}{4}x = \frac{1}{3}y$$
$$\frac{1}{4}x = \frac{1}{3}\left(\frac{10}{3}\right)$$
$$\frac{1}{4}x = \frac{10}{9}$$
$$x = \frac{40}{9}$$

We obtain $\left(\frac{40}{9}, \frac{10}{3}\right)$ as the solution.

13.
$$3x + y - z = -1, \quad (1)$$
$$2x - y + 4z = 2, \quad (2)$$
$$x - y + 3z = 3 \quad (3)$$

1., 2. The equations are already in standard form with no fractions or decimals.
3. Use Equations (1) and (2) to eliminate y.
$$\begin{aligned} 3x + y \ - z &= -1 \quad (1) \\ 2x - y + 4z &= 2 \quad (2) \\ \hline 5x \quad\ + 3z &= 1 \quad (4) \text{ Adding} \end{aligned}$$

4. Use a different pair of equations and eliminate y.
$$\begin{aligned} 3x + y \ - z &= -1 \quad (1) \\ x - y + 3z &= 3 \quad (3) \\ \hline 4x \quad\ + 2z &= 2 \quad (5) \text{ Adding} \end{aligned}$$

5. Now solve the system of Equations (4) and (5).
$$5x + 3z = 1 \quad (4)$$
$$4x + 2z = 2 \quad (5)$$

$$\begin{aligned} -10x - 6z &= -2 \quad \text{Multiplying (4) by } -2 \\ 12x + 6z &= 6 \quad \text{Multiplying (5) by 3} \\ \hline 2x \quad\ &= 4 \quad \text{Adding} \\ x &= 2 \end{aligned}$$

$$5 \cdot 2 + 3z = 1 \quad \text{Substituting 2 for } x \text{ in (4)}$$
$$10 + 3z = 1$$
$$3z = -9$$
$$z = -3$$

6. Substitute in one of the original equations to find y.
$$2 - y + 3(-3) = 3 \quad \text{Substituting 2 for } x \text{ and } -3 \text{ for } z \text{ in (3)}$$
$$2 - y - 9 = 3$$
$$-y = 10$$
$$y = -10$$

We obtain $(2, -10, -3)$. This checks, so it is the solution.

14.
$$2x + y - 3z = -4, \quad (1)$$
$$4x + y + 3z = -1, \quad (2)$$
$$2x - y + 6z = 7 \quad (3)$$

1., 2. The equations are already in standard form with no fractions or decimals.
3. Use Equations (1) and (2) to eliminate z.

$$\begin{aligned} 2x + y - 3z &= -4 \quad (1) \\ 4x + y + 3z &= -1 \quad (2) \\ \hline 6x + 2y \quad &= -5 \quad (4) \text{ Adding} \end{aligned}$$

4. Use a different pair of equations and eliminate z.
$$\begin{aligned} 4x + 2y - 6z &= -8 \quad \text{Multiplying (1) by 2} \\ 2x - y + 6z &= 7 \quad (3) \\ \hline 6x + y \quad &= -1 \quad (5) \text{ Adding} \end{aligned}$$

5. Now solve the system of Equations (4) and (5).
$$6x + 2y = -5 \quad (4)$$
$$6x + y = -1 \quad (5)$$

$$\begin{aligned} -6x - 2y &= 5 \quad \text{Multiplying (4) by } -1 \\ 6x + \ y &= -1 \quad (5) \\ \hline -y &= 4 \quad \text{Adding} \\ y &= -4 \end{aligned}$$

$$6x - 4 = -1 \quad \text{Substituting } -4 \text{ for } y \text{ in (5)}$$
$$6x = 3$$
$$x = \frac{1}{2}$$

6. Substitute in one of the original equations to find z.
$$2\left(\frac{1}{2}\right) - 4 - 3z = -4 \quad \text{Substituting } \frac{1}{2} \text{ for } x \text{ and } -4 \text{ for } y \text{ in (1)}$$
$$1 - 4 - 3z = -4$$
$$-3z = -1$$
$$z = \frac{1}{3}$$

We obtain $\left(\frac{1}{2}, -4, \frac{1}{3}\right)$. This checks, so it is the solution.

15.
$$3x + 5y - z = 8, \quad (1)$$
$$x + 6y \quad = 4, \quad (2)$$
$$x - 7y - z = 3 \quad (3)$$

1., 2. The equations are already in standard form with no fractions or decimals.
3., 4. We eliminate x from two different pairs of equations.
$$\begin{aligned} 3x + 5y - z &= 8 \quad (1) \\ -3x - 18y \quad &= -12 \quad \text{Multiplying (2) by } -3 \\ \hline -13y - z &= -4 \quad (4) \end{aligned}$$

$$\begin{aligned} 3x + 5y \ - z &= 8 \quad (1) \\ -3x + 21y + 3z &= -9 \quad \text{Multiplying (3) by } -3 \\ \hline 26y + 2z &= -1 \quad (4) \end{aligned}$$

5. Now solve the system of Equations (4) and (5).
$$-13y - z = -4 \quad (4)$$
$$26y + 2z = -1 \quad (5)$$

$$\begin{aligned} -26y - 2z &= -8 \quad \text{Multiplying (4) by 2} \\ 26y + 2z &= -1 \quad (5) \\ \hline 0 &= -9 \end{aligned}$$

We get a false equation, so there is no solution.

16.
$$x - y \quad = 4, \quad (1)$$
$$2x + y - z = 5, \quad (2)$$
$$3x \quad - z = 9 \quad (3)$$

1., 2. The equations are already in standard form with no fractions or decimals.
3. Note that there is no y in Equation (3). We will use Equations (1) and (2) to eliminate y:

$$x - y \quad = 4 \quad (1)$$
$$\underline{2x + y - z = 5} \quad (2)$$
$$3x \quad - z = 9 \quad \text{Adding}$$

Equations (3) and (4) are identical, so Equations (1), (2), and (3) are dependent.

17. *Familiarize*. Let $x =$ the amount of text messages received and $y =$ the amount of text messages sent.
Translate.

$\underbrace{\text{Total texts}}$ is $\underbrace{3853,}$

$\quad \downarrow \qquad \downarrow \quad \downarrow$

$\quad x + y \quad = \quad 3853$

$\underbrace{\text{texts sent}}$ is $\underbrace{191 \text{ more}}$ than $\underbrace{\text{texts received}}$

$\quad \downarrow \qquad \downarrow \qquad \downarrow \qquad \downarrow \qquad \downarrow$

$\quad y \quad = \quad 191 \quad + \quad x$

We have a system of equations:

$$x + y = 3853 \quad (1)$$
$$y = 191 + x \quad (2)$$

Carry out. We use the substitution method to solve the system of equations. Substitute $191 + x$ for y in Equation (1).

$$x + 191 + x = 3853$$
$$2x + 191 = 3853$$
$$2x = 3662$$
$$x = 1831$$

Substitute 1831 for x in Equation (1) and solve for x.

$$1831 + y = 3853$$
$$y = 2022$$

Check. The total amount is $1831 + 2022$, or 3853 texts. Texts received $191 + 1831 = 2022$. The answer checks.
State. There were 2022 text messages sent and 1831 text messages received.

18. Let $x =$ the number of 5-cent bottles or cans and $y =$ the number of 10-cent bottles or cans.
Solve: $\quad x + y = 430,$
$$0.05x + 0.10y = 26.20$$

The solution is (336, 94). So, 336 5-cent bottles or cans and 94 10-cent bottles or cans were collected.

19. *Familiarize*. Let $x =$ the number of pounds of Pecan Morning Granola and $y =$ the number of pounds of Oat Dream Granola to be used in the mixture. The amount of nuts and dried fruit in the mixture is 19%(20 lb), or $0.19(20 \text{ lb}) = 3.8 \text{ lb}$.
Translate. We organize the information in a table.

	Pecan Morning	Oat Dream	Mixture
Number of pounds	x	y	20
Percent of nuts and dried fruit	25%	10%	19%
Amount of nuts and dried fruit	$0.25x$	$0.10y$	3.8 lb

We get one equation from the "Number of pounds" row of the table:

$$x + y = 20$$

The last row of the table yields a second equation:

$$0.25x + 0.1y = 3.8$$

After clearing decimals, we have the problem translated to a system of equations:

$$x + y = 20, \quad (1)$$
$$25x + 10y = 380 \quad (2)$$

Carry out. We use the elimination method to solve the system of equations.

$$-10x - 10y = -200 \quad \text{Multiplying (1) by } -10$$
$$\underline{25x + 10y = 380}$$
$$15x \quad = 180$$
$$x = 12$$

Substitute 12 for x in (1) and solve for y.

$$12 + y = 20$$
$$y = 8$$

Check. The amount of the mixture is 12 lb + 8 lb, or 20 lb. The amount of nuts and dried fruit in the mixture is $0.25(12 \text{ lb}) + 0.1(8 \text{ lb}) = 3 \text{ lb} + 0.8 \text{ lb} = 3.8 \text{ lb}$. The answer checks.
State. 12 lb of Pecan Morning Granola and 8 lb of Oat Dream Granola should be mixed.

20. Let $r =$ the speed of the boat in still water.

	Distance	Rate	Time
With current	d	$r + 6$	1.5
Against current	d	$r - 6$	3

Solve: $\quad d = 1.5(r + 6)$
$$d = 3(r - 6)$$
We find that $r = 18$ mph.

Exercise Set 3.5

1. a

3. d

5. *Familiarize*. Let $x =$ the first number, $y =$ the second number, and $z =$ the third number.
Translate.

$\underbrace{\text{The sum of the three numbers}}$ is $\underbrace{85,}$

$\qquad\qquad \downarrow \qquad\qquad \downarrow \quad \downarrow$

$\qquad x + y + z \qquad = \quad 85$

$\underbrace{\text{The second}}$ is $\underbrace{\text{seven}}$ $\underbrace{\text{more than}}$ $\underbrace{\text{the first.}}$

$\quad \downarrow \qquad \downarrow \qquad \downarrow \qquad \downarrow \qquad \downarrow$

$\quad y \quad = \quad 7 \quad + \quad x$

$\underbrace{\text{The third}}$ is $\underbrace{\text{two}}$ $\underbrace{\text{more than}}$ $\underbrace{\text{four}}$ times $\underbrace{\text{the second.}}$

$\downarrow \quad \downarrow \quad \downarrow \quad \downarrow \quad \downarrow \quad \downarrow \quad \downarrow$

$z \;=\; 2 \;+\; 4 \;\times\; y$

We have a system of equations:

$$x + y + z = 85, \quad \text{or} \quad x + y + z = 85$$
$$y = 7 + x \qquad\quad -x + y = 7$$
$$z = 2 + 4y \qquad\quad -4y + z = 2$$

Carry out. Solving the system we get (8, 15, 62).
Check. The sum of the three numbers is $8 + 15 + 62$, or 85. The second number 15, is 7 more than the first number 8. The third number, 62 is 2 more than four times the second number, 15. The numbers check.
State. The numbers are 8, 15, and 62.

7. *Familiarize*. Let x = the first number, y = the second number, and z = the third number.
***Translate*.**

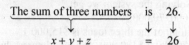

The sum of three numbers is 26.
$$x + y + z = 26$$

Twice the first minus the second is the third less 2.
$$2x - y = z - 2$$

The third is the second minus 3 times the first.
$$z = y - 3x$$

We now have a system of equations.
$$\begin{array}{ll} x + y + z = 26, & \text{or} \quad x + y + z = 26, \\ 2x - y = z - 2, & \quad\quad\; 2x - y - z = -2, \\ z = y - 3x & \quad\quad\; 3x - y + z = 0 \end{array}$$

***Carry out*.** Solving the system we get $(8, 21, -3)$.
***Check*.** The sum of the numbers is $8 + 21 - 3$, or 26. Twice the first minus the second is $2 \cdot 8 - 21$, or -5, which is 2 less than the third. The second minus three times the first is $21 - 3 \cdot 8$, or -3, which is the third. The numbers check.
***State*.** The numbers are 8, 21, and -3.

9. *Familiarize*. We first make a drawing.

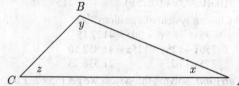

We let x, y, and z represent the measures of angles A, B, and C, respectively. The measures of the angles of a triangle add up to $180°$.
***Translate*.**

The sum of the measures is 180.
$$x + y + z = 180$$

The measure of angle B is three times the measure of angle A.
$$y = 3x$$

The measure of angle C is 20 more than the measure of angle A.
$$z = x + 20$$

We now have a system of equations.
$$\begin{array}{l} x + y + z = 180, \\ y = 3x, \\ z = x + 20 \end{array}$$

***Carry out*.** Solving the system we get $(32, 96, 52)$.
***Check*.** The sum of the measures is $32° + 96° + 52°$, or $180°$. Three times the measure of angle A is $3 \cdot 32°$, or

$96°$, the measure of angle B. 20 more than the measure of angle A is $32° + 20°$, or $52°$, the measure of angle C. The numbers check.
***State*.** The measures of angles A, B, and C are $32°$, $96°$, and $52°$, respectively.

11. *Familiarize*. Let x, y and z represent the GRE critical verbal score, quantitative score, and writing score, respectively.
***Translate*.**

The sum of three scores is 306.3.
$$x + y + z = 306.3$$

quantitative score is 1.6 more than verbal score.
$$y = 1.6 + x$$

verbal score is 147.1 more than writing score
$$x = 147.1 + z$$

We have a system of equations:
$$\begin{array}{ll} x + y + z = 306.3, & \text{or} \quad x + y + z = 306.3 \\ y = 1.6 + x & \quad\quad -x + y = 1.6 \\ x = 147.1 + z & \quad\quad x - z = 147.1 \end{array}$$

Carry out Solving the system we get $(150.6, 152.2, 3.5)$
***Check*.** The sum of the scores is $150.6 + 152.2 + 3.5$, or 306.3. The quantitative score, 152.2, is 1.6 more than the verbal score, 150.6. The verbal score, 150.6 is 147.1 more than the writing score, 3.5. The answer checks.
***State*.** The average score for verbal reasononing was 150.6, for quantitative reasoning 152.2, and for analytical writing 3.5.

13. *Familiarize*. Let x, y, and z represent the number of grams of fiber in 1 bran muffin, 1 banana, and a 1-cup serving of Wheaties, respectively.
***Translate*.**
Two bran muffins, 1 banana, and a 1-cup serving of Wheaties contain 9 g of fiber, so we have
$$2x + y + z = 9.$$
One bran muffin, 2 bananas, and a 1-cup serving of Wheaties contain 10.5 g of fiber, so we have
$$x + 2y + z = 10.5.$$
Two bran muffins and a 1-cup serving of Wheaties contain 6 g of fiber, so we have
$$2x + z = 6.$$
We now have a system of equations.
$$\begin{array}{l} 2x + y + z = 9, \\ x + 2y + z = 10.5, \\ 2x + z = 6 \end{array}$$

***Carry out*.** Solving the system, we get $(1.5, 3, 3)$.
***Check*.** Two bran muffins, 1 banana, and a 1-cup serving of Wheaties contain $2(1.5) + 3 + 3$, or 9 g of fiber. One bran muffin, 2 bananas, and a 1-cup serving of Wheaties contain $1.5 + 2 \cdot 3 + 3$, or 10.5 g of fiber. Two bran muffins and a 1-cup serving of Wheaties contain $2(1.5) + 3$, or 6 g of fiber. The answer checks.
***State*.** A bran muffin has 1.5 g of fiber, a banana has 3 g, and a 1-cup serving of Wheaties has 3 g.

15. Observe that the basic model plus tow package costs $24,290 and when a hard top is added the price rises to $25,285. This tells us that the price of a hard top is $25,285 – $24,290, or $995. Now observe that the base model and hard top costs $24,890 so the base model costs $24,890 – $995, or $23,895. Finally we know the tow package is $24,290 – $23,895, or $395.

17. *Familiarize*. Let x = the number of 12-oz cups, y = the number of 16-oz cups, and z = the number of 20-oz cups that Reba filled. Note that six 144-oz brewers contain $6 \cdot 144$, or 864 oz of coffee. Also, x 12-oz cups contain a total of $12x$ oz of coffee and bring in $1.75x$, y 16-oz cups contain $16y$ oz and bring in $1.95y$, and z 20-oz cups contain $20z$ oz and bring in $2.25z$.
Translate.

The total number of coffees served was 55.
$$x + y + z = 55$$

The total amount of coffee served was 864 oz.
$$12x + 16y + 20z = 864$$

The total amount collected was $107.75.
$$1.75x + 1.95y + 2.25z = 107.75$$

Now we have a system of equations.
$$x + y + z = 55,$$
$$12x + 16y + 20z = 864,$$
$$1.75x + 1.95y + 2.25z = 107.75$$

Carry out. Solving the system we get (17, 25, 13).
Check. The total number of coffees served was $17 + 25 + 13$, or 55. The total amount of coffee served was $12 \cdot 17 + 16 \cdot 25 + 20 \cdot 13 = 204 + 400 + 260 = 864$ oz. The total amount collected was $1.75(17) + 1.95(25) + 2.25(13) = 29.75 + 48.75 + 29.25 = 107.75$. The numbers check.
State. Reba filled 17 12-oz cups, 25 16-oz cups, and 13 20-oz cups.

19. *Familiarize*. Let x = the amount of the loan at 7%, y = the amount of the loan at 5%, and z = the amount of the loan at 3.2%.
Translate.

Total of the three loans is $120,000.
$$x + y + z = 120,000$$

Total interest due is $5040.
$$0.07x + 0.05y + 0.032z = 5040$$

home-equity interest is $1190 more than business-equipment loan interest.
$$0.032z = 1190 + 0.07x$$

We have a system of equations:

$$x + y + z = 120,000$$
$$0.07x + 0.05y + 0.032z = 5040$$
$$-0.07x + 0.032z = 1190$$

Carry out. Solving the system we get (15,000, 35,000, 70,000).
Check. The total of the three loans is $15,000 + $35,000 + $70,000 = $120,000. The total interest due is $0.07($15,000) + 0.05($35,000) + 0.032($70,000)$
$= $1050 + $1750 + $2240 = 5040. The interest from the home-equity loan, $0.032($70,000)$, or $2240 is $1190 more than the business-equipment interest $0.07($15,000)$, or $1050. The numbers check.
State. The business-equipment loan was $15,000, the small-business loan was $35,000, and the home-equity loan was $70,000.

21. *Familiarize*. Let x = the price of 1 g of gold, y = the the price of 1 g of silver, and z = the price of 1 g of copper.
Translate.

Cost of 100 g of red gold is $4177.15.
$$100(0.75x + 0.05y + 0.20z) = 4177.15$$

Cost of 100 g of yellow gold is $4185.25.
$$100(0.75x + 0.125y + 0.125z) = 4185.25$$

Cost of 100 g of white gold is $2153.875.
$$100(0.375x + 0.625y) = 2153.875$$

We have a system of equations:
$$75x + 5y + 20z = 4177.15$$
$$750x + 125y + 125z = 41,852.50$$
$$375x + 625y = 21,538.75$$

Carry out. Solving the system we get (55.62, 1.09, 0.01).
Check. The cost of 100 g of red gold is
$100[0.75($55.62) + 0.05($1.09) + 0.20($0.01)]$
$= $4171.5 + $5.45 + $0.20 = 4177.15. The cost of 100 g of yellow gold is
$100[0.75($55.62) + 0.125($1.09) + 0.125($0.01)]$
$= $4171.50 + $13.625 + $0.125 = 4185.25. The cost of 100 g of white gold is
$100[0.375($55.62) + 0.625($1.09)] = $2085.75 + 68.125
$= 2153.875. The numbers check.
State. The price of 1 g of gold is $55.62, of silver is $1.09 and of copper is $0.01.

23. *Familiarize*. Let r = the number of servings of roast beef, p = the number of baked potatoes, and b = the number of servings of broccoli. Then r servings of roast beef contain $300r$ Calories, $20r$ g of protein, and no vitamin C. In p baked potatoes there are $100p$ Calories, $5p$ g of protein, and $20p$ mg of vitamin C. And b servings of broccoli contain $50b$ Calories, $5b$ g of protein, and $100b$ mg of vitamin C. The patient requires 800 Calories, 55 g of protein, and 220 mg of vitamin C.
Translate. Write equations for the total number of

calories, the total amount of protein, and the total amount of vitamin C.

$$300r + 100p + 50b = 800 \quad \text{(Calories)}$$
$$20r + 5p + 5b = 55 \quad \text{(protein)}$$
$$20p + 100b = 220 \quad \text{(vitamin C)}$$

We now have a system of equations.

Carry out. Solving the system we get (2, 1, 2).

Check. Two servings of roast beef provide 600 Calories, 40 g of protein, and no vitamin C. One baked potato provides 100 Calories, 5 g of protein, and 20 mg of vitamin C. And 2 servings of broccoli provide 100 Calories, 10 g of protein, and 200 mg of vitamin C. Together, then, they provide 800 Calories, 55 g of protein, and 220 mg of vitamin C. The values check.

State. The dietician should prepare 2 servings of roast beef, 1 baked potato, and 2 servings of broccoli.

25. *Familiarize*. Let x, y, and z be the number of tickets sold for the first mezzanine, main floor and second mezzanine, respectively.

Translate.

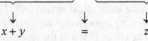

Total number of tickets is 40.

$$x + y + z = 40$$

Number of tickets for first mezzanine and main floor is number of tickets for second mezzanine.

$$x + y = z$$

Total cost of tickets is $1432.

$$52x + 38y + 28z = 1432$$

We have a system of equations:

$$x + y + z = 40$$
$$x + y = z$$
$$52x + 38y + 28z = 1432$$

Carry out. Solving the system we get (8, 12, 20).

Check. The total number of tickets is 8 + 12 + 20, or 40. The sum of first mezzanine and main floor is 8 + 12, is the number of second mezzanine, 20. The total cost is $52(8) + $38(12) + $28(20)

$= $416 + $456 + $560 = 1432. The numbers check.

State. There were 8 first mezzanine tickets, 12 main floor tickets and 20 second mezzanine tickets sold.

27. *Familiarize*. Let x, y, and z represent the populations of Asia, Africa, and the rest of the world, respectively, in billions, in 2050.

Translate.

The total world population will be 9.4 billion.

$$x + y + z = 9.4$$

Population of Asia will be 2.9 billion more than Population of Africa

$$x = 2.9 + y$$

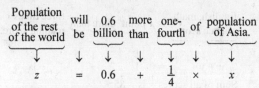

Population of the rest of the world will be 0.6 billion more than one-fourth of population of Asia.

$$z = 0.6 + \frac{1}{4} \times x$$

We have a system of equations.

$$x + y + z = 9.4$$
$$x = 2.9 + y$$
$$z = 0.6 + \frac{1}{4}x$$

Carry out. Solving the system we get (5.2, 2.3, 1.9).

Check. The total population will be 5.2 + 2.3 + 1.9, or 9.4 billion. The population of Asia, 5.2 billion, is 2.9 billion more than the population of Africa, 2.3 billion. Also, the rest of the world population, 1.9 billion is 0.6 billion more than $\frac{1}{4}$ of 5.2, or 1.3 billion. The numbers check.

State. In 2050, the population of Asia will be 5.2 billion, the population of Africa will be 2.3 billion and the population of the rest of the world will be 1.9 billion.

29. *Writing Exercise*.

31. Graph $y = 4$.

33. Graph $y - 3x = 3$.
    $$y = 3x + 3$$

35. Graph $f(x) = 2x - 1$.

37. *Writing Exercise*.

39. *Familiarize*. Let w, x, y, and z represent the ACT benchmarks for English, reading, mathematics, and science test scores, respectively.

Translate.

science score is 6 points more than English score.

$$z = 6 + w$$

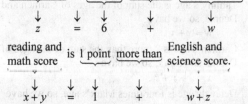

reading and math score is 1 point more than English and science score.

$$x + y = 1 + w + z$$

$$\underbrace{\text{English, math, science score}}\ \text{is}\ \underbrace{\text{1 point}}\ \underbrace{\text{more than}}\ \underbrace{\text{three times reading score.}}$$

$$w + y + z = 1 + 3x$$

$$\underline{\text{Sum of all scores}}\ \text{is}\ \underline{85.}$$

$$w + x + y + z = 85$$

We have a system of equations:

$$
\begin{aligned}
-w + z &= 6 \quad (1)\\
-w + x + y - z &= 1 \quad (2)\\
w - 3x + y + z &= 1 \quad (3)\\
w + x + y + z &= 85 \quad (4)
\end{aligned}
$$

Carry out. We solve the system of equations. First add Equations (2) and (4).

$$
\begin{aligned}
-w + x + y - z &= 1 \quad (2)\\
\underline{w - 3x + y + z} &= 1 \quad (3)\\
-2x + 2y &= 2 \quad (5)
\end{aligned}
$$

Next, add Equations (2) and (4)

$$
\begin{aligned}
-w + x + y - z &= 1 \quad (2)\\
\underline{w + x + y + z} &= 85 \quad (4)\\
2x + 2y &= 86 \quad (6)
\end{aligned}
$$

Then, add Equations (5) and (6).

$$
\begin{aligned}
-2x + 2y &= 2 \quad (5)\\
\underline{2x + 2y} &= 86 \quad (6)\\
4y &= 88\\
y &= 22
\end{aligned}
$$

Next substitute 22 for y in (6) and solve for x.

$$
\begin{aligned}
2x + 2(22) &= 86\\
2x &= 42\\
x &= 21
\end{aligned}
$$

Substitute 21 for x and 22 for y in (4).

$$
\begin{aligned}
w + 21 + 22 + z &= 85\\
w + z &= 42 \quad (7)
\end{aligned}
$$

Add Equations (1) and (7).

$$
\begin{aligned}
-w + z &= 6 \quad (1)\\
\underline{w + z} &= 42 \quad (7)\\
2z &= 48\\
z &= 24
\end{aligned}
$$

Finally, substitute 24 for z in (1) and solve for w.

$$
\begin{aligned}
-w + 24 &= 6\\
w &= 18
\end{aligned}
$$

The solution is (18, 21, 22, 24).
Check. The check is left to the student.
State. The benchmarks for English, reading, math, and science scores are 18, 21, 22, and 24, respectively.

41. *Familiarize*. Let $w, x, y,$ and z represent the ages of Tammy, Carmen, Dennis, and Mark respectively.
Translate.
Tammy's age is the sum of the ages of Carmen and Dennis, so we have
$$w = x + y.$$
Carmen's age is 2 more than the sum of the ages of Dennis and Mark, so we have
$$x = 2 + y + z.$$
Dennis's age is four times Mark's age, so we have

$$y = 4z.$$
The sum of all four ages is 42, so we have
$$w + x + y + z = 42.$$
Now we have a system of equations.

$$
\begin{aligned}
w &= x + y, \quad (1)\\
x &= 2 + y + z, \quad (2)\\
y &= 4z, \quad (3)\\
w + x + y + z &= 42 \quad (4)
\end{aligned}
$$

Carry out. We solve the system of equations. First we will express $w, x,$ and y in terms of z and then solve for z. From (3) we know that $y = 4z$. Substitute $4z$ for y in (2):
$$x = 2 + 4z + z = 2 + 5z.$$
Substitute $2 + 5z$ for x and $4z$ for y in (1):
$$w = 2 + 5z + 4z = 2 + 9z.$$
Now substitute $2 + 9z$ for w, $2 + 5z$ for x, and $4z$ for y in (4) and solve for z.

$$
\begin{aligned}
2 + 9z + 2 + 5z + 4z + z &= 42\\
19z + 4 &= 42\\
19z &= 38\\
z &= 2
\end{aligned}
$$

Then we have:
$$
\begin{aligned}
w &= 2 + 9z = 2 + 9 \cdot 2 = 20,\\
x &= 2 + 5z = 2 + 5 \cdot 2 = 12, \text{ and}\\
y &= 4z = 4 \cdot 2 = 8
\end{aligned}
$$
Although we were asked to find only Tammy's age, we found all of the ages so that we can check the result.
Check. The check is left to the student.
State. Tammy is 20 years old.

43. Let $T, G,$ and H represent the number of tickets Tom, Gary, and Hal begin with, respectively. After Hal gives tickets to Tom and Gary, each has the following number of tickets:

Tom: $T + T$, or $2T$,
Gary: $G + G$, or $2G$,
Hal: $H - T - G$.

After Tom gives tickets to Gary and Hal, each has the following number of tickets:

Gary: $2G + 2G$, or $4G$,
Hal: $(H - T - G) + (H - T - G)$, or $2(H - T - G)$,
Tom: $2T - 2G - (H - T - G)$, or $3T - H - G$

After Gary gives tickets to Hal and Tom, each has the following number of tickets:

Hal: $2(H - T - G) + 2(H - T - G)$, or $4(H - T - G)$
Tom: $(3T - H - G) + (3T - H - G)$, or $2(3T - H - G)$,
Gary: $4G - 2(H - T - G) - (3T - H - G)$, or $7G - H - T$.

Since Hal, Tom, and Gary each finish with 40 tickets, we write the following system of equations:

$$
\begin{aligned}
4(H - T - G) &= 40,\\
2(3T - H - G) &= 40,\\
7G - H - T &= 40
\end{aligned}
$$

Solving the system we find that $T = 35$, so Tom started with 35 tickets.

Exercise Set 3.6

1. matrix

3. entry

5. rows

7. $x + 2y = 11,$
$3x - y = 5$

Write a matrix using only the constants.

$$\begin{bmatrix} 1 & 2 & | & 11 \\ 3 & -1 & | & 5 \end{bmatrix}$$

Multiply the first row by -3 and add it to the second row.

$$\begin{bmatrix} 1 & 2 & | & 11 \\ 0 & -7 & | & -28 \end{bmatrix} \text{ New Row 2} = -3(\text{Row 1}) + \text{Row 2}$$

Reinserting the variables, we have

$x + 2y = 11,$ (1)
$-7y = -28$ (2)

Solve Equation (2) for y.
$-7y = -28$
$y = 4$

Substitute 4 for y in Equation (1) and solve for x.
$x + 2(4) = 11$
$x + 8 = 11$
$x = 3$

The solution is $(3, 4)$.

9. $3x + y = -1,$
$6x + 5y = 13$

We first write a matrix using only the constants.

$$\begin{bmatrix} 3 & 1 & | & -1 \\ 6 & 5 & | & 13 \end{bmatrix}$$

Multiply the first row by -2 and add it to the second row.

$$\begin{bmatrix} 3 & 1 & | & -1 \\ 0 & 3 & | & 15 \end{bmatrix} \text{ New Row 2} = -2(\text{Row 1}) + \text{Row 2}$$

Reinserting the variables, we have

$3x + y = -1,$ (1)
$3y = 15$ (2)

Solve Equation (2) for y.
$3y = 15$
$y = 5$

Substitute 5 for y in Equation (1) and solve for x.
$3x + 5 = -1$
$3x = -6$
$x = -2$

The solution is $(-2, 5)$.

11. $6x - 2y = 4,$
$7x + y = 13$

Write a matrix using only the constants.

$$\begin{bmatrix} 6 & -2 & | & 4 \\ 7 & 1 & | & 13 \end{bmatrix}$$

Multiply the second row by 6 to make the first number in row 2 a multiple of 6.

$$\begin{bmatrix} 6 & -2 & | & 4 \\ 42 & 6 & | & 78 \end{bmatrix} \text{ New Row 2} = 6(\text{Row 2})$$

Now multiply the first row by -7 and add it to the second row.

$$\begin{bmatrix} 6 & -2 & | & 4 \\ 0 & 20 & | & 50 \end{bmatrix} \text{ New Row 2} = -7(\text{Row 1}) + \text{Row 2}$$

Reinserting the variables, we have

$6x - 2y = 4,$ (1)
$20y = 50.$ (2)

Solve Equation (2) for y.
$20y = 50$
$y = \dfrac{5}{2}$

Substitute $\dfrac{5}{2}$ for y in Equation (1) and solve for x.

$6x - 2y = 4$
$6x - 2\left(\dfrac{5}{2}\right) = 4$
$6x - 5 = 4$
$6x = 9$
$x = \dfrac{3}{2}$

The solution is $\left(\dfrac{3}{2}, \dfrac{5}{2}\right)$.

13. $3x + 2y + 2z = 3,$
$x + 2y - z = 5,$
$2x - 4y + z = 0$

We first write a matrix using only the constants.

$$\begin{bmatrix} 3 & 2 & 2 & | & 3 \\ 1 & 2 & -1 & | & 5 \\ 2 & -4 & 1 & | & 0 \end{bmatrix}$$

First interchange rows 1 and 2 so that each number below the first number in the first row is a multiple of that number.

$$\begin{bmatrix} 1 & 2 & -1 & | & 5 \\ 3 & 2 & 2 & | & 3 \\ 2 & -4 & 1 & | & 0 \end{bmatrix}$$

Multiply row 1 by -3 and add it to row 2.
Multiply row 1 by -2 and add it to row 3.

$$\begin{bmatrix} 1 & 2 & -1 & | & 5 \\ 0 & -4 & 5 & | & -12 \\ 0 & -8 & 3 & | & -10 \end{bmatrix}$$

Multiply row 2 by -2 and add it to row 3.

$$\begin{bmatrix} 1 & 2 & -1 & | & 5 \\ 0 & -4 & 5 & | & -12 \\ 0 & 0 & -7 & | & 14 \end{bmatrix}$$

Reinserting the variables, we have

$x + 2y - z = 5,$ (1)
$-4y + 5z = -12,$ (2)
$-7z = 14.$ (3)

Solve (3) for z.
$-7z = 14$
$z = -2$

Substitute -2 for z in (2) and solve for y.

$$-4y + 5(-2) = -12$$
$$-4y - 10 = -12$$
$$-4y = -2$$
$$y = \frac{1}{2}$$

Substitute $\frac{1}{2}$ for y and -2 for z in (1) and solve for x.

$$x + 2 \cdot \frac{1}{2} - (-2) = 5$$
$$x + 1 + 2 = 5$$
$$x + 3 = 5$$
$$x = 2$$

The solution is $\left(2, \ \frac{1}{2}, \ -2\right)$.

15. $p - 2q - 3r = 3,$
$\quad\ 2p - q - 2r = 4,$
$\quad\ 4p + 5q + 6r = 4$

We first write a matrix using only the constants.

$$\begin{bmatrix} 1 & -2 & -3 & | & 3 \\ 2 & -1 & -2 & | & 4 \\ 4 & 5 & 6 & | & 4 \end{bmatrix}$$

$$\begin{bmatrix} 1 & -2 & -3 & | & 3 \\ 0 & 3 & 4 & | & -2 \\ 0 & 13 & 18 & | & -8 \end{bmatrix} \begin{matrix} \text{New Row } 2 = -2(\text{Row 1}) + \text{Row 2} \\ \text{New Row } 3 = -4(\text{Row 1}) + \text{Row 3} \end{matrix}$$

$$\begin{bmatrix} 1 & -2 & -3 & | & 3 \\ 0 & 3 & 4 & | & -2 \\ 0 & 39 & 54 & | & -24 \end{bmatrix} \quad \text{New Row 3} = 3(\text{Row 3})$$

$$\begin{bmatrix} 1 & -2 & -3 & | & 3 \\ 0 & 3 & 4 & | & -2 \\ 0 & 0 & 2 & | & 2 \end{bmatrix} \quad \text{New Row 3} = -13(\text{Row 2}) + \text{Row 3}$$

Reinserting the variables, we have

$$p - 2q - 3r = 3, \quad (1)$$
$$3q + 4r = -2, \quad (2)$$
$$2r = 2 \quad (3)$$

Solve (3) for r.
$$2r = 2$$
$$r = 1$$

Substitute 1 for r in (2) and solve for q.
$$3q + 4 \cdot 1 = -2$$
$$3q + 4 = -2$$
$$3q = -6$$
$$q = -2$$

Substitute -2 for q and 1 for r in (1) and solve for p.
$$p - 2(-2) - 3 \cdot 1 = 3$$
$$p + 4 - 3 = 3$$
$$p + 1 = 3$$
$$p = 2$$

The solution is $(2, -2, 1)$.

17. $3p + 2r = 11,$
$\quad\ q - 7r = 4,$
$\quad\ p - 6q = 1$

We first write a matrix using only the constants.

$$\begin{bmatrix} 3 & 0 & 2 & | & 11 \\ 0 & 1 & -7 & | & 4 \\ 1 & -6 & 0 & | & 1 \end{bmatrix}$$

$$\begin{bmatrix} 1 & -6 & 0 & | & 1 \\ 0 & 1 & -7 & | & 4 \\ 3 & 0 & 2 & | & 11 \end{bmatrix} \begin{matrix} \text{Interchange} \\ \text{Row 1 and Row 3} \end{matrix}$$

$$\begin{bmatrix} 1 & -6 & 0 & | & 1 \\ 0 & 1 & -7 & | & 4 \\ 0 & 18 & 2 & | & 8 \end{bmatrix} \quad \text{New Row 3} = -3(\text{Row 1}) + \text{Row 3}$$

$$\begin{bmatrix} 1 & -6 & 0 & | & 1 \\ 0 & 1 & -7 & | & 4 \\ 0 & 0 & 128 & | & -64 \end{bmatrix} \quad \text{New Row 3} = -18(\text{Row 2}) + \text{Row 3}$$

Reinserting the variables, we have

$$p - 6q = 1, \quad (1)$$
$$q - 7r = 4, \quad (2)$$
$$128r = -64. \quad (3)$$

Solve (3) for r.
$$128r = -64$$
$$r = -\frac{1}{2}$$

Substitute $-\frac{1}{2}$ for r in (2) and solve for q.
$$q - 7r = 4$$
$$q - 7\left(-\frac{1}{2}\right) = 4$$
$$q + \frac{7}{2} = 4$$
$$q = \frac{1}{2}$$

Substitute $\frac{1}{2}$ for q in (1) and solve for p.
$$p - 6 \cdot \frac{1}{2} = 1$$
$$p - 3 = 1$$
$$p = 4$$

The solution is $\left(4, \ \frac{1}{2}, -\frac{1}{2}\right)$.

19. We will rewrite the equations with the variables in alphabetical order:

$$-2w + 2x + 2y - 2z = -10,$$
$$w + x + y + z = -5,$$
$$3w + x - y + 4z = -2,$$
$$w + 3x - 2y + 2z = -6$$

Write a matrix using only the constants.

$$\begin{bmatrix} -2 & 2 & 2 & -2 & | & -10 \\ 1 & 1 & 1 & 1 & | & -5 \\ 3 & 1 & -1 & 4 & | & -2 \\ 1 & 3 & -2 & 2 & | & -6 \end{bmatrix}$$

$$\begin{bmatrix} -1 & 1 & 1 & -1 & | & -5 \\ 1 & 1 & 1 & 1 & | & -5 \\ 3 & 1 & -1 & 4 & | & -2 \\ 1 & 3 & -2 & 2 & | & -6 \end{bmatrix} \quad \text{New Row 1} = \frac{1}{2}(\text{Row 1})$$

$$\begin{bmatrix} -1 & 1 & 1 & -1 & | & -5 \\ 0 & 2 & 2 & 0 & | & -10 \\ 0 & 4 & 2 & 1 & | & -17 \\ 0 & 4 & -1 & 1 & | & -11 \end{bmatrix} \begin{matrix} \text{New Row 2} = \text{Row 1} + \text{Row 2} \\ \text{New Row 3} = 3(\text{Row 1}) + \text{Row 3} \\ \text{New Row 4} = \text{Row 1} + \text{Row 4} \end{matrix}$$

$$\begin{bmatrix} -1 & 1 & 1 & -1 & | & -5 \\ 0 & 2 & 2 & 0 & | & -10 \\ 0 & 0 & -2 & 1 & | & 3 \\ 0 & 0 & -5 & 1 & | & 9 \end{bmatrix} \begin{matrix} \\ \\ \text{New Row 3} = -2(\text{Row 2}) + \text{Row 3} \\ \text{New Row 4} = -2(\text{Row 2}) + \text{Row 4} \end{matrix}$$

$$\begin{bmatrix} -1 & 1 & 1 & -1 & | & -5 \\ 0 & 2 & 2 & 0 & | & -10 \\ 0 & 0 & -2 & 1 & | & 3 \\ 0 & 0 & -10 & 2 & | & 18 \end{bmatrix} \begin{matrix} \\ \\ \\ \text{New Row 4} = 2(\text{Row 4}) \end{matrix}$$

$$\begin{bmatrix} -1 & 1 & 1 & -1 & | & -5 \\ 0 & 2 & 2 & 0 & | & -10 \\ 0 & 0 & -2 & 1 & | & 3 \\ 0 & 0 & 0 & -3 & | & 3 \end{bmatrix} \begin{matrix} \\ \\ \\ \text{New Row 4} = -5(\text{Row 3}) + \text{Row 4} \end{matrix}$$

Reinserting the variables, we have

$$\begin{aligned} -w + x + y - z &= -5, \ (1) \\ 2x + 2y &= -10, \ (2) \\ -2y + z &= 3, \ (3) \\ -3z &= 3. \ (4) \end{aligned}$$

Solve (4) for z.

$$\begin{aligned} -3z &= 3 \\ z &= -1 \end{aligned}$$

Substitute -1 for z in (3) and solve for y.

$$\begin{aligned} -2y + (-1) &= 3 \\ -2y &= 4 \\ y &= -2 \end{aligned}$$

Substitute -2 for y in (2) and solve for x.

$$\begin{aligned} 2x + 2(-2) &= -10 \\ 2x - 4 &= -10 \\ 2x &= -6 \\ x &= -3 \end{aligned}$$

Substitute -3 for x, -2 for y, and -1 for z in (1) and solve for w.

$$\begin{aligned} -w + (-3) + (-2) - (-1) &= -5 \\ -w - 3 - 2 + 1 &= -5 \\ -w - 4 &= -5 \\ -w &= -1 \\ w &= 1 \end{aligned}$$

The solution is $(1, -3, -2, -1)$.

21. *Familiarize*. Let d = the number of dimes and n = the number of nickels. The value of d dimes is $\$0.10d$, and the value of n nickels is $\$0.05n$.

Translate.

Total number of coins is 42.

$$d + n = 42$$

Total value of coins is $3.

$$0.10d + 0.05n = 3$$

After clearing decimals, we have this system.

$$\begin{aligned} d + n &= 42, \\ 10d + 5n &= 300 \end{aligned}$$

Carry out. Solve using matrices.

$$\begin{bmatrix} 1 & 1 & | & 42 \\ 10 & 5 & | & 300 \end{bmatrix}$$

$$\begin{bmatrix} 1 & 1 & | & 42 \\ 0 & -5 & | & -120 \end{bmatrix} \ \text{New Row 2} = -10(\text{Row 1}) + \text{Row 2}$$

Reinserting the variables, we have

$$\begin{aligned} d + n &= 42, \quad (1) \\ -5n &= -120 \quad (2) \end{aligned}$$

Solve (2) for n.

$$\begin{aligned} -5n &= -120 \\ n &= 24 \\ d + 24 &= 42 \quad \text{Back-substituting} \\ d &= 18 \end{aligned}$$

Check. The sum of the two numbers is 42. The total value is $\$0.10(18) + \$0.05(24) = \$1.80 + \$1.20 = \$3$. The numbers check.

State. There are 18 dimes and 24 nickels.

23. *Familiarize*. Let x = the number of pounds of dried fruit and y = the number of pounds of macadamia nuts. We organize the information in a table.

	Dried fruit	Macadamia nuts	Mixture
Number of pounds	x	y	15
Price per pound	$5.80	$14.75	$9.38
Value of mixture	$5.80x$	$14.75y$	140.70

Translate.

The total number of pounds is 15.

$$x + y = 15$$

The total value of the mixture is $140.70.

$$5.80x + 14.75y = 140.70$$

After clearing decimals, we have this system:

$$\begin{aligned} x + y &= 15, \\ 580x + 1475y &= 14{,}070 \end{aligned}$$

Carry out. Solve using matrices.

$$\begin{bmatrix} 1 & 1 & | & 15 \\ 580 & 1475 & | & 14{,}070 \end{bmatrix}$$

Multiply row 1 by -580 and add it to the second row.

$$\begin{bmatrix} 1 & 1 & | & 15 \\ 0 & 895 & | & 5370 \end{bmatrix} \ \text{New Row 2} = -580(\text{Row 1}) + \text{Row 2}$$

Reinserting the variables, we have

$$\begin{aligned} x + y &= 15 \\ 895y &= 5370 \\ y &= 6 \end{aligned}$$

Back-substitute 6 for y in Equation (1) and solve for x.

$$\begin{aligned} x + 6 &= 15 \\ x &= 9 \end{aligned}$$

Check. The sum of the numbers, $6 + 9 = 15$. The total value is $\$5.90(9) + \$14.75(6)$, or $\$52.20 + \$88.50$, or $\$140.70$. The numbers check.

State. 9 pounds of dried fruit and 6 pounds of macadamia nuts should be used.

25. *Familiarize*. We let x, y, and z represent the amounts invested at 3%, 4%, and 5%, respectively. Recall the formula for simple interest:

Interest = Principal × Rate × Time

Translate. We organize the information in a table.

	First Investment	Second Investment	Third Investment	Total
P	x	y	z	$2500
R	3%	4%	5%	
T	1 yr	1 yr	1 yr	
I	$0.03x$	$0.04y$	$0.05z$	$112

The first row gives us one equation:

$x + y + z = 2500$

The last row gives a second equation:

$0.03x + 0.04y + 0.05z = 112$

Amount invested at 5% is $1100 more than amount invested at 4%.

$z = 1100 + y$

After clearing decimals, we have this system:

$x + y + z = 2500,$
$3x + 4y + 5z = 11,200,$
$-y + z = 1100$

Carry out. Solve using matrices.

$$\begin{bmatrix} 1 & 1 & 1 & | & 2500 \\ 3 & 4 & 5 & | & 11,200 \\ 0 & -1 & 1 & | & 1100 \end{bmatrix}$$

$$\begin{bmatrix} 1 & 1 & 1 & | & 2500 \\ 0 & 1 & 2 & | & 3700 \\ 0 & -1 & 1 & | & 1100 \end{bmatrix}$$ New Row 2 $= -3$(Row 1) + Row 2

$$\begin{bmatrix} 1 & 1 & 1 & | & 2500 \\ 0 & 1 & 2 & | & 3700 \\ 0 & 0 & 3 & | & 4800 \end{bmatrix}$$ New Row 3 = Row 2 + Row 3

Reinserting the variables, we have

$x + y + z = 2500,$ (1)
$y + 2z = 3700,$ (2)
$3z = 4800$ (3)

Solve (3) for z.

$3z = 4800$
$z = 1600$

Back-substitute 1600 for z in (2) and solve for y.

$y + 2 \cdot 1600 = 3700$
$y + 3200 = 3700$
$y = 500$

Back-substitute 500 for y and 1600 for z in (1) and solve for x.

$x + 500 + 1600 = 2500$
$x + 2100 = 2500$
$x = 400$

Check. The total investment is $400 + $500 + $1600, or $2500. The total interest is 0.03($400) + 0.04($500) + 0.05($1600) = $12 + $20 + $80 = $112. The amount invested at 5%, $1600, is $1100 more than the amount invested at 4%, $500. The numbers check.
State. $400 is invested at 3%, $500 is invested at 4%, and $1600 is invested at 5%.

27. *Writing Exercise.*

29. $(-7)^2 = 49$

31. $-7^2 = -49$

33. *Writing Exercise.*

35. *Familiarize.* Let $w, x, y,$ and z represent the thousand's, hundred's, ten's, and one's digits, respectively. *Translate.*

The sum of the digits is 10.

$w + x + y + z = 10$

Twice the sum of the thousand's and ten's digits is the sum of the hundred's and one's digits less one.

$2(w + y) = x + z - 1$

The ten's digit is twice the thousand's digit.

$y = 2 \cdot w$

The one's digit equals the sum of the thousand's and hundred's digits.

$z = w + x$

We have a system of equations which can be written as

$w + x + y + z = 10,$
$2w - x + 2y - z = -1,$
$-2w + y = 0,$
$w + x - z = 0.$

Carry out. We can use matrices to solve the system. We get (1, 3, 2, 4).
Check. The sum of the digits is 10. Twice the sum of 1 and 2 is 6. This is one less than the sum of 3 and 4. The ten's digit, 2, is twice the thousand's digit, 1. The one's digit, 4, equals 1 + 3. The numbers check.
State. The number is 1324.

Exercise Set 3.7

1. True

3. True

5. False; see page 205 in the text.

7. $\begin{vmatrix} 3 & 5 \\ 4 & 8 \end{vmatrix} = 3 \cdot 8 - 4 \cdot 5 = 24 - 20 = 4$

9. $\begin{vmatrix} 10 & 8 \\ -5 & -9 \end{vmatrix} = 10(-9) - 8(-5) = -90 + 40 = -50$

11. $\begin{vmatrix} 1 & 4 & 0 \\ 0 & -1 & 2 \\ 3 & -2 & 1 \end{vmatrix}$

$= 1\begin{vmatrix} -1 & 2 \\ -2 & 1 \end{vmatrix} - 0\begin{vmatrix} 4 & 0 \\ -2 & 1 \end{vmatrix} + 3\begin{vmatrix} 4 & 0 \\ -1 & 2 \end{vmatrix}$

$= 1[-1 \cdot 1 - (-2) \cdot 2] - 0 + 3[4 \cdot 2 - (-1) \cdot 0]$

$= 1 \cdot 3 - 0 + 3 \cdot 8$

$= 3 - 0 + 24$

$= 27$

13. $\begin{vmatrix} -1 & -2 & -3 \\ 3 & 4 & 2 \\ 0 & 1 & 2 \end{vmatrix}$

$= -1\begin{vmatrix} 4 & 2 \\ 1 & 2 \end{vmatrix} - 3\begin{vmatrix} -2 & -3 \\ 1 & 2 \end{vmatrix} + 0\begin{vmatrix} -2 & -3 \\ 4 & 2 \end{vmatrix}$

$= -1[4 \cdot 2 - 1 \cdot 2] - 3[-2 \cdot 2 - 1(-3)] + 0$

$= -1 \cdot 6 - 3 \cdot (-1) + 0$

$= -6 + 3 + 0$

$= -3$

15. $\begin{vmatrix} -4 & -2 & 3 \\ -3 & 1 & 2 \\ 3 & 4 & -2 \end{vmatrix}$

$= -4\begin{vmatrix} 1 & 2 \\ 4 & -2 \end{vmatrix} - (-3)\begin{vmatrix} -2 & 3 \\ 4 & -2 \end{vmatrix} + 3\begin{vmatrix} -2 & 3 \\ 1 & 2 \end{vmatrix}$

$= -4[1(-2) - 4 \cdot 2] + 3[-2(-2) - 4 \cdot 3]$
$\qquad + 3[-2 \cdot 2 - 1 \cdot 3]$

$= -4(-10) + 3(-8) + 3(-7)$

$= 40 - 24 - 21 = -5$

17. $5x + 8y = 1,$
$3x + 7y = 5$

We compute D, D_x, and D_y.

$D = \begin{vmatrix} 5 & 8 \\ 3 & 7 \end{vmatrix} = 35 - 24 = 11$

$D_x = \begin{vmatrix} 1 & 8 \\ 5 & 7 \end{vmatrix} = 7 - 40 = -33$

$D_y = \begin{vmatrix} 5 & 1 \\ 3 & 5 \end{vmatrix} = 25 - 3 = 22$

Then,

$\qquad x = \dfrac{D_x}{D} = \dfrac{-33}{11} = -3$ and $y = \dfrac{D_y}{D} = \dfrac{22}{11} = 2$.

The solution is $(-3, 2)$.

19. $5x - 4y = -3,$
$7x + 2y = 6$

We compute D, D_x, and D_y.

$D = \begin{vmatrix} 5 & -4 \\ 7 & 2 \end{vmatrix} = 10 - (-28) = 38$

$D_x = \begin{vmatrix} -3 & -4 \\ 6 & 2 \end{vmatrix} = -6 - (-24) = 18$

$D_y = \begin{vmatrix} 5 & -3 \\ 7 & 6 \end{vmatrix} = 30 - (-21) = 51$

Then,

$x = \dfrac{D_x}{D} = \dfrac{18}{38} = \dfrac{9}{19}$

and

$y = \dfrac{D_y}{D} = \dfrac{51}{38}$.

The solution is $\left(\dfrac{9}{19}, \dfrac{51}{38}\right)$.

21. $\begin{aligned} 3x - y + 2z &= 1, \\ x - y + 2z &= 3, \\ -2x + 3y + z &= 1 \end{aligned}$

We compute D, D_x, D_y and D_z.

$D = \begin{vmatrix} 3 & -1 & 2 \\ 1 & -1 & 2 \\ -2 & 3 & 1 \end{vmatrix}$

$= 3\begin{vmatrix} -1 & 2 \\ 3 & 1 \end{vmatrix} - 1\begin{vmatrix} -1 & 2 \\ 3 & 1 \end{vmatrix} - 2\begin{vmatrix} -1 & 2 \\ -1 & 2 \end{vmatrix}$

$= 3(-7) - 1(-7) - 2(0)$

$= -21 + 7 - 0$

$= -14$

$D_x = \begin{vmatrix} 1 & -1 & 2 \\ 3 & -1 & 2 \\ 1 & 3 & 1 \end{vmatrix}$

$= 1\begin{vmatrix} -1 & 2 \\ 3 & 1 \end{vmatrix} - 3\begin{vmatrix} -1 & 2 \\ 3 & 1 \end{vmatrix} + 1\begin{vmatrix} -1 & 2 \\ -1 & 2 \end{vmatrix}$

$= 1(-7) - 3(-7) + 1(0)$

$= -7 + 21 + 0$

$= 14$

$D_y = \begin{vmatrix} 3 & 1 & 2 \\ 1 & 3 & 2 \\ -2 & 1 & 1 \end{vmatrix}$

$= 3\begin{vmatrix} 3 & 2 \\ 1 & 1 \end{vmatrix} - 1\begin{vmatrix} 1 & 2 \\ 1 & 1 \end{vmatrix} - 2\begin{vmatrix} 1 & 2 \\ 3 & 2 \end{vmatrix}$

$= 3 \cdot 1 - 1(-1) - 2(-4)$

$= 3 + 1 + 8$

$= 12$

$D_z = \begin{vmatrix} 3 & -1 & 1 \\ 1 & -1 & 3 \\ -2 & 3 & 1 \end{vmatrix}$

$= 3\begin{vmatrix} -1 & 3 \\ 3 & 1 \end{vmatrix} - 1\begin{vmatrix} -1 & 1 \\ 3 & 1 \end{vmatrix} - 2\begin{vmatrix} -1 & 1 \\ -1 & 3 \end{vmatrix}$

$= 3(-10) - 1(-4) - 2(-2)$

$= -30 + 4 + 4$

$= -22$

Then,

$x = \dfrac{D_x}{D} = \dfrac{14}{-14} = -1$

and

$y = \dfrac{D_y}{D} = \dfrac{12}{-14} = -\dfrac{6}{7}$

and

$z = \dfrac{D_z}{D} = \dfrac{-22}{-14} = \dfrac{11}{7}$.

The solution is $\left(-1, -\dfrac{6}{7}, \dfrac{11}{7}\right)$.

23. $2x - 3y + 5z = 27,$
$\quad x + 2y - z = -4,$
$\quad 5x - y + 4z = 27$

We compute D, D_x, D_y and D_z.

$$D = \begin{vmatrix} 2 & -3 & 5 \\ 1 & 2 & -1 \\ 5 & -1 & 4 \end{vmatrix}$$

$$= 2\begin{vmatrix} 2 & -1 \\ -1 & 4 \end{vmatrix} - 1\begin{vmatrix} -3 & 5 \\ -1 & 4 \end{vmatrix} + 5\begin{vmatrix} -3 & 5 \\ 2 & -1 \end{vmatrix}$$

$$= 2(7) - 1(-7) + 5(-7)$$
$$= 14 + 7 - 35 = -14$$

$$D_x = \begin{vmatrix} 27 & -3 & 5 \\ -4 & 2 & -1 \\ 27 & -1 & 4 \end{vmatrix}$$

$$= 27\begin{vmatrix} 2 & -1 \\ -1 & 4 \end{vmatrix} - (-4)\begin{vmatrix} -3 & 5 \\ -1 & 4 \end{vmatrix} + 27\begin{vmatrix} -3 & 5 \\ 2 & -1 \end{vmatrix}$$

$$= 27(7) + 4(-7) + 27(-7)$$
$$= 189 - 28 - 189$$
$$= -28$$

$$D_y = \begin{vmatrix} 2 & 27 & 5 \\ 1 & -4 & -1 \\ 5 & 27 & 4 \end{vmatrix}$$

$$= 2\begin{vmatrix} -4 & -1 \\ 27 & 4 \end{vmatrix} - 1\begin{vmatrix} 27 & 5 \\ 27 & 4 \end{vmatrix} + 5\begin{vmatrix} 27 & 5 \\ -4 & -1 \end{vmatrix}$$

$$= 2(11) - 1(-27) + 5(-7)$$
$$= 22 + 27 - 35$$
$$= 14$$

$$D_z = \begin{vmatrix} 2 & -3 & 27 \\ 1 & 2 & -4 \\ 5 & -1 & 27 \end{vmatrix}$$

$$= 2\begin{vmatrix} 2 & -4 \\ -1 & 27 \end{vmatrix} - 1\begin{vmatrix} -3 & 27 \\ -1 & 27 \end{vmatrix} + 5\begin{vmatrix} -3 & 27 \\ 2 & -4 \end{vmatrix}$$

$$= 2(50) - 1(-54) + 5(-42)$$
$$= 100 + 54 - 210$$
$$= -56$$

Then,

$$x = \frac{D_x}{D} = \frac{-28}{-14} = 2$$

and

$$y = \frac{D_y}{D} = \frac{14}{-14} = -1$$

and

$$z = \frac{D_z}{D} = \frac{-56}{-14} = 4.$$

The solution is $(2, -1, 4)$.

25. $r - 2s + 3t = 6,$
$\quad 2r - s - t = -3,$
$\quad r + s + t = 6$

We compute D, D_r, D_s and D_t.

$$D = \begin{vmatrix} 1 & -2 & 3 \\ 2 & -1 & -1 \\ 1 & 1 & 1 \end{vmatrix}$$

$$= 1\begin{vmatrix} -1 & -1 \\ 1 & 1 \end{vmatrix} - 2\begin{vmatrix} -2 & 3 \\ 1 & 1 \end{vmatrix} + 1\begin{vmatrix} -2 & 3 \\ -1 & -1 \end{vmatrix}$$

$$= 1(0) - 2(-5) + 1(5)$$
$$= 0 + 10 + 5$$
$$= 15$$

$$D_r = \begin{vmatrix} 6 & -2 & 3 \\ -3 & -1 & -1 \\ 6 & 1 & 1 \end{vmatrix}$$

$$= 6\begin{vmatrix} -1 & -1 \\ 1 & 1 \end{vmatrix} - (-3)\begin{vmatrix} -2 & 3 \\ 1 & 1 \end{vmatrix} + 6\begin{vmatrix} -2 & 3 \\ -1 & -1 \end{vmatrix}$$

$$= 6(0) + 3(-5) + 6(5)$$
$$= 0 - 15 + 30$$
$$= 15$$

$$D_s = \begin{vmatrix} 1 & 6 & 3 \\ 2 & -3 & -1 \\ 1 & 6 & 1 \end{vmatrix}$$

$$= 1\begin{vmatrix} -3 & -1 \\ 6 & 1 \end{vmatrix} - 2\begin{vmatrix} 6 & 3 \\ 6 & 1 \end{vmatrix} + 1\begin{vmatrix} 6 & 3 \\ -3 & -1 \end{vmatrix}$$

$$= 1(3) - 2(-12) + 1(3)$$
$$= 3 + 24 + 3$$
$$= 30$$

$$D_t = \begin{vmatrix} 1 & -2 & 6 \\ 2 & -1 & -3 \\ 1 & 1 & 6 \end{vmatrix}$$

$$= 1\begin{vmatrix} -1 & -3 \\ 1 & 6 \end{vmatrix} - 2\begin{vmatrix} -2 & 6 \\ 1 & 6 \end{vmatrix} + 1\begin{vmatrix} -2 & 6 \\ -1 & -3 \end{vmatrix}$$

$$= 1(-3) - 2(-18) + 1(12)$$
$$= -3 + 36 + 12$$
$$= 45$$

Then,

$$r = \frac{D_r}{D} = \frac{15}{15} = 1$$

and

$$s = \frac{D_s}{D} = \frac{30}{15} = 2$$

and

$$t = \frac{D_t}{D} = \frac{45}{15} = 3.$$

The solution is $(1, 2, 3)$.

27. *Writing Exercise.*

29. $f(x) = \frac{1}{2}x - 10$

31. $m = \frac{8 - 0}{-2 - 3} = -\frac{8}{5}$

$$y - 0 = -\frac{8}{5}(x - 3)$$

$$y = -\frac{8}{5}x + \frac{24}{5}$$

$$f(x) = -\frac{8}{5}x + \frac{24}{5}$$

33. *Writing Exercise.*

35. $\begin{vmatrix} y & -2 \\ 4 & 3 \end{vmatrix} = 44$

$\quad y \cdot 3 - 4(-2) = 44$ Evaluating the determinant
$\quad\quad 3y + 8 = 44$
$\quad\quad\quad 3y = 36$
$\quad\quad\quad\quad y = 12$

37. $\begin{vmatrix} m+1 & -2 \\ m-2 & 1 \end{vmatrix} = 27$

$\quad (m+1)(1) - (m-2)(-2) = 27$ Evaluating
$\quad\quad\quad\quad\quad\quad\quad\quad\quad$ the determinant
$\quad\quad m+1+2m-4 = 27$
$\quad\quad\quad\quad\quad\quad 3m = 30$
$\quad\quad\quad\quad\quad\quad m = 10$

Exercise Set 3.8

1. b

3. h

5. e

7. c

9. $C(x) = 35x + 200,000 \quad R(x) = 55x$

  a. $P(x) = R(x) - C(x)$
$\quad\quad\quad\quad = 55x - (35x + 200,000)$
$\quad\quad\quad\quad = 55x - 35x - 200,000$
$\quad\quad\quad\quad = 20x - 200,000$

  b. Solve the system
$\quad\quad R(x) = 55x,$
$\quad\quad C(x) = 35x + 200,000.$

Since both $R(x)$ and $C(x)$ are in dollars and they are equal at the break-even point, we can rewrite the system:

$\quad\quad d = 55x, \quad\quad\quad\quad (1)$
$\quad\quad d = 35x + 200,000 \quad (2)$

We solve using substitution.

$\quad\quad 55x = 35x + 200,000$ Substituting 55x for
$\quad\quad\quad\quad\quad\quad\quad\quad\quad\quad\quad d$ in (2)
$\quad\quad 20x = 200,000$
$\quad\quad\quad x = 10,000$

Thus, 10,000 units must be produced and sold in order to break even.
The revenue will be
$R(10,000) = 55 \cdot 10,000 = 550,000$.
The break-even point is (10,000 units, \$550,000).

11. $C(x) = 15x + 3100 \quad R(x) = 40x$

  a. $P(x) = R(x) - C(x)$
$\quad\quad\quad\quad = 40x - (15x + 3100)$
$\quad\quad\quad\quad = 40x - 15x - 3100$
$\quad\quad\quad\quad = 25x - 3100$

  b. Solve the system
$\quad\quad R(x) = 40x,$
$\quad\quad C(x) = 15x + 3100.$

Since both $R(x)$ and $C(x)$ are in dollars and they are equal at the break-even point, we can rewrite the

system:
$\quad\quad d = 40x, \quad\quad\quad\quad (1)$
$\quad\quad d = 15x + 3100 \quad (2)$

We solve using substitution.
$\quad\quad 40x = 15x + 3100$ Substituting 40x for
$\quad\quad\quad\quad\quad\quad\quad\quad\quad\quad d$ in (2)
$\quad\quad 25x = 3100$
$\quad\quad\quad x = 124$

Thus, 124 units must be produced and sold in order to break even.
The revenue will be $R(124) = 40 \cdot 124 = 4960$.
The break-even point is (124 units, \$4960).

13. $C(x) = 40x + 22,500 \quad R(x) = 85x$

  a. $P(x) = R(x) - C(x)$
$\quad\quad\quad\quad = 85x - (40x + 22,500)$
$\quad\quad\quad\quad = 85x - 40x - 22,500$
$\quad\quad\quad\quad = 45x - 22,500$

  b. Solve the system
$\quad\quad R(x) = 85x,$
$\quad\quad C(x) = 40x + 22,500.$

Since both $R(x)$ and $C(x)$ are in dollars and they are equal at the break-even point, we can rewrite the system:

$\quad\quad d = 85x, \quad\quad\quad\quad (1)$
$\quad\quad d = 40x + 22,500 \quad (2)$

We solve using substitution.

$\quad\quad 85x = 40x + 22,500$ Substituting 85x for d in (2)
$\quad\quad 45x = 22,500$
$\quad\quad\quad x = 500$

Thus, 500 units must be produced and sold in order to break even.
The revenue will be $R(500) = 85 \cdot 500 = 42,500$.
The break-even point is (500 units, \$42,500).

15. $C(x) = 24x + 50,000 \quad R(x) = 40x$

  a. $P(x) = R(x) - C(x)$
$\quad\quad\quad\quad = 40x - (24x + 50,000)$
$\quad\quad\quad\quad = 40x - 24x - 50,000$
$\quad\quad\quad\quad = 16x - 50,000$

  b. Solve the system
$\quad\quad R(x) = 40x,$
$\quad\quad C(x) = 24x + 50,000.$

Since both $R(x)$ and $C(x)$ are in dollars and they are equal at the break-even point, we can rewrite the system:

$\quad\quad d = 40x, \quad\quad\quad\quad (1)$
$\quad\quad d = 24x + 50,000 \quad (2)$

We solve using substitution.

$\quad\quad 40x = 24x + 50,000$ Substituting 40x for d in (2)
$\quad\quad 16x = 50,000$
$\quad\quad\quad x = 3125$

Thus, 3125 units must be produced and sold in order to break even.
The revenue will be $R(3125) = 40 \cdot 3125 = \$125,000$.
The break-even point is (3125 units, \$125,000).

17. $C(x) = 75x + 100,000$ $R(x) = 125x$

a.
$$P(x) = R(x) - C(x)$$
$$= 125x - (75x + 100,000)$$
$$= 125x - 75x - 100,000$$
$$= 50x - 100,000$$

b.   Solve the system
$$R(x) = 125x,$$
$$C(x) = 75x + 100,000.$$

Since $R(x) = C(x)$ at the break-even point, we can rewrite the system:
$$R(x) = 125x,   \quad (1)$$
$$C(x) = 75x + 100,000 \quad (2)$$

We solve using substitution.
$$125x = 75x + 100,000 \quad \text{Substituting } 125x$$
$$50x = 100,000 \qquad\quad \text{for } R(x) \text{ in (2)}$$
$$x = 2000$$

To break even 2000 units must be produced and sold.
The revenue will be $R(2000) = 125 \cdot 2000 = 250,000$.
The break-even point is (2000 units, $250,000).

19. $D(p) = 2000 - 15p,$
$S(p) = 740 + 6p$

Rewrite the system:
$$q = 2000 - 15p, \quad (1)$$
$$q = 740 + 6p \qquad (2)$$

Substitute $2000 - 15p$ for q in (2) and solve.
$$2000 - 15p = 740 + 6p$$
$$1260 = 21p$$
$$60 = p$$

The equilibrium price is $60 per unit.
To find the equilibrium quantity we substitute $60 into either $D(p)$ or $S(p)$.
$$D(60) = 2000 - 15(60) = 2000 - 900 = 1100$$

The equilibrium quantity is 1100 units.
The equilibrium point is ($60, 1100).

21. $D(p) = 760 - 13p,$
$S(p) = 430 + 2p$

Rewrite the system:
$$q = 760 - 13p, \quad (1)$$
$$q = 430 + 2p \qquad (2)$$

Substitute $760 - 13p$ for q in (2) and solve.
$$760 - 13p = 430 + 2p$$
$$330 = 15p$$
$$22 = p$$

The equilibrium price is $22 per unit.
To find the equilibrium quantity we substitute $22 into either $D(p)$ or $S(p)$.
$$S(22) = 430 + 2(22) = 430 + 44 = 474$$

The equilibrium quantity is 474 units.
The equilibrium point is ($22, 474).

23. $D(p) = 7500 - 25p,$
$S(p) = 6000 + 5p$

Rewrite the system:
$$q = 7500 - 25p, \quad (1)$$
$$q = 6000 + 5p \qquad (2)$$

Substitute $7500 - 25p$ for q in (2) and solve.
$$7500 - 25p = 6000 + 5p$$
$$1500 = 30p$$
$$50 = p$$

The equilibrium price is $50 per unit.
To find the equilibrium quantity we substitute $50 into either $D(p)$ or $S(p)$.
$$D(50) = 7500 - 25(50) = 7500 - 1250 = 6250$$

The equilibrium quantity is 6250 units.
The equilibrium point is ($50, 6250).

25. $D(p) = 1600 - 53p,$
$S(p) = 320 + 75p$

Rewrite the system:
$$q = 1600 - 53p, \quad (1)$$
$$q = 320 + 75p \qquad (2)$$

Substitute $1600 - 53p$ for q in (2) and solve.
$$1600 - 53p = 320 + 75p$$
$$1280 = 128p$$
$$10 = p$$

The equilibrium price is $10 per unit.
To find the equilibrium quantity we substitute $10 into either $D(p)$ or $S(p)$.
$$S(10) = 320 + 75(10) = 320 + 750 = 1070$$

The equilibrium quantity is 1070 units.
The equilibrium point is ($10, 1070).

27. a. $C(x) = $ Fixed costs $+$ Variable costs
$$C(x) = 45,000 + 40x,$$
where x is the number of cell phones.

b.   Each cell phone sells for $130. The total revenue is 130 times the number of cell phone sold. We assume that all cell phones produced are sold.
$$R(x) = 130x$$

c.
$$P(x) = R(x) - C(x)$$
$$P(x) = 130x - (45,000 + 40x)$$
$$= 130x - 45,000 - 40x$$
$$= 90x - 45,000$$

d.
$$P(3000) = 90(3000) - 45,000$$
$$= 270,000 - 45,000$$
$$= \$225,000$$

The company will realize a profit of $225,000 when 3000 cell phones are produced and sold.
$$P(400) = 90(400) - 45,000$$
$$= 36,000 - 45,000$$
$$= -\$9000$$

The company will realize a $9000 loss when 400 cell phones are produced and sold.

e.   Solve the system
$$R(x) = 130x,$$
$$C(x) = 45,000 + 40x.$$

Since both $R(x)$ and $C(x)$ are in dollars and they are equal at the break-even point, we can rewrite the system:
$$d = 130x,   \quad (1)$$
$$d = 45,000 + 40x \quad (2)$$

We solve using substitution.

$130x = 45,000 + 40x$ Substituting $130x$ for d in (2)

$90x = 45,000$

$x = 500$

The firm will break even if it produces and sells 500 cell phones and takes in a total of $R(500) = 130 \cdot 500 = \$65,000$ in revenue. Thus, the break-even point is (500 cell phones, \$65,000).

29. a. $C(x) = $ Fixed costs $+$ Variable costs

$C(x) = 10,000 + 30x,$

where x is the number of pet car seats produced.

b.   Each pet car seat sells for \$80. The total revenue is 80 times the number of seats sold. We assume that all seats produced are sold.

$R(x) = 80x$

c. $P(x) = R(x) - C(x)$

$P(x) = 80x - (10,000 + 30x)$

$= 80x - 10,000 - 30x$

$= 50x - 10,000$

d. $P(2000) = 50(2000) - 10,000$

$= 100,000 - 10,000$

$= 90,000$

The company will realize a profit of \$90,000 when 2000 seats are produced and sold.

$P(50) = 50(50) - 10,000$

$= 2500 - 10,000$

$= -7500$

The company will realize a loss of \$7500 when 50 seats are produced and sold.

e.   Solve the system

$R(x) = 80x,$

$C(x) = 10,000 + 30x.$

Since both $R(x)$ and $C(x)$ are in dollars and they are equal at the break-even point, we can rewrite the system:

$d = 80x,$ (1)

$d = 10,000 + 30x$ (2)

We solve using substitution.

$80x = 10,000 + 30x$ Substituting $80x$ for d in (2)

$50x = 10,000$

$x = 200$

The firm will break even if it produces and sells 200 seats and takes in a total of $R(200) = 80 \cdot 200 = \$16,000$ in revenue. Thus, the break-even point is (200 seats, \$16,000).

31. *Writing Exercise.*

33. $(1.25 \times 10^{-15})(8 \times 10^4) = (1.25 \times 8)(10^{-15} \times 10^4)$

$= 10 \times 10^{-11}$

$= 1 \times 10^{-10}$

35. $C = \dfrac{2}{3}(x - y)$

$\dfrac{3}{2} C = \dfrac{3}{2} \cdot \dfrac{2}{3}(x - y)$

$\dfrac{3}{2} C = x - y$

$y = x - \dfrac{3}{2} C$

37. *Writing Exercise.*

39. The supply function contains the points (\$2, 100) and (\$8, 500). We find its equation:

$m = \dfrac{500 - 100}{8 - 2} = \dfrac{400}{6} = \dfrac{200}{3}$

$y - y_1 = m(x - x_1)$ Point-slope form

$y - 100 = \dfrac{200}{3}(x - 2)$

$y - 100 = \dfrac{200}{3} x - \dfrac{400}{3}$

$y = \dfrac{200}{3} x - \dfrac{100}{3}$

We can equivalently express supply S as a function of price p:

$S(p) = \dfrac{200}{3} p - \dfrac{100}{3}$

The demand function contains the points (\$1, 500) and (\$9, 100). We find its equation:

$m = \dfrac{100 - 500}{9 - 1} = \dfrac{-400}{8} = -50$

$y - y_1 = m(x - x_1)$

$y - 500 = -50(x - 1)$

$y - 500 = -50x + 50$

$y = -50x + 550$

We can equivalently express demand D as a function of price p:

$D(p) = -50p + 550$

We have a system of equations

$S(p) = \dfrac{200}{3} p - \dfrac{100}{3},$

$D(p) = -50p + 550.$

Rewrite the system:

$q = \dfrac{200}{3} p - \dfrac{100}{3},$ (1)

$q = -50p + 550$ (2)

Substitute $\dfrac{200}{3} p - \dfrac{100}{3}$ for q in (2) and solve.

$\dfrac{200}{3} p - \dfrac{100}{3} = -50p + 550$

$200p - 100 = -150p + 1650$ Multiplying by 3 to clear fractions

$350p - 100 = 1650$

$350p = 1750$

$p = 5$

The equilibrium price is \$5 per unit.

To find the equilibrium quantity, we substitute \$5 into either $S(p)$ or $D(p)$.

$D(5) = -50(5) + 550 = -250 + 550 = 300$

The equilibrium quantity is 300 yo-yo's.

The equilibrium point is (\$5, 300 yo-yo's).

41. a.   Use a graphing calculator to find the first coordinate of the point of intersection of $y_1 = -14.97x + 987.35$ and $y_2 = 98.55x - 5.13$, to the nearest hundredth. It is 8.74, so the price per unit that should be charged is \$8.74.

b.   Use a graphing calculator to find the first coordinate of the point of intersection of $y_1 = 87,985 + 5.13x$ and $y_2 = 8.74x$. It is about $24,508.4$, so 24,509 units must be sold in order to break even.

43. Convert the yearly savings of $175 to daily savings.

$$\frac{\$175}{1 \text{ year}} \cdot \frac{1 \text{ year}}{365 \text{ days}} \approx \$0.48 \text{ per day}$$

Find how many days to reach $8.14 at $0.48 per day:

$$\frac{\$8.14}{\$0.48 \text{ per day}} \approx 17 \text{ days}$$

It will take about 17 days to break even on the purchase.

Chapter 3 Review

1. substitution; see Section 3.2.

2. elimination, see Section 3.2.

3. graphical; see Sections 3.1 and 3.2.

4. dependent; see Section 3.1.

5. inconsistent; see Section 3.1.

6. contradiction; see Section 3.2.

7. parallel; see Section 3.1.

8. square; see Section 3.7.

9. determinant; see Section 3.7.

10. zero; see Section 3.8.

11. Graph the equations.

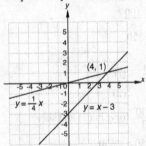

The solution (point of intersection) is (4, 1).

12. Graph the equations:

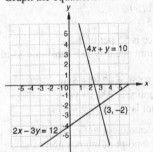

13. $5x - 2y = 4$,   (1)
    $x = y - 2$ (2)

We substitute $y - 2$ for x in Equation (1) and solve for y.

$$
\begin{aligned}
5x - 2y &= 4 &&(1)\\
5(y - 2) - 2y &= 4 &&\text{Substituting}\\
5y - 10 - 2y &= 4\\
3y - 10 &= 4\\
3y &= 14\\
y &= \frac{14}{3}
\end{aligned}
$$

Next we substitute $\frac{14}{3}$ for y in either equation of the original system and solve for x.

$$
\begin{aligned}
x &= y - 2 &&(2)\\
x &= \frac{14}{3} - 2 = \frac{8}{3}
\end{aligned}
$$

Since $\left(\frac{8}{3}, \frac{14}{3}\right)$ checks, it is the solution.

14. $y = x + 2$ (1)
    $y - x = 8$,   (2)

We substitute $x + 2$ for y in the second equation and solve for x.

$$
\begin{aligned}
y - x &= 8 &&(2)\\
x + 2 - x &= 8 &&\text{Substituting}\\
2 &= 8
\end{aligned}
$$

We have a contradiction, or an equation that is always false. Therefore, there is no solution.

15. $2x + 5y = 8$ (1)
    $\underline{6x - 5y = 10}$ (2)
    $\quad 8x \quad\;\; = 18$ Adding
    $\qquad\;\; x = \frac{9}{4}$

Substitute $\frac{9}{4}$ for x in Equation (1) and solve for y.

$$
\begin{aligned}
2x + 5y &= 8\\
2\left(\frac{9}{4}\right) + 5y &= 8 &&\text{Substituting}\\
\frac{9}{2} + 5y &= 8\\
5y &= \frac{7}{2}\\
y &= \frac{7}{10}
\end{aligned}
$$

We obtain $\left(\frac{9}{4}, \frac{7}{10}\right)$. This checks, so it is the solution.

16. $3x - 5y = 9$,   (1)
    $5x - 3y = -1$ (2)

We multiply Equation (1) by -3 and Equation (2) by 5.

$$
\begin{aligned}
-9x + 15y &= -27 &&\text{Multiplying (1) by } -3\\
\underline{25x - 15y} &= -5 &&\text{Multiplying (2) by 5}\\
16x \qquad\; &= -32 &&\text{Adding}\\
x &= -2
\end{aligned}
$$

Substitute -2 for x in Equation (1) and solve for y.

$$
\begin{aligned}
3x - 5y &= 9\\
3(-2) - 5y &= 9 &&\text{Substituting}\\
-6 - 5y &= 9\\
-5y &= 15\\
y &= -3
\end{aligned}
$$

We obtain $(-2, -3)$. This checks, so it is the solution.

17. $x - 3y = -2,$ (1)
$7y - 4x = 6$ (2)

We solve Equation (1) for x.

$x - 3y = -2$
$x = 3y - 2$ (3)

We substitute $3y - 2$ for x in the second equation and solve for y.

$7y - 4x = 6$ (2)
$7y - 4(3y - 2) = 6$ Substituting
$7y - 12y + 8 = 6$
$-5y + 8 = 6$
$-5y = -2$
$y = \frac{2}{5}$

We substitute $\frac{2}{5}$ for y in Equation (3) and solve for x.

$x = 3y - 2$ (3)
$x = 3 \cdot \frac{2}{5} - 2 = \frac{6}{5} - 2 = -\frac{4}{5}$

Since $\left(-\frac{4}{5}, \frac{2}{5}\right)$ checks, it is the solution.

18. $4x - 7y = 18,$ (1)
$9x + 14y = 40$ (2)

We multiply Equation (1) by 2.

$8x - 14y = 36$ Multiplying (1) by 2
$\underline{9x + 14y = 40}$ (2)
$17x \qquad = 76$ Adding
$x = \frac{76}{17}$

Substitute $\frac{76}{17}$ for x in Equation (1) and solve for y.

$4x - 7y = 18$
$4\left(\frac{76}{17}\right) - 7y = 18$ Substituting
$\frac{304}{17} - 7y = 18$
$-7y = \frac{2}{17}$
$y = -\frac{2}{119}$

We obtain $\left(\frac{76}{17}, -\frac{2}{119}\right)$. This checks, so it is the solution.

19. $1.5x - 3 = -2y,$ (1)
$3x + 4y = 6$ (2)

Rewriting the equations in standard form.

$1.5x + 2y = 3,$ (1)
$3x + 4y = 6$ (2)

Observe that, if we multiply Equation (1) by 2, we obtain Equation (2). Thus, any pair that is a solution of Equation (1) is also a solution of Equation (2). The equations are dependent and the solution set is infinite: $\{(x, y) \mid 3x + 4y = 6\}$.

20. $y = 2x - 5,$ (1)
$y - \frac{1}{2}x + 1$ (2)

We substitute $2x - 5$ for y in Equation (2) and solve for x.

$y = \frac{1}{2}x + 1$ (2)
$2x - 5 = \frac{1}{2}x + 1$ Substituting
$2(2x - 5) = 2\left(\frac{1}{2}x + 1\right)$
$4x - 10 = x + 2$
$3x - 10 = 2$
$3x = 12$
$x = 4$

Next we substitute 4 for x in either equation of the original system and solve for y.

$y = 2x - 5$ (1)
$y = 2(4) - 5 = 3$

Since (4, 3) checks, it is the solution.

21. *Familiarize*. Let $g =$ the number of students for group lessons and $p =$ the number of students for private lessons.

Translate. We organize the information in a table.

	Group	Private	Total
Number of students	g	p	12
Price per lesson	$18	$25	
Earnings	$18g$	$25p$	$265

The "Number of students" row of the table gives us one equation:

$g + p = 12$

The "Earnings" row yields a second equation:

$18g + 25p = 265$

We have translated to a system of equations:

$g + p = 12$ (1)
$18g + 25p = 265$ (2)

Carry out. We use the elimination method to solve the system of equations.

$-18g - 18p = -216$ Multiplying (1) by -18
$\underline{18g + 25p = 265}$ (2)
$7p = 49$
$p = 7$

Substitute 7 for p in Equation (1) and solve for g.

$g + 7 = 12$
$g = 5$

Check. The total number of students is $5 + 7$, or 12. The total earnings is $\$18 \cdot 5 + \$25 \cdot 7 = \$90 + \$175 = \$265$. The answer checks.

State. There were 7 students taking private lessons and 5 students taking group lessons.

22. *Familiarize*. Let $t =$ the number of hours for the passenger train, and $t + 1 =$ the number of hours for the freight train.

Now complete the chart.

$$d = r \cdot t$$

	Distance	Rate	Time	
Passenger train	d	55	t	$\rightarrow d = 55t$
Freight train	d	44	$t + 1$	$\rightarrow d = 44(t + 1)$

Translate. Using $d = rt$ in each row of the table, we get a system of equations:

$$d = 55t,$$
$$d = 44(t + 1)$$

Carry out. We solve the system of equations.

$$55t = 44(t + 1) \quad \text{Using substitution}$$
$$55t = 44t + 44$$
$$11t = 44$$
$$t = 4$$

Check. At 55 mph, in 4 hr the passenger train will travel $55 \cdot 4 = 220$ mi. At 44 mph, in 4 + 1, or 5 hr the freight train will travel $44 \cdot 5 = 220$ mi. The numbers check.

State. The passenger train overtakes the freight train after 4 hours.

23. Let x = the number of liters of 15% juice and y = the number of liters of 8% juice.

Solve:
$$x + y = 14$$
$$0.15x + 0.08y = 0.10(14)$$

The solution is (4, 10). So, 4 liters of 15% juice and 10 liters of 8% juice should be purchased.

24.
$$x + 4y + 3z = 2, \quad (1)$$
$$2x + y + z = 10, \quad (2)$$
$$-x + y + 2z = 8 \quad (3)$$

1., 2. The equations are already in standard form with no fractions or decimals.
3. Use Equations (1) and (3) to eliminate x.

$$\begin{array}{ll} x + 4y + 3z = 2 & (1) \\ -x + y + 2z = 8 & (3) \\ \hline 5y + 5z = 10 & (4) \text{ Adding} \end{array}$$

4. Use a different pair of equations and eliminate x.

$$\begin{array}{ll} 2x + y + z = 10 & (2) \\ -2x + 2y + 4z = 16 & \text{Multiplying (3) by 2} \\ \hline 3y + 5z = 26 & (5) \text{ Adding} \end{array}$$

5. Now solve the system of Equations (4) and (5).

$$5y + 5z = 10 \quad (4)$$
$$3y + 5z = 26 \quad (5)$$

$$\begin{array}{ll} 5y + 5z = 10 & (4) \\ -3y - 5z = -26 & \text{Multiplying (5) by } -1 \\ \hline 2y = -16 & \text{Adding} \\ y = -8 & \end{array}$$

$$5(-8) + 5z = 10 \quad \text{Substituting } -8 \text{ for } y \text{ in (4)}$$
$$-40 + 5z = 10$$
$$5z = 50$$
$$z = 10$$

6. Substitute in one of the original equations to find x.

$$x + 4(-8) + 3(10) = 2 \quad \text{Substituting } -8 \text{ for } y \text{ and}$$
$$\qquad\qquad\qquad\qquad 10 \text{ for } z \text{ in (1)}$$
$$x - 2 = 2$$
$$x = 4$$

We obtain (4, −8, 10). This checks, so it is the solution.

25.
$$4x + 2y - 6z = 34, \quad (1)$$
$$2x + y + 3z = 3, \quad (2)$$
$$6x + 3y - 3z = 37 \quad (3)$$

1., 2. The equations are already in standard form with no fractions or decimals.

3., 4. We eliminate z from two different pairs of equations.

$$\begin{array}{ll} 4x + 2y - 6z = 34 & (1) \\ 4x + 2y + 6z = 6 & \text{Multiplying (2) by 2} \\ \hline 8x + 4y = 40 & (4) \text{ Adding} \end{array}$$

$$\begin{array}{ll} 2x + y + 3z = 3 & (2) \\ 6x + 3y - 3z = 37 & (3) \\ \hline 8x + 4y = 40 & (5) \text{ Adding} \end{array}$$

5. Now solve the system of Equations (4) and (5).

$$8x + 4y = 40 \quad (4)$$
$$8x + 4y = 40 \quad (5)$$

$$\begin{array}{ll} 8x + 4y = 40 & (4) \\ -8x - 4y = -40 & \text{Multiplying (5) by } -1 \\ \hline 0 = 0 & (6) \end{array}$$

Equation (6) indicates Equations (1), (2), and (3) are dependent. (Note that if Equation (1) is added to Equation (2), the result is Equation (3).) We could also have concluded that the equations are dependent by observing that Equations (4) and (5) are identical.

26.
$$2x - 5y - 2z = -4, \quad (1)$$
$$7x + 2y - 5z = -6, \quad (2)$$
$$-2x + 3y + 2z = 4 \quad (3)$$

1., 2. The equations are already in standard form with no fractions or decimals.
3. Use Equations (1) and (2) to eliminate x.

$$\begin{array}{ll} 14x - 35y - 14z = -28 & \text{Multiplying (1) by 7} \\ -14x - 4y + 10z = 12 & \text{Multiplying (2) by } -2 \\ \hline -39y - 4z = -16 & (4) \text{ Adding} \end{array}$$

4. Use Equations (1) and (3) to eliminate x.

$$\begin{array}{ll} 2x - 5y - 2z = -4 & (1) \\ -2x + 3y + 2z = 4 & (3) \\ \hline -2y = 0 & (5) \text{ Adding} \\ y = 0 & \end{array}$$

5. When we used Equation (1) and (3) to eliminate x, we also eliminated z and found $y = 0$. Substitute 0 for y in Equation (4) to find z.

$$-39 \cdot 0 - 4z = -16 \quad \text{Substituting 0 for } y \text{ in (4)}$$
$$-4z = -16$$
$$z = 4$$

6. Substitute in one of the original equations to find x.

$$2x - 5 \cdot 0 - 2 \cdot 4 = -4$$
$$2x = 4$$
$$x = 2$$

We obtain (2, 0, 4). This checks, so it is the solution.

27.
$$3x + y = 2, \quad (1)$$
$$x + 3y + z = 0, \quad (2)$$
$$x + z = 2 \quad (3)$$

1., 2. The equations are already in standard form with no fractions or decimals.

3., 4. Note that there is no z in Equation (1). We will use Equations (2) and (3) to obtain another equation with no z-term.

$$\begin{array}{ll} x + 3y + z = 0 & (2) \\ -x - z = -2 & \text{Multiplying (3) by } -1 \\ \hline 3y = -2 & (4) \\ y = -\dfrac{2}{3} & \end{array}$$

5. Now substitute $-\frac{2}{3}$ for y in (1) to solve for x.

$$3x - \frac{2}{3} = 2 \quad \text{Substituting } -\frac{2}{3} \text{ for } y \text{ in (1)}$$
$$3x = \frac{8}{3}$$
$$x = \frac{8}{9}$$

6. Substitute in Equation (3) to find z.

$$\frac{8}{9} + z = 2 \quad \text{Substituting } \frac{8}{9} \text{ for } x \text{ in (3)}$$
$$z = \frac{10}{9}$$

We obtain $\left(\frac{8}{9}, -\frac{2}{3}, \frac{10}{9}\right)$. This checks, so it is the solution.

28.
$$2x - 3y + z = 1, \quad (1)$$
$$x - y + 2z = 5, \quad (2)$$
$$3x - 4y + 3z = -2 \quad (3)$$

1., 2. The equations are already in standard form with no fractions or decimals.

3., 4. We eliminate y from two different pairs of equations.

$$\begin{array}{ll} 2x - 3y + z = 1 & (1) \\ \underline{-3x + 3y - 6z = -15} & \text{Multiplying (2) by } -3 \\ -x - 5z = -14 & (4) \end{array}$$

$$\begin{array}{ll} -4x + 4y - 8z = -20 & \text{Multiplying (2) by } -4 \\ \underline{3x - 4y + 3z = -2} & (3) \\ -x - 5z = -22 & (4) \end{array}$$

5. Now solve the system of Equations (4) and (5).

$$-x - 5z = -14 \quad (4)$$
$$-x - 5z = -22 \quad (5)$$

$$\begin{array}{ll} -x - 5z = -14 & (4) \\ \underline{x + 5z = 22} & \text{Multiplying (5) by } -1 \\ 0 = 8 & \end{array}$$

We get a false equation, so there is no solution.

29. **Familiarize.** We let x, y, and z represent the measures of angles A, B, and C, respectively. The measures of the angles of a triangle add up to 180º.
Translate.

The sum of the measures is 180.
↓ ↓ ↓
$x + y + z$ = 180

The measure of angle A is four times the measure of angle C.
↓ ↓ ↓
x = $4z$

The measure of angle B is 45° more than the measure of angle C.
↓ ↓ ↓
y = $z + 45$

We now have a system of equations.
$$x + y + z = 180,$$
$$x = 4z,$$
$$y = z + 45$$

Carry out. Solving the system we get (90, 67.5, 22.5).

Check. The sum of the measures is
$90° + 67.5° + 22.5°$, or 180º. Four times the measure of angle C is $4 \cdot 22.5°$, or 90º, the measure of angle A. 45 more than the measure of angle C is $45° + 22.5°$, or 67.5º, the measure of angle B. The numbers check.
State. The measures of angles A, B, and C are 90º, 67.5º, and 22.5º, respectively.

30. **Familiarize.** Let x, y, and z represent the average number of cries for a man, woman and a one-year-old child, respectively.
Translate.
The sum for each month is 56.7, so we have
$$x + y + z = 56.7.$$
The number of cries for a woman is 3.9 more than the man, so we have
$$y = 3.9 + x.$$
The number of cries for the child is 43.3 more than the sum for the man and woman, so we have
$$z = 43.3 + x + y.$$
We now have a system of equations.
$$x + y + z = 56.7,$$
$$y = 3.9 + x,$$
$$z = 43.3 + x + y$$

Carry out. Solving the system, we get (1.4, 5.3, 50).
Check. The sum of the average number of cry times for a man, a woman, and a one-year-old child is
$1.4 + 5.3 + 50$, or 56.7. A woman cries $3.9 + 1.4$, or 5.3 times a month. A one-year-old child cries $43.3 + 1.4 + 5.3$, or 50 times a month. The answer checks.
State. The monthy average number of cries for a man is 1.4, for a woman is 5.3, and a one-year-old child is 50.

31.
$$3x + 4y = -13,$$
$$5x + 6y = 8$$
Write a matrix.
$$\begin{bmatrix} 3 & 4 & | & -13 \\ 5 & 6 & | & 8 \end{bmatrix}$$
Multiply the second row by 3
$$\begin{bmatrix} 3 & 4 & | & -13 \\ 15 & 18 & | & 24 \end{bmatrix} \quad \text{New Row 2} = 3(\text{Row 2})$$
Multiply row 1 by -5 and add it to row 2
$$\begin{bmatrix} 3 & 4 & | & -13 \\ 0 & -2 & | & 89 \end{bmatrix} \quad \text{New Row 2} = -5(\text{Row 1}) + \text{Row 2}$$
Reinserting the variables, we have
$$3x + 4y = -13, \quad (1)$$
$$-2y = 89 \quad (2)$$
Solve Equation (2) for y.
$$-2y = 89$$
$$y = -\frac{89}{2}$$
Substitute $-\frac{89}{2}$ for y in Equation (1) and solve for x.
$$3x + 4\left(-\frac{89}{2}\right) = -13$$
$$3x - 178 = -13$$
$$3x = 165$$
$$x = 55$$
The solution is $\left(55, -\frac{89}{2}\right)$.

32.　　$3x - y + z = -1,$
　　　　　$2x + 3y + z = 4,$
　　　　　$5x + 4y + 2z = 5$

We first write a matrix.

$$\begin{bmatrix} 3 & -1 & 1 & | & -1 \\ 2 & 3 & 1 & | & 4 \\ 5 & 4 & 2 & | & 5 \end{bmatrix}$$

$$\begin{bmatrix} 3 & -1 & 1 & | & -1 \\ 0 & 11 & 1 & | & 14 \\ 0 & 17 & 1 & | & 20 \end{bmatrix} \begin{matrix} \text{New Row 2} = -2(\text{Row 1}) + 3(\text{Row 2}) \\ \text{New Row 3} = -5(\text{Row 1}) + 3(\text{Row 3}) \end{matrix}$$

$$\begin{bmatrix} 3 & -1 & 1 & | & -1 \\ 0 & 11 & 1 & | & 14 \\ 0 & 187 & 11 & | & 220 \end{bmatrix} \quad \text{New Row 3} = 11(\text{Row 3})$$

$$\begin{bmatrix} 3 & -1 & 1 & | & -1 \\ 0 & 11 & 1 & | & 14 \\ 0 & 0 & -6 & | & -18 \end{bmatrix} \quad \text{New Row 3} = -17(\text{Row 2}) + \text{Row 3}$$

Reinserting the variables, we have

　　$3x - y + z = -1,$　　(1)
　　　　$11y + z = 14,$　　(2)
　　　　　　$-6z = -18$　　(3)

Solve (3) for z.

　　$-6z = -18$
　　　　$z = 3$

Substitute 3 for z in (2) and solve for y.

　　$11y + 3 = 14$
　　　　$11y = 11$
　　　　　$y = 1$

Substitute 1 for y and 3 for z in (1) and solve for x.

　　$3x - 1 + 3 = -1$
　　　　$3x + 2 = -1$
　　　　　$3x = -3$
　　　　　　$x = -1$

The solution is $(-1, 1, 3)$.

33.　$\begin{vmatrix} -2 & -5 \\ 3 & 10 \end{vmatrix} = -2(10) - (-5)(3) = -20 + 15 = -5$

34.　$\begin{vmatrix} 2 & 3 & 0 \\ 1 & 4 & -2 \\ 2 & -1 & 5 \end{vmatrix}$

$$= 2\begin{vmatrix} 4 & -2 \\ -1 & 5 \end{vmatrix} - 1\begin{vmatrix} 3 & 0 \\ -1 & 5 \end{vmatrix} + 2\begin{vmatrix} 3 & 0 \\ 4 & -2 \end{vmatrix}$$

$$= 2[4 \cdot 5 - (-2)(-1)] - 1[3 \cdot 5 - 0(-1)]$$
$$\quad + 2[3(-2) - 0 \cdot 4]$$
$$= 2(18) - 1(15) + 2(-6)$$
$$= 36 - 15 - 12$$
$$= 9$$

35.　$2x + 3y = 6,$
　　　　$x - 4y = 14$

We compute D, D_x, and D_y.

$$D = \begin{vmatrix} 2 & 3 \\ 1 & -4 \end{vmatrix} = -8 - 3 = -11$$

$$D_x = \begin{vmatrix} 6 & 3 \\ 14 & -4 \end{vmatrix} = -24 - 42 = -66$$

$$D_y = \begin{vmatrix} 2 & 6 \\ 1 & 14 \end{vmatrix} = 28 - 6 = 22$$

Then,

$$x = \frac{D_x}{D} = \frac{-66}{-11} = 6 \quad \text{and} \quad y = \frac{D_y}{D} = \frac{22}{-11} = -2.$$

The solution is $(6, -2)$.

36.　$2x + y + z = -2,$
　　　　$2x - y + 3z = 6,$
　　　　$3x - 5y + 4z = 7$

We compute D, D_x, D_y and D_z.

$$D = \begin{vmatrix} 2 & 1 & 1 \\ 2 & -1 & 3 \\ 3 & -5 & 4 \end{vmatrix}$$

$$= 2\begin{vmatrix} -1 & 3 \\ -5 & 4 \end{vmatrix} - 2\begin{vmatrix} 1 & 1 \\ -5 & 4 \end{vmatrix} + 3\begin{vmatrix} 1 & 1 \\ -1 & 3 \end{vmatrix}$$

$$= 2(11) - 2(9) + 3(4)$$
$$= 22 - 18 + 12 = 16$$

$$D_x = \begin{vmatrix} -2 & 1 & 1 \\ 6 & -1 & 3 \\ 7 & -5 & 4 \end{vmatrix}$$

$$= -2\begin{vmatrix} -1 & 3 \\ -5 & 4 \end{vmatrix} - 6\begin{vmatrix} 1 & 1 \\ -5 & 4 \end{vmatrix} + 7\begin{vmatrix} 1 & 1 \\ -1 & 3 \end{vmatrix}$$

$$= -2(11) - 6(9) + 7(4)$$
$$= -22 - 54 + 28 = -48$$

$$D_y = \begin{vmatrix} 2 & -2 & 1 \\ 2 & 6 & 3 \\ 3 & 7 & 4 \end{vmatrix} = 2\begin{vmatrix} 6 & 3 \\ 7 & 4 \end{vmatrix} - 2\begin{vmatrix} -2 & 1 \\ 7 & 4 \end{vmatrix} + 3\begin{vmatrix} -2 & 1 \\ 6 & 3 \end{vmatrix}$$

$$= 2(3) - 2(-15) + 3(-12)$$
$$= 6 + 30 - 36 = 0$$

$$D_z = \begin{vmatrix} 2 & 1 & -2 \\ 2 & -1 & 6 \\ 3 & -5 & 7 \end{vmatrix}$$

$$= 2\begin{vmatrix} -1 & 6 \\ -5 & 7 \end{vmatrix} - 2\begin{vmatrix} 1 & -2 \\ -5 & 7 \end{vmatrix} + 3\begin{vmatrix} 1 & -2 \\ -1 & 6 \end{vmatrix}$$

$$= 2(23) - 2(-3) + 3(4)$$
$$= 46 + 6 + 12 = 64$$

Then,

$$x = \frac{D_x}{D} = \frac{-48}{16} = -3$$

$$y = \frac{D_y}{D} = \frac{0}{16} = 0.$$

$$z = \frac{D_z}{D} = \frac{64}{16} = 4.$$

The solution is $(-3, 0, 4)$.

37.　$C(x) = 30x + 15,800$　$R(x) = 50x$

a.　$P(x) = R(x) - C(x)$
　　　　　$= 50x - (30x + 15,800)$
　　　　　$= 50x - 30x - 15,800$
　　　　　$= 20x - 15,800$

b.　Solve the system
　　　$R(x) = 50x,$
　　　$C(x) = 30x + 15,800.$

Since both $R(x)$ and $C(x)$ are in dollars and they are equal at the break-even point, we can rewrite the system:

$$d = 50x, \qquad (1)$$
$$d = 30x + 15,800 \quad (2)$$

We solve using substitution.

$$50x = 30x + 15,800 \quad \text{Substituting } 50x \text{ for}$$
$$\qquad\qquad\qquad\qquad d \text{ in (2)}$$
$$20x = 15,800$$
$$x = 790$$

Thus, 790 units must be produced and sold in order to break even.

The revenue will be $R(790) = 50 \cdot 790 = 39,500$.

The break-even point is (790 units, $39,500).

38. $60 + 7p = 120 - 13p$
$$20p = 60$$
$$p = 3$$

$$S(3) = 60 + 7 \cdot 3 = 81$$

The equilibrium point is ($3, 81).

39. **a.** $C(x) = 4.75x + 54,000$, where x is the number of pints of honey.

b.   Each pint of honey sells for $9.25. The total revenue is 9.25 times the number of pints of honey. We assume all the honey is sold. $R(x) = 9.25x$

c. $P(x) = R(x) - C(x)$
$$= 9.25x - (4.75x + 54,000)$$
$$= 9.25x - 4.75x - 54,000$$
$$= 4.5x - 54,000$$

d. $P(5000) = 4.5(5000) - 54,000$
$$= 22,500 - 54,000$$
$$= -31,500$$

Danae will realize a $31,500 loss when 5000 pints of honey are produced and sold.

$$P(15,000) = 4.5(15,000) - 54,000$$
$$= 67,500 - 54,000$$
$$= 13,500$$

Danae will realize a profit of $13,500 when 15,000 pints of honey are produced and sold.

e.   Solve the system
$$R(x) = 9.25x,$$
$$C(x) = 4.75x + 54,000.$$

Since both $R(x)$ and $C(x)$ are in dollars and they are equal at the break-even point, we can rewrite the system:

$$d = 9.25x, \qquad\qquad (1)$$
$$d = 4.75x + 54,000 \quad (2)$$

We solve using substitution.

$$9.25x = 4.75x + 54,000 \quad \text{Substituting } 9.25x \text{ for}$$
$$\qquad\qquad\qquad\qquad d \text{ in (2)}$$
$$4.5x = 54,000$$
$$x = 12,000$$

Thus, 12,000 units must be produced and sold in order to break even. The revenue will be

$R(12,000) = 9.25 \cdot 12,000 = \$111,000$.

The break-even point is (12,000 pints, $111,000).

40. *Writing Exercise.* To solve a problem involving four variables, go through the *Familiarize* and *Translate* steps as usual. The resulting system of equations can be solved using the elimination method just as for three variables but likely with more steps.

41. *Writing Exercise.* A system of equations can be both dependent and inconsistent if it is equivalent to a system with fewer equations that has no solution. An example is a system of three equations in three unknowns in which two of the equations represent the same plane, and the third represents a parallel plane.

42. From Exercise 39, we have
$$P(x) = 4.5x - 54,000.$$

Danae's salary was $36,000, or
$$S(x) = 36,000.$$

Since both $P(x)$ and $S(x)$ are in dollars and they are equal, we can rewrite the system:

$$d = 4.5x - 54,000 \quad (1)$$
$$d = 36,000 \qquad\qquad (2)$$

We solve using substitution.

$$4.5x - 54,000 = 36,000$$
$$4.5x = 90,000$$
$$x = 20,000$$

Danae's earnings will be the same if 20,000 pints of honey are produced and sold.

43. Graph both equations.

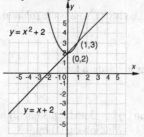

The solutions are apparently (0, 2) and (1, 3). Both pairs check.

Chapter 3 Test

1. Graph the equations.

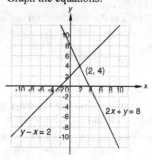

2. $x + 3y = -8, \quad (1)$
$$4x - 3y = 23 \quad (2)$$

Solve Equation (1) for x.
$$x + 3y = -8$$
$$x = -3y - 8$$

We substitute $-3y - 8$ for x in (2) and solve for y.

$$4x - 3y = 23 \quad (2)$$
$$4(-3y - 8) - 3y = 23 \quad \text{Substituting}$$
$$-12y - 32 - 3y = 23$$
$$-15y - 32 = 23$$
$$-15y = 55$$
$$y = -\frac{11}{3}$$

Next we substitute $-\frac{11}{3}$ for y in either equation of the original system and solve for x.

$$x + 3y = -8 \quad (1)$$
$$x + 3\left(-\frac{11}{3}\right) = -8$$
$$x - 11 = -8$$
$$x = 3$$

Since $\left(3, -\frac{11}{3}\right)$ checks, it is the solution.

3. $3x - y = 7 \quad (1)$
$\underline{x + y = 1} \quad (2)$
$4x = 8 \quad \text{Adding}$
$x = 2$

Substitute 2 for x in Equation (2) and solve for y.

$$x + y = 1$$
$$2 + y = 1 \quad \text{Substituting}$$
$$y = -1$$

We obtain $(2, -1)$. This checks, so it is the solution.

4. $4y + 2x = 18,$
$3x + 6y = 26$

Rewrite the equations in standard form.

$2x + 4y = 18, \quad (1)$
$3x + 6y = 26 \quad (2)$

We multiply Equation (1) by -3 and Equation (2) by 2.

$-6x - 12y = -54 \quad \text{Multiplying (1) by } -3$
$\underline{6x + 12y = 52} \quad \text{Multiplying (2) by 2}$
$0 = -2 \quad \text{Adding}$

We have a contradiction, or an equation that is always false. Therefore, there is no solution.

5. $2x - 4y = -6 \quad (1)$
$x = 2y - 3 \quad (2)$

We substitute $2y - 3$ for x in the first equation and solve for y.

$$2x - 4y = -6$$
$$2(2y - 3) - 4y = -6$$
$$4y - 6 - 4y = -6$$
$$-6 = -6$$

We have an identity, or an equation that is always true. The equations are dependent and the solution set is infinite: $\{(x, y)|2y - 3 = x\}$ or $\{(x, y)|2x - 4y = -6\}$.

6. $4x - 6y = 3, \quad (1)$
$6x - 4y = -3 \quad (2)$

We multiply Equation (1) by 6 and Equation (2) by -4.

$24x - 36y = 18 \quad \text{Multiplying (1) by 6}$
$\underline{-24x + 16y = 12} \quad \text{Multiplying (2) by } -4$
$-20y = 30 \quad \text{Adding}$
$y = -\frac{3}{2}$

Substitute $-\frac{3}{2}$ for y in Equation (1) and solve for x.

$$4x - 6y = 3$$
$$4x - 6\left(-\frac{3}{2}\right) = 3 \quad \text{Substituting}$$
$$4x + 9 = 3$$
$$4x = -6$$
$$x = -\frac{3}{2}$$

We obtain $\left(-\frac{3}{2}, -\frac{3}{2}\right)$. This checks, so it is the solution.

7. *Familiarize*. Let w = the width of the basketball court and let l = the length of the basketball court. Recall that the perimeter of a rectangle is given by the formula $P = 2w + 2l$.
Translate.
The perimeter of the court is 288 ft, so we have
$$2w + 2l = 288.$$
The length if 44 longer than the width, so we have
$$l = 44 + w.$$
We now have a system of equations.
$$2w + 2l = 288. \quad (1)$$
$$l = 44 + w \quad (2)$$
Carry out. Substitute $44 + w$ for l in the first equation and solve for w.
$$2w + 2l = 288$$
$$2w + 2(44 + w) = 288$$
$$2w + 88 + 2w = 288$$
$$4w + 88 = 288$$
$$4w = 200$$
$$w = 50$$
Finally substitute 50 for w in Equation (2) and solve for l.
$$l = 44 + w$$
$$l = 44 + 50 = 94$$
Check. The perimeter of the court is $2(94 \text{ ft})$
$+ 2(50 \text{ ft}) = 188 \text{ ft} + 100 \text{ ft} = 288 \text{ ft}$. The length, 94 ft, is 44 ft more than the width, or $44 + 50 = 94$.
State. The basketball court is 94 ft long and 50 ft wide.

8. Let x = the number of grams of Goldfish and y = the number of grams of Pretzels.
Solve: $x + y = 620,$
$0.40x + 0.09y = 93$
The solution is $(120, 500)$. So, the mixture contains 120 g of Goldfish and 500 g of Pretzels.

9. *Familiarize*. Let d = the distance and r = the speed of the boat in still water. Then when the boat travels downstream, its speed is $r + 5$, and its speed upstream is $r - 5$. The distances are the same.
Organize the information in a table.

	Distance	Rate	Time
Downstream	d	$r+5$	3
Upstream	d	$r-5$	5

Translate. Using $d = rt$ in each row of the table, we get a system of equations:

$d = 3(r + 5),$ $d = 3r + 15,$
$\qquad\qquad$ or
$d = 5(r - 5)$ $d = 5r - 25$

Carry out. Solve the system of equations.

$3r + 15 = 5r - 25$
$\quad\; 15 = 2r - 25$
$\quad\; 40 = 2r$
$\quad\; 20 = r$

Check. When $r = 20$, then $r + 5 = 20 + 5 = 25$, and the distance traveled in 3 hours is $3 \cdot 25 = 75$ mi. Also $r - 5 = 20 - 5 = 15$, and the distance traveled in 5 hours is $15 \cdot 5 = 75$ mi. The answer checks.

State. The speed of the boat in still water is 20 mph.

10. $-3x + y - 2z = 8,$ (1)
$\quad\; -x + 2y - z = 5,$ (2)
$\quad\;\; 2x + y + z = -3$ (3)

1., 2. The equations are already in standard form with no fractions or decimals.
3., 4. We eliminate x from two different pairs of equations.

$-3x + y - 2z = 8$ (1)
$\underline{\;\; 3x - 6y + 3z = -15}$ Multiplying (2) by -3
$\qquad -5y + z = -7$ (4)

$-2x + 4y - 2z = 10$ Multiplying (2) by 2
$\underline{\;\; 2x + y + z = -3}$ (3)
$\qquad 5y - z = 7$ (5)

5.   Now solve the system of Equations (4) and (5).

$-5y + z = -7$ (4)
$\underline{\;\; 5y - z = 7}$ (5)
$\qquad\; 0 = 0$ (6)

Equation (6) indicates that Equations (1), (2), and (3) are dependent.

11. $6x + 2y - 4z = 15,$ (1)
$\quad\; -3x - 4y + 2z = -6,$ (2)
$\quad\;\; 4x - 6y + 3z = 8$ (3)

1., 2. The equations are already in standard form with no fractions or decimals.
3., 4. We eliminate x from two different pairs of equations.

$6x + 2y - 4z = 15$ (1)
$\underline{-6x - 8y + 4z = -12}$ Multiplying (2) by 2
$\qquad -6y = 3$ (4)
$\qquad\qquad y = -\dfrac{1}{2}$

$-12x - 16y + 8z = -24$ Multiplying (2) by 4
$\underline{\;\; 12x - 18y + 9z = 24}$ Multiplying (3) by 3
$\qquad -34y + 17z = 0$ (5)

5. Now substitute $-\dfrac{1}{2}$ for y in (5) to solve for z.

$-34\left(-\dfrac{1}{2}\right) + 17z = 0$ Substituting $-\dfrac{1}{2}$ for y in (5)
$\qquad\; 17 + 17z = 0$
$\qquad\qquad\; z = -1$

6. Substitute in Equation (1) to find x.

$6x + 2\left(-\dfrac{1}{2}\right) - 4(-1) = 15$ Substituting
$\qquad\quad 6x - 1 + 4 = 15$
$\qquad\qquad\quad 6x = 12$
$\qquad\qquad\qquad x = 2$

We obtain $\left(2, -\dfrac{1}{2}, -1\right)$. This checks, so it is the solution.

12. $2x + 2y = 0,$ (1)
$\quad\; 4x + 4z = 4,$ (2)
$\quad\; 2x + y + z = 2$ (3)

1., 2. The equations are already in standard form with no fractions or decimals.
3., 4. Note that there is no z in Equation (1). We will use Equations (2) and (3) to obtain another equation with no z-term.

$4x + 4z = 4$ (2)
$\underline{-8x - 4y - 4z = -8}$ Multiplying (3) by -4
$-4x - 4y = -4$ (4)

5. Now solve the system of Equations (1) and (4).

$2x + 2y = 0$ (1)
$-4x - 4y = -4$ (4)

$4x + 4y = 0$ Multiplying (1) by 2
$\underline{-4x - 4y = -4}$ (4)
$\qquad\quad 0 = -4$ Adding

We get a false equation, or contradiction. There is no solution.

13. $3x + 3z = 0,$ (1)
$\quad\; 2x + 2y = 2,$ (2)
$\quad\qquad 3y + 3z = 3$ (3)

1., 2. The equations are already in standard form with no fractions or decimals.
3., 4. Note that there is no z in Equation (2). We will use Equations (1) and (3) to obtain another equation with no z-term.

$3x + 3z = 0$ (1)
$\underline{\;\; -3y - 3z = -3}$ Multiplying (3) by -1
$3x - 3y = -3$ (4)

5. Now solve the system of Equations (2) and (4).

$2x + 2y = 2$ (2)
$3x - 3y = -3$ (4)

$6x + 6y = 6$ Multiplying (2) by 3
$\underline{-6x + 6y = 6}$ Multiplying (4) by -2
$\qquad 12y = 12$
$\qquad\quad y = 1$

$2x + 2 \cdot 1 = 2$ Substituting in (2)
$\qquad 2x = 0$
$\qquad\; x = 0$

6. Substitute in Equation (1) or (3) to find z.

$3 \cdot 0 + 3z = 0$ Substituting in (1)
$\qquad 3z = 0$
$\qquad\; z = 0$

We obtain $(0, 1, 0)$. This checks, so it is the solution.

14. $\begin{bmatrix} 4 & 1 & | & 12 \\ 3 & 2 & | & 2 \end{bmatrix}$

Multiply row 2 by 4.

$\begin{bmatrix} 4 & 1 & | & 12 \\ 12 & 8 & | & 8 \end{bmatrix}$ New Row 2 = 4(Row 2)

Multiply row 1 by –3 and add it to row 2.

$\begin{bmatrix} 4 & 1 & | & 12 \\ 0 & 5 & | & -28 \end{bmatrix}$ New Row 2 = –3(Row 1) + Row 2

Then $4x + y = 12,$
$\qquad\quad 5y = -28$

and $y = -\dfrac{28}{5}$, $x = \dfrac{22}{5}$, or $\left(-\dfrac{28}{5}, \dfrac{22}{5}\right)$.

15. $x + 3y - 3z = 12,$
$\quad\ 3x - y + 4z = 0,$
$\quad -x + 2y - z = 1$

We first write a matrix.

$\begin{bmatrix} 1 & 3 & -3 & | & 12 \\ 3 & -1 & 4 & | & 0 \\ -1 & 2 & -1 & | & 1 \end{bmatrix}$

$\begin{bmatrix} 1 & 3 & -3 & | & 12 \\ 0 & -10 & 13 & | & -36 \\ 0 & 5 & -4 & | & 13 \end{bmatrix}$ New Row 2 = –3(Row 1) + Row 2
New Row 3 = (Row 1) + Row 3

$\begin{bmatrix} 1 & 3 & -3 & | & 12 \\ 0 & 5 & -4 & | & 13 \\ 0 & -10 & 13 & | & -36 \end{bmatrix}$ Interchange
Row 2 and Row 3

$\begin{bmatrix} 1 & 3 & -3 & | & 12 \\ 0 & 5 & -4 & | & 13 \\ 0 & 0 & 5 & | & -10 \end{bmatrix}$ New Row 3 = 2(Row 2) + Row 3

Reinserting the variables, we have

$\quad x + 3y - 3z = 12,$ (1)
$\qquad\quad 5y - 4z = 13,$ (2)
$\qquad\qquad\quad 5z = -10.$ (3)

Solve (3) for z.

$\quad 5z = -10$
$\qquad z = -2$

Substitute –2 for z in (2) and solve for y.

$\quad 5y - 4(-2) = 13$
$\qquad\qquad 5y = 5$
$\qquad\qquad\ y = 1$

Substitute 1 for y and –2 for z in (1) and solve for x.

$\quad x + 3 \cdot 1 - 3(-2) = 12$
$\qquad\quad x + 3 + 6 = 12$
$\qquad\qquad\qquad x = 3$

The solution is $(3, 1, -2)$.

16. $\begin{vmatrix} 4 & -2 \\ 3 & -5 \end{vmatrix} = 4(-5) - (-2)(3) = -20 + 6 = -14$

17. $\begin{vmatrix} 3 & 4 & 2 \\ -2 & -5 & 4 \\ 0 & 5 & -3 \end{vmatrix}$

$= 3\begin{vmatrix} -5 & 4 \\ 5 & -3 \end{vmatrix} - (-2)\begin{vmatrix} 4 & 2 \\ 5 & -3 \end{vmatrix} + 0\begin{vmatrix} 4 & 2 \\ -5 & 4 \end{vmatrix}$

$= 3[(-5)(-3) - 5 \cdot 4] + 2[4(-3) - 5 \cdot 2]$
$\qquad + 0[4 \cdot 4 - 2(-5)]$

$= 3(-5) + 2(-22) + 0$

$= -15 - 44 = -59$

18. $3x + 4y = -1,$
$\quad\ 5x - 2y = 4$

We compute D, D_x, and D_y.

$D = \begin{vmatrix} 3 & 4 \\ 5 & -2 \end{vmatrix} = -6 - 20 = -26$

$D_x = \begin{vmatrix} -1 & 4 \\ 4 & -2 \end{vmatrix} = 2 - 16 = -14$

$D_y = \begin{vmatrix} 3 & -1 \\ 5 & 4 \end{vmatrix} = 12 - (-5) = 17$

$x = \dfrac{D_x}{D} = \dfrac{-14}{-26} = \dfrac{7}{13}$ $y = \dfrac{D_y}{D} = \dfrac{-17}{26}$.

The solution is $\left(\dfrac{7}{13}, -\dfrac{17}{26}\right)$.

19. *Familiarize*. Let x, y, and z represent the number of hours for the electrician, carpenter and plumber, respectively.
Translate.
The total number of hours worked is 21.5, so we have
$\quad x + y + z = 21.5.$
The total earnings is $673, so we have
$\quad 30x + 28.50y + 34z = 673.$
The plumber worked 2 more hours than the carpenter, so we have
$\quad z = 2 + y.$
We have a system of equations:
$\qquad\qquad x + y + z = 21.5,$
$\qquad 30x + 28.50y + 34z = 673,$
$\qquad\qquad\qquad\quad z = 2 + y$

Carry out. Solving the system we get $(3.5, 8, 10)$.
Check. The total number of hours is $3.5 + 8 + 10$, or 21.5 h. The total amount earned is
$30(3.5) + 28.50(8) + 34(10) = 105 + 228 + 340$,
or $673. The plumber worked 10 h, which is 2 h more than the 8 h worked by the carpenter. The numbers check.
State. The electrician worked 3.5 h, the carpenter worked 8 h and the plumber worked 10 h.

20. $79 - 8p = 37 + 6p$
$\qquad\quad 42 = 14p$
$\qquad\qquad 3 = p$

$D(3) = 79 - 8 \cdot 3 = 55$

The equilibrium point is ($3, 55 units).

21. a. $C(x) = 25x + 44,000$, where x is the number of hammocks produced.

b. $R(x) = 80x$ We assume all hammocks are sold.

c. $P(x) = R(x) - C(x)$
$= 80x - (25x + 44,000)$
$= 80x - 25x - 44,000$
$= 55x - 44,000$

d. $P(300) = 55(300) - 44,000$
$= 16,500 - 44,000$
$= -27,500$

The company will realize a \$27,500 loss when 300 hammocks are produced and sold.
$P(900) = 55(900) - 44,000$
$= 49,500 - 44,000$
$= 5500$

The company will realize a profit of \$5500 when 900 hammocks are produced and sold.

e. Solve the system
$R(x) = 80x$,
$C(x) = 25x + 44,000$.

Since both $R(x)$ and $C(x)$ are in dollars and they are equal at the break-even point, we can rewrite the system:

$d = 80x$,   (1)
$d = 25x + 44,000$ (2)

We solve using substitution.
$80x = 25x + 44,000$ Substituting $80x$ for d in (2)
$55x = 44,000$
$x = 800$

Thus, 800 hammocks must be produced and sold in order to break even.

The revenue will be $R(800) = 80 \cdot 800 = \$64,000$.

The break-even point is (800 hammocks, \$64,000).

22. For $(-1, 3)$:
$f(-1) = m(-1) + b$
$= -m + b = 3$
For $(-2, -4)$:
$f(-2) = m(-2) + b$
$= -2m + b = -4$
We have a system of equations:
$-m + b = 3$ (1)
$-2m + b = -4$ (2)
Solving we get $m = 7$ and $b = 10$. Thus the function is $f(x) = 7x + 10$.

23. *Familiarize*. Let $x =$ the number of pounds of Kona coffee

Translate. We organize the information in a table.

	Kona	Mexican	Mixture
Number of pounds	x	40	$x + 40$
Percent Kona	100%	0%	30%
Amount of Kona	$1.00x$	0	$0.30(x + 40)$

The last row of the table gives us one equation:
$1.00x + 0 = 0.3(x + 40)$.

Carry out. After clearing decimals, we solve the equation.
$100x + 0 = 30(x + 40)$
$100x = 30x + 1200$
$70x = 1200$
$x = \dfrac{120}{7}$

Check. 30% of $\left(\dfrac{120}{7} + 40\right)$ lb is $0.3\left(\dfrac{400}{7}\right)$, or $\dfrac{120}{7}$ lb.

of Kona coffee. The answer checks.

State. At least $\dfrac{120}{7}$ lb of Kona coffee is added to the 40 lb of Mexican coffee to market the mixture as Kona Blend.

Chapter 4

Inequalities and Problem Solving

Exercise Set 4.1

1. Because $-8 < -1$ is true, -8 is a *solution* of $x < -1$.

3. The interval $(-7, 1]$ is a *half-open* interval.

5. If we add $3x$ to both sides of the equation $5x + 7 = 6 - 3x$, we get the equation $8x + 7 = 6$, so these are equivalent equations.

7. If we add 7 to both sides of the inequality $x - 7 > -2$, we get the inequality $x > 5$ so these are equivalent inequalities.

9. If we multiply the equation $\frac{3}{5}a + \frac{1}{5} = 2$ by 5 on both sides, we get the equation $3a + 1 = 10$, so these are equivalent equations.

11. $x - 4 \geq 1$
    a. -4: We substitute and get $-4 - 4 \geq 1$, or $-8 \geq 1$, a false sentence. Therefore, -4 is not a solution.
    b. 4: We substitute and get $4 - 4 \geq 1$, or $0 \geq 1$, a false sentence. Therefore, 4 is not a solution.
    c. 5: We substitute and get $5 - 4 \geq 1$, or $1 \geq 1$, a true sentence. Therefore, 5 is a solution.
    d. 8: We substitute and get $8 - 4 \geq 1$, or $4 \geq 1$, a true sentence. Therefore, 8 is a solution.

13. $2y + 3 < 6 - y$
    a. 0: We substitute and get $2(0) + 3 < 6 - 0$, or $3 < 6$, a true sentence. Therefore, 0 is a solution.
    b. 1: We substitute and get $2(1) + 3 < 6 - 1$, or $5 < 5$, a false sentence. Therefore, 1 is not a solution.
    c. -1: We substitute and get $2(-1) + 3 < 6 - (-1)$, or $1 < 7$, a true sentence. Therefore, -1 is a solution.
    d. 4: We substitute and get $2(4) + 3 < 6 - 4$, or $11 < 2$, a false sentence. Therefore, 4 is not a solution.

15. $y < 6$
    Graph: The solutions consist of all real numbers less than 6, so we shade all numbers to the left of 6 and use a parenthesis at 6 to indicate that it is not a solution.

    Set builder notation: $\{y \mid y < 6\}$

    Interval notation: $(-\infty, 6)$

17. $x \geq -4$
    Graph: We shade all numbers to the right of -4 and use a bracket at -4 to indicate that it is also a solution.

    Set builder notation: $\{x \mid x \geq -4\}$

    Interval notation: $[-4, \infty)$

19. $t > -3$
    Graph: We shade all numbers to the right of -3 and use a parenthesis at -3 to indicate that it is not a solution.

    Set builder notation: $\{t \mid t > -3\}$

    Interval notation: $(-3, \infty)$

21. $x \leq -7$
    Graph: We shade all numbers to the left of -7 and use a bracket at -7 to indicate that it is also a solution.

    Set builder notation: $\{x \mid x \leq -7\}$

    Interval notation: $(-\infty, -7]$

23. $\quad x + 2 > 1$
    $x + 2 + (-2) > 1 + (-2) \quad$ Adding -2
    $\quad x > -1$
    The solution set is $\{x \mid x > -1\}$, or $(-1, \infty)$.

25. $\quad t - 6 \leq 4$
    $t - 6 + 6 \leq 4 + 6 \quad$ Adding 6
    $\quad t \leq 10$
    The solution set is $\{t \mid t \leq 10\}$, or $(-\infty, 10]$.

27. $\quad x - 12 \geq -11$
    $x - 12 + 12 \geq -11 + 12 \quad$ Adding 12
    $\quad x \geq 1$
    The solution set is $\{x \mid x \geq 1\}$, or $[1, \infty)$.

29. $\quad 9t < -81$
    $\frac{1}{9} \cdot 9t < \frac{1}{9}(-81) \quad$ Multiplying by $\frac{1}{9}$
    $\quad t < -9$
    The solution set is $\{t \mid t < -9\}$, or $(-\infty, -9)$.

31. $\quad -0.3x > -15$
    $-\frac{1}{0.3}(-0.3x) < -\frac{1}{0.3}(-15) \quad$ Multiplying by $-\frac{1}{0.3}$ and reversing the inequality symbol
    $\quad x < 50$
    The solution set is $\{x \mid x < 50\}$, or $(-\infty, 50)$.

33. $-9x \geq 8.1$

$-\frac{1}{9}(-9x) \leq -\frac{1}{9}(8.1)$ Multiplying by $-\frac{1}{9}$ and reversing the inequality symbol

$x \leq -0.9$

The solution set is $\{x \mid x \leq -0.9\}$, or $(-\infty, -0.9]$.

35. $\frac{3}{4}y \geq -\frac{5}{8}$

$\frac{4}{3}\left(\frac{3}{4}y\right) \geq \frac{4}{3}\left(-\frac{5}{8}\right)$ Multiplying by $\frac{4}{3}$

$y \geq -\frac{5}{6}$

The solution set is $\left\{y \mid y \geq -\frac{5}{6}\right\}$, or $\left[-\frac{5}{6}, \infty\right)$.

37. $3x + 1 < 7$

$3x < 6$ Adding -1

$x < 2$ Dividing by 3

The solution set is $\{x \mid x < 2\}$, or $(-\infty, 2)$.

39. $3 - x \geq 12$

$-x \geq 9$ Adding -3

$x \leq -9$ Dividing by -1 and reversing the inequality symbol

The solution set is $\{x \mid x \leq -9\}$, or $(-\infty, -9]$.

41. $\frac{2x+7}{5} < -9$

$5 \cdot \frac{2x+7}{5} < 5(-9)$ Multiplying by 5

$2x + 7 < -45$

$2x < -52$ Adding -7

$x < -26$ Dividing by 2

The solution set is $\{x \mid x < -26\}$, or $(-\infty, -26)$.

43. $\frac{3t-7}{-4} \leq 5$

$-4 \cdot \frac{3t-7}{-4} \geq -4 \cdot 5$ Multiplying by -4 and reversing the inequality symbol

$3t - 7 \geq -20$

$3t \geq -13$ Adding 7

$t \geq -\frac{13}{3}$ Dividing by 3

The solution set is $\left\{t \mid t \geq -\frac{13}{3}\right\}$, or $\left[-\frac{13}{3}, \infty\right)$.

45. $\frac{9-x}{-2} \geq -6$

$9 - x \leq 12$ Multipling by -2 and reversing the inequality symbol

$-x \leq 3$ Adding -9

$x \geq -3$ Multipling by -1 and reversing the inequality symbol

The solution set is $\{x \mid x \geq -3\}$, or $[-3, \infty)$.

47. $f(x) = 7 - 3x$, $g(x) = 2x - 3$

$f(x) \leq g(x)$

$7 - 3x \leq 2x - 3$

$7 - 5x \leq -3$ Adding $-2x$

$-5x \leq -10$ Adding -7

$x \geq 2$ Multiplying by $-\frac{1}{5}$ and reversing the inequality symbol

The solution set is $\{x \mid x \geq 2\}$, or $[2, \infty)$.

49. $f(x) = 2x - 7$, $g(x) = 5x - 9$

$f(x) < g(x)$

$2x - 7 < 5x - 9$

$-3x - 7 < -9$ Adding $-5x$

$-3x < -2$ Adding 7

$x > \frac{2}{3}$ Dividing by -3

The solution set is $\left\{x \mid x > \frac{2}{3}\right\}$, or $\left(\frac{2}{3}, \infty\right)$.

51. $y_1 = \frac{3}{8} + 2x$, $y_2 = 3x - \frac{1}{8}$

$y_2 \geq y_1$

$3x - \frac{1}{8} \geq \frac{3}{8} + 2x$

$x - \frac{1}{8} \geq \frac{3}{8}$ Adding $-2x$

$x \geq \frac{1}{2}$ Adding $\frac{1}{8}$

The solution set is $\left\{x \mid x \geq \frac{1}{2}\right\}$, or $\left[\frac{1}{2}, \infty\right)$.

53. $3 - 8y \geq 9 - 4y$

$-4y + 3 \geq 9$

$-4y \geq 6$

$y \leq -\frac{3}{2}$

The solution set is $\left\{y \mid y \leq -\frac{3}{2}\right\}$, or $\left(-\infty, -\frac{3}{2}\right]$.

55.
$$5(t-3)+4t<2(7+2t)$$
$$5t-15+4t<14+4t$$
$$9t-15<14+4t$$
$$5t-15<14$$
$$5t<29$$
$$t<\frac{29}{5}$$

The solution set is $\left\{t\,\middle|\,t<\frac{29}{5}\right\}$, or $\left(-\infty,\,\frac{29}{5}\right)$.

57.
$$5\big[3m-(m+4)\big]>-2(m-4)$$
$$5(3m-m-4)>-2(m-4)$$
$$5(2m-4)>-2(m-4)$$
$$10m-20>-2m+8$$
$$12m-20>8$$
$$12m>28$$
$$m>\frac{28}{12}$$
$$m>\frac{7}{3}$$

The solution set is $\left\{m\,\middle|\,m>\frac{7}{3}\right\}$, or $\left(\frac{7}{3},\,\infty\right)$.

59.
$$19-(2x+3)\le 2(x+3)+x$$
$$19-2x-3\le 2x+6+x$$
$$16-2x\le 3x+6$$
$$16-5x\le 6$$
$$-5x\le -10$$
$$x\ge 2$$

The solution set is $\{x\,|\,x\ge 2\}$, or $[2,\,\infty)$.

61.
$$\frac{1}{4}(8y+4)-17<-\frac{1}{2}(4y-8)$$
$$2y+1-17<-2y+4$$
$$2y-16<-2y+4$$
$$4y-16<4$$
$$4y<20$$
$$y<5$$

The solution set is $\{y\,|\,y<5\}$, or $(-\infty,\,5)$.

63.
$$2[8-4(3-x)]-2\ge 8[2(4x-3)+7]-50$$
$$2[8-12+4x]-2\ge 8[8x-6+7]-50$$
$$2[-4+4x]-2\ge 8[8x+1]-50$$
$$-8+8x-2\ge 64x+8-50$$
$$8x-10\ge 64x-42$$
$$-56x-10\ge -42$$
$$-56x\ge -32$$
$$x\le \frac{32}{56}$$
$$x\le \frac{4}{7}$$

The solution set is $\left\{x\,\middle|\,x\le\frac{4}{7}\right\}$, or $\left(-\infty,\,\frac{4}{7}\right]$.

65. Let n represent the number. Then we have $n<10$.

67. Let t represent the temperature, in °C. Then we have $t\le -3$.

69. Let a represent the age of the Mayan altar, in years. Then we have $a>1200$.

71. Let f represent the focus group session, in hours. Then we have $f\le 2$.

73. Let d represent the number of years of driving experience. Then we have $d\ge 5$.

75. Let c represent the cost of production, in dollars. Then we have $c\le 12{,}500$.

77. ***Familiarize***. Let n = the number of hours. Then the total fee using the hourly plan is $120n$.
Translate. We write an inequality stating that the hourly plan costs less than the flat fee.
$$120n<900$$
Carry out.
$$120n<900$$
$$n<\frac{15}{2},\text{ or }7\frac{1}{2}$$
Check. We can do a partial check by substituting a value for n less than $\frac{15}{2}$. When $n=7$, the hourly plan costs $120(7)$, or \$840, so the hourly plan is less than the flat fee of \$900. When $n=8$, the hourly plan costs $120(8)$, or \$960, so the hourly plan is more expensive than the flat fee of \$900.
State. The hourly rate is less expensive for lengths of time less than $7\frac{1}{2}$ hours.

79. ***Familiarize***. Let q = Chloe's undergraduate grade point average. Unconditional acceptance is 500 plus 200 times the grade point average.
Translate. We write an inequality stating the score is at least 1020.
$$500+200q\ge 1020.$$
Carry out.
$$500+200q\ge 1020$$
$$200q\ge 520$$
$$q\ge 2.6$$
Check. As a partial check we show that the acceptance score is 1020.
$$500+200(2.6)=500+520=1020.$$
State. For unconditional acceptance, Chloe must have a GPA of at least 2.6.

81. ***Familiarize***. Let n = the number of correct answers. Then the points earned are $2n$, and the points deducted are $\frac{1}{2}$ of the rest of the questions, $80-n$, or $\frac{1}{2}(80-n)$.
Translate. We write an inequality stating the score is at least 100.
$$2n-\frac{1}{2}(80-n)\ge 100.$$
Carry out.
$$2n-\frac{1}{2}(80-n)\ge 100$$
$$2n-40+\frac{1}{2}n\ge 100$$
$$\frac{5}{2}n-40\ge 100$$
$$\frac{5}{2}n\ge 140$$
$$n\ge 56$$
Check. When $n=56$, the score earned is
$$2(56)-\frac{1}{2}(80-56),\text{ or }112-\frac{1}{2}(24),\text{ or }112-12,\text{ or}$$
100. When $n=58$, the score earned is

$2(58) - \frac{1}{2}(80-58)$, or $116 - \frac{1}{2}(22)$, or $116 - 11$, or 105. Since the score is exactly 100 when 56 questions are answered correctly and more than 100 when 58 questions are correct, we have performed a partial check.

State. At least 56 questions are correct for a score of at least 100.

83. **Familiarize.** Let d = the depth of the well, in feet. Then the cost on the pay-as-you-go plan is $\$500 + \$8d$. The cost of the guaranteed-water plan is $4000. We want to find the values of d for which the pay-as-you-go plan costs less than the guaranteed-water plan.

Translate. We write an inequality stating that the charge is at most $4000.

$$500 + 8d < 4000$$

Carry out.

$$500 + 8d < 4000$$
$$8d < 3500$$
$$d < 437.5$$

Check. We check to see that the solution is reasonable.

When $d = 437$, $\$500 + \$8 \cdot 437 = \$3996 < \$4000$

When $d = 437.5$, $\$500 + \$8(437.5) = \$4000$

When $d = 438$, $\$500 + \$8(438) = \$4004 > \$4000$

From these calculations, it appears that the solution is correct.

State. It would save a customer money to use the pay-as-you-go plan for a well of less than 437.5 ft.

85. **Familiarize.** We list the given information in a table.

Plan A: Monthly Income	Plan B: Monthly Income
$400 salary	$610
8% of sales	5% of sales
Total: 400+8% of sales	Total: 610+5% of sales

Suppose Toni had gross sales of $5000 one month. Then under plan A she would earn

$\$400 + 0.08(\$5000)$, or $800.

Under plan B she would earn

$\$610 + 0.05(\$5000)$, or $860.

This shows that, for gross sales of $5000, plan B is better.

If Toni had gross sales of $10,000 one month, then under plan A she would earn

$\$400 + 0.08(\$10,000)$, or $1200.

Under plan B she would earn

$\$610 + 0.05(\$10,000)$, or $1110.

This shows that, for gross sales of $10,000, plan A is better. To determine all values for which plan A is better we solve an inequality.

Translate.

Income from plan A	is greater than	Income from plan B.
↓	↓	↓
$400 + 0.08s$	$>$	$610 + 0.05s$

Carry out.

$$400 + 0.08s > 610 + 0.05s$$
$$400 + 0.03s > 610$$
$$0.03s > 210$$
$$s > 7000$$

Check. For $s = \$7000$, the income from plan A is

$\$400 + 0.08(\$7000)$, or $960

and the income from plan B is

$\$610 + 0.05(\$7000)$, or $960.

This shows that for sales of $7000 Toni's income is the same from each plan. In the Familiarize step we show that, for a value less than $7000, plan B is better and, for a value greater than $7000, plan A is better. Since we cannot check all possible values, we stop here.

State. Toni should select plan A for gross sales greater than $7000.

87. **Familiarize.** Let b = the number of bins collected. Then the Purple Plan will cost $5 + $3b per month and the Blue Plan will cost $15 + $1.75b per month.

Translate. We write an inequality stating that the Blue Plan costs less than the Purple Plan.

$$5 + 3b > 15 + 1.75b$$

Carry out.

$$5 + 3b > 15 + 1.75b$$
$$1.25b > 10$$
$$b > 8$$

Check. We do a partial check by substituting a value for b less than 8 and a value for b greater than 8.

When $b = 7$, the Purple Plan costs $5 + $3(7), or $26 and the Blue Plan costs $15 + $1.75(7), or $27.25. So the Purple Plan is less expensive.

When $b = 9$, the Purple Plan costs $5 + $3(9), or $32, and the Blue Plan costs $15 + $1.75(9), or $30.75. So the Blue Plan is less expensive.

State. The Blue Plan costs less for more than 8 bins collected per month.

89. **Familiarize.** Find the values of t for which $c(t) > 3.42$.

Translate. $-0.42t + 11 < 3.42$

Carry out.

$$-0.42t + 11 < 3.42$$
$$-0.42t < -7.58$$
$$t > 18.05$$

Check. $c(18.05) \approx 3.42$.

When $t = 18$, $c(18) = -0.42(18) + 11$, or 3.44.

When $t = 19$, $c(19) = -0.42(19) + 11$, or 3.02.

State. The cost will be less in the United States than the 2011 cost in Germany 19 or more years after 2011, or 2019 and later.

91. a. **Familiarize.** Find the values of d for which $F(d) > 25$.

Translate.

$$\left(\frac{4.95}{d} - 4.50\right) \times 100 > 25$$

Carry out.

$$\left(\frac{4.95}{d} - 4.50\right) \times 100 > 25$$

$$\frac{495}{d} - 450 > 25$$

$$\frac{495}{d} > 475$$

$$495 > 475d$$

$$\frac{495}{475} > d$$

$$\frac{99}{95} > d, \text{ or } d < 1.04$$

Check. When $d = 1$,

$$F(1) = \left(\frac{4.95}{1} - 4.50\right) \times 100, \text{ or } 45 \text{ percent}.$$

When $d = 1.05$,

$$F(1.05) = \left(\frac{4.95}{1.05} - 4.50\right) \times 100, \text{ or } 21 \text{ percent}.$$

State. A man is considered obese for body density less than $\frac{99}{95}$ kg/L, or about 1.04 kg/L.

b.   Familiarize. Find the values of d for which $F(d) > 32$.

Translate.

$$\left(\frac{4.95}{d} - 4.50\right) \times 100 > 32$$

Carry out.

$$\left(\frac{4.95}{d} - 4.50\right) \times 100 > 32$$

$$\frac{495}{d} - 450 > 32$$

$$\frac{495}{d} > 482$$

$$495 > 482d$$

$$\frac{495}{482} > d, \text{ or } d < 1.03$$

Check. Our check from part (a) leads to the result that $F(d) > 32$ when $d < 1.03$.

State. A woman is considered obese for body density less than $\frac{495}{482}$ kg/L, or about 1.03 kg/L.

93. Familiarize. Find the values of t for which $p(t) > f(t)$.

Translate.

$$27t + 325 > 16t + 500$$

Carry out.

$$27t + 325 > 16t + 500$$

$$11t > 175$$

$$t > 15.9$$

Check. $p(15.9) = 754.3 \approx r(15.9)$. Calculate $p(t)$ and $f(t)$ for some t greater than 15.9 and for some t less than 15.9.

Suppose $t = 15$:

$$p(15) = 27(15) + 325 = 730 \text{ and}$$
$$f(15) = 16(15) + 500 = 740.$$

In this case $p(t) < f(t)$.

Suppose $t = 16$:

$$p(16) = 27(16) + 325 = 757 \text{ and}$$
$$f(16) = 16(16) + 500 = 756.$$

In this case $p(t) > f(t)$..

Then for $t > 16$, $p(t) > f(t)$.

State. There will be more part-time faculty than full-time faculty in the year 2011 and later.

95. a.   Familiarize. Find the values of x for which $R(x) < C(x)$.

Translate.

$$48x < 90,000 + 25x$$

Carry out.

$$23x < 90,000$$

$$x < 3913\frac{1}{23}$$

Check. $R\left(3913\frac{1}{23}\right) = \$187,826.09 = C\left(3913\frac{1}{23}\right)$.

Calculate $R(x)$ and $C(x)$ for some x greater than $3913\frac{1}{23}$ and for some x less than $3913\frac{1}{23}$.

Suppose $x = 4000$:

$$R(x) = 48(4000) = 192,000 \text{ and}$$
$$C(x) = 90,000 + 25(4000) = 190,000.$$

In this case $R(x) > C(x)$.

Suppose $x = 3900$:

$$R(x) = 48(3900) = 187,200 \text{ and}$$
$$C(x) = 90,000 + 25(3900) = 187,500.$$

In this case $R(x) < C(x)$.

Then for $x < 3913\frac{1}{23}$, $R(x) < C(x)$.

State. We will state the result in terms of integers, since the company cannot sell a fraction of a lamp. For 3913 or fewer lamps the company loses money.

b.   Our check in part a) shows that for $x > 3913\frac{1}{23}$, $R(x) > C(x)$ and the company makes a profit. Again, we will state the result in terms of an integer. For more than 3913 lamps the company makes money.

97.   Writing Exercise.

99.
$$x - (9 - x) = -3(x + 5)$$
$$x - 9 + x = -3x - 15$$
$$2x - 9 = -3x - 15$$
$$5x = -6$$
$$x = -\frac{6}{5}$$

The solution is $-\frac{6}{5}$.

101.
$$\begin{array}{ll} 2x - 3y = 5 & (1) \\ \underline{x + 3y = -1} & (2) \\ 3x + 0 = 4 & \text{Adding} \end{array}$$

$$3x = 4$$

$$x = \frac{4}{3}$$

Substitute $\frac{4}{3}$ for x in Equation (2) and solve for y.

$$x + 3y = -1$$

$$\frac{4}{3} + 3y = -1 \qquad \text{Substituting}$$

$$3y = -\frac{7}{3}$$

$$y = -\frac{7}{9}$$

We obtain $\left(\frac{4}{3}, -\frac{7}{9}\right)$. This checks, so it is the solution.

103.
$$ar = b - cr$$
$$ar + cr = b$$
$$r(a + c) = b$$
$$r = \frac{b}{a+c}$$

105. *Writing Exercise.*

107. $3ax + 2x \geq 5ax - 4$
$$2x - 2ax \geq -4$$
$$2x(1 - a) \geq -4$$
$$x(1 - a) \geq -2$$
$$x \leq -\frac{2}{1-a}, \text{ or } \frac{2}{a-1}$$

We reversed the inequality symbol when we divided because when $a > 1$, then $1 - a < 0$.

The solution set is $\left\{ x \middle| x \leq \frac{2}{a-1} \right\}$.

109. $a(by - 2) \geq b(2y + 5)$
$$aby - 2a \geq 2by + 5b$$
$$aby - 2by \geq 2a + 5b$$
$$y(ab - 2b) \geq 2a + 5b$$
$$y \geq \frac{2a + 5b}{ab - 2b}, \text{ or } \frac{2a + 5b}{b(a - 2)}$$

The inequality symbol remained unchanged when we divided because when $a > 2$ and $b > 0$, then $ab - 2b > 0$.

The solution set is $\left\{ y \middle| y \geq \frac{2a + 5b}{b(a - 2)} \right\}$.

111. $c(2 - 5x) + dx > m(4 + 2x)$
$$2c - 5cx + dx > 4m + 2mx$$
$$-5cx + dx - 2mx > 4m - 2c$$
$$x(-5c + d - 2m) > 4m - 2c$$
$$x[d - (5c + 2m)] > 4m - 2c$$
$$x > \frac{4m - 2c}{d - (5c + 2m)}$$

The inequality symbol remained unchanged when we divided because when $5c + 2m < d$, then $d - (5c + 2m) > 0$.

The solution set is $\left\{ x \middle| x > \frac{4m - 2c}{d - (5c + 2m)} \right\}$.

113. False. If $a = 2$, $b = 3$, $c = 4$, and $d = 5$, then $2 < 3$ and $4 < 5$ but $2 - 4 = 3 - 5$.

115. *Writing Exercise.*

117. $x + 5 \leq 5 + x$
$$5 \leq 5 \qquad \text{Subtracting } x$$

We get an inequality that is true for all real numbers x. Thus the solution set is all real numbers.

119. $0^2 = 0$, $x^2 > 0$ for $x \neq 0$

The solution is $\{ x \mid x \text{ is a real number } and \ x \neq 0 \}$.

121. From Exercise 120, the first option, prepayment, was more economical if more than 7.8 gal of gas was used. If the car gets 30 mpg, then Abriana would need to drive at least $7.8 \text{ gal} \times 30 \text{ mpg}$, or about 234 miles to make the first option more economical.

123. a. The graph of y_1 lies above the graph of y_2 for x-values to the left of the point of intersection, or in the interval $(-\infty, 4)$.

    b. The graph of y_2 lies on or below the graph of y_3 for x-values at and to the right of the point of intersection, or in the interval $[2, \infty)$.

    c. The graph of y_3 lies on or above the graph of y_1 at and to the right of the point of intersection, or in the interval $[3.2, \infty)$.

Exercise Set 4.2

1. The *intersection* of two sets is the set of all elements that are in both sets.

3. The word "and" corresponds to *intersection*.

5. The solution of a disjunction is the *union* of the solution sets of the individual sentences.

7. h

9. f

11. e

13. b

15. c

17. $\{2, 4, 16\} \cap \{4, 16, 256\}$

The numbers 4 and 16 are common to both sets, so the intersection is $\{4, 16\}$.

19. $\{0, 5, 10, 15\} \cup \{5, 15, 20\}$

The numbers in either or both sets are 0, 5, 10, 15, and 20, so the union is $\{0, 5, 10, 15, 20\}$.

21. $\{a, b, c, d, e, f\} \cap \{b, d, f\}$

The letters b, d, and f are common to both sets, so the intersection is $\{b, d, f\}$.

23. $\{x, y, z\} \cup \{u, v, x, y, z\}$

The letters in either or both sets are u, v, x, y, and z, so the union is $\{u, v, x, y, z\}$.

25. $\{3, 6, 9, 12\} \cap \{5, 10, 15\}$

There are no numbers common to both sets, so the solution set has no members. It is $\varnothing$.

27. $\{1, 3, 5\} \cup \varnothing$

The numbers in either or both sets are 1, 3, and 5, so the union is $\{1, 3, 5\}$.

29. $1 < x < 3$

This inequality is an abbreviation for the conjunction $1 < x$ *and* $x < 3$. The graph is the intersection of two separate solution sets:

$\{x | 1 < x\} \cap \{x | x < 3\} = \{x | 1 < x < 3\}$.

Interval notation: $(1, 3)$

31. $-6 \leq y \leq 0$

This inequality is an abbreviation for the conjunction $-6 \leq y$ and $y \leq 0$.

Interval notation: $[-6, 0]$

33. $x < -1$ *or* $x > 4$

The graph of this disjunction is the union of the graphs of the individual solution sets $\{x | x < -1\}$ and $\{x | x > 4\}$.

Interval notation: $(-\infty, -1) \cup (4, \infty)$

35. $x \leq -2$ *or* $x > 1$

Interval notation: $(-\infty, -2] \cup (1, \infty)$

37. $-4 \leq -x < 2$

$\quad 4 \geq x > -2$ Multiplying by -1 and
$\qquad\qquad\qquad$ reversing the inequality symbols
$\quad -2 < x \leq 4$ Rewriting

Interval notation: $(-2, 4]$

39. $x > -2$ *and* $x < 4$

This conjunction can be abbreviated as $-2 < x < 4$.

Interval notation: $(-2, 4)$

41. $5 > a$ *or* $a > 7$

Interval notation: $(-\infty, 5) \cup (7, \infty)$

43. $x \geq 5$ *or* $-x \geq 4$

Multiplying the second inequality by -1 and reversing the inequality symbols, we get $x \geq 5$ or $x \leq -4$.

Interval notation: $(-\infty, -4] \cup [5, \infty)$

45. $7 > y$ *and* $y \geq -3$

This conjunction can be abbreviated as $-3 \leq y < 7$.

Interval notation: $[-3, 7)$

47. $-x < 7$ *and* $-x \geq 0$

Multiplying the inequalities by -1 and reversing the inequality symbols, we get $x > -7$ *and* $x \leq 0$.

Interval notation: $(-7, 0]$

49. $t < 2$ *or* $t < 5$

Observe that every number that is less than 2 is also less than 5. Then $t < 2$ *or* $t < 5$ is equivalent to $t < 5$ and the graph of this disjunction is the set $\{t | t < 5\}$.

Interval notation: $(-\infty, 5)$

51. $\quad -3 \leq x + 2 < 9$
$\quad -3 - 2 \leq x < 9 - 2$
$\qquad -5 \leq x < 7$

The solution set is $\{x | -5 \leq x < 7\}$ or $[-5, 7)$.

53. $\quad 0 < t - 4 \quad$ *and* $\quad t - 1 \leq 7$
$\qquad 4 < t \qquad$ *and* $\qquad t \leq 8$

We can abbreviate the answer as $4 < t \leq 8$.
The solution set is $\{t | 4 < t \leq 8\}$, or $(4, 8]$.

55. $\quad -7 \leq 2a - 3 \quad$ *and* $\quad 3a + 1 < 7$
$\qquad -4 \leq 2a \qquad$ *and* $\qquad 3a < 6$
$\qquad -2 \leq a \qquad$ *and* $\qquad a < 2$

We can abbreviate the answer as $-2 \leq a < 2$. The solution set is $\{a | -2 \leq a < 2\}$, or $[-2, 2)$.

57. $x + 3 \leq -1$ *or* $x + 3 > -2$

Observe that any real number is either less than or equal to -1 or greater than or equal to -2. Then the solution set is $\{x | x \text{ is a real number}\}$, or $(-\infty, \infty)$.

59. $\quad -10 \leq 3x - 1 \leq 5$
$\qquad -9 \leq 3x \leq 6$
$\qquad -3 \leq x \leq 2$

The solution set is $\{x | -3 \leq x \leq 2\}$, or $[-3, 2]$.

61. $\quad 5 > \dfrac{x - 3}{4} > 1$
$\quad 20 > x - 3 > 4$ Multiplying by 4
$\quad 23 > x > 7$, or
$\qquad 7 < x < 23$

The solution set is $\{x | 7 < x < 23\}$, or $(7, 23)$.

63. $\quad -2 \leq \dfrac{x + 2}{-5} \leq 6$
$\quad 10 \geq x + 2 \geq -30$ $\qquad$ Multiplying by -5
$\quad 8 \geq x \geq -32$, or
$\quad -32 \leq x \leq 8$

The solution set is $\{x | -32 \leq x \leq 8\}$, or $[-32, 8]$.

65. $2 \le 3x - 1 \le 8$

$\quad 3 \le 3x \le 9$

$\quad 1 \le x \le 3$

The solution set is $\{x \mid 1 \le x \le 3\}$, or $[1, 3]$.

67. $-21 \le -2x - 7 < 0$

$\quad -14 \le -2x < 7$

$\quad 7 \ge x > -\dfrac{7}{2}$, or

$\quad -\dfrac{7}{2} < x \le 7$

The solution set is $\left\{x \mid -\dfrac{7}{2} < x \le 7\right\}$, or $\left(-\dfrac{7}{2}, 7\right]$.

69. $5t + 3 < 3 \quad or \quad 5t + 3 > 8$

$\quad\quad 5t < 0 \quad or \quad\quad 5t > 5$

$\quad\quad\; t < 0 \quad or \quad\quad\; t > 1$

The solution set is $\{t \mid t < 0 \; or \; t > 1\}$, or

$(-\infty, 0) \cup (1, \infty)$.

71. $6 > 2a - 1 \quad or \quad -4 \le -3a + 2$

$\quad 7 > 2a \quad\quad or \quad -6 \le -3a$

$\quad \dfrac{7}{2} > a \quad\quad or \quad\quad 2 \ge a$

The solution set is $\left\{a \mid \dfrac{7}{2} > a\right\} \cup \{a \mid 2 \ge a\} = \left\{a \mid \dfrac{7}{2} > a\right\}$,

or $\left\{a \mid a < \dfrac{7}{2}\right\}$, or $\left(-\infty, \dfrac{7}{2}\right)$.

73. $a + 3 < -2 \quad and \quad 3a - 4 < 8$

$\quad\quad a < -5 \quad and \quad\quad 3a < 12$

$\quad\quad a < -5 \quad and \quad\quad\; a < 4$

The solution set is

$\{a \mid a < -5\} \cap \{a \mid a < 4\} = \{a \mid a < -5\}$, or $(-\infty, -5)$.

75. $3x + 2 < 2 \quad and \quad 3 - x < 1$

$\quad\quad 3x < 0 \quad and \quad\;\; -x < -2$

$\quad\quad\; x < 0 \quad and \quad\quad\;\; x > 2$

The solution set is $\{x \mid x < 0\} \cap \{x \mid x > 2\} = \varnothing$.

77. $2t - 7 \le 5 \quad or \quad 5 - 2t > 3$

$\quad\quad 2t \le 12 \quad or \quad\;\; -2t > -2$

$\quad\quad\; t \le 6 \quad or \quad\quad\;\; t < 1$

The solution set is $\{t \mid t \le 6\} \cup \{t \mid t < 1\} = \{t \mid t \le 6\}$, or

$(-\infty, 6]$.

79. $f(x) = \dfrac{9}{x + 6}$

$f(x)$ cannot be computed when the denominator is 0.

Since $x + 6 = 0$ is equivalent to $x = -6$, we have

Domain of $f = \{x \mid x$ is a real number and $x \ne -6\}$

$= (-\infty, -6) \cup (-6, \infty)$.

81. $f(x) = \dfrac{1}{x}$

$f(x)$ cannot be computed when the denominator is 0.

We have Domain of f

$= \{x \mid x$ is a real number and $x \ne 0\} = (-\infty, 0) \cup (0, \infty)$.

83. $f(x) = \dfrac{x + 3}{2x - 8}$

$f(x)$ cannot be computed when the denominator is 0.

Since $2x - 8 = 0$ is equivalent to $x = 4$, we have

Domain of $f = \{x \mid x$ is a real number and $x \ne 4\}$, or

$(-\infty, 4) \cup (4, \infty)$.

85. $f(x) = \sqrt{x - 10}$

$x - 10 \ge 0 \quad\quad x - 10$ must be nonnegative.

$\quad x \ge 10 \quad\quad$ Adding 10

When $x \ge 10$, the expression $x - 10$ is nonnegative.

Thus the domain of f is $\{x \mid x \ge 10\}$, or $[10, \infty)$.

87. $f(x) = \sqrt{3 - x}$

$3 - x \ge 0 \quad\quad 3 - x$ must be nonnegative.

$\;\; -x \ge -3 \quad\quad$ Adding -3

$\quad\;\; x \le 3 \quad\quad$ Multiplying by -1 and

$\quad\quad\quad\quad\quad\quad$ reversing the inequality symbol

When $x \le 3$, the expression $3 - x$ is nonnegative.

Thus the domain of f is $\{x \mid x \le 3\}$, or $(-\infty, 3]$.

89. $f(x) = \sqrt{2x + 7}$

$2x + 7 \ge 0 \quad\quad 2x + 7$ must be nonnegative.

$\quad 2x \ge -7 \quad\quad$ Adding -7

$\quad\;\; x \ge -\dfrac{7}{2} \quad\quad$ Dividing by 2

When $x \ge -\dfrac{7}{2}$, the expression $2x + 7$ is nonnegative.

Thus the domain of f is $\left\{x \mid x \ge -\dfrac{7}{2}\right\}$, or $\left[-\dfrac{7}{2}, \infty\right)$.

91. $f(x) = \sqrt{8 - 2x}$

$8 - 2x \ge 0 \quad\quad 8 - 2x$ must be nonnegative.

$\;\; -2x \ge -8 \quad\quad$ Adding -8

$\quad\;\; x \le 4 \quad\quad$ Dividing by -2 and reversing

$\quad\quad\quad\quad\quad\quad$ the inequality symbol

When $x \le 4$, the expression $8 - 2x$ is nonnegative.

Thus the domain of f is $\{x \mid x \le 4\}$, or $(-\infty, 4]$.

93. *Writing Exercise.*

95. $-\dfrac{2}{15}\left(-\dfrac{5}{8}\right) = \dfrac{1}{12}$

97. $-2 - 6^2 \div 4(-3) - (8 - 12)$
$= -2 - 36 \div 4(-3) - (-4)$
$= -2 - 9(-3) + 4$
$= -2 + 27 + 4$
$= 29$

99. *Writing Exercise.*

101. From the graph we observe that the values of x for which $2x - 5 > -7$ *and* $2x - 5 < 7$ are $\{x \mid -1 < x < 6\}$, or $(-1, 6)$.

103. Solve $1 \le P(d) \le 7$, or $1 \le 1 + \dfrac{d}{33} \le 7$.

$$1 \le 1 + \frac{d}{33} \le 7$$
$$0 \le \frac{d}{33} \le 6$$
$$0 \le d \le 198$$

Thus, $0 \text{ ft} \le d \le 198 \text{ ft}$.

105. Solve $25 \le F(d) \le 31$.
$$25 \le (4.95 / d - 4.50) \times 100 \le 31$$
$$25 \le \frac{495}{d} - 450 \le 31$$
$$475 \le \frac{495}{d} \le 481$$
$$\frac{1}{475} \ge \frac{d}{495} \ge \frac{1}{481}$$
$$\frac{495}{475} \ge d \ge \frac{495}{481},$$
or $1.03 \le d \le 1.04$.

Acceptable body densities are between 1.03 kg/L and 1.04 kg/L.

107. Let $c =$ the number of crossings in six months. Then at the $6 per crossing rate, the total cost of c crossings is $6c$. A six-month pass costs $50 and additional $2 per crossing toll brings the total cost of c crossings to $50 + $2c$. An unlimited crossing pass costs $300.

We write an inequality that states that the cost of c crossings using six-month passes is less than the cost using $6 per crossing toll and is less than the cost of using the unlimited-trip pass. Solve:
$$50 + 2c < 6c \quad and \quad 50 + 2c < 300.$$
We get $12.5 < c$ *and* $c < 125$.
For more than 12 crossings but less than 125 crossings in six months, the reduced-fare pass is more economical.

109. $4m - 8 > 6m + 5 \quad or \quad 5m - 8 < -2$
$\quad\quad -13 > 2m \quad\quad or \quad\quad 5m < 6$
$\quad\quad -\dfrac{13}{2} > m \quad\quad or \quad\quad m < \dfrac{6}{5}$

$\left\{ m \mid m < \dfrac{6}{5} \right\}$, or $\left(-\infty, \dfrac{6}{5} \right)$

111. $3x < 4 - 5x < 5 + 3x$
$\quad\quad 0 < 4 - 8x < 5$
$\quad\quad -4 < -8x < 1$
$\quad\quad \dfrac{1}{2} > x > -\dfrac{1}{8}$

The solution set is $\left\{ x \mid -\dfrac{1}{8} < x < \dfrac{1}{2} \right\}$, or $\left(-\dfrac{1}{8}, \dfrac{1}{2} \right)$.

113. Let $a = b = c = 2$. Then $a \le c$ and $c \le b$, but $b \not> a$. The given statement is false.

115. If $-a < c$, then $-1(-a) > -1 \cdot c$, or $a > -c$. Then if $a > -c$ and $-c > b$, we have $a > -c > b$, so $a > b$ and the given statement is true.

117. $f(x) = \dfrac{\sqrt{3 - 4x}}{x + 7}$

$3 - 4x \ge 0$ is equivalent to $x \le \dfrac{3}{4}$ and $x + 7 = 0$ is equivalent to $x = -7$. Then we have Domain of $f = \left\{ x \mid x \le \dfrac{3}{4} \text{ and } x \ne -7 \right\}$, or $(-\infty, -7) \cup \left(-7, \dfrac{3}{4} \right]$.

119. Observe that the graph of y_2 lies on or above the graph of y_1 and below the graph of y_3 for x in the interval $[-3, 4)$.

121. We substitute 1.2 for s in the formula.
$$r = 206.835 - 1.015n - 84.6s$$
$$r = 206.835 - 1.015n - 84.6(1.2)$$
$$r = 105.315 - 1.015n$$
5th graders score $90 \le r \le 100$. Substitute $105.315 - 1.015n$ for r in the inequality and solve for n.
$$90 \le 105.315 - 1.015n \le 100$$
$$-15.315 \le -1.015n \le -5.315$$
$$15.09 \ge n \ge 5.24$$
His average sentence should be between 5.24 and 15.09 words per sentence.

123. Let w represent the number of ounces in a bottle;
$$15.9 \le w \le 16.1, \text{ or } [15.9, 16.1]$$

125. *Graphing Calculator Exercise*

127. *Graphing Calculator Exercise*

Exercise Set 4.3

1. $|x| \geq 0$, so the statement is true.

3. True

5. True

7. False; the solution is all real numbers.

9. g

11. d

13. a

15. $|x| = 10$
$x = -10$ or $x = 10$　Using the absolute value principle
The solution set is $\{-10, 10\}$.

17. $|x| = -1$
The absolute value of a number is always nonnegative.
Therefore, the solution set is $\varnothing$.

19. $|p| = 0$
The only number whose absolute value is 0 is 0. The solution set is $\{0\}$.

21. $|2x - 3| = 4$

$$2x - 3 = -4 \quad or \quad 2x - 3 = 4 \qquad \text{Absolute-value}$$
$$\qquad\qquad\qquad\qquad\qquad\qquad\qquad\qquad \text{principle}$$
$$2x = -1 \quad or \qquad 2x = 7$$
$$x = -\frac{1}{2} \quad or \qquad x = \frac{7}{2}$$

The solution set is $\left\{-\frac{1}{2}, \frac{7}{2}\right\}$.

23. $|3x + 5| = -8$
Absolute value is always nonnegative, so the equation has no solution. The solution set is $\varnothing$.

25. $|x - 2| = 6$

$$x - 2 = -6 \quad or \quad x - 2 = 6 \qquad \text{Absolute-value}$$
$$\qquad\qquad\qquad\qquad\qquad\qquad\qquad \text{principle}$$
$$x = -4 \quad or \qquad x = 8$$

The solution set is $\{-4, 8\}$.

27. $|x - 7| = 1$
$$x - 7 = -1 \quad or \quad x - 7 = 1$$
$$x = 6 \quad or \qquad x = 8$$
The solution set is $\{6, 8\}$.

29. $|t| + 1.1 = 6.6$
$$|t| = 5.5 \qquad \text{Adding } -1.1$$
$t = -5.5$ or $t = 5.5$
The solution set is $\{-5.5, 5.5\}$.

31. $|5x| - 3 = 37$
$$|5x| = 40 \qquad \text{Adding 3}$$
$$5x = -40 \quad or \quad 5x = 40$$
$$x = -8 \quad or \qquad x = 8$$
The solution set is $\{-8, 8\}$.

33. $7|q| + 2 = 9$
$$7|q| = 7 \qquad \text{Adding } -2$$
$$|q| = 1 \qquad \text{Multiplying by } \frac{1}{7}$$
$q = -1$ or $q = 1$
The solution set is $\{-1, 1\}$.

35. $\left|\dfrac{2x - 1}{3}\right| = 4$

$$\frac{2x - 1}{3} = -4 \quad or \quad \frac{2x - 1}{3} = 4$$
$$2x - 1 = -12 \quad or \quad 2x - 1 = 12$$
$$2x = -11 \quad or \qquad 2x = 13$$
$$x = -\frac{11}{2} \quad or \qquad x = \frac{13}{2}$$

The solution set is $\left\{-\frac{11}{2}, \frac{13}{2}\right\}$.

37. $|5 - m| + 9 = 16$
$$|5 - m| = 7 \qquad \text{Adding } -9$$
$$5 - m = -7 \quad or \quad 5 - m = 7$$
$$-m = -12 \quad or \qquad -m = 2$$
$$m = 12 \quad or \qquad m = -2$$
The solution set is $\{-2, 12\}$.

39. $5 - 2|3x - 4| = -5$
$$-2|3x - 4| = -10$$
$$|3x - 4| = 5$$
$$3x - 4 = -5 \quad or \quad 3x - 4 = 5$$
$$3x = -1 \quad or \qquad 3x = 9$$
$$x = -\frac{1}{3} \quad or \qquad x = 3$$

The solution set is $\left\{-\frac{1}{3}, 3\right\}$.

41. $|2x + 6| = 8$
$$2x + 6 = -8 \quad or \quad 2x + 6 = 8$$
$$2x = -14 \quad or \qquad 2x = 2$$
$$x = -7 \quad or \qquad x = 1$$
The solution set is $\{-7, 1\}$.

43. $|x| - 3 = 5.7$
$$|x| = 8.7$$
$$x = -8.7 \quad or \quad x = 8.7$$
The solution set is $\{-8.7, 8.7\}$.

45. $\left|\dfrac{1 - 2x}{5}\right| = 2$

$$\frac{1 - 2x}{5} = -2 \quad or \quad \frac{1 - 2x}{5} = 2$$
$$1 - 2x = -10 \quad or \quad 1 - 2x = 10$$
$$-2x = -11 \quad or \qquad -2x = 9$$
$$x = \frac{11}{2} \quad or \qquad x = -\frac{9}{2}$$

The solution set is $\left\{-\frac{9}{2}, \frac{11}{2}\right\}$.

47. $|x-7| = |2x+1|$

$\qquad x-7 = 2x+1 \quad or \quad x-7 = -(2x+1)$
$\qquad -x = 8 \qquad\quad or \quad x-7 = -2x-1$
$\qquad x = -8 \qquad\quad or \qquad 3x = 6$
$\qquad\qquad\qquad\qquad\qquad\qquad x = 2$

The solution set is $\{-8, \ 2\}$.

49. $|x+4| = |x-3|$

$\qquad x+4 = x-3 \quad or \quad x+4 = -(x-3)$
$\qquad\quad 4 = -3 \qquad or \quad x+4 = -x+3$
$\qquad\quad$ False $\qquad\qquad\qquad 2x = -1$
$\qquad\qquad\qquad\qquad\qquad\qquad x = -\dfrac{1}{2}$

The solution set is $\left\{-\dfrac{1}{2}\right\}$.

51. $|3a-1| = |2a+4|$

$\qquad 3a-1 = 2a+4 \quad or \quad 3a-1 = -(2a+4)$
$\qquad\quad a-1 = 4 \qquad\quad or \quad 3a-1 = -2a-4$
$\qquad\qquad a = 5 \qquad\quad or \quad 5a-1 = -4$
$\qquad\qquad\qquad\qquad\qquad\qquad 5a = -3$
$\qquad\qquad\qquad\qquad\qquad\qquad a = -\dfrac{3}{5}$

The solution set is $\left\{-\dfrac{3}{5}, \ 5\right\}$.

53. Since $|n-3|$ and $|3-n|$ are equivalent expressions, the solution set of $|n-3| = |3-n|$ is the set of all real numbers.

55. $|7-4a| = |4a+5|$

$\qquad 7-4a = 4a+5 \quad or \quad 7-4a = -(4a+5)$
$\qquad\quad -8a = -2 \qquad or \quad 7-4a = -4a-5$
$\qquad\qquad a = \dfrac{1}{4} \qquad\quad or \qquad\quad 7 = -5$
$\qquad\qquad\qquad\qquad\qquad\qquad$ False

The solution set is $\left\{\dfrac{1}{4}\right\}$.

57. $|a| \le 3$

$\qquad -3 \le a \le 3 \qquad$ Part (b)

The solution set is $\{a| -3 \le a \le 3\}$, or $[-3, 3]$.

59. $|t| > 0$

$\qquad t < 0$ or $0 < t \quad$ Part (c)

The solution set is $\{t \,|\, t < 0 \ or \ t > 0\}$, or $\{t \,|\, t \ne 0\}$, or $(-\infty, \ 0) \cup (0, \ \infty)$.

61. $|x-1| < 4$

$\qquad -4 < x-1 < 4 \quad$ Part (b)
$\qquad -3 < x < 5$

The solution set is $\{x \,|\, -3 < x < 5\}$, or $(-3, \ 5)$.

63. $|n+2| \le 6$

$\qquad -6 \le n+2 \le 6 \qquad$ Part (b)
$\qquad -8 \le n \le 4$

The solution set is $\{n| -8 \le n \le 4\}$, or $[-8, 4]$.

65. $|x-3| + 2 > 7$

$\qquad |x-3| > 5 \qquad$ Adding -2

$\qquad x-3 < -5 \quad or \quad 5 < x-3 \quad$ Part(c)
$\qquad\quad x < -2 \quad or \quad 8 < x$

The solution set is $\{x \,|\, x < -2 \ or \ x > 8\}$, or $(-\infty, -2) \cup (8, \ \infty)$.

67. $|2y-9| > -5$

Since absolute value is never negative, any value of $2y-9$, and hence any value of y, will satisfy the inequality. The solution set is the set of all real numbers, or $(-\infty, \ \infty)$.

69. $|3a+4| + 2 \ge 8$

$\qquad |3a+4| \ge 6 \qquad$ Adding -2

$\qquad 3a+4 \le -6 \quad or \quad 6 \le 3a+4 \quad$ Part (c)
$\qquad\quad 3a \le -10 \quad or \quad 2 \le 3a$
$\qquad\quad a \le -\dfrac{10}{3} \quad or \quad \dfrac{2}{3} \le a$

The solution set is $\left\{a \Big| a \le -\dfrac{10}{3} \ or \ a \ge \dfrac{2}{3}\right\}$, or $\left(-\infty, -\dfrac{10}{3}\right] \cup \left[\dfrac{2}{3}, \ \infty\right)$.

71. $|y-3| < 12$

$\qquad -12 < y-3 < 12 \qquad$ Part (b)
$\qquad\quad -9 < y < 15 \qquad$ Adding 3

The solution set is $\{y| -9 < y < 15\}$, or $(-9, 15)$.

73. $9 - |x+4| \le 5$

$\qquad -|x+4| \le -4$
$\qquad\quad |x+4| \ge 4$

$\qquad x+4 \le -4 \quad or \quad 4 \le x+4 \quad$ Part (c)
$\qquad\quad x \le -8 \quad or \quad 0 \le x$

The solution set is $\{x| x \le -8 \ or \ x \ge 0\}$, or $(-\infty, -8] \cup [0, \ \infty)$.

75. $6 + |3 - 2x| > 10$

$\qquad |3 - 2x| > 4$

$\quad 3 - 2x < -4 \quad$ or $\quad 4 < 3 - 2x$

$\quad\ \ -2x < -7 \quad$ or $\quad 1 < -2x$

$\qquad\ \ x > \dfrac{7}{2} \quad$ or $\quad -\dfrac{1}{2} > x$

The solution set is $\left\{ x \middle| x < -\dfrac{1}{2} \text{ or } x > \dfrac{7}{2} \right\}$, or

$\left(-\infty, -\dfrac{1}{2}\right) \cup \left(\dfrac{7}{2}, \infty\right)$.

77. $|5 - 4x| < -6$

Absolute value is always nonnegative, so the inequality has no solution. The solution set is $\varnothing$.

79. $\left| \dfrac{1 + 3x}{5} \right| > \dfrac{7}{8}$

$\quad \dfrac{1 + 3x}{5} < -\dfrac{7}{8} \quad$ or $\quad \dfrac{7}{8} < \dfrac{1 + 3x}{5}$

$\quad\ \ 1 + 3x < -\dfrac{35}{8} \quad$ or $\quad \dfrac{35}{8} < 1 + 3x$

$\qquad\quad 3x < -\dfrac{43}{8} \quad$ or $\quad \dfrac{27}{8} < 3x$

$\qquad\quad\ x < -\dfrac{43}{24} \quad$ or $\quad \dfrac{9}{8} < x$

The solution set is $\left\{ x \middle| x < -\dfrac{43}{24} \text{ or } x > \dfrac{9}{8} \right\}$, or

$\left(-\infty, -\dfrac{43}{24}\right) \cup \left(\dfrac{9}{8}, \infty\right)$.

81. $|m + 3| + 8 \le 14$

$\quad |m + 3| \le 6 \quad$ Adding -8

$\quad -6 \le m + 3 \le 6$

$\quad -9 \le m \le 3$

The solution set is $\{ m \mid -9 \le m \le 3 \}$, or $[-9, 3]$.

83. $25 - 2|a + 3| > 19$

$\quad -2|a + 3| > -6$

$\qquad |a + 3| < 3 \quad$ Multiplying by $-\dfrac{1}{2}$

$\quad -3 < a + 3 < 3 \quad$ Part(b)

$\quad -6 < a < 0$

The solution set is $\{ a \mid -6 < a < 0 \}$, or $(-6, 0)$.

85. $|2x - 3| \le 4$

$\quad -4 \le 2x - 3 \le 4 \quad$ Part (b)

$\quad -1 \le 2x \le 7 \quad$ Adding 3

$\quad -\dfrac{1}{2} \le x \le \dfrac{7}{2} \quad$ Multiplying by $\dfrac{1}{2}$

The solution set is $\left\{ x \middle| -\dfrac{1}{2} \le x \le \dfrac{7}{2} \right\}$, or $\left[-\dfrac{1}{2}, \dfrac{7}{2} \right]$.

87. $5 + |3x - 4| \ge 16$

$\qquad |3x - 4| \ge 11$

$\quad 3x - 4 \le -11 \quad$ or $\quad 11 \le 3x - 4 \quad$ Part (c)

$\quad\ \ 3x \le -7 \quad$ or $\quad 15 \le 3x$

$\qquad\ \ x \le -\dfrac{7}{3} \quad$ or $\quad 5 \le x$

The solution set is $\left\{ x \middle| x \le -\dfrac{7}{3} \text{ or } x \ge 5 \right\}$, or

$\left(-\infty, -\dfrac{7}{3}\right] \cup [5, \infty)$.

89. $7 + |2x - 1| < 16$

$\qquad |2x - 1| < 9$

$\quad -9 < 2x - 1 < 9 \quad$ Part (b)

$\quad -8 < 2x < 10$

$\quad -4 < x < 5$

The solution set is $\{ x \mid -4 < x < 5 \}$, or $(-4, 5)$.

91. *Writing Exercise.*

93. $f(x) = \dfrac{1}{3}x - 2$

95. $m = \dfrac{3 - (-3)}{-1 - (-4)} = \dfrac{6}{3} = 2$

$\quad y - 3 = 2(x + 1)$

$\quad y - 3 = 2x + 2$

$\qquad\ \ y = 2x + 5$

$\quad f(x) = 2x + 5$

97. *Writing Exercise.*

99. $|3x - 5| = x$

$\quad 3x - 5 = -x \quad$ or $\quad 3x - 5 = x$

$\quad\ \ -5 = -4x \quad$ or $\quad -5 = -2x$

$\qquad \dfrac{5}{4} = x \quad$ or $\qquad \dfrac{5}{2} = x$

The solution set is $\left\{ \dfrac{5}{4}, \dfrac{5}{2} \right\}$.

101. $2 \le |x - 1| \le 5$

$\quad 2 \le |x - 1| \text{ and } |x - 1| \le 5.$

For $2 \le |x - 1|$:

$\quad x - 1 \le -2 \quad$ or $\quad 2 \le x - 1$

$\qquad\ \ x \le -1 \quad$ or $\quad 3 \le x$

The solution set of $2 \le |x - 1|$ is $\{ x \mid x \le -1 \text{ or } x \ge 3 \}$.

For $|x-1| \le 5$:
$$-5 \le x - 1 \le 5$$
$$-4 \le x \le 6$$

The solution set of $|x-1| \le 5$ is $\{x \mid -4 \le x \le 6\}$.

The solution set of $2 \le |x-1| \le 5$ is
$\{x \mid x \le -1 \text{ or } x \ge 3\} \cap \{x \mid -4 \le x \le 6\}$
$= \{x \mid -4 \le x \le -1 \text{ or } 3 \le x \le 6\}$, or $[-4, -1] \cup [3, 6]$.

103. $t - 2 \le |t-3|$

$$t - 3 \le -(t-2) \quad \text{or} \quad t - 2 \le t - 3$$
$$t - 3 \le -t + 2 \quad \text{or} \quad -2 \le -3$$
$$2t - 3 \le 2 \qquad\qquad\quad \text{False}$$
$$2t \le 5$$
$$t \le \frac{5}{2}$$

The solution set is $\left\{ t \mid t \le \frac{5}{2} \right\}$, or $\left(-\infty, \frac{5}{2} \right]$.

105. Using part (b), we find that $-3 < x < 3$ is equivalent to $|x| < 3$.

107. $x \le -6$ or $6 \le x$
$|x| \ge 6$ Using part (c)

109.
$$x < -8 \quad \text{or} \quad 2 < x$$
$$x + 3 < -5 \quad \text{or} \quad 5 < x + 3 \qquad \text{Adding 3}$$
$$|x+3| > 5 \qquad\qquad\qquad\qquad \text{Using part (c)}$$

111. The distance from x to 7 is $|x-7|$ or $|7-x|$, so we have $|x-7| < 2$, or $|7-x| < 2$.

113. The length of the segment from -1 to 7 is $|-1-7| = |-8| = 8$ units. The midpoint of the segment is $\frac{-1+7}{2} = \frac{6}{2} = 3$. Thus, the interval extends 8/2, or 4, units on each side of 3. An inequality for which the closed interval is the solution set is then $|x-3| \le 4$.

115. The length of the segment from -7 to -1 is $|-7-(-1)| = |-6| = 6$ units. The midpoint of the segment is $\frac{-7+(-1)}{2} = \frac{-8}{2} = -4$. Thus, the interval extends 6/2, or 3, units on each side of -4. An inequality for which the open interval is the solution set is $|x-(-4)| < 3$, or $|x+4| < 3$.

117. $|d - 60\text{ft}| \le 10\text{ft}$
$$-10\text{ft} \le d - 60\text{ft} \le 10\text{ft}$$
$$50\text{ft} \le d \le 70\text{ft}$$
When the bungee jumper is 50 ft above the river, she is $150 - 50$, or 100 ft, from the bridge. When she is 70 ft above the river, she is $150 - 70$, or 80 ft, from the bridge. Thus, at any given time, the bungee jumper is between 80 ft and 100 ft from the bridge.

119. a. Let $x = 4$ (in hundreds of kWh), and solve for y.
$$y = 7.2 - |x-5|$$
$$y = 7.2 - |4-5| = 6.2$$
Since y is the number of customers in hundreds,

there are 620 customers using 400 kWh per month.

b. Let $y = 5.2$ (in hundreds of customers), and solve for x.
$$y = 7.2 - |x-5|$$
$$5.2 = 7.2 - |x-5|$$
$$|x-5| = 2$$
$$x - 5 = -2 \quad \text{or} \quad x - 5 = 2$$
$$x = 3 \quad \text{or} \quad x = 7$$

Since x is the power used in hundreds of kWh, 520 customers draw 300 kWh and 700 kWh of power each month.

121. *Writing Exercise.*

Mid-Chapter Review

1. $-3 \le x - 5 \le 6$
$2 \le x \le 11$
The solution is [2, 11].

2. $|x-1| > 9$
$$x - 1 < -9 \quad \text{or} \quad 9 < x - 1$$
$$x < -8 \quad \text{or} \quad 10 < x$$
The solution is $(-\infty, -8) \cup (10, \infty)$.

3. $|x| = 15$
$$x = -15 \quad \text{or} \quad x = 15$$
The solution is $\{-15, 15\}$.

4. $|t| < 10$
$-10 < t < 10$
The solution is $\{t \mid -10 < t < 10\}$, or $(-10, 10)$.

5. $|p| > 15$
$$p < -15 \quad \text{or} \quad 15 < p$$
The solution is
$\{p \mid p < -15 \text{ or } p > 15\}$, or $(-\infty, -15) \cup (15, \infty)$.

6. $|2x+1| = 7$
$$2x + 1 = -7 \quad \text{or} \quad 2x + 1 = 7$$
$$x = -4 \quad \text{or} \quad x = 3$$
The solution is $\{-4, 3\}$.

7. $-1 < 10 - x < 8$
$-11 < -x < -2$
$11 > x > 2$ Reversing the inequality symbol
The solution is (2, 11).

8. $5|t| < 20$
$|t| < 4$
$-4 < t < 4$
The solution is $\{t \mid -4 < t < 4\}$, or $(-4, 4)$.

9. $x + 8 < 2 \quad \text{or} \quad x - 4 > 9$
$\quad\quad x < -6 \quad \text{or} \quad\quad x > 13$
The solution is $\{x \mid x < -6 \text{ or } x > 13\}$, or $(-\infty, -6) \cup (13, \infty)$.

10. $|x+2| \le 5$

$-5 \le x+2 \le 5$

$-7 \le x \le 3$

The solution is $\{x|-7 \le x \le 3\}$, or $[-7, 3]$.

11. $2+|3x|=10$

$|3x|=8$

$3x=-8 \quad or \quad 3x=8$

$x=-\frac{8}{3} \quad or \quad x=\frac{8}{3}$

The solution is $\left\{-\frac{8}{3}, \frac{8}{3}\right\}$.

12. $2(x-7)-5x > 4-(x+5)$

$2x-14-5x > 4-x-5$

$-3x-14 > -x-1$

$-2x > 13$

$x < -\frac{13}{2}$

The solution is $\left\{x\middle|x < -\frac{13}{2}\right\}$, or $\left(-\infty, -\frac{13}{2}\right)$.

13. $-12 < 2n+6 \quad and \quad 3n-1 \le 7$

$-18 < 2n \quad\quad and \quad\quad 3n \le 8$

$-9 < n \quad\quad and \quad\quad n \le \frac{8}{3}$

The solution is $\left\{n\middle|-9 < n \le \frac{8}{3}\right\}$, or $\left(-9, \frac{8}{3}\right]$.

14. $|x+5|+1 \ge 13$

$|2x+5| \ge 12$

$2x+5 \le -12 \quad or \quad 12 \le 2x+5$

$2x \le -17 \quad or \quad 7 \le 2x$

$x \le -\frac{17}{2} \quad or \quad \frac{7}{2} \le x$

The solution is $\left\{x\middle|x \le -\frac{17}{2} \text{ or } x \ge \frac{7}{2}\right\}$, or

$\left(-\infty, -\frac{17}{2}\right] \cup \left[\frac{7}{2}, \infty\right)$.

15. $\frac{1}{2}(2x-6) \le \frac{1}{3}(9x+3)$

$x-3 \le 3x+1$

$-2x \le 4$

$x \ge -2$

The solution is $\{x|x \ge -2\}$, or $[-2, \infty)$.

16. $\left|\frac{x+2}{5}\right|=8$

$\frac{x+2}{5}=-8 \quad or \quad \frac{x+2}{5}=8$

$x=-42 \quad or \quad x=38$

The solution is $\{-42, 38\}$.

17. $|8x-11|+6 < 2$

$|8x-11| < -4$

The absolute value of a number is always nonnegative. Therefore, the solution set is $\varnothing$.

18. $8-5|a+6| > 3$

$-5|a+6| > -5$

$|a+6| < 1$

$-1 < a+6 < 1$

$-7 < a < -5$

The solution is $\{a|-7 < a < -5\}$, or $(-7, -5)$.

19. $|5x+7|+9 \ge 4$

$|5x+7| \ge -5$

Since the absolute value of a quantity is always greater than zero, it is always greater than -5.

The solution set is $\mathbb{R}$, or $(-\infty, \infty)$.

20. $3x-7 < 5 \quad or \quad 2x+1 > 0$

$3x < 12 \quad or \quad\quad 2x > -1$

$x < 4 \quad or \quad\quad x > -\frac{1}{2}$

The solution set is $\mathbb{R}$, or $(-\infty, \infty)$.

Connecting the Concepts

1. $x+2=7$

$x=5$

2. $x+2 > 7$

$x > 5$

3. $x+2 \le 7$

$x \le 5$

4. $x+y=2$

$y=-x+2$

5. $x+y < 2$

$y < -x+2$

6. $x+y \ge 2$

$y \ge -x+2$

7. $x + 2 \le 7$
   $x \le 5$

$x + 2 \le 7$

8. $y = x - 1$,
   $y = -x + 1$

$(1, 0)$

9. $y \ge 1 - x$,
   $y \le x - 3$,
   $y \le 2$

Exercise Set 4.4

1. e

3. d

5. b

7. We replace x with –2 and y with 3.

$$\frac{2x - 3y > -4}{2(-2) - 3 \mid -4}$$
$$\overset{?}{-7 > -4} \quad \text{FALSE}$$

Since $-7 > -4$ is false, $(-2, 3)$ is not a solution.

9. We replace x with 5 and y with 8.

$$\frac{3y - 5x \le 0}{3 \cdot 8 - 5 \cdot 5 \mid 0}$$
$$24 - 25 \mid$$
$$\overset{?}{-1 \le 0} \quad \text{TRUE}$$

Since $-1 \le 0$ is true, $(5, 8)$ is a solution.

11. Graph: $y \ge \frac{1}{2}x$

We first graph the line $y = \frac{1}{2}x$. We draw the line solid since the inequality symbol is $\ge$. To determine which half-plane to shade, test a point not on the line, $(0, 1)$:

$$\frac{y \ge \frac{1}{2}x}{1 \mid \frac{1}{2} \cdot 0}$$
$$\overset{?}{1 \ge 0} \quad \text{TRUE}$$

Since $1 \ge 0$ is true, $(0, 1)$ is a solution as are all of the points in the half-plane containing $(0, 1)$. We shade that half-plane and obtain the graph.

$y \ge \frac{1}{2}x$

13. Graph: $y > x - 3$.
First graph the line $y = x - 3$. Draw it dashed since the inequality symbol is >. Test the point $(0, 0)$ to determine if it is a solution.

$$\frac{y > x - 3}{0 \mid 0 - 3}$$
$$\overset{?}{0 > -3} \quad \text{TRUE}$$

Since $0 > -3$ is true, we shade the half-plane that contains $(0, 0)$ and obtain the graph.

$y > x - 3$

15. Graph: $y \le x + 2$.
First graph the line $y = x + 2$. Draw it solid since the inequality symbol is $\le$. Test the point $(0, 0)$ to determine if it is a solution.

$$\frac{y \le x + 2}{0 \mid 0 + 2}$$
$$\overset{?}{0 \le 2} \quad \text{TRUE}$$

Since $0 \le 2$ is true, we shade the half-plane that contains $(0, 0)$ and obtain the graph.

$y \le x + 2$

17. Graph: $x - y \le 4$
First graph the line $x - y = 4$. Draw a solid line since the inequality symbol is $\le$. Test the point $(0, 0)$ to determine if it is a solution.

$$\frac{x - y \le 4}{0 - 0 \mid 4}$$
$$\overset{?}{0 \le 4} \quad \text{TRUE}$$

Since $0 \le 4$ is true, we shade the half-plane that contains $(0, 0)$ and obtain the graph.

$x - y \le 4$

19. Graph: $2x + 3y < 6$
First graph $2x + 3y = 6$. Draw the line dashed since the inequality symbol is <. Test the point $(0, 0)$ to determine if it is a solution.

$$\frac{2x + 3y < 6}{2 \cdot 0 + 3 \cdot 0 \mid 6}$$
$$\overset{?}{0 < 6} \quad \text{TRUE}$$

Since $0 < 6$ is true, we shade the half-plane containing

53. Graph: $8x + 5y \leq 40$,   (1)
         $x + 2y \leq 8$ (2)
              $x \geq 0$,   (3)
              $y \geq 0$ (4)

Graph the lines $8x + 5y = 40$, $x + 2y = 8$, $x = 0$, and $y = 0$ using solid lines. Determine the region for each inequality. Shade the region where they overlap.

To find the vertices we solve four different systems of equations.

From (1) and (2) we have $8x + 5y = 40$,
                         $x + 2y = 8$.

Solving, we obtain the vertex $\left(\frac{40}{11}, \frac{24}{11}\right)$.

From (1) and (4) we have $8x + 5y = 40$,
                         $y = 0$.

Solving, we obtain the vertex $(5, 0)$.

From (2) and (3) we have $x + 2y = 8$,
                         $x = 0$.

Solving, we obtain the vertex $(0, 4)$.

From (3) and (4) we have $x = 0$,
                         $y = 0$.

Solving, we obtain the vertex $(0, 0)$.

55. Graph: $y - x \geq 2$,   (1)
         $y - x \leq 4$,   (2)
           $2 \leq x \leq 5$ (3)

Think of (3) as two inequalities:
     $2 \leq x$,   (4)
     $x \leq 5$ (5)

Graph the lines $y - x = 2$, $y - x = 4$, $x = 2$, and $x = 5$, using solid lines. Determine the region for each inequality. Shade the region where they overlap.

To find the vertices we solve four different systems of equations.

From (1) and (4) we have $y - x = 2$,
                         $x = 2$.

Solving, we obtain the vertex $(2, 4)$.

From (1) and (5) we have $y - x = 2$,
                         $x = 5$.

Solving, we obtain the vertex $(5, 7)$.

From (2) and (4) we have $y - x = 4$,
                         $x = 2$.

Solving, we obtain the vertex $(2, 6)$.

From (2) and (5) we have $y - x = 4$,
                         $x = 5$.

Solving, we obtain the vertex $(5, 9)$.

57. *Writing Exercise.*

59. $(-152, 0)$ is on the x-axis.

61. Slope: $\frac{4}{3}$; y-intercept $(0, 15)$

63. $2x = 4 - 3y$
       $2x - 4 = -3y$
   $-\frac{2}{3}x + \frac{4}{3} = y$

     The slope is $-\frac{2}{3}$.

   $3x - 2y = 10$
     $-2y = -3x + 10$
        $y = \frac{3}{2}x - 5$.

     The slope is $\frac{3}{2}$.

Since $-\frac{2}{3} \cdot \frac{3}{2} = -1$, the lines are perpendicular.

65. *Writing Exercise.*

67. Graph: $x + y > 8$,
         $x + y \leq -2$

Graph the line $x + y = 8$ using a dashed line and graph $x + y = -2$, using a solid line. Indicate the region for each inequality by arrows. The regions do not overlap (the solution set is $\varnothing$), so we do not shade any portion of the graph.

69. Graph: $x - 2y \leq 0$,
         $-2x + y \leq 2$,
               $x \leq 2$,
               $y \leq 2$,
        $x + y \leq 4$

Graph the five inequalities above, and shade the region where they overlap.

71. Both the width and the height must be positive, so we have
     $w > 0$,
     $h > 0$.

To be checked as luggage, the sum of the width, height, and length cannot exceed 62 in., so we have
     $w + h + 30 \leq 62$, or
        $w + h \leq 32$.

The girth is represented by $2w + 2h$ and the length is 30 in. In order to meet postal regulations the sum of the girth and the length cannot exceed 130 in., so we have:

$$2w + 2h + 30 < 130, \text{or}$$
$$2w + 2h \le 100, \text{or}$$
$$w + h \le 50$$

Thus, have a system of inequalities:

$$w > 0,$$
$$h > 0,$$
$$w + h \le 32,$$
$$w + h \le 50$$

73. We graph the following inequalities:

$$q + v \ge 287$$
$$v \ge 145$$
$$q \le 170$$
$$v \le 170$$

75. Graph: $35c + 75a > 1000,$
$$c \ge 0,$$
$$a \ge 0$$

77. $h < 2w$
$w \le 1.5h$
$h \le 3200$
$h \ge 0$
$w \ge 0$

Crest width (in feet) / Height (in feet)

79. a. $3x + 6y > 2$

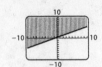

b. $x - 5y \le 10$

c. $13x - 25y + 10 \le 0$

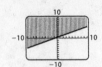

d. $2x + 5y > 0$

Exercise Set 4.5

1. In linear programming, the quantity we wish to maximize or minimize is represented by the *objective* function.

3. To solve a linear programming problem, we make use of the *corner* principle.

5. In linear programming, the corners of the shaded portion of the graph are referred to as *vertices*.

7. Find the maximum and minimum values of
$$F = 2x + 14y,$$
subject to
$$5x + 3y \le 34, \quad (1)$$
$$3x + 5y \le 30, \quad (2)$$
$$x \ge 0, \quad (3)$$
$$y \ge 0. \quad (4)$$

Graph the system of inequalities and find the coordinates of the vertices.

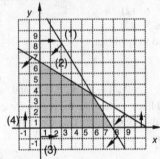

To find one vertex we solve the system
$$x = 0,$$
$$y = 0.$$

This vertex is $(0, 0)$.

To find a second vertex we solve the system
$$5x + 3y = 34,$$
$$y = 0.$$

This vertex is $\left(\frac{34}{5}, \ 0 \right)$.

To find a third vertex we solve the system
$$5x + 3y = 34,$$
$$3x + 5y = 30.$$

This vertex is $(5, \ 3)$.

To find the fourth vertex we solve the system
$$3x + 5y = 30,$$
$$x = 0.$$

This vertex is $(0, \ 6)$.

Now find the value of F at each of these points.

Vertex (x, y)	$F = 2x + 14y$	
(0, 0)	$2 \cdot 0 + 14 \cdot 0 = 0 + 0 = 0$	← Minimum
$\left(\frac{34}{5}, 0\right)$	$2 \cdot \frac{34}{5} + 14 \cdot 0 = \frac{68}{5} + 0 = 13\frac{3}{5}$	
(5, 3)	$2 \cdot 5 + 14 \cdot 3 = 10 + 42 = 52$	
(0, 6)	$2 \cdot 0 + 14 \cdot 6 = 0 + 84 = 84$	← Maximum

The maximum value of F is 84 when $x = 0$ and $y = 6$.

The minimum value of F is 0 when $x = 0$ and $y = 0$.

9. Find the maximum and minimum values of
$$P = 8x - y + 20,$$
subject to
$$6x + 8y \le 48, \quad (1)$$
$$0 \le y \le 4, \quad (2)$$
$$0 \le x \le 7. \quad (3)$$
Think of (2) as $0 \le y$, (4)
$$y \le 4. \quad (5)$$
Think of (3) as $0 \le x$, (6)
$$x \le 7. \quad (7)$$
Graph the system of inequalities.

To determine the coordinates of the vertices, we solve the following systems:

$$x = 0, \qquad x = 7, \qquad 6x + 8y = 48,$$
$$y = 0; \qquad y = 0; \qquad x = 7;$$

$$6x + 8y = 48, \qquad x = 0,$$
$$y = 4; \qquad y = 4$$

The vertices are $(0, 0)$, $(7, 0)$, $\left(7, \frac{3}{4}\right)$, $\left(\frac{8}{3}, 4\right)$, and $(0, 4)$, respectively. Compute the value of P at each of these points.

Vertex (x, y)	$P = 8x - y + 20$	
(0, 0)	$8 \cdot 0 - 0 + 20$ $= 0 - 0 + 20 = 20$	
(7, 0)	$8 \cdot 7 - 0 + 20$ $= 56 - 0 + 20 = 76$	← Maximum
$\left(7, \frac{3}{4}\right)$	$8 \cdot 7 - \frac{3}{4} + 20$ $= 56 - \frac{3}{4} + 20 = 75\frac{1}{4}$	
$\left(\frac{8}{3}, 4\right)$	$8 \cdot \frac{8}{3} - 4 + 20$ $= \frac{64}{3} - 4 + 20 = 37\frac{1}{3}$	
(0,4)	$8 \cdot 0 - 4 + 20$ $= 0 - 4 + 20 = 16$	← Minimum

The maximum is 76 when $x = 7$ and $y = 0$. The minimum is 16 when $x = 0$ and $y = 4$.

11. Find the maximum and minimum values of
$$F = 2y - 3x,$$
subject to
$$y \le 2x + 1, \quad (1)$$
$$y \ge -2x + 3, \quad (2)$$
$$x \le 3 \quad (3)$$
Graph the system of inequalities and find the coordinates of the vertices.

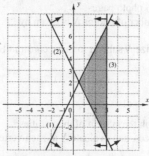

To determine the coordinates of the vertices, we solve the following systems:

$$y = 2x + 1, \qquad y = 2x + 1, \qquad y = -2x + 3,$$
$$y = -2x + 3; \qquad x = 3; \qquad x = 3$$

The solutions of the systems are $\left(\frac{1}{2}, 2\right)$, $(3, 7)$, and $(3, -3)$, respectively. Now find the value of F at each of these points.

Vertex (x, y)	$F = 2y - 3x$	
$\left(\frac{1}{2}, 2\right)$	$2 \cdot 2 - 3 \cdot \frac{1}{2} = \frac{5}{2}$	
(3, 7)	$2 \cdot 7 - 3 \cdot 3 = 5$	← Maximum
(3,−3)	$2(-3) - 3 \cdot 3 = -15$	← Minimum

The maximum value is 5 when $x = 3$ and $y = 7$. The minimum value is −15 when $x = 3$ and $y = -3$.

13. **Familiarize.** Let x = the number of train rides, and y = the number of bus rides.
Translate. The number of cost C is given by
$$C = 5x + 4y.$$
We wish to minimize C subject to these constraints.
$$x + y \ge 5$$
$$x + 1.5y \le 6$$
$$x \ge 0$$
$$y > 0.$$

Carry out. We graph the system of inequalities, determine the vertices, and evaluate N at each vertex.

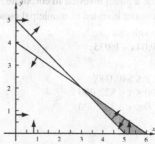

Vertex	$C = 5x + 4y$
(3, 2)	$5 \cdot 3 + 4 \cdot 2 = 23$
(5, 0)	$5 \cdot 5 + 6 \cdot 0 = 25$
(6, 0)	$5 \cdot 6 + 6 \cdot 0 = 30$

The smallest cost is $23, obtained when 3 train rides and 2 bus rides are taken.

Check. Go over the algebra and arithmetic.

State. The minimum cost is achieved by taking 3 train rides and 2 bus rides.

15. *Familiarize*. Let x = the number of 4 photo pages, and y = the number of 6 photo pages.

Translate. The number of photos N is given by
$$N = 4x + 6y.$$
We wish to maximize N subject to these constraints.
$$x + y \le 20$$
$$3x + 5y \le 90$$
$$x \ge 0$$
$$y \ge 0.$$

Carry out. We graph the system of inequalities, determine the vertices, and evaluate N at each vertex.

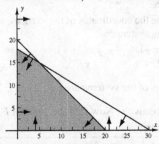

Vertex	$N = 4x + 6y$
(0, 0)	$4 \cdot 0 + 6 \cdot 0 = 0$
(0, 18)	$4 \cdot 0 + 6 \cdot 18 = 108$
(20, 0)	$4 \cdot 20 + 6 \cdot 0 = 80$
(5, 15)	$4 \cdot 5 + 6 \cdot 15 = 110$

The greatest number of photos is 110, obtained when 5 pages of 4-photos and 15 pages of 6-photos are used.

Check. Go over the algebra and arithmetic.

State. The maximum number of photos is achieved by using 5 pages or 4-photos and 15 pages of 6-photos.

17. In order to earn the most interest Rosa should invest the entire $40,000. She should also invest as much as possible in the type of investment that has the higher interest rate. Thus, she should invest $22,000 in corporate bonds and the remaining $18,000 in municipal bonds. The maximum income is
$$0.04(\$22,000) + 0.035(\$18,000) = \$1510.$$

We can also solve this problem as follows.

Let $x =$ the amount invested in corporate bonds and $y =$ the amount invested in municipal bonds. Find the maximum value of
$$I = 0.04x + 0.035y$$
subject to
$$x + y \le \$40,000,$$
$$\$6000 \le x \le \$22,000$$
$$0 \le y \le \$30,000.$$

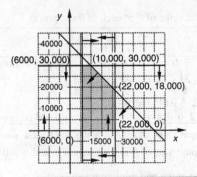

Vertex	$I = 0.04x + 0.035y$
($6000, $0)	$240
($6000, $30,000)	$1290
($10,000, $30,000)	$1450
($22,000, $18,000)	$1510
($22,000, $0)	$880

The maximum income of $1510 occurs when $22,000 is invested in corporate bonds and $18,000 is invested in municipal bonds.

19. *Familiarize*. Let x = the number of short-answer questions and y = the number of essay questions answered.

Translate. The score S is given by
$$S = 10x + 15y.$$
We wish to maximize S subject to these constraints:
$$x + y \le 16$$
$$3x + 6y \le 60$$
$$x \ge 0$$
$$y \ge 0.$$

Carry out. We graph the system of inequalities, determine the vertices, and evaluate S at each vertex.

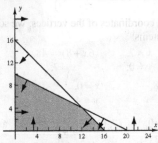

Vertex	$S = 10x + 15y$
(0, 0)	$10 \cdot 0 + 15 \cdot 0 = 0$
(0, 10)	$10 \cdot 0 + 15 \cdot 10 = 150$
(16, 0)	$10 \cdot 16 + 15 \cdot 0 = 160$
(12, 4)	$10 \cdot 12 + 15 \cdot 4 = 180$

The greatest score in the table is 180, obtained when 12 short-answer questions and 4 essay questions are answered.

Check. Go over the algebra and arithmetic.

State. The maximum score is 180 points when 12 short-answer questions and 4 essay questions are answered.

21. Familiarize. Let $x =$ the Merlot acreage and $y =$ the Cabernet acreage.

Translate. The profit P is given by

$P = \$400x + \$300y$.

We wish to maximize P subject to these constraints:

$$x + y \le 240,$$
$$2x + y \le 320,$$
$$x \ge 0,$$
$$y \ge 0.$$

Carry out. We graph the system of inequalities, determine the vertices, and evaluate P at each vertex.

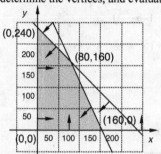

Vertex	$P = \$400x + \$300y$
(0, 0)	$0
(0, 240)	$72,000
(80, 160)	$80,000
(160, 0)	$64,000

Check. Go over the algebra and arithmetic.

State. The maximum profit occurs by planting 80 acres of Merlot grapes and 160 acres of Cabernet grapes.

23. Familiarize. Let $x =$ the number of servings of goat cheese and $y =$ the number of servings of hazelnuts.

Translate. The total number of calories is given by

$C = 264x + 628y$.

We wish to minimize C subject to these constraints.

$$15 \le x + 5y$$
$$x + 5y \le 45$$
$$1500 \le 500x + 100y$$
$$500x + 100y \le 2500$$

Carry out. We graph the system of inequalities, determine the vertices and evaluate C at each vertex.

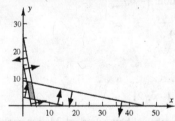

Vertex	$C = 264x + 628y$
(2.5, 2.5)	$264 \cdot 2.5 + 628 \cdot 2.5 = 2230$
(1.25, 8.75)	$264 \cdot 1.25 + 628 \cdot 8.75 = 5825$
$\left(\dfrac{55}{12}, \dfrac{25}{12}\right)$	$264 \cdot \dfrac{55}{12} + 628 \cdot \dfrac{25}{12} = 2518.3 = \dfrac{30,220}{12}$
$\left(\dfrac{10}{3}, \dfrac{25}{3}\right)$	$264 \cdot \dfrac{10}{3} + 628 \cdot \dfrac{25}{3} = 6113.3 = \dfrac{18,340}{3}$

The least number of calories in the table is 2230, obtained with 2.5 servings of each.

Check. Go over the algebra and arithmetic.

State. The minimum calories consumed is 2230 with 2.5 servings of each.

25. Writing Exercise.

27. $10^{-2} = \dfrac{1}{100}$

29. $\dfrac{-6x^2}{3x^{-10}} = -2x^{12}$

31. $\left(\dfrac{4c^2 d}{6cd^4}\right)^{-1} = \dfrac{6cd^4}{4c^2 d} = \dfrac{3d^3}{2c}$

33. Writing Exercise.

35. Familiarize. Let x represent the number of T3 planes and y represent the number of S5 planes. Organize the information in a table.

Plane	Number of planes	Passengers		
		First	Tourist	Economy
T3	x	$40x$	$40x$	$120x$
S5	y	$80y$	$30y$	$40y$

Plane	Cost per mile
T3	$30x$
S5	$25y$

Translate. Suppose C is the total cost per mile. Then $C = 30x + 25y$. We wish to minimize C subject to these facts (constraints) about x and y.

$$40x + 80y \ge 2000,$$
$$40x + 30y \ge 1500,$$
$$120x + 40y \ge 2400,$$
$$x \ge 0, \quad y \ge 0$$

Carry out. Graph the system of inequalities, determine the vertices, and evaluate C at each vertex.

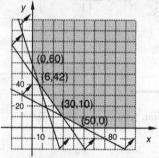

Vertex	$C = 30x + 25y$
(0, 60)	$30(0) + 25(60) = 1500$
(6, 42)	$30(6) + 25(42) = 1230$
(30, 10)	$30(30) + 25(10) = 1150$
(50, 0)	$30(50) + 25(0) = 1500$

Check. Go over the algebra and arithmetic.

State. In order to minimize the operating cost, 30 T3 planes and 10 S5 planes should be used.

Chapter 4 Review

1. True

2. False, if $c < 0$, then $ac < bc$.

3. True

4. False, the inequality $2 < 5x + 1 < 9$ is equivalent to $2 < 5x + 1$ *and* $5x + 1 < 9$.

5. True

6. True

7. True

8. True

9. False

10. False

11. Graph: $x \le -1$.

    Set builder notation: $\{x \mid x \le -1\}$

    Interval notation: $(-\infty, -1]$

12. $a + 3 \le 7$

    Graph: $a \le 4$.

    Set builder notation: $\{a \mid a \le 4\}$

    Interval notation: $(-\infty, 4]$

13. $4y > -15$

    Graph: $y > -\dfrac{15}{4}$.

    Set builder notation: $\left\{ y \middle| y > -\dfrac{15}{4} \right\}$

    Interval notation: $\left(-\dfrac{15}{4}, \infty \right)$

14. $-0.2y < 6$

    Graph: $y > -30$.

    Set builder notation: $\{y \mid y > -30\}$

    Interval notation: $(-30, \infty)$

15. $-6x - 5 < 4$

    $-6x < 9$

    Graph: $x > -\dfrac{3}{2}$

    Set builder notation: $\left\{ x \middle| x > -\dfrac{3}{2} \right\}$

    Interval notation: $\left(-\dfrac{3}{2}, \infty \right)$

16.
$$-\frac{1}{2}x - \frac{1}{4} > \frac{1}{2} - \frac{1}{4}x$$
$$-\frac{1}{2}x - \frac{1}{4} + \frac{1}{4}x > \frac{1}{2} - \frac{1}{4}x + \frac{1}{4}x$$
$$-\frac{1}{4}x - \frac{1}{4} > \frac{1}{2}$$
$$-\frac{1}{4}x > \frac{3}{4}$$
$$x < -3$$

    Graph: $x < -3$.

    Set builder notation: $\{x \mid x < -3\}$

    Interval notation: $(-\infty, -3)$

17. $0.3y - 7 < 2.6y + 15$

    $-22 < 2.3y$

    $-\dfrac{22}{2.3} < y$

    $-\dfrac{220}{23} < y$

    Graph: $-\dfrac{220}{23} < y$

    Set builder notation: $\left\{ y \middle| y > -\dfrac{220}{23} \right\}$

    Interval notation: $\left(-\dfrac{220}{23}, \infty \right)$

18. $-2(x - 5) \ge 6(x + 7) - 12$

    $-2x + 10 \ge 6x + 42 - 12$

    $-2x + 10 \ge 6x + 30$

    $-20 \ge 8x$

    $-\dfrac{5}{2} \ge x$

    Graph: $-\dfrac{5}{2} \ge x$

    Set builder notation: $\left\{ x \middle| x \le -\dfrac{5}{2} \right\}$

    Interval notation: $\left(-\infty, -\dfrac{5}{2} \right]$

19. $f(x) \le g(x)$

    $3x + 2 \le 10 - x$

    $4x \le 8$

    $x \le 2$

    $\{x \mid x \le 2\}$ or $(-\infty, 2]$

20. Let x = the number of hours worked. Mariah earns $\$8.40x$ at the sandwich shop and in carpentry she earns $\$16x$, but spends \$950, or $16x - 950$. We solve the inequality.

    $8.40x < 16x - 950$

    $950 < 7.6x$

    $125 < x$

    She must work more than 125 hours in carpentry to be more profitable.

21. Let $x =$ the amount invested at 3% and $9000 - x =$ the amount invested at 3.5%. The interest from the first investment is $0.03x$ and the interest from the second investment is $0.035(9000 - x)$. We solve the inequality.

$$0.03x + 0.035(9000 - x) \geq 300$$
$$0.03x + 315 - 0.035x \geq 300$$
$$-0.005x + 315 \geq 300$$
$$-0.005x \geq -15$$
$$x \leq 3000$$

Clay should invest at most $3000 at 3%.

22. $\{a, b, c, d\} \cap \{a, c, e, f, g\} = \{a, c\}$

23. $\{a, b, c, d\} \cup \{a, c, e, f, g\} = \{a, b, c, d, e, f, g\}$

24. Graph: $x \leq 2$ *and* $x > -3$

$(-3, 2]$

25. Graph: $x \leq 3$ *or* $x > -5$

$(-\infty, \infty)$

26. $-3 < x + 5 \leq 5$
$-8 < x \leq 0$
$\{x | -8 < x \leq 0\}$

$(-8, 0]$

27. $-15 < -4x - 5 < 0$
$-10 < -4x < 5$
$\dfrac{5}{2} > x > -\dfrac{5}{4}$, or $-\dfrac{5}{4} < x < \dfrac{5}{2}$

$\left\{ x \middle| -\dfrac{5}{4} < x < \dfrac{5}{2} \right\}$

$\left(-\dfrac{5}{4}, \dfrac{5}{2} \right)$

28. $3x < -9$ *or* $-5x < -5$
$x < -3$ *or* $x > 1$

$\{x | x < -3 \text{ or } x > 1\}$ or $(-\infty, -3) \cup (1, \infty)$

29. $2x + 5 < -17$ *or* $-4x + 10 \leq 34$
$2x < -22$ *or* $-4x \leq 24$
$x < -11$ *or* $x \geq -6$

$\{x | x < -11 \text{ or } x \geq -6\}$ or $(-\infty, -11) \cup [-6, \infty)$

30. $2x + 7 \leq -5$ *or* $x + 7 \geq 15$
$2x \leq -12$ *or* $x \geq 8$
$x \leq -6$

$\{x | x \leq -6 \text{ or } x \geq 8\}$ or $(-\infty, -6] \cup [8, \infty)$

31. $f(x) < -5$ *or* $f(x) > 5$
$3 - 5x < -5$ *or* $3 - 5x > 5$
$-5x < -8$ *or* $-5x > 2$
$x > \dfrac{8}{5}$ *or* $x < -\dfrac{2}{5}$

$\left\{ x \middle| x < -\dfrac{2}{5} \text{ or } x > \dfrac{8}{5} \right\}$ or $\left(-\infty, -\dfrac{2}{5} \right) \cup \left(\dfrac{8}{5}, \infty \right)$

32. $f(x) = \dfrac{2x}{x + 3}$
$x + 3 = 0$
$x = -3$
The domain of f is $(-\infty, -3) \cup (-3, \infty)$.

33. $f(x) = \sqrt{5x - 10}$
$5x - 10 \geq 0$
$5x \geq 10$
$x \geq 2$
The domain of f is $[2, \infty)$.

34. $f(x) = \sqrt{1 - 4x}$
$1 - 4x \geq 0$
$-4x \geq -1$
$x \leq \dfrac{1}{4}$
The domain of f is $\left(-\infty, \dfrac{1}{4} \right]$.

35. $|x| = 11$
$x = -11$ *or* $x = 11$
$\{-11, 11\}$

36. $|t| \geq 21$
$t \leq -21$ *or* $21 \leq t$
$\{t | t \leq -21 \text{ or } t \geq 21\}$ or $(-\infty, -21] \cup [21, \infty)$

37. $|x - 8| = 3$
$x - 8 = -3$ *or* $x - 8 = 3$
$x = 5$ *or* $x = 11$
$\{5, 11\}$

38. $|4a + 3| < 11$
$-11 < 4a + 3 < 11$
$-14 < 4a < 8$
$-\dfrac{7}{2} < a < 2$
$\left\{ a \middle| -\dfrac{7}{2} < a < 2 \right\}$ or $\left(-\dfrac{7}{2}, 2 \right)$

39. $|3x - 4| \geq 15$
$3x - 4 \leq -15$ *or* $15 \leq 3x - 4$
$3x \leq -11$ *or* $19 \leq 3x$
$x \leq -\dfrac{11}{3}$ *or* $\dfrac{19}{3} \leq x$
$\left\{ x \middle| x \leq -\dfrac{11}{3} \text{ or } x \geq \dfrac{19}{3} \right\}$ or $\left(-\infty, -\dfrac{11}{3} \right] \cup \left[\dfrac{19}{3}, \infty \right)$

40. $|2x+5| = |x-9|$

$2x+5 = x-9 \quad or \quad 2x+5 = -(x-9)$
$\qquad x = -14 \quad or \quad 2x+5 = -x+9$
$\qquad\qquad\qquad\qquad\qquad 3x = 4$
$\qquad\qquad\qquad\qquad\qquad x = \dfrac{4}{3}$

$\left\{-14, \dfrac{4}{3}\right\}$

41. $|5n+6| = -11$

Absolute value is never negative.

The solution is $\varnothing$.

42. $\left|\dfrac{x+4}{6}\right| \le 2$

$-2 \le \dfrac{x+4}{6} \le 2$
$-12 \le x+4 \le 12$
$-16 \le x \le 8$

$\{x | -16 \le x \le 8\}$ or $[-16, 8]$

43. $2|x-5| - 7 > 3$

$\qquad 2|x-5| > 10$
$\qquad\quad |x-5| > 5$

$x-5 < -5 \quad or \quad 5 < x-5$
$\quad x < 0 \quad or \quad 10 < x$

$\{x | x < 0 \; or \; x > 10\}$ or $(-\infty, 0) \cup (10, \infty)$

44. $19 - 3|x+1| \ge 4$

$\qquad -3|x+1| \ge -15$
$\qquad\quad |x+1| \le 5$

$-5 \le x+1 \le 5$
$-6 \le x \le 4$

$\{x | -6 \le x \le 4\}$ or $[-6, 4]$

45. $|8x-3| < 0$

Absolute value is never negative.

The solution is $\varnothing$.

46. Graph $x - 2y \ge 6$.

47. Graph $x + 3y > -1$,
$\qquad\qquad x + 3y < 4$

The lines are parallel, there are no vertices.

48. Graph $x - 3y \le 3$,
$\qquad\quad x + 3y \ge 9$,
$\qquad\quad y \le 6$

Vertices: (–9, 6), (6, 1) and (21, 6)

49. For $F = 3x + y + 4$, subject to
$\qquad y \le 2x+1$,
$\qquad x \le 7$,
$\qquad y \ge 3$.

Vertices	$F = 3x + y + 4$
(7, 3)	$3 \cdot 7 + 3 + 4 = 28$
(1, 3)	$3 \cdot 1 + 3 + 4 = 10$
(7, 15)	$3 \cdot 7 + 15 + 4 = 40$

The maximum value of F is 40 at $x = 7$, $y = 15$.
The minimum value of F is 10 at $x = 1$, $y = 3$.

50. Let x = the number of books ordered from the East coast supplier and y = the number of books ordered from the West coast supplier. Minimize the time $T = 5x + 2y$ subject to the constraints.
$\qquad 2x + 4y \le 320$,
$\qquad x + y \ge 100$,
$\qquad x \ge 0$
$\qquad y \ge 0$

Vertices	$T = 5x + 2y$
(160, 0)	$5 \cdot 160 + 2 \cdot 0 = 800$
(100, 0)	$5 \cdot 100 + 2 \cdot 0 = 500$
(40, 60)	$5 \cdot 40 + 2 \cdot 60 = 320$

The minimum is 320 when 40 books are ordered from the East coast supplier and 60 books are ordered from the West coast supplier.

51. *Writing Exercise.* The equation $|X| = p$ has two solutions when p is positive because X can be either p or $-p$. The same equation has no solution when p is negative because no number has a negative absolute value.

52. *Writing Exercise.* The solution set of a system of inequalities is all ordered pairs that make *all* the individual inequalities true. This consists of ordered pairs that are common to all the individual solution sets, or the intersection of the graphs.

53. $|2x+5| \le |x+3|$

$2x+5 \le x+3$ *and* $2x+5 \ge -(x+3)$

$\quad\quad x \le -2$ *and* $2x+5 \ge -x-3$

$\quad\quad\quad\quad\quad\quad\quad\quad\quad 3x \ge -8$

$\quad\quad\quad\quad\quad\quad\quad\quad\quad\quad x \ge -\dfrac{8}{3}$

$\left\{x \middle| -\dfrac{8}{3} \le x \le -2\right\}$, or $\left[-\dfrac{8}{3}, -2\right]$

54. False, $-5 < 3$ is true, but $(-5)^2 < 9$ is false.

55. $|d - 2.5| \le 0.003$

Chapter 4 Test

1. $x - 3 < 8$

$\quad x < 11$

$\{x \mid x < 11\}$ or $(-\infty, 11)$

2. $-\dfrac{1}{2}t < 12$

$\quad t > -24$

$\{t \mid t > -24\}$ or $(-24, \infty)$

3. $-4y - 3 \ge 5$

$\quad -4y \ge 8$

$\quad\quad y \le -2$

$\{y \mid y \le -2\}$ or $(-\infty, -2]$

4. $3a - 5 \le -2a + 6$

$\quad 5a - 5 \le 6$

$\quad\quad 5a \le 11$

$\quad\quad\quad a \le \dfrac{11}{5}$

$\left\{a \middle| a \le \dfrac{11}{5}\right\}$ or $\left(-\infty, \dfrac{11}{5}\right]$

5. $3(7-x) < 2x + 5$

$\quad 21 - 3x < 2x + 5$

$\quad\quad 21 < 5x + 5$

$\quad\quad 16 < 5x$

$\quad\quad \dfrac{16}{5} < x$

$\left\{x \middle| x > \dfrac{16}{5}\right\}$ or $\left(\dfrac{16}{5}, \infty\right)$

6. $-2(3x-1) - 5 \ge 6x - 4(3-x)$

$\quad -6x + 2 - 5 \ge 6x - 12 + 4x$

$\quad\quad -6x - 3 \ge 10x - 12$

$\quad\quad\quad -3 \ge 16x - 12$

$\quad\quad\quad\quad 9 \ge 16x$

$\quad\quad\quad\quad \dfrac{9}{16} \ge x$

$\left\{x \middle| x \le \dfrac{9}{16}\right\}$ or $\left(-\infty, \dfrac{9}{16}\right]$

7. $f(x) > g(x)$

$\quad -5x - 1 > -9x + 3$

$\quad\quad 4x - 1 > 3$

$\quad\quad 4x > 4$

$\quad\quad x > 1$

$\{x \mid x > 1\}$ or $(1, \infty)$

8. Let $x =$ the number of miles driven. The cost for unlimited is \$80 and the cost for the other plan is $\$45 + \$0.40x$. We solve the inequality.

$\quad 80 < 45 + 0.40x$

$\quad 35 < 0.4x$

$\quad 87.5 < x$

The unlimited mileage plan is less expensive for more than 87.5 miles.

9. Let $x =$ the number of additional hours. The cost is $\$80 + \$60x$ and \$200 is budgeted. We solve the inequality.

$\quad 80 + 60x \le 200$

$\quad\quad 60x \le 120$

$\quad\quad\quad x \le 2$

The time of service is $x + \dfrac{1}{2}$ hr or $2\dfrac{1}{2}$ hours or less.

10. $\{a, e, i, o, u\} \cap \{a, b, c, d, e\} = \{a, e\}$

11. $\{a, e, i, o, u\} \cup \{a, b, c, d, e\} = \{a, b, c, d, e, i, o, u\}$

12. $f(x) = \sqrt{6 - 3x}$

$\quad 6 - 3x \ge 0$

$\quad\quad 6 \ge 3x$

$\quad\quad 2 \ge x$

The domain of f is $(-\infty, 2]$.

13. $f(x) = \dfrac{x}{x - 7}$

$\quad x - 7 = 0$

$\quad\quad x = 7$

The domain of f is $(-\infty, 7) \cup (7, \infty)$.

14. $-5 < 4x + 1 \le 3$

$\quad -6 < 4x \le 2$

$\quad -\dfrac{3}{2} < x \le \dfrac{1}{2}$

$\left\{x \middle| -\dfrac{3}{2} < x \le \dfrac{1}{2}\right\}$ or $\left(-\dfrac{3}{2}, \dfrac{1}{2}\right]$

15. $3x - 2 < 7$ *or* $x - 2 > 4$
  $\quad 3x < 9$ *or* $\quad x > 6$
  $\quad\quad x < 3$

The solution set is $\{x | x < 3 \ or \ x > 6\}$ or

$(-\infty, \ 3) \cup (6, \ \infty)$.

$$\underset{0\quad 3\quad 6}{\longleftrightarrow}$$

16. $-3x > 12$ *or* $4x \ge -10$
  $\quad x < -4$ *or* $\quad x \ge -\dfrac{5}{2}$

The solution set is $\left\{x \middle| x < -4 \ or \ x \ge -\dfrac{5}{2}\right\}$ or

$(-\infty, -4) \cup \left[-\dfrac{5}{2}, \ \infty\right)$.

$$\underset{-4\quad -\frac{5}{2}\quad 0}{\longleftrightarrow}$$

17. $1 \le 3 - 2x \le 9$
  $-2 \le -2x \le 6$
  $\quad 1 \ge x \ge -3$

The solution set is $\{x | -3 \le x \le 1\}$ or $[-3, \ 1]$.

$$\underset{-3\quad 0\quad 1}{\longleftrightarrow}$$

18. $|n| = 15$
  $n = -15$ *or* $n = 15$
  $\{-15, \ 15\}$

$$\underset{-15\quad 0\quad 15}{\longleftrightarrow}$$

19. $|a| > 5$
  $a < -5$ *or* $5 < a$
  $\{a \ | \ a < -5 \ or \ a > 5\}$ or $(-\infty, -5) \cup (5, \ \infty)$

$$\underset{-5\quad 0\quad 5}{\longleftrightarrow}$$

20. $|3x - 1| < 7$
  $-7 < 3x - 1 < 7$
  $\quad -6 < 3x < 8$
  $\quad\quad -2 < x < \dfrac{8}{3}$

$\left\{x \middle| -2 < x < \dfrac{8}{3}\right\}$, or $\left(-2, \ \dfrac{8}{3}\right)$

$$\underset{-2\quad 0\quad 2}{\overset{\frac{8}{3}}{\longleftrightarrow}}$$

21. $|-5t - 3| \ge 10$
  $-5t - 3 \le -10$ *or* $10 \le -5t - 3$
  $\quad -5t \le -7$ *or* $\quad 13 \le -5t$
  $\quad\quad t \ge \dfrac{7}{5}$ *or* $-\dfrac{13}{5} \ge t$

$\left\{t \middle| t \le -\dfrac{13}{5} \ or \ t \ge \dfrac{7}{5}\right\}$ or $\left(-\infty, -\dfrac{13}{5}\right] \cup \left[\dfrac{7}{5}, \ \infty\right)$

$$\underset{-\frac{13}{5}\quad 0\quad \frac{7}{5}}{\longleftrightarrow}$$

22. $|2 - 5x| = -12$
Absolute value is never negative.
The solution is $\varnothing$.

23. $g(x) < -3$ *or* $g(x) > 3$
  $4 - 2x < -3$ *or* $4 - 2x > 3$
  $\quad -2x < -7$ *or* $\quad -2x > -1$
  $\quad\quad x > \dfrac{7}{2}$ *or* $\quad\quad x < \dfrac{1}{2}$

$\left\{x \middle| x < \dfrac{1}{2} \ or \ x > \dfrac{7}{2}\right\}$ or $\left(-\infty, \ \dfrac{1}{2}\right) \cup \left(\dfrac{7}{2}, \ \infty\right)$

$$\underset{0\ \frac{1}{2}\qquad\quad \frac{7}{2}}{\longleftrightarrow}$$

24. $f(x) = g(x)$
  $|2x - 1| = |2x + 7|$

  $2x - 1 = 2x + 7$ *or* $2x - 1 = -(2x + 7)$
  $\quad\quad -1 = 7$ $\quad\quad\quad 2x - 1 = -2x - 7$
  $\quad\quad$ FALSE $\quad\quad\quad\quad\quad 4x = -6$
  $\quad\quad\quad\quad\quad\quad\quad\quad\quad\quad x = -\dfrac{3}{2}$

$\left\{-\dfrac{3}{2}\right\}$

25. $y \le 2x + 1$

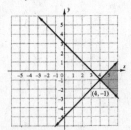

26. $x + y \ge 3$,
  $x - y \ge 5$

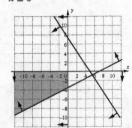

Vertex: $(4, -1)$

27. $2y - x \ge -7$,
  $2y + 3x \le 15$,
  $y \le 0$,
  $x \le 0$

Vertices: $(0, 0)$ and $\left(0, -\dfrac{7}{2}\right)$

28.

Vertices	$F = 5x + 3y$
(1, 0)	$5 \cdot 1 + 3 \cdot 0 = 5$
(6, 0)	$5 \cdot 6 + 3 \cdot 0 = 30$
(1, 12)	$5 \cdot 1 + 3 \cdot 12 = 41$
(3, 12)	$5 \cdot 3 + 3 \cdot 12 = 51$
(6, 9)	$5 \cdot 6 + 3 \cdot 9 = 57$

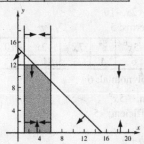

The maximum is 57 when $x = 6$, $y = 9$.
The minimum is 5 when $x = 1$, $y = 0$.

29. Let x = the number of manicures and y = the number of haircuts. The profit is $P = 12x + 18y$. We maximize the profit subject to the constraints:

$$30x + 50y \le 5(6)(60)$$
$$x + y \le 50$$
$$x \ge 0$$
$$y \ge 0$$

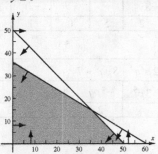

Vertices	$P = 12x + 18y$
(0, 0)	$12 \cdot 0 + 18 \cdot 0 = 0$
(50, 0)	$12 \cdot 50 + 18 \cdot 0 = 600$
(0, 36)	$12 \cdot 0 + 18 \cdot 36 = 648$
(35, 15)	$12 \cdot 35 + 18 \cdot 15 = 690$

The maximum profit is $690 when there are 35 manicures and 15 haircuts.

30. $|2x - 5| \le 7$ and $|x - 2| \ge 2$

$-7 \le 2x - 5 \le 7$ *and* $x - 2 \le -2$ *or* $2 \le x - 2$

$-2 \le 2x \le 12$ *and* $x \le 0$ *or* $4 \le x$

$-1 \le x \le 6$

$\{x | -1 \le x \le 0 \ or \ 4 \le x \le 6\}$ or $[-1, 0] \cup [4, 6]$

31. $7x < 8 - 3x < 6 + 7x$

$0 < 8 - 10x < 6$

$-8 < -10x < -2$

$\dfrac{4}{5} > x > \dfrac{1}{5}$

$\left\{ x \left| \dfrac{1}{5} < x < \dfrac{4}{5} \right. \right\}$ or $\left(\dfrac{1}{5}, \dfrac{4}{5} \right)$

32. $\dfrac{-8 + 2}{2} = -3$

$\dfrac{2 - (-8)}{2} = 5$

$|x - (-3)| \le 5$ or $|x + 3| \le 5$

Chapter 5

Polynomials and Polynomial Functions

Exercise Set 5.1

1. g

3. a

5. b

7. j

9. f

11. $7x^4$, x^3, $-5x$, 8

13. $-t^6$, $7t^3$, $-3t^2$, 6

15. $x^2 - 23x + 17$ is a trinomial.

17. $x^3 - 7x^2 + 2x - 4$ is a polynomial with no special name.

19. $y + 5$ is a binomial.

21. 17 is a monomial.

23.

Term	Degree
$3x^2$	2
$-5x$	1

25.

Term	Degree
$2t^5$	5
$-t^2$	2
1	0

27.

Term	Degree
$8x^2y$	3
$-3x^4y^3$	7
y^4	4

29.

Term	Coefficient
$4x^5$	4
$7x$	7
-3	-3

31.

Term	Coefficient
x^4	1
$-x^3$	-1
$4x$	4

33.

Term	Coefficient
a^2b^3	1
$-5ab$	5
$7b^2$	7
1	1

35. $-5x^6 + x^4 + 7x^3 - 2x - 10$

   a. Number of terms: 5

   b.

Term	$-5x^6$	x^4	$7x^3$	$-2x$	-10
Degree	6	4	3	1	0

   c. Degree of polynomial: 6

   d. Leading term: $-5x^6$

   e. Leading coefficient: -5

37. $7a^4 + a^3b^2 - 5a^2b + 3$

   a. Number of terms: 4

   b.

Term	$7a^4$	a^3b^2	$-5a^2b$	3
Degree	4	5	3	0

   c. Degree of polynomial: 5

   d. Leading term: a^3b^2

   e. Leading coefficient: 1

39. $8y^2 + y^5 - 9 - 2y + 3y^4$

Term	$8y^2$	y^5	-9	$-2y$	$3y^4$
Degree	2	5	0	1	4
Degree of polynomial	5				

41. $3p^4 - 5pq + 2p^3q^3 + 8pq^2 - 7$

Term	$3p^4$	$-5pq$	$2p^3q^3$	$8pq^2$	-7
Degree	4	2	6	3	0
Degree of polynomial	6				

43. $-15t^4 + 2t^3 + 5t^2 - 8t + 4$; $-15t^4$; -15

45. $-x^6 + 6x^5 + 7x^2 + 3x - 5$; $-x^6$; -1

47. $-9 + 4x + 5x^3 - x^6$

49. $8y + 5xy^3 + 2x^2y - x^3$

51. $g(x) = x - 5x^2 + 4$

   $g(3) = 3 - 5(3)^2 + 4 = -38$

53. $f(x) = -3x^4 + 5x^3 + 6x - 2$

   $f(x) = -3(-1)^4 + 5(-1)^3 + 6(-1) - 2$

      $= -3 - 5 - 6 - 2$

      $= -16$

55. $F(x) = 2x^2 - 6x - 9$

   $F(2) = 2 \cdot 2^2 - 6 \cdot 2 - 9 = 8 - 12 - 9 = -13$

   $F(5) = 2 \cdot 5^2 - 6 \cdot 5 - 9 = 50 - 30 - 9 = 11$

57. $Q(y) = -8y^3 + 7y^2 - 4y - 9$

$Q(-3) = -8(-3)^3 + 7(-3)^2 - 4(-3) - 9$
$= 216 + 63 + 12 - 9 = 282$

$Q(0) = -8 \cdot 0^3 + 7 \cdot 0^2 - 4 \cdot 0 - 9$
$= 0 + 0 + 0 - 9 = -9$

59. $N(p) = p^3 - 3p^2 + 2p$

$N(20) = 20^3 - 3 \cdot 20^2 + 2 \cdot 20$
$= 8000 - 1200 + 40$
$= 6840$

A president, vice president, and treasurer can be elected in 6840 ways.

61. Evaluate the polynomial function for $v = 180$.

$h(v) = \dfrac{0.354}{8250} v^3$

$h(180) = \dfrac{0.354}{8250} \cdot (180)^3 \approx 250$

The race car needs about 250 horsepower.

63. Locate 10 on the horizontal axis. From there, move vertically to the function and then horizontally to the vertical axis. This locates a value of about 20 W.

65. To approximate $P(20)$, locate 20 on the horizontal axis. From there, move vertically to the function and then horizontally to the vertical axis. This locates the value of about 150. Thus, $P(20) \approx 150$.

67. Using this function, we find $N(3)$.

$N(x) = \dfrac{1}{3}x^3 + \dfrac{1}{2}x^2 + \dfrac{1}{6}x$

$N(3) = \dfrac{1}{3} \cdot 3^3 + \dfrac{1}{2} \cdot 3^2 + \dfrac{1}{6} \cdot 3$

$= \dfrac{1}{3} \cdot 27 + \dfrac{1}{2} \cdot 9 + \dfrac{1}{6} \cdot 3$

$= 9 + \dfrac{9}{2} + \dfrac{1}{2} = 14$

From the figure we see that the bottom layer has 9 spheres, the second layer has 4, and the third layer has 1. Thus there are 9 + 4 + 1, or 14 spheres.

Using either the function or the figure, we find that $N(3) = 14$.

To calculate the number of oranges in a pyramid with 5 layers, we evaluate the function for $x = 5$.

$N(5) = \dfrac{1}{3} \cdot 5^3 + \dfrac{1}{2} \cdot 5^2 + \dfrac{1}{6} \cdot 5$

$= \dfrac{1}{3} \cdot 125 + \dfrac{1}{2} \cdot 25 + \dfrac{1}{6} \cdot 5$

$= \dfrac{125}{3} + \dfrac{25}{2} + \dfrac{5}{6}$

$= \dfrac{250}{6} + \dfrac{75}{6} + \dfrac{5}{6} = \dfrac{330}{6}$

$= 55$ oranges

69. Locate 1 on the horizontal axis. From there, move vertically to the graph and then horizontally to the $M(t)$-axis. This locates a value of about 260. Thus, about 260 mg of ibuprofen is in the bloodstream 1 hr after swallowed.

71. Locate 2 on the horizontal axis. From there, move vertically to the graph and then horizontally to the

$M(t)$-axis. This locates a value of about 340. Thus, $M(2) \approx 340$.

73. We evaluate the polynomial for $h = 6.3$ and $r = 1.2$:

$2\pi rh + 2\pi r^2 = 2\pi(1.2)(6.3) + 2\pi(1.2)^2 \approx 56.5$

The surface area is about 56.5 in^2.

75. Evaluate the polynomial function for $x = 75$:

$R(x) = 280x - 0.4x^2$

$R(75) = 280 \cdot 75 - 0.4(75)^2$
$= 21,000 - 0.4(5625)$
$= 21,000 - 2250 = 18,750$

The total revenue is $18,750.

77. Evaluate the polynomial function for $x = 75$:

$C(x) = 5000 + 0.6x^2$

$C(75) = 5000 + 0.6(75)^2$
$= 5000 + 0.6(5625)$
$= 5000 + 3375$
$= 8375$

The total cost is $8375.

79. $8x + 2 - 5x + 3x^3 - 4x - 1$
$= 3x^3 + (8 - 5 - 4)x + 2 - 1$
$= 3x^3 - x + 1$

81. $3a^2b + 4b^2 - 9a^2b - 7b^2$
$= (3 - 9)a^2b + (4 - 7)b^2$
$= -6a^2b - 3b^2$

83. $9x^2 - 3xy + 12y^2 + x^2 - y^2 + 5xy + 4y^2$
$= (9 + 1)x^2 + (-3 + 5)xy + (12 - 1 + 4)y^2$
$= 10x^2 + 2xy + 15y^2$

85. $\left(5t^4 - 2t^3 + t\right) + \left(-t^4 - t^3 + 6t^2\right)$
$= (5 - 1)t^4 + (-2 - 1)t^3 + 6t^2 + t$
$= 4t^4 - 3t^3 + 6t^2 + t$

87. $(x^2 + 2x - 3xy - 7) + (-3x^2 - x + 2y^2 + 6)$
$= (1 - 3)x^2 + (2 - 1)x - 3xy + 2y^2 + (-7 + 6)$
$= -2x^2 + x - 3xy + 2y^2 - 1$

89. $(8x^2y - 3xy^2 + 4xy) + (-2x^2y - xy^2 + xy)$
$= (8 - 2)x^2y + (-3 - 1)xy^2 + (4 + 1)xy$
$= 6x^2y - 4xy^2 + 5xy$

91. $(2r^2 + 12r - 11) + (6r^2 - 2r + 4) + (r^2 - r - 2)$
$= (2 + 6 + 1)r^2 + (12 - 2 - 1)r + (-11 + 4 - 2)$
$= 9r^2 + 9r - 9$

93. $\left(\frac{1}{8}xy - \frac{3}{5}x^3y^2 + 4.3y^3\right) + \left(-\frac{1}{3}xy - \frac{3}{4}x^3y^2 - 2.9y^3\right)$

$= \left(\frac{1}{8} - \frac{1}{3}\right)xy + \left(-\frac{3}{5} - \frac{3}{4}\right)x^3y^2 + (4.3 - 2.9)y^3$

$= \left(\frac{3}{24} - \frac{8}{24}\right)xy + \left(-\frac{12}{20} - \frac{15}{20}\right)x^3y^2 + 1.4y^3$

$= -\frac{5}{24}xy - \frac{27}{20}x^3y^2 + 1.4y^3$

95. $3t^4 + 8t^2 - 7t - 1$

   a. $-(3t^4 + 8t^2 - 7t - 1)$ Writing the opposite of P as $-P$

   b. $-3t^4 - 8t^2 + 7t + 1$ Changing the sign of every term

97. $-12y^5 + 4ay^4 - 7by^2$

   a. $-(-12y^5 + 4ay^4 - 7by^2)$

   b. $12y^5 - 4ay^4 + 7by^2$

99. $(-3x^2 + 2x + 9) - (x^2 + 5x - 4)$

$= (-3x^2 + 2x + 9) + (-x^2 - 5x + 4)$

$= -4x^2 - 3x + 13$

101. $(8a - 3b + c) - (2a + 3b - 4c)$

$= (8a - 3b + c) + (-2a - 3b + 4c)$

$= 6a - 6b + 5c$

103. $(6a^2 + 5ab - 4b^2) - (8a^2 - 7ab + 3b^2)$

$= (6a^2 + 5ab - 4b^2) + (-8a^2 + 7ab - 3b^2)$

$= -2a^2 + 12ab - 7b^2$

105. $(6ab - 4a^2b + 6ab^2) - (3ab^2 - 10ab - 12a^2b)$

$= (6ab - 4a^2b + 6ab^2) + (-3ab^2 + 10ab + 12a^2b)$

$= 8a^2b + 16ab + 3ab^2$

107. $\left(\frac{5}{8}x^4 - \frac{1}{4}x^2 - \frac{1}{2}\right) - \left(-\frac{3}{8}x^4 + \frac{3}{4}x^2 + \frac{1}{2}\right)$

$= \left(\frac{5}{8}x^4 - \frac{1}{4}x^2 - \frac{1}{2}\right) + \left(\frac{3}{8}x^4 - \frac{3}{4}x^2 - \frac{1}{2}\right)$

$= x^4 - x^2 - 1$

109. $(6t^2 + 7) - (2t^2 + 3) + (t^2 + t)$

$= (6t^2 + 7) + (-2t^2 - 3) + t^2 + t$

$= (6 - 2 + 1)t^2 + t + 7 - 3$

$= 5t^2 + t + 4$

111. $(8r^2 - 6r) - (2r - 6) + (5r^2 - 7)$

$= (8r^2 - 6r) + (-2r + 6) + (5r^2 - 7)$

$= (8 + 5)r^2 + (-6 - 2)r + (6 - 7)$

$= 13r^2 - 8r - 1$

113. $(x^2 - 4x + 7) + (3x^2 - 9) - (x^2 - 4x + 7)$

Note that $x^2 - 4x + 7$ and $-(x^2 - 4x + 7)$ are opposites

so their sum is 0. Then the result is $3x^2 - 9$.

115. $P(x) = R(x) - C(x)$

$P(x) = (280x - 0.4x^2) - (5000 + 0.6x^2)$

$P(x) = (280x - 0.4x^2) + (-5000 - 0.6x^2)$

$P(x) = 280x - x^2 - 5000$

$P(70) = 280 \cdot 70 - 70^2 - 5000$

$= 19,600 - 4900 - 5000 = 9700$

The profit is \$9700.

117. *Writing Exercise.*

119. $-\frac{3}{20} - \frac{1}{8} = -\frac{6}{40} - \frac{5}{40} = -\frac{11}{40}$

121. $(-120)(-2) = 240$

123. $3 - (4 - 10)^2 \div 3(2 - 4)$

$= 3 - (-6)^2 \div 3(2 - 4)$

$= 3 - 36 \div 3(2 - 4)$

$= 3 - 12(2 - 4)$

$= 3 - 12(-2)$

$= 3 + 24$

$= 27$

125. *Writing Exercise.*

127. $2[P(x)] = 2(13x^5 - 22x^4 - 36x^3 + 40x^2 - 16x + 75)$

$= 26x^5 - 44x^4 - 72x^3 + 80x^2 - 32x + 150$

Use columns to add:

$26x^5 - 44x^4 - 72x^3 + 80x^2 - 32x + 150$
$\underline{42x^5 - 37x^4 + 50x^3 - 28x^2 + 34x + 100}$
$68x^5 - 81x^4 - 22x^3 + 52x^2 + 2x + 250$

129. $2[Q(x)] = 2(42x^5 - 37x^4 + 50x^3 - 28x^2 + 34x + 100)$

$= 84x^5 - 74x^4 + 100x^3 - 56x^2 + 68x + 200$

$3[P(x)] = 3(13x^5 - 22x^4 - 36x^3 + 40x^2 - 16x + 75)$

$= 39x^5 - 66x^4 - 108x^3 + 120x^2 - 48x + 225$

Use columns to subtract, adding the opposite of $3[P(x)]$:

$84x^5 - 74x^4 + 100x^3 - 56x^2 + 68x + 200$
$\underline{- 39x^5 + 66x^4 + 108x^3 - 120x^2 + 48x - 225}$
$45x^5 - 8x^4 + 208x^3 - 176x^2 + 116x - 25$

131. First we find the number of truffles in the display.

$N(x) = \frac{1}{6}x^3 + \frac{1}{2}x^2 + \frac{1}{3}x$

$N(5) = \frac{1}{6} \cdot 5^3 + \frac{1}{2} \cdot 5^2 + \frac{1}{3} \cdot 5$

$= \frac{1}{6} \cdot 125 + \frac{1}{2} \cdot 25 + \frac{5}{3}$

$= \frac{125}{6} + \frac{25}{2} + \frac{5}{3}$

$= \frac{125}{6} + \frac{75}{6} + \frac{10}{6}$

$= \frac{210}{6} = 35$

There are 35 truffles in the display. Now find the volume of one truffle. Each truffle's diameter is 3 cm, so the radius is $\frac{3}{2}$, or 1.5 cm.

$$V(r) = \frac{4}{3}\pi r^3$$

$$V(1.5) \approx \frac{4}{3}(3.14)(1.5)^3 \approx 14.13 \text{ cm}^3$$

Finally, multiply the number of truffles and the volume of a truffle to find the total volume of chocolate.

$$35(14.13 \text{ cm}^3) = 494.55 \text{ cm}^3$$

The display contains about 494.55 cm^3 of chocolate.

133. The area of the base is $x \cdot x$, or x^2.

The area of each side is $x \cdot (x-2)$.

The total area of all four sides is $4x(x-2)$.

The surface area of this box can be expressed as a polynomial function.

$$\begin{aligned} S(x) &= x^2 + 4x(x-2) \\ &= x^2 + 4x^2 - 8x \\ &= 5x^2 - 8x \end{aligned}$$

135. $(2x^{5b} + 4x^{4b} + 3x^{3b} + 8) - (x^{5b} + 2x^{3b} + 6x^{2b} + 9x^b + 8)$

$$= (2-1)x^{5b} + 4x^{4b} + (3-2)x^{3b} - 6x^{2b} - 9x^b + (8-8)$$

$$= x^{5b} + 4x^{4b} + x^{3b} - 6x^{2b} - 9x^b$$

137. *Writing Exercise.*

Exercise Set 5.2

1. False; the coefficient of $3x^5$ is 3.

3. True; see Example 2.

5. False; FOIL is intended to be used to multiply two binomials.

7. True

9. $3x^4 \cdot 5x = (3 \cdot 5)(x^4 \cdot x) = 15x^5$

11. $6a^2(-8ab^2) = 6(-8)(a^2 \cdot a)b^2 = -48a^3b^2$

13. $(-4x^3y^2)(-9x^2y^4) = (-4)(-9)(x^3 \cdot x^2)(y^2 \cdot y^4)$

$$= 36x^5y^6$$

15. $7x(3-x) = 7x \cdot 3 - 7x \cdot x$

$$= 21x - 7x^2$$

17. $5cd(4c^2d - 5cd^2)$

$$= 5cd \cdot 4c^2d - 5cd \cdot 5cd^2$$

$$= 20c^3d^2 - 25c^2d^3$$

19. $(x+3)(x+5)$

$$= x^2 + 5x + 3x + 15 \quad \text{FOIL}$$

$$= x^2 + 8x + 15$$

21. $(2a+3)(4a-1)$

$$= 8a^2 - 2a + 12a - 3 \quad \text{FOIL}$$

$$= 8a^2 + 10a - 3$$

23. $(x+2)(x^2 - 3x + 1)$

$$= x(x^2 - 3x + 1) + 2(x^2 - 3x + 1)$$

$$= x^3 - 3x^2 + x + 2x^2 - 6x + 2$$

$$= x^3 - x^2 - 5x + 2$$

25. $(t-5)(t^2 + 2t - 3)$

$$= t(t^2 + 2t - 3) - 5(t^2 + 2t - 3)$$

$$= t^3 + 2t^2 - 3t - 5t^2 - 10t + 15$$

$$= t^3 - 3t^2 - 13t + 15$$

27.

$$\begin{array}{rl} a^2 + \ a - 1 & \\ a^2 + 4a - 5 & \\ \hline -5a^2 - 5a + 5 & \text{Multiplying by } -5 \\ 4a^3 + \ 4a^2 - 4a & \text{Multiplying by } 4a \\ a^4 + \ a^3 - \ a^2 & \text{Multiplying by } a^2 \\ \hline a^4 + 5a^3 - 2a^2 - 9a + 5 & \text{Adding} \end{array}$$

29. $(x+3)(x^2 - 3x + 9)$

$$= (x+3)(x^2) + (x+3)(-3x) + (x+3)(9)$$

$$= x^3 + 3x^2 - 3x^2 - 9x + 9x + 27$$

$$= x^3 + 27$$

31. $(a-b)(a^2 + ab + b^2)$

$$= (a-b)(a^2) + (a-b)(ab) + (a-b)(b^2)$$

$$= a^3 - a^2b + a^2b - ab^2 + ab^2 - b^3$$

$$= a^3 - b^3$$

33. $(t-3)(t+2)$

$$= t^2 + 2t - 3t - 6$$

$$= t^2 - t - 6$$

35. $(5x+2y)(4x+y)$

$$= 20x^2 + 5xy + 8xy + 2y^2$$

$$= 20x^2 + 13xy + 2y^2$$

37. $\left(t - \frac{1}{3}\right)\left(t - \frac{1}{4}\right)$

$$= t^2 - \frac{1}{4}t - \frac{1}{3}t + \frac{1}{12} \quad \text{FOIL}$$

$$= t^2 - \frac{3}{12}t - \frac{4}{12}t + \frac{1}{12}$$

$$= t^2 - \frac{7}{12}t + \frac{1}{12}$$

39. $(1.2t + 3s)(2.5t - 5s)$

$$= 3t^2 - 6st + 7.5st - 15s^2 \quad \text{FOIL}$$

$$= 3t^2 + 1.5st - 15s^2$$

41. $(r+3)(r+2)(r-1)$

$$= (r^2 + 2r + 3r + 6)(r-1) \quad \text{FOIL}$$

$$= (r^2 + 5r + 6)(r-1)$$

$$= (r^2 + 5r + 6) \cdot r + (r^2 + 5r + 6)(-1)$$

$$= r^3 + 5r^2 + 6r - r^2 - 5r - 6$$

$$= r^3 + 4r^2 + r - 6$$

43. $(x+5)^2$

$\quad = x^2 + 2 \cdot x \cdot 5 + 5^2 \qquad (A+B)^2 = A^2 + 2AB + B^2$

$\quad = x^2 + 10x + 25 \cdot$

45. $(2y-7)^2$

$\quad = (2y)^2 - 2 \cdot 2y \cdot 7 + 7^2 \qquad (A-B)^2 = A^2 - 2AB + B^2$

$\quad = 4y^2 - 28y + 49$

47. $(5c-2d)^2$

$\quad = (5c)^2 - 2 \cdot 5c \cdot 2d + (2d)^2$

$\qquad\qquad (A-B)^2 = A^2 - 2AB + B^2$

$\quad = 25c^2 - 20cd + 4d^2$

49. $\left(3a^3 - 10b^2\right)^2$

$\quad = \left(3a^3\right)^2 - 2 \cdot 3a^3 \cdot 10b^2 + \left(10b^2\right)^2$

$\quad = 9a^6 - 60a^3b^2 + 100b^4$

51. $(x^3 y^4 + 5)^2$

$\quad = (x^3 y^4)^2 + 2 \cdot x^3 y^4 \cdot 5 + 5^2$

$\qquad\qquad (A+B)^2 = A^2 + 2AB + B^2$

$\quad = x^6 y^8 + 10x^3 y^4 + 25$

53. $(c+7)(c-7)$

$\quad = c^2 - 7^2 \qquad (A+B)(A-B) = A^2 - B^2$

$\quad = c^2 - 49$

55. $(1-4x)(1+4x)$

$\quad = 1^2 - (4x)^2 \qquad (A+B)(A-B) = A^2 - B^2$

$\quad = 1 - 16x^2$

57. $\left(3m - \dfrac{1}{2}n\right)\left(3m + \dfrac{1}{2}n\right)$

$\quad = (3m)^2 - \left(\dfrac{1}{2}n\right)^2 \qquad (A+B)(A-B) = A^2 - B^2$

$\quad = 9m^2 - \dfrac{1}{4}n^2$

59. $(x^3 + yz)(x^3 - yz)$

$\quad = \left(x^3\right)^2 - (yz)^2 \qquad (A+B)(A-B) = A^2 - B^2$

$\quad = x^6 - y^2 z^2$

61. $\left(-mn + 3m^2\right)\left(mn + 3m^2\right)$

$\quad = \left(3m^2 - mn\right)\left(3m^2 + mn\right)$

$\quad = \left(3m^2\right)^2 - (mn)^2 \qquad (A+B)(A-B) = A^2 - B^2$

$\quad = 9m^4 - m^2 n^2, \text{ or } -m^2 n^2 + 9m^4$

63. $(x+7)^2 - (x+3)(x-3)$

$\quad = x^2 + 2 \cdot x \cdot 7 + 7^2 - (x^2 - 3^2)$

$\quad = x^2 + 14x + 49 - (x^2 - 9)$

$\quad = x^2 + 14x + 49 - x^2 + 9$

$\quad = 14x + 58$

65. $(2m-n)(2m+n) - (m-2n)^2$

$\quad = \left[(2m)^2 - n^2\right] - \left[m^2 - 2 \cdot m \cdot 2n + (2n)^2\right]$

$\quad = 4m^2 - n^2 - (m^2 - 4mn + 4n^2)$

$\quad = 4m^2 - n^2 - m^2 + 4mn - 4n^2$

$\quad = 3m^2 + 4mn - 5n^2$

67. $(a+b+1)(a+b-1)$

$\quad = [(a+b)+1][(a+b)-1]$

$\quad = (a+b)^2 - 1^2$

$\quad = a^2 + 2ab + b^2 - 1$

69. $(2x+3y+4)(2x+3y-4)$

$\quad = [(2x+3y)+4][(2x+3y)-4]$

$\quad = (2x+3y)^2 - 4^2$

$\quad = 4x^2 + 12xy + 9y^2 - 16$

71. $A = P(1+r)^2$

$\quad A = P\left(1 + 2r + r^2\right)$

$\quad A = P + 2Pr + Pr^2$

73. $P(x) \cdot Q(x)$

$\quad = (3x^2 - 5)(4x^2 - 7x + 1)$

$\quad = (3x^2 - 5)(4x^2) + (3x^2 - 5)(-7x) + (3x^2 - 5)(1)$

$\quad = 12x^4 - 20x^2 - 21x^3 + 35x + 3x^2 - 5$

$\quad = 12x^4 - 21x^3 - 17x^2 + 35x - 5$

75. $P(x) \cdot P(x)$

$\quad = (5x-2)(5x-2)$

$\quad = (5x)^2 - 2 \cdot 5x \cdot 2 + 2^2$

$\qquad\qquad (A-B)^2 = A^2 - 2AB + B^2$

$\quad = 25x^2 - 20x + 4$

77. $\left[F(x)\right]^2$

$\quad = \left(2x - \dfrac{1}{3}\right)^2$

$\quad = (2x)^2 - 2 \cdot 2x \cdot \dfrac{1}{3} + \left(\dfrac{1}{3}\right)^2$

$\qquad\qquad (A-B)^2 = A^2 - 2AB + B^2$

$\quad = 4x^2 - \dfrac{4}{3}x + \dfrac{1}{9}$

79. a. Replace x with $t-1$.

$\quad f(t-1) = (t-1)^2 + 5$

$\qquad\qquad = t^2 - 2t + 1 + 5$

$\qquad\qquad = t^2 - 2t + 6$

b. $f(a+h) - f(a)$

$\quad = [(a+h)^2 + 5] - (a^2 + 5)$

$\quad = a^2 + 2ah + h^2 + 5 - a^2 - 5$

$\quad = 2ah + h^2$

c. $f(a) - f(a-h)$

$\quad = (a^2 + 5) - [(a-h)^2 + 5]$

$\quad = a^2 + 5 - (a^2 - 2ah + h^2 + 5)$

$\quad = a^2 + 5 - a^2 + 2ah - h^2 - 5$

$\quad = 2ah - h^2$

81. a. $f(a)+f(-a)$
$$=\left(a^2+a\right)+\left[(-a)^2+(-a)\right]$$
$$=a^2+a+a^2-a$$
$$=2a^2$$

b. $f(a+h)$
$$=(a+h)^2+(a+h)$$
$$=a^2+2ah+h^2+a+h$$

c. $f(a+h)-f(a)$
$$=(a+h)^2+(a+h)-\left(a^2+a\right)$$
$$=a^2+2ah+h^2+a+h-a^2-a$$
$$=2ah+h^2+h$$

83. *Writing Exercise.*

85. $\dfrac{x}{3}-7=\dfrac{1}{4}$
$$\frac{x}{3}=\frac{29}{4}$$
$$x=\frac{87}{4}$$

87. $|3x-6|>8$

$3x-6<-8$ *or* $8<3x-6$
$3x<-2$ *or* $14<3x$
$x<-\dfrac{2}{3}$ *or* $\dfrac{14}{3}<x$

The solution set is $\left\{x\middle|x<-\dfrac{2}{3}\ or\ x>\dfrac{14}{3}\right\}$, or

$$\left(-\infty,-\frac{2}{3}\right)\cup\left(\frac{14}{3},\infty\right).$$

89. $2x-3y=4,$ (1)
     $x+2y=5$ (2)

We solve the second equation for x.
$$2x-3y=4, \qquad (1)$$
$$x=-2y+5 \quad (3)$$

We substitute $-2y+5$ for x in the first equation and solve for y.
$$2(-2y+5)-3y=4$$
$$-4y+10-3y=4$$
$$-7y+10=4$$
$$-7y=-6$$
$$y=\frac{6}{7}$$

Now we substitute $\dfrac{6}{7}$ for y in Equation (3).

$$x=-2\left(\frac{6}{7}\right)+5=-\frac{12}{7}+5=\frac{23}{7}$$

The solution is $\left(\dfrac{23}{7},\dfrac{6}{7}\right)$.

91. *Writing Exercise.*

93. $\left(x^2+y^n\right)\left(x^2-y^n\right)=\left(x^2\right)^2-\left(y^n\right)^2=x^4-y^{2n}$

95. $x^2y^3(5x^n+4y^n)=x^2y^3\cdot5x^n+x^2y^3\cdot4y^n$
$$=5x^{n+2}y^3+4x^2y^{n+3}$$

97. $(a-b+c-d)(a+b+c+d)$
$$=[(a+c)-(b+d)][(a+c)+(b+d)]$$
$$=(a+c)^2-(b+d)^2$$
$$=\left(a^2+2ac+c^2\right)-\left(b^2+2bd+d^2\right)$$
$$=a^2+2ac+c^2-b^2-2bd-d^2$$

99. $(x^2-3x+5)(x^2+3x+5)$
$$=[(x^2+5)-3x][(x^2+5)+3x]$$
$$=(x^2+5)^2-(3x)^2$$
$$=x^4+10x^2+25-9x^2$$
$$=x^4+x^2+25$$

101. $(x-1)(x^2+x+1)(x^3+1)$
$$=(x^3+x^2+x-x^2-x-1)(x^3+1)$$
$$=(x^3-1)(x^3+1)$$
$$=x^6-1$$

103. $\left(x^{a-b}\right)^{a+b}=x^{(a-b)(a+b)}=x^{a^2-b^2}$

105. $(x-a)(x-b)(x-c)\cdots(x-z)$
$$=(x-a)(x-b)\cdots(x-x)(x-y)(x-z)$$
$$=(x-a)(x-b)\cdots0\cdot(x-y)(x-z)$$
$$=0$$

107. $(2x^{-2}+3x^{-1})(5x^{-3}-x^2)$
$$=10x^{-5}-2x^0+15x^{-4}-3x^1$$
$$=10x^{-5}+15x^{-4}-2-3x$$

109. $\dfrac{g(a+h)-g(a)}{h}$
$$=\frac{(a+h)^2-9-(a^2-9)}{h}$$
$$=\frac{a^2+2ah+h^2-9-a^2+9}{h}$$
$$=\frac{2ah+h^2}{h}=\frac{h(2a+h)}{h}$$
$$=2a+h$$

111. $(A-B)^2 = A^2 - 2AB + B^2$

113. One method is as follows. For each equation, let y_1 represent the left-hand side and y_2 represent the right-hand side, and let $y_3 = y_2 - y_1$. Then use a graphing calculator to view the graph of y_3 and/or a table of values for y_3. If $y_3 = 0$, the equation is an identity. If $y_3 \neq 0$, the equation is not an identity.

   a. Not an identity
   b. Identity
   c. Identity
   d. Not an identity
   e. Not an identity

Exercise Set 5.3

1. True

3. True

5. True

7. True; $-(a-b) = -a+b = b-a;$
    $-1(a-b) = -a+b = b-a.$

9. $10x^2 + 35 = 5 \cdot 2x^2 + 5 \cdot 7$
             $= 5(2x^2 + 7)$

11. $2y^2 - 18y = 2y \cdot y - 2y \cdot 9$
             $= 2y(y-9)$

13. $5t^3 - 15t + 5 = 5 \cdot t^3 - 5 \cdot 3t + 5 \cdot 1$
                $= 5(t^3 - 3t + 1)$

15. $a^6 + 2a^4 - a^3 = a^3 \cdot a^3 + a^3 \cdot 2a - a^3 \cdot 1$
                $= a^3(a^3 + 2a - 1)$

17. $12x^4 - 30x^3 + 42x = 6x \cdot 2x^3 - 6x \cdot 5x^2 + 6x \cdot 7$
                $= 6x(2x^3 - 5x^2 + 7)$

19. $6a^2b - 2ab - 9b = b \cdot 6a^2 - b \cdot 2a - b \cdot 9$
                $= b(6a^2 - 2a - 9)$

21. $15m^4n + 30m^5n^2 + 25m^3n^3$
    $= 5m^3n \cdot 3m + 5m^3n \cdot 6m^2n + 5m^3n \cdot 5n^2$
    $= 5m^3n(3m + 6m^2n + 5n^2)$

23. $9x^3y^6z^2 - 12x^4y^4z^4 + 15x^2y^5z^3$
    $= 3x^2y^4z^2 \cdot 3xy^2 - 3x^2y^4z^2 \cdot 4x^2z^2 + 3x^2y^4z^2 \cdot 5yz$
    $= 3x^2y^4z^2(3xy^2 - 4x^2z^2 + 5yz)$

25. $-5x - 40 = -5(x+8)$

27. $-16t^2 + 96 = -16(t^2 - 6)$

29. $-2x^2 + 12x + 40 = -2(x^2 - 6x - 20)$

31. $5 - 10y = -5(-1 + 2y)$, or $-5(2y - 1)$

33. $8d^2 - 12cd = -4d(-2d + 3c)$, or $-4d(3c - 2d)$

35. $-m^3 + 8 = -1(m^3 - 8)$

37. $-p^3 - 2p^2 - 5p + 2 = -1(p^3 + 2p^2 + 5p - 2)$

39. $a(b-5) + c(b-5) = (b-5)(a+c)$

41. $(x+7)(x-1) + (x+7)(x-2)$
    $= (x+7)(x-1+x-2)$
    $= (x+7)(2x-3)$

43. $a^2(x-y) + 5(y-x)$
    $= a^2(x-y) + 5(-1)(x-y)$ Factoring out -1
                   to reverse the second subtraction
    $= a^2(x-y) - 5(x-y)$ Simplifying
    $= (x-y)(a^2 - 5)$

45. $xy + xz + wy + wz = x(y+z) + w(y+z)$
                      $= (y+z)(x+w)$

47. $y^3 - y^2 + 3y - 3 = y^2(y-1) + 3(y-1)$
                    $= (y-1)(y^2 + 3)$

49. $t^3 + 6t^2 - 2t - 12 = t^2(t+6) - 2(t+6)$
                      $= (t+6)(t^2 - 2)$

51. $12a^4 - 21a^3 - 9a^2$
    $= 3a^2 \cdot 4a^2 - 3a^2 \cdot 7a - 3a^2 \cdot 3$
    $= 3a^2(4a^2 - 7a - 3)$

53. $y^8 - 1 - y^7 + y = y^8 - y^7 + y - 1$
                  $= y^7(y-1) + 1(y-1)$
                  $= (y-1)(y^7 + 1)$

55. $2xy + 3x - x^2y - 6 = 2xy - x^2y - 6 + 3x$
                    $= xy(2-x) - 3(2-x)$
                    $= (2-x)(xy-3)$, or $(x-2)(3-xy)$

57. a. $h(t) = -16t^2 + 72t$
         $h(t) = -8t(2t - 9)$

   b.   Using $h(t) = -16t^2 + 72t$:
         $h(1) = -16 \cdot 1^2 + 72 \cdot 1 = -16 \cdot 1 + 72$
               $= -16 + 72 = 56$ ft
        Using $h(t) = -8t(2t-9)$:
         $h(1) = -8(1)(2 \cdot 1 - 9) = -8(1)(-7) = 56$ ft
    The expressions have the same value for $t = 1$, so the factorization is probably correct.

59. $2\pi rh + \pi r^2 = \pi r(2h + r)$

61. $P(t) = t^2 - 5t$
$P(t) = t(t - 5)$

63. $R(x) = 280x - 0.4x^2$
$R(x) = 0.4x(700 - x)$

65. $P(n) = \frac{1}{2}n^2 - \frac{3}{2}n$
$P(n) = \frac{1}{2}(n^2 - 3n)$

67. $N(x) = \frac{1}{6}x^3 + \frac{1}{2}x^2 + \frac{1}{3}x$
$N(x) = \frac{1}{6}(x^3 + 3x^2 + 2x)$ Factoring out $\frac{1}{6}$

69. *Writing Exercise.*

71. Graph $f(x) = -\frac{1}{2}x + 3$.

Slope is $-\frac{1}{2}$; y-intercept is $(0, 3)$
From the y-intercept we go down 1 unit and to the right 2 units. This gives us $(2, 2)$. We can now draw the graph.

73. Graph $y - 1 = 2(x + 3)$.
From the point-slope equation, $m = 2$, and a point is $(-3, 1)$.

75. Graph $6x = 3$.
Since y does not appear, we solve for x.
$6x = 3$
$x = \frac{1}{2}$

This is a vertical line that crosses the x-axis at $\left(\frac{1}{2}, 0\right)$.

77. *Writing Exercise.*

79. $x^5 y^4 + \underline{\quad} = x^3 y(\underline{\quad} + xy^5)$
The term that goes in the first blank is the product of $x^3 y$ and xy^5, or $x^4 y^6$.
The term that goes in the second blank is the expression that is multiplied with $x^3 y$ to obtain $x^5 y^4$, or $x^2 y^3$.
Thus, we have $x^5 y^4 + x^4 y^6 = x^3 y(x^2 y^3 + xy^5)$.

81. $rx^2 - rx + 5r + sx^2 - sx + 5s$
$= r(x^2 - x + 5) + s(x^2 - x + 5)$
$= (x^2 - x + 5)(r + s)$

83. $a^4 x^4 + a^4 x^2 + 5a^4 + a^2 x^4 + a^2 x^2 + 5a^2 + 5x^4 + 5x^2 + 25$
$= a^4(x^4 + x^2 + 5) + a^2(x^4 + x^2 + 5) + 5(x^4 + x^2 + 5)$
$= (x^4 + x^2 + 5)(a^4 + a^2 + 5)$

85. $x^{-6} + x^{-9} + x^{-3}$
$= x^{-9} \cdot x^3 + x^{-9} \cdot 1 + x^{-9} \cdot x^6$
$= x^{-9}(x^3 + 1 + x^6)$

87. $x^{1/3} - 5x^{1/2} + 3x^{3/4}$
$= x^{4/12} - 5x^{6/12} + 3x^{9/12}$
$= x^{4/12}(1 - 5x^{2/12} + 3x^{5/12})$
$= x^{1/3}(1 - 5x^{1/6} + 3x^{5/12})$

89. $x^{-5/2} + x^{-3/2}$
$= x^{-5/2} \cdot 1 + x^{-5/2} \cdot x$
$= x^{-5/2}(1 + x)$

91. $x^{-4/5} - x^{-7/5} + x^{-1/3}$
$= x^{-7/5} \cdot x^{3/5} - x^{-7/5} \cdot 1 + x^{-7/5} \cdot x^{16/15}$
$= x^{-7/5}(x^{3/5} - 1 + x^{16/15})$

93. $3a^{n+1} + 6a^n - 15a^{n+2}$
$= 3a^n \cdot a + 3a^n \cdot 2 - 3a^n(5a^2)$
$= 3a^n(a + 2 - 5a^2)$

95. $7y^{2a+b} - 5y^{a+b} + 3y^{a+2b}$
$= y^{a+b} \cdot 7y^a - y^{a+b}(5) + y^{a+b} \cdot 3y^b$
$= y^{a+b}(7y^a - 5 + 3y^b)$

97. One method is to let $y_1 = (x^2 - 3x + 2)^4$ and let $y_2 = x^8 + 81x^4 + 16$. Then use a table to show that $y_1 \neq y_2$ for all values of x.

Exercise Set 5.4

1. True

3. False

5. False; whenever the product of a pair of factors is negative, the factors have different signs.

7. True; see Example 8.

9. $x^2 + 5x + 4$
We look for two numbers whose product is 4 and whose sum is 5. Since 4 and 5 are both positive, we need only consider positive factors. The only positive pair is 1 and 4. They are the numbers we need. The factorization is $(x+1)(x+4)$.

11. $y^2 - 12y + 27$

Since the constant term is positive and the coefficient of the middle term is negative, we look for a factorization of 27 in which both factors are negative. Their sum must be –12.

Pair of Factors	Sum of Factors
–1, –27	–28
–3, –9	–12

The numbers we need are –3 and –9. The factorization is $(y-3)(y-9)$.

13. $t^2 - 2t - 8$

Since the constant term is negative, we look for a factorization of –8 in which one factor is positive and one factor is negative. Their sum must be –2, so the negative factor must have the larger absolute value. Thus we consider only pairs of factors in which the negative factor has the larger absolute value.

Pair of Factors	Sum of Factors
–8, 1	–7
–4, 2	–2

The numbers we need are –4 and 2. The factorization is $(t-4)(t+2)$.

15. $a^2 + a - 2$

Since the constant term is negative, we look for a factorization of –2 in which one factor is positive and one factor is negative. Their sum must be 1, so the positive factor must have the larger absolute value. The only pair is 2 and –1. The factorization is $(a+2)(a-1)$.

17. $2x^2 + 6x - 108$

$= 2(x^2 + 3x - 54)$ Removing the common factor

We now factor $x^2 + 3x - 54$. Since the constant term is negative, we look for a factorization of –54 in which one factor is positive and one factor is negative. We consider only pairs of factors in which the positive factor has the larger absolute value, since the sum of the factors, 3, is positive.

Pair of Factors	Sum of Factors
–1, 54	53
–2, 27	25
–3, 18	15
–6, 9	3

The numbers we need are –6 and 9.

$$x^2 + 3x - 54 = (x-6)(x+9)$$

We must not forget to include the common factor 2.

$$2x^2 + 6x - 108 = 2(x-6)(x+9)$$

19. $14a + a^2 + 45 = a^2 + 14a + 45$

Since the constant term and the middle term are both positive, we look for a factorization of 45 in which both factors are positive. Their sum must be 14.

Pair of Factors	Sum of Factors
45, 1	46
15, 3	18
9, 5	14

The numbers we need are 9 and 5. The factorization is $(a+9)(a+5)$.

21. $p^3 - p^2 - 72p$

$= p(p^2 - p - 72)$ Removing the common factor

We now factor $p^2 - p - 72$. Since the constant term is negative, we look for a factorization of –72 in which one factor is positive and one factor is negative. We consider only pairs of factors in which the negative factor has the larger absolute value, since the sum of the factors, –1, is negative.

Pair of Factors	Sum of Factors
–72, 1	–71
–36, 2	–34
–24, 3	–21
–18, 4	–14
–12, 6	–6
–9, 8	–1

The numbers we need are –9 and 8.

$$p^2 - p - 72 = (p-9)(p+8)$$

We must not forget to include the common factor p.

$$p^3 - p^2 - 72p = p(p-9)(p+8)$$

23. $a^2 - 11a + 28$

Since the constant term is positive and the coefficient of the middle term is negative, we look for a factorization of 28 in which both factors are negative. Their sum must be -11.

Pair of Factors	Sum of Factors
–1, –28	–29
–2, –14	–16
–4, –7	–11

The numbers we need are –4 and –7. The factorization is $(a-4)(a-7)$.

25. $x + x^2 - 6 = x^2 + x - 6$

Since the constant term is negative, we look for a factorization of –6 in which one factor is positive and one factor is negative. We consider only pairs of factors in which the positive factor has the larger absolute value, since the sum of the factors, 1, is positive.

Pair of Factors	Sum of Factors
6, –1	5
3, –2	1

The numbers we need are 3 and –2. The factorization is $(x+3)(x-2)$.

27. $5y^2 + 40y + 35$

$= 5(y^2 + 8y + 7)$ Removing the common factor

We now factor $y^2 + 8y + 7$. We look for two numbers whose product is 7 and whose sum is 8. Since 7 and 8 are both positive, we need consider only positive factors. The only possible pair is 1 and 7. They are the numbers we need.

$$y^2 + 8y + 7 = (y+1)(y+7)$$

We must not forget to include the common factor 5.

$$5y^2 + 40y + 35 = 5(y+1)(y+7)$$

29. $32 + 4y - y^2 = -y^2 + 4y + 32 = -(y^2 - 4y - 32)$

We now factor $y^2 - 4y - 32$. Since the constant term is negative, we look for a factorization of -32 in which one factor is positive and one factor is negative. We consider only pairs of factors in which the negative factor has the larger absolute value, since the sum of the factors, -4, is negative.

Pair of Factors	Sum of Factors
$-32, 1$	-31
$-16, 2$	-14
$-8, 4$	-4

The numbers we need are -8 and 4. Thus,

$y^2 - 4y - 32 = (y - 8)(y + 4)$. We must not forget to include the factor that was factored out earlier:

$32 + 4y - y^2 = -(y - 8)(y + 4)$,

or $(-y + 8)(y + 4)$, or $(8 - y)(4 + y)$

31. $56x + x^2 - x^3$

There is a common factor, x. We also factor out -1 in order to make the leading coefficient positive.

$56x + x^2 - x^3 = -x(-56 - x + x^2)$

$= -x(x^2 - x - 56)$

Now we factor $x^2 - x - 56$. Since the constant term is negative, we look for a factorization of -56 in which one factor is positive and one factor is negative. We consider only pairs of factors in which the negative factor has the larger absolute value, since the sum of the factors, -1, is negative.

Pair of Factors	Sum of Factors
$-56, 1$	-55
$-28, 2$	-26
$-14, 4$	-10
$-8, 7$	-1

The numbers we need are -8 and 7. Thus,

$x^2 - x - 56 = (x - 8)(x + 7)$. We must not forget to include the common factor:

$56x + x^2 - x^3 = -x(x - 8)(x + 7)$,

or $x(-x + 8)(x + 7)$, or $x(8 - x)(7 + x)$

33. $y^4 + 5y^3 - 84y^2$

$= y^2(y^2 + 5y - 84)$ Removing the common factor

We now factor $y^2 + 5y - 84$. We look for pairs of factors of -84, one positive and one negative, such that the positive factor has the larger absolute value and the sum of the factors is 5.

Pair of Factors	Sum of Factors
$84, -1$	83
$42, -2$	40
$28, -3$	25
$21, -4$	17
$14, -6$	8
$12, -7$	5

The numbers we need are 12 and -7. Then

$y^2 + 5y - 84 = (y + 12)(y - 7)$.

We must not forget to include the common factor:

$y^4 + 5y^3 - 84y^2 = y^2(y + 12)(y - 7)$

35. $x^2 - 3x + 5$

There are no factors of 5 whose sum is -3. This trinomial is not factorable into binomials with integer coefficients. The polynomial is prime.

37. $x^2 + 12xy + 27y^2$

We look for numbers p and q such that

$x^2 + 12xy + 27y^2 = (x + py)(x + qy)$. Our thinking is

much the same as if we were factoring $x^2 + 12x + 27$. Since the constant term is positive and the coefficient of the middle term is positive, we look for a factorization of 27 in which both factors are positive. Their sum must be 12.

Pair of Factors	Sum of Factors
$1, 27$	28
$3, 9$	12

The numbers we need are 3 and 9. The factorization is $(x + 3y)(x + 9y)$.

39. $x^2 - 14xy + 49y^2$

We look for numbers p and q such that

$x^2 - 14xy + 49y^2 = (x + py)(x + qy)$. Our thinking is

much the same as if we were factoring $x^2 - 14x + 49$. We look for factors of 49 whose sum is -14. Since the constant term is positive and the coefficient of the middle term is negative, both factors must be negative.

Pair of Factors	Sum of Factors
$-49, -1$	-50
$-7, -7$	-14

The numbers we need are -7 and -7. The factorization is $(x - 7y)(x - 7y)$, or $(x - 7y)^2$.

41. $n^5 - 80n^4 + 79n^3$

$= n^3(n^2 - 80n + 79)$ Removing the common factor

Now we factor $n^2 - 80n + 79$. We look for a pair of factors of 79 whose sum is -80. The numbers we need are -1 and -79. Then $n^2 - 80n + 79 = (n - 79)(n - 1)$. We must not forget to include the common factor.

$n^5 - 80n^4 + 79n^3 = n^3(n - 79)(n - 1)$

43. $x^6 + 2x^5 - 63x^4$

$= x^4(x^2 + 2x - 63)$ Removing the common factor

We now factor $x^2 + 2x - 63$. We look for a pair of factors of -63 whose sum is 2. The numbers we need are 9 and -7. Then $x^2 + 2x - 63 = (x + 9)(x - 7)$. We must not forget to include the common factor:

$x^6 + 2x^5 - 63x^4 = x^4(x + 9)(x - 7)$

45. $3x^2 - 4x - 4$

We will use the FOIL method.

1. There is no common factor (other than 1 or −1.)
2. Factor the first term, $3x^2$. The factors are $3x$, x. We have this possibility:

$$(3x + \quad)(x + \quad).$$

3. Factor the last term, −4. The possibilities are $4(-1)$, $-4 \cdot 1$, and $2(-2)$.
4. We need factors for which the sum of the products (the "outer" and "inner" parts of FOIL) is the middle term, −4x. Try some possibilities and check by multiplying.

$$(3x + 1)(x - 4) = 3x^2 - 3x - 4$$

We try again.

$$(3x + 2)(x - 2) = 3x^2 - 4x - 4$$

The factorization is $(3x + 2)(x - 2)$.

47. $6t^2 + t - 15$

We will use the grouping method.

1. Factor the trinomial $6t^2 + t - 15$. Multiply the leading coefficient, 6, and the constant, −15.

$$6(-15) = -90$$

2. Try to factor −90 so the sum of the factors is 1. We need only consider pairs of factors in which the positive factor has the larger absolute value, since their sum is positive.

Pair of Factors	Sum of Factors
90, −1	89
45, −2	43
30, −3	27
18, −5	13
15, −6	9
10, −9	1

3. Split the middle term, t, using the results of step (2).

$$t = 10t - 9t$$

4. Factor by grouping.

$$6t^2 + t - 15 = 6t^2 + 10t - 9t - 15$$
$$= 2t(3t + 5) - 3(3t + 5)$$
$$= (3t + 5)(2t - 3)$$

49. $6p^2 - 20p + 16$

We will use the FOIL method.

1. Factor out the common factor 2.

$$2(3p^2 - 10p + 8)$$

2. Factor the first term $3p^2$. The factors are $3p$, p. We have this possibility:

$$(3p + \quad)(p + \quad).$$

3. Factor the last term, 8. The possibilities are $8 \cdot 1$, $-8(-1)$, $4 \cdot 2$, and $-4(-2)$.
4. Look for factors such that the sum of the products is the middle term, −10p. Trial and error leads us to the correct factorization.

$$3p^2 - 10p + 8 = (3p - 4)(p - 2).$$

We must include the common factor.

$$6p^2 - 20p + 16 = 2(3p - 4)(p - 2)$$

51. $9a^2 + 18a + 8$

We will use the grouping method.

1. There is no common factor (other than 1 or −1).
2. Multiply the leading coefficient, 9, and the constant, 8: $9(8) = 72$
3. Try to factor 72 so the sum of the factors is 18. We need only consider pairs of positive factors since 72 and 18 are both positive.

Pair of Factors	Sum of Factors
72, 1	73
36, 2	38
24, 3	27
18, 4	22
12, 6	18
9, 8	17

4. Split 18a using the results of step (3):

$$18a = 12a + 6a$$

5. Factor by grouping:

$$9a^2 + 18a + 8 = 9a^2 + 12a + 6a + 8$$
$$= 3a(3a + 4) + 2(3a + 4)$$
$$= (3a + 4)(3a + 2)$$

53. $8y^2 + 30y^3 - 6y = 30y^3 + 8y^2 - 6y$

We will use the FOIL method.

1. Factor out the common factor 2y.

$$2y(15y^2 + 4y - 3)$$

2. Now we factor the trinomial $15y^2 + 4y - 3$.

Factor the first term, $15y^2$. The factors are $15y$, y and $5y$, $3y$. We have these possibilities:

$$(15y + \quad)(y + \quad) \text{ and } (5y + \quad)(3y + \quad).$$

3. Factor the last term, −3. The possibilities are $(1)(-3)$ and $(-1)3$ as well as $(-3)(1)$ and $3(-1)$.
4. Look for factors such that the sum of the products is the middle term, 4y. Trial and error leads us to the correct factorization.

$$15y^2 + 4y - 3 = (5y + 3)(3y - 1)$$

We must include the common factor to get a factorization of the original trinomial.

$$8y^2 + 30y^3 - 6y = 2y(5y + 3)(3y - 1)$$

55. $18x^2 - 24 - 6x = 18x^2 - 6x - 24$

We will use the grouping method.

1. Factor out the common factor, 6:

$$6(3x^2 - x - 4)$$

2. Now we factor the trinomial $3x^2 - x - 4$. Multiply the leading coefficient, 3, and the constant, −4:

$$3(-4) = -12$$

3. Factor −12 so the sum of the factors is −1. We need only consider pairs of factors in which the negative factor has the larger absolute value, since their sum is negative.

Pair of Factors	Sum of Factors
−12, 1	−11
−6, 2	−4
−4, 3	−1

4. Split −x using the results of step (3):
$$-x = -4x + 3x$$

5. Factor by grouping:
$$3x^2 - x - 4 = 3x^2 - 4x + 3x - 4$$
$$= x(3x - 4) + (3x - 4)$$
$$= (3x - 4)(x + 1)$$

We must include the common factor to get a factorization of the original trinomial:
$$18x^2 - 24 - 6x = 6(3x - 4)(x + 1)$$

57. $t^8 + 5t^7 - 14t^6$
$$= t^6(t^2 + 5t - 14) \quad \text{Removing the common factor}$$

We now factor $t^2 + 5t - 14$. We look for a pair of factors of −14 whose sum is 5. The numbers we need are 7 and −2. Then we have
$t^2 + 5t - 14 = (t + 7)(t - 2)$. We must not forget to include the common factor:
$$t^8 + 5t^7 - 14t^6 = t^6(t + 7)(t - 2)$$

59. $70x^4 - 68x^3 + 16x^2$
We will use the grouping method.

1. Factor out the common factor, $2x^2$:
$$2x^2(35x^2 - 34x + 8)$$

2. Now we factor the trinomial $35x^2 - 34x + 8$. Multiply the leading coefficient, 35, and the constant, 8: $35 \cdot 8 = 280$

3. Factor 280 so the sum of the factors is −34. We need only consider pairs of negative factors since the sum is negative.

Pair of Factors	Sum of Factors
−280, −1	−281
−140, −2	−142
−70, −4	−74
−56, −5	−61
−40, −7	−47
−35, −8	−43
−28, −10	−38
−20, −14	−34

4. Split −34x using the results of step (3):
$$-34x = -20x - 14x$$

5. Factor by grouping:
$$35x^2 - 34x + 8 = 35x^2 - 20x - 14x + 8$$
$$= 5x(7x - 4) - 2(7x - 4)$$
$$= (7x - 4)(5x - 2)$$

We must include the common factor to get a factorization of the original trinomial:
$$70x^4 - 68x^3 + 16x^2 = 2x^2(7x - 4)(5x - 2)$$

61. $18y^2 - 9y - 20$
We will use the FOIL method.

1. There is no common factor (other than 1 or −1).

2. Factor the first term, $18y^2$. The possibilities are
$(18y + \ )(y + \ )$, $(9y + \ )(2y + \ )$ and
$(6y + \ )(3y + \ )$.

3. Factor the last term, −20. The possibilities are
$-20 \cdot 1$, $20(-1)$, $-10 \cdot 2$, $10(-2)$, $-5 \cdot 4$, $5(-4)$.

4. We need factors for which the sum of the products is the middle term, −9y. Trial and error leads us to the correct factorization.
$$18y^2 - 9y - 20 = (3y - 4)(6y + 5).$$

63. $16x^2 + 24x + 5$
We will use the grouping method.
1. There is no common factor (other than 1 or −1).
2. Multiply the leading coefficient and constant:
$16(5) = 80$.
3. Factor 80 so the sum of the factors is 24. We need only consider pairs of positive factors since 80 and 24 are both positive.

Pair of Factors	Sum of Factors
80, 1	81
40, 2	42
20, 4	24
16, 5	21
10, 8	18

4. Split 24x using the results of step (3).
$$24x = 20x + 4x$$
5. Factor by grouping.
$$16x^2 + 24x + 5 = 16x^2 + 20x + 4x + 5$$
$$= 4x(4x + 5) + 1(4x + 5)$$
$$= (4x + 5)(4x + 1)$$

65. $5x^2 + 24x + 16$
We will use the FOIL method.
1. There is no common factor (other than 1 or −1).
2. Factor the first term, $5x^2$. The factors are $5x, x$. We have this possibility:
$$(5x + \ )(x + \ ).$$
3. Factor the last term, 16. We consider only positive factors since both the middle term and the last term are positive. The possibilities are $16 \cdot 1$, $8 \cdot 2$, and $4 \cdot 4$.
4. We need factors for which the sum of products is the middle term, 24x. Trial and error leads us to the correct factorization.
$$5x^2 + 24x + 16 = (5x + 4)(x + 4)$$

67. $-8t^2 - 8t + 30$
We will use the grouping method.

1. Factor out −2: $-2(4t^2 + 4t - 15)$

2. Now we factor the trinomial $4t^2 + 4t - 15$. Multiply the leading coefficient and the constant:
$4(-15) = -60$

3. Factor −60 so the sum of the factors is 4. The desired factorization is $10(-6)$.

4. Split $4t$ using the results of step (3): $4t = 10t - 6t$
5. Factor by grouping:
$$4t^2 + 4t - 15 = 4t^2 + 10t - 6t - 15$$
$$= 2t(2t + 5) - 3(2t + 5)$$
$$= (2t + 5)(2t - 3)$$

We must include the common factor to get a factorization of the original trinomial:
$$-8t^2 - 8t + 30 = -2(2t + 5)(2t - 3)$$

69. $18xy^3 + 3xy^2 - 10xy$

We will use the FOIL method.

1. Factor out the common factor, xy.
$$xy(18y^2 + 3y - 10)$$

2. We now factor the trinomial $18y^2 + 3y - 10$.

    Factor the first term, $18y^2$. The possibilities are $(18y + \ )(y + \ )$, $(9y + \ )(2y + \ )$, and $(6y + \ )(3y + \ )$.

3. Factor the last term, -10. The possibilities are $-10 \cdot 1$, $-5 \cdot 2$, $10(-1)$ and $5(-2)$.

4. We need factors for which the sum of the products is the middle term, $3y$. Trial and error leads us to the correct factorization.
$$18y^2 + 3y - 10 = (6y + 5)(3y - 2)$$

We must include the common factor to get a factorization of the original trinomial:
$$18xy^3 + 3xy^2 - 10xy = xy(6y + 5)(3y - 2)$$

71. $24x^2 - 2 - 47x = 24x^2 - 47x - 2$

We will use the grouping method.

1. There is no common factor (other than 1 or –1).
2. Multiply the leading coefficient and the constant: $24(-2) = -48$
3. Factor -48 so the sum of the factors is -47. The desired factorization is $-48 \cdot 1$.
4. Split $-47x$ using the results of step (3):
$$-47x = -48x + x$$
5. Factor by grouping:
$$24x^2 - 47x - 2 = 24x^2 - 48x + x - 2$$
$$= 24x(x - 2) + (x - 2)$$
$$= (x - 2)(24x + 1)$$

73. $63x^3 + 111x^2 + 36x$

We will use the FOIL method.

1. Factor out the common factor, $3x$.
$$3x(21x^2 + 37x + 12)$$

2. Now we will factor the trinomial $21x^2 + 37x + 12$.

    Factor the first term, $21x^2$. The factors are $21x$, x and $7x$, $3x$. We have these possibilities:
$(21x + \ )(x + \ )$ and $(7x + \ )(3x + \ )$.

3. Factor the last term, 12. The possibilities are $12 \cdot 1$, $(-12)(-1)$, $6 \cdot 2$, $(-6)(-2)$, $4 \cdot 3$, and $(-4)(-3)$ as well as $1 \cdot 12$, $(-1)(-12)$, $2 \cdot 6$, $(-2)(-6)$, $3 \cdot 4$, and $(-3)(-4)$.

4. Look for factors such that the sum of the products

is the middle term, $37x$. Trial and error leads us to the correct factorization:
$$(7x + 3)(3x + 4)$$

We must include the common factor to get a factorization of the original trinomial:
$$63x^3 + 111x^2 + 36x = 3x(7x + 3)(3x + 4)$$

75. $48x^4 + 4x^3 - 30x^2$

We will use the grouping method.

1. We factor out the common factor, $2x^2$:
$$2x^2(24x^2 + 2x - 15)$$

2. We now factor $24x^2 + 2x - 15$. Multiply the leading coefficient and the constant:
$$24(-15) = -360$$

3. Factor -360 so the sum of the factors is 2. The desired factorization is $-18 \cdot 20$.

4. Split $2x$ using the results of step (3):
$$2x = -18x + 20x$$

5. Factor by grouping:
$$24x^2 + 2x - 15 = 24x^2 - 18x + 20x - 15$$
$$= 6x(4x - 3) + 5(4x - 3)$$
$$= (4x - 3)(6x + 5)$$

We must not forget to include the common factor:
$$48x^4 + 4x^3 - 30x^2 = 2x^2(4x - 3)(6x + 5)$$

77. $12a^2 - 17ab + 6b^2$

We will use the FOIL method. (Our thinking is much the same as if we were factoring $12a^2 - 17a + 6$.)

1. There is no common factor (other than 1 or –1).

2. Factor the first term, $12a^2$. The factors are $12a$, a and $6a$, $2a$ and $4a$, $3a$. We have these possibilities: $(12a + \ )(a + \ )$ and $(6a + \ )(2a + \ )$ and $(4a + \ )(3a + \ )$.

3. Factor the last term, $6b^2$. The possibilities are $6b \cdot b$, $(-6b)(-b)$, $3b \cdot 2b$, and $(-3b)(-2b)$ as well as $b \cdot 6b$, $(-b)(-6b)$, $2b \cdot 3b$, and $(-2b)(-3b)$.

4. Look for factors such that the sum of the products is the middle term, $-17ab$. Trial and error leads us to the correct factorization:
$$(4a - 3b)(3a - 2b)$$

79. $2x^2 + xy - 6y^2$

We will use the grouping method.

1. There is no common factor (other than 1 or –1).
2. Multiply the coefficients of the first and last terms:
$$2(-6) = -12$$
3. Factor -12 so the sum of the factors is 1. The desired factorization is $4(-3)$.
4. Split xy using the results of step (3):
$$xy = 4xy - 3xy$$
5. Factor by grouping:
$$2x^2 + xy - 6y^2 = 2x^2 + 4xy - 3xy - 6y^2$$
$$= 2x(x + 2y) - 3y(x + 2y)$$
$$= (x + 2y)(2x - 3y)$$

81. $6x^2 - 29xy + 28y^2$

We will use the FOIL method.

1. There is no common factor (other than 1 or –1).
2. Factor the first term, $6x^2$. The factors are $6x$, x and $3x$, $2x$. We have these possibilities: $(6x+\ )(x+\ )$ and $(3x+\ )(2x+\ )$.
3. Factor the last term, $28y^2$. The possibilities are $28y \cdot y$, $(-28y)(-y)$, $14y \cdot 2y$, $(-14y)(-2y)$, $7y \cdot 4y$, and $(-7y)(-4y)$ as well as $y \cdot 28y$, $(-y)(-28y)$, $2y \cdot 14y$, $(-2y)(-14y)$, $4y \cdot 7y$, and $(-4y)(-7y)$.
4. Look for factors such that the sum of the products is the middle term, $-29xy$. Trial and error leads us to the correct factorization: $(3x - 4y)(2x - 7y)$

83. $9x^2 - 30xy + 25y^2$

We will use the grouping method.

1. There is no common factor (other than 1 or –1).
2. Multiply the coefficients of the first and last terms: $9(25) = 225$
3. Factor 225 so the sum of the factors is –30. The desired factorization is $-15(-15)$.
4. Split $-30xy$ using the results of step (3): $-30xy = -15xy - 15xy$
5. Factor by grouping:

$$9x^2 - 30xy + 25y^2 = 9x^2 - 15xy - 15xy + 25y^2$$
$$= 3x(3x - 5y) - 5y(3x - 5y)$$
$$= (3x - 5y)(3x - 5y), \text{ or } (3x - 5y)^2$$

85. $9x^2y^2 + 5xy - 4$

Let $u = xy$ and $u^2 = x^2y^2$. Factor $9u^2 + 5u - 4$. We will use the FOIL method.

1. There is no common factor (other than 1 or –1).
2. Factor the first term, $9u^2$. The factors are $9u$, u and $3u$, $3u$. We have these possibilities: $(9u+\ )(u+\ )$ and $(3u+\ )(3u+\ )$.
3. Factor the last term, –4. The possibilities are $-4 \cdot 1$, $-2 \cdot 2$, and $-1 \cdot 4$.
4. We need factors for which the sum of the products is the middle term, $5u$. Trial and error leads us to the factorization: $(9u - 4)(u + 1)$. Replace u by xy.

We have $9x^2y^2 + 5xy - 4 = (9xy - 4)(xy + 1)$.

87. *Writing Exercise.*

89. $(2a^{-6}b)^{-3} = 2^{-3}a^{18}b^{-3} = \dfrac{a^{18}}{8b^3}$

91. $\dfrac{12t^{-11}}{8t^6} = \dfrac{3}{2t^{17}}$

93. $0.000607 = 6.07 \times 10^{-4}$

95. *Writing Exercise.*

97. $60x^8y^6 + 35x^4y^3 + 5$

$= 5(12x^8y^6 + 7x^4y^3 + 1)$ Removing the common factor

To factor the trinomial $12x^8y^6 + 7x^4y^3 + 1$, first note that $(x^4y^3)^2 = x^8y^6$, so the trinomial is of the form $12u^2 + 7u + 1$. Trial and error leads us to the factorization:

$$12x^8y^6 + 7x^4y^3 + 1 = (4x^4y^3 + 1)(3x^4y^3 + 1)$$

Then the factorization of the original trinomial is

$5(4x^4y^3 + 1)(3x^4y^3 + 1)$.

99. $y^2 - \dfrac{8}{49} + \dfrac{2}{7}y = y^2 + \dfrac{2}{7}y - \dfrac{8}{49}$

We look for factors of $-\dfrac{8}{49}$ whose sum is $\dfrac{2}{7}$. The factors are $\dfrac{4}{7}$ and $-\dfrac{2}{7}$. The factorization is

$\left(y + \dfrac{4}{7}\right)\left(y - \dfrac{2}{7}\right)$.

101. $20a^3b^6 - 3a^2b^4 - 2ab^2$

Factor the common factor, ab^2.

$$20a^3b^6 - 3a^2b^4 - 2ab^2 = ab^2\left(20a^2b^4 - 3ab^2 - 2\right)$$

Substitute u for ab^2 (and u^2 for $\left(ab^2\right)^2 = a^2b^4$). We factor $20u^2 - 3u - 2$. Trial and error leads us to the factorization: $20u^2 - 3u - 2 = (4u + 1)(5u - 2)$.

Replace u with ab^2:

$$20a^2b^4 - 3ab^2 - 2 = \left(4ab^2 + 1\right)\left(5ab^2 - 2\right).$$

Include the common factor:

$$20a^3b^6 - 3a^2b^4 - 2ab^2 = ab^2\left(4ab^2 + 1\right)\left(5ab^2 - 2\right)$$

103. $x^{2a} + 5x^a - 24$

Substitute u for x^a (and u^2 for x^{2a}). We factor $u^2 + 5u - 24$. We look for factors of –24 whose sum is 5. The factors are 8 and –3. We have

$u^2 + 5u - 24 = (u + 8)(u - 3)$. Replace u with x^a:

$x^{2a} + 5x^a - 24 = (x^a + 8)(x^a - 3)$.

105. $2ar^2 + 4asr + as^2 - asr$

$= 2ar^2 + 3asr + as^2$
$= a(2r^2 + 3sr + s^2)$
$= a(2r + s)(r + s)$

107. $(x + 3)^2 - 2(x + 3) - 35$

Substitute u for $x + 3$ (and u^2 for $(x + 3)^2$). We factor $u^2 - 2u - 35$. Look for factors of –35 whose sum is –2. The factors are –7 and 5. We have

$u^2 - 2u - 35 = (u - 7)(u + 5)$. Replace u with $x + 3$:

$(x + 3)^2 - 2(x + 3) - 35 = [(x + 3) - 7][(x + 3) + 5]$, or $(x - 4)(x + 8)$

109. $x^2 + mx + 75$

All such m are the sums of the factors of 75.

Pair of Factors	Sum of Factors
75, 1	76
−75, −1	−76
25, 3	28
−25, −3	−28
15, 5	20
−15, −5	−20

m can be 76, −76, 28, −28, 20, or −20.

111. Since $ax^2 + bx + c = (mx + r)(nx + s)$, from FOIL we know that $a = mn$, $c = rs$, and $b = ms + rn$. If $P = ms$ and $Q = rn$, then $b = P + Q$. Since $ac = mnrs = msrn$, we have $ac = PQ$.

113. *Graphing Calculator Exercise*

115. *Writing Exercise.*

Mid-Chapter Review

1. $(2x - 3)(x + 4) = 2x^2 + 8x - 3x - 12$
$= 2x^2 + 5x - 12$

2. $3x^3 + 7x^2 + 2x = x(3x^2 + 7x + 2)$
$= x(3x + 1)(x + 2)$

3. $(4t^3 - 2t + 6) + (8t^2 - 11t - 7)$
$= 4t^3 + 8t^2 - 2t - 11t + 6 - 7$
$= 4t^3 + 8t^2 - 13t - 1$

4. $4x^2 y(3xy - 2x^3 + 6y^2) = 12x^3y^2 - 8x^5y + 24x^2y^3$

5. $(8n^2 + 5n - 2) - (-n^2 + 6n - 2)$
$= 8n^2 + 5n - 2 + n^2 - 6n + 2$
$= 9n^2 - n$

6. $(x + 1)(x + 7)$
$= x^2 + 7x + x + 7$
$= x^2 + 8x + 7$

7. $(2x - 3)(5x - 1) = 10x^2 - 17x + 3$

8. $\left(\frac{1}{2}x^2 + \frac{1}{3}x - \frac{3}{2}\right) + \left(\frac{2}{3}x^2 - \frac{1}{2}x - \frac{1}{3}\right) = \frac{7}{6}x^2 - \frac{1}{6}x - \frac{11}{6}$

9. $(3m - 10)^2 = (3m - 10)(3m - 10)$
$= 9m^2 - 30m - 30m + 100$
$= 9m^2 - 60m + 100$

10. $(1.2x^2 - 3.7x) - (2.8x^2 - x + 1.4) = -1.6x^2 - 2.7x - 1.4$

11. $(a + 2)(a^2 - a - 6)$
$= a^3 - a^2 - 6a + 2a^2 - 2a - 12$
$= a^3 + a^2 - 8a - 12$

12. $(c + 9)(c - 9) = c^2 - 81$

13. $8x^2y^3z + 12x^3y^2 - 16x^2yz^3 = 4x^2y(2y^2z + 3xy - 4z^3)$

14. $3t^3 - 3t^2 - 1 + t$
$= 3t^2(t - 1) + 1(t - 1)$
$= (t - 1)(3t^2 + 1)$

15. $x^2 - x - 90 = (x - 10)(x + 9)$

16. $6x^3 + 60x^2 + 126x = 6x(x^2 + 10x + 21)$
$= 6x(x + 7)(x + 3)$

17. $5x^2 + 7x - 6 = (5x - 3)(x + 2)$

18. $2x + 2y + ax + ay = (x + y)(2 + a)$

Exercise Set 5.5

1. $x^2 - 100 = (x)^2 - 10^2$
This is a difference of squares.

3. $36x^2 - 12x + 1 = (6x)^2 - 2 \cdot 6x \cdot 1 + 1^2$
This is a perfect-square trinomial.

5. $4r^2$ and 9 are squares but $8r \neq 2 \cdot 2r \cdot 3$ and $8r \neq -2 \cdot 2r \cdot 3$, so this trinomial is classified as none of these.

7. $4x^2 + 8x + 10 = 2(2x^2 + 4x + 5)$ and $2x^2 + 4x + 5$ cannot be factored, so this is a polynomial having a common factor.

9. $4t^2 + 9s^2 + 12st = 4t^2 + 12st + 9s^2$
$= (2t)^2 + 2 \cdot 2t \cdot 3s + (3s)^2$

This is a perfect-square trinomial.

11. $x^2 + 20x + 100 = (x + 10)^2$
Find the square terms and write the square roots with a plus sign between them.

13. $t^2 - 2t + 1 = (t - 1)^2$
Find the square terms and write the square roots with a minus sign between them.

15. $4a^2 - 24a + 36$
$= 4(a^2 - 6a + 9)$ Factoring out the common factor
$= 4(a - 3)^2$ Factoring the perfect-square trinomial

17. $y^2 + 36 + 12y$
$= y^2 + 12y + 36$ Changing order
$= (y + 6)^2$ Factoring the perfect-square trinomial

19. $-18y^2 + y^3 + 81y$
$= y^3 - 18y^2 + 81y$ Changing order
$= y(y^2 - 18y + 81)$ Factoring out the common factor
$= y(y - 9)^2$

21. $2x^2 - 40x + 200$

$= 2(x^2 - 20x + 100)$ Factoring out the common factor

$= 2(x - 10)^2$ Factoring the perfect–square
                                   trinomial

23. $1 - 8d + 16d^2$

$= (1 - 4d)^2$ Factoring the perfect-square trinomial

25. $-y^3 - 8y^2 - 16y$

$= -y(y^2 + 8y + 16)$

$= -y(y + 4)^2$

27. $0.25x^2 + 0.30x + 0.09 = (0.5x + 0.3)^2$

Find the square terms and write the square
roots with a plus sign between them.

29. $p^2 - 2pq + q^2 = (p - q)^2$

31. $25a^2 + 30ab + 9b^2 = (5a + 3b)^2$

33. $5a^2 + 10ab + 5b^2$

$= 5(a^2 + 2ab + b^2)$

$= 5(a + b)^2$

35. $x^2 - 25 = x^2 - 5^2 = (x + 5)(x - 5)$

37. $m^2 - 64 = m^2 - 8^2 = (m + 8)(m - 8)$

39. $4a^2 - 81 = (2a)^2 - 9^2 = (2a + 9)(2a - 9)$

41. $12c^2 - 12d^2 = 12(c^2 - d^2) = 12(c + d)(c - d)$

43. $7xy^4 - 7xz^4$

$= 7x(y^4 - z^4)$

$= 7x\left[(y^2)^2 - (z^2)^2\right]$

$= 7x(y^2 + z^2)(y^2 - z^2)$

$= 7x(y^2 + z^2)(y + z)(y - z)$

45. $4a^3 - 49a = a(4a^2 - 49)$

$= a\left[(2a)^2 - 7^2\right]$

$= a(2a + 7)(2a - 7)$

47. $3x^8 - 3y^8$

$= 3(x^8 - y^8)$

$= 3\left[(x^4)^2 - (y^4)^2\right]$

$= 3(x^4 + y^4)(x^4 - y^4)$

$= 3(x^4 + y^4)[(x^2)^2 - (y^2)^2]$

$= 3(x^4 + y^4)(x^2 + y^2)(x^2 - y^2)$

$= 3(x^4 + y^4)(x^2 + y^2)(x + y)(x - y)$

49. $p^2q^2 - 100 = (pq + 10)(pq - 10)$

51. $9a^4 - 25a^2b^4 = a^2(9a^2 - 25b^4)$

$= a^2\left[(3a)^2 - (5b^2)^2\right]$

$= a^2(3a + 5b^2)(3a - 5b^2)$

53. $y^2 - \dfrac{1}{4} = \left(y + \dfrac{1}{2}\right)\left(y - \dfrac{1}{2}\right)$

55. $\dfrac{1}{100} - x^2 = \left(\dfrac{1}{10} + x\right)\left(\dfrac{1}{10} - x\right)$

57. $(a + b)^2 - 36 = (a + b + 6)(a + b - 6)$

59. $x^2 - 6x + 9 - y^2$

$= (x^2 - 6x + 9) - y^2$ Grouping as a difference of squares

$= (x - 3)^2 - y^2$

$= (x - 3 + y)(x - 3 - y)$

61. $t^3 + 8t^2 - t - 8$

$= t^2(t + 8) - (t + 8)$ Factoring by

$= (t + 8)(t^2 - 1)$ grouping

$= (t + 8)(t + 1)(t - 1)$ Factoring the difference
                                    of squares

63. $r^3 - 3r^2 - 9r + 27$

$= r^2(r - 3) - 9(r - 3)$ Factoring by

$= (r - 3)(r^2 - 9)$ grouping

$= (r - 3)(r + 3)(r - 3)$,   Factoring the difference
                                    of squares

or $(r - 3)^2(r + 3)$

65. $m^2 - 2mn + n^2 - 25$

$= (m^2 - 2mn + n^2) - 25$ Grouping as a difference
                                         of squares

$= (m - n)^2 - 5^2$

$= (m - n + 5)(m - n - 5)$

67. $81 - (x + y)^2 = \left[9 + (x + y)\right]\left[9 - (x + y)\right]$

$= (9 + x + y)(9 - x - y)$

69. $r^2 - 2r + 1 - 4s^2$

$= (r^2 - 2r + 1) - 4s^2$ Grouping as a difference
                                      of squares

$= (r - 1)^2 - (2s)^2$

$= (r - 1 + 2s)(r - 1 - 2s)$

71. $16 - a^2 - 2ab - b^2$

$= 16 - (a^2 + 2ab + b^2)$ Grouping as a difference
                                        of squares

$= 4^2 - (a + b)^2$

$= [4 + (a + b)][4 - (a + b)]$

$= (4 + a + b)(4 - a - b)$

73. $x^3 + 5x^2 - 4x - 20$
$= x^2(x+5) - 4(x+5)$
$= (x+5)(x^2-4)$
$= (x+5)(x+2)(x-2)$

75. $a^3 - ab^2 - 2a^2 + 2b^2$
$= a(a^2 - b^2) - 2(a^2 - b^2)$ Factoring by
$= (a^2 - b^2)(a-2)$ grouping
$= (a+b)(a-b)(a-2)$ Factoring the
                              difference of squares

77. *Writing Exercise.*

79. $-(-16) = 16$

81. $3x = y - ax$
          $3x + ax = y$
          $x(3+a) = y$
              $x = \dfrac{y}{3+a}$

83. $\{1,\ 2,\ 3\} \cap \{1,\ 3,\ 5,\ 7\} = \{1,\ 3\}$

85. *Writing Exercise.*

87. $-\dfrac{8}{27}r^2 - \dfrac{10}{9}rs - \dfrac{1}{6}s^2 + \dfrac{2}{3}rs$
$= -\dfrac{8}{27}r^2 - \dfrac{4}{9}rs - \dfrac{1}{6}s^2$
$= -\dfrac{1}{54}(16r^2 + 24rs + 9s^2)$
$= -\dfrac{1}{54}(4r + 3s)^2$

89. $0.09x^8 + 0.48x^4 + 0.64 = (0.3x^4 + 0.8)^2$, or
$\dfrac{1}{100}(3x^4 + 8)^2$

91. $r^2 - 8r - 25 - s^2 - 10s + 16$
$= (r^2 - 8r + 16) - (s^2 + 10s + 25)$
$= (r-4)^2 - (s+5)^2$
$= [(r-4) + (s+5)][(r-4) - (s+5)]$
$= (r-4+s+5)(r-4-s-5)$
$= (r+s+1)(r-s-9)$

93. $25y^{2a} - (x^{2b} - 2x^b + 1)$
$= (5y^a)^2 - (x^b - 1)^2$
$= [5y^a + (x^b - 1)][5y^a - (x^b - 1)]$
$= (5y^a + x^b - 1)(5y^a - x^b + 1)$

95. $3(x+1)^2 + 12(x+1) + 12 = 3\left[(x+1)^2 + 4(x+1) + 4\right]$
$= 3(x+1+2)^2$, or $3(x+3)^2$

97. $s^2 - 4st + 4t^2 + 4s - 8t + 4$
$= (s - 2t)^2 + 4(s - 2t) + 4$
$= (s - 2t + 2)^2$

99. $9x^{2n} - 6x^n + 1 = (3x^n)^2 - 6x^n + 1$
$= (3x^n - 1)^2$

101. If $P(x) = x^2$, then
$P(a+h) - P(a)$
$= (a+h)^2 - a^2$
$= [(a+h) + a][(a+h) - a]$
$= (2a+h)h$, or $h(2a+h)$

103. a. $\pi R^2 h - \pi r^2 h = \pi h(R^2 - r^2)$
$= \pi h(R+r)(R-r)$

b.   Note that 4 m = 400 cm.
$\pi R^2 h - \pi r^2 h$
$= \pi(50)^2(400) - \pi(10)^2(400)$
$= 1,000,000\pi - 40,000\pi$
$= 960,000\pi$ cm^3 (or 0.96π m^3)
$\approx 3,014,400$ cm^3 Using 3.14 for π
$\pi h(R+r)(R-r)$
$= \pi(400)(50+10)(50-10)$
$= \pi(400)(60)(40)$
$= 960,000\pi$ cm^3 (or $0.96\ \pi$ m^3)
$\approx 3,014,400$ cm^3 Using 3.14 for π
If we use the π key on a calculator, the result is
approximately 3,015,929 cm^3.

105. *Graphing Calculator Exercise*

Exercise Set 5.6

1. $x^3 - 1 = (x)^3 - 1^3$
This is a difference of two cubes.

3. $9x^4 - 25 = (3x^2)^2 - 5^2$
This is a difference of two squares.

5. $1000t^3 + 1 = (10t)^3 + 1^3$
This is a sum of two cubes.

7. $25x^2 + 8x$ has a common factor of x so it is not prime, but it does not fall into any of the other categories. It is classified as none of these.

9. $s^{21} - t^{15} = (s^7)^3 - (t^5)^3$
This is a difference of two cubes.

11. $x^3 - 64 = x^3 - 4^3$
$= (x-4)(x^2 + 4x + 16)$
$A^3 - B^3 = (A - B)(A^2 + AB + B^2)$

13. $z^3 + 1 = z^3 + 1^3$
$= (z+1)(z^2 - z + 1)$
$A^3 + B^3 = (A + B)(A^2 - AB + B^2)$

15. $t^3 - 1000 = t^3 - 10^3$
$= (t - 10)(t^2 + 10t + 100)$
$A^3 - B^3 = (A - B)(A^2 + AB + B^2)$

17. $27x^3 + 1 = (3x)^3 + 1^3$
$= (3x + 1)(9x^2 - 3x + 1)$
$A^3 + B^3 = (A + B)(A^2 - AB + B^2)$

19. $64 - 125x^3 = 4^3 - (5x)^3 = (4 - 5x)(16 + 20x + 25x^2)$

21. $x^3 - y^3 = (x - y)(x^2 + xy + y^2)$

23. $a^3 + \dfrac{1}{8} = a^3 + \left(\dfrac{1}{2}\right)^3 = \left(a + \dfrac{1}{2}\right)\left(a^2 - \dfrac{1}{2}a + \dfrac{1}{4}\right)$

25. $8t^3 - 8 = 8(t^3 - 1) = 8(t^3 - 1^3) = 8(t - 1)(t^2 + t + 1)$

27. $54x^3 + 2 = x(27x^3 + 1) = 2\left[(3x)^3 + 1^3\right]$
$= 2(3x + 1)(9x^2 - 3x + 1)$

29. $rs^4 + 64rs = rs(s^3 + 64)$
$= rs(s^3 + 4^3)$
$= rs(s + 4)(s^2 - 4s + 16)$

31. $5x^3 - 40z^3 = 5(x^3 - 8z^3)$
$= 5\left[x^3 - (2z)^3\right]$
$= 5(x - 2z)(x^2 + 2xz + 4z^2)$

33. $y^3 - \dfrac{1}{1000} = y^3 - \left(\dfrac{1}{10}\right)^3 = \left(y - \dfrac{1}{10}\right)\left(y^2 + \dfrac{1}{10}y + \dfrac{1}{100}\right)$

35. $x^3 + 0.001 = x^3 + (0.1)^3 = (x + 0.1)(x^2 - 0.1x + 0.01)$

37. $64x^6 - 8t^6 = 8(8x^6 - t^6)$
$= 8\left[(2x^2)^3 - (t^2)^3\right]$
$= 8(2x^2 - t^2)(4x^4 + 2x^2t^2 + t^4)$

39. $54y^4 - 128y = 2y(27y^3 - 64)$
$= 2y\left[(3y)^3 - 4^3\right]$
$= 2y(3y - 4)(9y^2 + 12y + 16)$

41. $z^6 - 1$
$= (z^3)^2 - 1^2$ Writing as a difference
                               of squares
$= (z^3 + 1)(z^3 - 1)$ Factoring a difference
                               of squares
$= (z + 1)(z^2 - z + 1)(z - 1)(z^2 + z + 1)$
                               Factoring a sum and
                               a difference of cubes

43. $t^6 + 64y^6 = (t^2)^3 + (4y^2)^3$
$= (t^2 + 4y^2)(t^4 - 4t^2y^2 + 16y^4)$

45. $x^{12} - y^3z^{12} = (x^4)^3 - (yz^4)^3$
$= (x^4 - yz^4)(x^8 + x^4yz^4 + y^2z^8)$

47. *Writing Exercise.*

49. *Familiarize.* Let x = the length of a side of a regular octagon and $2x - 1$ = the length of a regular pentagon.
Translate. The perimeter of a regular pentagon has the same length as the perimeter of a regular octagon, so we have one equation:
$$8x = 5(2x - 1).$$
Carry out. We solve the equation.
$$8x = 5(2x - 1)$$
$$8x = 10x - 5$$
$$-2x = -5$$
$$x = 2.5$$
When $x = 2.5$, $8x = 20$.
Check. For a regular pentagon $5(2 \cdot 2.5 - 1) = 20$. For a regular octagon $8x = 20$. Our answer checks.
State. So, the perimeter is 20 cm.

51. *Familiarize.* Let x = the lab time for Friday.
Translate. Averaging the time over 5 days, we have an inequality:
$$\frac{30 + 0 + 50 + 80 + x}{5} \geq 45.$$
Carry out. We solve the inequality.
$$\frac{30 + 0 + 50 + 80 + x}{5} \geq 45$$
$$x + 160 \geq 225$$
$$x \geq 65$$
Check. Averaging the five times, we have
$$\frac{30 + 0 + 50 + 80 + 65}{5} = \frac{225}{5} = 45.$$
State. Kyle must spend 65 minutes or more.

53. *Writing Exercise.*

55. $x^{6a} - y^{3b} = (x^{2a})^3 - (y^b)^3$
$= (x^{2a} - y^b)(x^{4a} + x^{2a}y^b + y^{2b})$

57. $(x + 5)^3 + (x - 5)^3$ Sum of cubes
$= [(x + 5) + (x - 5)][(x + 5)^2 - (x + 5)(x - 5) + (x - 5)^2]$
$= 2x[(x^2 + 10x + 25) - (x^2 - 25) + (x^2 - 10x + 25)]$
$= 2x(x^2 + 10x + 25 - x^2 + 25 + x^2 - 10x + 25)$
$= 2x(x^2 + 75)$

59. $5x^3y^6 - \frac{5}{8}$

$= 5\left(x^3y^6 - \frac{1}{8}\right)$

$= 5\left(xy^2 - \frac{1}{2}\right)\left(x^2y^4 + \frac{1}{2}xy^2 + \frac{1}{4}\right)$

61. $x^{6a} - (x^{2a} + 1)^3$

$= \left[x^{2a} - (x^{2a}+1)\right]\left[x^{4a} + x^{2a}(x^{2a}+1) + (x^{2a}+1)^2\right]$

$= (x^{2a} - x^{2a} - 1)(x^{4a} + x^{4a} + x^{2a} + x^{4a} + 2x^{2a} + 1)$

$= -(3x^{4a} + 3x^{2a} + 1)$

63. $t^4 - 8t^3 - t + 8 = t^3(t-8) - (t-8)$

$= (t-8)(t^3 - 1)$

$= (t-8)(t-1)(t^2 + t + 1)$

65. If $Q(x) = x^6$, then

$Q(a+h) - Q(a)$

$= (a+h)^6 - a^6$

$= \left[(a+h)^3 + a^3\right]\left[(a+h)^3 - a^3\right]$

$= \left[(a+h)+a\right] \cdot \left[(a+h)^2 - (a+h)a + a^2\right]$

$\quad \cdot \left[(a+h)-a\right] \cdot \left[(a+h)^2 + (a+h)a + a^2\right]$

$= (2a+h) \cdot (a^2 + 2ah + h^2 - a^2 - ah + a^2) \cdot (h)$

$\quad \cdot (a^2 + 2ah + h^2 + a^2 + ah + a^2)$

$= h(2a+h)(a^2 + ah + h^2)(3a^2 + 3ah + h^2)$

67. *Graphing Calculator Exercise*

Exercise Set 5.7

1. b

3. f

5. c

7. a

9. $x^2 - 3x - 4$ Factor a trinomial

$= (x-4)(x+1)$ FOIL or grouping method

11. $4x^3 - 10x^2 - 2x + 5$

$= 2x^2(2x-5) - 1(2x-5)$ Factoring by grouping

$= (2x-5)(2x^2 - 1)$ Difference of squares

13. $24a^2 - 16a - 8$

$= 8(3a^2 - 2a - 1)$ Factor out a common factor

15. $x^2 - 81$

$= x^2 - 9^2$ Difference of squares

$= (x+9)(x-9)$

17. $9m^4 - 900$

$= 9(m^4 - 100)$

$= 9\left[(m^2)^2 - 10^2\right]$ Difference of squares

$= 9(m^2 + 10)(m^2 - 10)$

19. $2x^3 + 12x^2 + 16x$

$= 2x(x^2 + 6x + 8)$

$= 2x(x+4)(x+2)$ Trial and error

21. $a^2 + 25 + 10a$

$= a^2 + 10a + 25$ Perfect-square trinomial

$= (a+5)^2$

23. $2y^2 - 11y + 12$

$= (2y-3)(y-4)$ FOIL or grouping method

25. $3x^2 + 15x - 252$

$= 3(x^2 + 5x - 84)$

$= 3(x+12)(x-7)$ FOIL or grouping method

27. $25x^2 - 9y^2$

$= (5x)^2 - (3y)^2$ Difference of squares

$= (5x+3y)(5x-3y)$

29. $t^6 + 1$

$= (t^2)^3 + 1^3$ Sum of cubes

$= (t^2 + 1)(t^4 - t^2 + 1)$

31. $x^2 + 6x - y^2 + 9$

$= x^2 + 6x + 9 - y^2$

$= (x+3)^2 - y^2$ Difference of squares

$= \left[(x+3)+y\right]\left[(x+3)-y\right]$

$= (x+y+3)(x-y+3)$

33. $128a^3 + 250b^3$

$= 2(64a^3 + 125b^3)$ Sum of cubes

$= 2(4a+5b)(16a^2 - 20ab + 25b^2)$

35. $7x^3 - 14x^2 - 105x$

$= 7x(x^2 - 2x - 15)$

$= 7x(x-5)(x+3)$ Trial and error

37. $-9t^2 + 16t^4$

$= t^2(-9 + 16t^2)$

$= t^2(16t^2 - 9)$ Difference of squares

$= t^2(4t+3)(4t-3)$

39. $8m^3 + m^6 - 20$

$= (m^3)^2 + 8m^3 - 20$

$= (m^3 - 2)(m^3 + 10)$ Trial and error

41. $ac + cd - ab - bd$

$= c(a+d) - b(a+d)$ Factoring by grouping

$= (a+d)(c-b)$

43. $4c^2 - 4cd + d^2$ Perfect-square trinomial

$= (2c-d)^2$

45. $40x^2 + 3xy - y^2$
$= (5x + y)(8x - y)$ FOIL or grouping method

47. $4a - 5a^2 - 10 + 2a^3$
$= 2a^3 - 5a^2 + 4a - 10$ Factoring by grouping
$= a^2(2a - 5) + 2(2a - 5)$
$= (2a - 5)(a^2 + 2)$

49. $2x^3 + 6x^2 - 8x - 24$
$= 2(x^3 + 3x^2 - 4x - 12)$
$= 2[x^2(x + 3) - 4(x + 3)]$ Factoring by grouping
$= 2(x + 3)(x^2 - 4)$ Difference of squares
$= 2(x + 3)(x + 2)(x - 2)$

51. $54a^3 - 16b^3$
$= 2(27a^3 - 8b^3)$
$= 2[(3a)^3 - (2b)^3]$ Difference of cubes
$= 2(3a - 2b)(9a^2 + 6ab + 4b^2)$

53. $36y^2 - 35 + 12y$
$= 36y^2 + 12y - 35$
$= (6y - 5)(6y + 7)$ FOIL or grouping method

55. $4m^4 - 64n^4$
$= 4(m^4 - 16n^4)$ Difference of squares
$= 4(m^2 + 4n^2)(m^2 - 4n^2)$ Difference of squares
$= 4(m^2 + 4n^2)(m + 2n)(m - 2n)$

57. $a^5b - 16ab^5$
$= ab(a^4 - 16b^4)$
$= ab[(a^2)^2 - (4b^2)^2]$ Difference of squares
$= ab(a^2 + 4b^2)(a^2 - 4b^2)$
$= ab(a^2 + 4b^2)[a^2 - (2b)^2]$ Difference of squares
$= ab(a^2 + 4b^2)(a + 2b)(a - 2b)$

59. $34t^3 - 6t = 2t(17t^2 - 3)$

61. $(a - 3)(a + 7) + (a - 3)(a - 1)$
$= (a - 3)(a + 7 + a - 1)$
$= (a - 3)(2a + 6)$
$= (a - 3)(2)(a + 3)$
$= 2(a - 3)(a + 3)$

63. $7a^4 - 14a^3 + 21a^2 - 7a$
$= 7a(a^3 - 2a^2 + 3a - 1)$ Removing a common factor

65. $42ab + 27a^2b^2 + 8$
$= 27a^2b^2 + 42ab + 8$
$= (9ab + 2)(3ab + 4)$ FOIL or grouping method

67. $-10t^3 + 15t = -5t(2t^2 - 3)$

69. $-6x^4 + 8x^3 - 12x = -2x(3x^3 - 4x^2 + 6)$

71. $p - 64p^4$
$= p(1 - 64p^3)$ Sum of cubes
$= p(1 - 4p)(1 + 4p + 16p^2)$

73. $a^2 - b^2 - 6b - 9$
$= a^2 - (b^2 + 6b + 9)$ Factoring out -1
$= a^2 - (b + 3)^2$ Difference of squares
$= [a + (b + 3)][a - (b + 3)]$
$= (a + b + 3)(a - b - 3)$

75. *Writing Exercise.*

77. $g(-10) = (-10)^2 - 2 = 100 - 2 = 98$

79. $(f + g)(5) = f(5) + g(5)$
$= 3(5) + 1 + (5)^2 - 2$
$= 15 + 1 + 25 - 2$
$= 39$

81. The domain is $\mathbb{R}$.

83. *Writing Exercise.*

85. $28a^3 - 25a^2bc + 3ab^2c^2$
$= a(28a^2 - 25abc + 3b^2c^2)$
$= a(7a - bc)(4a - 3bc)$

87. $(x - p)^2 - p^2$
$= (x - p + p)(x - p - p)$
$= x(x - 2p)$

89. $(y - 1)^4 - (y - 1)^2$
$= (y - 1)^2[(y - 1)^2 - 1]$
$= (y - 1)^2[(y - 1) + 1][(y - 1) - 1]$
$= (y - 1)^2(y)(y - 2), \text{ or } y(y - 1)^2(y - 2)$

91. $4x^2 + 4xy + y^2 - r^2 + 6rs - 9s^2$
$= (4x^2 + 4xy + y^2) - (r^2 - 6rs + 9s^2)$ Grouping
$= (2x + y)^2 - (r - 3s)^2$ Difference of squares
$= [(2x + y) + (r - 3s)][(2x + y) - (r - 3s)]$
$= (2x + y + r - 3s)(2x + y - r + 3s)$

93. $\dfrac{x^{27}}{1000} - 1 = \left(\dfrac{x^9}{10}\right)^3 - 1^3 = \left(\dfrac{x^9}{10} - 1\right)\left(\dfrac{x^{18}}{100} + \dfrac{x^9}{10} + 1\right)$

95. $3(x+1)^2 - 9(x+1) - 12$

Substitute u for $x+1$ (and u^2 for $(x+1)^2$.)

$$3u^2 - 9u - 12 = 3(u^2 - 3u - 4)$$
$$= 3(u-4)(u+1)$$

Now replace u with $x+1$.

$$3(x+1-4)(x+1+1) = 3(x-3)(x+2)$$

97. $3(a+2)^2 + 30(a+2) + 75$

Substitute u for $a+2$ (and u^2 for $(a+2)^2$.)

$$3u^2 + 30u + 75 = 3(u^2 + 10u + 25)$$
$$= 3(u+5)^2$$

Now replace u with $a+2$.

$$3(a+2+5)^2 = 3(a+7)^2$$

99. $2x^{-1} - 2x^{-3} - 12x^{-5} = 2x^{-5}(x^4 - x^2 - 6)$
$$= 2x^{-5}(x^2 + 2)(x^2 - 3)$$

101. $a^{2w+1} + 2a^{w+1} + a = a(a^{2w} + 2a^w + 1)$
$$= a(a^w + 1)^2$$

Connecting the Concepts

1. $x^2 + 5x + 6 = (x+2)(x+3)$

2. $\quad x^2 + 5x + 6 = 0$
$(x+3)(x+2) = 0$
$x+3 = 0 \quad or \quad x+2 = 0$
$\quad x = -3 \quad or \quad\quad x = -2$
The solution set is $\{-3, -2\}$.

3. $\quad\quad x^2 + 6 = 5x$
$\quad x^2 - 5x + 6 = 0$
$(x-2)(x-3) = 0$
$x-2 = 0 \quad or \quad x-3 = 0$
$\quad x = 2 \quad or \quad\quad x = 3$
The solution set is $\{2, 3\}$.

4. $3x^2 - x + x^2 - 5 = 4x^2 - x - 5$

5. $(3x^2 - x) - (x^2 - 5) = 3x^2 - x - x^2 + 5$
$$= 2x^2 - x + 5$$

6. $a^2 - 1 = (a+1)(a-1)$

7. $(a+1)(a-1) = a^2 - 1$

8. $(a+1)(a-1) = 24$
$\quad\quad a^2 - 1 = 24$
$\quad\quad a^2 - 25 = 0$
$(a+5)(a-5) = 0$
$a+5 = 0 \quad or \quad a-5 = 0$
$\quad a = -5 \quad or \quad\quad a = 5$
The solution set is $\{-5, 5\}$.

Exercise Set 5.8

1. True

3. False; see Example 2(b).

5. False; hypotenuse is defined only for a right triangle.

7. $(x-2)(x-5) = 0$
$x-2 = 0 \quad or \quad x-5 = 0 \quad$ Principle of zero products
$\quad x = 2 \quad or \quad\quad x = 5$
The solutions are 2 and 5. The solution set is $\{2, 5\}$.

9. $\quad x^2 + 8x + 7 = 0$
$(x+7)(x+1) = 0 \quad$ Factoring
$x+7 = 0 \quad or \quad x+1 = 0 \quad$ Principle of zero products
$\quad x = -7 \quad or \quad\quad x = -1$
The solutions are -7 and -1. The solution set is $\{-7, -1\}$.

11. $9t(2t+1) = 0$
$9t = 0 \quad or \quad 2t+1 = 0 \quad$ Principle of zero products
$t = 0 \quad or \quad\quad 2t = -1$
$t = 0 \quad or \quad\quad t = -\dfrac{1}{2}$
The solutions are 0 and $-\dfrac{1}{2}$. The solution set is $\left\{0, -\dfrac{1}{2}\right\}$.

13. $15t^2 - 12t = 0$
$3t(5t-4) = 0 \quad$ Factoring
$3t = 0 \quad or \quad 5t-4 = 0 \quad$ Principle of zero products
$t = 0 \quad or \quad\quad 5t = 4$
$t = 0 \quad or \quad\quad t = \dfrac{4}{5}$
The solutions are 0 and $\dfrac{4}{5}$. The solution set is $\left\{0, \dfrac{4}{5}\right\}$.

15. $(2t+5)(t-7) = 0$
$2t+5 = 0 \quad or \quad t-7 = 0 \quad$ Principle of zero products
$\quad 2t = -5 \quad or \quad\quad t = 7$
$\quad t = -\dfrac{5}{2} \quad or \quad\quad t = 7$
The solutions are $-\dfrac{5}{2}$ and 7. The solution set is $\left\{-\dfrac{5}{2}, 7\right\}$.

17. $\quad x^2 - 3x - 18 = 0$
$(x+3)(x-6) = 0 \quad$ Factoring
$x+3 = 0 \quad or \quad x-6 = 0 \quad$ Principle of zero products
$\quad x = -3 \quad or \quad\quad x = 6$
The solutions are -3 and 6. The solution set is $\{-3, 6\}$.

19. $t^2 - 10t = 0$
  $t(t - 10) = 0$ Factoring
  $t = 0$ or $t - 10 = 0$ Principle of zero products
  $t = 0$ or $t = 10$

The solutions are 0 and 10. The solution set is $\{0, 10\}$.

21. $(3x - 1)(4x - 5) = 0$
  $3x - 1 = 0$ or $4x - 5 = 0$ Principle of zero products
    $3x = 1$ or $4x = 5$
     $x = \dfrac{1}{3}$ or $x = \dfrac{5}{4}$

The solutions are $\dfrac{1}{3}$ and $\dfrac{5}{4}$. The solution set is

$\left\{\dfrac{1}{3}, \dfrac{5}{4}\right\}$.

23. $4a^2 = 10a$
  $4a^2 - 10a = 0$ Getting 0 on one side
  $2a(2a - 5) = 0$ Factoring
  $2a = 0$ or $2a - 5 = 0$ Principle of zero products
   $a = 0$ or $2a = 5$
   $a = 0$ or $a = \dfrac{5}{2}$

The solutions are 0 and $\dfrac{5}{2}$. The solution set is $\left\{0, \dfrac{5}{2}\right\}$.

25. $t^2 - 6t - 16 = 0$
  $(t - 8)(t + 2) = 0$ Factoring
  $t - 8 = 0$ or $t + 2 = 0$ Principle of zero products
   $t = 8$ or $t = -2$

The solutions are 8 and –2. The solution set is $\{-2, 8\}$.

27. $t^2 - 3t = 28$
  $t^2 - 3t - 28 = 0$ Getting 0 on one side
  $(t - 7)(t + 4) = 0$ Factoring
  $t - 7 = 0$ or $t + 4 = 0$ Principle of zero products
   $t = 7$ or $t = -4$

The solutions are 7 and –4. The solution set is $\{-4, 7\}$.

29. $r^2 + 16 = 8r$
  $r^2 - 8r + 16 = 0$ Getting 0 on one side
  $(r - 4)(r - 4) = 0$ Factoring
  $r - 4 = 0$ or $r - 4 = 0$ Principle of zero products
   $r = 4$ or $r = 4$

There is only one solution, 4. The solution set is $\{4\}$.

31. $a^2 + 20a + 100 = 0$
  $(a + 10)(a + 10) = 0$ Factoring
  $a + 10 = 0$ or $a + 10 = 0$ Principle of zero products
   $a = -10$ or $a = -10$

The solution is –10. The solution set is $\{-10\}$.

33. $8y + y^2 + 15 = 0$
  $y^2 + 8y + 15 = 0$ Changing order
  $(y + 5)(y + 3) = 0$ Factoring
  $y + 5 = 0$ or $y + 3 = 0$ Principle of zero products
   $y = -5$ or $y = -3$

The solutions are –5 and –3. The solution set is
$\{-5, -3\}$.

35. $n^2 - 81 = 0$

Observe that we can write this equation as $n^2 = 81$.
Then the solutions are the numbers which, when
squared are 81. These are the square roots of 81, –9 or
9. The solution set is $\{-9, 9\}$.
We could also use the principle of zero products to
solve this equation, as shown below.

$$n^2 - 81 = 0$$
$$(n + 9)(n - 9) = 0$$
$$n + 9 = 0 \quad \text{or} \quad n - 9 = 0$$
$$n = -9 \quad \text{or} \quad n = 9$$

The solutions are –9 and 9. The solution set is $\{-9, 9\}$.

37. $x^3 - 2x^2 = 63x$
  $x^3 - 2x^2 - 63x = 0$ Getting 0 on one side
  $x(x^2 - 2x - 63) = 0$
  $x(x - 9)(x + 7) = 0$
  $x = 0$ or $x - 9 = 0$ or $x + 7 = 0$ Principle of
                                         zero products
  $x = 0$ or $x = 9$ or $x = -7$

The solutions are 0, 9, and –7. The solution set is
$\{-7, 0, 9\}$.

39. $t^2 = 25$

Using the reasoning in Exercise 35, we see that the
solution set is composed of the square roots of 25,
$\{-5, 5\}$. We could also do this exercise as follows:

$$t^2 = 25$$
$$t^2 - 25 = 0$$
$$(t + 5)(t - 5) = 0$$
$$t + 5 = 0 \quad \text{or} \quad t - 5 = 0$$
$$t = -5 \quad \text{or} \quad t = 5$$

The solutions are –5 and 5. The solution set is $\{-5, 5\}$.

41. $(a - 4)(a + 4) = 20$
      $a^2 - 16 = 20$
      $a^2 - 36 = 0$
  $(a + 6)(a - 6) = 0$
  $a + 6 = 0$ or $a - 6 = 0$
   $a = -6$ or $a = 6$

The solutions are –6 and 6. The solution set is $\{-6, 6\}$.

43. $-9x^2 + 15x - 4 = 0$
  $9x^2 - 15x + 4 = 0$ Multiplying by -1
  $(3x - 4)(3x - 1) = 0$
  $3x - 4 = 0$ or $3x - 1 = 0$
    $3x = 4$ or $3x = 1$
     $x = \dfrac{4}{3}$ or $x = \dfrac{1}{3}$

The solutions are $\frac{4}{3}$ and $\frac{1}{3}$. The solution set is

$\left\{ \frac{1}{3}, \frac{4}{3} \right\}$.

45. $-8y^3 - 10y^2 - 3y = 0$

$-y(8y^2 + 10y + 3) = 0$

$-y(2y + 1)(4y + 3) = 0$

$-y = 0$ or $2y + 1 = 0$ or $4y + 3 = 0$

$y = 0$ or $2y = -1$ or $4y = -3$

$y = 0$ or $y = -\frac{1}{2}$ or $y = -\frac{3}{4}$

The solutions are 0, $-\frac{1}{2}$, and $-\frac{3}{4}$. The solution set is

$\left\{ -\frac{3}{4}, -\frac{1}{2}, 0 \right\}$.

47. $(z + 4)(z - 2) = -5$

$z^2 + 2z - 8 = -5$ Multiplying

$z^2 + 2z - 3 = 0$

$(z + 3)(z - 1) = 0$

$z + 3 = 0$ or $z - 1 = 0$

$z = -3$ or $z = 1$

The solutions are -3 and 1. The solution set is $\{-3, 1\}$.

49. $x(5 + 12x) = 28$

$5x + 12x^2 = 28$ Multiplying

$5x + 12x^2 - 28 = 0$

$12x^2 + 5x - 28 = 0$ Rearranging

$(4x + 7)(3x - 4) = 0$

$4x + 7 = 0$ or $3x - 4 = 0$

$4x = -7$ or $3x = 4$

$x = -\frac{7}{4}$ or $x = \frac{4}{3}$

The solutions are $-\frac{7}{4}$ and $\frac{4}{3}$. The solution set is

$\left\{ -\frac{7}{4}, \frac{4}{3} \right\}$.

51. $a^2 - \frac{1}{100} = 0$

$\left(a + \frac{1}{10} \right)\left(a - \frac{1}{10} \right) = 0$

$a + \frac{1}{10} = 0$ or $a - \frac{1}{10} = 0$

$a = -\frac{1}{10}$ or $a = \frac{1}{10}$

The solutions are $-\frac{1}{10}$ and $\frac{1}{10}$. The solution set is

$\left\{ -\frac{1}{10}, \frac{1}{10} \right\}$.

53. $t^4 - 26t^2 + 25 = 0$

$(t^2 - 1)(t^2 - 25) = 0$

$(t + 1)(t - 1)(t + 5)(t - 5) = 0$

$t + 1 = 0$ or $t - 1 = 0$ or $t + 5 = 0$ or $t - 5 = 0$

$t = -1$ or $t = 1$ or $t = -5$ or $t = 5$

The solutions are -1, 1, -5, and 5. The solution set is $\{-5, -1, 1, 5\}$.

55. We set $f(a)$ equal to 8.

$a^2 + 12a + 40 = 8$

$a^2 + 12a + 32 = 0$

$(a + 8)(a + 4) = 0$

$a + 8 = 0$ or $a + 4 = 0$

$a = -8$ or $a = -4$

The values of a for which $f(a) = 8$ are -8 and -4.

57. We set $g(a)$ equal to 12.

$2a^2 + 5a = 12$

$2a^2 + 5a - 12 = 0$

$(2a - 3)(a + 4) = 0$

$2a - 3 = 0$ or $a + 4 = 0$

$2a = 3$ or $a = -4$

$a = \frac{3}{2}$ or $a = -4$

The values of a for which $g(a) = 12$ are $\frac{3}{2}$ and -4.

59. We set $h(a)$ equal to -27.

$12a + a^2 = -27$

$12a + a^2 + 27 = 0$

$a^2 + 12a + 27 = 0$ Rearranging

$(a + 3)(a + 9) = 0$

$a + 3 = 0$ or $a + 9 = 0$

$a = -3$ or $a = -9$

The values of a for which $h(a) = -27$ are -3 and -9.

61. $f(x) = g(x)$

$12x^2 - 15x = 8x - 5$

$12x^2 - 23x + 5 = 0$

$(4x - 1)(3x - 5) = 0$

$4x - 1 = 0$ or $3x - 5 = 0$

$4x = 1$ or $3x = 5$

$x = \frac{1}{4}$ or $x = \frac{5}{3}$

The values of x for which $f(x) = g(x)$ are $\frac{1}{4}$ and $\frac{5}{3}$.

63. $f(x) = g(x)$

$2x^3 - 5x = 10x - 7x^2$

$2x^3 + 7x^2 - 15x = 0$

$x(2x^2 + 7x - 15) = 0$

$x(2x - 3)(x + 5) = 0$

$x = 0$ or $2x - 3 = 0$ or $x + 5 = 0$

$x = 0$ or $2x = 3$ or $x = -5$

$x = 0$ or $x = \frac{3}{2}$ or $x = -5$

The values of x for which $f(x) = g(x)$ are 0, $\frac{3}{2}$,

and -5.

65. $f(x) = \dfrac{3}{x^2 - 3x - 4}$

$f(x)$ cannot be calculated for any x-value for which the denominator, $x^2 - 3x - 4$, is 0. To find the excluded values, we solve:

$$x^2 - 3x - 4 = 0$$
$$(x - 4)(x + 1) = 0$$
$$x - 4 = 0 \quad or \quad x + 1 = 0$$
$$x = 4 \quad or \quad x = -1$$

The domain of f is

$\{x \mid x \text{ is a real number } and \ x \neq 4 \ and \ x \neq -1\}$, or

$(-\infty, -1) \cup (-1, 4) \cup (4, \infty)$.

67. $f(x) = \dfrac{x}{6x^2 - 54}$

$f(x)$ cannot be calculated for any x-value for which the denominator, $6x^2 - 54$, is 0. To find the excluded values, we solve:

$$6x^2 - 54 = 0$$
$$6(x^2 - 9) = 0$$
$$6(x + 3)(x - 3) = 0$$
$$x + 3 = 0 \quad or \quad x - 3 = 0$$
$$x = -3 \quad or \quad x = 3$$

The domain of f is

$\{x \mid x \text{ is a real number } and \ x \neq -3 \ and \ x \neq 3\}$, or

$(-\infty, -3) \cup (-3, 3) \cup (3, \infty)$.

69. $f(x) = \dfrac{x - 5}{9x - 18x^2}$

$f(x)$ cannot be calculated for any x-value for which the denominator, $9x - 18x^2$, is 0. To find the excluded values, we solve:

$$9x - 18x^2 = 0$$
$$9x(1 - 2x) = 0$$
$$9x = 0 \quad or \quad 1 - 2x = 0$$
$$x = 0 \quad or \quad -2x = -1$$
$$x = 0 \quad or \quad x = \frac{1}{2}$$

The domain of f is

$\left\{x \mid x \text{ is a real number } and \ x \neq 0 \ and \ x \neq \dfrac{1}{2}\right\}$, or

$(-\infty, 0) \cup \left(0, \dfrac{1}{2}\right) \cup \left(\dfrac{1}{2}, \infty\right)$.

71. $f(x) = \dfrac{7}{5x^3 - 35x^2 + 50x}$

$f(x)$ cannot be calculated for any x-value for which the denominator, $5x^3 - 35x^2 + 50x$, is 0. To find the excluded values, we solve:

$$5x^3 - 35x^2 + 50x = 0$$
$$5x(x^2 - 7x + 10) = 0$$
$$5x(x - 2)(x - 5) = 0$$
$$5x = 0 \quad or \quad x - 2 = 0 \quad or \quad x - 5 = 0$$
$$x = 0 \quad or \quad x = 2 \quad or \quad x = 5$$

The domain of f is

$\{x \mid x \text{ is a real number } and \ x \neq 0 \ and \ x \neq 2 \ and \ x \neq 5\}$,

or $(-\infty, 0) \cup (0, 2) \cup (2, 5) \cup (5, \infty)$.

73. *Familiarize*. We let w represent the width and $w + 3$ represent the length. We make a drawing and label it.

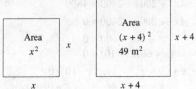

Recall that the formula for the area of a rectangle is

$A = \text{length} \times \text{width}$.

***Translate*.**

$$\underbrace{\text{Area}}_{\downarrow} \quad \text{is} \quad \underbrace{180 \text{ in}^2}_{\downarrow},$$
$$w(w + 3) \quad = \quad 180$$

***Carry out*.** We solve the equation.

$$w(w + 3) = 180$$
$$w^2 + 3w = 180$$
$$w^2 + 3w - 180 = 0$$
$$(w + 15)(w - 12) = 0$$
$$w + 15 = 0 \quad or \quad w - 12 = 0$$
$$w = -15 \quad or \quad w = 12$$

***Check*.** The number -15 is not a solution, because width cannot be negative. If the width is 12 in. and the length is 3 in. more, or 15 in., then the area is $12 \cdot 15$, or 180 in^2. This is a solution.

***State*.** The length is 15 in. and the width is 12 in.

75. *Familiarize*. We make a drawing and label it. We let x represent the length of a side of the original square, in meters.

***Translate*.**

$$\underbrace{\text{Area of new square}}_{\downarrow} \quad \text{is} \quad \underbrace{49 \text{ m}^2}_{\downarrow}$$
$$(x + 4)^2 \quad = \quad 49$$

***Carry out*.** We solve the equation:

$$(x + 4)^2 = 49$$
$$x^2 + 8x + 16 = 49$$
$$x^2 + 8x - 33 = 0$$
$$(x - 3)(x + 11) = 0$$
$$x - 3 = 0 \quad or \quad x + 11 = 0$$
$$x = 3 \quad or \quad x = -11$$

***Check*.** We check only 3 since the length of a side cannot be negative. If we increase the length by 4, the new length is $3 + 4$, or 7 m. Then the new area is $7 \cdot 7$, or 49 m^2. We have a solution.

***State*.** The length of a side of the original square is 3 m.

77. *Familiarize*. We let x represent the width of the frame. The length and width of the picture that shows are represented by $20 - 2x$ and $14 - 2x$. The area of the picture that shows is 160 cm^2.

Translate. Using the formula for the area of a rectangle, $A = l \cdot w$, we have
$$160 = (20 - 2x)(14 - 2x).$$

Carry out. We solve the equation:
$$160 = 280 - 68x + 4x^2$$
$$160 = 4(70 - 17x + x^2)$$
$$40 = 70 - 17x + x^2 \qquad \text{Dividing by 4}$$
$$0 = x^2 - 17x + 30$$
$$0 = (x - 2)(x - 15)$$
$$x - 2 = 0 \quad or \quad x - 15 = 0$$
$$x = 2 \quad or \quad x = 15$$

Check. We see that 15 is not a solution because when $x = 15$, $20 - 2x = -10$ and $14 - 2x = -16$, and the length and width of the frame cannot be negative. We check 2. When $x = 2$, $20 - 2x = 16$ and $14 - 2x = 10$ and $16 \cdot 10 = 160$. The area is 160. The value checks.

State. The width of the frame is 2 cm.

79. *Familiarize*. Let $x =$ the length on each side. The length of the table cloth is $2x + 60$, and the width is $2x + 40$. The area of a rectangle is $A = $ length $\times$ width.

Translate.

Area of table cloth	is	Twice the area of the table,
$\downarrow$	$\downarrow$	$\downarrow$
$(2x + 60)(2x + 40)$	$=$	$2(60 \cdot 40)$

Carry out. We solve the equation.
$$(2x + 60)(2x + 40) = 2(60 \cdot 40)$$
$$4x^2 + 200x + 2400 = 4800$$
$$4x^2 + 200x - 2400 = 0$$
$$x^2 + 50x - 600 = 0$$
$$(x + 60)(x - 10) = 0$$
$$x + 60 = 0 \quad or \quad x - 10 = 0$$
$$x = -60 \quad or \quad x = 10$$

Check. We check only 10 since the length cannot be negative. The length of the table cloth is $2(10) + 60$, or 80 in. and the width is $2(10) + 40$, or 60 in. The area is $80 \cdot 60$, or 4800 in^2 which is twice 2400 in^2, the area of the table. The answer checks.

State. On each side, the table cloth hangs down 10 in.

81. *Familiarize*. Let x represent the first integer, $x + 2$ the second, and $x + 4$ the third.

Translate.

Square of the third	is	76	more than	square of the second.
$\downarrow$	$\downarrow$	$\downarrow$	$\downarrow$	$\downarrow$
$(x + 4)^2$	$=$	76	$+$	$(x + 2)^2$

Carry out. We solve the equation:

$$(x + 4)^2 = 76 + (x + 2)^2$$
$$x^2 + 8x + 16 = 76 + x^2 + 4x + 4$$
$$x^2 + 8x + 16 = x^2 + 4x + 80$$
$$4x = 64$$
$$x = 16$$

Check. We check the integers 16, 18, and 20. The square of 20, or 400, is 76 more than 324, the square of 18. The answer checks.

State. The integers are 16, 18, and 20.

83. *Familiarize*. Using the labels on the drawing in the text, we let x represent the height of the triangle and $x + 20$ represent the base. Recall that the formula for the area of the triangle with base b and height h is $\frac{1}{2}bh$.

Translate.

Area	is	750 in^2.
$\downarrow$	$\downarrow$	$\downarrow$
$\frac{1}{2}x(x + 20)$	$=$	750

Carry out. We solve the equation.
$$\frac{1}{2}x(x + 20) = 750$$
$$x(x + 20) = 1500$$
$$x^2 + 20x = 1500$$
$$x^2 + 20x - 1500 = 0$$
$$(x + 50)(x - 30) = 0$$
$$x + 50 = 0 \quad or \quad x - 30 = 0$$
$$x = -50 \quad or \quad x = 30$$

Check. We check only 30 since the height cannot be negative. If the height is 30 in., then the base is $30 + 20$, or 50 in., and the area is $\frac{1}{2}(30)(50)$, or 750 in^2. The answer checks.

State. The height is 30 in. and the base is 50 in.

85. *Familiarize*. We make a drawing. Let $x =$ one side and $x + 5$ is the other side of the triangle.

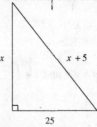

Translate. We use the Pythagorean theorem.
$$25^2 + x^2 = (x + 5)^2$$

Carry out. We solve the equation:
$$625 + x^2 = x^2 + 10x + 25$$
$$600 = 10x$$
$$60 = x$$

Check. If $x = 60$, then $x + 5 = 65$;
$25^2 + 60^2 = 625 + 3600 = 4225 = 65^2$, so the answer checks.

State. One side is 60 ft and the other side is 65 ft.

87. *Familiarize*. We make a drawing. Let h = the height the ladder reaches on the wall. Then the length of the ladder is $h + 1$.

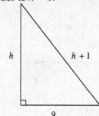

Translate. We use the Pythagorean theorem.

$$9^2 + h^2 = (h+1)^2$$

Carry out. We solve the equation.

$$81 + h^2 = h^2 + 2h + 1$$
$$80 = 2h$$
$$40 = h$$

Check. If $h = 40$, then $h + 1 = 41$;

$9^2 + 40^2 = 81 + 1600 = 1681 = 41^2$, so the answer checks.

State. The ladder is 41 ft long.

89. *Familiarize*. Let w represent the width and $w + 25$ represent the length, in meters. Make a drawing.

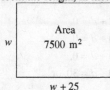

Recall that the formula for the area of a rectangle is $A =$ length $\times$ width.

Translate.

$$\underset{\downarrow}{\underline{\text{Area}}} \quad \underset{\downarrow}{\text{is}} \quad \underset{\downarrow}{\underline{7500 \text{ m}^2}};$$

$$w(w + 25) = 7500$$

Carry out. We solve the equation.

$$w(w + 25) = 7500$$
$$w^2 + 25w = 7500$$
$$w^2 + 25w - 7500 = 0$$
$$(w + 100)(w - 75) = 0$$
$$w + 100 = 0 \quad or \quad w - 75 = 0$$
$$w = -100 \quad or \quad w = 75$$

Check. The number -100 is not a solution because width cannot be negative. If the width is 75 m and the length is 25 m more, or 100 m, then the area will be $75 \cdot 100$, or 7500 m^2. This is a solution.

State. The dimensions will be 100 m by 75 m.

91. *Familiarize*. The company breaks even when the cost and the revenue are the same. We use the functions given in the text.

Translate.

$$\underset{\downarrow}{\underline{\text{Cost}}} \quad \underset{\downarrow}{\text{equals}} \quad \underset{\downarrow}{\underline{\text{revenue}}},$$

$$x^2 - 2x + 10 = 2x^2 + x$$

Carry out. We solve the equation.

$$x^2 - 2x + 10 = 2x^2 + x$$
$$x^2 + 3x - 10 = 0$$
$$(x + 5)(x - 2) = 0$$
$$x + 5 = 0 \quad or \quad x - 2 = 0$$
$$x = -5 \quad or \quad x = 2$$

Check. We check only 2 since the number of audio systems cannot be negative. If 2 systems are produced, the cost is $C(2) = (2)^2 - 2(2) + 10 = \$10$ thousand. If 2 systems are sold, the revenue is

$R(2) = 2(2)^2 + 2 = 8 + 2 = \$10$ thousand. The answer checks.

State. The company breaks even when 2 systems are produced and sold.

93. *Familiarize*. We will use the formula in Example 6,

$$h(t) = -15t^2 + 75t + 10$$

Translate. We need to find the value of t for which $h(t) = 100$.

$$-15t^2 + 75t + 10 = 100$$

Carry out. We solve the equation.

$$-15t^2 + 75t + 10 = 100$$
$$15t^2 - 75t + 90 = 0$$
$$15(t^2 - 5t + 6) = 0$$
$$15(t - 2)(t - 3) = 0$$
$$t - 2 = 0 \quad or \quad t - 3 = 0$$
$$t = 2 \quad or \quad t = 3$$

Check. We have

$$h(2) = -15(2)^2 + 75(2) + 10 = 100$$
$$h(3) = -15(3)^2 + 75(3) + 10 = 100$$

The solutions of the equation are 2 and 3. We reject 3 since it indicates when the height of the tee shirt was 100 ft on the way down.

State. The tee shirt was airborne for 2 sec before it was caught.

95. *Familiarize*. We will use the formula

$$h(t) = -16t^2 + 64t + 80.$$

Translate. We find the value of t for which $h(t) = 0$. We have

$$-16t^2 + 64t + 80 = 0.$$

Carry out. We solve the equation.

$$-16t^2 + 64t + 80 = 0$$
$$t^2 - 4t - 5 = 0 \quad \text{Diving by } -16$$
$$(t + 1)(t - 5) = 0$$
$$t + 1 = 0 \quad or \quad t - 5 = 0$$
$$t = -1 \quad or \quad t = 5$$

Check. Time cannot be negative in this application.

$h(5) = -16(5)^2 + 64(5) + 80 = 0$ The answer checks.

State. The cardboard will reach the ground 5 sec after it is launched.

97. *Familiarize.* We will use the formula

$$p(t) = 0.006t^2 - 0.4t + 20.$$

Translate. We find the value of t for which $p(t) = 15$.
We have

$$0.006t^2 - 0.4t + 20 = 15.$$

Carry out. We solve the equation.

$$0.006t^2 - 0.4t + 20 = 15$$
$$0.006t^2 - 0.4t + 5 = 0$$
$$6t^2 - 400t + 5000 = 0 \quad \text{Multiplying by 1000}$$
$$(6t - 100)(t - 50) = 0$$
$$6t - 100 = 0 \quad or \quad t - 50 = 0$$
$$t = 16\tfrac{2}{3} \quad or \quad t = 50$$

Check. We have

$$p\left(16\tfrac{2}{3}\right) = 0.006\left(16\tfrac{2}{3}\right)^2 - 0.4\left(16\tfrac{2}{3}\right) + 20 = 15$$

$$p(50) = 0.006(50)^2 - 0.4(50) + 20 = 15$$

The answers check.

State. So $16\tfrac{2}{3}$ years after 1950 and 50 years after

1950, or in 1966 and in 2000, there were 15% of
Americans living in multigenerational households.

99. *Writing Exercise.*

101. $m = \dfrac{-6 - 10}{1 - 3} = \dfrac{-16}{-2} = 8$

103. $x - 5y = 20$

For x-intercept, set $y = 0$.
$$x = 20$$
The x-intercept is $(20, 0)$.
For y-intercept, set $x = 0$.
$$-5y = 20$$
$$y = -4$$
The y-intercept is $(0, -4)$.

105. $m = \dfrac{-10 - (-5)}{6 - 4} = -\dfrac{5}{2}$

$$y - (-5) = -\frac{5}{2}(x - 4)$$
$$y + 5 = -\frac{5}{2}x + 10$$
$$y = -\frac{5}{2}x + 5$$
$$f(x) = -\frac{5}{2}x + 5$$

107. *Writing Exercise.*

109. $(8x + 11)(12x^2 - 5x - 2) = 0$
$$(8x + 11)(3x - 2)(4x + 1) = 0$$
$$8x + 11 = 0 \quad or \quad 3x - 2 = 0 \quad or \quad 4x + 1 = 0$$
$$8x = -11 \quad or \quad 3x = 2 \quad or \quad 4x = -1$$
$$x = -\frac{11}{8} \quad or \quad x = \frac{2}{3} \quad or \quad x = -\frac{1}{4}$$

The solution set is $\left\{-\dfrac{11}{8}, -\dfrac{1}{4}, \dfrac{2}{3}\right\}$.

111.

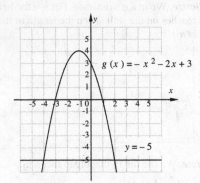

The solutions of $-x^2 - 2x + 3 = 0$ are the first
coordinates of the x-intercepts. From the graph we see
that these are -3 and 1. The solution set is $\{-3, 1\}$.

To solve $-x^2 - 2x + 3 \geq -5$ we find the x-values for
which $g(x) \geq -5$. From the graph we see that these are
the values in the interval $[-4, 2]$. The solution set can
also be expressed as $\{x \mid -4 \leq x \leq 2\}$.

113. Answers may vary. A polynomial function of lowest
degree that meets the given criteria is of the form
$f(x) = ax^3 + bx^2 + cx + d$. Substituting, we have

$$a \cdot 2^3 + b \cdot 2^2 + c \cdot 2 + d = 0,$$
$$a(-1)^3 + b(-1)^2 + c(-1) + d = 0,$$
$$a \cdot 3^3 + b \cdot 3^2 + c \cdot 3 + d = 0,$$
$$a \cdot 0^3 + b \cdot 0^2 + c \cdot 0 + d = 30, \text{ or}$$
$$8a + 4b + 2c + d = 0,$$
$$-a + b - c + d = 0,$$
$$27a + 9b + 3c + d = 0,$$
$$d = 30.$$

Solving the system of equations we get $(5, -20, 5, 30)$,
so the corresponding function is

$$f(x) = 5x^3 - 20x^2 + 5x + 30.$$

115. *Familiarize.* Using the labels on the drawing in the
text, we let x represent the width of the piece of tin and
$2x$ represent the length. Then the width and length of
the base of the box are represented by $x - 4$ and
$2x - 4$, respectively. Recall that the formula for the
volume of a rectangular solid with length l, width w,
and height h is $l \cdot w \cdot h$.

    Translate.

The volume	is	480 cm^3.
↓	↓	↓
$(2x-4)(x-4)(2)$	$=$	480

Carry out. We solve the equation:
$$(2x - 4)(x - 4)(2) = 480$$
$$(2x - 4)(x - 4) = 240 \quad \text{Dividing by 2}$$
$$2x^2 - 12x + 16 = 240$$
$$2x^2 - 12x - 224 = 0$$
$$x^2 - 6x - 112 = 0 \quad \text{Dividing by 2}$$
$$(x + 8)(x - 14) = 0$$
$$x + 8 = 0 \quad or \quad x - 14 = 0$$
$$x = -8 \quad or \quad x = 14$$

Check. We check only 14 since the width cannot be negative. If the width of the piece of tin is 14 cm, then its length is $2 \cdot 14$, or 28 cm, and the dimensions of the base of the box are $14 - 4$, or 10 cm by $28 - 4$, or 24 cm. The volume of the box is $24 \cdot 10 \cdot 2$, or 480 cm^3. The answer checks.

State. The dimensions of the piece of tin are 14 cm by 28 cm.

117. Graph $y_1 = 11.12(x+1)^2$ and $y_2 = 15.4x^2$ in a window that shows the point of intersection of the graphs. The window $[0, 10, 0, 600]$, Xscl = 1, Yscl = 100 is one good choice. Then find the first coordinate of the point of intersection. It is approximately 5.7, so it will take the camera about 5.7 sec to catch up to the skydiver.

119. *Graphing Calculator Exercise*

121. $\{6.90\}$

123. $\{3.48\}$

125. Since $n^2 + m^2$ represents the largest number, let $c = n^2 + m^2$. Then $a = n^2 - m^2$ and $b = 2mn$.

$$a^2 + b^2 = (n^2 - m^2)^2 + (2mn)^2$$
$$= n^4 - 2n^2m^2 + m^4 + 4m^2n^2$$
$$= n^4 + 2m^2n^2 + m^4;$$
$$c^2 = (n^2 + m^2)^2$$
$$= n^4 + 2m^2n^4 + m^4$$

Since $a^2 + b^2 = c^2$, the expressions satisfy the Pythagorean equation.

Chapter 5 Review

1. g

2. b

3. a

4. d

5. e

6. j

7. h

8. c

9. i

10. f

11.

Term	$2xy^6$	$-7x^8y^3$	$2x^3$	9
Degree	7	11	3	0

Degree of polynomial: 11

12. $-5x^3 + 2x^2 + 3x + 9$; $-5x^3$; -5

13. $-3x^2 + 2x^3 + 8x^6y - 7x^8y^3$

14. $P(x) = x^3 - x^2 + 4x$
$P(0) = 0^3 - 0^2 + 4 \cdot 0 = 0$
$P(-1) = (-1)^3 - (-1)^2 + 4(-1) = -6$

15. $P(x) = x^2 + 10x$
$P(a+h) - P(a) = (a+h)^2 + 10(a+h) - (a^2 + 10a)$
$\qquad = a^2 + 2ah + h^2 + 10a + 10h - a^2 - 10a$
$\qquad = 2ah + h^2 + 10h$

16. $6 - 4a + a^2 - 2a^3 - 10 + a = -2a^3 + a^2 - 3a - 4$

17. $4x^2y - 3xy^2 - 5x^2y + xy^2$
$= (4-5)x^2y + (-3+1)xy^2$
$= -x^2y - 2xy^2$

18. $(-7x^3 - 4x^2 + 3x + 2) + (5x^3 + 2x + 6x^2 + 1)$
$= -2x^3 + 2x^2 + 5x + 3$

19. $(4n^3 + 2n^2 - 12n + 7) + (-6n^3 + 9n + 4 + n)$
$= (4-6)n^3 + 2n^2 + (-12+9+1)n + 7 + 4$
$= -2n^3 + 2n^2 - 2n + 11$

20. $(-9xy^2 - xy - 6x^2y) + (-5x^2y - xy + 4xy^2)$
$= -5xy^2 - 2xy - 11x^2y$

21. $(8x - 5) - (-6x + 2)$
$= 8x - 5 + 6x - 2$
$= 14x - 7$

22. $(4a - b - 3c) - (6a - 7b - 3c) = -2a + 6b$

23. $(8x^2 - 4xy + y^2) - (2x^2 - 3y^2 - 9y)$
$= 8x^2 - 4xy + y^2 - 2x^2 + 3y^2 + 9y$
$= 6x^2 - 4xy + 4y^2 + 9y$

24. $(3x^2y)(-6xy^3) = -18x^3y^4$

25.
$$
\begin{array}{r}
x^4 \; - 2x^2 \; + 3 \\
x^4 \; + x^2 \; - 1 \\
\hline
-\,x^4 \; + 2x^2 \; - 3 \\
x^6 \; - 2x^4 \; + 3x^2 \\
x^8 - 2x^6 + 3x^4 \\
\hline
x^8 \; - \; x^6 \qquad + 5x^2 \; - 3
\end{array}
$$

26. $(4ab + 3c)(2ab - c) = 8a^2b^2 + 2abc - 3c^2$

27. $(7t + 1)(7t - 1) = 49t^2 - 1$ Difference of squares

28. $(3x - 4y)^2 = 9x^2 - 24xy + 16y^2$

29. $(x + 3)(2x - 1)$
$= 2x^2 - x + 6x - 3$ FOIL
$= 2x^2 + 5x - 3$

30. $(x^2 + 4y^3)^2 = x^4 + 8x^2y^3 + 16y^6$

31. $(3t-5)^2 - (2t+3)^2$
$= 9t^2 - 30t + 25 - (4t^2 + 12t + 9)$
$= 5t^2 - 42t + 16$

32. $\left(x - \dfrac{1}{3}\right)\left(x - \dfrac{1}{6}\right) = x^2 - \dfrac{1}{2}x + \dfrac{1}{18}$

33. $-3y^4 - 9y^2 + 12y = -3y\left(y^3 + 3y - 4\right)$

34. $a^2 - 12a + 27 = (a-9)(a-3)$

35. $3m^2 - 10m - 8$
We will use the FOIL method.
1. There is no common factor (other than 1 or –1.)
2. Factor the first term, $3m^2$. The factors are $3m, m$. We have this possibility:
    $(3m + \quad)(m + \quad).$
3. Factor the last term, –8. The possibilities are $8(-1)$, $-8 \cdot 1$, $4(-2)$, and $-4 \cdot 2$.
4. We need factors for which the sum of the products (the "outer" and "inner" parts of FOIL) is the middle term, $-10m$. Try some possibilities and check by multiplying. Trial and error leads us to
    $(3m + 2)(m - 4) = 3m^2 - 10m - 8$
The factorization is $(3m + 2)(m - 4)$.

36. $25x^2 + 20x + 4 = (5x+2)(5x+2) = (5x+2)^2$

37. $4y^2 - 16$
$= 4\left(y^2 - 4\right)$ Difference of squares
$= 4(y+2)(y-2)$

38. $5x^2 + x^3 - 14x = x^3 + 5x^2 - 14x$
$= x\left(x^2 + 5x - 14\right)$
$= x(x-2)(x+7)$

39. $ax + 2bx - ay - 2by$
$= x(a+2b) - y(a+2b)$ Grouping
$= (a+2b)(x-y)$

40. $3y^3 + 6y^2 - 5y - 10$
$= 3y^2(y+2) - 5(y+2)$
$= (y+2)\left(3y^2 - 5\right)$

41. $a^4 - 81 = a^4 - 3^4$ Difference of squares
$= \left(a^2 + 3^2\right)\left(a^2 - 3^2\right)$ Difference of squares
$= \left(a^2 + 9\right)(a+3)(a-3)$

42. $4x^4 + 4x^2 + 20 = 4\left(x^4 + x^2 + 5\right)$

43. $27x^3 + 8 = (3x)^3 + 2^3$ Sum of cubes
$= (3x+2)\left(9x^2 - 6x + 4\right)$

44. $\dfrac{1}{125}b^3 - \dfrac{1}{8}c^6 = \left(\dfrac{1}{5}b\right)^3 - \left(\dfrac{1}{2}c^2\right)^3$ Difference of cubes
$= \left(\dfrac{1}{5}b - \dfrac{1}{2}c^2\right)\left(\dfrac{1}{25}b^2 + \dfrac{1}{10}bc^2 + \dfrac{1}{4}c^4\right)$

45. $a^2b^4 - 64$
$= (ab^2 + 8)(ab^2 - 8)$ Difference of squares

46. Prime

47. $0.01x^4 - 1.44y^6$
$= \left(0.1x^2\right)^2 - \left(1.2y^3\right)^2$ Difference of squares
$= \left(0.1x^2 + 1.2y^3\right)\left(0.1x^2 - 1.2y^3\right)$

48. $4x^2y + 100y - 40xy$
$= 4y\left(x^2 - 10x + 25\right)$
$= 4y(x-5)^2$

49. $6t^2 + 17pt + 5p^2$
We will use the grouping method.
1. There is no common factor (other than 1 or –1).
2. Multiply the leading coefficient, 6, and the last coefficient, 5: $6(5) = 30$
3. Try to factor 30 so the sum of the factors is 17. We need only consider pairs of positive factors since 30 and 17 are both positive.

Pair of Factors	Sum of Factors
30, 1	31
15, 2	17
10, 3	13
6, 5	11

4. Split $17pt$ using the results of step (3):
    $17pt = 15pt + 2pt$
5. Factor by grouping:
    $6t^2 + 17pt + 5p^2 = 6t^2 + 15pt + 2pt + 5p^2$
    $= 3t(2t + 5p) + p(2t + 5p)$
    $= (2t + 5p)(3t + p)$

50. $x^3 + 2x^2 - 9x - 18$
$= x^2(x+2) - 9(x+2)$
$= (x+2)\left(x^2 - 9\right)$
$= (x+2)(x+3)(x-3)$

51. $a^2 - 2ab + b^2 - 4t^2$
$= \left(a^2 - 2ab + b^2\right) - 4t^2$ Grouping as a difference
$= (a-b)^2 - 4t^2$ of squares
$= (a-b+2t)(a-b-2t)$

52. $x^2 - 12x + 36 = 0$
$(x-6)(x-6) = 0$
$x - 6 = 0$ or $x - 6 = 0$
$x = 6$ or $x = 6$
The solution set is $\{6\}$.

53.
$$6b^2 + 6 = 13b$$
$$6b^2 - 13b + 6 = 0$$
$$(3b - 2)(2b - 3) = 0$$
$$3b - 2 = 0 \quad \text{or} \quad 2b - 3 = 0$$
$$b = \frac{2}{3} \quad \text{or} \quad b = \frac{3}{2}$$
The solution set is $\left\{\frac{2}{3}, \frac{3}{2}\right\}$.

54.
$$8y^2 = 14y$$
$$8y^2 - 14y = 0$$
$$2y(4y - 7) = 0$$
$$2y = 0 \quad \text{or} \quad 4y - 7 = 0$$
$$y = 0 \quad \text{or} \quad y = \frac{7}{4}$$
The solution set is $\left\{0, \frac{7}{4}\right\}$.

55.
$$3r^2 = 12$$
$$3r^2 - 12 = 0$$
$$3(r^2 - 4) = 0$$
$$3(r + 2)(r - 2) = 0$$
$$r + 2 = 0 \quad \text{or} \quad r - 2 = 0$$
$$r = -2 \quad \text{or} \quad r = 2$$
The solution set is $\{-2, 2\}$.

56.
$$a^3 = 4a^2 + 21a$$
$$a^3 - 4a^2 - 21a = 0$$
$$a(a^2 - 4a - 21) = 0$$
$$a(a + 3)(a - 7) = 0$$
$$a = 0 \quad \text{or} \quad a + 3 = 0 \quad \text{or} \quad a - 7 = 0$$
$$a = 0 \quad \text{or} \quad a = -3 \quad \text{or} \quad a = 7$$
The solution set is $\{-3, 0, 7\}$.

57.
$$(y - 1)(y - 4) = 10$$
$$y^2 - 5y + 4 = 10$$
$$y^2 - 5y - 6 = 0$$
$$(y + 1)(y - 6) = 0$$
$$y + 1 = 0 \quad \text{or} \quad y - 6 = 0$$
$$y = -1 \quad \text{or} \quad y = 6$$
The solution set is $\{-1, 6\}$.

58.
$$a^2 - 7a - 40 = 4$$
$$a^2 - 7a - 44 = 0$$
$$(a - 11)(a + 4) = 0$$
$$a - 11 = 0 \quad \text{or} \quad a + 4 = 0$$
$$a = 11 \quad \text{or} \quad a = -4$$
The values of a for which $f(a) = 4$ are –4 and 11.

59. $f(x) = \dfrac{x - 5}{x^2 - x - 56}$

$f(x)$ cannot be calculated for any x-value for which the denominator, $x^2 - x - 56$, is 0. To find the excluded values, we solve:

$$x^2 - x - 56 = 0$$
$$(x - 8)(x + 7) = 0$$
$$x - 8 = 0 \quad \text{or} \quad x + 7 = 0$$
$$x = 8 \quad \text{or} \quad x = -7$$

The domain of f is
$\{x \mid x \text{ is a real number } and \ x \neq -7 \ and \ x \neq 8\}$, or
$(-\infty, -7) \cup (-7, \ 8) \cup (8, \ \infty)$.

60. Let $x =$ the base of the triangle and the height is $\dfrac{3}{4}x$.

Solve $\dfrac{1}{2}x\left(\dfrac{3}{4}x\right) = 216$.

$x = -24$ or $x = 24$.
The base cannot be negative. The base is 24 m and the height is 18 m.

61. ***Familiarize.*** We let x represent the width of the photograph. The length and width of the picture that shows are represented by $(x + 3) + 4$ and $x + 4$. The area of the picture that shows is 108 in^2.

Translate. Using the formula for the area of a rectangle, $A = l \cdot w$, we have
$$108 = ((x + 3) + 4)(x + 4).$$

Carry out. We solve the equation:
$$108 = (x + 7)(x + 4)$$
$$108 = x^2 + 11x + 28$$
$$0 = x^2 + 11x - 80$$
$$0 = (x + 16)(x - 5)$$
$$x + 16 = 0 \quad \text{or} \quad x - 5 = 0$$
$$x = -16 \quad \text{or} \quad x = 5$$

Check. We see that –16 is not a solution because the length and width of the frame cannot be negative. We check 5. When $x = 5$, $x + 7 = 12$ and $x + 4 = 9$ and $12 \cdot 9 = 108$. The area is 108. The value checks.

State. The length is 8 in. and the width is 5 in.

62. Let $x =$ second leg and $x + 2$ is the hypotenuse of the triangle. The first leg is 8 ft. Solve $8^2 + x^2 = (x + 2)^2$. The solution of the equation is 15. If $x = 15$, then $x + 2 = 17$. The lengths of the other sides are 15 ft and 17 ft.

63. ***Familiarize.*** We will use the formula
$$p(t) = -\frac{1}{50}t^2 + \frac{4}{5}t + 38.$$

Translate. We find the value of t for which $p(t) = 44$. We have
$$-\frac{1}{50}t^2 + \frac{4}{5}t + 38 = 44.$$

Carry out. We solve the equation.
$$-\frac{1}{50}t^2 + \frac{4}{5}t + 38 = 44$$
$$-\frac{1}{50}t^2 + \frac{4}{5}t - 6 = 0$$
$$t^2 - 40t + 300 = 0 \qquad \text{Multiplying by } -50$$
$$(t - 10)(t - 30) = 0$$

$$t - 10 = 0 \quad or \quad t - 30 = 0$$
$$t = 10 \quad or \quad t = 30$$

Check. We check 10: $p(10) = -\frac{1}{50}(10)^2 + \frac{4}{5}(10) + 38$

$= -2 + 8 + 38 = 44.$

We check 30: $p(30) = -\frac{1}{50}(30)^2 + \frac{4}{5}(30) + 38$

$= -18 + 24 + 38 = 44.$ The answers check.

State. In the years 1994 + 10, or 2004, and 1994 + 30, or 2024, there were 44% of the males ages 55 and older in the labor force.

64. *Writing Exercise*. When multiplying polynomials, we begin with a product and carry out the multiplication to write a sum of terms. When factoring a polynomial, we write an equivalent expression that is a product.

65. *Writing Exercise*. The principle of zero products states that if a product is equal to 0, at least one of the factors must be 0. If a product is nonzero, we cannot conclude that any one of the factors is a particular value.

66. $128x^6 - 2y^6$

$= 2(64x^6 - y^6) \qquad 64 = 2^6$

$= 2[(2x)^6 - y^6] \qquad$ Difference of squares

$= 2[(2x)^3 - y^3][(2x)^3 + y^3] \quad$ Difference/sum of cubes

$= 2(2x - y)(4x^2 + 2xy + y^2)(2x + y)(4x^2 - 2xy + y^2)$

67. $(x-1)^3 - (x+1)^3$

$= x^3 - 3x^2 + 3x - 1 - (x^3 + 3x^2 + 3x + 1)$

$= -6x^2 - 2$

$= -2(3x^2 + 1)$

68. $3x^{-6} - 12x^{-4} + 15x^{-3} = 3x^{-6}(1 - 4x^2 + 5x^3)$

69. $(x+1)^3 = x^2(x+1)$

$x^3 + 3x^2 + 3x + 1 = x^3 + x^2$

$2x^2 + 3x + 1 = 0$

$(x+1)(2x+1) = 0$

$x + 1 = 0 \quad or \quad 2x + 1 = 0$

$x = -1 \quad or \quad x = -\frac{1}{2}$

The solution set is $\left\{-1, -\frac{1}{2}\right\}$.

70. $x^2 + 100 = 0$

$x^2 = -100$

No solution

71. **a.**   Area of base = πr^2

Area of cylinder = $2\pi r(h-r) = 2\pi rh - 2\pi r^2$

Area of half-sphere = $\frac{1}{2}(4\pi r^2) = 2\pi r^2$

Area of silo = $\pi r^2 + 2\pi rh - 2\pi r^2 + 2\pi r^2$

$= 2\pi hr + \pi r^2$

b.   Area of base = πr^2

Area of cylinder = $2\pi rx$

Area of half-sphere = $\frac{1}{2}(4\pi r^2) = 2\pi r^2$

Area of silo = $\pi r^2 + 2\pi rx + 2\pi r^2$

$= 2\pi rx + 3\pi r^2$

c.   Since $x = h - r$, then

$2\pi rx + 3\pi r^2 = 2\pi r(h-r) + 3\pi r^2$

$= 2\pi rh - 2\pi r^2 + 3\pi r^2$

$= 2\pi rh + \pi r^2$

Chapter 5 Test

1.

Term	$8xy^3$	$-14x^2y$	$5x^5y^4$	$-9x^4y$
Degree	4	3	9	5

Degree of polynomial: 9

2. $5x^5y^4 - 9x^4y - 14x^2y + 8xy^3$

3. Leading term: $-5a^3$

4. $P(x) = 2x^3 + 3x^2 - x + 4$

$P(0) = 2(0)^3 + 3(0)^2 - 0 + 4 = 4$

$P(-2) = 2(-2)^3 + 3(-2)^2 - (-2) + 4 = 2$

5. $P(x) = x^2 - 3x$

$P(a+h) - P(a) = (a+h)^2 - 3(a+h) - (a^2 - 3a)$

$= a^2 + 2ah + h^2 - 3a - 3h - a^2 + 3a$

$= 2ah + h^2 - 3h$

6. $6xy - 2xy^2 - 2xy + 5xy^2 = 4xy + 3xy^2$

7. $(-4y^3 + 6y^2 - y) + (3y^3 - 9y - 7)$

$= (-4 + 3)y^3 + 6y^2 + (-1 - 9)y - 7$

$= -y^3 + 6y^2 - 10y - 7$

8. $(2m^3 - 4m^2n - 5n^2) + (8m^3 - 3mn^2 + 6n^2)$

$= 10m^3 - 4m^2n - 3mn^2 + n^2$

9. $(8a - 4b) - (3a + 4b)$

$= 8a - 4b - 3a - 4b$

$= 5a - 8b$

10. $(9y^2 - 2y - 5y^3) - (4y^2 - 2y - 6y^3) = 5y^2 + y^3$

11. $(-4x^2y^3)(-16xy^5) = (-4)(-16)x^{2+1}y^{3+5} = 64x^3y^8$

12. $(6a - 5b)(2a + b) = 12a^2 - 4ab - 5b^2$

13. $(x - y)(x^2 - xy - y^2)$

$= x(x^2 - xy - y^2) - y(x^2 - xy - y^2)$

$= x^3 - x^2y - xy^2 - x^2y + xy^2 + y^3$

$= x^3 - 2x^2y + y^3$

14. $(4t-3)^2 = 16t^2 - 24t + 9$

15. $\left(5a^3 + 9\right)^2$

$= \left(5a^3\right)^2 + 2 \cdot 5a^3 \cdot 9 + 9^2 \quad (A+B)^2 = A^2 + 2AB + B^2$

$= 25a^6 + 90a^3 + 81$

16. $(x-2y)(x+2y) = x^2 - 4y^2$

17. $x^2 - 10x + 25 = (x-5)^2$

18. $y^3 + 5y^2 - 4y - 20$

$= y^2(y+5) - 4(y+5)$

$= (y+5)\left(y^2 - 4\right)$

$= (y+5)(y+2)(y-2)$

19. $p^2 - 12p - 28$

Since the constant term is negative, we look for a factorization of -28 in which one factor is positive and one factor is negative. Their sum must be -12, so the negative factor must have the larger absolute value. Thus we consider only pairs of factors in which the negative factor has the larger absolute value.

Pair of Factors	Sum of Factors
$-28, 1$	-27
$-14, 2$	-12
$-7, 4$	-3

The numbers we need are -14 and 2. The factorization is $(p-14)(p+2)$.

20. $t^7 - 3t^5 = t^5(t^2 - 3)$

21. $12m^2 + 20m + 3 = (6m+1)(2m+3)$

22. $9y^2 - 25$ Difference of squares

$= (3y+5)(3y-5)$

23. $3r^3 - 3 = 3(r^3 - 1)$ Difference of cubes

$= 3(r-1)(r^2 + r + 1)$

24. $45x^2 + 20 + 60x$

$= 45x^2 + 60x + 20$

$= 5(9x^2 + 12x + 4)$ Perfect-square trinomial

$= 5(3x+2)(3x+2)$

$= 5(3x+2)^2$

25. $3x^4 - 48y^4 = 3(x^4 - 16y^4)$

$= 3\left(x^2 + 4y^2\right)\left(x^2 - 4y^2\right)$

$= 3\left(x^2 + 4y^2\right)(x+2y)(x-2y)$

26. $y^2 + 8y + 16 - 100t^2$

$= \left(y^2 + 8y + 16\right) - 100t^2$ Grouping as a difference

$= (y+4)^2 - 100t^2$ of squares

$= (y+4+10t)(y+4-10t)$

27. Prime

28. $20a^2 - 5b^2 = 5\left(4a^2 - b^2\right) = 5(2a+b)(2a-b)$

29. $24x^2 - 46x + 10$

$= 2\left(12x^2 - 23x + 5\right)$

$= 2(4x-1)(3x-5)$

30. $16a^7b + 54ab^7 = 2ab\left(8a^6 + 27b^6\right)$

$= 2ab\left(2a^2 + 3b^2\right)\left(4a^4 - 6a^2b^2 + 9b^4\right)$

31. $x^2 - 3x - 18 = 0$

$(x+3)(x-6) = 0$

$x+3 = 0 \quad$ or $\quad x-6 = 0$

$x = -3 \quad$ or $\quad x = 6$

The solution set is $\{-3, 6\}$.

32. $\qquad 5t^2 = 125$

$\quad 5t^2 - 125 = 0$

$\quad 5(t^2 - 25) = 0$

$5(t+5)(t-5) = 0$

$t+5 = 0 \quad$ or $\quad t-5 = 0$

$t = -5 \quad$ or $\quad t = 5$

The solution set is $\{-5, 5\}$.

33. $\qquad 2x^2 + 21 = -17x$

$2x^2 + 17x + 21 = 0$

$(2x+3)(x+7) = 0$

$2x+3 = 0 \quad$ or $\quad x+7 = 0$

$x = -\dfrac{3}{2} \quad$ or $\quad x = -7$

The solution set is $\left\{-7, -\dfrac{3}{2}\right\}$.

34. $9x^2 + 3x = 0$

$3x(3x+1) = 0$

$3x = 0 \quad$ or $\quad 3x+1 = 0$

$x = 0 \quad$ or $\quad x = -\dfrac{1}{3}$

The solution set is $\left\{-\dfrac{1}{3}, 0\right\}$.

35. $\qquad x^2 + 81 = 18x$

$x^2 - 18x + 81 = 0$

$\quad (x-9)^2 = 0$

$\qquad x = 9$

The solution set is $\{9\}$.

36. $3a^2 - 15a + 11 = 11$

$\quad 3a^2 - 15a = 0$

$\quad 3a(a-5) = 0$

$3a = 0 \quad$ or $\quad a-5 = 0$

$a = 0 \quad$ or $\quad a = 5$

The values of a for which $f(a) = 11$ are 0 and 5.

37. $f(x) = \dfrac{8-x}{x^2 + 2x + 1}$

$f(x)$ cannot be calculated for any x-value for which

the denominator, $x^2 + 2x + 1$, is 0. To find the
excluded values, we solve:

$$x^2 + 2x + 1 = 0$$
$$(x+1)(x+1) = 0$$
$$x + 1 = 0 \quad or \quad x + 1 = 0$$
$$x = -1 \quad or \quad\quad x = -1$$

The domain of f is

$\{x \mid x \text{ is a real number } and \; x \neq -1\}$, or

$(-\infty, -1) \cup (-1, \infty)$.

38. Let w = the width.
Solve $w(w+3) = 40$.
$w = -8$ or $w = 5$
Width cannot be negative.
The length is 8 cm, and the width is 5 cm.

39. *Familiarize*. We will use the formula

$h(t) = -16t^2 + 64t + 36$.

Translate. We find the value of t for which $h(t) = 0$.
We have

$$-16t^2 + 64t + 36 = 0.$$

Carry out. We solve the equation.

$$-16t^2 + 64t + 36 = 0$$
$$4t^2 - 16t - 9 = 0 \quad \text{Diving by } -4$$
$$(2t+1)(2t-9) = 0$$
$$2t + 1 = 0 \quad or \quad 2t - 9 = 0$$
$$t = -\frac{1}{2} \quad or \quad\quad t = 4\frac{1}{2}$$

Check. Time cannot be negative in this application.

$h\left(4\frac{1}{2}\right) = -16\left(4\frac{1}{2}\right)^2 + 64\left(4\frac{1}{2}\right) + 36 = 0$ The answer

checks.

State. The fireworks will reach the water in $4\frac{1}{2}$ sec.

40. Let h = the height that the ladder reaches on the wall.

Solve $10^2 + h^2 = (h+2)^2$
$h = 24$
The ladder reaches 24 ft on the wall.

41. $(a+3)^2 - 2(a+3) - 35$

Substitute u for $a+3$ (and u^2 for $(a+3)^2$). We factor

$u^2 - 2u - 35$. Look for factors of -35 whose sum is -2.
The factors are -7 and 5. We have

$u^2 - 2u - 35 = (u-7)(u+5)$. Replace u with $a+3$:

$(a+3)^2 - 2(a+3) - 35 = [(a+3) - 7][(a+3) + 5]$, or

$(a-4)(a+8)$

42.
$$20x(x+2)(x-1) = 5x^3 - 24x - 14x^2$$
$$20x(x^2 + x - 2) = 5x^3 - 14x^2 - 24x$$
$$20x^3 + 20x^2 - 40x = 5x^3 - 14x^2 - 24x$$
$$15x^3 + 34x^2 - 16x = 0$$
$$x(15x^2 + 34x - 16) = 0$$
$$x(3x+8)(5x-2) = 0$$
$$x = 0 \quad or \quad 3x + 8 = 0 \quad or \quad 5x - 2 = 0$$
$$x = 0 \quad or \quad 3x = -8 \quad or \quad\quad 5x = 2$$
$$x = 0 \quad or \quad\quad x = -\frac{8}{3} \quad or \quad\quad x = \frac{2}{5}$$

The solution set is $\left\{-\dfrac{8}{3},\; 0,\; \dfrac{2}{5}\right\}$.

Chapter 6

Rational Expressions, Equations, and Functions

Exercise Set 6.1

1. The expression $\frac{x-4}{5x}$ is an example of a *rational* expression.

3. When simplifying rational expressions, remove a *factor* equal to 1.

5. Since $x - 5 = 0$ when $x = 5$, choice (c) is correct.

7. Since $(x-2)(x-5) = 0$ when $x = 2$ or $x = 5$, choice (f) is correct.

9. Since $x + 3 = 0$ when $x = -3$, choice (b) is correct.

11. $f(x) = \frac{2x^2 - x - 5}{x - 1}$

   a. $f(0) = \frac{2 \cdot 0^2 - 0 - 5}{0 - 1} = \frac{-5}{-1} = 5$

   b. $f(-1) = \frac{2 \cdot (-1)^2 - (-1) - 5}{(-1) - 1} = \frac{2 + 1 - 5}{-2} = 1$

   c. $f(3) = \frac{2 \cdot 3^2 - 3 - 5}{3 - 1} = \frac{18 - 3 - 5}{2} = 5$

13. $r(t) = \frac{t^2 - 8t - 9}{t^2 - 4}$

   a. $r(0) = \frac{0^2 - 8 \cdot 0 - 9}{0^2 - 4} = \frac{-9}{-4} = \frac{9}{4}$

   b. $r(2) = \frac{2^2 - 8 \cdot 2 - 9}{2^2 - 4} = \frac{4 - 16 - 9}{4 - 4}$

       $= \frac{-21}{0}$ does not exist

   c. $r(-1) = \frac{(-1)^2 - 8(-1) - 9}{(-1)^2 - 4} = \frac{1 + 8 - 9}{1 - 4} = \frac{0}{-3} = 0$

15. $H(t) = \frac{t^2 + t}{2t + 1}$

   $H(5) = \frac{5^2 + 5}{2 \cdot 5 + 1} = \frac{25 + 5}{10 + 1} = \frac{30}{11}$ hr, or $2\frac{8}{11}$ hr

17. $\frac{8t^4}{40t}$

   $= \frac{8t \cdot t^3}{8t \cdot 5}$ Factoring; the greatest common factor is $8t$.

   $= \frac{8t}{8t} \cdot \frac{t^3}{5}$ Factoring the rational expression

   $= 1 \cdot \frac{t^3}{5}$ $\frac{8t}{8t} = 1$

   $= \frac{t^3}{5}$ Removing a factor equal to 1

19. $\frac{24x^3y}{30x^5y^8}$

   $= \frac{6x^3y \cdot 4}{6x^3y \cdot 5x^2y^7}$ Factoring the numerator and the denominator

   $= \frac{6x^3y}{6x^3y} \cdot \frac{4}{5x^2y^7}$ Factoring the rational expression

   $= \frac{4}{5x^2y^7}$ Removing a factor equal to 1

21. $\frac{2a - 10}{2} = \frac{2(a-5)}{2 \cdot 1} = \frac{2}{2} \cdot \frac{a-5}{1} = a - 5$

23. $\frac{5}{25y - 30} = \frac{5 \cdot 1}{5(5y - 6)} = \frac{5}{5} \cdot \frac{1}{5y - 6} = \frac{1}{5y - 6}$

25. $\frac{3x - 12}{3x + 15} = \frac{3(x - 4)}{3(x + 5)} = \frac{3}{3} \cdot \frac{x-4}{x+5} = \frac{x-4}{x+5}$

27. $f(x) = \frac{5x + 30}{x^2 + 6x}$

   $= \frac{5(x + 6)}{x(x + 6)}$ Note that $x \neq 0$ and $x \neq -6$

   $= \frac{5}{x} \cdot \frac{x + 6}{x + 6}$

   $= \frac{5}{x}$

   Thus, $f(x) = \frac{5}{x}$, $x \neq 0, -6$.

29. $g(x) = \frac{x^2 - 9}{5x + 15}$

   $= \frac{(x + 3)(x - 3)}{5(x + 3)}$ Note that $x \neq -3$.

   $= \frac{x + 3}{x + 3} \cdot \frac{x - 3}{5}$

   $= \frac{x - 3}{5}$

   Thus, $g(x) = \frac{x-3}{5}$, $x \neq -3$.

31. $h(x) = \frac{2 - x}{7x - 14}$

   $= \frac{-1(x - 2)}{7(x - 2)}$ Note that $x \neq 2$

   $= \frac{-1}{7} \cdot \frac{x - 2}{x - 2}$

   $= -\frac{1}{7}$

   Thus, $h(x) = -\frac{1}{7}$, $x \neq 2$.

33. $f(t) = \dfrac{t^2 - 16}{t^2 - 8t + 16}$

$\quad = \dfrac{(t+4)(t-4)}{(t-4)(t-4)}$ Note that $t \neq 4$.

$\quad = \dfrac{t+4}{t-4} \cdot \dfrac{t-4}{t-4}$

$\quad = \dfrac{t+4}{t-4}$

Thus, $f(t) = \dfrac{t+4}{t-4}$, $t \neq 4$.

35. $g(t) = \dfrac{21 - 7t}{3t - 9}$

$\quad = \dfrac{-7(t-3)}{3(t-3)}$ Note that $t \neq 3$.

$\quad = \dfrac{-7}{3} \cdot \dfrac{t-3}{t-3}$

$\quad = -\dfrac{7}{3}$

Thus, $g(t) = -\dfrac{7}{3}$, $t \neq 3$.

37. $h(t) = \dfrac{t^2 + 5t + 4}{t^2 - 8t - 9}$

$\quad = \dfrac{(t+1)(t+4)}{(t+1)(t-9)}$ Note that $t \neq -1$ and $t \neq 9$.

$\quad = \dfrac{t+1}{t+1} \cdot \dfrac{t+4}{t-9}$

$\quad = \dfrac{t+4}{t-9}$

Thus, $h(t) = \dfrac{t+4}{t-9}$, $t \neq -1, 9$.

39. $f(x) = \dfrac{9x^2 - 4}{3x - 2}$

$\quad = \dfrac{(3x+2)(3x-2)}{3x-2}$ Note that $x \neq \dfrac{2}{3}$.

$\quad = \dfrac{3x+2}{1} \cdot \dfrac{3x-2}{3x-2}$

$\quad = 3x + 2$

Thus, $f(x) = 3x + 2$, $x \neq \dfrac{2}{3}$.

41. $g(t) = \dfrac{16 - t^2}{t^2 - 8t + 16}$

$\quad = \dfrac{(4+t)(4-t)}{(t-4)(t-4)}$ Note that $t \neq 4$.

$\quad = \dfrac{(4+t)(4-t)}{-1(4-t)(t-4)}$

$\quad = \dfrac{4-t}{4-t} \cdot \dfrac{4+t}{-1(t-4)}$

$\quad = \dfrac{4+t}{4-t}$

Thus, $g(t) = \dfrac{4+t}{4-t}$, $t \neq 4$. (We could also write this as

$g(t) = \dfrac{-t-4}{t-4}$, $t \neq 4$.)

43. $\dfrac{3y^3}{5z} \cdot \dfrac{10z^4}{7y^6}$

$\quad = \dfrac{3y^3 \cdot 10z^4}{5z \cdot 7y^6}$ Multiplying the numerators and also the denominators

$\quad = \dfrac{3 \cdot y^3 \cdot 2 \cdot 5 \cdot z \cdot z^3}{5 \cdot z \cdot 7 \cdot y^3 \cdot y^3}$ Factoring the numerator and denominator

$\quad = \dfrac{3 \cdot y^3 \cdot 2 \cdot 5 \cdot z \cdot z^3}{5 \cdot z \cdot 7 \cdot y^3 \cdot y^3}$ Removing a factor equal to 1

$\quad = \dfrac{6z^3}{7y^3}$

45. $\dfrac{8x - 16}{5x} \cdot \dfrac{x^3}{5x - 10} = \dfrac{(8x-16)(x^3)}{5x(5x-10)}$

$\qquad\qquad = \dfrac{8(x-2)(x)(x^2)}{5 \cdot x \cdot 5(x-2)}$

$\qquad\qquad = \dfrac{8(x-2)(x)(x^2)}{5 \cdot x \cdot 5(x-2)}$

$\qquad\qquad = \dfrac{8x^2}{25}$

47. $\dfrac{y^2 - 9}{y^2} \cdot \dfrac{y^2 - 3y}{y^2 - y - 6} = \dfrac{(y^2-9)(y^2-3y)}{y^2(y^2-y-6)}$

$\qquad\qquad = \dfrac{(y+3)(y-3)(y)(y-3)}{y \cdot y(y-3)(y+2)}$

$\qquad\qquad = \dfrac{(y+3)(y-3)(y)(y-3)}{y \cdot y(y-3)(y+2)}$

$\qquad\qquad = \dfrac{(y+3)(y-3)}{y(y+2)}$

49. $\dfrac{7a - 14}{4 - a^2} \cdot \dfrac{5a^2 + 6a + 1}{35a + 7}$

$\quad = \dfrac{(7a-14)(5a^2+6a+1)}{(4-a^2)(35a+7)}$

$\quad = \dfrac{7(a-2)(5a+1)(a+1)}{(2+a)(2-a)(7)(5a+1)}$

$\quad = \dfrac{7(-1)(2-a)(5a+1)(a+1)}{(2+a)(2-a)(7)(5a+1)}$

$\quad = \dfrac{7(-1)(2-a)(5a+1)(a+1)}{(2+a)(2-a)(7)(5a+1)}$

$\quad = \dfrac{-1(a+1)}{2+a}$

$\quad = \dfrac{-a-1}{2+a}$, or $-\dfrac{a+1}{2+a}$

51. $\dfrac{t^3 - 4t}{t - t^4} \cdot \dfrac{t^4 - t}{4t - t^3}$

$\quad = \dfrac{t^3 - 4t}{t - t^4} \cdot \dfrac{-1(t - t^4)}{-1(t^3 - 4t)}$

$\quad = \dfrac{(t^3-4t)(-1)(t-t^4)}{(t-t^4)(-1)(t^3-4t)}$

$\quad = 1$

53. $\dfrac{c^3+8}{c^5-4c^3} \cdot \dfrac{c^6-4c^5+4c^4}{c^2-2c+4}$

$= \dfrac{(c^3+8)(c^6-4c^5+4c^4)}{(c^5-4c^3)(c^2-2c+4)}$

$= \dfrac{(c+2)(c^2-2c+4)(c^4)(c-2)(c-2)}{c^3(c+2)(c-2)(c^2-2c+4)}$

$= \dfrac{c^3(c+2)(c^2-2c+4)(c-2)}{c^3(c+2)(c^2-2c+4)(c-2)} \cdot \dfrac{c(c-2)}{1}$

$= c(c-2)$

55. $\dfrac{a^3-b^3}{3a^2+9ab+6b^2} \cdot \dfrac{a^2+2ab+b^2}{a^2-b^2}$

$= \dfrac{(a^3-b^3)(a^2+2ab+b^2)}{(3a^2+9ab+6b^2)(a^2-b^2)}$

$= \dfrac{(a-b)(a^2+ab+b^2)(a+b)(a+b)}{3(a+b)(a+2b)(a+b)(a-b)}$

$= \dfrac{\cancel{(a-b)}\,(a^2+ab+b^2)\,\cancel{(a+b)}\,\cancel{(a+b)}}{3\,\cancel{(a+b)}\,(a+2b)\,\cancel{(a+b)}\,\cancel{(a-b)}}$

$= \dfrac{a^2+ab+b^2}{3(a+2b)}$

57. $\dfrac{12a^3}{5b^2} \div \dfrac{4a^2}{15b}$

$= \dfrac{12a^3}{5b^2} \cdot \dfrac{15b}{4a^2}$ Multiplying by the reciprocal of the divisor

$= \dfrac{12a^3(15b)}{5b^2(4a^2)}$

$= \dfrac{3 \cdot 4 \cdot a^2 \cdot a \cdot 5 \cdot 3 \cdot b}{5 \cdot b \cdot b \cdot 4 \cdot a^2}$

$= \dfrac{3 \cdot \cancel{4} \cdot \cancel{a^2} \cdot a \cdot \cancel{5} \cdot 3 \cdot \cancel{b}}{\cancel{5} \cdot \cancel{b} \cdot b \cdot \cancel{4} \cdot \cancel{a^2}}$

$= \dfrac{9a}{b}$

59. $\dfrac{5x+20}{x^6} \div \dfrac{x+4}{x^2} = \dfrac{5x+20}{x^6} \cdot \dfrac{x^2}{x+4}$

$= \dfrac{(5x+20)(x^2)}{x^6(x+4)}$

$= \dfrac{5\,\cancel{(x+4)}\,\cancel{(x^2)}}{x^2(x^4)\,\cancel{(x+4)}}$

$= \dfrac{5}{x^4}$

61. $\dfrac{25x^2-4}{x^2-9} \div \dfrac{2-5x}{x+3} = \dfrac{25x^2-4}{x^2-9} \cdot \dfrac{x+3}{2-5x}$

$= \dfrac{(25x^2-4)(x+3)}{(x^2-9)(2-5x)}$

$= \dfrac{(5x+2)(5x-2)(x+3)}{(x+3)(x-3)(-1)(5x-2)}$

$= \dfrac{(5x+2)\,\cancel{(5x-2)}\,\cancel{(x+3)}}{\cancel{(x+3)}\,(x-3)(-1)\,\cancel{(5x-2)}}$

$= \dfrac{5x+2}{-x+3}$, or $-\dfrac{5x+2}{x-3}$

63. $\dfrac{5y-5x}{15y^3} \div \dfrac{x^2-y^2}{3x+3y} = \dfrac{5y-5x}{15y^3} \cdot \dfrac{3x+3y}{x^2-y^2}$

$= \dfrac{(5y-5x)(3x+3y)}{15y^3(x^2-y^2)}$

$= \dfrac{5(y-x)(3)(x+y)}{5 \cdot 3 \cdot y^3(x+y)(x-y)}$

$= \dfrac{5(-1)(x-y)(3)(x+y)}{5 \cdot 3 \cdot y^3(x+y)(x-y)}$

$= \dfrac{\cancel{5}(-1)\,\cancel{(x-y)}\,\cancel{(3)}\,\cancel{(x+y)}}{\cancel{5} \cdot \cancel{3} \cdot y^3\,\cancel{(x+y)}\,\cancel{(x-y)}}$

$= \dfrac{-1}{y^3}$, or $-\dfrac{1}{y^3}$

65. $\dfrac{y^2-36}{y^2-8y+16} \div \dfrac{3y-18}{y^2-y-12}$

$= \dfrac{y^2-36}{y^2-8y+16} \cdot \dfrac{y^2-y-12}{3y-18}$

$= \dfrac{(y+6)(y-6)(y-4)(y+3)}{(y-4)(y-4)3(y-6)}$

$= \dfrac{(y+6)\,\cancel{(y-6)}\,\cancel{(y-4)}\,(y+3)}{\cancel{(y-4)}\,(y-4)(3)\,\cancel{(y-6)}}$

$= \dfrac{(y+6)(y+3)}{3(y-4)}$

67. $\dfrac{x^3-64}{x^3+64} \div \dfrac{x^2-16}{x^2-4x+16}$

$= \dfrac{x^3-64}{x^3+64} \cdot \dfrac{x^2-4x+16}{x^2-16}$

$= \dfrac{(x^3-64)(x^2-4x+16)}{(x^3+64)(x^2-16)}$

$= \dfrac{(x-4)(x^2+4x+16)(x^2-4x+16)}{(x+4)(x^2-4x+16)(x+4)(x-4)}$

$= \dfrac{(x-4)(x^2-4x+16)}{(x-4)(x^2-4x+16)} \cdot \dfrac{x^2+4x+16}{(x+4)(x+4)}$

$= \dfrac{x^2+4x+16}{(x+4)(x+4)}$, or $\dfrac{x^2+4x+16}{(x+4)^2}$

69. $f(t) = \dfrac{t^2 - 100}{5t + 20} \cdot \dfrac{t + 4}{t - 10}$

The denominators of the rational expressions are $5t + 20$ and $t - 10$, so the domain of f cannot contain -4 or 10.

$f(t) = \dfrac{t^2 - 100}{5t + 20} \cdot \dfrac{t + 4}{t - 10}$

$= \dfrac{(t^2 - 100)(t + 4)}{(5t + 20)(t - 10)}$

$= \dfrac{(t + 10)(t - 10)(t + 4)}{5(t + 4)(t - 10)}$

$= \dfrac{(t + 10)\,\cancel{(t - 10)}\,\cancel{(t + 4)}}{5\,\cancel{(t + 4)}\,\cancel{(t - 10)}}$

$= \dfrac{t + 10}{5}, \quad t \neq -4,\ 10$

71. $g(x) = \dfrac{x^2 - 2x - 35}{2x^3 - 3x^2} \cdot \dfrac{4x^3 - 9x}{7x - 49}$

First we find the values of x for which the denominator $2x^3 - 3x^2 = 0$.

$2x^3 - 3x^2 = 0$

$x^2(2x - 3) = 0$

$x \cdot x \cdot (2x - 3) = 0$

$x = 0 \quad \text{or} \quad x = 0 \quad \text{or} \quad 2x - 3 = 0$

$x = 0 \quad \text{or} \quad x = 0 \quad \text{or} \quad x = \dfrac{3}{2}$

We see that the domain cannot contain 0 or $\dfrac{3}{2}$. Since $7x - 49 = 0$ when $x = 7$, the number 7 must also be excluded from the domain.

$g(x) = \dfrac{x^2 - 2x - 35}{2x^3 - 3x^2} \cdot \dfrac{4x^3 - 9x}{7x - 49}$

$= \dfrac{(x^2 - 2x - 35)(4x^3 - 9x)}{(2x^3 - 3x^2)(7x - 49)}$

$= \dfrac{(x - 7)(x + 5)(x)(2x + 3)(2x - 3)}{x \cdot x \cdot (2x - 3)(7)(x - 7)}$

$= \dfrac{\cancel{(x - 7)}(x + 5)\,\cancel{(x)}(2x + 3)\,\cancel{(2x - 3)}}{\cancel{x} \cdot x \cdot \cancel{(2x - 3)}(7)\,\cancel{(x - 7)}}$

$= \dfrac{(x + 5)(2x + 3)}{7x}, \quad x \neq 0,\ \dfrac{3}{2},\ 7$

73. $f(x) = \dfrac{x^2 - 4}{x^3} \div \dfrac{x^5 - 2x^4}{x + 4}$

$x^3 = 0$ when $x = 0$ and $x + 4 = 0$ when $x = -4$, so the domain cannot contain 0 or -4. Also, division by $(x^5 - 2x^4)/(x + 4)$ is defined only when this expression is nonzero. We find the values of x for which $x^5 - 2x^4$ is zero.

$x^5 - 2x^4 = 0$

$x^4(x - 2) = 0$

The solutions of this equation are 0 and 2 so, in addition to 0 and -4, we must also exclude 2 from the domain.

$f(x) = \dfrac{x^2 - 4}{x^3} \div \dfrac{x^5 - 2x^4}{x + 4}$

$= \dfrac{x^2 - 4}{x^3} \cdot \dfrac{x + 4}{x^5 - 2x^4}$

$= \dfrac{(x^2 - 4)(x + 4)}{x^3(x^5 - 2x^4)}$

$= \dfrac{(x + 2)(x - 2)(x + 4)}{x^3 \cdot x^4(x - 2)}$

$= \dfrac{(x + 2)\,\cancel{(x - 2)}\,(x + 4)}{x^7\,\cancel{(x - 2)}}$

$= \dfrac{(x + 2)(x + 4)}{x^7}, \quad x \neq -4,\ 0,\ 2$

75. $h(n) = \dfrac{n^3 + 3n}{n^2 - 9} \div \dfrac{n^2 + 5n - 14}{n^2 + 4n - 21}$

First we find the values of n for which the denominators of the rational expressions are zero.

$n^2 - 9 = 0$

$(n + 3)(n - 3) = 0$

$n + 3 = 0 \quad \text{or} \quad n - 3 = 0$

$n = -3 \quad \text{or} \quad n = 3$

$n^2 + 4n - 21 = 0$

$(n + 7)(n - 3) = 0$

$n + 7 = 0 \quad \text{or} \quad n - 3 = 0$

$n = -7 \quad \text{or} \quad n = 3$

We see that the domain cannot contain -7, -3, or 3. Now we find the values of n for which $(n^2 + 5n - 14)/(n^2 + 4n - 21)$ is zero.

$n^2 + 5n - 14 = 0$

$(n + 7)(n - 2) = 0$

$n + 7 = 0 \quad \text{or} \quad n - 2 = 0$

$n = -7 \quad \text{or} \quad n = 2$

We must also exclude 2 from the domain.

$h(n) = \dfrac{n^3 + 3n}{n^2 - 9} \div \dfrac{n^2 + 5n - 14}{n^2 + 4n - 21}$

$= \dfrac{n^3 + 3n}{n^2 - 9} \cdot \dfrac{n^2 + 4n - 21}{n^2 + 5n - 14}$

$= \dfrac{(n^3 + 3n)(n^2 + 4n - 21)}{(n^2 - 9)(n^2 + 5n - 14)}$

$= \dfrac{n(n^2 + 3)(n + 7)(n - 3)}{(n + 3)(n - 3)(n + 7)(n - 2)}$

$= \dfrac{n(n^2 + 3)\,\cancel{(n + 7)}\,\cancel{(n - 3)}}{(n + 3)\,\cancel{(n - 3)}\,\cancel{(n + 7)}(n - 2)}$

$= \dfrac{n(n^2 + 3)}{(n + 3)(n - 2)}, \quad n \neq -7,\ -3,\ 2,\ 3$

77. $\dfrac{4x^2 - 9y^2}{8x^3 - 27y^3} \div \dfrac{4x + 6y}{3x - 9y} \cdot \dfrac{4x^2 + 6xy + 9y^2}{4x^2 - 8xy + 3y^2}$

$= \dfrac{4x^2 - 9y^2}{8x^3 - 27y^3} \cdot \dfrac{3x - 9y}{4x + 6y} \cdot \dfrac{4x^2 + 6xy + 9y^2}{4x^2 - 8xy + 3y^2}$

$= \dfrac{(4x^2 - 9y^2)(3x - 9y)(4x^2 + 6xy + 9y^2)}{(8x^3 - 27y^3)(4x + 6y)(4x^2 - 8xy + 3y^2)}$

$$= \frac{(2x+3y)(2x-3y)(3)(x-3y)(4x^2+6xy+9y^2)}{(2x-3y)(4x^2+6xy+9y^2)(2)(2x+3y)(2x-y)(2x-3y)}$$

$$= \frac{(2x+3y)(2x-3y)(4x^2+6xy+9y^2)}{(2x+3y)(2x-3y)(4x^2+6xy+9y^2)} \cdot \frac{3(x-3y)}{2(2x-y)(2x-3y)}$$

$$= \frac{3(x-3y)}{2(2x-y)(2x-3y)}$$

79. $\dfrac{a^3-ab^2}{2a^2+3ab+b^2} \cdot \dfrac{4a^2-b^2}{a^2-2ab+b^2} \div \dfrac{a^2+a}{a-1}$

$$= \frac{a^3-ab^2}{2a^2+3ab+b^2} \cdot \frac{4a^2-b^2}{a^2-2ab+b^2} \cdot \frac{a-1}{a^2+a}$$

$$= \frac{(a^3-ab^2)(4a^2-b^2)(a-1)}{(2a^2+3ab+b^2)(a^2-2ab+b^2)(a^2+a)}$$

$$= \frac{a(a+b)(a-b)(2a+b)(2a-b)(a-1)}{(2a+b)(a+b)(a-b)(a-b)(a)(a+1)}$$

$$= \frac{a(a+b)(a-b)(2a+b)}{a(a+b)(a-b)(2a+b)} \cdot \frac{(2a-b)(a-1)}{(a-b)(a+1)}$$

$$= \frac{(2a-b)(a-1)}{(a-b)(a+1)}$$

81. *Writing Exercise.*

83. $2n^2-11n+12 = (2n-3)(n-4)$

85. $8x^3+125 = (2x+5)(4x^2-10x+25)$

87. $t^3+8t^2-33t = t(t^2+8t-33) = t(t+11)(t-3)$

89. *Writing Exercise.*

91. $m = \dfrac{f(a+h)-f(a)}{(a+h)-a} = \dfrac{(a+h)^2+5-(a^2+5)}{a+h-a}$

$$= \frac{a^2+2ah+h^2+5-a^2-5}{h}$$

$$= \frac{2ah+h^2}{h} = \frac{h(2a+h)}{h}$$

$$= 2a+h$$

93. To find the domain of f we set
$$x-3 = 0$$
$$x = 3.$$
The domain of $f = \{x \,|\, x \text{ is a real number and } x \neq 3\}$.

Simplify: $f(x) = \dfrac{x^2-9}{x-3} = \dfrac{(x+3)(x-3)}{(x-3)\cdot 1}$

$$= \frac{x-3}{x-3} \cdot \frac{x+3}{1} = x+3$$

Graph $f(x) = x+3$ using the domain found above.

95. $\dfrac{d^2-d}{d^2-6d+8} \cdot \dfrac{d-2}{d^2+5d} \div \left(\dfrac{5d^2}{d^2-9d+20}\right)^2$

$$= \frac{d^2-d}{d^2-6d+8} \cdot \frac{d-2}{d^2+5d} \cdot \frac{(d^2-9d+20)^2}{(5d^2)^2}$$

$$= \frac{d(d-1)(d-2)[(d-4)(d-5)]^2}{(d-2)(d-4)(d)(d+5)(25d^4)}$$

$$= \frac{d(d-1)(d-2)(d-4)(d-4)(d-5)^2}{(d-2)(d-4)(d)(d+5)(25d^4)}$$

$$= \frac{(d-1)(d-4)(d-5)^2}{25d^4(d+5)}$$

97. $\dfrac{m^2-t^2}{m^2+t^2+m+t+2mt}$

$$= \frac{m^2-t^2}{(m^2+2mt+t^2)+(m+t)}$$

$$= \frac{(m+t)(m-t)}{(m+t)^2+(m+t)}$$

$$= \frac{(m+t)(m-t)}{(m+t)[(m+t)+1]}$$

$$= \frac{(m+t)(m-t)}{(m+t)(m+t+1)}$$

$$= \frac{m-t}{m+t+1}$$

99. $\dfrac{x^3+x^2-y^3-y^2}{x^2-2xy+y^2}$

$$= \frac{(x^3-y^3)+(x^2-y^2)}{x^2-2xy+y^2}$$

$$= \frac{(x-y)(x^2+xy+y^2)+(x+y)(x-y)}{(x-y)^2}$$

$$= \frac{(x-y)(x^2+xy+y^2+x+y)}{(x-y)(x-y)}$$

$$= \frac{x^2+xy+y^2+x+y}{x-y}$$

101. $\dfrac{x^5-x^3+x^2-1-(x^3-1)(x+1)^2}{(x^2-1)^2}$

$$= \frac{x^5-x^3+(x^2-1)-[(x^3-1)(x+1)^2]}{(x^2-1)^2}$$

$$= \frac{x^3(x^2-1)+(x^2-1)-[(x-1)(x^2+x+1)(x+1)(x+1)]}{(x^2-1)^2}$$

$$= \frac{x^3(x^2-1)+(x^2-1)-[(x^2-1)(x+1)(x^2+x+1)]}{(x^2-1)^2}$$

$$= \frac{(x^2-1)[x^3+1-(x^3+x^2+x+x^2+x+1)]}{(x^2-1)(x^2-1)}$$

$$= \frac{(x^2-1)(-2x^2-2x)}{(x^2-1)(x^2-1)}$$

$$= \frac{-2x^2 - 2x}{x^2 - 1}$$

$$= \frac{-2x(x+1)}{(x+1)(x-1)}$$

$$= \frac{-2x\,\cancel{(x+1)}}{\cancel{(x+1)}(x-1)}$$

$$= \frac{-2x}{x-1}, \text{ or } -\frac{2x}{x-1}$$

103. a. $(f \cdot g)(x) = \dfrac{4}{x^2 - 1} \cdot \dfrac{4x^2 + 8x + 4}{x^3 - 1}$

$$= \frac{4(4x^2 + 8x + 4)}{(x^2 - 1)(x^3 - 1)}$$

$$= \frac{4 \cdot 4(x+1)(x+1)}{(x+1)(x-1)(x-1)(x^2 + x + 1)}$$

$$= \frac{4 \cdot 4\,\cancel{(x+1)}(x+1)}{\cancel{(x+1)}(x-1)(x-1)(x^2 + x + 1)}$$

$$= \frac{16(x+1)}{(x-1)^2(x^2 + x + 1)}$$

(Note that $x \neq -1$ is an additional restriction, since -1 is not in the domain of f.)

b. $(f/g)(x) = \dfrac{4}{x^2 - 1} \div \dfrac{4x^2 + 8x + 4}{x^3 - 1}$

$$= \frac{4}{x^2 - 1} \cdot \frac{x^3 - 1}{4x^2 + 8x + 4}$$

$$= \frac{4(x^3 - 1)}{(x^2 - 1)(4x^2 + 8x + 4)}$$

$$= \frac{4(x-1)(x^2 + x + 1)}{(x+1)(x-1)(4)(x+1)(x+1)}$$

$$= \frac{\cancel{4}\,\cancel{(x-1)}(x^2 + x + 1)}{(x+1)\cancel{(x-1)}\,\cancel{(4)}(x+1)(x+1)}$$

$$= \frac{x^2 + x + 1}{(x+1)^3}$$

(Note that $x \neq 1$ is an additional restriction, since 1 is not in the domain of either f or g.)

c. $(g/f)(x) = \dfrac{1}{(f/g)(x)}$

$$= \frac{(x+1)^3}{x^2 + x + 1} \quad \text{(See part (b) above)}$$

(Note that $x \neq -1$ and $x \neq 1$ are restrictions, since -1 is not in the domain of f and 1 is not in the domain of either f or g.)

105. *Writing Exercise.*

Exercise Set 6.2

1. True

3. False

5. False; see Example 4.

7. False

9. $\dfrac{4}{3a} + \dfrac{11}{3a} = \dfrac{15}{3a}$ Adding the numerators. The denominator is unchanged.

$$= \frac{3 \cdot 5}{3 \cdot a}$$

$$= \frac{\cancel{3} \cdot 5}{\cancel{3} \cdot a}$$

$$= \frac{5}{a}$$

11. $\dfrac{5}{3m^2 n^2} - \dfrac{4}{3m^2 n^2} = \dfrac{1}{3m^2 n^2}$

13. $\dfrac{x - 3y}{x + y} + \dfrac{x + 5y}{x + y} = \dfrac{2x + 2y}{x + y} = \dfrac{2(x + y)}{x + y} = \dfrac{2\cancel{(x+y)}}{1\cancel{(x+y)}} = 2$

15. $\dfrac{3t + 2}{t - 4} - \dfrac{t - 2}{t - 4} = \dfrac{3t + 2 - (t - 2)}{t - 4} = \dfrac{3t + 2 - t + 2}{t - 4} = \dfrac{2t + 4}{t - 4}$

17. $\dfrac{5 - 7x}{x^2 - 3x - 10} + \dfrac{8x - 3}{x^2 - 3x - 10} = \dfrac{x + 2}{x^2 - 3x - 10}$

$$= \frac{x + 2}{(x + 2)(x - 5)}$$

$$= \frac{\cancel{x + 2}}{\cancel{(x + 2)}(x - 5)}$$

$$= \frac{1}{x - 5}$$

19. $\dfrac{a - 2}{a^2 - 25} - \dfrac{2a - 7}{a^2 - 25} = \dfrac{a - 2 - (2a - 7)}{a^2 - 25}$

$$= \frac{a - 2 - 2a + 7}{a^2 - 25}$$

$$= \frac{-a + 5}{a^2 - 25}$$

$$= \frac{-1(a - 5)}{(a + 5)(a - 5)}$$

$$= \frac{-1\cancel{(a - 5)}}{(a + 5)\cancel{(a - 5)}}$$

$$= \frac{-1}{a + 5}$$

21. $f(x) = \dfrac{2x + 1}{x^2 + 6x + 5} + \dfrac{x - 2}{x^2 + 6x + 5}$ Note that $x \neq -5, -1$

$$= \frac{3x - 1}{(x + 5)(x + 1)}, \; x \neq -5, -1$$

23. $f(x) = \dfrac{x - 4}{x^2 - 1} - \dfrac{2x + 1}{x^2 - 1}$ Note that $x \neq -1, 1$

$$= \frac{x - 4 - (2x + 1)}{(x + 1)(x - 1)}$$

$$= \frac{x - 4 - 2x - 1}{(x + 1)(x - 1)}$$

$$= \frac{-x - 5}{(x + 1)(x - 1)}, \; x \neq -1, 1$$

25. $8x^2 = 2^3 x^2$, $12x^5 = 2^2 \cdot 3x^5$

$\text{LCM} = 2^3 \cdot 3x^5 = 24x^5$

27. $x^2 - 9 = (x + 3)(x - 3)$, $x^2 - 6x + 9 = (x - 3)^2$

$\text{LCM} = (x + 3)(x - 3)^2$

29. $\dfrac{2}{15x^2}+\dfrac{3}{5x}$ LCD is $15x^2$

$=\dfrac{2}{15x^2}+\dfrac{3}{5x}\cdot\dfrac{3x}{3x}$

$=\dfrac{2}{15x^2}+\dfrac{9x}{15x^2}$

$=\dfrac{9x+2}{15x^2}$

31. $\dfrac{y+1}{y-2}-\dfrac{y-1}{2y-4}=\dfrac{y+1}{y-2}-\dfrac{y-1}{2(y-2)}$ LCD is $2(y-2)$

$=\dfrac{y+1}{y-2}\cdot\dfrac{2}{2}-\dfrac{y-1}{2(y-2)}$

$=\dfrac{2(y+1)-(y-1)}{2(y-2)}$

$=\dfrac{2y+2-y+1}{2(y-2)}$

$=\dfrac{y+3}{2(y-2)}$

33. $\dfrac{4xy}{x^2-y^2}+\dfrac{x-y}{x+y}$

$=\dfrac{4xy}{(x+y)(x-y)}+\dfrac{x-y}{x+y}$ LCD is $(x+y)(x-y)$.

$=\dfrac{4xy}{(x+y)(x-y)}+\dfrac{x-y}{x+y}\cdot\dfrac{x-y}{x-y}$

$=\dfrac{4xy+x^2-2xy+y^2}{(x+y)(x-y)}$

$=\dfrac{x^2+2xy+y^2}{(x+y)(x-y)}=\dfrac{(x+y)(x+y)}{(x+y)(x-y)}$

$=\dfrac{\cancel{(x+y)}(x+y)}{\cancel{(x+y)}(x-y)}=\dfrac{x+y}{x-y}$

35. $\dfrac{8}{2x^2-7x+5}+\dfrac{3x+2}{2x^2-x-10}$

$=\dfrac{8}{(2x-5)(x-1)}+\dfrac{3x+2}{(2x-5)(x+2)}$

LCD is $(2x-5)(x-1)(x+2)$.

$=\dfrac{8}{(2x-5)(x-1)}\cdot\dfrac{x+2}{x+2}+\dfrac{3x+2}{(2x-5)(x+2)}\cdot\dfrac{x-1}{x-1}$

$=\dfrac{8x+16+3x^2-x-2}{(2x-5)(x-1)(x+2)}$

$=\dfrac{3x^2+7x+14}{(2x-5)(x-1)(x+2)}$

37. $\dfrac{5ab}{a^2-b^2}-\dfrac{a-b}{a+b}$

$=\dfrac{5ab}{(a+b)(a-b)}-\dfrac{a-b}{a+b}$ LCD is $(a+b)(a-b)$.

$=\dfrac{5ab}{(a+b)(a-b)}-\dfrac{a-b}{a+b}\cdot\dfrac{a-b}{a-b}$

$=\dfrac{5ab-\left(a^2-2ab+b^2\right)}{(a+b)(a-b)}$

$=\dfrac{5ab-a^2+2ab-b^2}{(a+b)(a-b)}$

$=\dfrac{-a^2+7ab-b^2}{(a+b)(a-b)}$

39. $\dfrac{x}{x^2+9x+20}-\dfrac{4}{x^2+7x+12}$

$=\dfrac{x}{(x+5)(x+4)}-\dfrac{4}{(x+3)(x+4)}$

[LCD is $(x+5)(x+4)(x+3)$.]

$=\dfrac{x}{(x+5)(x+4)}\cdot\dfrac{x+3}{x+3}-\dfrac{4}{(x+3)(x+4)}\cdot\dfrac{x+5}{x+5}$

$=\dfrac{x^2+3x-(4x+20)}{(x+5)(x+4)(x+3)}$

$=\dfrac{x^2+3x-4x-20}{(x+5)(x+4)(x+3)}$

$=\dfrac{x^2-x-20}{(x+5)(x+4)(x+3)}$

$=\dfrac{(x-5)(x+4)}{(x+5)(x+4)(x+3)}$

$=\dfrac{(x-5)\cancel{(x+4)}}{(x+5)\cancel{(x+4)}(x+3)}$

$=\dfrac{x-5}{(x+5)(x+3)}$

41. $\dfrac{3}{t}-\dfrac{6}{-t}=\dfrac{3}{t}-\dfrac{-1}{-1}\cdot\dfrac{6}{-t}=\dfrac{3}{t}+\dfrac{6}{t}=\dfrac{9}{t}$

43. $\dfrac{s^2}{r-s}+\dfrac{r^2}{s-r}=\dfrac{s^2}{r-s}+\dfrac{-1}{-1}\cdot\dfrac{r^2}{s-r}$

$=\dfrac{s^2}{r-s}+\dfrac{-r^2}{r-s}$

$=\dfrac{s^2-r^2}{r-s}$

$=\dfrac{(s-r)(s+r)}{-1(s-r)}$

$=\dfrac{\cancel{(s-r)}(s+r)}{-1\cancel{(s-r)}}$

$=-(s+r)$

45. $\dfrac{a+2}{a-4}+\dfrac{a-2}{a+3}$ LCD is $(a-4)(a+3)$.

$=\dfrac{a+2}{a-4}\cdot\dfrac{a+3}{a+3}+\dfrac{a-2}{a+3}\cdot\dfrac{a-4}{a-4}$

$=\dfrac{(a^2+5a+6)+(a^2-6a+8)}{(a-4)(a+3)}$

$=\dfrac{2a^2-a+14}{(a-4)(a+3)}$

47. $4+\dfrac{x-3}{x+1}=\dfrac{4}{1}+\dfrac{x-3}{x+1}$ [LCD is $x+1$.]

$=\dfrac{4}{1}\cdot\dfrac{x+1}{x+1}+\dfrac{x-3}{x+1}$

$=\dfrac{(4x+4)+(x-3)}{x+1}$

$=\dfrac{5x+1}{x+1}$

49. $\dfrac{x+6}{5x+10}-\dfrac{x-2}{4x+8}$

$=\dfrac{x+6}{5(x+2)}-\dfrac{x-2}{4(x+2)}$ [LCD is $5\cdot4(x+2)$.]

$=\dfrac{x+6}{5(x+2)}\cdot\dfrac{4}{4}-\dfrac{x-2}{4(x+2)}\cdot\dfrac{5}{5}$

$=\dfrac{4(x+6)-5(x-2)}{5\cdot4(x+2)}$

$$= \frac{4x+24-5x+10}{5 \cdot 4(x+2)}$$

$$= \frac{-x+34}{5 \cdot 4(x+2)}, \text{ or } \frac{-x+34}{20(x+2)}$$

51. $\dfrac{4}{x+1} + \dfrac{x+2}{x^2-1} + \dfrac{3}{x-1}$

$$= \frac{4}{x+1} + \frac{x+2}{(x+1)(x-1)} + \frac{3}{x-1} \qquad [\text{LCD is } (x+1)(x-1).]$$

$$= \frac{4}{x+1} \cdot \frac{x-1}{x-1} + \frac{x+2}{(x+1)(x-1)} + \frac{3}{x-1} \cdot \frac{x+1}{x+1}$$

$$= \frac{4x-4+x+2+3x+3}{(x+1)(x-1)}$$

$$= \frac{8x+1}{(x+1)(x-1)}$$

53. $\dfrac{y-4}{y^2-25} - \dfrac{9-2y}{25-y^2} = \dfrac{y-4}{y^2-25} - \dfrac{-1}{-1} \cdot \dfrac{9-2y}{25-y^2}$

$$= \frac{y-4}{y^2-25} - \frac{2y-9}{y^2-25}$$

$$= \frac{y-4-(2y-9)}{y^2-25}$$

$$= \frac{y-4-2y+9}{y^2-25}$$

$$= \frac{-y+5}{y^2-25}$$

$$= \frac{-1(y-5)}{(y+5)(y-5)}$$

$$= \frac{-1\cancel{(y-5)}}{(y+5)\cancel{(y-5)}}$$

$$= \frac{-1}{y+5}, \text{ or } -\frac{1}{y+5}$$

55. $\dfrac{y^2-5}{y^4-81} + \dfrac{4}{81-y^4} = \dfrac{y^2-5}{y^4-81} + \dfrac{-1}{-1} \cdot \dfrac{4}{81-y^4}$

$$= \frac{y^2-5}{y^4-81} + \frac{-4}{y^4-81}$$

$$= \frac{y^2-5+(-4)}{y^4-81}$$

$$= \frac{y^2-9}{y^4-81} = \frac{y^2-9}{(y^2+9)(y^2-9)}$$

$$= \frac{y^2-9}{y^2-9} \cdot \frac{1}{y^2+9}$$

$$= \frac{1}{y^2+9}$$

57. $\dfrac{r-6s}{r^3-s^3} - \dfrac{5s}{s^3-r^3} = \dfrac{r-6s}{r^3-s^3} - \dfrac{-1}{-1} \cdot \dfrac{5s}{s^3-r^3}$

$$= \frac{r-6s}{r^3-s^3} - \frac{-5s}{r^3-s^3}$$

$$= \frac{r-6s-(-5s)}{r^3-s^3}$$

$$= \frac{r-s}{(r-s)(r^2+rs+s^2)}$$

$$= \frac{\cancel{r-s}}{\cancel{(r-s)}\,(r^2+rs+s^2)}$$

$$= \frac{1}{r^2+rs+s^2}$$

59. $\dfrac{3y}{y^2-7y+10} - \dfrac{2y}{y^2-8y+15}$

$$= \frac{3y}{(y-5)(y-2)} - \frac{2y}{(y-5)(y-3)}$$
$$[\text{LCD is } (y-5)(y-2)(y-3).]$$

$$= \frac{3y}{(y-5)(y-2)} \cdot \frac{y-3}{y-3} - \frac{2y}{(y-5)(y-3)} \cdot \frac{y-2}{y-2}$$

$$= \frac{3y^2-9y-(2y^2-4y)}{(y-5)(y-2)(y-3)}$$

$$= \frac{3y^2-9y-2y^2+4y}{(y-5)(y-2)(y-3)}$$

$$= \frac{y^2-5y}{(y-5)(y-2)(y-3)} = \frac{y(y-5)}{(y-5)(y-2)(y-3)}$$

$$= \frac{y\cancel{(y-5)}}{\cancel{(y-5)}(y-2)(y-3)} = \frac{y}{(y-2)(y-3)}$$

61. $\dfrac{2x+1}{x-y} + \dfrac{5x^2-5xy}{x^2-2xy+y^2} = \dfrac{2x+1}{x-y} + \dfrac{5x(x-y)}{(x-y)(x-y)}$

$$= \frac{2x+1}{x-y} + \frac{5x\cancel{(x-y)}}{\cancel{(x-y)}(x-y)}$$

$$= \frac{2x+1}{x-y} + \frac{5x}{x-y}$$

$$= \frac{7x+1}{x-y}$$

63. $\dfrac{2y-6}{y^2-9} - \dfrac{y}{y-1} + \dfrac{y^2+2}{y^2+2y-3}$

$$= \frac{2\cancel{(y-3)}}{(y+3)\cancel{(y-3)}} - \frac{y}{y-1} + \frac{y^2+2}{(y+3)(y-1)}$$

$$= \frac{2}{y+3} - \frac{y}{y-1} + \frac{y^2+2}{(y+3)(y-1)}$$
$$[\text{LCD is } (y+3)(y-1).]$$

$$= \frac{2}{y+3} \cdot \frac{y-1}{y-1} - \frac{y}{y-1} \cdot \frac{y+3}{y+3} + \frac{y^2+2}{(y+3)(y-1)}$$

$$= \frac{2y-2-y^2-3y+y^2+2}{(y+3)(y-1)}$$

$$= \frac{-y}{(y+3)(y-1)}, \text{ or } -\frac{y}{(y+3)(y-1)}$$

65. $\dfrac{5y}{1-4y^2} - \dfrac{2y}{2y+1} + \dfrac{5y}{4y^2-1}$

Observe that $\dfrac{5y}{1-4y^2}$ and $\dfrac{5y}{4y^2-1}$ are opposites, so

their sum is 0. Then the result is the remaining

expression, $-\dfrac{2y}{2y+1}$.

67. $f(x) = 2 + \dfrac{x}{x-3} - \dfrac{18}{x^2-9}$

$= \dfrac{2}{1} + \dfrac{x}{x-3} - \dfrac{18}{(x+3)(x-3)}$ Note that $x \neq -3,\ 3$.

LCD is $(x+3)(x-3)$.

$= \dfrac{2}{1} \cdot \dfrac{(x+3)(x-3)}{(x+3)(x-3)} + \dfrac{x}{x-3} \cdot \dfrac{x+3}{x+3} - \dfrac{18}{(x+3)(x-3)}$

$= \dfrac{2(x+3)(x-3) + x(x+3) - 18}{(x+3)(x-3)}$

$= \dfrac{2x^2 - 18 + x^2 + 3x - 18}{(x+3)(x-3)}$

$= \dfrac{3x^2 + 3x - 36}{(x+3)(x-3)}$

$= \dfrac{3(x+4)(x-3)}{(x+3)(x-3)}$

$= \dfrac{3(x+4)}{x+3},\ x \neq -3,\ 3$

69. $f(x) = \dfrac{3x-1}{x^2+2x-3} - \dfrac{x+4}{x^2-16}$

$= \dfrac{3x-1}{(x+3)(x-1)} - \dfrac{x+4}{(x+4)(x-4)}$

Note that $x \neq -3,\ 1, -4,\ 4$.

$= \dfrac{3x-1}{(x+3)(x-1)} - \dfrac{1}{x-4}$ Removing a factor

equal to 1; LCD is

$(x+3)(x-1)(x-4)$.

$= \dfrac{3x-1}{(x+3)(x-1)} \cdot \dfrac{x-4}{x-4} - \dfrac{1}{x-4} \cdot \dfrac{(x+3)(x-1)}{(x+3)(x-1)}$

$= \dfrac{(3x-1)(x-4) - (x+3)(x-1)}{(x+3)(x-1)(x-4)}$

$= \dfrac{3x^2 - 13x + 4 - (x^2 + 2x - 3)}{(x+3)(x-1)(x-4)}$

$= \dfrac{3x^2 - 13x + 4 - x^2 - 2x + 3}{(x+3)(x-1)(x-4)}$

$= \dfrac{2x^2 - 15x + 7}{(x+3)(x-1)(x-4)}$

$= \dfrac{(2x-1)(x-7)}{(x+3)(x-1)(x-4)},\ x \neq -4, -3,\ 1,\ 4$

71. $f(x) = \dfrac{1}{x^2+5x+6} - \dfrac{2}{x^2+3x+2} - \dfrac{1}{x^2+5x+6}$

$= \dfrac{1}{(x+3)(x+2)} - \dfrac{2}{(x+2)(x+1)} - \dfrac{1}{(x+3)(x+2)}$

Note that $x \neq -3,\ -2,\ -1$.

Observe that the sum of the first and third terms is 0, so the result is the remaining term.

$f(x) = -\dfrac{2}{(x+2)(x+1)},\ \text{or}\ \dfrac{-2}{(x+2)(x+1)},$

$x \neq -3,\ -2, -1$

73. *Writing Exercise.*

75. $x^2 = 6 + x$

$x^2 - x - 6 = 0$

$(x+2)(x-3) = 0$

$x + 2 = 0 \quad or \quad x - 3 = 0$

$x = -2 \quad or \quad\quad x = 3$

77. $-5 < 2x + 1 < 0$

$-6 < 2x < -1$

$-3 < x < -\dfrac{1}{2}$

$\left\{ x \middle| -3 < x < -\dfrac{1}{2} \right\},\ \text{or}\ \left(-3, -\dfrac{1}{2} \right)$

79. $y = x + 2,\quad$ (1)

$2x = y - 4\quad$ (2)

Use substitution. Substitute $x + 2$ for y in Equation (2).

$2x = y - 4\quad$ (2)

$2x = x + 2 - 4$

$2x = x - 2$

$x = -2$

Solve for y. Use Equation (1).

$y = -2 + 2 = 0$

The solution set is $(-2, 0)$.

81. *Writing Exercise.*

83. We find the least common multiple of 14 (2 weeks = 14 days), 20, and 30.

$14 = 2 \cdot 7$

$20 = 2 \cdot 2 \cdot 5$

$30 = 2 \cdot 3 \cdot 5$

$\text{LCM} = 2 \cdot 2 \cdot 3 \cdot 5 \cdot 7 = 420$

It will be 420 days until Jinney can refill all three prescriptions on the same day.

85. The smallest number of parts possible is the least common multiple of 6 and 4.

$6 = 2 \cdot 3$

$4 = 2 \cdot 2$

$\text{LCM} = 2 \cdot 3 \cdot 2,\ \text{or}\ 12$

A measure should be divided into 12 parts.

87. $x^8 - x^4 = x^4(x^2+1)(x+1)(x-1)$

$x^5 - x^2 = x^2(x-1)(x^2+x+1)$

$x^5 - x^3 = x^3(x+1)(x-1)$

$x^5 + x^2 = x^2(x+1)(x^2-x+1)$

The LCM is

$x^4(x^2+1)(x+1)(x-1)(x^2+x+1)(x^2-x+1)$.

89. The LCM is $8a^4b^7$.

One expression is $2a^3b^7$.

Then the other expression must contain 8, a^4, and one of the following:

no factor of b, b, b^2, b^3, b^4, b^5, b^6, or b^7.

Thus, all the possibilities for the other expression are

$8a^4$, $8a^4b$, $8a^4b^2$, $8a^4b^3$, $8a^4b^4$, $8a^4b^5$, $8a^4b^6$,

$8a^4b^7$.

91. $(f+g)(x) = \dfrac{x^3}{x^2-4} + \dfrac{x^2}{x^2+3x-10}$

$= \dfrac{x^3}{(x+2)(x-2)} + \dfrac{x^2}{(x+5)(x-2)}$

$= \dfrac{x^3(x+5) + x^2(x+2)}{(x+2)(x-2)(x+5)}$

$$= \frac{x^4 + 5x^3 + x^3 + 2x^2}{(x+2)(x-2)(x+5)}$$

$$= \frac{x^4 + 6x^3 + 2x^2}{(x+2)(x-2)(x+5)}$$

93. $(f \cdot g)(x) = \dfrac{x^3}{x^2-4} \cdot \dfrac{x^2}{x^2+3x-10}$

$$= \frac{x^5}{(x^2-4)(x^2+3x-10)}$$

95. $x^{-2} + 2x^{-1} = \dfrac{1}{x^2} + \dfrac{2}{x}$ LCD is x^2

$$= \frac{1}{x^2} + \frac{2}{x} \cdot \frac{x}{x}$$

$$= \frac{1}{x^2} + \frac{2x}{x^2}$$

$$= \frac{2x+1}{x^2}$$

97. $5(x-3)^{-1} + 4(x+3)^{-1} - 2(x+3)^{-2}$

$$= \frac{5}{x-3} + \frac{4}{x+3} - \frac{2}{(x+3)^2}$$

[LCD is $(x-3)(x+3)^2$.]

$$= \frac{5(x+3)^2 + 4(x-3)(x+3) - 2(x-3)}{(x-3)(x+3)^2}$$

$$= \frac{5x^2 + 30x + 45 + 4x^2 - 36 - 2x + 6}{(x-3)(x+3)^2}$$

$$= \frac{9x^2 + 28x + 15}{(x-3)(x+3)^2}$$

99. $\dfrac{x+4}{6x^2-20x}\left(\dfrac{x}{x^2-x-20} + \dfrac{2}{x+4}\right)$

$$= \frac{x+4}{2x(3x-10)}\left(\frac{x}{(x-5)(x+4)} + \frac{2}{x+4}\right)$$

$$= \frac{x+4}{2x(3x-10)}\left(\frac{x+2(x-5)}{(x-5)(x+4)}\right)$$

$$= \frac{x+4}{2x(3x-10)}\left(\frac{x+2x-10}{(x-5)(x+4)}\right)$$

$$= \frac{(x+4)(3x-10)}{2x(3x-10)(x-5)(x+4)}$$

$$= \frac{(x+4)(3x-10)(1)}{2x(3x-10)(x-5)(x+4)}$$

$$= \frac{1}{2x(x-5)}$$

101. $\dfrac{8t^5}{2t^2-10t+12} \div \left(\dfrac{2t}{t^2-8t+15} - \dfrac{3t}{t^2-7t+10}\right)$

$$= \frac{8t^5}{2t^2-10t+12} \div \left(\frac{2t}{(t-5)(t-3)} - \frac{3t}{(t-5)(t-2)}\right)$$

$$= \frac{8t^5}{2t^2-10t+12} \div \left(\frac{2t(t-2) - 3t(t-3)}{(t-5)(t-3)(t-2)}\right)$$

$$= \frac{8t^5}{2t^2-10t+12} \div \left(\frac{2t^2-4t-3t^2+9t}{(t-5)(t-3)(t-2)}\right)$$

$$= \frac{8t^5}{2t^2-10t+12} \div \frac{-t^2+5t}{(t-5)(t-3)(t-2)}$$

$$= \frac{8t^5}{2(t-3)(t-2)} \cdot \frac{(t-5)(t-3)(t-2)}{-t(t-5)}$$

$$= \frac{2 \cdot 4 \cdot t \cdot t^4 \, (t-5) \, (t-3) \, (t-2)}{2 \, (t-3) \, (t-2)(-1)(t)(t-5)}$$

$$= -4t^4$$

103. *Graphing Calculator Exercise*

Exercise Set 6.3

1. (b)

3. (f)

5. (d)

7. $\dfrac{\frac{1}{2}+\frac{1}{3}}{\frac{1}{4}-\frac{1}{6}} = \dfrac{\frac{1}{2}+\frac{1}{3}}{\frac{1}{4}-\frac{1}{6}} \cdot \dfrac{12}{12}$ Multiplying by 1 using the LCD

$$= \frac{\frac{1}{2}\cdot 12 + \frac{1}{3}\cdot 12}{\frac{1}{4}\cdot 12 - \frac{1}{6}\cdot 12}$$

$$= \frac{6+4}{3-2}$$

$$= \frac{10}{1} = 10$$

9. $\dfrac{1+\frac{1}{4}}{2+\frac{3}{4}} = \dfrac{1+\frac{1}{4}}{2+\frac{3}{4}} \cdot \dfrac{4}{4}$ Multiplying by 1 using the LCD

$$= \frac{1\cdot 4 + \frac{1}{4}\cdot 4}{2\cdot 4 + \frac{3}{4}\cdot 4}$$

$$= \frac{4+1}{8+3}$$

$$= \frac{5}{11}$$

11. $\dfrac{\frac{x}{4}+x}{\frac{4}{x}+x} = \dfrac{\frac{x}{4}+x}{\frac{4}{x}+x} \cdot \dfrac{4x}{4x}$ Multiplying by 1 using the LCD

$$= \frac{\frac{x}{4}\cdot 4x + x\cdot 4x}{\frac{4}{x}\cdot 4x + x\cdot 4x}$$

$$= \frac{x^2 + 4x^2}{16 + 4x^2}$$

$$= \frac{5x^2}{4(x^2+4)}$$

13. $\dfrac{\frac{x+5}{x-3}}{\frac{x-2}{x+1}} = \dfrac{x+5}{x-3} \div \dfrac{x-2}{x+1}$

$$= \frac{x+5}{x-3} \cdot \frac{x+1}{x-2}$$

$$= \frac{(x+1)(x+5)}{(x-3)(x-2)}$$

15. $\dfrac{\dfrac{3}{x}+\dfrac{2}{x^3}}{\dfrac{5}{x}-\dfrac{3}{x^2}}=\dfrac{\dfrac{3}{x}+\dfrac{2}{x^3}}{\dfrac{5}{x}-\dfrac{3}{x^2}}\cdot\dfrac{x^3}{x^3}$ Multiplying by 1 using the LCD

$=\dfrac{\dfrac{3}{x}\cdot x^3+\dfrac{2}{x^3}\cdot x^3}{\dfrac{5}{x}\cdot x^3-\dfrac{3}{x^2}\cdot x^3}$ Multiplying the numerators and the denominators

$=\dfrac{3x^2+2}{5x^2-3x}$

$=\dfrac{3x^2+2}{x(5x-3)}$

17. $\dfrac{\dfrac{6}{r}-\dfrac{1}{s}}{\dfrac{2}{r}+\dfrac{3}{s}}=\dfrac{\dfrac{6}{r}-\dfrac{1}{s}}{\dfrac{2}{r}+\dfrac{3}{s}}\cdot\dfrac{rs}{rs}$ Multiplying by 1 using the LCD

$=\dfrac{\dfrac{6}{r}\cdot rs-\dfrac{1}{s}\cdot rs}{\dfrac{2}{r}\cdot rs+\dfrac{3}{s}\cdot rs}$

$=\dfrac{6s-r}{2s+3r}$

19. $\dfrac{\dfrac{3}{z^2}+\dfrac{2}{yz}}{\dfrac{4}{zy^2}-\dfrac{1}{y}}=\dfrac{\dfrac{3}{z^2}+\dfrac{2}{yz}}{\dfrac{4}{zy^2}-\dfrac{1}{y}}\cdot\dfrac{y^2z^2}{y^2z^2}$ Multiplying by 1 using the LCD

$=\dfrac{\dfrac{3}{z^2}\cdot y^2z^2+\dfrac{2}{yz}\cdot y^2z^2}{\dfrac{4}{zy^2}\cdot y^2z^2-\dfrac{1}{y}\cdot y^2z^2}$

$=\dfrac{3y^2+2yz}{4z-yz^2}$, or $\dfrac{y(3y+2z)}{z(4-yz)}$

21. $\dfrac{\dfrac{a^2-b^2}{ab}}{\dfrac{a-b}{b}}=\dfrac{a^2-b^2}{ab}\div\dfrac{a-b}{b}$

$=\dfrac{a^2-b^2}{ab}\cdot\dfrac{b}{a-b}$

$=\dfrac{(a+b)(a-b)(b)}{a\cdot b\cdot(a-b)}$

$=\dfrac{(a+b)\,\cancel{(a-b)}\,\cancel{(b)}}{a\cdot\cancel{b}\cdot\cancel{(a-b)}}$

$=\dfrac{a+b}{a}$

23. $\dfrac{1-\dfrac{2}{3x}}{x-\dfrac{4}{9x}}=\dfrac{1-\dfrac{2}{3x}}{x-\dfrac{4}{9x}}\cdot\dfrac{9x}{9x}$ Multiplying by 1, using the LCD

$=\dfrac{1\cdot 9x-\dfrac{2}{3x}\cdot 9x}{x\cdot 9x-\dfrac{4}{9x}\cdot 9x}$

$=\dfrac{9x-6}{9x^2-4}$

$=\dfrac{3(3x-2)}{(3x+2)(3x-2)}$

$=\dfrac{3\,(3x\!-\!\cancel{2})}{(3x+2)\,(3x\!-\!\cancel{2})}$

$=\dfrac{3}{3x+2}$

25. $\dfrac{y^{-1}-x^{-1}}{\dfrac{x^2-y^2}{xy}}=\dfrac{\dfrac{1}{y}-\dfrac{1}{x}}{\dfrac{x^2-y^2}{xy}}$ Rewriting with positive exponents

$=\dfrac{\dfrac{1}{y}-\dfrac{1}{x}}{\dfrac{x^2-y^2}{xy}}\cdot\dfrac{xy}{xy}$ Multiplying by 1, using the LCD

$=\dfrac{\dfrac{1}{y}\cdot xy-\dfrac{1}{x}\cdot xy}{\dfrac{x^2-y^2}{xy}\cdot xy}$

$=\dfrac{x-y}{x^2-y^2}=\dfrac{x-y}{(x+y)(x-y)}$

$=\dfrac{\cancel{x-y}}{(x+y)\,\cancel{(x-y)}}$

$=\dfrac{1}{x+y}$

27. $\dfrac{\dfrac{1}{x+h}-\dfrac{1}{x}}{h}=\dfrac{\dfrac{1}{x+h}\cdot\dfrac{x}{x}-\dfrac{1}{x}\cdot\dfrac{x+h}{x+h}}{h}$ Adding in the numerator

$=\dfrac{\dfrac{x-x-h}{x(x+h)}}{h}=\dfrac{\dfrac{-h}{x(x+h)}}{h}$

$=\dfrac{-h}{x(x+h)}\cdot\dfrac{1}{h}$ Multiplying by the reciprocal of the divisor

$=\dfrac{-1\cdot\cancel{h}\cdot 1}{x(x+h)(\cancel{h})}$ $(-h=-1\cdot h)$

$=-\dfrac{1}{x(x+h)}$

29. $\dfrac{\dfrac{a^2-4}{a^2+3a+2}}{\dfrac{a^2-5a-6}{a^2-6a-7}}$

$=\dfrac{a^2-4}{a^2+3a+2}\cdot\dfrac{a^2-6a-7}{a^2-5a-6}$ Multiplying by the reciprocal of the divisor

$=\dfrac{(a+2)(a-2)}{(a+2)(a+1)}\cdot\dfrac{(a+1)(a-7)}{(a+1)(a-6)}$

$=\dfrac{(a+2)(a-2)(a+1)(a-7)}{(a+2)(a+1)(a+1)(a-6)}$

$=\dfrac{\cancel{(a+2)}\,(a-2)\,\cancel{(a+1)}\,(a-7)}{\cancel{(a+2)}\,\cancel{(a+1)}\,(a+1)(a-6)}$

$=\dfrac{(a-2)(a-7)}{(a+1)(a-6)}$

31. $\dfrac{\dfrac{x}{x^2+3x-4}-\dfrac{1}{x^2+3x-4}}{\dfrac{x}{x^2+6x+8}+\dfrac{3}{x^2+6x+8}}$

$=\dfrac{\dfrac{x-1}{x^2+3x-4}}{\dfrac{x+3}{x^2+6x+8}}$ Adding in the numerator and the denominator

$=\dfrac{x-1}{x^2+3x-4}\cdot\dfrac{x^2+6x+8}{x+3}$

$=\dfrac{(x-1)(x+4)(x+2)}{(x+4)(x-1)(x+3)}$

$=\dfrac{\cancel{(x-1)}\,\cancel{(x+4)}\,(x+2)}{\cancel{(x+4)}\,\cancel{(x-1)}\,(x+3)}=\dfrac{x+2}{x+3}$

33. $\dfrac{\dfrac{1}{y}+2}{\dfrac{1}{y}-3}=\dfrac{\dfrac{1}{y}+2}{\dfrac{1}{y}-3}\cdot\dfrac{y}{y}$ Multiplying by 1, using the LCD

$=\dfrac{\left(\dfrac{1}{y}+2\right)y}{\left(\dfrac{1}{y}-3\right)y}$

$=\dfrac{\dfrac{1}{y}\cdot y+2\cdot y}{\dfrac{1}{y}\cdot y-3\cdot y}$

$=\dfrac{1+2y}{1-3y}$

35. $\dfrac{y+y^{-2}}{y-y^{-2}}=\dfrac{y+\dfrac{1}{y^2}}{y-\dfrac{1}{y^2}}$ Rewriting with positive exponents

$=\dfrac{y+\dfrac{1}{y^2}}{y-\dfrac{1}{y^2}}\cdot\dfrac{y^2}{y^2}$ Multiplying by 1, using the LCD

$=\dfrac{y\cdot y^2+\dfrac{1}{y^2}\cdot y^2}{y\cdot y^2-\dfrac{1}{y^2}\cdot y^2}$

$=\dfrac{y^3+1}{y^3-1}$

$=\dfrac{(y+1)(y^2-y+1)}{(y-1)(y^2+y+1)}$

37. $\dfrac{\dfrac{3}{ab^4}+\dfrac{4}{a^3b}}{ab}=\dfrac{\dfrac{3}{ab^4}+\dfrac{4}{a^3b}}{ab}\cdot\dfrac{a^3b^4}{a^3b^4}$ Multiplying by 1 using the LCD

$=\dfrac{\dfrac{3}{ab^4}\cdot a^3b^4+\dfrac{4}{a^3b}\cdot a^3b^4}{ab\cdot a^3b^4}$

$=\dfrac{3a^2+4b^3}{a^4b^5}$

39. $\dfrac{x-y}{x^{-3}-y^{-3}}=\dfrac{x-y}{\dfrac{1}{x^3}-\dfrac{1}{y^3}}$ Rewriting with positive exponents

$=\dfrac{x-y}{\dfrac{1}{x^3}-\dfrac{1}{y^3}}\cdot\dfrac{x^3y^3}{x^3y^3}$ Multiplying by 1, using the LCD

$=\dfrac{x\cdot x^3y^3-y\cdot x^3y^3}{\dfrac{1}{x^3}\cdot x^3y^3-\dfrac{1}{y^3}\cdot x^3y^3}$

$=\dfrac{x^4y^3-x^3y^4}{y^3-x^3}$

$=\dfrac{x^3y^3(x-y)}{(y-x)(y^2+xy+x^2)}$

$=\dfrac{-x^3y^3}{y^2+xy+x^2}$

41. $\dfrac{\dfrac{1}{x-2}+\dfrac{3}{x-1}}{\dfrac{2}{x-1}+\dfrac{5}{x-2}}$

$=\dfrac{\dfrac{1}{x-2}+\dfrac{3}{x-1}}{\dfrac{2}{x-1}+\dfrac{5}{x-2}}\cdot\dfrac{(x-2)(x-1)}{(x-2)(x-1)}$ Multiplying by 1, using the LCD

$=\dfrac{\dfrac{1}{x-2}\cdot(x-2)(x-1)+\dfrac{3}{x-1}\cdot(x-2)(x-1)}{\dfrac{2}{x-1}\cdot(x-2)(x-1)+\dfrac{5}{x-2}\cdot(x-2)(x-1)}$

$=\dfrac{x-1+3(x-2)}{2(x-2)+5(x-1)}$

$=\dfrac{x-1+3x-6}{2x-4+5x-5}$

$=\dfrac{4x-7}{7x-9}$

43. $\dfrac{a(a+3)^{-1}-2(a-1)^{-1}}{a(a+3)^{-1}-(a-1)^{-1}}$

$=\dfrac{\dfrac{a}{a+3}-\dfrac{2}{a-1}}{\dfrac{a}{a+3}-\dfrac{1}{a-1}}$

$=\dfrac{\dfrac{a}{a+3}-\dfrac{2}{a-1}}{\dfrac{a}{a+3}-\dfrac{1}{a-1}}\cdot\dfrac{(a+3)(a-1)}{(a+3)(a-1)}$ Multiplying by 1, using the LCD

$=\dfrac{\dfrac{a}{a+3}\cdot(a+3)(a-1)-\dfrac{2}{a-1}\cdot(a+3)(a-1)}{\dfrac{a}{a+3}\cdot(a+3)(a-1)-\dfrac{1}{a-1}\cdot(a+3)(a-1)}$

$=\dfrac{a(a-1)-2(a+3)}{a(a-1)-(a+3)}$

$=\dfrac{a^2-a-2a-6}{a^2-a-a-3}=\dfrac{a^2-3a-6}{a^2-2a-3}$

(Although the denominator can be factored, doing so does not lead to further simplification.)

45. $\dfrac{\dfrac{2}{a^2-1}+\dfrac{1}{a+1}}{\dfrac{3}{a^2-1}+\dfrac{2}{a-1}}$

$=\dfrac{\dfrac{2}{(a+1)(a-1)}+\dfrac{1}{a+1}}{\dfrac{3}{(a+1)(a-1)}+\dfrac{2}{a-1}}$

$=\dfrac{\dfrac{2}{(a+1)(a-1)}+\dfrac{1}{a+1}}{\dfrac{3}{(a+1)(a-1)}+\dfrac{2}{a-1}}\cdot\dfrac{(a+1)(a-1)}{(a+1)(a-1)}$ Multiplying by 1, using the LCD

$=\dfrac{\dfrac{2}{(a+1)(a-1)}\cdot(a+1)(a-1)+\dfrac{1}{a+1}\cdot(a+1)(a-1)}{\dfrac{3}{(a+1)(a-1)}\cdot(a+1)(a-1)+\dfrac{2}{a-1}\cdot(a+1)(a-1)}$

$=\dfrac{2+a-1}{3+2(a+1)}=\dfrac{a+1}{3+2a+2}=\dfrac{a+1}{2a+5}$

47. $\dfrac{\dfrac{5}{x^2-4}-\dfrac{3}{x-2}}{\dfrac{4}{x^2-4}-\dfrac{2}{x+2}}$

$=\dfrac{\dfrac{5}{(x+2)(x-2)}-\dfrac{3}{x-2}}{\dfrac{4}{(x+2)(x-2)}-\dfrac{2}{x+2}}$

$=\dfrac{\dfrac{5}{(x+2)(x-2)}-\dfrac{3}{x-2}}{\dfrac{4}{(x+2)(x-2)}-\dfrac{2}{x+2}}\cdot\dfrac{(x+2)(x-2)}{(x+2)(x-2)}$ Multiplying by 1, using the LCD

$=\dfrac{\dfrac{5}{(x+2)(x-2)}\cdot(x+2)(x-2)-\dfrac{3}{x-2}\cdot(x+2)(x-2)}{\dfrac{4}{(x+2)(x-2)}\cdot(x+2)(x-2)-\dfrac{2}{x+2}\cdot(x+2)(x-2)}$

$=\dfrac{5-3(x+2)}{4-2(x-2)}=\dfrac{5-3x-6}{4-2x+4}=\dfrac{-1-3x}{8-2x}$, or $\dfrac{3x+1}{2x-8}$

49. $\dfrac{\dfrac{y^3}{y^2-4}+\dfrac{125}{4-y^2}}{\dfrac{y}{y^2-4}+\dfrac{5}{4-y^2}}$

$=\dfrac{\dfrac{y^3}{y^2-4}+\dfrac{-1}{-1}\cdot\dfrac{125}{4-y^2}}{\dfrac{y}{y^2-4}+\dfrac{-1}{-1}\cdot\dfrac{5}{4-y^2}}$

$=\dfrac{\dfrac{y^3}{y^2-4}-\dfrac{125}{y^2-4}}{\dfrac{y}{y^2-4}-\dfrac{5}{y^2-4}}$

$=\dfrac{\dfrac{y^3-125}{y^2-4}}{\dfrac{y-5}{y^2-4}}$ Adding the numerator and the denominator

$=\dfrac{y^3-125}{y^2-4}\cdot\dfrac{y^2-4}{y-5}$ Multiplying by the reciprocal of the divisor

$=\dfrac{(y-5)(y^2+5y+25)(y^2-4)}{(y^2-4)(y-5)}$

$=y^2+5y+25$

51. $\dfrac{\dfrac{y^2}{y^2-25}-\dfrac{y}{y-5}}{\dfrac{y}{y^2-25}-\dfrac{1}{y+5}}$

$=\dfrac{\dfrac{y^2}{(y+5)(y-5)}-\dfrac{y}{y-5}}{\dfrac{y}{(y+5)(y-5)}-\dfrac{1}{y+5}}\cdot\dfrac{(y+5)(y-5)}{(y+5)(y-5)}$ Multiplying by 1, using the LCD

$=\dfrac{\dfrac{y^2}{(y+5)(y-5)}\cdot(y+5)(y-5)-\dfrac{y}{y-5}\cdot(y+5)(y-5)}{\dfrac{y}{(y+5)(y-5)}\cdot(y+5)(y-5)-\dfrac{1}{y+5}\cdot(y+5)(y-5)}$

$=\dfrac{y^2-y(y+5)}{y-(y-5)}$

$=\dfrac{y^2-y^2-5y}{y-y+5}=\dfrac{-5y}{5}=-y$

53. $\dfrac{\dfrac{a}{a+2}+\dfrac{5}{a}}{\dfrac{a}{2a+4}+\dfrac{1}{3a}}$

$=\dfrac{\dfrac{a}{a+2}+\dfrac{5}{a}}{\dfrac{a}{2(a+2)}+\dfrac{1}{3a}}$

$=\dfrac{\dfrac{a}{a+2}+\dfrac{5}{a}}{\dfrac{a}{2(a+2)}+\dfrac{1}{3a}}\cdot\dfrac{6a(a+2)}{6a(a+2)}$ Multiplying by 1, using the LCD

$=\dfrac{\dfrac{a}{a+2}\cdot6a(a+2)+\dfrac{5}{a}\cdot6a(a+2)}{\dfrac{a}{2(a+2)}\cdot6a(a+2)+\dfrac{1}{3a}\cdot6a(a+2)}$

$=\dfrac{6a^2+30(a+2)}{3a^2+2(a+2)}$

$=\dfrac{6a^2+30a+60}{3a^2+2a+4}=\dfrac{6(a^2+5a+10)}{3a^2+2a+4}$

55. $\dfrac{\dfrac{1}{x^2-3x+2}+\dfrac{1}{x^2-4}}{\dfrac{1}{x^2+4x+4}+\dfrac{1}{x^2-4}}$

$=\dfrac{\dfrac{1}{(x-1)(x-2)}+\dfrac{1}{(x+2)(x-2)}}{\dfrac{1}{(x+2)(x+2)}+\dfrac{1}{(x+2)(x-2)}}$

$=\dfrac{\dfrac{1}{(x-1)(x-2)}+\dfrac{1}{(x+2)(x-2)}}{\dfrac{1}{(x+2)(x+2)}+\dfrac{1}{(x+2)(x-2)}}$

$\cdot\dfrac{(x-1)(x-2)(x+2)(x+2)}{(x-1)(x-2)(x+2)(x+2)}$ Multiplying by 1, using the LCD

$=\dfrac{(x+2)(x+2)+(x-1)(x+2)}{(x-1)(x-2)+(x-1)(x+2)}$

$$= \frac{x^2 + 4x + 4 + x^2 + x - 2}{x^2 - 3x + 2 + x^2 + x - 2}$$

$$= \frac{2x^2 + 5x + 2}{2x^2 - 2x}$$

$$= \frac{(2x + 1)(x + 2)}{2x(x - 1)}$$

57. $$\frac{\dfrac{3}{a^2 - 4a + 3} + \dfrac{3}{a^2 - 5a + 6}}{\dfrac{3}{a^2 - 3a + 2} + \dfrac{3}{a^2 + 3a - 10}}$$

$$= \frac{\dfrac{3}{(a-1)(a-3)} + \dfrac{3}{(a-2)(a-3)}}{\dfrac{3}{(a-1)(a-2)} + \dfrac{3}{(a+5)(a-2)}}$$

$$= \frac{\dfrac{3}{(a-1)(a-3)} + \dfrac{3}{(a-2)(a-3)}}{\dfrac{3}{(a-1)(a-2)} + \dfrac{3}{(a+5)(a-2)}}$$

$$\cdot \frac{(a-1)(a-3)(a-2)(a+5)}{(a-1)(a-3)(a-2)(a+5)} \qquad \text{Multiplying by 1, using the LCD}$$

$$= \frac{3(a-2)(a+5) + 3(a-1)(a+5)}{3(a-3)(a+5) + 3(a-1)(a-3)}$$

$$= \frac{3[(a-2)(a+5) + (a-1)(a+5)]}{3[(a-3)(a+5) + (a-1)(a-3)]}$$

$$= \frac{a^2 + 3a - 10 + a^2 + 4a - 5}{a^2 + 2a - 15 + a^2 - 4a + 3}$$

$$= \frac{2a^2 + 7a - 15}{2a^2 - 2a - 12}$$

$$= \frac{(2a-3)(a+5)}{2(a^2 - a - 6)}$$

$$= \frac{(2a-3)(a+5)}{2(a-3)(a+2)}$$

59. $$\frac{\dfrac{y}{y^2 - 4} - \dfrac{2y}{y^2 + y - 6}}{\dfrac{2y}{y^2 + y - 6} - \dfrac{y}{y^2 - 4}}$$

Observe that

$$\frac{y}{y^2 - 4} - \frac{2y}{y^2 + y - 6} = -\left(\frac{2y}{y^2 + y - 6} - \frac{y}{y^2 - 4} \right). \text{ Then,}$$

the numerator and denominator are opposites and thus their quotient is –1.

61. $$\frac{t + 5 + \dfrac{3}{t}}{t + 2 + \dfrac{1}{t}} = \frac{t + 5 + \dfrac{3}{t}}{t + 2 + \dfrac{1}{t}} \cdot \frac{t}{t} \qquad \text{LCD is } t$$

$$= \frac{t \cdot t + 5 \cdot t + \dfrac{3}{t} \cdot t}{t \cdot t + 2 \cdot t + \dfrac{1}{t} \cdot t}$$

$$= \frac{t^2 + 5t + 3}{t^2 + 2t + 1}$$

$$= \frac{t^2 + 5t + 3}{(t+1)^2}$$

63. *Writing Exercise.*

65. $(2x^2 - x - 7) - (x^3 - x + 16)$

$= 2x^2 - x - 7 - x^3 + x - 16$

$= -x^3 + 2x^2 - 23$

67. $(m - 6)^2 = m^2 - 12m + 36$

69. *Writing Exercise.*

71. $$\frac{5x^{-2} + 10x^{-1}y^{-1} + 5y^{-2}}{3x^{-2} - 3y^{-2}}$$

$$= \frac{\dfrac{5}{x^2} + \dfrac{10}{xy} + \dfrac{5}{y^2}}{\dfrac{3}{x^2} - \dfrac{3}{y^2}}$$

$$= \frac{\dfrac{5}{x^2} + \dfrac{10}{xy} + \dfrac{5}{y^2}}{\dfrac{3}{x^2} - \dfrac{3}{y^2}} \cdot \frac{x^2 y^2}{x^2 y^2}$$

$$= \frac{5y^2 + 10xy + 5x^2}{3y^2 - 3x^2}$$

$$= \frac{5(y^2 + 2xy + x^2)}{3(y^2 - x^2)}$$

$$= \frac{5(y+x)(y+x)}{3(y+x)(y-x)}$$

$$= \frac{5\cancel{(y+x)}(y+x)}{3\cancel{(y+x)}(y-x)}$$

$$= \frac{5(y+x)}{3(y-x)}$$

73. *Writing Exercise.*

75. $$\frac{30,000 \cdot \dfrac{0.075}{12}}{\left(1 + \dfrac{0.075}{12}\right)^{120} - 1} = \frac{30,000(0.00625)}{(1 + 0.00625)^{120} - 1}$$

$$= \frac{187.5}{(1.00625)^{120} - 1}$$

$$\approx \frac{187.5}{2.112064637 - 1}$$

$$\approx \frac{187.5}{1.112064637}$$

$$\approx 168.61$$

Michael's monthly investment is \$168.61.

77. The reciprocal is $\dfrac{1}{x^2 + x + 1 + \dfrac{1}{x} + \dfrac{1}{x^2}}$.

We simplify.

$$\frac{1}{x^2 + x + 1 + \dfrac{1}{x} + \dfrac{1}{x^2}}$$

$$= \frac{1}{\dfrac{x^4 + x^3 + x^2 + x + 1}{x^2}} \qquad \text{Adding in the denominator}$$

$$= 1 \cdot \frac{x^2}{x^4 + x^3 + x^2 + x + 1}$$

$$= \frac{x^2}{x^4 + x^3 + x^2 + x + 1}$$

79. $f(x) = \dfrac{3}{x}$, $f(x+h) = \dfrac{3}{x+h}$

$$\frac{f(x+h) - f(x)}{h} = \frac{\dfrac{3}{x+h} - \dfrac{3}{x}}{h}$$

$$= \frac{\dfrac{3x - 3(x+h)}{x(x+h)}}{h}$$

$$= \frac{3x - 3(x+h)}{x(x+h)} \cdot \frac{1}{h}$$

$$= \frac{3x - 3x - 3h}{xh(x+h)}$$

$$= \frac{-3h}{xh(x+h)}$$

$$= \frac{-3\cancel{h}}{x\cancel{h}(x+h)}$$

$$= \frac{-3}{x(x+h)}$$

81. $f(x) = \dfrac{2}{2+x}$

$\qquad f(a) = \dfrac{2}{2+a}$

$\qquad f(f(a)) = \dfrac{2}{2 + \dfrac{2}{2+a}}$ Note that $a \ne -2, -3$.

$$= \frac{2}{2 + \dfrac{2}{2+a}} \cdot \frac{2+a}{2+a}$$

$$= \frac{2(2+a)}{2(2+a) + \dfrac{2}{2+a} \cdot 2 + a}$$

$$= \frac{4 + 2a}{4 + 2a + 2}$$

$$= \frac{4 + 2a}{6 + 2a} = \frac{2(2+a)}{2(3+a)}$$

$$= \frac{\cancel{2}(2+a)}{\cancel{2}(3+a)} = \frac{2+a}{3+a}, \quad a \ne -2, -3$$

83. $\left[\dfrac{\dfrac{x+3}{x-3} + 1}{\dfrac{x+3}{x-3} - 1} \right]^4 = \left[\dfrac{\dfrac{x+3}{x-3} + 1}{\dfrac{x+3}{x-3} - 1} \cdot \dfrac{x-3}{x-3} \right]^4$

$$= \left[\frac{x+3 + x - 3}{x+3 - x + 3} \right]^4$$

$$= \left(\frac{2x}{6} \right)^4 = \left(\frac{x}{3} \right)^4 = \frac{x^4}{81}$$

Division by zero occurs in both the numerator and the denominator of the original fraction when $x = 3$. To avoid division by zero in the complex fraction we solve:

$$\frac{x+3}{x-3} - 1 = 0$$

$$\frac{x+3}{x-3} = 1$$

$$x + 3 = x - 3$$

$$3 = -3$$

The equation has no solution, so the denominator of the complex fraction cannot be zero. Thus, the domain of $f = \{x \mid x \text{ is a real number } and \ x \ne 3\}$.

85. *Writing Exercise.*

Connecting the Concepts

1. $\dfrac{5x^2 - 10x}{5x^2 + 5x} = \dfrac{5x(x-2)}{5x(x+1)}$ Factoring

$\qquad\qquad = \dfrac{5x}{5x} \cdot \dfrac{x-2}{x+1}$

$\qquad\qquad = \dfrac{x-2}{x+1}$ Removing a factor of 1

2. $\dfrac{5}{3t} + \dfrac{1}{2t-1} = \dfrac{2t-1}{2t-1} \cdot \dfrac{5}{3t} + \dfrac{3t}{3t} \cdot \dfrac{1}{2t-1}$

$\qquad\qquad = \dfrac{5(2t-1)}{3t(2t-1)} + \dfrac{3t}{3t(2t-1)}$

$\qquad\qquad = \dfrac{10t - 5 + 3t}{3t(2t-1)}$

$\qquad\qquad = \dfrac{13t - 5}{3t(2t-1)}$

3. $\dfrac{t}{2} + \dfrac{t}{3} = 5$,   LCD is 6

$\qquad 6\left(\dfrac{t}{2} + \dfrac{t}{3} \right) = 6 \cdot 5$

$\qquad\qquad 3t + 2t = 30$

$\qquad\qquad\quad 5t = 30$

$\qquad\qquad\quad\ t = 6$

The solution is 6.

4. $\dfrac{1}{y} - \dfrac{1}{2} = \dfrac{5}{6y}$ Note that $y \ne 0$.

$\qquad 6y\left(\dfrac{1}{y} - \dfrac{1}{2} \right) = 6y \cdot \dfrac{5}{6y}$

$\qquad\qquad 6 - 3y = 5$

$\qquad\qquad\quad -3y = -1$

$\qquad\qquad\qquad y = \dfrac{1}{3}$

5. $\dfrac{\dfrac{1}{z} + 1}{\dfrac{1}{z^2} - 1} = \dfrac{\dfrac{1}{z} + 1}{\dfrac{1}{z^2} - 1} \cdot \dfrac{z^2}{z^2}$

$\qquad = \dfrac{\dfrac{1}{z} \cdot z^2 + 1 \cdot z^2}{\dfrac{1}{z^2} \cdot z^2 - 1 \cdot z^2}$

$\qquad = \dfrac{z + z^2}{1 - z^2}$

$\qquad = \dfrac{z(1+z)}{(1+z)(1-z)}$

$\qquad = \dfrac{z}{1-z}$

6. $\dfrac{5}{x+3} = \dfrac{3}{x+2}$,   LCD is $(x+2)(x+3)$

$\qquad\qquad$ Note that $x \ne -3$ and $x \ne -2$.

$(x+2)(x+3)\dfrac{5}{x+3} = (x+2)(x+3)\dfrac{3}{x+2}$

$\qquad\qquad 5(x+2) = 3(x+3)$

$\qquad\qquad 5x + 10 = 3x + 9$

$\qquad\qquad\qquad 2x = -1$

$\qquad\qquad\qquad\ x = -\dfrac{1}{2}$

The solution is $-\dfrac{1}{2}$.

Exercise Set 6.4

1. False

3. True

5. Equation

7. Expression

9. Expression

11. Note that there is no value of t that makes a denominator 0.

$$\frac{t}{10} + \frac{t}{15} = 1, \quad \text{LCD is } 30$$

$$30\left(\frac{t}{10} + \frac{t}{15}\right) = 30 \cdot 1$$

$$3t + 2t = 30$$

$$5t = 30$$

$$t = 6$$

Check: $\dfrac{\dfrac{t}{10} + \dfrac{t}{15} = 1}{\begin{array}{c|c} \dfrac{6}{10} + \dfrac{6}{15} & 1 \\ \dfrac{18}{30} + \dfrac{12}{30} & \\ & 1 \overset{?}{=} 1 \quad \text{TRUE} \end{array}}$

The solution is 6.

13. Note that t cannot be 0.

$$\frac{1}{8} + \frac{1}{10} = \frac{1}{t}, \quad \text{LCD} = 40t$$

$$40t\left(\frac{1}{8} + \frac{1}{10}\right) = 40t \cdot \frac{1}{t}$$

$$40t \cdot \frac{1}{8} + 40t \cdot \frac{1}{10} = 40t \cdot \frac{1}{t}$$

$$5t + 4t = 40$$

$$9t = 40$$

$$t = \frac{40}{9}$$

Check:

$$\frac{\dfrac{1}{8} + \dfrac{1}{10} = \dfrac{1}{t}}{\begin{array}{c|c} \dfrac{1}{8} + \dfrac{1}{10} & \dfrac{1}{\frac{40}{9}} \\ \dfrac{5}{40} + \dfrac{4}{40} & 1 \cdot \dfrac{9}{40} \\ \dfrac{9}{40} \overset{?}{=} \dfrac{9}{40} & \text{TRUE} \end{array}}$$

This checks, so the solution is $\frac{40}{9}$.

15. Note that x cannot be 0.

$$\frac{d}{7} - \frac{7}{d} = 0, \quad \text{LCD} = 7d$$

$$7d\left(\frac{d}{7} - \frac{7}{d}\right) = 7d \cdot 0$$

$$7d \cdot \frac{d}{7} - 7d \cdot \frac{7}{d} = 7d \cdot 0$$

$$d^2 - 49 = 0$$

$$(d+7)(d-7) = 0$$

$d + 7 = 0 \quad or \quad d - 7 = 0$

$d = -7 \quad or \qquad d = 7$

Check:

$$\frac{\dfrac{d}{7} - \dfrac{7}{d} = 0}{\begin{array}{c|c} \dfrac{-7}{7} - \dfrac{7}{-7} & 0 \\ -1 + 1 & \\ & 0 \overset{?}{=} 0 \quad \text{TRUE} \end{array}} \qquad \frac{\dfrac{d}{7} - \dfrac{7}{d} = 0}{\begin{array}{c|c} \dfrac{7}{7} - \dfrac{7}{7} & 0 \\ 1 - 1 & \\ & 0 \overset{?}{=} 0 \quad \text{TRUE} \end{array}}$$

Both of these check, so the two solutions are −7 and 7.

17. Because $\frac{1}{x}$ is undefined when x is 0, at the outset we state the restriction that $x \neq 0$.

$$\frac{3}{4} - \frac{1}{x} = \frac{7}{8}, \quad \text{LCD is } 8x$$

$$8x\left(\frac{3}{4} - \frac{1}{x}\right) = 8x\left(\frac{7}{8}\right)$$

$$8x \cdot \frac{3}{4} - 8x \cdot \frac{1}{x} = 8x \cdot \frac{7}{8}$$

$$6x - 8 = 7x$$

$$-8 = x$$

Check: $\dfrac{\dfrac{3}{4} - \dfrac{1}{x} = \dfrac{7}{8}}{\begin{array}{c|c} \dfrac{3}{4} - \dfrac{1}{-8} & \dfrac{7}{8} \\ \dfrac{6}{8} + \dfrac{1}{8} & \\ & \dfrac{7}{8} \overset{?}{=} \dfrac{7}{8} \quad \text{TRUE} \end{array}}$

The solution is −8.

19. To avoid the division by 0, we must have $n - 5 \neq 0$, or $n \neq 5$.

$$\frac{n+3}{n-5} = \frac{1}{2}, \quad \text{LCD} = 2(n-5)$$

$$2(n-5) \cdot \frac{n+3}{n-5} = 2(n-5) \cdot \frac{1}{2}$$

$$2(n+3) = n-5$$

$$2n + 6 = n-5$$

$$n = -11$$

Check:

$$\frac{\dfrac{n+3}{n-5} = \dfrac{1}{2}}{\begin{array}{c|c} \dfrac{-11+3}{-11-5} & \dfrac{1}{2} \\ \dfrac{-8}{-16} & \\ \dfrac{1}{2} \overset{?}{=} \dfrac{1}{2} & \text{TRUE} \end{array}}$$

This checks, so the solution is −11.

21. Note that x cannot be 0.

$$\frac{9}{x} = \frac{x}{4}, \quad \text{LCD is } 4x$$

$$4x \cdot \frac{9}{x} = 4x \cdot \frac{x}{4}$$

$$36 = x^2$$

$$0 = x^2 - 36$$

$$0 = (x+6)(x-6)$$

$x + 6 = 0 \quad or \quad x - 6 = 0$

$x = -6 \quad or \quad x = 6$

These check, so the solutions are −6 and 6.

23. Because $\dfrac{1}{3t}$ and $\dfrac{1}{t}$ are undefined when t is 0, at the outset we state the restriction that $t \neq 0$.

$$\dfrac{1}{3t} + \dfrac{1}{t} = \dfrac{1}{2}, \quad \text{LCD is } 6t$$

$$6t\left(\dfrac{1}{3t} + \dfrac{1}{t}\right) = 6t\left(\dfrac{1}{2}\right)$$

$$6t \cdot \dfrac{1}{3t} + 6t \cdot \dfrac{1}{t} = 6t \cdot \dfrac{1}{2}$$

$$2 + 6 = 3t$$

$$8 = 3t$$

$$\dfrac{8}{3} = t$$

Check: $\quad \dfrac{1}{3t} + \dfrac{1}{t} = \dfrac{1}{2}$

$$\begin{array}{c|c} \dfrac{1}{3 \cdot \frac{8}{3}} + \dfrac{1}{\frac{8}{3}} & \dfrac{1}{2} \\[2mm] \dfrac{1}{8} + \dfrac{3}{8} & \\[2mm] \dfrac{1}{2} \overset{?}{=} \dfrac{1}{2} & \text{TRUE} \end{array}$$

The solution is $\dfrac{8}{3}$.

25. To ensure that no denominator is 0, at the outset we state the restriction that $x \neq 1$.

$$\dfrac{3}{x-1} + \dfrac{3}{10} = \dfrac{5}{2x-2}$$

$$\dfrac{3}{x-1} + \dfrac{3}{10} = \dfrac{5}{2(x-1)},$$

LCD is $10(x-1)$

$$10(x-1)\left(\dfrac{3}{x-1} + \dfrac{3}{10}\right) = 10(x-1) \cdot \dfrac{5}{2(x-1)}$$

$$10(x-1) \cdot \dfrac{3}{x-1} + 10(x-1) \cdot \dfrac{3}{10} = 10(x-1) \cdot \dfrac{5}{2(x-1)}$$

$$30 + 3(x-1) = 5 \cdot 5$$

$$30 + 3x - 3 = 25$$

$$3x + 27 = 25$$

$$3x = -2$$

$$x = -\dfrac{2}{3}$$

Check: $\quad \dfrac{3}{x-1} + \dfrac{3}{10} = \dfrac{5}{2x-2}$

$$\begin{array}{c|c} \dfrac{3}{-\frac{2}{3}-1} + \dfrac{3}{10} & \dfrac{5}{2\left(-\frac{2}{3}\right)-2} \\[3mm] \dfrac{3}{-\frac{5}{3}} + \dfrac{3}{10} & \dfrac{5}{-\frac{4}{3}-2} \\[3mm] -\dfrac{9}{5} + \dfrac{3}{10} & \dfrac{5 \cdot -3}{1 \quad 10} \\[3mm] -\dfrac{18}{10} + \dfrac{3}{10} & \\[3mm] -\dfrac{15}{10} \overset{?}{=} -\dfrac{15}{10} & \text{TRUE} \end{array}$$

The solution is $-\dfrac{2}{3}$.

27. $\dfrac{2}{6} + \dfrac{1}{2x} = \dfrac{1}{3}$

Because $\dfrac{1}{2x}$ is undefined when x is 0, at the outset we state the restriction that $x \neq 0$. Observe that $\dfrac{2}{6}$ is equivalent to $\dfrac{1}{3}$. This means that $\dfrac{1}{2x}$ must be 0 in order for the equation to be true. But there is no value of x for which $\dfrac{1}{2x} = 0$, so the equation has no solution.

29. $y + \dfrac{4}{y} = -5$

Because $\dfrac{4}{y}$ is undefined when y is 0, we note at the outset that $y \neq 0$. Then we multiply both sides by the LCD, y.

$$y\left(y + \dfrac{4}{y}\right) = y(-5)$$

$$y \cdot y + y \cdot \dfrac{4}{y} = -5y$$

$$y^2 + 4 = -5y$$

$$y^2 + 5y + 4 = 0$$

$$(y+1)(y+4) = 0$$

$$y + 1 = 0 \quad or \quad y + 4 = 0$$

$$y = -1 \quad or \quad y = -4$$

Both values check. The solutions are −1 and −4.

31. Because $\dfrac{12}{x}$ is undefined when x is 0, at the outset we state the restriction that $x \neq 0$.

$$x - \dfrac{12}{x} = 4, \quad \text{LCD is } x$$

$$x\left(x - \dfrac{12}{x}\right) = x \cdot 4$$

$$x \cdot x - x \cdot \dfrac{12}{x} = x \cdot 4$$

$$x^2 - 12 = 4x$$

$$x^2 - 4x - 12 = 0$$

$$(x-6)(x+2) = 0$$

$$x = 6 \quad or \quad x = -2$$

Both numbers check. The solutions are −2 and 6.

33. Because $\dfrac{1}{y}$ is undefined when y is 0, at the outset we state the restriction that $y \neq 0$.

$$\dfrac{9}{10} = \dfrac{1}{y}, \quad \text{LCD is } 10y$$

$$10y\left(\dfrac{9}{10}\right) = 10y \cdot \dfrac{1}{y}$$

$$9y = 10$$

$$y = \dfrac{10}{9}$$

This number checks. The solution is $\dfrac{10}{9}$.

35. $\dfrac{t-1}{t-3} = \dfrac{2}{t-3}$

To ensure that neither denominator is 0, we note at the outset that $t \neq 3$. Then we multiply both sides by the LCD, $t - 3$.

$$(t-3) \cdot \frac{t-1}{t-3} = (t-3) \cdot \frac{2}{t-3}$$
$$t-1 = 2$$
$$t = 3$$

Recall that, because of the restriction above, 3 cannot be a solution. A check confirms this.

Check: $\dfrac{t-1}{t-3} = \dfrac{2}{t-3}$

$$\dfrac{\dfrac{3-1}{3-3}}{} \Bigg| \dfrac{\dfrac{2}{3-3}}{}$$

$$\frac{2}{0} \overset{?}{=} \frac{2}{0} \qquad \text{UNDEFINED}$$

The equation has no solution.

37. $\dfrac{x}{x-5} = \dfrac{25}{x^2-5x}$

$$\dfrac{x}{x-5} = \dfrac{25}{x(x-5)}$$

To ensure that neither denominator is 0, we note at the outset that $x \neq 0$ and $x \neq 5$. Then we multiply both sides by the LCD, $x(x-5)$.

$$x(x-5) \cdot \frac{x}{x-5} = x(x-5) \cdot \frac{25}{x(x-5)}$$
$$x^2 = 25$$
$$x^2 - 25 = 0$$
$$(x+5)(x-5) = 0$$
$$x = -5 \ or \ x = 5$$

Recall that, because of the restrictions above, 5 cannot be a solution. The number -5 checks and is the solution.

39. To avoid division by 0, we must have $n+2 \neq 0$ and $n+1 \neq 0$, or $n \neq -2$ and $n \neq -1$.

$$\frac{n+1}{n+2} = \frac{n-3}{n+1}, \text{ LCD} = (n+2)(n+1)$$
$$(n+2)(n+1) \cdot \frac{n+1}{n+2} = (n+2)(n+1) \cdot \frac{n-3}{n+1}$$
$$(n+1)(n+1) = (n+2)(n-3)$$
$$n^2 + 2n + 1 = n^2 - n - 6$$
$$3n = -7$$
$$n = -\frac{7}{3}$$

This checks, so the solution is $-\dfrac{7}{3}$.

41. $\dfrac{x^2+4}{x-1} = \dfrac{5}{x-1}$

To ensure that neither denominator is 0, we note at the outset that $x \neq 1$. Then we multiply both sides by the LCD, $x-1$.

$$(x-1) \cdot \frac{x^2+4}{x-1} = (x-1) \cdot \frac{5}{x-1}$$
$$x^2 + 4 = 5$$
$$x^2 - 1 = 0$$
$$(x+1)(x-1) = 0$$
$$x+1 = 0 \quad or \quad x-1 = 0$$
$$x = -1 \quad or \qquad x = 1$$

Recall that, because of the restriction above, 1 cannot be a solution. The number -1 checks and is the solution.

We might also observe that since the denominators are the same, the numerators must be the same. Solving $x^2 + 4 = 5$, we get $x = -1$ or $x = 1$ as shown above. Again, because of the restriction $x \neq 1$, only -1 is a solution of the equation.

43. $\dfrac{6}{a+1} = \dfrac{a}{a-1}$

To ensure that neither denominator is 0, we note at the outset that $a \neq -1$ and $a \neq 1$. Then we multiply both sides by the LCD, $(a+1)(a-1)$.

$$(a+1)(a-1) \cdot \frac{6}{a+1} = (a+1)(a-1) \cdot \frac{a}{a-1}$$
$$6(a-1) = a(a+1)$$
$$6a - 6 = a^2 + a$$
$$0 = a^2 - 5a + 6$$
$$0 = (a-2)(a-3)$$
$$a-2 = 0 \quad or \quad a-3 = 0$$
$$a = 2 \quad or \qquad a = 3$$

Both values check. The solutions are 2 and 3.

45. $\dfrac{60}{t-5} - \dfrac{18}{t} = \dfrac{40}{t}$

Listing the restrictions, we note at the outset that $t \neq 5$ and $t \neq 0$. Then we multiply both sides by the LCD, $t(t-5)$.

$$t(t-5) \left(\frac{60}{t-5} - \frac{18}{t} \right) = t(t-5) \cdot \frac{40}{t}$$
$$60t - 18(t-5) = 40(t-5)$$
$$60t - 18t + 90 = 40t - 200$$
$$2t = -290$$
$$t = -145$$

This value checks. The solution is -145.

47. $\dfrac{4}{y^2+y-12} = \dfrac{1}{y+4} - \dfrac{2}{y-3}$

$$\dfrac{4}{(y+4)(y-3)} = \dfrac{1}{y+4} - \dfrac{2}{y-3}$$

Listing the restrictions, we note at the outset that $y \neq -4$ and $y \neq 3$. Then we multiply both sides by the LCD, $(y+4)(y-3)$.

$$(y+4)(y-3) \cdot \frac{4}{(y+4)(y-3)} = (y+4)(y-3) \left(\frac{1}{y+4} - \frac{2}{y-3} \right)$$
$$4 = y-3-2(y+4)$$
$$4 = y-3-2y-8$$
$$4 = -y-11$$
$$15 = -y$$
$$-15 = y$$

This value checks. The solution is -15.

49. $\dfrac{3}{x-3} + \dfrac{5}{x+2} = \dfrac{5x}{x^2-x-6}$

$$\dfrac{3}{x-3} + \dfrac{5}{x+2} = \dfrac{5x}{(x-3)(x+2)}$$

Listing the restrictions, we note at the outset that $x \neq 3$ and $x \neq -2$. Then we multiply both sides by the LCD, $(x-3)(x+2)$.

$(x-3)(x+2)\left(\dfrac{3}{x-3}+\dfrac{5}{x+2}\right)=(x-3)(x+2)\cdot\dfrac{5x}{(x-3)(x+2)}$

$3(x+2)+5(x-3)=5x$

$3x+6+5x-15=5x$

$8x-9=5x$

$-9=-3x$

$3=x$

Recall that, because of the restriction above, 3 cannot be a solution. Thus, the equation has no solution.

51. $\dfrac{3}{x}+\dfrac{x}{x+2}=\dfrac{4}{x^2+2x}$

$\dfrac{3}{x}+\dfrac{x}{x+2}=\dfrac{4}{x(x+2)}$

Listing the restrictions, we note at the outset that $x\neq 0$ and $x\neq -2$. Then we multiply both sides by the LCD, $x(x+2)$.

$x(x+2)\left(\dfrac{3}{x}+\dfrac{x}{x+2}\right)=x(x+2)\cdot\dfrac{4}{x(x+2)}$

$3(x+2)+x\cdot x=4$

$3x+6+x^2=4$

$x^2+3x+2=0$

$(x+1)(x+2)=0$

$x+1=0\quad or\quad x+2=0$

$x=-1\quad or\quad\quad x=-2$

Recall that, because of the restrictions above, -2 cannot be a solution. The number -1 checks. The solution is -1.

53. $\dfrac{2}{t-4}+\dfrac{1}{t}=\dfrac{t}{4-t}$

$\dfrac{2}{t-4}+\dfrac{1}{t}=\dfrac{-t}{t-4}$

Listing the restrictions, we note at the outset that $t\neq 0$ and $t\neq 4$. Then we multiply both sides by the LCD, $t(t-4)$.

$t(t-4)\left(\dfrac{2}{t-4}+\dfrac{1}{t}\right)=t(t-4)\cdot\dfrac{-t}{t-4}$

$t\cdot 2+t-4=t(-t)$

$2t+t-4=-t^2$

$t^2+3t-4=0$

$(t+4)(t-1)=0$

$t+4=0\quad or\quad t-1=0$

$t=-4\quad or\quad\quad t=1$

Both values check. The solutions are -4 and 1.

55. $\dfrac{5}{x+2}-\dfrac{3}{x-2}=\dfrac{2x}{4-x^2}$

$\dfrac{5}{x+2}-\dfrac{3}{x-2}=\dfrac{2x}{(2+x)(2-x)}$

$\dfrac{5}{x+2}+\dfrac{3}{2-x}=\dfrac{2x}{(2+x)(2-x)}\quad\left(-\dfrac{3}{x-2}=\dfrac{3}{2-x}\right)$

First note that $x\neq -2$ and $x\neq 2$. Then multiply both sides by the LCD, $(2+x)(2-x)$.

$(2+x)(2-x)\left(\dfrac{5}{x+2}+\dfrac{3}{2-x}\right)=(2+x)(2-x)\cdot\dfrac{2x}{(2+x)(2-x)}$

$5(2-x)+3(2+x)=2x$

$10-5x+6+3x=2x$

$16-2x=2x$

$16=4x$

$4=x$

This value checks. The solution is 4.

57. $\dfrac{1}{x^2+2x+1}=\dfrac{x-1}{3x+3}+\dfrac{x+2}{5x+5}$

$\dfrac{1}{(x+1)(x+1)}=\dfrac{x-1}{3(x+1)}+\dfrac{x+2}{5(x+1)}$

Note that $x\neq -1$. Then multiply both sides by the LCD, $15(x+1)^2$.

$15(x+1)^2\cdot\dfrac{1}{(x+1)(x+1)}=15(x+1)^2\left(\dfrac{x-1}{3(x+1)}+\dfrac{x+2}{5(x+1)}\right)$

$15=5(x+1)(x-1)+3(x+1)(x+2)$

$15=5x^2-5+3x^2+9x+6$

$15=8x^2+9x+1$

$0=8x^2+9x-14$

$0=(x+2)(8x-7)$

$x=-2\quad or\quad x=\dfrac{7}{8}$

Both values check. The solutions are -2 and $\dfrac{7}{8}$.

59. $\dfrac{3-2y}{y+1}-\dfrac{10}{y^2-1}=\dfrac{2y+3}{1-y}$

$\dfrac{3-2y}{y+1}-\dfrac{10}{(y+1)(y-1)}=\dfrac{2y+3}{1-y}$

$\dfrac{3-2y}{y+1}+\dfrac{10}{(y+1)(1-y)}=\dfrac{2y+3}{1-y}$

Note that $y\neq -1$ and $y\neq 1$.

$(y+1)(1-y)\left(\dfrac{3-2y}{y+1}+\dfrac{10}{(y+1)(1-y)}\right)=(y+1)(1-y)\cdot\dfrac{2y+3}{1-y}$

$(1-y)(3-2y)+10=(y+1)(2y+3)$

$3-5y+2y^2+10=2y^2+5y+3$

$10=10y$

$1=y$

Recall that because of the restriction above, 1 is not a solution. Thus, the equation has no solution.

61. We find all values of a for which $2a-\dfrac{15}{a}=7$. First note that $a\neq 0$. Then multiply both sides by the LCD, a.

$a\left(2a-\dfrac{15}{a}\right)=a\cdot 7$

$a\cdot 2a-a\cdot\dfrac{15}{a}=7a$

$2a^2-15=7a$

$2a^2-7a-15=0$

$(2a+3)(a-5)=0$

$a=-\dfrac{3}{2}\ or\ a=5$

Both values check. The solutions are $-\dfrac{3}{2}$ and 5.

63. We find all values of a for which $\frac{a-5}{a+1} = \frac{3}{5}$. First note

that $a \neq -1$. Then multiply both sides by the LCD, $5(a+1)$.

$$5(a+1) \cdot \frac{a-5}{a+1} = 5(a+1) \cdot \frac{3}{5}$$
$$5(a-5) = 3(a+1)$$
$$5a-25 = 3a+3$$
$$2a = 28$$
$$a = 14$$

This value checks. The solution is 14.

65. We find all values of a for which $\frac{12}{a} - \frac{12}{2a} = 8$. First

note that $a \neq 0$. Then multiply both sides by the LCD, $2a$.

$$2a\left(\frac{12}{a} - \frac{12}{2a}\right) = 2a \cdot 8$$
$$2a \cdot \frac{12}{a} - 2a \cdot \frac{12}{2a} = 16a$$
$$24 - 12 = 16a$$
$$12 = 16a$$
$$\frac{3}{4} = a$$

This value checks. The solution is $\frac{3}{4}$.

67.
$$f(a) = g(a)$$
$$\frac{3a-1}{a^2-7a+10} = \frac{a-1}{a^2-4} + \frac{2a+1}{a^2-3a-10}$$
$$\frac{3a-1}{(a-2)(a-5)} = \frac{a-1}{(a+2)(a-2)} + \frac{2a+1}{(a-5)(a+2)}$$

First note that $a \neq 2$, $a \neq 5$, and $a \neq -2$. Then multiply both sides by the LCD, $(a-2)(a-5)(a+2)$.

$$(a-2)(a-5)(a+2) \cdot \frac{3a-1}{(a-2)(a-5)}$$
$$= (a-2)(a-5)(a+2)\left(\frac{a-1}{(a+2)(a-2)} + \frac{2a+1}{(a-5)(a+2)}\right)$$
$$(a+2)(3a-1) = (a-5)(a-1) + (a-2)(2a+1)$$
$$3a^2+5a-2 = a^2-6a+5+2a^2-3a-2$$
$$3a^2+5a-2 = 3a^2-9a+3$$
$$5a-2 = -9a+3$$
$$14a-2 = 3$$
$$14a = 5$$
$$a = \frac{5}{14}$$

This number checks. Then $f(a) = g(a)$ for $a = \frac{5}{14}$.

69.
$$f(a) = g(a)$$
$$\frac{2}{a^2-8a+7} = \frac{3}{a^2-2a-3} - \frac{1}{a^2-1}$$
$$\frac{2}{(a-1)(a-7)} = \frac{3}{(a+1)(a-3)} - \frac{1}{(a+1)(a-1)}$$

First note that $a \neq 1$, $a \neq 7$, $a \neq -1$, and $a \neq 3$. Then multiply both sides by the LCD, $(a-1)(a-7)(a+1)(a-3)$.

$$(a-1)(a-7)(a+1)(a-3) \cdot \frac{2}{(a-1)(a-7)}$$
$$= (a-1)(a-7)(a+1)(a-3)\left(\frac{3}{(a+1)(a-3)} - \frac{1}{(a+1)(a-1)}\right)$$

$$2(a+1)(a-3) = 3(a-1)(a-7) - (a-7)(a-3)$$
$$2(a^2-2a-3) = 3(a^2-8a+7) - (a^2-10a+21)$$
$$2a^2-4a-6 = 3a^2-24a+21-a^2+10a-21$$
$$2a^2-4a-6 = 2a^2-14a$$
$$-4a-6 = -14a$$
$$-6 = -10a$$
$$\frac{3}{5} = a$$

This number checks. Then $f(a) = g(a)$ for $a = \frac{3}{5}$.

71. *Writing Exercise.*

73. $2x - y^2 \div 3x$ for $x = 3$ and $y = -6$
$$2(3) - (-6)^2 \div 3 \cdot 3$$
$$= 2(3) - 36 \div 3 \cdot 3$$
$$= 6 - 12 \cdot 3$$
$$= 6 - 36$$
$$= -30$$

75. $-3^{-2} = -\frac{1}{3^2} = -\frac{1}{9}$

77. $\frac{24a^{-4}c^{-8}}{16a^5 c^{-7}} = \frac{3}{2a^{5+4}c^{8-7}} = \frac{3}{2a^9 c}$

79. *Writing Exercise.*

81.
$$\frac{2 - \frac{a}{4}}{2} = \frac{\frac{a}{4} - 2}{\frac{a}{2} + 2}$$
$$\frac{\frac{8-a}{4}}{2} = \frac{\frac{a-8}{4}}{\frac{a+4}{2}}$$
$$\frac{8-a}{4} \cdot \frac{1}{2} = \frac{a-8}{4} \cdot \frac{2}{a+4}$$
$$\frac{8-a}{8} = \frac{2a-16}{4(a+4)}$$

To ensure that the denominator on the right side of the equation is not 0, we note at the outset that $a \neq -4$. Then we multiply both sides by the LCD, $8(a+4)$.

$$8(a+4) \cdot \frac{8-a}{8} = 8(a+4) \cdot \frac{2a-16}{4(a+4)}$$
$$(a+4)(8-a) = 2(2a-16)$$
$$-a^2+4a+32 = 4a-32$$
$$0 = a^2 - 64$$
$$0 = (a+8)(a-8)$$
$$a = -8 \ \ or \ \ a = 8$$

Both numbers check.

83. Set $f(a)$ equal to $g(a)$ and solve for a.

$$\frac{1}{1+a} + \frac{a}{1-a} = \frac{1}{1-a} - \frac{a}{1+a}$$

Note that $a \neq -1$ and $a \neq 1$. Then multiply by the LCD, $(1+a)(1-a)$.

$$(1+a)(1-a)\left(\frac{1}{1+a} + \frac{a}{1-a}\right) = (1+a)(1-a)\left(\frac{1}{1-a} - \frac{a}{1+a}\right)$$
$$1-a+a(1+a) = 1+a-a(1-a)$$
$$1-a+a+a^2 = 1+a-a+a^2$$
$$1+a^2 = 1+a^2$$
$$1 = 1$$

We get an equation that is true for all real numbers. Recall that, because of the restrictions above, -1 and 1 are not solutions. Thus, $f(a) = g(a)$ for

$\{a \mid a$ is a real number and $a \neq -1$ and $a \neq 1\}$.

85. Note that x cannot be 0.

$$\frac{\frac{1}{x}+1}{x} = \frac{\frac{1}{x}}{2}$$

$$\left(\frac{1}{x}+1\right) \cdot \frac{1}{x} = \frac{1}{x} \cdot \frac{1}{2}$$

$$\frac{1}{x^2} + \frac{1}{x} = \frac{1}{2x}, \ \ LCD = 2x^2$$

$$2 + 2x = x$$

$$2 = -x$$

$$-2 = x$$

This checks so the solution is -2.

87. *Graphing Calculator Exercise*

Mid-Chapter Review

1. $\dfrac{2}{x} + \dfrac{1}{x^2 + x} = \dfrac{2}{x} + \dfrac{1}{x(x+1)}$

$= \dfrac{2}{x} \cdot \dfrac{x+1}{x+1} + \dfrac{1}{x(x+1)}$

$= \dfrac{2x+2}{x(x+1)} + \dfrac{1}{x(x+1)}$

$= \dfrac{2x+3}{x(x+1)}$

2.

$$\frac{10}{(t+1)(t-1)} = \frac{t+4}{t(t-1)} \quad \begin{matrix} t \neq -1, t \neq 1, \\ t \neq 0 \end{matrix}$$

$$t(t+1)(t-1)\left(\frac{10}{(t+1)(t-1)}\right) = t(t+1)(t-1)\left(\frac{t+4}{t(t-1)}\right)$$

$$10t = (t+1)(t+4)$$

$$10t = t^2 + 5t + 4$$

$$0 = t^2 - 5t + 4$$

$$0 = (t-1)(t-4)$$

$$t - 1 = 0 \quad or \quad t - 4 = 0$$

$$t = 1 \quad or \quad t = 4$$

Since $t \neq 1$, the solution is 4.

3. $\dfrac{2x-6}{5x+10} \cdot \dfrac{x+2}{6x-12} = \dfrac{2(x-3)(x+2)}{5(x+2) \cdot 2 \cdot 3(x-2)}$

$= \dfrac{(x-3)}{15(x-2)}$

4. $\dfrac{2}{x-5} \div \dfrac{6}{x-5} = \dfrac{2}{x-5} \cdot \dfrac{x-5}{6}$

$= \dfrac{1}{3} \cdot \dfrac{2(x-5)}{2(x-5)}$

$= \dfrac{1}{3}$

5. $\dfrac{x}{x+2} - \dfrac{1}{x-1}$

$= \dfrac{x}{x+2} \cdot \dfrac{x-1}{x-1} - \dfrac{1}{x-1} \cdot \dfrac{x+2}{x+2} \quad LCD = (x+2)(x-1)$

$= \dfrac{x(x-1) - (x+2)}{(x+2)(x-1)}$

$= \dfrac{x^2 - x - x - 2}{(x+2)(x-1)}$

$= \dfrac{x^2 - 2x - 2}{(x+2)(x-1)}$

6. $\dfrac{2}{x+3} + \dfrac{3}{x+4}$

$= \dfrac{2}{x+3} \cdot \dfrac{x+4}{x+4} + \dfrac{3}{x+4} \cdot \dfrac{x+3}{x+3} \quad LCD = (x+3)(x-4)$

$= \dfrac{2(x+4) + 3(x+3)}{(x+3)(x+4)}$

$= \dfrac{2x+8 + 3x+9}{(x+3)(x+4)}$

$= \dfrac{5x+17}{(x+3)(x+4)}$

7. $\dfrac{3}{x-4} - \dfrac{2}{4-x} = \dfrac{3}{x-4} + \dfrac{2}{x-4}$

$= \dfrac{5}{x-4}$

8. $\dfrac{x^2-16}{x^2-x} \cdot \dfrac{x^2}{x^2-5x+4} = \dfrac{(x+4)(x-4)x^2}{x(x-1)(x-4)(x-1)}$

$= \dfrac{x(x+4)}{(x-1)^2}$

9. $\dfrac{x+1}{x^2-7x+10} + \dfrac{3}{x^2-x-2} \quad LCD = (x+1)(x-2)(x-5)$

$= \dfrac{x+1}{(x-5)(x-2)} \cdot \dfrac{x+1}{x+1} + \dfrac{3}{(x-2)(x+1)} \cdot \dfrac{x-5}{x-5}$

$= \dfrac{x^2 + 2x + 1 + 3x - 15}{(x+1)(x-2)(x-5)}$

$= \dfrac{x^2 + 5x - 14}{(x+1)(x-2)(x-5)}$

$= \dfrac{(x+7)(x-2)}{(x+1)(x-2)(x-5)}$

$= \dfrac{x+7}{(x+1)(x-5)}$

10. $(t^2 + t - 20) \cdot \dfrac{t+5}{t-4} = \dfrac{(t+5)(t-4)(t+5)}{t-4} = (t+5)^2$

11. $\dfrac{a^2 - 2a + 1}{a^2 - 4} \div (a^2 - 3a + 2)$

$= \dfrac{(a-1)(a-1)}{(a+2)(a-2)} \cdot \dfrac{1}{(a-2)(a-1)}$

$= \dfrac{a-1}{(a+2)(a-2)^2}$

12. $\dfrac{\dfrac{3}{z}+\dfrac{2}{y}}{\dfrac{4}{z}-\dfrac{1}{y}}=\dfrac{\dfrac{3}{z}+\dfrac{2}{y}}{\dfrac{4}{z}-\dfrac{1}{y}}\cdot\dfrac{yz}{yz}=\dfrac{3y+2z}{4y-z}$

13. $\dfrac{xy^{-1}+x^{-1}}{2x^{-1}+4y^{-1}}=\dfrac{\dfrac{x}{y}+\dfrac{1}{x}}{\dfrac{2}{x}+\dfrac{4}{y}}=\dfrac{\dfrac{x}{y}+\dfrac{1}{x}}{\dfrac{2}{x}+\dfrac{4}{y}}\cdot\dfrac{xy}{xy}$

$\qquad=\dfrac{x^2+y}{2y+4x}$

$\qquad=\dfrac{x^2+y}{2(y+2x)}$

14. $\dfrac{\dfrac{y}{y^2-4}+\dfrac{5}{4-y^2}}{\dfrac{y^2}{y^2-4}+\dfrac{25}{4-y^2}}=\dfrac{\dfrac{y}{y^2-4}+\dfrac{5}{4-y^2}}{\dfrac{y^2}{y^2-4}+\dfrac{25}{4-y^2}}\cdot\dfrac{y^2-4}{y^2-4}$

$\qquad=\dfrac{y-5}{y^2-25}$

$\qquad=\dfrac{y-5}{(y+5)(y-5)}$

$\qquad=\dfrac{1}{y+5}$

15. $\dfrac{\dfrac{1}{a}-\dfrac{1}{b}}{\dfrac{1}{a^3}-\dfrac{1}{b^3}}=\dfrac{\dfrac{1}{a}-\dfrac{1}{b}}{\dfrac{1}{a^3}-\dfrac{1}{b^3}}\cdot\dfrac{a^3b^3}{a^3b^3}=\dfrac{a^2b^3-a^3b^2}{b^3-a^3}$

$\qquad=\dfrac{a^2b^2(b-a)}{(b-a)(b^2+ab+a^2)}$

$\qquad=\dfrac{a^2b^2}{b^2+ab+a^2}$

16. Note that there is no value of a that makes a denominator 0.

$\qquad\dfrac{a+1}{3}+\dfrac{a-4}{5}=\dfrac{2a}{9},\quad$ LCD is 45

$\qquad 45\left(\dfrac{a+1}{3}+\dfrac{a-4}{5}\right)=45\cdot\dfrac{2a}{9}$

$\qquad 45\cdot\dfrac{a+1}{3}+45\cdot\dfrac{a-4}{5}=45\cdot\dfrac{2a}{9}$

$\qquad 15(a+1)+9(a-4)=5\cdot 2a$

$\qquad 15a+15+9a-36=10a$

$\qquad 24a-21=10a$

$\qquad -21=-14a$

$\qquad \dfrac{3}{2}=a$

This number checks. The solution is $\dfrac{3}{2}$.

17. $\dfrac{5}{4t}=\dfrac{7}{5t-2}$

Listing the restrictions, we note at the outset that $t\neq 0$ and $t\neq\dfrac{2}{5}$. Then we multiply both sides by the LCD, $4t(5t-2)$.

$4t(5t-2)\cdot\dfrac{5}{4t}=4t(5t-2)\cdot\dfrac{7}{5t-2}$

$\qquad 5(5t-2)=4t\cdot 7$

$\qquad 25t-10=28t$

$\qquad -10=3t$

$\qquad -\dfrac{10}{3}=t$

This value checks. The solution is $-\dfrac{10}{3}$.

18.

$\qquad\dfrac{2}{1-x}=\dfrac{-4}{x^2-1}$

$\qquad\dfrac{-2}{x-1}=\dfrac{-4}{(x+1)(x-1)}$

$\qquad\qquad$ Note that $x\neq -1$ and $x\neq 1$.

$(x+1)(x-1)\dfrac{-2}{x-1}=(x+1)(x-1)\dfrac{-4}{(x+1)(x-1)}$

$\qquad -2x-2=-4$

$\qquad -2x=-2$

$\qquad x=1$

Recall that because of the restriction, 1 is not a solution. Thus, the equation has no solution.

19. $\qquad\dfrac{3}{x}+\dfrac{2}{x-2}=1,\quad$ LCD is $x(x-2)$

$\qquad\qquad$ Note that $x\neq 0$ and $x\neq 2$.

$\qquad x(x-2)\left(\dfrac{3}{x}+\dfrac{2}{x-2}\right)=x(x-2)\cdot 1$

$\qquad 3(x-2)+2x=x(x-2)$

$\qquad 3x-6+2x=x^2-2x$

$\qquad 0=x^2-7x+6$

$\qquad 0=(x-1)(x-6)$

$\qquad x=1\ \ or\ \ x=6$

20.

$\qquad\dfrac{t-1}{t^2-3t}-\dfrac{4}{t^2-9}=0\qquad$ Note that

$\qquad\dfrac{t-1}{t(t-3)}-\dfrac{4}{(t+3)(t-3)}=0\qquad t\neq 0,\ t\neq 3,$ and $t\neq -3$.

$t(t+3)(t-3)\left(\dfrac{t-1}{t(t-3)}-\dfrac{4}{(t+3)(t-3)}\right)=t(t+3)(t-3)\cdot 0$

$\qquad (t+3)(t-1)-4t=0$

$\qquad t^2+2t-3-4t=0$

$\qquad t^2-2t-3=0$

$\qquad (t+1)(t-3)=0$

$\qquad\qquad t=-1\ \ or\ \ t=3$

Recall that because of the restriction, 3 is not a solution. Thus, the solution is -1.

Exercise Set 6.5

1. False. To find the time that it would take two people working together, we need to solve $\dfrac{1}{a}+\dfrac{1}{b}=\dfrac{1}{t}$ for t, where a and b represent the time needed for each person to complete the work alone.

3. True

5. True

7. Let n = the number.

The reciprocal of 3   plus   the reciprocal of 6   is   the reciprocal of the number.

$$\frac{1}{3} + \frac{1}{6} = \frac{1}{n}$$

9. Let n = the number.

A number   plus   6   times   its reciprocal   is   −5.

$$n + 6 \cdot \frac{1}{n} = -5$$

11. Let x = the first integer. Then $x + 1$ = the second, and their product = $x(x + 1)$.

Reciprocal of the product   is $\frac{1}{90}$.

$$\frac{1}{x(x+1)} = \frac{1}{90}$$

13. *Familiarize.* Let n = the number.
Translate.

$$\frac{1}{3} + \frac{1}{6} = \frac{1}{n}$$

Carry out. We solve the equation.

$$\frac{1}{3} + \frac{1}{6} = \frac{1}{n}, \quad \text{LCD is } 6x$$
$$6n\left(\frac{1}{3} + \frac{1}{6}\right) = 6n \cdot \frac{1}{n}$$
$$2n + n = 6$$
$$3n = 6$$
$$n = 2$$

Check. $\frac{1}{3} + \frac{1}{6} = \frac{2}{6} + \frac{1}{6} = \frac{3}{6} = \frac{1}{2}$. This is the reciprocal of 2, so the result checks.
State. The number is 2.

15. *Familiarize.* We let n = the number.
Translate.

$$n + \frac{6}{n} = -5$$

Carry out. We solve the equation.

$$n + \frac{6}{n} = -5, \quad \text{LCD is } n$$
$$n\left(n + \frac{6}{n}\right) = n(-5)$$
$$n^2 + 6 = -5n$$
$$n^2 + 5n + 6 = 0$$
$$(n + 3)(n + 2) = 0$$
$$n = -3 \quad or \quad n = -2$$

Check. The possible solutions are −3 and −2. We check −3 in the conditions of the problem.

Number :	−3
6 times the reciprocal of the number :	$6\left(-\frac{1}{3}\right) = -2$
Sum of the number and 6 times its reciprocal :	$-3 + (-2) = -5$

The number −3 checks.

Now we check −2:

Number :	−2
6 times the reciprocal of the number :	$6\left(-\frac{1}{2}\right) = -3$
Sum of the number and 6 times its reciprocal :	$-2 + (-3) = -5$

The number −2 also checks.
State. The number is −3 or −2.

17. *Familiarize.* Let x = the first integer. Then $x + 1$ = the second, and their product = $x(x + 1)$.
Translate.

$$\frac{1}{x(x+1)} = \frac{1}{90}$$

Carry out. We solve the equation.

$$\frac{1}{x(x+1)} = \frac{1}{90}, \quad \text{LCD is } 90x(x+1)$$
$$90x(x+1)\left(\frac{1}{x(x+1)}\right) = 90x(x+1) \cdot \frac{1}{90}$$
$$90 = x(x+1)$$
$$90 = x^2 + x$$
$$0 = x^2 + x - 90$$
$$0 = (x + 10)(x - 9)$$
$$x = -10 \quad or \quad x = 9$$

Check. The possible solutions are −10 and 9.
If the first is −10, then the second is −9. $\frac{1}{-10(-9)} = \frac{1}{90}$.

If the first is 9, then the second is 10. $\frac{1}{9(10)} = \frac{1}{90}$.

Both values check.
State. The integers are −10 and −9, or 9 and 10.

19. *Familiarize.* Let t represent the number of hours it takes Chandra and Traci to complete the job working together.
Translate. Chandra takes 6 hr and Traci takes 9 hr to complete the job, so we have

$$\frac{t}{6} + \frac{t}{9} = 1$$

Carry out. We solve the equation. Multiply on both sides by the LCD, 18.

$$18\left(\frac{t}{6} + \frac{t}{9}\right) = 18 \cdot 1$$
$$3t + 2t = 18$$
$$5t = 18$$
$$t = \frac{18}{5}, \text{ or } 3\frac{3}{5}$$

Check. If Chandra does the job alone in 6 hr, then in $3\frac{3}{5}$ hr she does $\frac{18/5}{6}$, or $\frac{3}{5}$ of the job. If Traci does the job alone in 9 hr, then in $3\frac{3}{5}$ hr she does $\frac{18/5}{9}$, or $\frac{2}{5}$ of the job. Together, they do $\frac{3}{5} + \frac{2}{5}$, or 1 entire job. The result checks.

State. It would take Chandra and Traci $3\frac{3}{5}$ hr to finish the job working together.

21. *Familiarize*. In 1 minute the Gempler pump does $\frac{1}{48}$

of the job and the Liberty pump does $\frac{1}{30}$ of the job.

Working together, they do $\frac{1}{48} + \frac{1}{30}$ of the job in 1

minute. Suppose it takes t minutes to do the job
working together.

Translate. We find t such that

$$t\left(\frac{1}{48}\right) + t\left(\frac{1}{30}\right) = 1, \text{ or } \frac{t}{48} + \frac{t}{30} = 1.$$

Carry out. We solve the equation. We multiply both
sides by the LCD, 240.

$$240\left(\frac{t}{48} + \frac{t}{30}\right) = 240 \cdot 1$$
$$5t + 8t = 240$$
$$13t = 240$$
$$t = \frac{240}{13}$$

Check. In $\frac{240}{13}$ min the Gempler pump does $\frac{240}{13} \cdot \frac{1}{48}$,

or $\frac{5}{13}$ of the job and the Liberty pump does $\frac{240}{13} \cdot \frac{1}{30}$,

or $\frac{8}{13}$ of the job. Together they do $\frac{5}{13} + \frac{8}{13}$, or 1 entire

job. The answer checks.

State. The two pumps can pump out the basement in

$\frac{240}{13}$ min, or $18\frac{6}{13}$ min, working together.

23. *Familiarize*. Let t represent the time, in minutes, that it
takes the DS-860 to scan the manuscript working alone.
Then $2t$ represents the time it takes the DS-530 to do

the job, working alone. In 1 min the DS-860 does $\frac{1}{t}$ of

the job and the DS-530 does $\frac{1}{2t}$ of the job.

Translate. Working together, they can do the entire job
in 5 min, so we want to find t such that

$$5\left(\frac{1}{t}\right) + 5\left(\frac{1}{2t}\right) = 1, \text{ or } \frac{5}{t} + \frac{5}{2t} = 1.$$

Carry out. We solve the equation. We multiply both
sides by the LCD, $2t$.

$$2t\left(\frac{5}{t} + \frac{5}{2t}\right) = 2t \cdot 1$$
$$10 + 5 = 2t$$
$$15 = 2t$$
$$\frac{15}{2} = t, \text{ or } 7\frac{1}{2}$$

Check. If the DS-860 can do the job in $\frac{15}{2}$ min, then in

5 min it does $5 \cdot \frac{1}{15/2}$, or $\frac{2}{3}$ of the job. If it takes the

DS-530 $2 \cdot \frac{15}{2}$, or 15 min, to do the job, then in 5 min

it does $5 \cdot \frac{1}{15}$, or $\frac{1}{3}$ of the job. Working together, the

two machines do $\frac{2}{3} + \frac{1}{3}$, or 1 entire job, in 5 min.

State. Working alone, it takes the DS-860 $7\frac{1}{2}$ min and

the DS-530 15 min to scan the manuscript.

25. *Familiarize*. Let t represent the number of days it takes
Tori to mulch the gardens working alone.
Then $t - 3$ represents the time it takes Anita to
mulch the gardens, working alone. In 1 day, Tori does

$\frac{1}{t}$ of the job and Anita does $\frac{1}{t-3}$ of the job.

Translate. Working together, Tori and Anita can mulch
the gardens in 2 days.

$$2\left(\frac{1}{t}\right) + 2\left(\frac{1}{t-3}\right) = 1, \text{ or } \frac{2}{t} + \frac{2}{t-3} = 1.$$

Carry out. We solve the equation. Multiply on both
sides by the LCD, $t(t-3)$.

$$t(t-3)\left(\frac{2}{t} + \frac{2}{t-3}\right) = t(t-3) \cdot 1$$
$$2(t-3) + 2t = t(t-3)$$
$$2t - 6 + 2t = t^2 - 3t$$
$$0 = t^2 - 7t + 6$$
$$0 = (t-1)(t-6)$$
$$t = 1 \text{ or } t = 6$$

Check. If $t = 1$, then $t - 3 = 1 - 3 = -2$. Since negative
time has no meaning in this application, 1 cannot be a
solution. If $t = 6$, then $t - 3 = 6 - 3 = 3$. In 2 days Anita

does $2 \cdot \frac{1}{3}$, or $\frac{2}{3}$ of the job. In 2 days Tori does

$2 \cdot \frac{1}{6}$, or $\frac{1}{3}$ of the job. Together they do $\frac{2}{3} + \frac{1}{3}$, or 1

entire job. The answer checks.

State. It would take Anita 3 days and Tori 6 days to do
the job, working alone.

27. *Familiarize*. Let t represent the number of months it
takes Tristan to program alone. Then $3t$ represents the
number of months it takes Sara program alone.

Translate. In 1 month Tristan and Sara will do one
entire job, so we have

$$1\left(\frac{1}{t}\right) + 1\left(\frac{1}{3t}\right) = 1, \text{ or } \frac{1}{t} + \frac{1}{3t} = 1$$

Carry out. We solve the equation. Multiply on both
sides by the LCD, $3t$.

$$3t\left(\frac{1}{t} + \frac{1}{3t}\right) = 3t \cdot 1$$
$$3 + 1 = 3t$$
$$4 = 3t$$
$$\frac{4}{3} = t$$

Check. If Tristan does the job alone in $\frac{4}{3}$ months, then

in 1 month he does $\frac{1}{4/3}$, or $\frac{3}{4}$ of the job. If Sara does

the job alone in $3 \cdot \frac{4}{3}$, or 4 months, then in 1 month she

does $\frac{1}{4}$ of the job. Together, they do $\frac{3}{4} + \frac{1}{4}$, or 1 entire

job, in 1 month. The result checks.

State. It would take Tristan $\frac{4}{3}$ months and it would

take Sara 4 months to program alone.

29. *Familiarize*. Let t represent the number of minutes it
takes Lia to decorate the cupcakes working alone. Then
$t + 20$ represents the time it takes Zeno to decorate the
cupcakes working alone.

Translate. In 10.5 min Lia and Zeno will do one entire

job, so we have

$$10.5\left(\frac{1}{t}\right) + 10.5\left(\frac{1}{t+20}\right) = 1, \text{ or } \frac{10.5}{t} + \frac{10.5}{t+20} = 1$$

Carry out. We solve the equation. Multiply on both sides by the LCD, $t(t+20)$.

$$t(t+20)\left(\frac{10.5}{t} + \frac{10.5}{t+20}\right) = t(t+20)\cdot 1$$
$$10.5(t+20) + 10.5t = t(t+20)$$
$$10.5t + 210 + 10.5t = t^2 + 20t$$
$$0 = t^2 - t - 210$$
$$0 = (t-15)(t+14)$$
$$t = 15 \text{ or } t = -14$$

Check. If Lia does the job alone in 15 min, then in 10.5 min she does $\frac{7}{10}$ of the job. If Zeno does the job alone in $15 + 20$, or 35 min, then in 10.5 min he does $\frac{10.5}{35}$, or $\frac{3}{10}$ of the job. Together, they do $\frac{7}{10} + \frac{3}{10}$, or 1 entire job, in 10.5 min. The result checks.

State. It would take Lia 15 min and it would take Zeno 35 min to decorate the cupcakes alone.

31. **Familiarize**. Let t represent the number of minutes it takes Chris to do the job working alone. Then $t + 120$ represents the time it takes Kim to do the job working alone.

We will convert hours to minutes:

2 hr = $2 \cdot 60$ min = 120 min

2 hr 55 min = 120 min + 55 min = 175 min

Translate. In 175 min Chris and Kim will do one entire job, so we have

$$175\left(\frac{1}{t}\right) + 175\left(\frac{1}{t+120}\right) = 1, \text{ or } \frac{175}{t} + \frac{175}{t+120} = 1$$

Carry out. We solve the equation. Multiply on both sides by the LCD, $t(t+120)$.

$$t(t+120)\left(\frac{175}{t} + \frac{175}{t+120}\right) = t(t+120)\cdot 1$$
$$175(t+120) + 175t = t(t+120)$$
$$175t + 21,000 + 175t = t^2 + 120t$$
$$0 = t^2 - 230t - 21,000$$
$$0 = (t-300)(t+70)$$
$$t = 300 \text{ or } t = -70$$

Check. Since negative time has no meaning in this problem -70 is not a solution of the original problem. If Chris does the job alone in 300 min, then in 175 min he does $\frac{175}{300} = \frac{7}{12}$ of the job. If Kim does the job alone in $300 + 120$, or 420 min, then in 175 min she does $\frac{175}{420} = \frac{5}{12}$ of the job. Together, they do $\frac{7}{12} + \frac{5}{12}$, or 1 entire job, in 175 min. The result checks.

State. It would take Chris 300 min, or 5 hours to do the job alone.

33. **Familiarize**. We first make a drawing. Let r = the kayak's speed in still water in mph. Then $r - 3$ = the speed upstream and $r + 3$ = the speed downstream.

Upstream   4 miles $r - 3$ mph

10 miles $r + 3$ mph   Downstream

We organize the information in a table. The time is the same both upstream and downstream so we use t for each time.

	Distance	Speed	Time
Upstream	4	$r - 3$	t
Downstream	10	$r + 3$	t

Translate. Using the formula Time = Distance/Rate in each row of the table and the fact that the times are the same, we can write an equation.

$$\frac{4}{r-3} = \frac{10}{r+3}$$

Carry out. We solve the equation.

$$\frac{4}{r-3} = \frac{10}{r+3}, \quad \text{LCD is } (r-3)(r+3)$$
$$(r-3)(r+3)\cdot\frac{4}{r-3} = (r-3)(r+3)\cdot\frac{10}{r+3}$$
$$4(r+3) = 10(r-3)$$
$$4r + 12 = 10r - 30$$
$$42 = 6r$$
$$7 = r$$

Check. If $r = 7$ mph, then $r - 3$ is 4 mph and $r + 3$ is 10 mph. The time upstream is $\frac{4}{4}$, or 1 hour. The time downstream is $\frac{10}{10}$, or 1 hour. Since the times are the same, the answer checks.

State. The speed of the kayak in still water is 7 mph.

35. **Familiarize**. We first make a drawing. Let r = Drew's speed on a nonmoving sidewalk in ft/sec. Then his speed moving forward (120 ft) on the moving sidewalk is $r + 1.7$, and his speed in the opposite direction (52 ft) is $r - 1.7$.

We organize the information in a table. The time is the same both forward and in the opposite direction, so we use t for each time.

	Distance	Speed	Time
Forward	120	$r + 1.7$	t
Opposite direction	52	$r - 1.7$	t

Translate. Using the formula Time = Distance/Rate in each row of the table and the fact that the times are the same, we can write an equation.

$$\frac{120}{r+1.7} = \frac{52}{r-1.7}$$

Carry out. We solve the equation.

$$\frac{120}{r+1.7} = \frac{52}{r-1.7}$$
$$\text{LCD is } (r+1.7)(r-1.7)$$
$$(r+1.7)(r-1.7)\frac{120}{r+1.7} = (r+1.7)(r-1.7)\frac{52}{r-1.7}$$
$$120(r-1.7) = 52(r+1.7)$$
$$120r - 204 = 52r + 88.4$$
$$68r = 292.4$$
$$r = 4.3$$

Check. If Drew's speed on a nonmoving sidewalk is 4.3 ft/sec, then his speed moving forward on the moving sidewalk is $4.3 + 1.7$, or 6 ft/sec, and his speed moving in the opposite direction on the sidewalk is $4.3 - 1.7$, or 2.6 ft/sec. Moving 120 ft at 6 ft/sec takes

$\frac{120}{6} = 29$ sec . Moving 52 ft at 2.6 ft/sec takes

$\frac{52}{2.6} = 29$ sec . Since the times are the same, the answer checks.

State. Drew would be walking 4.3 ft/sec on a nonmoving sidewalk.

37. Familiarize. Let r = the speed of the passenger train in mph. Then $r - 14$ = the speed of the freight train in mph. We organize the information in a table. The time is the same for both trains so we use t for each time.

	Distance	Speed	Time
Passenger train	400	r	t
Freight train	330	$r-14$	t

Translate. Using the formula Time = Distance/Rate in each row of the table and the fact that the times are the same, we can write an equation.

$$\frac{400}{r} = \frac{330}{r-14}$$

Carry out. We solve the equation.

$$\frac{400}{r} = \frac{330}{r-14}, \quad \text{LCD is } r(r-14)$$

$$r(r-14) \cdot \frac{400}{r} = r(r-14) \cdot \frac{330}{r-14}$$

$$400(r-14) = 330r$$

$$400r - 5600 = 330r$$

$$-5600 = -70r$$

$$80 = r$$

Check. If the passenger train's speed is 80 mph, then the freight train's speed is $80 - 14$, or 66 mph. Traveling 400 mi at 80 mph takes $\frac{400}{80} = 5$ hr .

Traveling 330 mi at 66 mph takes $\frac{330}{66} = 5$ hr . Since the times are the same, the answer checks.

State. The speed of the passenger train is 80 mph; the speed of the freight train is 66 mph.

39. Note that 38 mi is 7 mi less than 45 mi and that the local bus travels 7 mph slower than the express. Then the express travels 45 mi in one hr, or 45 mph, and the local bus travels 38 mi in one hr, or 38 mph.

41. Familiarize. Let c = the speed of the current, in km/h. Then $2 + c$ = the speed downriver and $2 - c$ = the speed upriver. We organize the information in a table.

	Distance	Speed	Time
Downriver	4	$2+c$	t
Upriver	1	$2-c$	t

Translate. Using the formula Time = Distance/Rate in each row of the table and the fact that the times are the same, we can write an equation.

$$\frac{4}{2+c} = \frac{1}{2-c}$$

Carry out. We solve the equation. Multiply both sides by the LCD, $(2+c)(2-c)$.

$$\frac{4}{2+c} = \frac{1}{2-c}, \quad \text{LCD is } (2+c)(2-c)$$

$$(2+c)(2-c) \cdot \frac{4}{2+c} = (2+c)(2-c) \cdot \frac{1}{2-c}$$

$$4(2-c) = 2+c$$

$$8 - 4c = 2+c$$

$$6 = 5c$$

$$1\frac{1}{5} = c$$

Check. If the speed of the current is $1\frac{1}{5}$ km/h, the boat travels upriver at $2 - 1\frac{1}{5}$, or $\frac{4}{5}$ km/h and downriver at $2 + 1\frac{1}{5}$, or $3\frac{1}{5}$ km/h. Traveling 4 km at $3\frac{1}{5}$ km/h takes $\frac{4}{16/5} = 1\frac{1}{4}$ hr . Traveling 1 km at $\frac{4}{5}$ km/h takes $\frac{1}{4/5} = 1\frac{1}{4}$ hr . Since the times are the same, the answer checks.

State. The speed of the current is $1\frac{1}{5}$ km/h.

43. Familiarize. Let w = the wind speed, in mph. Then the speed into the wind is $460 - w$, and the speed with the wind is $460 + w$. We organize the information in a table.

	Distance	Speed	Time
Into the wind	525	$460-w$	t_1
With the wind	525	$460+w$	t_2

Translate. Using the formula Time = Distance/Rate, we see that $t_1 = \frac{525}{460-w}$ and $t_2 = \frac{525}{460+w}$. The total time into the wind and back is 2.3 hr, so $t_1 + t_2 = 2.3$, or

$$\frac{525}{460-w} + \frac{525}{460+w} = 2.3 .$$

Carry out. We solve the equation. Multiply both sides by the LCD, $(460+w)(460-w)$, or $460^2 - w^2$.

$$(460^2 - w^2)\left(\frac{525}{460-w} + \frac{525}{460+w}\right) = (460^2 - w^2)2.3$$

$$525(460-w) + 525(460+w) = 2.3(211,600 - w^2)$$

$$241,500 - 525w + 241,500 + 525w = 486,680 - 2.3w^2$$

$$483,000 = 486,680 - 2.3w^2$$

$$2.3w^2 - 3680 = 0$$

$$2.3(w^2 - 1600) = 0$$

$$2.3(w+40)(w-40) = 0$$

$$w = -40 \ \text{ or } \ w = 40$$

Check. We check only 40 since the wind speed cannot be negative. If the wind speed is 40 mph, then the plane's speed into the wind is $460 - 40$, or 420 mph, and the speed with the wind is $460 + 40$, or 500 mph. Flying 525 mi into the wind takes $\frac{525}{420} = 1.25$ hr. Flying 525 mi with the wind takes $\frac{525}{500} = 1.05$ hr. The total time is 1.25 + 1.05, or 2.3 hr. The answer checks.

State. The wind speed is 40 mph.

45. *Familiarize*. Let r = the speed at which the train actually traveled in mph, and let t = the actual travel time in hours. We organize the information in a table.

	Distance	Speed	Time
Actual speed	120	r	t
Faster speed	120	$r+10$	$t-2$

Translate. From the first row of the table we have $120 = rt$, and from the second row we have $120 = (r+10)(t-2)$. Solving the first equation for t, we have $t = \dfrac{120}{r}$. Substituting for t in the second equation, we have

$$120 = (r+10)\left(\dfrac{120}{r} - 2\right).$$

Carry out. We solve the equation.

$$120 = (r+10)\left(\dfrac{120}{r} - 2\right)$$
$$120 = 120 - 2r + \dfrac{1200}{r} - 20$$
$$20 = -2r + \dfrac{1200}{r}$$
$$r \cdot 20 = r\left(-2r + \dfrac{1200}{r}\right)$$
$$20r = -2r^2 + 1200$$
$$2r^2 + 20r - 1200 = 0$$
$$2(r^2 + 10r - 600) = 0$$
$$2(r+30)(r-20) = 0$$
$$r = -30 \ or \ r = 20$$

Check. Since speed cannot be negative in this problem, -30 cannot be a solution of the original problem. If the speed is 20 mph, it takes $\frac{120}{20}$, or 6 hr, to travel 120 mi. If the speed is 10 mph faster, or 30 mph, it takes $\frac{120}{30}$, or 4 hr, to travel 120 mi. Since 4 hr is 2 hr less time than 6 hr, the answer checks.
State. The speed was 20 mph.

47. *Writing Exercise*.

49. The interest formula is $I = Prt$, where $t = 1$ year. The amount in the 4% account is $2200. Then the amount in the 5% account is $5000 − $2200 = $2800. Find the interest from both accounts.

$$I = 2200 \cdot 0.04 \cdot 1 + 2800 \cdot 0.05 \cdot 1$$
$$I = 88 + 140$$
$$I = 228$$

Omar earned $228 in interest from both accounts.

51. *Familiarize*. Let x = the amount of lemon juice and y = the amount of linseed oil, in ounces.
Translate.

Amount of oil	is	Twice the amount of lemon juice
↓	↓	↓
y	$=$	$2x$

Total amount	is	32 oz
↓	↓	↓
$x + y$	$=$	32

We have a system of equations:

$$y = 2x \quad (1)$$
$$x + y = 32 \quad (2)$$

Carry out. We use the substitution method to solve the system of equations. Substitute $2x$ for y in Equation (2) and solve for x.

$$x + y = 32 \quad (2)$$
$$x + 2x = 32$$
$$3x = 32$$
$$x = 10\tfrac{2}{3}$$
$$y = 2x = 2\left(10\tfrac{2}{3}\right) = 21\tfrac{1}{3}$$

Check. The total amount is $10\tfrac{2}{3} + 21\tfrac{1}{3} = 32$. Since $2 \cdot 10\tfrac{2}{3} = 21\tfrac{1}{3}$, the amount of oil, is twice as much as the amount of lemon juice.
State. There is $21\tfrac{1}{3}$ oz of oil and $10\tfrac{2}{3}$ oz of lemon juice.

53. *Writing Exercise*.

55. *Familiarize*. If the drainage gate is closed, $\frac{1}{9}$ of the bog is filled in 1 hr. If the bog is not being filled, $\frac{1}{11}$ of the bog is drained in 1 hr. If the bog is being filled with the drainage gate left open, $\frac{1}{9} - \frac{1}{11}$ of the bog is filled in 1 hr. Let t = the time it takes to fill the bog with the drainage gate left open.
Translate. We want to find t such that

$$t\left(\tfrac{1}{9} - \tfrac{1}{11}\right) = 1, \ or \ \tfrac{t}{9} - \tfrac{t}{11} = 1.$$

Carry out. We solve the equation. First we multiply by the LCD, 99.

$$99\left(\tfrac{t}{9} - \tfrac{t}{11}\right) = 99 \cdot 1$$
$$11t - 9t = 99$$
$$2t = 99$$
$$t = \dfrac{99}{2}$$

Check. In $\frac{99}{2}$ hr, we have

$$\dfrac{99}{2}\left(\tfrac{1}{9} - \tfrac{1}{11}\right) = \tfrac{11}{2} - \tfrac{9}{2} = \tfrac{2}{2} = 1 \ \text{full bog.}$$

State. It will take $\frac{99}{2}$, or $49\tfrac{1}{2}$ hr, to fill the bog.

57. Sean's speed downstream is 7 + 3, or 10 mph. Using Time = Distance/Rate, we find that the time it will take Sean to kayak 5 mi downstream is 5/10, or 1/2 hr, or 30 min.

59. *Familiarize*. Let p = the number of people per hour moved by the 60 cm-wide escalator. Then $2p$ = the number of people per hour moved by the 100 cm-wide escalator. We convert 1575 people per 14 minutes to people per hour:

$$\frac{1575 \text{ people}}{14 \text{ min}} \cdot \frac{60 \text{ min}}{1 \text{ hr}} = 6750 \text{ people / hr}$$

Translate. We use the information that together the escalators move 6750 people per hour to write an equation.

$$p + 2p = 6750$$

Carry out. We solve the equation.

$$p + 2p = 6750$$
$$3p = 6750$$
$$p = 2250$$

Check. If the 60 cm-wide escalator moves 2250 people per hour, then the 100 cm-wide escalator moves $2 \cdot 2250$, or 4500 people per hour. Together, they move $2250 + 4500$, or 6750 people per hour. The answer checks.

State. The 60 cm-wide escalator moves 2250 people per hour.

61. Familiarize. Let $d =$ the distance, in miles, the paddleboat can cruise upriver before it is time to turn around. The boat's speed upriver is $12 - 5$, or 7 mph, and its speed downriver is $12 + 5$, or 17 mph. We organize the information in a table.

	Distance	Speed	Time
Upriver	d	7	t_1
Downriver	d	17	t_2

Translate. Using the formula Time = Distance/Rate we see that $t_1 = \frac{d}{7}$ and $t_2 = \frac{d}{17}$. The time upriver and back is 3 hr, so $t_1 + t_2 = 3$, or $\frac{d}{7} + \frac{d}{17} = 3$.

Carry out. We solve the equation.

$$7 \cdot 17 \left(\frac{d}{7} + \frac{d}{17} \right) = 7 \cdot 17 \cdot 3$$
$$17d + 7d = 357$$
$$24d = 357$$
$$d = \frac{119}{8}$$

Check. Traveling $\frac{119}{8}$ mi upriver at a speed of 7 mph takes $\frac{119/8}{7} = \frac{17}{8}$ hr. Traveling $\frac{119}{8}$ mi downriver at a speed of 17 mph takes $\frac{119/8}{17} = \frac{7}{8}$ hr. The total time is $\frac{17}{8} + \frac{7}{8} = \frac{24}{8} = 3$ hr. The answer checks.

State. The pilot can go $\frac{119}{8}$, or $14\frac{7}{8}$ mi upriver before it is time to turn around.

63. Familiarize. Let t represent the time it takes the printers to print 500 pages working together. Translate. The faster machine can print 500 pages in 4 min, and it takes the slower printer 5 min to do the same job. Then we have

$$\frac{t}{4} + \frac{t}{5} = 1.$$

Carry out. We solve the equation.

$$\frac{t}{4} + \frac{t}{5} = 1, \quad \text{LCD is 20}$$
$$20 \left(\frac{t}{4} + \frac{t}{5} \right) = 20 \cdot 1$$
$$5t + 4t = 20 \cdot 1$$
$$9t = 20$$
$$t = \frac{20}{9}$$

In $\frac{20}{9}$ min, the faster printer does $\frac{20/9}{4}$, or $\frac{20}{9} \cdot \frac{1}{4}$, or $\frac{5}{9}$ of the job. Then starting at page 1, it would print $\frac{5}{9} \cdot 500$, or $277\frac{7}{9}$ pages. Thus, in $\frac{20}{9}$ min, the two machines will meet on page 278.

Check. We can check to see that the slower machine is also printing page 278 after $\frac{20}{9}$ min. In $\frac{20}{9}$ min, the slower machine does $\frac{20/9}{5}$, or $\frac{20}{9} \cdot \frac{1}{5}$, or $\frac{4}{9}$ of the job. Then it would print $\frac{4}{9} \cdot 500$, or $222\frac{2}{9}$ pages. Working backward from page 500, this machine would be on page $500 - 222\frac{2}{9}$, or $277\frac{7}{9}$. Thus, both machines are printing page 278 after $\frac{20}{9}$ min. The answer checks.

State. The two machines will meet on page 278.

65. Familiarize. Express the position of the hands in terms of minute units on the face of the clock. At 10:30 the hour hand is at $\frac{10.5}{12}$ hr $\times \frac{60 \text{ min}}{1 \text{ hr}}$, or 52.5 minutes, and the minute hand is at 30 minutes. The rate of the minute hand is 12 times the rate of the hour hand. (When the minute hand moves 60 minutes, the hour hand moves 5 minutes.) Let $t =$ the number of minutes after 10:30 that the hands will first be perpendicular. After t minutes the minute hand has moved t units, and the hour hand has moved $\frac{t}{12}$ units. The position of the hour hand will be 15 units "ahead" of the position of the minute hand when they are first perpendicular.

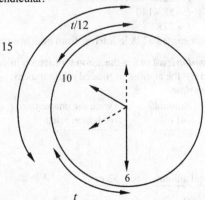

Translate.

Position of hour hand after t min	is	position of minute hand after t min	plus	15 min.
$\downarrow$	$\downarrow$	$\downarrow$	$\downarrow$	$\downarrow$
$52.5 + \dfrac{t}{12}$	$=$	$30 + t$	$+$	15

Carry out. We solve the equation.

$$52.5 + \frac{t}{12} = 30 + t + 15$$

$$52.5 + \frac{t}{12} = 45 + t, \quad \text{LCM is } 12$$

$$12\left(52.5 + \frac{t}{12}\right) = 12(45 + t)$$

$$630 + t = 540 + 12t$$

$$90 = 11t$$

$$\frac{90}{11} = t, \text{ or } 8\frac{2}{11} = t$$

Check. At $\dfrac{90}{11}$ min after 10:30, the position of the hour hand is at $52.5 + \dfrac{90/11}{12}$, or $53\dfrac{2}{11}$ min. The minute hand is at $30 + \dfrac{90}{11}$, or $38\dfrac{2}{11}$ min. The hour hand is 15 minutes ahead of the minute hand so the hands are perpendicular. The answer checks.

State. After 10:30 the hands of a clock will first be perpendicular in $8\dfrac{2}{11}$ min. The time is $10{:}38\dfrac{2}{11}$, or $21\dfrac{9}{11}$ min before 11:00.

67. Familiarize. Let $r =$ the speed in mph Garry would have to travel for the last half of the trip in order to average a speed of 45 mph for the entire trip. We organize the information in a table.

	Distance	Speed	Time
First half	50	40	t_1
Last half	50	r	t_2

The total distance is $50 + 50$, or 100 mi.

The total time is $t_1 + t_2$, or $\dfrac{50}{40} + \dfrac{50}{r}$, or $\dfrac{5}{4} + \dfrac{50}{r}$. The average speed is 45 mph.

Translate.

$$\text{Average speed} = \frac{\text{Total distance}}{\text{Total time}}$$

$$45 = \frac{100}{\dfrac{5}{4} + \dfrac{50}{r}}$$

Carry out. We solve the equation.

$$45 = \frac{100}{\dfrac{5}{4} + \dfrac{50}{r}}$$

$$45 = \frac{100}{\dfrac{5r + 200}{4r}}$$

$$45 = 100 \cdot \frac{4r}{5r + 200}$$

$$45 = \frac{400r}{5r + 200}$$

$$(5r + 200)(45) = (5r + 200) \cdot \frac{400r}{5r + 200}$$

$$225r + 9000 = 400r$$

$$9000 = 175r$$

$$\frac{360}{7} = r$$

Check. Traveling 50 mi at 40 mph takes $\dfrac{50}{40}$, or $\dfrac{5}{4}$ hr. Traveling 50 mi at $\dfrac{360}{7}$ mph takes $\dfrac{50}{360/7}$, or $\dfrac{35}{36}$ hr. Then the total time is $\dfrac{5}{4} + \dfrac{35}{36} = \dfrac{80}{36} = \dfrac{20}{9}$ hr.

The average speed when traveling 100 mi for $\dfrac{20}{9}$ hr is

$$\frac{100}{20/9} = 45 \text{ mph. The answer checks.}$$

State. Garry would have to travel at a speed of $\dfrac{360}{7}$, or $51\dfrac{3}{7}$ mph for the last half of the trip so that the average speed for the entire trip would be 45 mph.

Exercise Set 6.6

1. The divisor is $x - 3$.

3. The quotient is $x + 2$.

5. The degree of the divisor, $x - 3$, is 1.

7. $\dfrac{36x^6 + 18x^5 - 27x^2}{9x^2} = \dfrac{36x^6}{9x^2} + \dfrac{18x^5}{9x^2} - \dfrac{27x^2}{9x^2}$
$\qquad = 4x^4 + 2x^3 - 3$

9. $\dfrac{21a^3 + 7a^2 - 3a - 14}{-7a}$
$= \dfrac{21a^3}{-7a} + \dfrac{7a^2}{-7a} - \dfrac{3a}{-7a} - \dfrac{14}{-7a}$
$= -3a^2 - a + \dfrac{3}{7} + \dfrac{2}{a}$

11. $\dfrac{16y^4z^2 - 8y^6z^4 + 12y^8z^3}{-4y^4z}$
$= \dfrac{16y^4z^2}{-4y^4z} - \dfrac{8y^6z^4}{-4y^4z} + \dfrac{12y^8z^3}{-4y^4z}$
$= -4z + 2y^2z^3 - 3y^4z^2$

13. $\dfrac{16y^3 - 9y^2 - 8y}{2y^2}$
$= \dfrac{16y^3}{2y^2} - \dfrac{9y^2}{2y^2} - \dfrac{8y}{2y^2}$
$= 8y - \dfrac{9}{2} - \dfrac{4}{y}$

15. $\dfrac{15x^7 - 21x^4 - 3x^2}{-3x^2}$
$= \dfrac{15x^7}{-3x^2} + \dfrac{-21x^4}{-3x^2} + \dfrac{-3x^2}{-3x^2}$
$= -5x^5 + 7x^2 + 1$

17. $(a^2b - a^3b^3 - a^5b^5) \div (a^2b)$

$= \dfrac{a^2b}{a^2b} - \dfrac{a^3b^3}{a^2b} - \dfrac{a^5b^5}{a^2b}$

$= 1 - ab^2 - a^3b^4$

19. $(x^2 + 10x + 21) \div (x + 7)$

$= \dfrac{(x+7)(x+3)}{x+7}$

$= \dfrac{\cancel{(x+7)}(x+3)}{\cancel{x+7}}$

$= x + 3$

The answer is $x + 3$.

21.
$$
\begin{array}{r}
y - 5 \\
y-5\overline{)y^2 - 10y - 25} \\
\underline{y^2 - 5y} \\
-5y - 25 \\
\underline{-5y + 25} \\
-50
\end{array}
$$

$(y^2 - 10y) - (y^2 - 5y) = -5y$

$(-5y - 25) - (-5y + 25) = -50$

The answer is $y - 5$, R -50, or $y - 5 + \dfrac{-50}{y-5}$.

23.
$$
\begin{array}{r}
x - 5 \\
x-4\overline{)x^2 - 9x + 21} \\
\underline{x^2 - 4x} \\
-5x + 21 \\
\underline{-5x + 20} \\
1
\end{array}
$$

The answer is $x - 5$, R 1, or $x - 5 + \dfrac{1}{x-4}$.

25. $(y^2 - 25) \div (y + 5) = \dfrac{y^2 - 25}{y + 5}$

$= \dfrac{(y+5)(y-5)}{y+5}$

$= \dfrac{\cancel{(y+5)}(y-5)}{\cancel{y+5}}$

$= y - 5$

We could also find this quotient as follows.

$$
\begin{array}{r}
y - 5 \\
y+5\overline{)y^2 + 0y - 25} \\
\underline{y^2 + 5y} \\
-5y - 25 \\
\underline{-5y - 25} \\
0
\end{array}
$$
Writing in the missing term

The answer is $y - 5$.

27. $\dfrac{a^3 + 8}{a + 2} = \dfrac{(a+2)(a^2 - 2a + 4)}{a + 2}$

$= \dfrac{\cancel{(a+2)}(a^2 - 2a + 4)}{\cancel{a+2}}$

$= a^2 - 2a + 4$

We could also find this quotient as follows.

$$
\begin{array}{r}
a^2 - 2a + 4 \\
a+2\overline{)a^3 + 0a^2 + 0a + 8} \\
\underline{a^3 + 2a^2} \\
-2a^2 + 0a \\
\underline{-2a^2 + 4a} \\
-4a + 8 \\
\underline{-4a + 8} \\
0
\end{array}
$$
Writing in the missing terms

The answer is $a^2 - 2a + 4$.

29.
$$
\begin{array}{r}
x - 3 \\
5x+1\overline{)5x^2 - 14x + 0} \\
\underline{5x^2 + x} \\
-15x + 0 \\
\underline{-15x - 3} \\
3
\end{array}
$$

The answer is $x - 3$, R 3, or $x - 3 + \dfrac{3}{5x+1}$.

31.
$$
\begin{array}{r}
y^2 - 2y - 1 \\
y-2\overline{)y^3 - 4y^2 + 3y - 6} \\
\underline{y^3 - 2y^2} \\
-2y^2 + 3y \\
\underline{-2y^2 + 4y} \\
-y - 6 \\
\underline{-y + 2} \\
-8
\end{array}
$$

The answer is $y^2 - 2y - 1$, R -8, or

$y^2 - 2y - 1 + \dfrac{-8}{y-2}$.

33.
$$
\begin{array}{r}
2x^2 - x + 1 \\
x+2\overline{)2x^3 + 3x^2 - x - 3} \\
\underline{2x^3 + 4x^2} \\
-x^2 - x \\
\underline{-x^2 - 2x} \\
x - 3 \\
\underline{x + 2} \\
-5
\end{array}
$$

The answer is $2x^2 - x + 1$, R -5, or

$2x^2 - x + 1 + \dfrac{-5}{x+2}$.

35.
$$
\begin{array}{r}
a^2 - 4a + 6 \\
a+4\overline{)a^3 + 0a^2 - 10a + 24} \\
\underline{a^3 + 4a^2} \\
-4a^2 - 10a \\
\underline{-4a^2 - 16a} \\
6a + 24 \\
\underline{6a + 24} \\
0
\end{array}
$$

The answer is $a^2 - 4a + 6$.

37.

$$5y - 2 \overline{\smash{\big)}\, 10y^3 + 6y^2 - 9y + 10}$$

quotient $2y^2 + 2y - 1$

$$
\begin{array}{r}
2y^2 + 2y - 1 \\
5y - 2 \overline{\smash{\big)}\, 10y^3 + 6y^2 - 9y + 10} \\
\underline{10y^3 - 4y^2} \\
10y^2 - 9y \\
\underline{10y^2 - 4y} \\
-5y + 10 \\
\underline{-5y + 2} \\
8
\end{array}
$$

The answer is $2y^2 + 2y - 1$, R 8, or

$$2y^2 + 2y - 1 + \frac{8}{5y - 2}.$$

39.

$$
\begin{array}{r}
3x^2 + x + 1 \\
x^2 - 3 \overline{\smash{\big)}\, 3x^4 + x^3 - 8x^2 - 3x - 3} \\
\underline{3x^4 - 9x^2} \\
x^3 + x^2 - 3x \\
\underline{x^3 - 3x} \\
x^2 - 3 \\
\underline{x^2 - 3} \\
0
\end{array}
$$

The answer is $3x^2 + x + 1$.

41.

$$
\begin{array}{r}
2x^2 - x - 9 \\
x^2 + 2 \overline{\smash{\big)}\, 2x^4 - x^3 - 5x^2 + x - 6} \\
\underline{2x^4 + 4x^2} \\
-x^3 - 9x^2 + x \\
\underline{-x^3 - 2x} \\
-9x^2 + 3x - 6 \\
\underline{-9x^2 - 18} \\
3x + 12
\end{array}
$$

The answer is $2x^2 - x - 9$, R $3x + 12$, or

$$2x^2 - x - 9 + \frac{3x + 12}{x^2 + 2}.$$

43. $F(x) = \dfrac{f(x)}{g(x)} = \dfrac{6x^2 - 11x - 10}{3x + 2}$

$$
\begin{array}{r}
2x - 5 \\
3x + 2 \overline{\smash{\big)}\, 6x^2 - 11x - 10} \\
\underline{6x^2 + 4x} \\
-15x - 10 \\
\underline{-15x - 10} \\
0
\end{array}
$$

Since $g(x)$ is 0 for $x = -\dfrac{2}{3}$, we have

$F(x) = 2x - 5$, provided $x \neq -\dfrac{2}{3}$.

45. $F(x) = \dfrac{f(x)}{g(x)} = \dfrac{8x^3 - 27}{2x - 3}$

$$
\begin{array}{r}
4x^2 + 6x + 9 \\
2x - 3 \overline{\smash{\big)}\, 8x^3 - 27} \\
\underline{8x^3 - 12x^2} \\
12x^2 + 0x \\
\underline{12x^2 - 18x} \\
18x - 27 \\
\underline{18x - 27} \\
0
\end{array}
$$

Since $g(x)$ is 0 for $x = \dfrac{3}{2}$, we have

$F(x) = 4x^2 + 6x + 9$, provided $x \neq \dfrac{3}{2}$.

47. $F(x) = \dfrac{f(x)}{g(x)} = \dfrac{x^4 - 24x^2 - 25}{x^2 - 25}$

$$
\begin{array}{r}
x^2 + 1 \\
x^2 - 25 \overline{\smash{\big)}\, 5x^4 - 24x^2 - 25} \\
\underline{x^4 - 25x^2} \\
x^2 - 25 \\
\underline{x^2 - 25} \\
0
\end{array}
$$

Since $g(x)$ is 0 for $x = -5$ or $x = 5$, we have

$F(x) = x^2 + 1$, provided $x \neq -5$ and $x \neq 5$.

49. We rewrite $f(x)$ in descending order.

$$F(x) = \frac{f(x)}{g(x)} = \frac{2x^5 - 3x^4 - 2x^3 + 8x^2 - 5}{x^2 - 1}$$

$$
\begin{array}{r}
2x^3 - 3x^2 + 5 \\
x^2 - 1 \overline{\smash{\big)}\, 2x^5 - 3x^4 - 2x^3 + 8x^2 - 5} \\
\underline{2x^5 - 2x^3} \\
-3x^4 + 8x^2 \\
\underline{-3x^4 + 3x^2} \\
5x^2 - 5 \\
\underline{5x^2 - 5} \\
0
\end{array}
$$

Since $g(x)$ is 0 for $x = -1$ or $x = 1$, we have

$F(x) = 2x^3 - 3x^2 + 5$, provided $x \neq -1$ and $x \neq 1$.

51. *Writing Exercise.*

53. Graph $3x - y = 9$

$y = 3x - 9$

This graph is a line with y-intercept $(0, -9)$ and slope 3.

55. Graph $y < \frac{5}{2}x$

The graph is a dotted line with y-intercept $(0, 0)$ and shading below the line.

57. Graph $y = -\frac{3}{4}x + 1$

This graph is a line with slope $-\frac{3}{4}$ and y-intercept $(0, 1)$.

59. *Writing Exercise.*

61.
$$
\begin{array}{r}
a^2 + ab \\
a^2 + 3ab + 2b^2 \overline{\smash{\big)}\, a^4 + 4a^3b + 5a^2b^2 + 2ab^3} \\
\underline{a^4 + 3a^3b + 2a^2b^2} \\
a^3b + 3a^2b^2 + 2ab^3 \\
\underline{a^3b + 3a^2b^2 + 2ab^3} \\
0
\end{array}
$$

The answer is $a^2 + ab$.

63.
$$
\begin{array}{r}
a^6 - a^5b + a^4b^2 - a^3b^3 + a^2b^4 - ab^5 + b^6 \\
a+b \overline{\smash{\big)}\, a^7 \qquad\qquad\qquad\qquad\qquad + b^7} \\
\underline{a^7 + a^6b} \\
-a^6b \\
\underline{-a^6b - a^5b^2} \\
a^5b^2 \\
\underline{a^5b^2 + a^4b^3} \\
-a^4b^3 \\
\underline{-a^4b^3 - a^3b^4} \\
a^3b^4 \\
\underline{a^3b^4 + a^2b^5} \\
-a^2b^5 \\
\underline{-a^2b^5 - ab^6} \\
ab^6 + b^7 \\
\underline{ab^6 + b^7} \\
0
\end{array}
$$

The answer is
$a^6 - a^5b + a^4b^2 - a^3b^3 + a^2b^4 - ab^5 + b^6$.

65.
$$
\begin{array}{r}
x - 5 \\
x + 2 \overline{\smash{\big)}\, x^2 - 3x + 2k} \\
\underline{x^2 + 2x} \\
-5x + 2k \\
\underline{-5x \quad -10} \\
2k + 10
\end{array}
$$

The remainder is 7. Thus, we solve the following

equation for k.
$$2k + 10 = 7$$
$$2k = -3$$
$$k = -\frac{3}{2}$$

67. *Writing Exercise.*

69. *Graphing Calculator Exercise*

Exercise Set 6.7

1. True

3. False; synthetic division is a streamlined version of long division so it cannot be used where long division cannot be used.

5. True

7. $\left(x^3 - 4x^2 - 2x + 5\right) \div (x - 1)$

$$
\begin{array}{r|rrrr}
1 & 1 & -4 & -2 & 5 \\
  &   & 1 & -3 & -5 \\
\hline
  & 1 & -3 & -5 & \,|\,0
\end{array}
$$

The answer is $x^2 - 3x - 5$.

9. $\left(a^2 + 8a + 11\right) \div (a + 3)$

$= \left(a^2 + 8a + 11\right) \div [a - (-3)]$

$$
\begin{array}{r|rrr}
-3 & 1 & 8 & 11 \\
   &   & -3 & -15 \\
\hline
   & 1 & 5 & \,|\,-4
\end{array}
$$

The answer is $a + 5$, R -4, or $a + 5 + \dfrac{-4}{a+3}$.

11. $\left(2x^3 - x^2 - 7x + 14\right) \div (x + 2)$

$= \left(2x^3 - x^2 - 7x + 14\right) \div [x - (-2)]$

$$
\begin{array}{r|rrrr}
-2 & 2 & -1 & -7 & 14 \\
   &   & -4 & 10 & -6 \\
\hline
   & 2 & -5 & 3 & \,|\,8
\end{array}
$$

The answer is $2x^2 - 5x + 3$, R 8, or

$2x^2 - 5x + 3 + \dfrac{8}{x+2}$.

13. $\left(a^3 - 10a + 12\right) \div (a - 2)$

$= \left(a^3 + 0a^2 - 10a + 12\right) \div (a - 2)$

$$
\begin{array}{r|rrrr}
2 & 1 & 0 & -10 & 12 \\
  &   & 2 & 4 & -12 \\
\hline
  & 1 & 2 & -6 & \,|\,0
\end{array}
$$

The answer is $a^2 + 2a - 6$.

15. $\left(3y^3 - 7y^2 - 20\right) \div (y - 3)$

$= \left(3y^3 - 7y^2 + 0y - 20\right) \div (y - 3)$

$$
\begin{array}{r|rrrr}
3 & 3 & -7 & 0 & -20 \\
  &   & 9 & 6 & 18 \\
\hline
  & 3 & 2 & 6 & \,|\,-2
\end{array}
$$

The answer is $3y^2 + 2y + 6$, R -2, or

$3y^2 + 2y + 6 + \dfrac{-2}{y-3}$.

17. $(x^5 - 32) \div (x - 2)$

$= (x^5 + 0x^4 + 0x^3 + 0x^2 + 0x - 32) \div (x - 2)$

$$\underline{2|} \quad \begin{array}{rrrrrr} 1 & 0 & 0 & 0 & 0 & -32 \\ & 2 & 4 & 8 & 16 & 32 \\ \hline 1 & 2 & 4 & 8 & 16 & |\,0 \end{array}$$

The answer is $x^4 + 2x^3 + 4x^2 + 8x + 16$.

19. $\left(3x^3 + 1 - x + 7x^2\right) \div \left(x + \dfrac{1}{3}\right)$

$= \left(3x^3 + 7x^2 - x + 1\right) \div \left[x - \left(-\dfrac{1}{3}\right)\right]$

$$\underline{-\tfrac{1}{3}|} \quad \begin{array}{rrrr} 3 & 7 & -1 & 1 \\ & -1 & -2 & 1 \\ \hline 3 & 6 & -3 & |\,2 \end{array}$$

The answer is $3x^2 + 6x - 3$, R 2, or

$3x^2 + 6x - 3 + \dfrac{2}{x + \frac{1}{3}}$

21. $$\underline{-3|} \quad \begin{array}{rrrrr} 5 & 12 & 0 & 28 & 9 \\ & -15 & 9 & -27 & -3 \\ \hline 5 & -3 & 9 & 1 & |\,6 \end{array}$$

The remainder tells us that $f(-3) = 6$.

23. $$\underline{-3|} \quad \begin{array}{rrrrr} 2 & -1 & -7 & 1 & 2 \\ & -6 & 21 & -42 & 123 \\ \hline 2 & -7 & 14 & -41 & |\,125 \end{array}$$

The remainder tells us that $P(-3) = 125$.

25. $$\underline{4|} \quad \begin{array}{rrrrr} 1 & -6 & 11 & -17 & 20 \\ & 4 & -8 & 12 & -20 \\ \hline 1 & -2 & 3 & -5 & |\,0 \end{array}$$

The remainder tells us that $f(4) = 0$.

27. *Writing Exercise.*

29. $f(x) = 3x - 4$

31. $m = \dfrac{2 - 7}{4 - 1} = -\dfrac{5}{3}$

$y - 2 = -\dfrac{5}{3}(x - 4)$

$y - 2 = -\dfrac{5}{3}x + \dfrac{20}{3}$

$f(x) = -\dfrac{5}{3}x + \dfrac{26}{3}$

33. The line $2x - y = 7$ can be written as $y = 2x - 7$ which has slope 2. Now find a linear function perpendicular to this line containing the point (0, 6), the y-intercept.

$f(x) = -\dfrac{1}{2}x + 6$

35. *Writing Exercise.*

37. a. The degree of the remainder must be less than the degree of the divisor. Thus, the degree of the remainder must be 0, so R must be a constant.

b. $P(x) = (x - r) \cdot Q(x) + R$

$P(r) = (r - r) \cdot Q(r) + R = 0 \cdot Q(r) + R = R$

39. $$\underline{-3|} \quad \begin{array}{rrrr} 4 & 16 & -3 & -45 \\ & -12 & -12 & 45 \\ \hline 4 & 4 & -15 & |\,0 \end{array}$$

The remainder tells us that $f(-3) = 0$.

$f(x) = (x + 3)(4x^2 + 4x - 15) = (x + 3)(2x + 5)(2x - 3)$

Solve $f(x) = 0$:

$(x + 3)(2x + 5)(2x - 3) = 0$

$x + 3 = 0 \quad or \quad 2x + 5 = 0 \quad or \quad 2x - 3 = 0$

$x = -3 \quad or \qquad x = -\dfrac{5}{2} \quad or \qquad x = \dfrac{3}{2}$

The solutions are -3, $-\dfrac{5}{2}$, and $\dfrac{3}{2}$.

41. *Graphing Calculator Exercise*

43. $f(x) = 4x^3 + 16x^2 - 3x - 45$

$= x(4x^2 + 16x - 3) - 45$

$= x(x(4x + 16) - 3) - 45$

$f(-3) = -3(-3(4(-3) + 16) - 3) - 45$

$= -3(-3(-12 + 16) - 3) - 45$

$= -3(-3 \cdot 4 - 3) - 45$

$= -3(-12 - 3) - 45$

$= -3(-15) - 45$

$= 45 - 45$

$= 0$

Exercise Set 6.8

1. (d) LCM

3. (e) Product

5. (a) Directly

7. As the number of painters increases, the time required to scrape the house decreases, so we have inverse variation.

9. As the number of laps increases, the time required to swim them increases, so we have direct variation.

11. As the number of volunteers increases, the time required to wrap the toys decreases, so we have inverse variation.

13. $f = \dfrac{L}{d}$

$df = L$ Multiplying by d

$d = \dfrac{L}{f}$ Dividing by f

15. $s = \dfrac{(v_1 + v_2)t}{2}$

$2s = (v_1 + v_2)t$ Multiplying by 2

$\dfrac{2s}{t} = v_1 + v_2$ Dividing by t

$\dfrac{2s}{t} - v_2 = v_1$

This result can also be expressed as $v_1 = \dfrac{2s - tv_2}{t}$.

17.
$$\frac{t}{a} + \frac{t}{b} = 1$$
$$ab\left(\frac{t}{a} + \frac{t}{b}\right) = ab \cdot 1 \qquad \text{Multiplying by the LCD}$$
$$ab \cdot \frac{t}{a} + ab \cdot \frac{t}{b} = ab$$
$$bt + at = ab$$
$$at = ab - bt$$
$$at = b(a - t) \qquad \text{Factoring}$$
$$\frac{at}{a - t} = b$$

19.
$$R = \frac{gs}{g + s}$$
$$(g + s) \cdot R = (g + s) \cdot \frac{gs}{g + s} \qquad \text{Multiplying by the LCD}$$
$$Rg + Rs = gs$$
$$Rs = gs - Rg$$
$$Rs = g(s - R) \qquad \text{Factoring out } g$$
$$\frac{Rs}{s - R} = g \qquad \text{Multiplying by } \frac{1}{s - R}$$

21.
$$I = \frac{nE}{R + nr}$$
$$I(R + nr) = \frac{nE}{R + nr} \cdot (R + nr) \qquad \begin{array}{l}\text{Multiplying}\\\text{by the LCD}\end{array}$$
$$IR + Inr = nE$$
$$IR = nE - Inr$$
$$IR = n(E - Ir)$$
$$\frac{IR}{E - Ir} = n$$

23.
$$\frac{1}{p} + \frac{1}{q} = \frac{1}{f}$$
$$pqf\left(\frac{1}{p} + \frac{1}{q}\right) = pqf \cdot \frac{1}{f} \qquad \begin{array}{l}\text{Multiplying by}\\\text{the LCD}\end{array}$$
$$qf + pf = pq$$
$$pf = pq - qf$$
$$pf = q(p - f)$$
$$\frac{pf}{p - f} = q$$

25.
$$S = \frac{H}{m(t_1 - t_2)}$$
$$(t_1 - t_2)S = \frac{H}{m} \qquad \text{Multiplying by } t_1 - t_2$$
$$t_1 - t_2 = \frac{H}{Sm} \qquad \text{Dividing by } S$$
$$t_1 = \frac{H}{Sm} + t_2, \text{ or } \frac{H + Smt_2}{Sm}$$

27.
$$\frac{E}{e} = \frac{R + r}{r}$$
$$er \cdot \frac{E}{e} = er \cdot \frac{R + r}{r} \qquad \text{Multiplying by the LCD}$$
$$Er = e(R + r)$$
$$Er = eR + er$$
$$Er - er = eR$$
$$r(E - e) = eR$$
$$r = \frac{eR}{E - e}$$

29.
$$S = \frac{a}{1 - r}$$
$$(1 - r)S = a \qquad \text{Multiplying by the LCD, } 1 - r$$
$$1 - r = \frac{a}{S} \qquad \text{Dividing by } S$$
$$1 - \frac{a}{S} = r \qquad \text{Adding } r \text{ and } -\frac{a}{S}$$

This result can also be expressed as $r = \frac{S - a}{S}$.

31.
$$c = \frac{f}{(a + b)c}$$
$$\frac{a + b}{c} \cdot c = \frac{a + b}{c} \cdot \frac{f}{(a + b)c}$$
$$a + b = \frac{f}{c^2}$$

33.
$$P = \frac{A}{1 + r}$$
$$P(1 + r) = \frac{A}{1 + r} \cdot (1 + r)$$
$$P(1 + r) = A$$
$$1 + r = \frac{A}{P}$$
$$r = \frac{A}{P} - 1, \text{ or } \frac{A - P}{P}$$

35.
$$v = \frac{d_2 - d_1}{t_2 - t_1}$$
$$(t_2 - t_1)v = (t_2 - t_1) \cdot \frac{d_2 - d_1}{t_2 - t_1}$$
$$(t_2 - t_1)v = d_2 - d_1$$
$$t_2 - t_1 = \frac{d_2 - d_1}{v}$$
$$-t_1 = -t_2 + \frac{d_2 - d_1}{v}$$
$$t_1 = t_2 - \frac{d_2 - d_1}{v}, \text{ or } \frac{t_2 v - d_2 + d_1}{v}$$

37.
$$\frac{1}{t} = \frac{1}{a} + \frac{1}{b}$$
$$tab \cdot \frac{1}{t} = tab\left(\frac{1}{a} + \frac{1}{b}\right)$$
$$ab = tb + ta$$
$$ab = t(b + a)$$
$$\frac{ab}{b + a} = t$$

39.
$$A = \frac{2Tt + Qq}{2T + Q}$$
$$(2T + Q) \cdot A = (2T + Q) \cdot \frac{2Tt + Qq}{2T + Q}$$
$$2AT + AQ = 2Tt + Qq$$
$$AQ - Qq = 2Tt - 2AT \qquad \text{Adding } -2AT \text{ and } -Qq$$
$$Q(A - q) = 2Tt - 2AT$$
$$Q = \frac{2Tt - 2AT}{A - q}$$

41.
$$p = \frac{-98.42 + 4.15c - 0.082w}{w}$$
$$pw = -98.42 + 4.15c - 0.082w$$
$$pw + 0.082w = -98.42 + 4.15c$$
$$w(p + 0.082) = -98.42 + 4.15c$$
$$w = \frac{-98.42 + 4.15c}{p + 0.082}$$

43. $y = kx$

$30 = k \cdot 5$ Substituting

$6 = k$

The variation constant is 6.

The equation of variation is $y = 6x$.

45. $y = kx$

$3.4 = k \cdot 2$ Substituting

$1.7 = k$

The variation constant is 1.7.

The equation of variation is $y = 1.7x$.

47. $y = kx$

$2 = k \cdot \frac{1}{5}$ Substituting

$10 = k$ Multiplying by 5

The variation constant is 10.

The equation of variation is $y = 10x$.

49. $y = \frac{k}{x}$

$5 = \frac{k}{20}$ Substituting

$100 = k$

The variation constant is 100.

The equation of variation is $y = \frac{100}{x}$.

51. $y = \frac{k}{x}$

$11 = \frac{k}{4}$ Substituting

$44 = k$

The variation constant is 44.

The equation of variation is $y = \frac{44}{x}$.

53. $y = \frac{k}{x}$

$27 = \frac{k}{\frac{1}{3}}$ Substituting

$9 = k$

The variation constant is 9.

The equation of variation is $y = \frac{9}{x}$.

55. *Familiarize*. Because of the phrase "*d* … varies directly as … *m*," we express the distance as a function of the mass. Thus we have $d(m) = km$. We know that $d(3) = 20$.

Translate. We find the variation constant and then find the equation of variation.

$d(m) = km$

$d(3) = k \cdot 3$ Replacing m with 3

$20 = k \cdot 3$ Replacing $d(3)$ with 20

$\frac{20}{3} = k$ Variation constant

The equation of variation is $d(m) = \frac{20}{3}m$.

Carry out. We compute $d(5)$.

$d(m) = \frac{20}{3}m$

$d(5) = \frac{20}{3} \cdot 5$ Replacing m with 5

$= \frac{100}{3}$, or $33\frac{1}{3}$

Check. Reexamine the calculations.

State. The distance is $33\frac{1}{3}$ cm.

57. *Familiarize*. Because T varies inversely as P, we write $T(P) = k / P$. We know that $T(7) = 5$.

Translate. We find the variation constant and the equation of variation.

$T(P) = \frac{k}{P}$

$T(7) = \frac{k}{7}$ Replacing P with 7

$5 = \frac{k}{7}$ Replacing $T(P)$ with 5

$35 = k$ Variation constant

$T(P) = \frac{35}{P}$ Equation of variation

Carry out. We find $T(10)$.

$T(10) = \frac{35}{10} = 3.5$

Check. Reexamine the calculations.

State. It would take 3.5 hr for 10 volunteers to complete the job.

59. *Familiarize*. Because cost C varies directly as people fed P, we write $C(P) = kP$. We know that $C(7) = 15.75$.

Translate. We find the variation constant and the equation of variation.

$C(P) = kP$

$C(7) = k \cdot 7$ Replacing P with 7

$15.75 = k \cdot 7$ Replacing $C(7)$ with 15.75

$2.25 = k$ Variation constant

$C(P) = 2.25P$ Equation of variation

Carry out. We find P when $C(P) = 27.00$.

$27 = 2.25P$

$12 = P$

Check. Reexamine the calculations.

State. 12 people could be fed with a gift of $27.00.

61. *Familiarize*. Because the amount of salt A, in tons, varies directly as the number of storms n, we write $A(n) = kn$. We know that $A(8) = 1200$.

Translate. We find the variation constant and the equation of variation.

$A(n) = kn$

$W(8) = k \cdot 8$ Replacing n with 8

$1200 = k \cdot 8$ Replacing $W(8)$ with 1200

$150 = k$ Variation constant

$A(n) = 150n$ Equation of variation

Carry out. We find $A(4)$.

$A(4) = 150 \cdot 4 = 600$

Check. Reexamine the calculations.

State. For 4 storms, they would need 600 tons of salt.

63. Familiarize. Because the frequency, f varies inversely as length L, we write $f(L) = k/L$. We know that $f(33) = 260$.

Translate. We find the variation constant and the equation of variation.

$$f(L) = \frac{k}{L}$$

$$f(33) = \frac{k}{33} \quad \text{Replacing } L \text{ with } 33$$

$$260 = \frac{k}{33} \quad \text{Replacing } f(33) \text{ with } 260$$

$$8580 = k \quad \text{Variation constant}$$

$$f(L) = \frac{8580}{L} \quad \text{Equation of variation}$$

Carry out. We find $f(30)$.

$$f(30) = \frac{8580}{30} = 286$$

Check. Reexamine the calculations.

State. If the string was shortened to 30 cm the new frequency would be 286 Hz.

65. Familiarize. Because of the phrase "t varies inversely as …u," we write $t(u) = k/u$. We know that $t(4) = 75$.

Translate. We find the variation constant and then we find the equation of variation.

$$t(u) = \frac{k}{u}$$

$$t(4) = \frac{k}{4} \quad \text{Replacing } u \text{ with } 4$$

$$75 = \frac{k}{4} \quad \text{Replacing } t(4) \text{ with } 75$$

$$300 = k \quad \text{Variation constant}$$

$$t(u) = \frac{300}{u} \quad \text{Equation of variation}$$

Carry out. We find $t(14)$.

$$t(14) = \frac{300}{14} \approx 21$$

Check. Reexamine the calculations. Note that, as expected, as the UV rating increases, the time it takes to burn goes down.

State. It will take about 21 min to burn when the UV rating is 14.

67. Familiarize. The CPI c, varies inversely to the stitch length l. We write c as a function of l: $c(l) = \frac{k}{l}$. We know that $c(0.166) = 34.85$.

Translate.

$$c(l) = \frac{k}{l}.$$

$$c(0.166) = \frac{k}{0.166} \quad \text{Replacing } l \text{ with } 0.166$$

$$34.85 = \frac{k}{0.166} \quad \text{Replacing } c(0.166) \text{ with } 34.85$$

$$5.7851 = k \quad \text{Variation constant}$$

$$c(l) = \frac{5.7851}{l} \quad \text{Equation of variation}$$

Carry out. Find $c(0.175)$.

$$c(l) = \frac{5.7851}{l}$$

$$c(0.175) = \frac{5.7851}{0.175}$$

$$\approx 33.06$$

Check. Reexamine the calculations. Answers may vary slightly due to rounding differences.

State. The CPI would be about 33.06 for a stitch length of 0.175 in.

69.
$$y = kx^2$$
$$50 = k(10)^2 \quad \text{Substituting}$$
$$50 = k \cdot 100$$
$$\frac{1}{2} = k \quad \text{Variation constant}$$

The equation of variation is $y = \frac{1}{2}x^2$.

71.
$$y = \frac{k}{x^2}$$
$$50 = \frac{k}{(10)^2} \quad \text{Substituting}$$
$$50 = \frac{k}{100}$$
$$5000 = k \quad \text{Variation constant}$$

The equation of variation is $y = \frac{5000}{x^2}$.

73.
$$y = kxz$$
$$105 = k \cdot 14 \cdot 5 \quad \text{Substituting}$$
$$105 = k \cdot 70$$
$$1.5 = k \quad \text{Variation constant}$$

The equation of variation is $y = 1.5xz$.

75.
$$y = k \cdot \frac{wx^2}{z}$$
$$49 = k \cdot \frac{3 \cdot 7^2}{12} \quad \text{Substituting}$$
$$4 = k \quad \text{Variation constant}$$

The equation of variation is $y = \frac{4wx^2}{z}$.

77. Familiarize. Because the stopping distance d, in feet varies directly as the square of the speed r, in mph, we write $d = kr^2$. We know that $d = 138$ when $r = 60$.

Translate. Find k and the equation of variation.

$$d = kr^2$$
$$138 = k(60)^2$$
$$\frac{23}{600} = k$$
$$d = \frac{23}{600}r^2 \quad \text{Equation of variation}$$

Carry out. We find the value of d when r is 40.

$$d = \frac{23}{600}(40)^2 \approx 61.3 \text{ ft}$$

Check. Reexamine the calculations.

State. It would take a car going 40 mph about 61.3 ft to stop.

79. Familiarize. Because the wind power P, in megawatts varies directly as the cube of the wind speed v, in m/s, we write $P = kv^3$. We know that $P = 400$ when $v = 12$.

Translate. Find k and the equation of variation.

$$P = kv^3$$
$$400 = k(12)^3$$
$$k = \frac{25}{108} \approx 0.231$$
$$P = 0.231v^3 \quad \text{Equation of variation}$$

Carry out. We find the value of P when v is 15.

$$P = 0.231(15)^3 \approx 780 \text{ MW}$$

Check. Reexamine the calculations.
State. At 15 m/s, the generator would create about 780 MW of power.

81. Familiarize. Because I varies inversely as the square of d, we write $I = \dfrac{k}{d^2}$. We know that $I = 400$ when $d = 1$.

Translate. Find k and the equation of variation.

$$I = \frac{k}{d^2}$$
$$400 = \frac{k}{1^2}$$
$$400 = k$$
$$I = \frac{400}{d^2} \quad \text{Equation of variation}$$

Carry out. Substitute 2.5 for d and find I.

$$I = \frac{400}{(2.5)^2} = 64$$

Check. Reexamine the calculations.
State. The illumination is 64 foot-candles 2.5 ft from the source.

83. Familiarize. The drag W varies jointly as the surface area A and velocity v, so we write $W = kAv$. We know that $W = 222$ when $A = 37.8$ and $v = 40$.

Translate. Find k.

$$W = kAv$$
$$222 = k(37.8)(40)$$
$$\frac{222}{37.8(40)} = k$$
$$\frac{37}{252} = k$$
$$W = \frac{37}{252} Av \quad \text{Equation of variation}$$

Carry out. Substitute 51 for A and 430 for W and solve for v.

$$430 = \frac{37}{252} \cdot 51 \cdot v$$
$$57 \text{ mph} \approx v$$

Check. Reexamine the calculations.
State. The car must travel about 57 mph.

85. Writing Exercise.

87.
$$f(x) = 4x - 7$$
$$f(a) + h = 4a - 7 + h$$

89.
$$f(x) = \frac{x-5}{2x+1}$$
$$2x + 1 = 0$$
$$x = -\frac{1}{2}$$

The domain is $\left\{ x \middle| x \text{ is a real number } and \ x \neq -\frac{1}{2} \right\}$ or $\left(-\infty, -\frac{1}{2} \right) \cup \left(-\frac{1}{2}, \infty \right)$.

91.
$$f(x) = \sqrt{2x+8}$$
$$2x + 8 \geq 0$$
$$2x \geq -8$$
$$x \geq -4$$

The domain is $\{ x | x \text{ is a real number } and \ x \geq -4 \}$ or $[-4, \infty)$.

93. Writing Exercise.

95. Use the result of Example 2.

$$h = \frac{2R^2 g}{V^2} - R$$

We have $V = 6.5$ mi/sec, $R = 3960$ mi, and $g = 32.2$ ft/sec^2. We must convert 32.2 ft/sec^2 to mi/sec^2 so all units of length are the same.

$$32.2 \frac{\cancel{\text{ft}}}{\sec^2} \cdot \frac{1 \text{ mi}}{5280 \ \cancel{\text{ft}}} \approx 0.0060984 \frac{\text{mi}}{\sec^2}$$

Now we substitute and compute.

$$h = \frac{2(3960)^2 (0.0060984)}{(6.5)^2} - 3960$$
$$h \approx 567$$

The satellite is about 567 mi from the surface of Earth.

97.
$$c = \frac{a}{a+12} \cdot d$$
$$c = \frac{2a}{2a+12} \cdot d \quad \text{Doubling } a$$
$$= \frac{2a}{2(a+6)} \cdot d$$
$$= \frac{a}{a+6} \cdot d \quad \text{Simplifying}$$

The ratio of the larger dose to the smaller dose is

$$\frac{\dfrac{a}{a+6} \cdot d}{\dfrac{a}{a+12} \cdot d} = \frac{\dfrac{ad}{a+6}}{\dfrac{ad}{a+12}}$$
$$= \frac{ad}{a+6} \cdot \frac{a+12}{ad}$$
$$= \frac{\cancel{ad}(a+12)}{(a+6)\cancel{ad}}$$
$$= \frac{a+12}{a+6}$$

The amount by which the dosage increases is

$$\frac{a}{a+6} \cdot d - \frac{a}{a+12} \cdot d$$
$$= \frac{ad}{a+6} - \frac{ad}{a+12}$$
$$= \frac{ad}{a+6} \cdot \frac{a+12}{a+12} - \frac{ad}{a+12} \cdot \frac{a+6}{a+6}$$
$$= \frac{ad(a+12) - ad(a+6)}{(a+6)(a+12)}$$
$$= \frac{a^2 d + 12ad - a^2 d - 6ad}{(a+6)(a+12)}$$
$$= \frac{6ad}{(a+6)(a+12)}$$

Then the percent by which the dosage increases is

$$\dfrac{\dfrac{6ad}{(a+6)(a+12)}}{\dfrac{a}{a+12}\cdot d}=\dfrac{\dfrac{6ad}{(a+6)(a+12)}}{\dfrac{ad}{a+12}}$$

$$=\dfrac{6ad}{(a+6)(a+12)}\cdot\dfrac{a+12}{ad}$$

$$=\dfrac{6\cdot ad\cdot(a+12)}{(a+6)(a+12)\cdot ad}$$

$$=\dfrac{6}{a+6}$$

This is a decimal representation for the percent of increase. To give the result in percent notation we multiply by 100 and use a percent symbol. We have

$$\dfrac{6}{a+6}\cdot100\%,\text{ or }\dfrac{600}{a+6}\%.$$

99.
$$a=\dfrac{\dfrac{d_4-d_3}{t_4-t_3}-\dfrac{d_2-d_1}{t_2-t_1}}{t_4-t_2}$$

$$a(t_4-t_2)=\dfrac{d_4-d_3}{t_4-t_3}-\dfrac{d_2-d_1}{t_2-t_1}\quad\begin{array}{l}\text{Multiplying}\\\text{by }t_4-t_2\end{array}$$

$$a(t_4-t_2)(t_4-t_3)(t_2-t_1)=(d_4-d_3)(t_2-t_1)-(d_2-d_1)(t_4-t_3)$$

$$\text{Multiplying by }(t_4-t_3)(t_2-t_1)$$

$$a(t_4-t_2)(t_4-t_3)(t_2-t_1)-(d_4-d_3)(t_2-t_1)$$
$$=-(d_2-d_1)(t_4-t_3)$$

$$(t_2-t_1)[a(t_4-t_2)(t_4-t_3)-(d_4-d_3)]$$
$$=-(d_2-d_1)(t_4-t_3)$$

$$t_2-t_1=\dfrac{-(d_2-d_1)(t_4-t_3)}{a(t_4-t_2)(t_4-t_3)-(d_4-d_3)}$$

$$t_2+\dfrac{(d_2-d_1)(t_4-t_3)}{a(t_4-t_2)(t_4-t_3)+d_3-d_4}=t_1$$

101. Let w = the wattage of the bulb. Then we have $I=\dfrac{kw}{d^2}$.

Now substitute $2w$ for w and $2d$ for d.

$$I=\dfrac{k(2w)}{(2d)^2}=\dfrac{2kw}{4d^2}=\dfrac{kw}{2d^2}=\dfrac{1}{2}\cdot\dfrac{kw}{d^2}$$

We see that the intensity is halved.

103. *Familiarize.* We write $T=kml^2f^2$. We know that $T=100$ when $m=5$, $l=2$, and $f=80$.
Translate. Find k.

$$T=kml^2f^2$$
$$100=k(5)(2)^2(80)^2$$
$$0.00078125=k$$
$$T=0.00078125ml^2f^2$$

Carry out. Substitute 72 for T, 5 for m, and 80 for f and solve for l.

$$72=0.00078125(5)(l^2)(80)^2$$
$$2.88=l^2$$
$$1.7\approx l$$

Check. Recheck the calculations.
State. The string should be about 1 7 m long.

105. *Familiarize.* Because d varies inversely as s, we write $d(s)=k/s$. We know that $d(0.56)=50$.
Translate.

$$d(s)=\dfrac{k}{s}$$
$$d(0.56)=\dfrac{k}{0.56}\quad\text{Replacing }s\text{ with }0.56$$
$$50=\dfrac{k}{0.56}\quad\text{Replacing }d(0.56)\text{ with }50$$
$$28=k$$
$$d(s)=\dfrac{28}{s}\quad\text{Equation of variation}$$

Carry out. Find $d(0.40)$.

$$d(0.40)=\dfrac{28}{0.40}=70$$

Check. Reexamine the calculations. Also observe that, as expected, when d decreases, then s increases.

State. The equation of variation is $d(s)=\dfrac{28}{s}$. The distance is 70 yd.

Chapter 6 Review

1. True; when $x=-2$ or $x=2$, the denominator is zero, so these values are not included in the domain of f.

2. True

3. False; $3-x$ can be written as $-1(x-3)$, which is the LCM.

4. False

5. False

6. True

7. True

8. False

9. True

10. True

11. $f(t)=\dfrac{t^2-3t+2}{t^2-9}$

a. $f(0)=\dfrac{0^2-3\cdot0+2}{0^2-9}=-\dfrac{2}{9}$

b. $f(-1)=\dfrac{(-1)^2-3(-1)+2}{(-1)^2-9}=\dfrac{1+3+2}{1-9}=\dfrac{6}{-8}=-\dfrac{3}{4}$

c. $f(1)=\dfrac{1^2-3\cdot1+2}{1^2-9}=\dfrac{1-3+2}{1-9}=0$

12. $20x^3=2^2\cdot5x^3$, $24x^2=2^3\cdot3x^2$
LCM $=2^3\cdot3\cdot5x^3=120x^3$

13. $x^2+8x-20=(x+10)(x-2)$
$x^2+7x-30=(x+10)(x-3)$
LCM $=(x+10)(x-2)(x-3)$

14. $\dfrac{x^2}{x-8}-\dfrac{64}{x-8}=\dfrac{x^2-64}{x-8}=\dfrac{(x+8)(x-8)}{x-8}=x+8$

15. $\dfrac{12a^2b^3}{5c^3d^2} \cdot \dfrac{25c^9d^4}{9a^7b} = \dfrac{3\cdot 4\cdot 5\cdot 5a^2b^3c^9d^4}{5\cdot 3\cdot 3a^7bc^3d^2}$

$\qquad = \dfrac{20}{3}a^{2-7}b^{3-1}c^{9-3}d^{4-2}$

$\qquad = \dfrac{20}{3}a^{-5}b^2c^6d^2,$ or $\dfrac{20b^2c^6d^2}{3a^5}$

16. $\dfrac{5}{6m^2n^3p} + \dfrac{7}{9mn^4p^2}\qquad$ LCD is $18m^2n^4p^2$

$\qquad = \dfrac{5}{6m^2n^3p}\cdot\dfrac{3np}{3np} + \dfrac{7}{9mn^4p^2}\cdot\dfrac{2m}{2m}$

$\qquad = \dfrac{15np}{18m^2n^4p^2} + \dfrac{14m}{18m^2n^4p^2} = \dfrac{15np+14m}{18m^2n^4p^2}$

17. $\dfrac{x^3-8}{x^2-25} \cdot \dfrac{x^2+10x+25}{x^2+2x+4}$

$\qquad = \dfrac{(x-2)(x^2+2x+4)}{(x+5)(x-5)} \cdot \dfrac{(x+5)(x+5)}{x^2+2x+4}$

$\qquad = \dfrac{(x-2)(x^2+2x+4)(x+5)(x+5)}{(x+5)(x-5)(x^2+2x+4)}$

$\qquad = \dfrac{(x+5)(x^2+2x+4)}{(x+5)(x^2+2x+4)} \cdot \dfrac{(x-2)(x+5)}{x-5}$

$\qquad = \dfrac{(x-2)(x+5)}{x-5}$

18. $\dfrac{x^2-4x-12}{x^2-6x+8} \div \dfrac{x^2-4}{x^3-64}$

$\qquad = \dfrac{x^2-4x-12}{x^2-6x+8} \cdot \dfrac{x^3-64}{x^2-4}$

$\qquad = \dfrac{(x-6)(x+2)}{(x-4)(x-2)} \cdot \dfrac{(x-4)(x^2+4x+16)}{(x+2)(x-2)}$

$\qquad = \dfrac{(x+2)(x-4)}{(x+2)(x-4)} \cdot \dfrac{(x-6)(x^2+4x+16)}{(x-2)(x-2)}$

$\qquad = \dfrac{(x-6)(x^2+4x+16)}{(x-2)(x-2)},$ or $\dfrac{(x-6)(x^2+4x+16)}{(x-2)^2}$

19. $\dfrac{x}{x^2+5x+6} - \dfrac{2}{x^2+3x+2}$

$\qquad = \dfrac{x}{(x+2)(x+3)} - \dfrac{2}{(x+1)(x+2)}$

$\qquad$ [LCD is $(x+1)(x+2)(x+3)$.]

$\qquad = \dfrac{x}{(x+2)(x+3)}\cdot\dfrac{x+1}{x+1} - \dfrac{2}{(x+1)(x+2)}\cdot\dfrac{x+3}{x+3}$

$\qquad = \dfrac{x^2+x-(2x+6)}{(x+1)(x+2)(x+3)}$

$\qquad = \dfrac{x^2+x-2x-6}{(x+1)(x+2)(x+3)}$

$\qquad = \dfrac{x^2-x-6}{(x+1)(x+2)(x+3)}$

$\qquad = \dfrac{(x-3)(x+2)}{(x+1)(x+2)(x+3)}$

$\qquad = \dfrac{(x-3)(x+2)}{(x+1)(x+2)(x+3)}$

$\qquad = \dfrac{x-3}{(x+1)(x+3)}$

20. $\dfrac{-4xy}{x^2-y^2} + \dfrac{x+y}{x-y} = \dfrac{-4xy}{(x+y)(x-y)} + \dfrac{x+y}{x-y}$

$\qquad = \dfrac{-4xy}{(x+y)(x-y)} + \dfrac{x+y}{x-y}\cdot\dfrac{x+y}{x+y}$

$\qquad = \dfrac{-4xy+x^2+2xy+y^2}{(x+y)(x-y)}$

$\qquad = \dfrac{x^2-2xy+y^2}{(x+y)(x-y)} = \dfrac{(x-y)(x-y)}{(x+y)(x-y)}$

$\qquad = \dfrac{x-y}{x+y}$

21. $\dfrac{5a^2}{a-b} + \dfrac{5b^2}{b-a} = \dfrac{5a^2}{a-b} + \dfrac{-1}{-1}\cdot\dfrac{5b^2}{b-a}$

$\qquad = \dfrac{5a^2}{a-b} + \dfrac{-5b^2}{a-b}$

$\qquad = \dfrac{5a^2-5b^2}{a-b} = \dfrac{5(a^2-b^2)}{a-b}$

$\qquad = \dfrac{5(a+b)(a-b)}{a-b}$

$\qquad = \dfrac{5(a+b)(a-b)}{1\cdot(a-b)} = 5(a+b)$

22. $\dfrac{3}{y+4} - \dfrac{y}{y-1} + \dfrac{y^2+3}{y^2+3y-4}$

$\qquad = \dfrac{3}{y+4} - \dfrac{y}{y-1} + \dfrac{y^2+3}{(y-1)(y+4)}$

$\qquad = \dfrac{3(y-1)-y(y+4)+y^2+3}{(y-1)(y+4)}$

$\qquad = \dfrac{3y-3-y^2-4y+y^2+3}{(y-1)(y+4)}$

$\qquad = \dfrac{-y}{(y-1)(y+4)}$

23. $f(x) = \dfrac{4x-2}{x^2-5x+4} - \dfrac{3x+2}{x^2-5x+4}\qquad$ Note that $x\neq 1,\ 4$

$\qquad = \dfrac{x-4}{(x-1)(x-4)}$

$\qquad = \dfrac{1}{x-1},\ x\neq 1,\ 4$

24. $f(x) = \dfrac{x+8}{x+5}\cdot\dfrac{2x+10}{x^2-64}$

$\qquad = \dfrac{(x+8)\cdot 2(x+5)}{(x+5)(x+8)(x-8)}\qquad$ Note that $x\neq -5,-8,\ 8.$

$\qquad = \dfrac{2}{x-8}\cdot\dfrac{(x+5)(x+8)}{(x+5)(x+8)}$

$\qquad = \dfrac{2}{x-8}$

Thus, $f(x) = \dfrac{2}{x-8},\ x\neq -5,-8,\ 8.$

25. $f(x) = \dfrac{9x^2-1}{x^2-9} \div \dfrac{3x+1}{x+3}$

$x^2-9 = (x+3)(x-3)$ is zero when $x=-3$ or $x=3$;

$x+3$ is zero when $x=-3$. Also, $3x+1$ is zero when

$x = -\frac{1}{3}$. Thus, the domain cannot contain -3, $-\frac{1}{3}$, or 3.

$$f(x) = \frac{9x^2 - 1}{x^2 - 9} \div \frac{3x + 1}{x + 3}$$

$$= \frac{9x^2 - 1}{x^2 - 9} \cdot \frac{x + 3}{3x + 1}$$

$$= \frac{(3x + 1)(3x - 1)(x + 3)}{(x + 3)(x - 3)(3x + 1)}$$

$$= \frac{3x - 1}{x - 3}, \quad x \neq -3, \ -\frac{1}{3}, \ 3$$

26. $\dfrac{\dfrac{4}{x} - 4}{\dfrac{9}{x} - 9} = \dfrac{\dfrac{4}{x} - 4}{\dfrac{9}{x} - 9} \cdot \dfrac{x}{x}$ Multiplying by 1, using the LCD

$$= \frac{\dfrac{4}{x} \cdot x - 4 \cdot x}{\dfrac{9}{x} \cdot x - 9 \cdot x}$$

$$= \frac{4 - 4x}{9 - 9x}$$

$$= \frac{4(1 - x)}{9(1 - x)}$$

$$= \frac{4}{9}$$

27. $\dfrac{\dfrac{3}{a} + \dfrac{3}{b}}{\dfrac{6}{a^3} + \dfrac{6}{b^3}} = \dfrac{\dfrac{3}{a} + \dfrac{3}{b}}{\dfrac{6}{a^3} + \dfrac{6}{b^3}} \cdot \dfrac{a^3 b^3}{a^3 b^3}$ Multiplying by 1, using the LCD

$$= \frac{\dfrac{3}{a} \cdot a^3 b^3 + \dfrac{3}{b} \cdot a^3 b^3}{\dfrac{6}{a^3} \cdot a^3 b^3 + \dfrac{6}{b^3} \cdot a^3 b^3}$$

$$= \frac{3a^2 b^3 + 3a^3 b^2}{6b^3 + 6a^3} = \frac{3a^2 b^2 (a + b)}{6(b + a)(b^2 - ab + a^2)}$$

$$= \frac{a^2 b^2}{2(b^2 - ab + a^2)}$$

28. $\dfrac{\dfrac{y^2 + 4y - 77}{y^2 - 10y + 25}}{\dfrac{y^2 - 5y - 14}{y^2 - 25}}$

$$= \frac{y^2 + 4y - 77}{y^2 - 10y + 25} \cdot \frac{y^2 - 25}{y^2 - 5y - 14} \quad \text{Multiplying by the reciprocal of the divisor}$$

$$= \frac{(y - 7)(y + 11)}{(y - 5)(y - 5)} \cdot \frac{(y + 5)(y - 5)}{(y + 2)(y - 7)}$$

$$= \frac{(y - 7)(y + 11)(y + 5)(y - 5)}{(y - 5)(y - 5)(y + 2)(y - 7)}$$

$$= \frac{(y + 11)(y + 5)}{(y - 5)(y + 2)}$$

29. $\dfrac{\dfrac{5}{x^2 - 9} - \dfrac{3}{x + 3}}{\dfrac{4}{x^2 + 6x + 9} + \dfrac{2}{x - 3}}$

$$= \frac{\dfrac{5}{(x + 3)(x - 3)} - \dfrac{3}{x + 3}}{\dfrac{4}{(x + 3)(x + 3)} + \dfrac{2}{x - 3}}$$

$$= \frac{\dfrac{5}{(x + 3)(x - 3)} - \dfrac{3}{x + 3}}{\dfrac{4}{(x + 3)(x + 3)} + \dfrac{2}{x - 3}} \cdot \frac{(x + 3)^2 (x - 3)}{(x + 3)^2 (x - 3)} \quad \text{Multiplying by 1, using the LCD}$$

$$= \frac{\dfrac{5}{(x + 3)(x - 3)} \cdot (x + 3)^2 (x - 3) - \dfrac{3}{x + 3} \cdot (x + 3)^2 (x - 3)}{\dfrac{4}{(x + 3)(x + 3)} \cdot (x + 3)^2 (x - 3) + \dfrac{2}{x - 3} \cdot (x + 3)^2 (x - 3)}$$

$$= \frac{5(x + 3) - 3(x - 3)(x + 3)}{4(x - 3) + 2(x + 3)^2} = \frac{5x + 15 - 3x^2 + 27}{4x - 12 + 2x^2 + 12x + 18}$$

$$= \frac{-3x^2 + 5x + 42}{2x^2 + 16x + 6} = \frac{(14 - 3x)(x + 3)}{2x^2 + 16x + 6}$$

30. Because $\dfrac{1}{x}$ is undefined when x is 0, at the outset we state the restriction that $x \neq 0$.

$$\frac{3}{x} + \frac{7}{x} = 5, \quad \text{LCD is } x$$

$$x\left(\frac{3}{x} + \frac{7}{x}\right) = x(5)$$

$$x \cdot \frac{3}{x} + x \cdot \frac{7}{x} = x \cdot 5$$

$$3 + 7 = 5x$$

$$10 = 5x$$

$$2 = x$$

31. $\dfrac{5}{3x + 2} = \dfrac{3}{2x}$

To ensure that neither denominator is 0, we note at the outset that $x \neq -\dfrac{2}{3}$ and $x \neq 0$. Then we multiply both sides by the LCD, $2x(3x + 2)$.

$$2x(3x + 2) \cdot \frac{5}{3x + 2} = 2x(3x + 2) \cdot \frac{3}{2x}$$

$$2x(5) = (3x + 2)(3)$$

$$10x = 9x + 6$$

$$x = 6$$

The solution is 6.

32. $\dfrac{4x}{x + 1} + \dfrac{4}{x} + 9 = \dfrac{4}{x^2 + x}$

$$\frac{4x}{x + 1} + \frac{4}{x} + 9 = \frac{4}{x(x + 1)}$$

Listing the restrictions, we note at the outset that $x \neq -1$ and $x \neq 0$. Then we multiply both sides by the LCD, $x(x + 1)$.

$$\frac{4x}{x + 1} + \frac{4}{x} + 9 = \frac{4}{x(x + 1)}$$

$$x(x + 1)\left(\frac{4x}{x + 1} + \frac{4}{x} + 9\right) = x(x + 1) \cdot \frac{4}{x(x + 1)}$$

$$4x^2 + 4(x + 1) + 9x(x + 1) = 4$$

$$4x^2 + 4x + 4 + 9x^2 + 9x = 4$$

$$13x^2 + 13x = 0$$

$$13x(x + 1) = 0$$

$$x = 0 \ \text{ or } \ x = -1$$

Because of the restriction above, neither -1 nor 0 can be a solution. Thus, the equation has no solution.

33.

$$\frac{x+6}{x^2+x-6}+\frac{x}{x^2+4x+3}=\frac{x+2}{x^2-x-2}$$

$$\frac{x+6}{(x+3)(x-2)}+\frac{x}{(x+1)(x+3)}=\frac{x+2}{(x+1)(x-2)}$$

Listing the restrictions, we note at the outset that $x\neq-3$, $x\neq-1$ and $x\neq2$. Then we multiply both sides by the LCD, $(x+1)(x-2)(x+3)$.

$$(x+1)(x-2)(x+3)\cdot\left(\frac{x+6}{(x+3)(x-2)}+\frac{x}{(x+1)(x+3)}\right)$$
$$=(x+1)(x-2)(x+3)\cdot\frac{x+2}{(x+1)(x-2)}$$

$$(x+1)(x+6)+x(x-2)=(x+2)(x+3)$$
$$x^2+7x+6+x^2-2x=x^2+5x+6$$
$$x^2=0$$
$$x=0$$

The solution is 0.

34.

$$\frac{x}{x-3}-\frac{3x}{x+2}=\frac{5}{x^2-x-6}$$

$$\frac{x}{x-3}-\frac{3x}{x+2}=\frac{5}{(x-3)(x+2)}$$

First note that $x\neq-2$ and $x\neq3$. Then multiply both sides by the LCD $(x-3)(x+2)$.

$$(x-3)(x+2)\cdot\left(\frac{x}{x-3}-\frac{3x}{x+2}\right)=(x-3)(x+2)\cdot\frac{5}{(x-3)(x+2)}$$

$$x(x+2)-3x(x-3)=5$$
$$x^2+2x-3x^2+9x=5$$
$$-2x^2+11x-5=0$$
$$-(2x^2-11x+5)=0$$
$$-(2x-1)(x-5)=0$$

$$2x-1=0\quad or\quad x-5=0$$
$$x=\frac{1}{2}\quad or\quad x=5$$

35. We find all values of a for which $\dfrac{2}{a-1}+\dfrac{2}{a+2}=1$.

First note that $a\neq1$ and $a\neq-2$. Then multiply both sides by the LCD, $(a-1)(a+2)$.

$$(a-1)(a+2)\left(\frac{2}{a-1}+\frac{2}{a+2}\right)=(a-1)(a+2)\cdot1$$

$$2(a+2)+2(a-1)=(a-1)(a+2)$$
$$2a+4+2a-2=a^2+a-2$$
$$0=a^2-3a-4$$
$$0=(a+1)(a-4)$$
$$a=-1\ or\ a=4$$

Both values check. The solutions are -1 and 4.

36. Let t represent the number of hours it takes Meg and Kelly to complete the job working together.

Solve $\dfrac{t}{9}+\dfrac{t}{12}=1$.

The solution is $\dfrac{36}{7}$ hr, or $5\dfrac{1}{7}$ hr. The value checks.

37. *Familiarize.* Let t represent the number of hours it takes the Jon to build one section of trail working alone. Then $t-15$ represents the time it takes Ben to build one section of trail, working alone. In 1 hour Jon does $\dfrac{1}{t}$ of the job and Ben does $\dfrac{1}{t-15}$ of the job.

Translate. Working together, Jon and Ben can build the section of trail in 18 hr to find t such that

$$18\left(\frac{1}{t}\right)+18\left(\frac{1}{t-15}\right)=1,\ or\ \frac{18}{t}+\frac{18}{t-15}=1.$$

Carry out. We solve the equation. First we multiply both sides by the LCD, $t(t-15)$.

$$t(t-15)\left(\frac{18}{t}+\frac{18}{t-15}\right)=t(t-15)\cdot1$$
$$18(t-15)+18t=t(t-15)$$
$$18t-270+18t=t^2-15t$$
$$0=t^2-51t+270$$
$$0=(t-6)(t-45)$$
$$t=6\ or\ t=45$$

Check. If $t=6$, then $t-15=6-15=-9$. Since negative time has no meaning in this application, 6 cannot be a solution. If $t=45$, then $t-15=45-15=30$. In 18 hr Jon does $18\cdot\dfrac{1}{45}$, or $\dfrac{2}{5}$ of the job. In 18 hr Ben does $18\cdot\dfrac{1}{30}$, or $\dfrac{3}{5}$ of the job. Together they do $\dfrac{2}{5}+\dfrac{3}{5}$, or 1 entire job. The answer checks.

State. Working alone, Jon can build a section of trail in 45 hr and Ben can build a section of trail in 30 hr.

38. Let $r=$ the speed of the boat in still water in mph.

Solve $\dfrac{30}{r-6}=\dfrac{50}{r+6}$.

The solution is 24 mph. This answer checks.

39. *Familiarize.* Let $r=$ Elizabeth's speed in mph. Then $r+8=$ Jennifer's speed in mph. We organize the information in a table. The time is the same for both so we use t for each time.

	Distance	Speed	Time
Elizabeth	93	r	t
Jennifer	105	$r+8$	t

Translate. Using the formula Time = Distance/Rate in each row of the table and the fact that the times are the same, we can write an equation.

$$\frac{93}{r}=\frac{105}{r+8}$$

Carry out. We solve the equation.

$$\frac{93}{r}=\frac{105}{r+8},\quad LCD\ is\ r(r+8)$$
$$r(r+8)\cdot\frac{93}{r}=r(r+8)\cdot\frac{105}{r+8}$$
$$93(r+8)=105r$$
$$93r+744=105r$$
$$744=12r$$
$$62=r$$

Check. If Elizabeth's speed is 62 mph, then Jennifer's speed is $62+8$, or 70 mph. Traveling 93 mi at 62 mph takes $\dfrac{93}{62}=1.5$ hr. Traveling 105 mi at 70 mph takes $\dfrac{105}{70}=1.5$ hr. Since the times are the same, the answer checks.

State. Elizabeth's speed is 62 mph; Jennifer's speed is 70 mph.

40. $\left(30r^2s^3 + 25r^2s^2 - 20r^3s^3\right) \div \left(10r^2s\right)$

$= \dfrac{30r^2s^3 + 25r^2s^2 - 20r^3s^3}{10r^2s}$

$= \dfrac{30r^2s^3}{10r^2s} + \dfrac{25r^2s^2}{10r^2s} - \dfrac{20r^3s^3}{10r^2s}$

$= 3s^2 + \dfrac{5s}{2} - 2rs^2$

41. $\dfrac{y^3 + 8}{y + 2} = \dfrac{(y+2)(y^2 - 2y + 4)}{y + 2} = y^2 - 2y + 4$

42.

$$\begin{array}{r} 4x + 3 \\ x^2 + 1 \overline{)4x^3 + 3x^2 - 5x - 2} \\ \underline{4x^3 \qquad + 4x} \\ 3x^2 - 9x - 2 \\ \underline{3x^2 \qquad + 3} \\ -9x - 5 \end{array}$$

The answer is $4x + 3$, R $-9x - 5$, or $4x + 3 + \dfrac{-9x - 5}{x^2 + 1}$.

43. $\left(x^3 + 3x^2 + 2x - 6\right) \div (x - 3)$

$$\begin{array}{r|rrrr} 3 & 1 & 3 & 2 & -6 \\ & & 3 & 18 & 60 \\ \hline & 1 & 6 & 20 & 54 \end{array}$$

$x^2 + 6x + 20$, R 54, or $x^2 + 6x + 20 + \dfrac{54}{x - 3}$

44.

$$\begin{array}{r|rrrr} 5 & 4 & -6 & 0 & -9 \\ & & 20 & 70 & 350 \\ \hline & 4 & 14 & 70 & 341 \end{array}$$

The remainder tells us that $f(5) = 341$.

45. $I = \dfrac{2V}{R + 2r}$

$(R + 2r)I = (R + 2r) \cdot \dfrac{2V}{R + 2r}$

$IR + 2rI = 2V$

$2rI = 2V - IR$

$r = \dfrac{2V - IR}{2I}$, or $\dfrac{V}{I} - \dfrac{R}{2}$

46. $S = \dfrac{H}{m(t_1 - t_2)}$

$m(t_1 - t_2) \cdot S = m(t_1 - t_2) \cdot \dfrac{H}{m(t_1 - t_2)}$

$mS(t_1 - t_2) = H$

$m = \dfrac{H}{S(t_1 - t_2)}$

47. $\dfrac{1}{ac} = \dfrac{2}{ab} - \dfrac{3}{bc}$

$abc \cdot \dfrac{1}{ac} = abc \cdot \dfrac{2}{ab} - abc \cdot \dfrac{3}{bc}$

$b = 2c - 3a$

$b + 3a = 2c$

$\dfrac{b + 3a}{2} = c$

48. $T = \dfrac{A}{v(t_2 - t_1)}$

$v(t_2 - t_1) \cdot T = v(t_2 - t_1) \cdot \dfrac{A}{v(t_2 - t_1)}$

$vTt_2 - vTt_1 = A$

$vTt_2 - A = vTt_1$

$\dfrac{vTt_2 - A}{vT} = t_1$, or $t_1 = t_2 - \dfrac{A}{vT}$

49. *Familiarize.* Because of the phrase "the base varies inversely as the height," we write $b(h) = \dfrac{k}{h}$.

Since $b = 8$ when $h = 10$, we know that $b(10) = 8$.

Translate. We find the variation constant and then we find the equation of variation.

$b(h) = \dfrac{k}{h}$

$b(10) = \dfrac{k}{10}$ Replacing h with 10

$8 = \dfrac{k}{10}$ Replacing $b(10)$ with 8

$80 = k$ Variation constant

$b(h) = \dfrac{80}{h}$ Equation of variation

Carry out. We find $b(4)$.

$b(4) = \dfrac{80}{4} = 20$

Check. Reexamine the calculations.

State. The base will be 20 cm when the height is 4 cm.

50. $W(S) = kS$

$16.8 = k \cdot 150$

$0.112 = k$ Variation constant

$W(S) = 0.112S$ Equation of variation

$W(500) = 0.112(500)$

$= 56$ in.

51. *Familiarize.* Because the time t, in seconds varies inversely as the current I, in amperes, we write $t = \dfrac{k}{I^2}$.

We know that $t = 3.4$ when $I = 0.089$.

Translate. Find k and the equation of variation.

$t = \dfrac{k}{I^2}$

$3.4 = \dfrac{k}{(0.089)^2}$

$0.0269314 = k$

$t = \dfrac{0.0269314}{I^2}$ Equation of variation

Carry out. We find the value of t when I is 0.096.

$t = \dfrac{0.0269314}{(0.096)^2} \approx 2.9$ sec

Check. Reexamine the calculations.

State. A 0.096-amp current would be deadly after about 2.9 sec.

52. *Writing Exercise.* The least common denominator was used to add and subtract rational expressions, to simplify complex rational expressions, and to solve rational equations.

53. *Writing Exercise.* A rational *expression* is a quotient of two polynomials. Expressions can be simplified, multiplied, or added, but they cannot be solved for a variable. A rational *equation* is an equation containing rational expressions. In a rational equation, we often can solve for a variable.

54.
$$\frac{5}{x-13} - \frac{5}{x} = \frac{65}{x^2 - 13x}$$
$$\frac{5}{x-13} - \frac{5}{x} = \frac{65}{x(x-13)}$$

Listing the restrictions, we note at the outset that $x \neq 0$ and $x \neq 13$. Then we multiply both sides by the LCD, $x(x-13)$.

$$x(x-13)\left(\frac{5}{x-13} - \frac{5}{x}\right) = x(x-13) \cdot \frac{65}{x(x-13)}$$
$$5x - 5(x-13) = 65$$
$$5x - 5x + 65 = 65$$
$$0 = 0$$

This is a true statement. Recall that because of the restrictions above, 0 and 13 cannot be a solution, however all other real numbers *can* be a solution. The solution is $\{x | x \text{ is a real number } and\ x \neq 0\ and\ x \neq 13\}$.

55. There is more than one approach to solving this equation. Listing the restrictions, we note at the outset that $x \neq -5$ and $x \neq 5$.

$$\frac{\dfrac{x}{x^2-25} + \dfrac{2}{x-5}}{\dfrac{3}{x-5} - \dfrac{4}{x^2-10x+25}} = 1$$

$$\frac{\dfrac{x}{(x+5)(x-5)} + \dfrac{2}{x-5}}{\dfrac{3}{x-5} - \dfrac{4}{(x-5)(x-5)}} = 1$$

$$\frac{\dfrac{x}{(x+5)(x-5)} + \dfrac{2}{x-5}}{\dfrac{3}{x-5} - \dfrac{4}{(x-5)(x-5)}} \cdot \frac{(x-5)^2(x+5)}{(x-5)^2(x+5)} = 1$$

$$\frac{x(x-5) + 2(x-5)(x+5)}{3(x-5)(x+5) - 4(x+5)} = 1$$

$$\frac{x^2 - 5x + 2x^2 - 50}{3x^2 - 75 - 4x - 20} = 1$$

$$\frac{3x^2 - 5x - 50}{3x^2 - 4x - 95} = 1$$

$$3x^2 - 5x - 50 = 3x^2 - 4x - 95$$
$$45 = x$$

The solution is 45.

56. Let t represent the number of hours it takes the Andrew, Jon, and Ben working together to build the section of trail.

Solve $t\left(\dfrac{1}{20}\right) + t\left(\dfrac{1}{18}\right) = 1$.

The solution is $\dfrac{180}{19}$ hr, or $9\dfrac{9}{19}$ hr.

Chapter 6 Test

1. $\dfrac{t+1}{t+3} \cdot \dfrac{5t+15}{4t^2-4} = \dfrac{t+1}{t+3} \cdot \dfrac{5(t+3)}{4(t+1)(t-1)}$

$$= \frac{(t+1) \cdot 5(t+3)}{(t+3) \cdot 4(t+1)(t-1)}$$

$$= \frac{(t+1)(t+3)}{(t+1)(t+3)} \cdot \frac{5}{4(t-1)}$$

$$= \frac{5}{4(t-1)}$$

2. $\dfrac{x^3+27}{x^2-16} \div \dfrac{x^2+8x+15}{x^2+x-20} = \dfrac{x^3+27}{x^2-16} \cdot \dfrac{x^2+x-20}{x^2+8x+15}$

$$= \frac{(x+3)(x^2-3x+9)}{(x+4)(x-4)} \cdot \frac{(x+5)(x-4)}{(x+3)(x+5)}$$

$$= \frac{(x+3)(x-4)(x+5)}{(x+3)(x-4)(x+5)} \cdot \frac{x^2-3x+9}{x+4}$$

$$= \frac{x^2-3x+9}{x+4}$$

3. $\dfrac{25x}{x+5} + \dfrac{x^3}{x+5} = \dfrac{25x+x^3}{x+5} = \dfrac{x(25+x^2)}{x+5}$

4. $\dfrac{3a^2}{a-b} - \dfrac{3b^2-6ab}{b-a} = \dfrac{3a^2}{a-b} - \dfrac{-1}{-1} \cdot \dfrac{3b^2-6ab}{b-a}$

$$= \frac{3a^2}{a-b} + \frac{3b^2-6ab}{a-b}$$

$$= \frac{3a^2+3b^2-6ab}{a-b}$$

$$= \frac{3(a^2-2ab+b^2)}{a-b} = \frac{3(a-b)(a-b)}{a-b}$$

$$= 3(a-b)$$

5. $\dfrac{4ab}{a^2-b^2} + \dfrac{a^2+b^2}{a+b} = \dfrac{4ab}{(a+b)(a-b)} + \dfrac{a^2+b^2}{a+b}$

$$= \frac{4ab}{(a+b)(a-b)} + \frac{a^2+b^2}{a+b} \cdot \frac{a-b}{a-b}$$

$$= \frac{4ab + (a^2+b^2)(a-b)}{(a+b)(a-b)}$$

$$= \frac{4ab + a^3 - a^2b + ab^2 - b^3}{(a+b)(a-b)}$$

$$= \frac{a^3 - a^2b + 4ab + ab^2 - b^3}{(a+b)(a-b)}$$

6. $\dfrac{6}{x^3-64} - \dfrac{4}{x^2-16}$

$$= \frac{6}{(x-4)(x^2+4x+16)} - \frac{4}{(x+4)(x-4)}$$

LCD is $(x+4)(x-4)(x^2+4x+16)$.

$$= \frac{6}{(x-4)(x^2+4x+16)} \cdot \frac{x+4}{x+4}$$

$$\quad - \frac{4}{(x+4)(x-4)} \cdot \frac{x^2+4x+16}{x^2+4x+16}$$

$$= \frac{6x+24 - 4x^2 - 16x - 64}{(x+4)(x-4)(x^2+4x+16)}$$

$$= \frac{-4x^2 - 10x - 40}{(x+4)(x-4)(x^2+4x+16)}$$

$$= \frac{-2(2x^2+5x+20)}{(x+4)(x-4)(x^2+4x+16)}$$

7. $\dfrac{4}{x+3} - \dfrac{x}{x-2} + \dfrac{x^2+4}{x^2+x-6}$

$= \dfrac{4}{x+3} - \dfrac{x}{x-2} + \dfrac{x^2+4}{(x+3)(x-2)}$

$\qquad$ Note that $x \neq -3,\ 2$.

$= \dfrac{4}{x+3} \cdot \dfrac{x-2}{x-2} - \dfrac{x}{x-2} \cdot \dfrac{x+3}{x+3} + \dfrac{x^2+4}{(x+3)(x-2)}$

$= \dfrac{4(x-2) - x(x+3) + x^2+4}{(x+3)(x-2)} = \dfrac{4x-8-x^2-3x+x^2+4}{(x+3)(x-2)}$

$= \dfrac{x-4}{(x+3)(x-2)},\ \ x \neq -3,\ 2$

8. $f(x) = \dfrac{x^2-1}{x+2} \div \dfrac{x^2-2x}{x^2+x-2}$

$x + 2$ is zero when $x = -2$; $x^2 - 2x = x(x-2)$ is zero

when $x = 0$ or $x = 2$. Also, $x^2+x-2 = (x+2)(x-1)$ is

zero when $x = -2$ or $x = 1$. Thus, the domain cannot

contain -2, 0, 1, or 2.

$f(x) = \dfrac{x^2-1}{x+2} \div \dfrac{x^2-2x}{x^2+x-2}$

$= \dfrac{x^2-1}{x+2} \cdot \dfrac{x^2+x-2}{x^2-2x}$

$= \dfrac{(x+1)(x-1)\,(x+2)\,(x-1)}{(x+2)\,x(x-2)}$

$= \dfrac{(x+1)(x-1)^2}{x(x-2)}\ \ x \neq -2,\ 0,\ 1,\ 2$

9. $\dfrac{\frac{2}{a} + \frac{3}{b}}{\frac{5}{ab} + \frac{1}{a^2}} = \dfrac{\frac{2}{a} + \frac{3}{b}}{\frac{5}{ab} + \frac{1}{a^2}} \cdot \dfrac{a^2 b}{a^2 b}$ Multiplying by 1, using the LCD

$= \dfrac{\frac{2}{a} \cdot a^2 b + \frac{3}{b} \cdot a^2 b}{\frac{5}{ab} \cdot a^2 b + \frac{1}{a^2} \cdot a^2 b}$

$= \dfrac{2ab + 3a^2}{5a + b}$

$= \dfrac{a(2b + 3a)}{5a + b}$

10. $\dfrac{\dfrac{x^2-5x-36}{x^2-36}}{\dfrac{x^2+x-12}{x^2-12x+36}}$

$= \dfrac{x^2-5x-36}{x^2-36} \cdot \dfrac{x^2-12x+36}{x^2+x-12}$ Multiplying by the reciprocal of the divisor

$= \dfrac{(x-9)(x+4)}{(x+6)(x-6)} \cdot \dfrac{(x-6)(x-6)}{(x+4)(x-3)}$

$= \dfrac{(x-9)(x+4)(x-6)(x-6)}{(x+6)(x-6)(x+4)(x-3)}$

$= \dfrac{(x-9)\,(x+4)\,(x-6)\,(x-6)}{(x+6)\,(x-6)\,(x+4)\,(x-3)}$

$= \dfrac{(x-9)(x-6)}{(x+6)(x-3)}$

11. $\dfrac{\dfrac{x}{8} - \dfrac{8}{x}}{\dfrac{1}{8} + \dfrac{1}{x}}$ LCD is $8x$

$= \dfrac{8x}{8x} \cdot \dfrac{\frac{x}{8} - \frac{8}{x}}{\frac{1}{8} + \frac{1}{x}} = \dfrac{\frac{8x^2}{8} - \frac{64x}{x}}{\frac{8x}{8} + \frac{8x}{x}}$

$= \dfrac{x^2-64}{x+8} = \dfrac{(x+8)(x-8)}{x+8} = \dfrac{(x+8)(x-8)}{x+8}$

$= x - 8$

12. Note that $t \neq 0$.

$\dfrac{1}{t} + \dfrac{1}{3t} = \dfrac{1}{2},$ LCD $= 6t$

$6t\left(\dfrac{1}{t} + \dfrac{1}{3t}\right) = 6t\left(\dfrac{1}{2}\right)$

$6t \cdot \dfrac{1}{t} + 6t \cdot \dfrac{1}{3t} = 6t \cdot \dfrac{1}{2}$

$6 + 2 = 3t$

$8 = 3t$

$\dfrac{8}{3} = t$

The solution is $\dfrac{8}{3}$.

13. $\dfrac{t+11}{t^2-t-12} + \dfrac{1}{t-4} = \dfrac{4}{t+3}$

$\dfrac{t+11}{(t-4)(t+3)} + \dfrac{1}{t-4} = \dfrac{4}{t+3}$

Listing the restrictions, we note at the outset that $t \neq 4$

and $t \neq -3$. Then we multiply both sides by the LCD,

$(t-4)(t+3)$.

$(t-4)(t+3) \cdot \left(\dfrac{t+11}{(t-4)(t+3)} + \dfrac{1}{t-4}\right) = (t-4)(t+3) \cdot \dfrac{4}{t+3}$

$t + 11 + t + 3 = 4(t-4)$

$2t + 14 = 4t - 16$

$30 = 2t$

$15 = t$

The solution is 15.

14. $\qquad \dfrac{15}{x} - \dfrac{15}{x-2} = -2,$ LCD is $x(x-2)$

$\qquad\qquad$ Note that $x \neq 0$ and $x \neq 2$.

$x(x-2)\left(\dfrac{15}{x} - \dfrac{15}{x-2}\right) = x(x-2) \cdot (-2)$

$15(x-2) - 15x = -2x(x-2)$

$15x - 30 - 15x = -2x^2 + 4x$

$0 = -2x^2 + 4x + 30$

$0 = -2(x^2 - 2x - 15)$

$0 = (x+3)(x-5)$

$x = -3\ \ or\ \ x = 5$

15. $f(x) = \dfrac{x+5}{x-1}$

$f(0) = \dfrac{0+5}{0-1} = -5$

$f(-3) = \dfrac{-3+5}{-3-1} = \dfrac{2}{-4} = -\dfrac{1}{2}$

16. $\dfrac{a+5}{a-1}=10$

Note that $a \neq 1$.

$$(a-1)\cdot\dfrac{a+5}{a-1}=(a-1)\cdot 10$$
$$a+5=10a-10$$
$$15=9a$$
$$\dfrac{5}{3}=a$$

17. $\left(16a^4b^3c-10a^5b^2c^2+12a^2b^2c\right)\div\left(4a^2b\right)$

$$=\dfrac{16a^4b^3c-10a^5b^2c^2+12a^2b^2c}{4a^2b}$$
$$=\dfrac{16a^4b^3c}{4a^2b}-\dfrac{10a^5b^2c^2}{4a^2b}+\dfrac{12a^2b^2c}{4a^2b}$$
$$=4a^2b^2c-\dfrac{5a^3bc^2}{2}+3bc$$

18.

$$
\begin{array}{r}
y-14 \\
y-6\overline{\smash{\big)}\,y^2-20y+64} \\
\underline{y^2-6y} \\
-14y+64 \\
\underline{-14y+84} \\
-20
\end{array}
$$

The answer is $y-14$, R -20, or $y-14+\dfrac{-20}{y-6}$.

19.

$$
\begin{array}{r}
6x^2-9 \\
x^2+2\overline{\smash{\big)}\,6x^4+3x^2+5x+4} \\
\underline{6x^4+12x^2} \\
-9x^2+5x+4 \\
\underline{-9x^2-18} \\
5x+22
\end{array}
$$

The answer is $6x^2-9$, R $5x+22$, or

$$6x^2-9+\dfrac{5x+22}{x^2+2}.$$

20. $\left(x^3+5x^2+4x-7\right)\div(x-2)$

$$
\begin{array}{r|rrrr}
2 & 1 & 5 & 4 & -7 \\
  &   & 2 & 14 & 36 \\
\hline
  & 1 & 7 & 18 & 29
\end{array}
$$

$x^2+7x+18$, R 29, or $x^2+7x+18+\dfrac{29}{x-2}$

21.

$$
\begin{array}{r|rrrrr}
4 & 3 & -5 & 0 & 2 & -7 \\
  &   & 12 & 28 & 112 & 456 \\
\hline
  & 3 & 7 & 28 & 114 & 449
\end{array}
$$

The remainder tells us that $f(4)=449$.

22. $R=\dfrac{gs}{g+s}$

$$(g+s)R=(g+s)\cdot\dfrac{gs}{g+s}$$
$$gR+sR=gs$$
$$gr=gs-sR$$
$$gR=s(g-R)$$
$$\dfrac{gR}{g-R}=s$$

23. *Familiarize*. Let t represent the number of hours it takes Ella and Sari to install a countertop, working together.

Translate. Ella takes 5 hr and Sari takes 4 hr to complete the job, so we have

$$\dfrac{t}{5}+\dfrac{t}{4}=1$$

Carry out. We solve the equation. Multiply on both sides by the LCD, 20.

$$20\left(\dfrac{t}{5}+\dfrac{t}{4}\right)=20\cdot 1$$
$$4t+5t=20$$
$$9t=20$$
$$t=\dfrac{20}{9},\text{ or }2\dfrac{2}{9}$$

Check. If Ella does the job alone in 5 hr, then in $2\dfrac{2}{9}$ hr she does $\dfrac{20/9}{5}$, or $\dfrac{4}{9}$ of the job. If Sari does the job alone in 4 hr, then in $2\dfrac{2}{9}$ hr she does $\dfrac{20/9}{4}$, or $\dfrac{5}{9}$ of the job. Together, they do $\dfrac{4}{9}+\dfrac{5}{9}$, or 1 entire job. The result checks.

State. It would take Ella and Sari $2\dfrac{2}{9}$ hr to finish the job working together.

24. *Familiarize*. Let $w=$ the speed of the wind, in mph. Then $12+w=$ the speed with the wind and $12-w=$ the speed against the wind. We organize the information in a table.

	Distance	Speed	Time
With	14	$12+w$	t
Against	8	$12-w$	t

Translate. Using the formula Time = Distance/Rate in each row of the table and the fact that the times are the same, we can write an equation.

$$\dfrac{14}{12+w}=\dfrac{8}{12-w}$$

Carry out. We solve the equation. Multiply both sides by the LCD, $(12+w)(12-w)$.

$$(12+w)(12-w)\cdot\dfrac{14}{12+w}=(12+w)(12-w)\cdot\dfrac{8}{12-w}$$
$$14(12-w)=8(12+w)$$
$$168-14w=96+8w$$
$$72=22w$$
$$\dfrac{36}{11}=r,\text{ or }r=3\dfrac{3}{11}$$

Check. If $w=3\dfrac{3}{11}$, then the speed with the wind is $12+3\dfrac{3}{11}=15\dfrac{3}{11}$ km/h and the speed against the wind is $12-3\dfrac{3}{11}=8\dfrac{8}{11}$ km/h. The time for the trip against the wind is $\dfrac{8}{8\frac{8}{11}}$, or $\dfrac{11}{12}$ hours. The time for the trip with the wind is $\dfrac{14}{15\frac{3}{11}}$, or $\dfrac{11}{12}$ hours. The times are the same. The values check.

State. The speed of the wind is $3\dfrac{3}{11}$ mph.

25. *Familiarize*. Let t represent the number of hours it takes Tyler to prepare the meal, working alone. Then $t + 6$ represents the time it takes Katie to prepare the meal, working alone. In 1 hour Tyler does $\frac{1}{t}$ of the job and Katie does $\frac{1}{t+6}$ of the job.

Translate. Working together, Tyler and Katie can prepare the meal in $2\frac{6}{7}$ hr.

$$2\frac{6}{7}\left(\frac{1}{t}\right) + 2\frac{6}{7}\left(\frac{1}{t+6}\right) = 1, \text{ or } \frac{20}{7t} + \frac{20}{7(t+6)} = 1.$$

Carry out. We solve the equation. First we multiply both sides by the LCD, $7t(t+6)$.

$$7t(t+6)\left(\frac{20}{7t} + \frac{20}{7(t+6)}\right) = 7t(t+6)\cdot 1$$
$$20(t+6) + 20t = 7t(t+6)$$
$$20t + 120 + 20t = 7t^2 + 42t$$
$$0 = 7t^2 + 2t - 120$$
$$0 = (7t+30)(t-4)$$
$$t = -\frac{30}{7} \text{ or } t = 4$$

Check. Since negative time has no meaning in this application, $-\frac{30}{7}$ cannot be a solution. If $t = 4$, then $t + 6 = 4 + 6 = 10$. In $2\frac{6}{7}$ hr Tyler does $2\frac{6}{7}\cdot\frac{1}{4}$, or $\frac{5}{7}$ of the job. In $2\frac{6}{7}$ hr Katie does $2\frac{6}{7}\cdot\frac{1}{10}$, or $\frac{2}{7}$ of the job. Together they do $\frac{5}{7} + \frac{2}{7}$, or 1 entire job. The answer checks.

State. Working alone, Tyler can prepare the meal in 4 hr and Katie can prepare the meal in 10 hr.

26.
$$n(t) = \frac{k}{t}$$
$$25 = \frac{k}{6}$$
$$150 = k \qquad \text{Variation constant}$$
$$n(t) = \frac{150}{t} \qquad \text{Equation of variation}$$
$$n(5) = \frac{150}{5}$$
$$= 30 \qquad \text{people}$$

27. *Familiarize*. Because the surface area A, in square inches varies directly as the square of the radius r, in inches, we write $A = kr^2$. We know that $A = 325$ when $r = 5$.

Translate. Find k and the equation of variation.
$$A = kr^2$$
$$325 = k(5)^2$$
$$13 = k$$
$$A = 13r^2 \quad \text{Equation of variation}$$

Carry out. We find the value of A when r is 7.
$$A = 13(7)^2 = 637 \text{ in}^2$$

Check. Reexamine the calculations.

State. The area would be 637 in^2 when the radius is 7 in.

28.
$$f(a) = \frac{1}{a+3} + \frac{5}{a-2}$$
$$f(a+5) = \frac{1}{a+5+3} + \frac{5}{a+5-2} = \frac{1}{a+8} + \frac{5}{a+3}$$
$$f(a) = f(a+5)$$
$$\frac{1}{a+3} + \frac{5}{a-2} = \frac{1}{a+8} + \frac{5}{a+3}$$

First note that $a \neq -8$, $a \neq -3$, and $a \neq 2$. Then multiply both sides by the LCD, $(a+8)(a+3)(a-2)$.

$$(a+8)(a+3)(a-2)\cdot\left(\frac{1}{a+3} + \frac{5}{a-2}\right)$$
$$= (a+8)(a+3)(a-2)\left(\frac{1}{a+8} + \frac{5}{a+3}\right)$$
$$(a+8)(a-2)+5(a+8)(a+3)=(a+3)(a-2)+5(a+8)(a-2)$$
$$a^2+6a-16+5(a^2+11a+24)=a^2+a-6+5(a^2+6a-16)$$
$$6a^2 + 61a + 104 = 6a^2 + 31a - 86$$
$$30a = -190$$
$$a = -\frac{19}{3}$$

This number checks. Then $f(a) = f(a+5)$ for $a = -\frac{19}{3}$.

29.
$$\frac{6}{x-15} - \frac{6}{x} = \frac{90}{x^2-15x}$$
$$\frac{6}{x-15} - \frac{6}{x} = \frac{90}{x(x-15)}$$

Listing the restrictions, we note at the outset that $x \neq 0$ and $x \neq 15$. Then we multiply both sides by the LCD, $x(x-15)$.

$$x(x-15)\left(\frac{6}{x-15} - \frac{6}{x}\right) = x(x-15)\cdot\frac{90}{x(x-15)}$$
$$6x - 6(x-15) = 90$$
$$6x - 6x + 90 = 90$$
$$0 = 0$$

This is a true statement. Recall that because of the restrictions above, 0 and 15 cannot be a solution, however all other real numbers *can* be a solution. The solution is

$\{x|x \text{ is a real number } and\ x \neq 0\ and\ x \neq 15\}$.

30.
$$1 - \cfrac{1}{1 - \cfrac{1}{1 - \cfrac{1}{a}}} = 1 - \cfrac{1}{1 - \cfrac{1}{\frac{a-1}{a}}}$$
$$= 1 - \cfrac{1}{1 - \cfrac{a}{a-1}}$$
$$= 1 - \cfrac{1}{\frac{a-1-a}{a-1}}$$
$$= 1 - \cfrac{1}{\frac{-1}{a-1}}$$
$$= 1 + a - 1$$
$$= a$$

31. The ratio of the number of lawns Alex mowed to the number of lawns Ryan mowed is $\frac{4}{3}$.

Let $x = $ the number of lawns Alex mowed, then $98 - x = $ the number of lawns Ryan mowed. Set the ratios equal to one another and solve for x.

$$\frac{4}{3} = \frac{x}{98-x}, \quad \text{LCD is } 3(98-x)$$

$$3(98-x) \cdot \frac{4}{3} = 3(98-x) \cdot \frac{x}{98-x}$$

$$4(98-x) = 3x$$

$$392 - 4x = 3x$$

$$392 = 7x$$

$$56 = x$$

Then $98 - x = 98 - 56 = 42$.

$\frac{56}{42} = \frac{4}{3}$ and $56 + 42 = 98$, so the number checks.

Alex mowed 56 lawns and Ryan mowed 42 lawns.

Chapter 7

Exponents and Radicals

Exercise Set 7.1

1. Every positive number has *two* square roots.

3. Even if a represents a negative number, $\sqrt{a^2}$ is *positive*.

5. If a is a whole number that is not a perfect square, then $\sqrt{a}$ is an *irrational* number.

7. If $\sqrt[4]{x}$ is a real number, then x must be *nonnegative*.

9. The square roots of 64 are 8 and –8, because $8^2 = 64$ and $(-8)^2 = 64$.

11. The square roots of 100 are 10 and –10 because $10^2 = 100$ and $(-10)^2 = 100$.

13. The square roots of 400 are 20 and –20 because $20^2 = 400$ and $(-20)^2 = 400$.

15. The square roots of 625 are 25 and –25 because $25^2 = 625$ and $(-25)^2 = 625$.

17. $\sqrt{49} = 7$ \quad Remember, $\sqrt{}$ indicates the principle square root.

19. $-\sqrt{16} = -4$ \quad Since, $\sqrt{16} = 4$, $-\sqrt{16} = -4$

21. $\sqrt{\dfrac{36}{49}} = \dfrac{6}{7}$

23. $-\sqrt{\dfrac{16}{81}} = -\dfrac{4}{9}$ \quad Since, $\sqrt{\dfrac{16}{81}} = \dfrac{4}{9}$, $-\sqrt{\dfrac{16}{81}} = -\dfrac{4}{9}$

25. $\sqrt{0.04} = 0.2$

27. $\sqrt{0.0081} = 0.09$

29. $f(t) = \sqrt{5t - 10}$
$f(3) = \sqrt{5(3) - 10} = \sqrt{5}$
$f(2) = \sqrt{5(2) - 10} = \sqrt{0} = 0$
$f(1) = \sqrt{5(1) - 10} = \sqrt{-5}$
Since negative numbers do not have real-number square roots, $f(1)$ does not exist.
$f(-1) = \sqrt{5(-1) - 10} = \sqrt{-15}$
Since negative numbers do not have real-number square roots, $f(-1)$ does not exist.

31. $t(x) = -\sqrt{2x^2 - 1}$
$t(5) = -\sqrt{2 \cdot 5^2 - 1} = -\sqrt{49} = -7$
$t(0) = -\sqrt{2 \cdot 0^2 - 1} = \sqrt{-1}$
$\quad t(0)$ does not exist
$t(-1) = -\sqrt{2(-1)^2 - 1} = -\sqrt{1} = -1$
$t\left(-\dfrac{1}{2}\right) = -\sqrt{2\left(-\dfrac{1}{2}\right)^2 - 1} = -\sqrt{-\dfrac{1}{2}}$ does not exist

33. $f(t) = \sqrt{t^2 + 1}$
$f(0) = \sqrt{0^2 + 1} = \sqrt{1} = 1$
$f(-1) = \sqrt{(-1)^2 + 1} = \sqrt{2}$
$f(-10) = \sqrt{(-10)^2 + 1} = \sqrt{101}$

35. $\sqrt{100x^2} = \sqrt{(10x)^2} = |10x| = 10|x|$
Since x might be negative, absolute-value notation is necessary.

37. $\sqrt{(-4b)^2} = |-4b| = 4|b|$
Since b might be negative, absolute-value notation is necessary.

39. $\sqrt{(8-t)^2} = |8 - t|$
Since $8 - t$ might be negative, absolute-value notation is necessary.

41. $\sqrt{y^2 + 16y + 64} = \sqrt{(y+8)^2} = |y + 8|$
Since $y + 8$ might be negative, absolute-value notation is necessary.

43. $\sqrt{4x^2 + 28x + 49} = \sqrt{(2x+7)^2} = |2x + 7|$
Since $2x + 7$ might be negative, absolute-value notation is necessary.

45. $\sqrt{a^{22}} = |a^{11}|$ \quad Note that $\left(a^{11}\right)^2 = a^{22}$; a could have a negative value.

47. $\sqrt{-25}$ is not a real number, so $\sqrt{-25}$ cannot be simplified.

49. $\sqrt[3]{-1} = -1$ \quad Since $(-1)^3 = -1$

51. $-\sqrt[3]{64} = -4$ \quad $\left(4^3 = 64\right)$

53. $-\sqrt[3]{-125y^3} = -(-5y)$ \quad $\left[\left(-5y^3\right) = -125y^3\right]$
$\qquad\qquad\qquad = 5y$

55. radicand: $p^2 + 4$; index: 2

57. radicand: $\frac{x}{y+4}$; index: 5

59. $-\sqrt[4]{256} = -4$ Since $4^4 = 256$

61. $-\sqrt[5]{-\frac{32}{243}} = \frac{2}{3}$ Since $\left(-\frac{2}{3}\right)^5 = -\frac{32}{243}$

63. $\sqrt[6]{x^6} = |x|$

The index is even. Use absolute-value notation since x could have a negative value.

65. $\sqrt[9]{t^9} = t$

The index is odd. Absolute-value signs are not necessary.

67. $\sqrt[4]{(6a)^4} = |6a| = 6|a|$

The index is even. Use absolute-value notation since a could have a negative value.

69. $\sqrt[10]{(-6)^{10}} = |-6| = 6$

71. $\sqrt[414]{(a+b)^{414}} = |a+b|$

The index is even. Use absolute-value notation since $a + b$ could have a negative value.

73. $\sqrt{16x^2} = \sqrt{(4x)^2} = 4x$ Assuming x is nonnegative

75. $-\sqrt{(3t)^2} = -3t$ Assuming t is nonnegative

77. $\sqrt{(-5b)^2} = 5b$

79. $\sqrt{a^2 + 2a + 1} = \sqrt{(a+1)^2} = a+1$

81. $\sqrt[4]{16x^4} = \sqrt[4]{(2x)^4} = 2x$

83. $\sqrt[3]{(x-1)^3} = x-1$

85. $\sqrt{t^{18}} = \sqrt{(t^9)^2} = t^9$

87. $\sqrt{(x-2)^8} = \sqrt{\left[(x-2)^4\right]^2} = (x-2)^4$

89. $f(x) = \sqrt[3]{x+1}$
$f(7) = \sqrt[3]{7+1} = \sqrt[3]{8} = 2$
$f(26) = \sqrt[3]{26+1} = \sqrt[3]{27} = 3$
$f(-9) = \sqrt[3]{-9+1} = \sqrt[3]{-8} = -2$
$f(-65) = \sqrt[3]{-65+1} = \sqrt[3]{-64} = -4$

91. $g(t) = \sqrt[4]{t-3}$
$g(19) = \sqrt[4]{19-3} = \sqrt[4]{16} = 2$
$g(-13) = \sqrt[4]{-13-3} = \sqrt[4]{-16}$
$\qquad g(-13)$ does not exist
$g(1) = \sqrt[4]{1-3} = \sqrt[4]{-2}$
$\qquad g(1)$ does not exist
$g(84) = \sqrt[4]{84-3} = \sqrt[4]{81} = 3$

93. $f(x) = \sqrt{x-6}$

Since the index is even, the radicand, $x - 6$, must be non-negative. We solve the inequality:
$$x - 6 \geq 0$$
$$x \geq 6$$
Domain of $f = \{x | x \geq 6\}$, or $[6, \infty)$

95. $g(t) = \sqrt[4]{t+8}$

Since the index is even, the radicand, $t + 8$, must be non-negative. We solve the inequality:
$$t + 8 \geq 0$$
$$t \geq -8$$
Domain of $g = \{t | t \geq -8\}$, or $[-8, \infty)$

97. $g(x) = \sqrt[4]{10 - 2x}$

Since the index is even, the radicand, $10 - 2x$, must be nonnegative. We solve the inequality:
$$10 - 2x \geq 0$$
$$-2x \geq -10$$
$$x \leq 5$$
Domain of $g = \{x | x \leq 5\}$, or $(-\infty, 5]$

99. $f(t) = \sqrt[5]{2t + 7}$

Since the index is odd, the radicand can be any real number.
Domain of $f = \{t | t$ is a real number$\}$, or $(-\infty, \infty)$

101. $h(z) = -\sqrt[6]{5z + 2}$

Since the index is even, the radicand, $5z + 2$, must be nonnegative. We solve the inequality:
$$5z + 2 \geq 0$$
$$5z \geq -2$$
$$z \geq -\frac{2}{5}$$
Domain of $h = \left\{z \middle| z \geq -\frac{2}{5}\right\}$, or $\left[-\frac{2}{5}, \infty\right)$

103. $f(t) = 7 + \sqrt[8]{t^8}$

Since we can compute $7 + \sqrt[8]{t^8}$ for any real number t, the domain is the set of real numbers, or
$\{t | t$ is a real number$\}$ or $(-\infty, \infty)$.

105. *Writing Exercise.*

107. $f\left(\frac{1}{3}\right) = 3\left(\frac{1}{3}\right) - 1 = 1 - 1 = 0$

109. $\{x | x \neq 0\}$, or $(-\infty, 0) \cup (0, \infty)$

111. $(fg)(x) = (3x - 1)\left(\frac{1}{x}\right) = 3 - \frac{1}{x}$

113. *Writing Exercise.*

115. $f(p) = 118.8\sqrt{p}$

Substitute 50 for p.
$f(50) = 118.8\sqrt{50}$
$\qquad \approx 840$
The water flow is about 840 GPM.

117. $S = 88.63 \sqrt[4]{A}$

Substitute 63,000 for A.

$S = 88.63 \sqrt[4]{63{,}000}$

$S \approx 1404$

There are about 1404 species of plants.

119. $f(x) = \sqrt{x+5}$

Since the index is even, the radicand, $x + 5$, must be non-negative. We solve the inequality:

$x + 5 \geq 0$

$x \geq -5$

Domain of $f = \{x | x \geq -5\}$, or $[-5, \infty)$

Make a table of values, keeping in mind that x must be -5 or greater. Plot these points and draw the graph.

x	$f(x)$
-5	0
-4	1
-1	2
1	2.4
3	2.8
4	3

121. $g(x) = \sqrt{x} - 2$

Since the index is even, the radicand, x, must be non-negative, so we have $x \geq 0$.

Domain of $g = \{x | x \geq 0\}$, or $[0, \infty)$

Make a table of values, keeping in mind that x must be 0 or greater. Plot these points and draw the graph.

x	$g(x)$
0	-2
1	-1
4	0
6	0.4
8	0.8

123. $f(x) = \dfrac{\sqrt{x+3}}{\sqrt[4]{2-x}}$

In the numerator we must have $x + 3 \geq 0$, or $x \geq -3$, and in the denominator we must have $2 - x > 0$, or $x < 2$, so

Domain of $f = \{x | -3 \leq x < 2\}$, or $[-3, 2)$.

125. $F(x) = \dfrac{x}{\sqrt{x^2 - 5x - 6}}$

Since the radical expression in the denominator has an even index, so the radicand, $x^2 - 5x - 6$, must be nonnegative in order for $\sqrt{x^2 - 5x - 6}$ to exist. In addition, the denominator cannot be zero, so the radicand must be positive. We solve the inequality:

$x^2 - 5x - 6 > 0$

$(x + 1)(x - 6) > 0$

We have $x < -1$ *and* $x > 6$, so

Domain of $F = \{x | x < -1 \text{ or } x > 6\}$, or

$(-\infty, -1) \cup (6, \infty)$.

127. $P = 50 \sqrt[15]{\dfrac{\text{NYT}(Ah + Aw)}{\text{ENQ}(Sc + 5)} \cdot Md \cdot \left[\dfrac{Md}{(Md + 2)}\right]^{T^2}}$

Substitute 258 for NYT, 29 for Ah, 29 for Aw, 44 for ENQ, 0 for Sc, 120 for Md, and 5 for T.

$P = 50 \sqrt[15]{\dfrac{258(29 + 29)}{44(0 + 5)} \cdot 120 \cdot \left[\dfrac{120}{(120 + 2)}\right]^{5^2}}$

$P \approx 88.7$

The probability that the marriage will last 5 years is about 89%.

Exercise Set 7.2

1. The expression $\sqrt{3x}$ is an example of a *radical* expression.

3. The expressions $\sqrt[3]{5mn}$ and $(5mn)^{1/3}$ are *equivalent*.

5. Choice (g) is correct because $a^{m/n} = \sqrt[n]{a^m}$.

7. $x^{-5/2} = \dfrac{1}{x^{5/2}} = \dfrac{1}{(\sqrt{x})^5}$, so choice (e) is correct.

9. $x^{1/5} \cdot x^{2/5} = x^{1/5+2/5} = x^{3/5}$, so choice (a) is correct.

11. Choice (b) is correct because $\sqrt[n]{a^m}$ and $(\sqrt[n]{a})^m$ are equivalent.

13. $y^{1/3} = \sqrt[3]{y}$

15. $36^{1/2} = \sqrt{36} = 6$

17. $32^{1/5} = \sqrt[5]{32} = 2$

19. $64^{1/2} = \sqrt{64} = 8$

21. $(xyz)^{1/2} = \sqrt{xyz}$

23. $(a^2 b^2)^{1/5} = \sqrt[5]{a^2 b^2}$

25. $t^{5/6} = \sqrt[6]{t^5}$

27. $16^{3/4} = \sqrt[4]{16^3} = (\sqrt[4]{16})^3 = 2^3 = 8$

29. $125^{4/3} = \sqrt[3]{125^4} = (\sqrt[3]{125})^4 = 5^4 = 625$

31. $(81x)^{3/4} = \sqrt[4]{(81x)^3} = \sqrt[4]{81^3 x^3}$, or

$\sqrt[4]{81^3} \cdot \sqrt[4]{x^3} = (\sqrt[4]{81})^3 \cdot (\sqrt[4]{x^3}) = 3^3 \sqrt[4]{x^3} = 27\sqrt[4]{x^3}$

33. $(25x^4)^{3/2} = \sqrt{(25x^4)^3} = \sqrt{25^3 \cdot x^{12}} = \sqrt{25^3} \cdot \sqrt{x^{12}}$

$= (\sqrt{25})^3 x^6 = 5^3 x^6 = 125x^6$

35. $\sqrt[3]{18} = 18^{1/3}$

37. $\sqrt{30} = 30^{1/2}$

39. $\sqrt{x^7} = x^{7/2}$

41. $\sqrt[5]{m^2} = m^{2/5}$

43. $\sqrt[4]{xy} = (xy)^{1/4}$

45. $\sqrt[5]{xy^2z} = (xy^2z)^{1/5}$

47. $\left(\sqrt{3mn}\right)^3 = (3mn)^{3/2}$

49. $\left(\sqrt[7]{8x^2y}\right)^5 = (8x^2y)^{5/7}$

51. $\dfrac{2x}{\sqrt[3]{z^2}} = \dfrac{2x}{z^{2/3}}$

53. $8^{-1/3} = \dfrac{1}{8^{1/3}} = \dfrac{1}{(2^3)^{1/3}} = \dfrac{1}{2^{3/3}} = \dfrac{1}{2}$

55. $(2rs)^{-3/4} = \dfrac{1}{(2rs)^{3/4}}$

57. $\left(\dfrac{1}{16}\right)^{-3/4} = \left(\dfrac{16}{1}\right)^{3/4} = (2^4)^{3/4} = 2^{4(3/4)} = 2^3 = 8$

59. $\dfrac{8c}{a^{-3/5}} = 8a^{3/5}c$

61. $2a^{3/4}b^{-1/2}c^{2/3} = 2 \cdot a^{3/4} \cdot \dfrac{1}{b^{1/2}} \cdot c^{2/3} = \dfrac{2a^{3/4}c^{2/3}}{b^{1/2}}$

63. $3^{-5/2}a^3b^{-7/3} = \dfrac{1}{3^{5/2}} \cdot a^3 \cdot \dfrac{1}{b^{7/3}} = \dfrac{a^3}{3^{5/2}b^{7/3}}$

65. $\left(\dfrac{2ab}{3c}\right)^{-5/6} = \left(\dfrac{3c}{2ab}\right)^{5/6}$ Finding the reciprocal of the base and changing the sign of the exponent

67. $xy^{-1/4} = \dfrac{x}{y^{1/4}}$

69. $11^{1/2} \cdot 11^{1/3} = 11^{1/2+1/3} = 11^{3/6+2/6} = 11^{5/6}$
We added exponents after finding a common denominator.

71. $\dfrac{3^{5/8}}{3^{-1/8}} = 3^{5/8-(-1/8)} = 3^{5/8+1/8} = 3^{6/8} = 3^{3/4}$
We subtracted exponents and simplified.

73. $\dfrac{4.3^{-1/5}}{4.3^{-7/10}} = 4.3^{-1/5-(-7/10)} = 4.3^{-1/5+7/10}$
$\qquad = 4.3^{-2/10+7/10} = 4.3^{5/10} = 4.3^{1/2}$
We subtracted exponents after finding a common denominator. Then we simplified.

75. $\left(10^{3/5}\right)^{2/5} = 10^{3/5 \cdot 2/5} = 10^{6/25}$
We multiplied exponents.

77. $a^{2/3} \cdot a^{5/4} = a^{2/3+5/4} = a^{8/12+15/12} = a^{23/12}$
We added exponents after finding a common denominator.

79. $\left(64^{3/4}\right)^{4/3} = 64^{\frac{3}{4} \cdot \frac{4}{3}} = 64^1 = 64$

81. $\left(m^{2/3}n^{-1/4}\right)^{1/2} = m^{2/3 \cdot 1/2}n^{-1/4 \cdot 1/2} = m^{1/3}n^{-1/8}$
$\qquad = m^{1/3} \cdot \dfrac{1}{n^{1/8}} = \dfrac{m^{1/3}}{n^{1/8}}$

83. $\sqrt[9]{x^3} = x^{3/9}$ Converting to exponential notation
$\qquad = x^{1/3}$ Simplifying the exponent
$\qquad = \sqrt[3]{x}$ Returning to radical notation

85. $\sqrt[3]{y^{15}} = y^{15/3}$ Converting to exponential notation
$\qquad = y^5$ Simplifying

87. $\sqrt[12]{a^6} = a^{6/12}$ Converting to exponential notation
$\qquad = a^{1/2}$ Simplifying the exponent
$\qquad = \sqrt{a}$ Returning to radical notation

89. $\left(\sqrt[7]{xy}\right)^{14} = (xy)^{14/7}$ Converting to exponential notation
$\qquad = (xy)^2$ Simplifying the exponent
$\qquad = x^2y^2$ Using the laws of exponents

91. $\sqrt[4]{(7a)^2} = (7a)^{2/4}$ Converting to exponential notation
$\qquad = (7a)^{1/2}$ Simplifying the exponent
$\qquad = \sqrt{7a}$ Returning to radical notation

93. $\sqrt[8]{(2x)^6} = (2x)^{6/8}$ Converting to exponential notation
$\qquad = (2x)^{3/4}$ Simplifying the exponent
$\qquad = \sqrt[4]{(2x)^3}$ Returning to radical notation
$\qquad = \sqrt[4]{8x^3}$ Using the laws of exponents

95. $\sqrt{\sqrt[5]{m}} = \sqrt{m^{1/5}}$ Converting to exponential notation
$\qquad = \left(m^{1/5}\right)^{1/2}$
$\qquad = m^{1/10}$ Using the laws of exponents
$\qquad = \sqrt[10]{m}$ Returning to radical notation

97. $\sqrt[4]{(xy)^{12}} = (xy)^{12/4}$ Converting to exponential notation
$\qquad = (xy)^3$ Simplifying the exponent
$\qquad = x^3y^3$ Using the laws of exponents

99. $\left(\sqrt[5]{a^2b^4}\right)^{15}$

$= \left(a^2b^4\right)^{15/5}$ Converting to exponential notation

$= \left(a^2b^4\right)^3$ Simplifying the exponent

$= a^6b^{12}$ Using the laws of exponents

101. $\sqrt[3]{\sqrt[4]{xy}} = \sqrt[3]{(xy)^{1/4}}$ Converting to

$= \left[(xy)^{1/4}\right]^{1/3}$ exponential notation

$= (xy)^{1/12}$ Using the laws of exponents

$= \sqrt[12]{xy}$ Returning to radical notation

103. *Writing Exercise.*

105. $2(t+3)-5=1-(6-t)$

$2t+6-5=1-6+t$

$2t+1=-5+t$

$t+1=-5$

$t=-6$

The solution is -6.

109. $\dfrac{15}{x} - \dfrac{15}{x+2} = 2$

To ensure that none of the denominators is 0, we note at the outset that $x \neq 0$ and $x \neq -2$. Then we multiply both sides by the LCD, $x(x+2)$.

$x(x+2)\left(\dfrac{15}{x} - \dfrac{15}{x+2}\right) = x(x+2) \cdot 2$

$15(x+2) - 15x = 2x(x+2)$

$15x + 30 - 15x = 2x^2 + 4x$

$30 = 2x^2 + 4x$

$0 = 2x^2 + 4x - 30$

$0 = 2(x^2 + 2x - 15)$

$0 = 2(x+5)(x-3)$

$x+5=0 \quad or \quad x-3=0$

$x=-5 \quad or \quad x=3$

The numbers -5 and 3 check. The solution is -5 or 3.

111. *Writing Exercise.*

113. $\sqrt{x \sqrt[3]{x^2}} = \sqrt{x \cdot x^{2/3}} = \left(x^{5/3}\right)^{1/2} = x^{5/6} = \sqrt[6]{x^5}$

115. $\sqrt[14]{c^2 - 2cd + d^2} = \sqrt[14]{(c-d)^2} = \left[(c-d)^2\right]^{1/14}$

$= (c-d)^{2/14} = (c-d)^{1/7}$

$= \sqrt[7]{c-d}, \ c \geq d$

117. $2^{7/12} \approx 1.498 \approx 1.5$ so the G that is 7 half steps above middle C has a frequency that is about 1.5 times that of middle C.

119. a. $L = \dfrac{(0.000169)60^{2.27}}{1} \approx 1.8$ m

b. $L = \dfrac{(0.000169)75^{2.27}}{0.9906} \approx 3.1$ m

c. $L = \dfrac{(0.000169)80^{2.27}}{2.4} \approx 1.5$ m

d. $L = \dfrac{(0.000169)100^{2.27}}{1.1} \approx 5.3$ m

121. $T = 0.936d^{1.97}h^{0.85}$

$= 0.936(3)^{1.97}(80)^{0.85}$

≈ 338 cubic feet

123. $BSA = 0.007184w^{0.425}h^{0.725}$

$= 0.007184(29.5)^{0.425}(122)^{0.725}$

≈ 0.99 m^2

125. *Graphing Calculator Exercise*

Exercise Set 7.3

1. True

3. False; for instance, for $x = 4$, $\sqrt{x^2 - 9} = \sqrt{4^2 - 9} = \sqrt{7}$, but $x - 3 = 4 - 3 = 1$.

5. True

7. $\sqrt{3}\sqrt{10} = \sqrt{3 \cdot 10} = \sqrt{30}$

9. $\sqrt[3]{7}\sqrt[3]{5} = \sqrt[3]{7 \cdot 5} = \sqrt[3]{35}$

11. $\sqrt[4]{6}\sqrt[4]{9} = \sqrt[4]{6 \cdot 9} = \sqrt[4]{54}$

13. $\sqrt{2x}\sqrt{13y} = \sqrt{2x \cdot 13y} = \sqrt{26xy}$

15. $\sqrt[5]{8y^3}\sqrt[5]{10y} = \sqrt[5]{8y^3 \cdot 10y} = \sqrt[5]{80y^4}$

17. $\sqrt{y-b}\sqrt{y+b} = \sqrt{(y-b)(y+b)} = \sqrt{y^2 - b^2}$

19. $\sqrt[3]{0.7y}\sqrt[3]{0.3y} = \sqrt[3]{0.7y \cdot 0.3y} = \sqrt[3]{0.21y^2}$

21. $\sqrt[5]{x-2}\sqrt[5]{(x-2)^2} = \sqrt[5]{(x-2)(x-2)^2} = \sqrt[5]{(x-2)^3}$

23. $\sqrt{\dfrac{2}{t}}\sqrt{\dfrac{3s}{11}} = \sqrt{\dfrac{2}{t} \cdot \dfrac{3s}{11}} = \sqrt{\dfrac{6s}{11t}}$

25. $\sqrt[7]{\dfrac{x-3}{4}}\sqrt[7]{\dfrac{5}{x+2}} = \sqrt[7]{\dfrac{x-3}{4} \cdot \dfrac{5}{x+2}} = \sqrt[7]{\dfrac{5x-15}{4x+8}}$

27. $\sqrt{12}$

$= \sqrt{4 \cdot 3}$ 4 is the largest perfect square factor of 12

$= \sqrt{4} \cdot \sqrt{3}$

$= 2\sqrt{3}$

29. $\sqrt{45}$

$= \sqrt{9 \cdot 5}$ 9 is the largest perfect square factor of 45

$= \sqrt{9} \cdot \sqrt{5}$

$= 3\sqrt{5}$

31. $\sqrt{8x^9}$

$= \sqrt{4x^8 \cdot 2x}$ $4x^8$ is a perfect square

$= \sqrt{4x^8} \cdot \sqrt{2x}$ Factoring into two radicals

$= 2x^4\sqrt{2x}$ Taking the square root of $4x^8$

33. $\sqrt{120} = \sqrt{4 \cdot 30} = \sqrt{4} \cdot \sqrt{30} = 2\sqrt{30}$

35. $\sqrt{36a^4b}$

$= \sqrt{36a^4 \cdot b}$ $36a^4$ is a perfect square

$= \sqrt{36a^4} \cdot \sqrt{b}$ Factoring into two radicals

$= 6a^2\sqrt{b}$ Taking the square root of $36a^4$

37. $\sqrt[3]{8x^3y^2}$

$= \sqrt[3]{8x^3 \cdot y^2}$ $8x^3$ is a perfect cube

$= \sqrt[3]{8x^3} \cdot \sqrt[3]{y^2}$ Factoring into two radicals

$= 2x\sqrt[3]{y^2}$ Taking the cube root of $8x^3$

39. $\sqrt[3]{-16x^6}$

$= \sqrt[3]{-8x^6 \cdot 2}$ $-8x^6$ is a perfect cube

$= \sqrt[3]{-8x^6} \cdot \sqrt[3]{2}$ Factoring into two radicals

$= -2x^2\sqrt[3]{2}$ Taking the cube root of $-8x^6$

41. $f(x) = \sqrt[3]{40x^6}$

$= \sqrt[3]{8x^6 \cdot 5}$

$= \sqrt[3]{8x^6} \cdot \sqrt[3]{5}$

$= 2x^2\sqrt[3]{5}$

43. $f(x) = \sqrt{49(x-3)^2}$ $49(x-3)^2$ is a perfect square.

$= \left|7(x-3)\right|$, or $7|x-3|$

45. $f(x) = \sqrt{5x^2 - 10x + 5}$

$= \sqrt{5(x^2 - 2x + 1)}$

$= \sqrt{5(x-1)^2}$

$= \sqrt{(x-1)^2} \cdot \sqrt{5}$

$= |x-1|\sqrt{5}$

47. $\sqrt{a^{10}b^{11}}$

$= \sqrt{a^{10} \cdot b^{10} \cdot b}$ Identifying the largest even powers of a and b

$= \sqrt{a^{10}}\sqrt{b^{10}}\sqrt{b}$ Factoring into several radicals

$= a^5b^5\sqrt{b}$

49. $\sqrt[3]{x^5y^6z^{10}}$

$= \sqrt[3]{x^3 \cdot x^2 \cdot y^6 \cdot z^9 \cdot z}$ Identifying the largest perfect-cube powers of x, y and z

$= \sqrt[3]{x^3} \cdot \sqrt[3]{y^6} \cdot \sqrt[3]{z^9} \cdot \sqrt[3]{x^2z}$ Factoring into several radicals

$= xy^2z^3\sqrt[3]{x^2z}$

51. $\sqrt[4]{16x^5y^{11}} = \sqrt[4]{2^4 \cdot x^4 \cdot x \cdot y^8 \cdot y^3}$

$= \sqrt[4]{2^4} \cdot \sqrt[4]{x^4} \cdot \sqrt[4]{y^8} \cdot \sqrt[4]{xy^3}$

$= 2xy^2\sqrt[4]{xy^3}$

53. $\sqrt[5]{x^{13}y^8z^{17}} = \sqrt[5]{x^{10} \cdot x^3 \cdot y^5 \cdot y^3 \cdot z^{15} \cdot z^2}$

$= \sqrt[5]{x^{10}} \cdot \sqrt[5]{y^5} \cdot \sqrt[5]{z^{15}} \cdot \sqrt[5]{x^3y^3z^2}$

$= x^2yz^3\sqrt[5]{x^3y^3z^2}$

55. $\sqrt[3]{-80a^{14}} = \sqrt[3]{-8 \cdot 10 \cdot a^{12} \cdot a^2}$

$= \sqrt[3]{-8} \cdot \sqrt[3]{a^{12}} \cdot \sqrt[3]{10a^2}$

$= -2a^4\sqrt[3]{10a^2}$

57. $\sqrt{5}\sqrt{10} = \sqrt{5 \cdot 10} = \sqrt{50} = \sqrt{25 \cdot 2} = 5\sqrt{2}$

59. $\sqrt{6}\sqrt{33} = \sqrt{6 \cdot 33} = \sqrt{198} = \sqrt{9 \cdot 22} = 3\sqrt{22}$

61. $\sqrt[3]{9}\sqrt[3]{3} = \sqrt[3]{9 \cdot 3} = \sqrt[3]{27} = 3$

63. $\sqrt{24y^5}\sqrt{24y^5} = \sqrt{\left(24y^5\right)^2} = 24y^5$

65. $\sqrt[3]{5a^2}\sqrt[3]{2a} = \sqrt[3]{5a^2 \cdot 2a} = \sqrt[3]{10a^3} = \sqrt[3]{a^3 \cdot 10} = a\sqrt[3]{10}$

67. $3\sqrt{2x^5} \cdot 4\sqrt{10x^2} = 12\sqrt{20x^7} = 12\sqrt{4x^6 \cdot 5x} = 24x^3\sqrt{5x}$

69. $\sqrt[3]{s^2t^4}\sqrt[3]{s^4t^6} = \sqrt[3]{s^6t^{10}} = \sqrt[3]{s^6t^9 \cdot t} = s^2t^3\sqrt[3]{t}$

71. $\sqrt[3]{(x-y)^2}\sqrt[3]{(x-y)^{10}} = \sqrt[3]{(x-y)^{12}} = (x-y)^4$

73. $\sqrt[4]{20a^3b^7}\sqrt[4]{4a^2b^5} = \sqrt[4]{80a^5b^{12}} = \sqrt[4]{16a^4b^{12} \cdot 5a}$

$= 2ab^3\sqrt[4]{5a}$

75. $\sqrt[5]{x^3(y+z)^6}\sqrt[5]{x^3(y+z)^4} = \sqrt[5]{x^6(y+z)^{10}}$

$= \sqrt[5]{x^5(y+z)^{10} \cdot x} = x(y+z)^2\sqrt[5]{x}$

77. *Writing Exercise.*

79. $\dfrac{15a^2x}{8b} \cdot \dfrac{24b^2x}{5a} = \dfrac{(5a \cdot 3ax)(8b \cdot 3bx)}{8b \cdot 5a} = 9abx^2$

81. $\dfrac{x-3}{2x-10} - \dfrac{3x-5}{x^2-25} = \dfrac{x-3}{2(x-5)} - \dfrac{3x-5}{(x+5)(x-5)}$

$= \dfrac{x-3}{2(x-5)} \cdot \dfrac{x+5}{x+5} - \dfrac{3x-5}{(x+5)(x-5)} \cdot \dfrac{2}{2}$

$= \dfrac{x^2 + 2x - 15 - (6x - 10)}{2(x+5)(x-5)}$

$= \dfrac{x^2 - 4x - 5}{2(x+5)(x-5)}$

$= \dfrac{(x-5)(x+1)}{2(x+5)(x-5)}$

$= \dfrac{x+1}{2(x+5)}$

83. $\dfrac{a^{-1} + b^{-1}}{ab} = \dfrac{\dfrac{1}{a} + \dfrac{1}{b}}{ab} = \dfrac{\dfrac{1}{a} + \dfrac{1}{b}}{ab} \cdot \dfrac{ab}{ab} = \dfrac{b+a}{a^2b^2}$

85. *Writing Exercise.*

87.
$$R(x) = \frac{1}{2}\sqrt[4]{\frac{x \cdot 3.0 \times 10^6}{\pi^2}}$$
$$R(5 \times 10^4) = \frac{1}{2}\sqrt[4]{\frac{5 \times 10^4 \cdot 3.0 \times 10^6}{\pi^2}}$$
$$= \frac{1}{2}\sqrt[4]{\frac{15 \times 10^{10}}{\pi^2}}$$
$$\approx 175.6 \text{ mi}$$

89. a. $T_w = 33 - \dfrac{(10.45 + 10\sqrt{8} - 8)(33 - 7)}{22}$
$$\approx -3.3 \text{ °C}$$

  b. $T_w = 33 - \dfrac{(10.45 + 10\sqrt{12} - 12)(33 - 0)}{22}$
$$\approx -16.6 \text{ °C}$$

  c. $T_w = 33 - \dfrac{(10.45 + 10\sqrt{14} - 14)(33 - (-5))}{22}$
$$\approx -25.5 \text{ °C}$$

  d. $T_w = 33 - \dfrac{(10.45 + 10\sqrt{15} - 15)(33 - (-23))}{22}$
$$\approx -54.0 \text{ °C}$$

91. $\left(\sqrt[3]{25x^4}\right)^4 = \sqrt[3]{(25x^4)^4} = \sqrt[3]{25^4 x^{16}}$
$$= \sqrt[3]{25^3 \cdot 25 \cdot x^{15} \cdot x} = \sqrt[3]{25^3} \sqrt[3]{x^{15}} \sqrt[3]{25x}$$
$$= 25x^5 \sqrt[3]{25x}$$

93. $\left(\sqrt{a^3 b^5}\right)^7 = \sqrt{(a^3 b^5)^7} = \sqrt{a^{21} b^{35}}$
$$= \sqrt{a^{20} \cdot a \cdot b^{34} \cdot b} = \sqrt{a^{20}} \sqrt{b^{34}} \sqrt{ab} = a^{10} b^{17} \sqrt{ab}$$

95.

We see that $f(x) = h(x)$ and $f(x) \neq g(x)$.

97. $g(x) = x^2 - 6x + 8$

We must have $x^2 - 6x + 8 \geq 0$, or $(x - 2)(x - 4) \geq 0$.

We graph $y = x^2 - 6x + 8$.

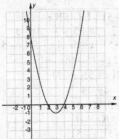

From the graph we see that $y \geq 0$ for $x \leq 2$ or $x \geq 4$, so the domain of g is $\{x \mid x \leq 2 \text{ or } x \geq 4\}$, or $(-\infty, 2] \cup [4, \infty)$.

99. $\sqrt[5]{4a^{3k+2}} \sqrt[5]{8a^{6-k}} = 2a^4$
$$\sqrt[5]{32a^{2k+8}} = 2a^4$$
$$2\sqrt[5]{a^{2k+8}} = 2a^4$$
$$\sqrt[5]{a^{2k+8}} = a^4$$
$$a^{\frac{2k+8}{5}} = a^4$$

Since the base is the same, the exponents must be equal. We have:
$$\frac{2k + 8}{5} = 4$$
$$2k + 8 = 20$$
$$2k = 12$$
$$k = 6$$

101. *Writing Exercise.*

Exercise Set 7.4

1. Quotient rule for radicals

3. Multiplying by 1

5. $\sqrt[4]{\dfrac{16a^6}{a^2}} = \sqrt[4]{16a^4} = 2a$, so choice (f) is correct.

7. $\sqrt[5]{\dfrac{a^6}{b^4}} = \sqrt[5]{\dfrac{a^6}{b^4} \cdot \dfrac{b}{b}} = \sqrt[5]{\dfrac{a^6 b}{b^4 \cdot b}}$, so choice (e) is correct.

9. $\dfrac{\sqrt{5a^4}}{\sqrt{5a^3}} = \sqrt{\dfrac{5a^4}{5a^3}} = \sqrt{a}$, so choice (c) is correct.

11. $\sqrt{\dfrac{49}{100}} = \dfrac{\sqrt{49}}{\sqrt{100}} = \dfrac{7}{10}$

13. $\sqrt[3]{\dfrac{125}{8}} = \dfrac{\sqrt[3]{125}}{\sqrt[3]{8}} = \dfrac{5}{2}$

15. $\sqrt{\dfrac{121}{t^2}} = \dfrac{\sqrt{121}}{\sqrt{t^2}} = \dfrac{11}{t}$

17. $\sqrt{\dfrac{36y^3}{x^4}} = \dfrac{\sqrt{36y^3}}{\sqrt{x^4}} = \dfrac{\sqrt{36y^2 \cdot y}}{\sqrt{x^4}} = \dfrac{\sqrt{36y^2}\sqrt{y}}{\sqrt{x^4}} = \dfrac{6y\sqrt{y}}{x^2}$

19. $\sqrt[3]{\dfrac{27a^4}{8b^3}} = \dfrac{\sqrt[3]{27a^4}}{\sqrt[3]{8b^3}} = \dfrac{\sqrt[3]{27a^3 \cdot a}}{\sqrt[3]{8b^3}} = \dfrac{\sqrt[3]{27a^3}\sqrt[3]{a}}{\sqrt[3]{8b^3}} = \dfrac{3a\sqrt[3]{a}}{2b}$

21. $\sqrt[4]{\dfrac{32a^4}{2b^4 c^8}} = \sqrt[4]{\dfrac{16a^4}{b^4 c^8}} = \dfrac{\sqrt[4]{16a^4}}{\sqrt[4]{b^4 c^8}} = \dfrac{2a}{bc^2}$

23. $\sqrt[4]{\dfrac{a^5 b^8}{c^{10}}} = \dfrac{\sqrt[4]{a^5 b^8}}{\sqrt[4]{c^{10}}} = \dfrac{\sqrt[4]{a^4 b^8 \cdot a}}{\sqrt[4]{c^8 \cdot c^2}} = \dfrac{\sqrt[4]{a^4 b^8}\sqrt[4]{a}}{\sqrt[4]{c^8}\sqrt[4]{c^2}} = \dfrac{ab^2 \sqrt[4]{a}}{c^2 \sqrt[4]{c^2}}$,

or $\dfrac{ab^2}{c^2}\sqrt[4]{\dfrac{a}{c^2}}$

25. $\sqrt[5]{\dfrac{32x^6}{y^{11}}} = \dfrac{\sqrt[5]{32x^6}}{\sqrt[5]{y^{11}}} = \dfrac{\sqrt[5]{32x^5 \cdot x}}{\sqrt[5]{y^{10} \cdot y}}$

$= \dfrac{\sqrt[5]{32x^5} \cdot \sqrt[5]{x}}{\sqrt[5]{y^{10}} \sqrt[5]{y}} = \dfrac{2x\sqrt[5]{x}}{y^2\sqrt[5]{y}}$, or $\dfrac{2x}{y^2}\sqrt[5]{\dfrac{x}{y}}$

27. $\sqrt[6]{\dfrac{x^6 y^8}{z^{15}}} = \dfrac{\sqrt[6]{x^6 y^8}}{\sqrt[6]{z^{15}}} = \dfrac{\sqrt[6]{x^6 y^6 \cdot y^2}}{\sqrt[6]{z^{12} \cdot z^3}} = \dfrac{\sqrt[6]{x^6 y^6} \sqrt[6]{y^2}}{\sqrt[6]{z^{12}} \sqrt[6]{z^3}}$,

$= \dfrac{xy \sqrt[6]{y^2}}{z^2 \sqrt[6]{z^3}}$, or $\dfrac{xy}{z^2}\sqrt[6]{\dfrac{y^2}{z^3}}$

29. $\dfrac{\sqrt{18y}}{\sqrt{2y}} = \sqrt{\dfrac{18y}{2y}} = \sqrt{9} = 3$

31. $\dfrac{\sqrt[3]{26}}{\sqrt[3]{13}} = \sqrt[3]{\dfrac{26}{13}} = \sqrt[3]{2}$

33. $\dfrac{\sqrt{40xy^3}}{\sqrt{8x}} = \sqrt{\dfrac{40xy^3}{8x}} = \sqrt{5y^3} = \sqrt{y^2 \cdot 5y}$

$= \sqrt{y^2}\sqrt{5y} = y\sqrt{5y}$

35. $\dfrac{\sqrt[3]{96a^4 b^2}}{\sqrt[3]{12a^2 b}} = \sqrt[3]{\dfrac{96a^4 b^2}{12a^2 b}} = \sqrt[3]{8a^2 b} = \sqrt[3]{8}\sqrt[3]{a^2 b} = 2\sqrt[3]{a^2 b}$

37. $\dfrac{\sqrt{100ab}}{5\sqrt{2}} = \dfrac{1}{5} \dfrac{\sqrt{100ab}}{\sqrt{2}} = \dfrac{1}{5}\sqrt{\dfrac{100ab}{2}} = \dfrac{1}{5}\sqrt{50ab}$

$= \dfrac{1}{5}\sqrt{25 \cdot 2ab} = \dfrac{1}{5} \cdot 5\sqrt{2ab} = \sqrt{2ab}$

39. $\dfrac{\sqrt[4]{48x^9 y^{13}}}{\sqrt[4]{3xy^{-2}}} = \sqrt[4]{\dfrac{48x^9 y^{13}}{3xy^{-2}}} = \sqrt[4]{16x^8 y^{15}} = \sqrt[4]{16x^8 y^{12}}\sqrt[4]{y^3}$

$= 2x^2 y^3 \sqrt[4]{y^3}$

41. $\dfrac{\sqrt[3]{x^3 - y^3}}{\sqrt[3]{x - y}} = \sqrt[3]{\dfrac{x^3 - y^3}{x - y}} = \sqrt[3]{\dfrac{(x-y)(x^2 + xy + y^2)}{x - y}}$

$= \sqrt[3]{x^2 + xy + y^2}$

43. $\sqrt{\dfrac{2}{5}} = \sqrt{\dfrac{2}{5} \cdot \dfrac{5}{5}} = \sqrt{\dfrac{10}{25}} = \dfrac{\sqrt{10}}{\sqrt{25}} = \dfrac{\sqrt{10}}{5}$

45. $\dfrac{2\sqrt{5}}{7\sqrt{3}} = \dfrac{2\sqrt{5}}{7\sqrt{3}} \cdot \dfrac{\sqrt{3}}{\sqrt{3}} = \dfrac{2\sqrt{15}}{21}$

47. $\sqrt[3]{\dfrac{5}{4}} = \sqrt[3]{\dfrac{5}{4} \cdot \dfrac{2}{2}} = \sqrt[3]{\dfrac{10}{8}} = \dfrac{\sqrt[3]{10}}{\sqrt[3]{8}} = \dfrac{\sqrt[3]{10}}{2}$

49. $\dfrac{\sqrt[3]{3a}}{\sqrt[3]{5c}} = \dfrac{\sqrt[3]{3a}}{\sqrt[3]{5c}} \cdot \dfrac{\sqrt[3]{5^2 c^2}}{\sqrt[3]{5^2 c^2}} = \dfrac{\sqrt[3]{75ac^2}}{\sqrt[3]{5^3 c^3}} = \dfrac{\sqrt[3]{75ac^2}}{5c}$

51. $\dfrac{\sqrt[4]{5y^6}}{\sqrt[4]{9x}} = \dfrac{\sqrt[4]{5y^6}}{\sqrt[4]{9x}} \cdot \dfrac{\sqrt[4]{9x^3}}{\sqrt[4]{9x^3}} = \dfrac{\sqrt[4]{5y^6 \cdot 9x^3}}{\sqrt[4]{81x^4}} = \dfrac{y\sqrt[4]{45x^3 y^2}}{3x}$

53. $\sqrt[3]{\dfrac{2}{x^2 y}} = \sqrt[3]{\dfrac{2}{x^2 y} \cdot \dfrac{xy^2}{xy^2}} = \sqrt[3]{\dfrac{2xy^2}{x^3 y^3}} = \dfrac{\sqrt[3]{2xy^2}}{\sqrt[3]{x^3 y^3}} = \dfrac{\sqrt[3]{2xy^2}}{xy}$

55. $\sqrt{\dfrac{7a}{18}} = \sqrt{\dfrac{7a}{18} \cdot \dfrac{2}{2}} = \sqrt{\dfrac{14a}{36}} = \dfrac{\sqrt{14a}}{\sqrt{36}} = \dfrac{\sqrt{14a}}{6}$

57. $\sqrt[5]{\dfrac{9}{32x^5 y}} = \sqrt[5]{\dfrac{9}{32x^5 y} \cdot \dfrac{y^4}{y^4}} = \dfrac{\sqrt[5]{9y^4}}{\sqrt[5]{32x^5 y^5}} = \dfrac{\sqrt[5]{9y^4}}{2xy}$

59. $\sqrt{\dfrac{10ab^2}{72a^3 b}} = \sqrt{\dfrac{5b}{36a^2}} = \dfrac{\sqrt{5b}}{6a}$

61. $\sqrt{\dfrac{5}{11}} = \sqrt{\dfrac{5}{11} \cdot \dfrac{5}{5}} = \sqrt{\dfrac{25}{55}} = \dfrac{\sqrt{25}}{\sqrt{55}} = \dfrac{5}{\sqrt{55}}$

63. $\dfrac{2\sqrt{6}}{5\sqrt{7}} = \dfrac{2\sqrt{6}}{5\sqrt{7}} \cdot \dfrac{\sqrt{6}}{\sqrt{6}} = \dfrac{2\sqrt{36}}{5\sqrt{42}} = \dfrac{2 \cdot 6}{5\sqrt{42}} = \dfrac{12}{5\sqrt{42}}$

65. $\dfrac{\sqrt{8}}{2\sqrt{3x}} = \dfrac{\sqrt{8}}{2\sqrt{3x}} \cdot \dfrac{\sqrt{2}}{\sqrt{2}} = \dfrac{\sqrt{16}}{2\sqrt{6x}} = \dfrac{4}{2\sqrt{6x}}$

$= \dfrac{2}{\sqrt{6x}}$

67. $\dfrac{\sqrt[3]{7}}{\sqrt[3]{2}} = \dfrac{\sqrt[3]{7}}{\sqrt[3]{2}} \cdot \dfrac{\sqrt[3]{7^2}}{\sqrt[3]{7^2}} = \dfrac{\sqrt[3]{7^3}}{\sqrt[3]{98}} = \dfrac{7}{\sqrt[3]{98}}$

69. $\sqrt{\dfrac{7x}{3y}} = \sqrt{\dfrac{7x}{3y} \cdot \dfrac{7x}{7x}} = \sqrt{\dfrac{(7x)^2}{21xy}} = \dfrac{7x}{\sqrt{21xy}}$

71. $\sqrt[3]{\dfrac{2a^5}{5b}} = \sqrt[3]{\dfrac{2a^5}{5b} \cdot \dfrac{4a}{4a}} = \sqrt[3]{\dfrac{8a^6}{20ab}} = \dfrac{2a^2}{\sqrt[3]{20ab}}$

73. $\sqrt{\dfrac{x^3 y}{2}} = \sqrt{\dfrac{x^3 y}{2} \cdot \dfrac{xy}{xy}} = \sqrt{\dfrac{x^4 y^2}{2xy}} = \dfrac{\sqrt{x^4 y^2}}{\sqrt{2xy}} = \dfrac{x^2 y}{\sqrt{2xy}}$

75. *Writing Exercise.*

77. $-\dfrac{2}{9} \div \dfrac{4}{6} = -\dfrac{2}{9} \cdot \dfrac{6}{4} = -\dfrac{12}{36} = -\dfrac{1}{3}$

79. $12 - 100 \div 5 \cdot (-2)^2 - 3(6 - 7)$

$= 12 - 100 \div 5 \cdot 4 - 3(-1)$

$= 12 - 20 \cdot 4 + 3$

$= 12 - 80 + 3$

$= -65$

81. $(12x^3 - 6x - 8) \div (x + 1)$

$= (12x^3 + 0x^2 - 6x - 8) \div (x + 1)$

$\underline{-1|}\ \ \begin{array}{rrrr} 12 & 0 & -6 & -8 \\ & -12 & 12 & -6 \\ \hline 12 & -12 & 6 & |-14 \end{array}$

$(12x^3 - 6x - 8) \div (x + 1) = 12x^2 - 12x + 6 + \dfrac{-14}{x + 1}$

83. *Writing Exercise.*

85. a. $T = 2\pi\sqrt{\dfrac{65}{980}} \approx 1.62$ sec

b. $T = 2\pi\sqrt{\dfrac{98}{980}} \approx 1.99$ sec

c. $T = 2\pi\sqrt{\dfrac{120}{980}} \approx 2.20$ sec

87. $\dfrac{\left(\sqrt[3]{81mn^2}\right)^2}{\left(\sqrt[3]{mn}\right)^2} = \dfrac{\sqrt[3]{(81mn^2)^2}}{\sqrt[3]{(mn)^2}}$

$\qquad = \dfrac{\sqrt[3]{6561m^2n^4}}{\sqrt[3]{m^2n^2}}$

$\qquad = \sqrt[3]{\dfrac{6561m^2n^4}{m^2n^2}}$

$\qquad = \sqrt[3]{6561n^2}$

$\qquad = \sqrt[3]{729\cdot 9n^2}$

$\qquad = \sqrt[3]{729}\sqrt[3]{9n^2}$

$\qquad = 9\sqrt[3]{9n^2}$

89. $\sqrt{a^2-3} - \dfrac{a^2}{\sqrt{a^2-3}} = \sqrt{a^2-3} - \dfrac{a^2}{\sqrt{a^2-3}}\cdot\dfrac{\sqrt{a^2-3}}{\sqrt{a^2-3}}$

$\qquad = \sqrt{a^2-3} - \dfrac{a^2\sqrt{a^2-3}}{a^2-3}$

$\qquad = \sqrt{a^2-3}\cdot\dfrac{a^2-3}{a^2-3} - \dfrac{a^2\sqrt{a^2-3}}{a^2-3}$

$\qquad = \dfrac{a^2\sqrt{a^2-3} - 3\sqrt{a^2-3} - a^2\sqrt{a^2-3}}{a^2-3}$

$\qquad = \dfrac{-3\sqrt{a^2-3}}{a^2-3}$, or $\dfrac{-3}{\sqrt{a^2-3}}$

91. Step 1: $\sqrt[n]{a} = a^{1/n}$, by definition;

Step 2: $\left(\dfrac{a}{b}\right)^n = \dfrac{a^n}{b^n}$, raising a quotient to a power;

Step 3: $a^{1/n} = \sqrt[n]{a}$, by definition

93. $f(x) = \sqrt{18x^3}$, $g(x) = \sqrt{2x}$

$(f/g)(x) = \dfrac{f(x)}{g(x)} = \dfrac{\sqrt{18x^3}}{\sqrt{2x}} = \sqrt{\dfrac{18x^3}{2x}} = \sqrt{9x^2} = 3x$

$\sqrt{2x}$ is defined for $2x \geq 0$, or $x \geq 0$. To avoid division by 0, we must exclude 0 from the domain. Thus, the domain of

$f/g = \{x\,|\,x \text{ is a real number and } x > 0\}$, or $(0, \infty)$.

95. $f(x) = \sqrt{x^2-9}$, $g(x) = \sqrt{x-3}$

$(f/g)(x) = \dfrac{f(x)}{g(x)} = \dfrac{\sqrt{x^2-9}}{\sqrt{x-3}} = \sqrt{\dfrac{x^2-9}{x-3}}$

$\qquad = \sqrt{\dfrac{(x+3)(x-3)}{x-3}} = \sqrt{x+3}$

$\sqrt{x-3}$ is defined for $x-3 \geq 0$, or $x \geq 3$. To avoid division by 0, we must exclude 3 from the domain. Thus, the domain of

$f/g = \{x\,|\,x \text{ is a real number and } x > 3\}$, or $(3, \infty)$.

Connecting the Concepts

1. $\dfrac{6}{\sqrt{7}} = \dfrac{6}{\sqrt{7}}\cdot\dfrac{\sqrt{7}}{\sqrt{7}} = \dfrac{6\sqrt{7}}{7}$

2. $\dfrac{1}{3-\sqrt{2}} = \dfrac{1}{3-\sqrt{2}}\cdot\dfrac{3+\sqrt{2}}{3+\sqrt{2}} = \dfrac{\sqrt{3}+2}{7}$

3. $\dfrac{2}{\sqrt{xy}} = \dfrac{2}{\sqrt{xy}}\cdot\dfrac{\sqrt{xy}}{\sqrt{xy}} = \dfrac{2\sqrt{xy}}{xy}$

4. $\dfrac{5}{\sqrt{8}} = \dfrac{5}{\sqrt{4\cdot 2}} = \dfrac{5}{2\sqrt{2}}\cdot\dfrac{\sqrt{2}}{\sqrt{2}} = \dfrac{5\sqrt{2}}{4}$

5. $\dfrac{\sqrt{2}}{\sqrt{5}+\sqrt{3}} = \dfrac{\sqrt{2}}{\sqrt{5}+\sqrt{3}}\cdot\dfrac{\sqrt{5}-\sqrt{3}}{\sqrt{5}-\sqrt{3}} = \dfrac{\sqrt{10}-\sqrt{6}}{2}$

6. $\dfrac{2}{1-\sqrt{5}} = \dfrac{2}{1-\sqrt{5}}\cdot\dfrac{1+\sqrt{5}}{1+\sqrt{5}} = \dfrac{2(1+\sqrt{5})}{-4} = \dfrac{-1-\sqrt{5}}{2}$

7. $\dfrac{1}{\sqrt[3]{x^2y}} = \dfrac{1}{\sqrt[3]{x^2y}}\cdot\dfrac{\sqrt[3]{xy^2}}{\sqrt[3]{xy^2}} = \dfrac{\sqrt[3]{xy^2}}{xy}$

8. $\dfrac{a}{\sqrt[4]{a^3b^2}} = \dfrac{a}{\sqrt[4]{a^3b^2}}\cdot\dfrac{\sqrt[4]{ab^2}}{\sqrt[4]{ab^2}} = \dfrac{a\sqrt[4]{ab^2}}{ab} = \dfrac{\sqrt[4]{ab^2}}{b}$

Exercise Set 7.5

1. To add radical expressions, both the *radicands* and the *indices* must be the same.

3. To find a product by adding exponents, the *bases* must be the same.

5. To rationalize the *numerator* of $\dfrac{\sqrt{c}-\sqrt{a}}{5}$, we multiply by a form of 1, using the *conjugate* of $\sqrt{c}-\sqrt{a}$, or $\sqrt{c}+\sqrt{a}$, to write 1.

7. $4\sqrt{3} + 7\sqrt{3} = (4+7)\sqrt{3} = 11\sqrt{3}$

9. $7\sqrt[3]{4} - 5\sqrt[3]{4} = (7-5)\sqrt[3]{4} = 2\sqrt[3]{4}$

11. $\sqrt[3]{y} + 9\sqrt[3]{y} = (1+9)\sqrt[3]{y} = 10\sqrt[3]{y}$

13. $8\sqrt{2} - \sqrt{2} + 5\sqrt{2} = (8-1+5)\sqrt{2} = 12\sqrt{2}$

15. $9\sqrt[3]{7} - \sqrt{3} + 4\sqrt[3]{7} + 2\sqrt{3}$
$\qquad = (9+4)\sqrt[3]{7} + (-1+2)\sqrt{3} = 13\sqrt[3]{7} + \sqrt{3}$

17. $4\sqrt{27} - 3\sqrt{3}$
$\qquad = 4\sqrt{9\cdot 3} - 3\sqrt{3}$ Factoring the
$\qquad = 4\sqrt{9}\cdot\sqrt{3} - 3\sqrt{3}$ first radical
$\qquad = 4\cdot 3\sqrt{3} - 3\sqrt{3}$ Taking the square root of 9
$\qquad = 12\sqrt{3} - 3\sqrt{3}$
$\qquad = 9\sqrt{3}$ Combining the radicals

19. $3\sqrt{45} - 8\sqrt{20}$

$= 3\sqrt{9 \cdot 5} - 8\sqrt{4 \cdot 5}$ Factoring the

$= 3\sqrt{9} \cdot \sqrt{5} - 8\sqrt{4} \cdot \sqrt{5}$ radicals

$= 3 \cdot 3\sqrt{5} - 8 \cdot 2\sqrt{5}$ Taking the square roots

$= 9\sqrt{5} - 16\sqrt{5}$

$= -7\sqrt{5}$ Combining like radicals

21. $3\sqrt[3]{16} + \sqrt[3]{54} = 3\sqrt[3]{8 \cdot 2} + \sqrt[3]{27 \cdot 2}$

$= 3\sqrt[3]{8} \cdot \sqrt[3]{2} + \sqrt[3]{27} \cdot \sqrt[3]{2} = 3 \cdot 2\sqrt[3]{2} + 3\sqrt[3]{2}$

$= 6\sqrt[3]{2} + 3\sqrt[3]{2} = 9\sqrt[3]{2}$

23. $\sqrt{a} + 3\sqrt{16a^3} = \sqrt{a} + 3\sqrt{16a^2 \cdot a} = \sqrt{a} + 3\sqrt{16a^2} \cdot \sqrt{a}$

$= \sqrt{a} + 3 \cdot 4a\sqrt{a} = \sqrt{a} + 12a\sqrt{a}$

$= (1 + 12a)\sqrt{a}$

25. $\sqrt[3]{6x^4} - \sqrt[3]{48x} = \sqrt[3]{x^3 \cdot 6x} - \sqrt[3]{8 \cdot 6x}$

$= \sqrt[3]{x^3} \cdot \sqrt[3]{6x} - \sqrt[3]{8} \cdot \sqrt[3]{6x} = x\sqrt[3]{6x} - 2\sqrt[3]{6x}$

$= (x - 2)\sqrt[3]{6x}$

27. $\sqrt{4a - 4} + \sqrt{a - 1} = \sqrt{4(a-1)} + \sqrt{a - 1}$

$= \sqrt{4}\sqrt{a - 1} + \sqrt{a - 1} = 2\sqrt{a - 1} + \sqrt{a - 1} = 3\sqrt{a - 1}$

29. $\sqrt{x^3 - x^2} + \sqrt{9x - 9} = \sqrt{x^2(x-1)} + \sqrt{9(x-1)}$

$= \sqrt{x^2} \cdot \sqrt{x - 1} + \sqrt{9} \cdot \sqrt{x - 1}$

$= x\sqrt{x - 1} + 3\sqrt{x - 1} = (x + 3)\sqrt{x - 1}$

31. $\sqrt{2}(5 + \sqrt{2}) = \sqrt{2} \cdot 5 + \sqrt{2} \cdot \sqrt{2} = 5\sqrt{2} + 2$

33. $3\sqrt{5}(\sqrt{6} - \sqrt{7}) = 3\sqrt{5} \cdot \sqrt{6} - 3\sqrt{5} \cdot \sqrt{7} = 3\sqrt{30} - 3\sqrt{35}$

35. $\sqrt{2}(3\sqrt{10} - \sqrt{8}) = \sqrt{2} \cdot 3\sqrt{10} - \sqrt{2} \cdot \sqrt{8} = 3\sqrt{20} - \sqrt{16}$

$= 3\sqrt{4 \cdot 5} - \sqrt{16} = 3 \cdot 2\sqrt{5} - 4$

$= 6\sqrt{5} - 4$

37. $\sqrt[3]{3}\left(\sqrt[3]{9} - 4\sqrt[3]{21}\right) = \sqrt[3]{3} \cdot \sqrt[3]{9} - \sqrt[3]{3} \cdot 4\sqrt[3]{21}$

$= \sqrt[3]{27} - 4\sqrt[3]{63}$

$= 3 - 4\sqrt[3]{63}$

39. $\sqrt[3]{a}\left(\sqrt[3]{a^2} + \sqrt[3]{24a^2}\right) = \sqrt[3]{a} \cdot \sqrt[3]{a^2} + \sqrt[3]{a}\sqrt[3]{24a^2}$

$= \sqrt[3]{a^3} + \sqrt[3]{24a^3}$

$= \sqrt[3]{a^3} + \sqrt[3]{8a^3 \cdot 3}$

$= a + 2a\sqrt[3]{3}$

41. $(2 + \sqrt{6})(5 - \sqrt{6}) = 2 \cdot 5 - 2\sqrt{6} + 5\sqrt{6} - \sqrt{6} \cdot \sqrt{6}$

$= 10 + 3\sqrt{6} - 6 = 4 + 3\sqrt{6}$

43. $(\sqrt{2} + \sqrt{7})(\sqrt{3} - \sqrt{7})$

$= \sqrt{2} \cdot \sqrt{3} - \sqrt{2} \cdot \sqrt{7} + \sqrt{7} \cdot \sqrt{3} - \sqrt{7} \cdot \sqrt{7}$

$= \sqrt{6} - \sqrt{14} + \sqrt{21} - 7$

45. $(2 - \sqrt{3})(2 + \sqrt{3}) = 2^2 - (\sqrt{3})^2 = 4 - 3 = 1$

47. $(\sqrt{10} - \sqrt{15})(\sqrt{10} + \sqrt{15}) = (\sqrt{10})^2 - (\sqrt{15})^2$

$= 10 - 15 = -5$

49. $(3\sqrt{7} + 2\sqrt{5})(2\sqrt{7} - 4\sqrt{5})$

$= 3\sqrt{7} \cdot 2\sqrt{7} - 3\sqrt{7} \cdot 4\sqrt{5} + 2\sqrt{5} \cdot 2\sqrt{7} - 2\sqrt{5} \cdot 4\sqrt{5}$

$= 6 \cdot 7 - 12\sqrt{35} + 4\sqrt{35} - 8 \cdot 5 = 42 - 8\sqrt{35} - 40$

$= 2 - 8\sqrt{35}$

51. $(4 + \sqrt{7})^2 = 4^2 + 2 \cdot 4 \cdot \sqrt{7} + (\sqrt{7})^2 = 16 + 8\sqrt{7} + 7$

$= 23 + 8\sqrt{7}$

53. $(\sqrt{3} - \sqrt{2})^2 = (\sqrt{3})^2 - 2 \cdot \sqrt{3} \cdot \sqrt{2} + (\sqrt{2})^2$

$= 3 - 2\sqrt{6} + 2 = 5 - 2\sqrt{6}$

55. $(\sqrt{2t} + \sqrt{5})^2 = (\sqrt{2t})^2 + 2 \cdot \sqrt{2t} \cdot \sqrt{5} + (\sqrt{5})^2$

$= 2t + 2\sqrt{10t} + 5$

57. $(3 - \sqrt{x + 5})^2 = 3^2 - 2 \cdot 3 \cdot \sqrt{x + 5} + (\sqrt{x + 5})^2$

$= 9 - 6\sqrt{x + 5} + x + 5$

$= 14 - 6\sqrt{x + 5} + x$

59. $(2\sqrt[4]{7} - \sqrt[4]{6})(3\sqrt[4]{9} + 2\sqrt[4]{5})$

$= 2\sqrt[4]{7} \cdot 3\sqrt[4]{9} + 2\sqrt[4]{7} \cdot 2\sqrt[4]{5} - \sqrt[4]{6} \cdot 3\sqrt[4]{9} - \sqrt[4]{6} \cdot 2\sqrt[4]{5}$

$= 6\sqrt[4]{63} + 4\sqrt[4]{35} - 3\sqrt[4]{54} - 2\sqrt[4]{30}$

61. $\dfrac{6}{3 - \sqrt{2}} = \dfrac{6}{3 - \sqrt{2}} \cdot \dfrac{3 + \sqrt{2}}{3 + \sqrt{2}} = \dfrac{6(3 + \sqrt{2})}{(3 - \sqrt{2})(3 + \sqrt{2})}$

$= \dfrac{18 + 6\sqrt{2}}{3^2 - (\sqrt{2})^2} = \dfrac{18 + 6\sqrt{2}}{9 - 2} = \dfrac{18 + 6\sqrt{2}}{7}$

63. $\dfrac{2 + \sqrt{5}}{6 + \sqrt{3}} = \dfrac{2 + \sqrt{5}}{6 + \sqrt{3}} \cdot \dfrac{6 - \sqrt{3}}{6 - \sqrt{3}}$

$= \dfrac{(2 + \sqrt{5})(6 - \sqrt{3})}{(6 + \sqrt{3})(6 - \sqrt{3})} = \dfrac{12 - 2\sqrt{3} + 6\sqrt{5} - \sqrt{15}}{36 - 3}$

$= \dfrac{12 - 2\sqrt{3} + 6\sqrt{5} - \sqrt{15}}{33}$

65. $\dfrac{\sqrt{a}}{\sqrt{a} + \sqrt{b}} = \dfrac{\sqrt{a}}{\sqrt{a} + \sqrt{b}} \cdot \dfrac{\sqrt{a} - \sqrt{b}}{\sqrt{a} - \sqrt{b}}$

$= \dfrac{\sqrt{a}(\sqrt{a} - \sqrt{b})}{(\sqrt{a} + \sqrt{b})(\sqrt{a} - \sqrt{b})} = \dfrac{a - \sqrt{ab}}{a - b}$

67. $\dfrac{\sqrt{7} - \sqrt{3}}{\sqrt{3} - \sqrt{7}} = \dfrac{-1(\sqrt{3} - \sqrt{7})}{\sqrt{3} - \sqrt{7}} = -1 \cdot \dfrac{\sqrt{3} - \sqrt{7}}{\sqrt{3} - \sqrt{7}} = -1 \cdot 1 = -1$

69. $\dfrac{3\sqrt{2} - \sqrt{7}}{4\sqrt{2} + 2\sqrt{5}} = \dfrac{3\sqrt{2} - \sqrt{7}}{4\sqrt{2} + 2\sqrt{5}} \cdot \dfrac{4\sqrt{2} - 2\sqrt{5}}{4\sqrt{2} - 2\sqrt{5}}$

$= \dfrac{(3\sqrt{2} - \sqrt{7})(4\sqrt{2} - 2\sqrt{5})}{(4\sqrt{2} + 2\sqrt{5})(4\sqrt{2} - 2\sqrt{5})}$

$= \dfrac{12 \cdot 2 - 6\sqrt{10} - 4\sqrt{14} + 2\sqrt{35}}{16 \cdot 2 - 4 \cdot 5}$

$= \dfrac{24 - 6\sqrt{10} - 4\sqrt{14} + 2\sqrt{35}}{32 - 20}$

$= \dfrac{24 - 6\sqrt{10} - 4\sqrt{14} + 2\sqrt{35}}{12}$

$= \dfrac{2(12 - 3\sqrt{10} - 2\sqrt{14} + \sqrt{35})}{2 \cdot 6}$

$= \dfrac{12 - 3\sqrt{10} - 2\sqrt{14} + \sqrt{35}}{6}$

71. $\dfrac{\sqrt{5}+1}{4} = \dfrac{\sqrt{5}+1}{4} \cdot \dfrac{\sqrt{5}-1}{\sqrt{5}-1} = \dfrac{(\sqrt{5}+1)(\sqrt{5}-1)}{4(\sqrt{5}-1)}$

$\qquad = \dfrac{(\sqrt{5})^2 - 1}{4(\sqrt{5}-1)} = \dfrac{5-1}{4(\sqrt{5}-1)} = \dfrac{4}{4(\sqrt{5}-1)}$

$\qquad = \dfrac{1}{\sqrt{5}-1}$

73. $\dfrac{\sqrt{6}-2}{\sqrt{3}+7} = \dfrac{\sqrt{6}-2}{\sqrt{3}+7} \cdot \dfrac{\sqrt{6}+2}{\sqrt{6}+2} = \dfrac{(\sqrt{6}-2)(\sqrt{6}+2)}{(\sqrt{3}+7)(\sqrt{6}+2)}$

$\qquad = \dfrac{6-4}{\sqrt{18}+2\sqrt{3}+7\sqrt{6}+14} = \dfrac{2}{3\sqrt{2}+2\sqrt{3}+7\sqrt{6}+14}$

75. $\dfrac{\sqrt{x}-\sqrt{y}}{\sqrt{x}+\sqrt{y}} = \dfrac{\sqrt{x}-\sqrt{y}}{\sqrt{x}+\sqrt{y}} \cdot \dfrac{\sqrt{x}+\sqrt{y}}{\sqrt{x}+\sqrt{y}}$

$\qquad = \dfrac{(\sqrt{x}-\sqrt{y})(\sqrt{x}+\sqrt{y})}{(\sqrt{x}+\sqrt{y})(\sqrt{x}+\sqrt{y})} = \dfrac{x-y}{x+2\sqrt{xy}+y}$

77. $\dfrac{\sqrt{a+h}-\sqrt{a}}{h} = \dfrac{\sqrt{a+h}-\sqrt{a}}{h} \cdot \dfrac{\sqrt{a+h}+\sqrt{a}}{\sqrt{a+h}+\sqrt{a}}$

$\qquad = \dfrac{(\sqrt{a+h}-\sqrt{a})(\sqrt{a+h}+\sqrt{a})}{h(\sqrt{a+h}+\sqrt{a})} = \dfrac{a+h-a}{h(\sqrt{a+h}+\sqrt{a})}$

$\qquad = \dfrac{h}{h(\sqrt{a+h}+\sqrt{a})} = \dfrac{1}{\sqrt{a+h}+\sqrt{a}}$

79. $\sqrt[3]{a}\,\sqrt[6]{a}$

$\qquad = a^{1/3} \cdot a^{1/6}$ Converting to exponential notation

$\qquad = a^{1/2}$ Adding exponents

$\qquad = \sqrt{a}$ Returning to radical notation

81. $\sqrt{b^3}\,\sqrt[5]{b^4}$

$\qquad = b^{3/2} \cdot b^{4/5}$ Converting to exponential notation

$\qquad = b^{23/10}$ Adding exponents

$\qquad = b^{2+3/10}$ Writing $\frac{23}{10}$ as a mixed number

$\qquad = b^2 b^{3/10}$ Factoring

$\qquad = b^2\,\sqrt[10]{b^3}$ Returning to radical notation

83. $\sqrt{xy^3}\,\sqrt[3]{x^2 y} = (xy^3)^{1/2}(x^2 y)^{1/3}$

$\qquad = (xy^3)^{3/6}(x^2 y)^{2/6}$

$\qquad = \left[(xy^3)^3(x^2 y)^2\right]^{1/6}$

$\qquad = \sqrt[6]{x^3 y^9 \cdot x^4 y^2}$

$\qquad = \sqrt[6]{x^7 y^{11}}$

$\qquad = \sqrt[6]{x^6 y^6 \cdot xy^5}$

$\qquad = xy\,\sqrt[6]{xy^5}$

85. $\sqrt[4]{9ab^3}\,\sqrt{3a^4 b} = (9ab^3)^{1/4}(3a^4 b)^{1/2}$

$\qquad = (9ab^3)^{1/4}(3a^4 b)^{2/4}$

$\qquad = \left[(9ab^3)(3a^4 b)^2\right]^{1/4}$

$\qquad = \sqrt[4]{9ab^3 \cdot 9a^8 b^2}$

$\qquad = \sqrt[4]{81a^9 b^5}$

$\qquad = \sqrt[4]{81a^8 b^4 \cdot ab}$

$\qquad = 3a^2 b\,\sqrt[4]{ab}$

87. $\sqrt{a^4 b^3 c^4}\,\sqrt[3]{ab^2 c} = (a^4 b^3 c^4)^{1/2}(ab^2 c)^{1/3}$

$\qquad = (a^4 b^3 c^4)^{3/6}(ab^2 c)^{2/6}$

$\qquad = \left[(a^4 b^3 c^4)^3(ab^2 c)^2\right]^{1/6}$

$\qquad = \sqrt[6]{a^{12} b^9 c^{12} \cdot a^2 b^4 c^2}$

$\qquad = \sqrt[6]{a^{14} b^{13} c^{14}}$

$\qquad = \sqrt[6]{a^{12} b^{12} c^{12} \cdot a^2 bc^2}$

$\qquad = a^2 b^2 c^2\,\sqrt[6]{a^2 bc^2}$

89. $\dfrac{\sqrt[3]{a^2}}{\sqrt[4]{a}}$

$\qquad = \dfrac{a^{2/3}}{a^{1/4}}$ Converting to exponential notation

$\qquad = a^{2/3 - 1/4}$ Subtracting exponents

$\qquad = a^{5/12}$ Converting back

$\qquad = \sqrt[12]{a^5}$ to radical notation

91. $\dfrac{\sqrt[4]{x^2 y^3}}{\sqrt[3]{xy}}$

$\qquad = \dfrac{(x^2 y^3)^{1/4}}{(xy)^{1/3}}$ Converting to exponential notation

$\qquad = \dfrac{x^{2/4} y^{3/4}}{x^{1/3} y^{1/3}}$ Using the power and product rules

$\qquad = x^{2/4 - 1/3} y^{3/4 - 1/3}$ Subtracting exponents

$\qquad = x^{2/12} y^{5/12}$

$\qquad = (x^2 y^5)^{1/12}$ Converting back to radical notation

$\qquad = \sqrt[12]{x^2 y^5}$

93. $\dfrac{\sqrt{ab^3}}{\sqrt[5]{a^2 b^3}}$

$\qquad = \dfrac{(ab^3)^{1/2}}{(a^2 b^3)^{1/5}}$ Converting to exponential notation

$\qquad = \dfrac{a^{1/2} b^{3/2}}{a^{2/5} b^{3/5}}$

$\qquad = a^{1/10} b^{9/10}$ Subtracting exponents

$\qquad = (ab^9)^{1/10}$ Converting back to radical notation

$\qquad = \sqrt[10]{ab^9}$

95. $\dfrac{\sqrt{(7-y)^3}}{\sqrt[3]{(7-y)^2}}$

$\qquad = \dfrac{(7-y)^{3/2}}{(7-y)^{2/3}}$ Converting to exponential notation

$\qquad = (7-y)^{3/2 - 2/3}$ Subtracting exponents

$\qquad = (7-y)^{5/6}$

$\qquad = \sqrt[6]{(7-y)^5}$ Returning to radical notation

97. $\dfrac{\sqrt[4]{(5+3x)^3}}{\sqrt[3]{(5+3x)^2}}$

$= \dfrac{(5+3x)^{3/4}}{(5+3x)^{2/3}}$ Converting to exponential notation

$= (5+3x)^{3/4-2/3}$ Subtracting exponents

$= (5+3x)^{1/12}$ Converting back to radical notation

$= \sqrt[12]{5+3x}$

99. $\sqrt[3]{x^2 y}\left(\sqrt{xy} - \sqrt[5]{xy^3}\right)$

$= \left(x^2 y\right)^{1/3}\left[(xy)^{1/2} - \left(xy^3\right)^{1/5}\right]$

$= x^{2/3} y^{1/3}\left(x^{1/2} y^{1/2} - x^{1/5} y^{3/5}\right)$

$= x^{2/3} y^{1/3} x^{1/2} y^{1/2} - x^{2/3} y^{1/3} x^{1/5} y^{3/5}$

$= x^{2/3+1/2} y^{1/3+1/2} - x^{2/3+1/5} y^{1/3+3/5}$

$= x^{7/6} y^{5/6} - x^{13/15} y^{14/15}$ Writing as a mixed numeral

$= x \cdot x^{1/6} y^{5/6} - x^{13/15} y^{14/15}$

$= x\left(xy^5\right)^{1/6} - \left(x^{13} y^{14}\right)^{1/15}$

$= x\sqrt[6]{xy^5} - \sqrt[15]{x^{13} y^{14}}$

101. $\left(m + \sqrt[3]{n^2}\right)\left(2m + \sqrt[4]{n}\right)$

$= \left(m + n^{2/3}\right)\left(2m + n^{1/4}\right)$ Converting to exponential notation

$= 2m^2 + mn^{1/4} + 2mn^{2/3} + n^{2/3} n^{1/4}$ Using FOIL

$= 2m^2 + mn^{1/4} + 2mn^{2/3} + n^{2/3+1/4}$ Adding exponents

$= 2m^2 + mn^{1/4} + 2mn^{2/3} + n^{11/12}$

$= 2m^2 + m\sqrt[4]{n} + 2m\sqrt[3]{n^2} + \sqrt[12]{n^{11}}$ Converting back to radical notation

103. $f(x) = \sqrt[4]{x},\; g(x) = 2\sqrt{x} - \sqrt[3]{x^2}$

$(f \cdot g)(x) = \sqrt[4]{x}\left(2\sqrt{x} - \sqrt[3]{x^2}\right)$

$= x^{1/4} \cdot 2x^{1/2} - x^{1/4} \cdot x^{2/3}$

$= 2x^{1/4+1/2} - x^{1/4+2/3}$

$= 2x^{1/4+2/4} - x^{3/12+8/12}$

$= 2x^{3/4} - x^{11/12}$

$= 2\sqrt[4]{x^3} - \sqrt[12]{x^{11}}$

105. $f(x) = x + \sqrt{7},\; g(x) = x - \sqrt{7}$

$(f \cdot g)(x) = \left(x + \sqrt{7}\right)\left(x - \sqrt{7}\right)$

$= x^2 - \left(\sqrt{7}\right)^2$

$= x^2 - 7$

107. $f(x) = x^2$

$f\left(3 - \sqrt{2}\right) = \left(3 - \sqrt{2}\right)^2 = 3^2 - 2 \cdot 3 \cdot \sqrt{2} + \left(\sqrt{2}\right)^2$

$= 9 - 6\sqrt{2} + 2 = 11 - 6\sqrt{2}$

109. $f(x) = x^2$

$f\left(\sqrt{6} + \sqrt{21}\right) = \left(\sqrt{6} + \sqrt{21}\right)^2$

$= \left(\sqrt{6}\right)^2 + 2 \cdot \sqrt{6} \cdot \sqrt{21} + \left(\sqrt{21}\right)^2$

$= 6 + 2\sqrt{126} + 21 = 27 + 2\sqrt{9 \cdot 14}$

$= 27 + 6\sqrt{14}$

111. *Writing Exercise.*

113. IV

115. Let $y = 0$, then $x = 10$. Then the x-intercept is $(10, 0)$.
Let $x = 0$, then $y = -10$. Then the y-intercept is $(0, -10)$.

117. A line perpendicular to $y = \frac{1}{2}x - 7$ has slope -2.

The line is $y = -2x + 12$.

119. *Writing Exercise.*

121. $f(x) = \sqrt{x^3 - x^2} + \sqrt{9x^3 - 9x^2} - \sqrt{4x^3 - 4x^2}$

$= \sqrt{x^2(x-1)} + \sqrt{9x^2(x-1)} - \sqrt{4x^2(x-1)}$

$= x\sqrt{x-1} + 3x\sqrt{x-1} - 2x\sqrt{x-1}$

$= 2x\sqrt{x-1}$

123. $f(x) = \sqrt[4]{x^5 - x^4} + 3\sqrt[4]{x^9 - x^8}$

$= \sqrt[4]{x^4(x-1)} + 3\sqrt[4]{x^8(x-1)}$

$= \sqrt[4]{x^4} \cdot \sqrt[4]{x-1} + 3\sqrt[4]{x^8}\sqrt[4]{x-1}$

$= x\sqrt[4]{x-1} + 3x^2\sqrt[4]{x-1}$

$= (x + 3x^2)\sqrt[4]{x-1}$

125. $7x\sqrt{(x+y)^3} - 5xy\sqrt{x+y} - 2y\sqrt{(x+y)^3}$

$= 7x\sqrt{(x+y)^2(x+y)} - 5xy\sqrt{x+y}$
$\qquad - 2y\sqrt{(x+y)^2(x+y)}$

$= 7x(x+y)\sqrt{x+y} - 5xy\sqrt{x+y} - 2y(x+y)\sqrt{x+y}$

$= [7x(x+y) - 5xy - 2y(x+y)]\sqrt{x+y}$

$= \left(7x^2 + 7xy - 5xy - 2xy - 2y^2\right)\sqrt{x+y}$

$= \left(7x^2 - 2y^2\right)\sqrt{x+y}$

127. $\sqrt{8x(y+z)^5}\sqrt[3]{4x^2(y+z)^2}$

$= \left[8x(y+z)^5\right]^{1/2}\left[4x^2(y+z)^2\right]^{1/3}$

$= \left[8x(y+z)^5\right]^{3/6}\left[4x^2(y+z)^2\right]^{2/6}$

$= \left\{\left[2^3 x(y+z)^5\right]^3\left[2^2 x^2(y+z)^2\right]^2\right\}^{1/6}$

$= \sqrt[6]{2^9 x^3(y+z)^{15} \cdot 2^4 x^4(y+z)^4}$

$= \sqrt[6]{2^{13} x^7(y+z)^{19}}$

$= \sqrt[6]{2^{12} x^6(y+z)^{18} \cdot 2x(y+z)}$

$= 2^2 x(y+z)^3\sqrt[6]{2x(y+z)}$, or

$4x(y+z)^3\sqrt[6]{2x(y+z)}$

129. $\dfrac{\dfrac{1}{\sqrt{w}} - \sqrt{w}}{\dfrac{\sqrt{w}+1}{\sqrt{w}}} = \dfrac{\dfrac{1}{\sqrt{w}} - \sqrt{w}}{\dfrac{\sqrt{w}+1}{\sqrt{w}}} \cdot \dfrac{\sqrt{w}}{\sqrt{w}} = \dfrac{1-w}{\sqrt{w}+1}$

$= \dfrac{1-w}{\sqrt{w}+1} \cdot \dfrac{\sqrt{w}-1}{\sqrt{w}-1} = \dfrac{\sqrt{w}-1 - w\sqrt{w} + w}{w-1}$

$= \dfrac{(w-1) - \sqrt{w}(w-1)}{w-1} = \dfrac{(w-1)(1-\sqrt{w})}{w-1}$

$= 1 - \sqrt{w}$

131. $x - 5 = \left(\sqrt{x}\right)^2 - \left(\sqrt{5}\right)^2 = \left(\sqrt{x} + \sqrt{5}\right)\left(\sqrt{x} - \sqrt{5}\right)$

133. $x - a = \left(\sqrt{x}\right)^2 - \left(\sqrt{a}\right)^2 = \left(\sqrt{x} + \sqrt{a}\right)\left(\sqrt{x} - \sqrt{a}\right)$

135. $\left(\sqrt{x+2} - \sqrt{x-2}\right)^2 = x + 2 - 2\sqrt{(x+2)(x-2)} + x - 2$
$$= x + 2 - 2\sqrt{x^2 - 4} + x - 2$$
$$= 2x - 2\sqrt{x^2 - 4}$$

137. $A^2 + B^2 = \left(A + B + \sqrt{2AB}\right)\left(A + B - \sqrt{2AB}\right)$

  a. $\left(A + B + \sqrt{2AB}\right)\left(A + B - \sqrt{2AB}\right)$
$$= (A + B)^2 - \left(\sqrt{2AB}\right)^2$$
$$= A^2 + 2AB + B^2 - 2AB$$
$$= A^2 + B^2$$

  b. $2AB$ must be a perfect square.

Mid-Chapter Review

1. $\sqrt{6x^9} \cdot \sqrt{2xy} = \sqrt{6x^9 \cdot 2xy}$
$$= \sqrt{12x^{10}y}$$
$$= \sqrt{4x^{10} \cdot 3y}$$
$$= \sqrt{4x^{10}} \cdot \sqrt{3y}$$
$$= 2x^5\sqrt{3y}$$

2. $\sqrt{12} - 3\sqrt{75} + \sqrt{8}$
$$= 2\sqrt{3} - 3 \cdot 5\sqrt{3} + 2\sqrt{2}$$
$$= 2\sqrt{3} - 15\sqrt{3} + 2\sqrt{2}$$
$$= -13\sqrt{3} + 2\sqrt{2}$$

3. $\sqrt{81} = 9$

4. $-\sqrt{\dfrac{9}{100}} = -\dfrac{3}{10}$

5. $\sqrt{64t^2} = |8t|$, or $8|t|$

6. $\sqrt[5]{x^5} = x$

7. $f(x) = \sqrt[3]{12x - 4}$
$$f(-5) = \sqrt[3]{12(-5) - 4} = \sqrt[3]{-64} = -4$$

8. $g(x) = \sqrt[4]{10 - x}$

Since the index is even, the radicand, $10 - x$, must be nonnegative. We solve the inequality:
$$10 - x \geq 0$$
$$-x \geq -10$$
$$x \leq 10$$
Domain of $g = \{x | x \leq 10\}$, or $(-\infty, 10]$

9. $8^{2/3} = \left(2^3\right)^{2/3} = 2^{3 \cdot \frac{2}{3}} = 2^2 = 4$

10. $\sqrt[6]{\sqrt{a}} = \sqrt[6]{a^{1/2}} = \left(a^{1/2}\right)^{1/6} = a^{\frac{1}{2} \cdot \frac{1}{6}} = a^{1/12} = \sqrt[12]{a}$

11. $\sqrt[3]{y^{24}} = y^8$

12. $\sqrt{(t+5)^2} = t + 5$

13. $\sqrt[3]{-27a^{12}} = \sqrt[3]{\left(-3a^4\right)^3} = -3a^4$

14. $\sqrt{6x}\sqrt{15x} = \sqrt{6x \cdot 15x} = \sqrt{90x^2} = \sqrt{9x^2 \cdot 10} = 3x\sqrt{10}$

15. $\dfrac{\sqrt{20y}}{\sqrt{45y}} = \sqrt{\dfrac{20y}{45y}} = \sqrt{\dfrac{4}{9}} = \dfrac{\sqrt{4}}{\sqrt{9}} = \dfrac{2}{3}$

16. $\sqrt{6}\left(\sqrt{10} - \sqrt{33}\right) = \sqrt{6}\sqrt{10} - \sqrt{6}\sqrt{33} = \sqrt{6 \cdot 10} - \sqrt{6 \cdot 33}$
$$= \sqrt{60} - \sqrt{198} = \sqrt{4 \cdot 15} - \sqrt{9 \cdot 22}$$
$$= 2\sqrt{15} - 3\sqrt{22}$$

17. $\dfrac{\sqrt{t}}{\sqrt[8]{t^3}} = \dfrac{t^{1/2}}{t^{3/8}} = t^{1/2 - 3/8} = t^{1/8} = \sqrt[8]{t}$

18. $\dfrac{\sqrt[5]{3a^{12}}}{\sqrt[5]{96a^2}} = \sqrt[5]{\dfrac{3a^{12}}{96a^2}} = \sqrt[5]{\dfrac{a^{10}}{32}} = \dfrac{a^2}{2}$

19. $2\sqrt{3} - 5\sqrt{12} = 2\sqrt{3} - 5\sqrt{4 \cdot 3} = 2\sqrt{3} - 5 \cdot 2\sqrt{3}$
$$= 2\sqrt{3} - 10\sqrt{3} = -8\sqrt{3}$$

20. $\left(\sqrt{5} + 3\right)\left(\sqrt{5} - 3\right) = \left(\sqrt{5}\right)^2 - 3^2 = 5 - 9 = -4$

21. $\left(\sqrt{15} + \sqrt{10}\right)^2 = \left(\sqrt{15}\right)^2 + 2 \cdot \sqrt{10}\sqrt{15} + \left(\sqrt{10}\right)^2$
$$= 15 + 2\sqrt{25 \cdot 6} + 10 = 25 + 2 \cdot 5\sqrt{6}$$
$$= 25 + 10\sqrt{6}$$

22. $\sqrt{25x - 25} - \sqrt{9x - 9} = \sqrt{25(x-1)} - \sqrt{9(x-1)}$
$$= \sqrt{25}\sqrt{x-1} - \sqrt{9}\sqrt{x-1}$$
$$= 5\sqrt{x-1} - 3\sqrt{x-1}$$
$$= 2\sqrt{x-1}$$

23. $\sqrt{x^3 y}\sqrt[5]{xy^4} = x^{3/2}y^{1/2}x^{1/5}y^{4/5} = x^{3/2 + 1/5}y^{1/2 + 4/5}$
$$= x^{17/10}y^{13/10} = x^{1 + 7/10}y^{1 + 3/10}$$
$$= xy\sqrt[10]{x^7 y^3}$$

24. $\sqrt[3]{5000} + \sqrt[3]{625} = \sqrt[3]{1000 \cdot 5} + \sqrt[3]{125 \cdot 5}$
$$= \sqrt[3]{1000}\sqrt[3]{5} + \sqrt[3]{125}\sqrt[3]{5}$$
$$= 10\sqrt[3]{5} + 5\sqrt[3]{5} = 15\sqrt[3]{5}$$

25. $\sqrt[3]{12x^2 y^5}\sqrt[3]{18x^7 y} = \sqrt[3]{216x^9 y^6} = \sqrt[3]{6^3 x^9 y^6} = 6x^3 y^2$

Exercise Set 7.6

1. When we "square both sides" of an equation, we are using the principle of *powers*.

3. To solve an equation with a radical term, we first *isolate* the radical term on one side of the equation.

5. True by the principle of powers

7. False; if $x^2 = 36$, then $x = 6$, or $x = -6$.

9. $\sqrt{5x+1} = 4$

$\left(\sqrt{5x+1}\right)^2 = 4^2$ Principle of powers (squaring)

$5x+1 = 16$

$5x = 15$

$x = 3$

Check: $\sqrt{5x+1} = 4$

$\begin{array}{c|c}\sqrt{5\cdot3+1} & 4 \\ \sqrt{16} & \end{array}$

$\overset{?}{4=4}$ TRUE

The solution is 3.

11. $\sqrt{3x}+1 = 5$

$\sqrt{3x} = 4$ Adding to isolate the radical

$\left(\sqrt{3x}\right)^2 = 4^2$ Principle of powers (squaring)

$3x = 16$

$x = \frac{16}{3}$

Check: $\sqrt{3x}+1 = 5$

$\begin{array}{c|c}\sqrt{3\cdot\frac{16}{3}+1} & 5 \\ \sqrt{16}+1 & \\ 4+1 & \end{array}$

$\overset{?}{5=5}$ TRUE

The solution is $\frac{16}{3}$.

13. $\sqrt{y+5}-4 = 1$

$\sqrt{y+5} = 5$ Adding to isolate the radical

$\left(\sqrt{y+5}\right)^2 = 5^2$ Principle of powers (squaring)

$y+5 = 25$

$y = 20$

Check: $\sqrt{y+5}-4 = 1$

$\begin{array}{c|c}\sqrt{20+5}-4 & 1 \\ \sqrt{25}-4 & \\ 5-4 & \end{array}$

$\overset{?}{1=1}$ TRUE

The solution is 20.

15. $\sqrt{8-x}+7 = 10$

$\sqrt{8-x} = 3$ Adding to isolate the radical

$\left(\sqrt{8-x}\right)^2 = 3^2$ Principle of powers (squaring)

$8-x = 9$

$x = -1$

Check: $\sqrt{8-x}+7 = 10$

$\begin{array}{c|c}\sqrt{8-(-1)}+7 & 10 \\ \sqrt{9}+7 & \\ 3+7 & \end{array}$

$\overset{?}{10=10}$ TRUE

The solution is -1.

17. $\sqrt[3]{y+3} = 2$

$\left(\sqrt[3]{y+3}\right)^3 = 2^3$ Principle of powers (cubing)

$y+3 = 8$

$y = 5$

Check: $\sqrt[3]{y+3} = 2$

$\begin{array}{c|c}\sqrt[3]{5+3} & 2 \\ \sqrt[3]{8} & \end{array}$

$\overset{?}{2=2}$ TRUE

The solution is 5.

19. $\sqrt[4]{t-10} = 3$

$\left(\sqrt[4]{t-10}\right)^4 = 3^4$

$t-10 = 81$

$t = 91$

Check: $\sqrt[4]{t-10} = 3$

$\begin{array}{c|c}\sqrt[4]{91-10} & 3 \\ \sqrt[4]{81} & \end{array}$

$\overset{?}{3=3}$ TRUE

The solution is 91.

21. $6\sqrt{x} = x$

$\left(6\sqrt{x}\right)^2 = x^2$

$36x = x^2$

$0 = x^2 - 36x$

$0 = x(x-36)$

$x = 0$ or $x = 36$

Check:

For $x = 0$: $6\sqrt{x} = x$

$\begin{array}{c|c}6\sqrt{0} & 0 \\ \end{array}$

$\overset{?}{0=0}$ TRUE

For $x = 36$: $6\sqrt{x} = x$

$\begin{array}{c|c}6\sqrt{36} & 36 \\ 6\cdot6 & \end{array}$

$\overset{?}{36=36}$ TRUE

The solutions are 0 and 36.

23. $2y^{1/2}-13 = 7$

$2\sqrt{y}-13 = 7$

$2\sqrt{y} = 20$

$\sqrt{y} = 10$

$\left(\sqrt{y}\right)^2 = 10^2$

$y = 100$

Check: $2y^{1/2}-13 = 7$

$\begin{array}{c|c}2\cdot100^{1/2}-13 & 7 \\ 2\cdot10-13 & \\ 20-13 & \end{array}$

$\overset{?}{7=7}$ TRUE

The solution is 100.

25. $\sqrt[3]{x} = -5$

$\left(\sqrt[3]{x}\right)^3 = (-5)^3$

$x = -125$

Check: $\dfrac{\sqrt[3]{x} = -5}{\begin{array}{c|c} \sqrt[3]{-125} & -5 \\ \sqrt[3]{(-5)^3} & \end{array}}$

$\overset{?}{-5 = -5}$ TRUE

The solution is -125.

27. $z^{1/4} + 8 = 10$

$z^{1/4} = 2$

$\left(z^{1/4}\right)^4 = 2^4$

$z = 16$

Check: $\dfrac{z^{1/4} + 8 = 10}{\begin{array}{c|c} 16^{1/4} + 8 & 10 \\ 2 + 8 & \end{array}}$

$\overset{?}{10 = 10}$ TRUE

The solution is 16.

29. $\sqrt{n} = -2$

This equation has no solution, since the principal square root is never negative.

31. $\sqrt[4]{3x+1} - 4 = -1$

$\sqrt[4]{3x+1} = 3$

$\left(\sqrt[4]{3x+1}\right)^4 = 3^4$

$3x + 1 = 81$

$3x = 80$

$x = \dfrac{80}{3}$

Check: $\dfrac{\sqrt[4]{3x+1} - 4 = -1}{\begin{array}{c|c} \sqrt[4]{3 \cdot \frac{80}{3} + 1} - 4 & -1 \\ \sqrt[4]{81} - 4 & \\ 3 - 4 & \end{array}}$

$\overset{?}{-1 = -1}$ TRUE

The solution is $\dfrac{80}{3}$.

33. $(21x + 55)^{1/3} = 10$

$\left[(21x+55)^{1/3}\right]^3 = 10^3$

$21x + 55 = 1000$

$21x = 945$

$x = 45$

Check: $\dfrac{(21x+55)^{1/3} = 10}{\begin{array}{c|c} (21 \cdot 45 + 55)^{1/3} & 10 \\ (945 + 55)^{1/3} & \end{array}}$

$\overset{?}{10 = 10}$ TRUE

The solution is 45.

35. $\sqrt[3]{3y+6} + 7 = 8$

$\sqrt[3]{3y+6} = 1$

$\left(\sqrt[3]{3y+6}\right)^3 = 1^3$

$3y + 6 = 1$

$3y = -5$

$y = -\dfrac{5}{3}$

Check: $\dfrac{\sqrt[3]{3y+6} + 7 = 8}{\begin{array}{c|c} \sqrt[3]{3\left(-\frac{5}{3}\right) + 6} + 7 & 8 \\ \sqrt[3]{1} + 7 & \\ 1 + 7 & \end{array}}$

$\overset{?}{8 = 8}$ TRUE

The solution is $-\dfrac{5}{3}$.

37. $3 + \sqrt{5-x} = x$

$\sqrt{5-x} = x - 3$

$\left(\sqrt{5-x}\right)^2 = (x-3)^2$

$5 - x = x^2 - 6x + 9$

$0 = x^2 - 5x + 4$

$0 = (x-1)(x-4)$

$x - 1 = 0 \quad or \quad x - 4 = 0$

$x = 1 \quad or \quad x = 4$

Check:

For 1: $\dfrac{3 + \sqrt{5-x} = x}{\begin{array}{c|c} 3 + \sqrt{5-1} & 1 \\ 3 + \sqrt{4} & \\ 3 + 2 & \end{array}}$

$\overset{?}{5 = 1}$ FALSE

For 4: $\dfrac{3 + \sqrt{5-x} = x}{\begin{array}{c|c} 3 + \sqrt{5-4} & 4 \\ 3 + \sqrt{1} & \\ 3 + 1 & \end{array}}$

$\overset{?}{4 = 4}$ TRUE

Since 4 checks but 1 does not, the solution is 4.

39. $\sqrt{3t+4} = \sqrt{4t+3}$

$\left(\sqrt{3t+4}\right)^2 = \left(\sqrt{4t+3}\right)^2$

$3t + 4 = 4t + 3$

$4 = t + 3$

$1 = t$

Check: $\dfrac{\sqrt{3t+4} = \sqrt{4t+3}}{\begin{array}{c|c} \sqrt{3 \cdot 1 + 4} & \sqrt{4 \cdot 1 + 3} \end{array}}$

$\overset{?}{\sqrt{7} = \sqrt{7}}$ TRUE

The solution is 1.

41.

$$3(4-t)^{1/4} = 6^{1/4}$$

$$\left[3(4-t)^{1/4}\right]^4 = \left(6^{1/4}\right)^4$$

$$81(4-t) = 6$$

$$324 - 81t = 6$$

$$-81t = -318$$

$$t = \frac{106}{27}$$

The number $\frac{106}{27}$ checks and is the solution.

43. $\sqrt{4x-3} = 2 + \sqrt{2x-5}$ One radical is already isolated.

$$\left(\sqrt{4x-3}\right)^2 = \left(2 + \sqrt{2x-5}\right)^2 \quad \text{Squaring both sides}$$

$$4x - 3 = 4 + 4\sqrt{2x-5} + 2x - 5$$

$$2x - 2 = 4\sqrt{2x-5}$$

$$x - 1 = 2\sqrt{2x-5}$$

$$x^2 - 2x + 1 = 8x - 20$$

$$x^2 - 10x + 21 = 0$$

$$(x-7)(x-3) = 0$$

$$x - 7 = 0 \quad or \quad x - 3 = 0$$

$$x = 7 \quad or \quad x = 3$$

Both numbers check. The solutions are 7 and 3.

45. $\sqrt{20-x} + 8 = \sqrt{9-x} + 11$

$$\sqrt{20-x} = \sqrt{9-x} + 3 \quad \text{Isolating one radical}$$

$$\left(\sqrt{20-x}\right)^2 = \left(\sqrt{9-x} + 3\right)^2 \quad \text{Squaring both sides}$$

$$20 - x = 9 - x + 6\sqrt{9-x} + 9$$

$$2 = 6\sqrt{9-x} \quad \text{Isolating the remaining radical}$$

$$1 = 3\sqrt{9-x} \quad \text{Multiplying by } \frac{1}{2}$$

$$1^2 = \left(3\sqrt{9-x}\right)^2 \quad \text{Squaring both sides}$$

$$1 = 9(9-x)$$

$$1 = 81 - 9x$$

$$-80 = -9x$$

$$\frac{80}{9} = x$$

The number $\frac{80}{9}$ checks and is the solution.

47. $\sqrt{x+2} + \sqrt{3x+4} = 2$

$$\sqrt{x+2} = 2 - \sqrt{3x+4} \quad \text{Isolating one radical}$$

$$\left(\sqrt{x+2}\right)^2 = \left(2 - \sqrt{3x+4}\right)^2$$

$$x + 2 = 4 - 4\sqrt{3x+4} + 3x + 4$$

$$-2x - 6 = -4\sqrt{3x+4} \quad \text{Isolating the remaining radical}$$

$$x + 3 = 2\sqrt{3x+4} \quad \text{Multiplying by } -\frac{1}{2}$$

$$(x+3)^2 = \left(2\sqrt{3x+4}\right)^2$$

$$x^2 + 6x + 9 = 4(3x+4)$$

$$x^2 + 6x + 9 = 12x + 16$$

$$x^2 - 6x - 7 = 0$$

$$(x-7)(x+1) = 0$$

$$x - 7 = 0 \quad or \quad x + 1 = 0$$

$$x = 7 \quad or \quad x = -1$$

Check:

For 7:

$$\frac{\sqrt{x+2} + \sqrt{3x+4} = 2}{\begin{array}{c|c} \sqrt{7+2} + \sqrt{3\cdot 7 + 4} & 2 \\ \sqrt{9} + \sqrt{25} & \end{array}}$$

$$8 \overset{?}{=} 2 \quad \text{FALSE}$$

For -1:

$$\frac{\sqrt{x+2} + \sqrt{3x+4} = 2}{\begin{array}{c|c} \sqrt{-1+2} + \sqrt{3\cdot(-1)+4} & \\ \sqrt{1} + \sqrt{1} & \end{array}}$$

$$2 \overset{?}{=} 2 \quad \text{TRUE}$$

Since -1 checks but 7 does not, the solution is -1.

49. We must have $f(x) = 1$, or $\sqrt{x} + \sqrt{x-9} = 1$.

$$\sqrt{x} + \sqrt{x-9} = 1$$

$$\sqrt{x-9} = 1 - \sqrt{x} \quad \text{Isolating one radical term}$$

$$\left(\sqrt{x-9}\right)^2 = \left(1 - \sqrt{x}\right)^2$$

$$x - 9 = 1 - 2\sqrt{x} + x$$

$$-10 = -2\sqrt{x} \quad \text{Isolating the remaining radical term}$$

$$5 = \sqrt{x}$$

$$25 = x$$

This value does not check. There is no solution, so there is no value of x for which $f(x) = 1$.

51. $\sqrt{t-2} - \sqrt{4t+1} = -3$

$$\sqrt{t-2} = \sqrt{4t+1} - 3$$

$$\left(\sqrt{t-2}\right)^2 = \left(\sqrt{4t+1} - 3\right)^2$$

$$t - 2 = 4t + 1 - 6\sqrt{4t+1} + 9$$

$$-3t - 12 = -6\sqrt{4t+1}$$

$$t + 4 = 2\sqrt{4t+1}$$

$$(t+4)^2 = \left(2\sqrt{4t+1}\right)^2$$

$$t^2 + 8t + 16 = 4(4t+1)$$

$$t^2 + 8t + 16 = 16t + 4$$

$$t^2 - 8t + 12 = 0$$

$$(t-2)(t-6) = 0$$

$$t - 2 = 0 \quad or \quad t - 6 = 0$$

$$t = 2 \quad or \quad t = 6$$

Both numbers check, so we have $f(t) = -3$ when $t = 2$ and when $t = 6$.

53. We must have $\sqrt{2x-3} = \sqrt{x+7} - 2$.

$$\sqrt{2x-3} = \sqrt{x+7} - 2$$
$$\left(\sqrt{2x-3}\right)^2 = \left(\sqrt{x+7} - 2\right)^2$$
$$2x - 3 = x + 7 - 4\sqrt{x+7} + 4$$
$$x - 14 = -4\sqrt{x+7}$$
$$(x-14)^2 = \left(-4\sqrt{x+7}\right)^2$$
$$x^2 - 28x + 196 = 16(x+7)$$
$$x^2 - 28x + 196 = 16x + 112$$
$$x^2 - 44x + 84 = 0$$
$$(x-2)(x-42) = 0$$
$$x = 2 \quad or \quad x = 42$$

Since 2 checks but 42 does not, we have $f(x) = g(x)$ when $x = 2$.

55. We must have $4 - \sqrt{t-3} = (t+5)^{1/2}$.

$$4 - \sqrt{t-3} = (t+5)^{1/2}$$
$$\left(4 - \sqrt{t-3}\right)^2 = \left[(t+5)^{1/2}\right]^2$$
$$16 - 8\sqrt{t-3} + t - 3 = t + 5$$
$$-8\sqrt{t-3} = -8$$
$$\sqrt{t-3} = 1$$
$$\left(\sqrt{t-3}\right)^2 = 1^2$$
$$t - 3 = 1$$
$$t = 4$$

The number 4 checks, so we have $f(t) = g(t)$ when $t = 4$.

57. *Writing Exercise.*

59. *Familiarize.* Let t = the score of Taylor's last test.

Translate. Write an inequality stating the average score is at least 80.

$$\frac{74 + 88 + 76 + 78 + t}{5} \geq 80$$

Carry out. We solve the inequality.

$$\frac{74 + 88 + 76 + 78 + t}{5} \geq 80$$
$$\frac{316 + t}{5} \geq 80$$
$$316 + t \geq 400$$
$$t \geq 84$$

Check. If the score is 84%, then the average score is $\frac{74 + 88 + 76 + 78 + 84}{5} = 80$. The answer checks.

State. Taylor needs to score at least 84% on the last test to earn at least a B in the course.

61. *Familiarize.* Let c = the speed of the current, in mph. Then $10 + c$ = the speed downriver and $10 - c$ = the speed upriver. We organize the information in a table.

	Distance	Speed	Time
Downriver	7	$10+c$	t_1
Upriver	7	$10-c$	t_2

Translate. Using the formula Time = Distance/Rate we see that $t_1 = \frac{7}{10+c}$ and $t_2 = \frac{7}{10-c}$. The total time upriver and back is $1\frac{2}{3}$ hr, so $t_1 + t_2 = 1\frac{2}{3}$, or

$$\frac{7}{10+c} + \frac{7}{10-c} = \frac{5}{3}.$$

Carry out. We solve the equation. Multiply both sides by the LCD, $3(10+c)(10-c)$.

$$3(10+c)(10-c)\left(\frac{7}{10+c} + \frac{7}{10-c}\right) = 3(10+c)(10-c)\frac{5}{3}$$
$$21(10-c) + 21(10+c) = 5(100 - c^2)$$
$$210 - 21c + 210 + 21c = 500 - 5c^2$$
$$5c^2 - 80 = 0$$
$$5(c+4)(c-4) = 0$$
$$c + 4 = 0 \quad or \quad c - 4 = 0$$
$$c = -4 \quad or \quad c = 4$$

Check. Since speed cannot be negative in this problem, -4 cannot be a solution of the original problem. If the speed of the current is 4 mph, the boat travels upriver at $10 - 4$, or 6 mph. At this rate it takes $\frac{7}{6}$, or $1\frac{1}{6}$ hr, to travel 7 mi. The boat travels downriver at $10 + 4$, or 14 mph. At this rate it takes $\frac{7}{14}$, or $\frac{1}{2}$ hr, to travel 7 mi. The total travel time is $1\frac{1}{6} + \frac{1}{2}$, or $1\frac{2}{3}$ hr. The answer checks.

State. The speed of the current is 4 mph.

63. *Writing Exercise.*

65. Substitute 100 for $v(p)$ and solve for p.

$$v(p) = 12.1\sqrt{p}$$
$$100 = 12.1\sqrt{p}$$
$$8.2645 \approx \sqrt{p}$$
$$(8.2645)^2 \approx \left(\sqrt{p}\right)^2$$
$$68.3013 \approx p$$

The nozzle pressure is about 68 psi.

67. Let f be the frequency of the string and t be the tension of the string. Substitute 260 for f, 28 for t, and solve for k, the constant of variation.

$$f = k\sqrt{t}$$
$$260 = k\sqrt{28}$$
$$k = \frac{260}{\sqrt{28}} \approx 49.135$$

Then substitute 32 for t and solve for f.

$$f = 49.135\sqrt{t}$$
$$f = 49.135\sqrt{32} \approx 277.952$$

The frequency is about 278 Hz.

69. Substitute 1880 for $S(t)$ and solve for t.

$$1880 = 1087.7\sqrt{\frac{9t + 2617}{2457}}$$

$$1.7284 \approx \sqrt{\frac{9t + 2617}{2457}} \qquad \text{Dividing by 1087.7}$$

$$(1.7284)^2 \approx \left(\sqrt{\frac{9t + 2617}{2457}}\right)^2$$

$$2.9874 \approx \frac{9t + 2617}{2457}$$

$$7340.0418 \approx 9t + 2617$$

$$4723.0418 \approx 9t$$

$$524.7824 \approx t$$

The temperature is about 524.8°C.

71.
$$S = 1087.7\sqrt{\frac{9t + 2617}{2457}}$$

$$\frac{S}{1087.7} = \sqrt{\frac{9t + 2617}{2457}}$$

$$\left(\frac{S}{1087.7}\right)^2 = \left(\sqrt{\frac{9t + 2617}{2457}}\right)^2$$

$$\frac{S^2}{1087.7^2} = \frac{9t + 2617}{2457}$$

$$\frac{2457S^2}{1087.7^2} = 9t + 2617$$

$$\frac{2457S^2}{1087.7^2} - 2617 = 9t$$

$$\frac{1}{9}\left(\frac{2457S^2}{1087.7^2} - 2617\right) = t$$

73.
$$v = \sqrt{2gr}\sqrt{\frac{h}{r + h}}$$

$$v^2 = 2gr \cdot \frac{h}{r + h} \qquad \text{Squaring both sides}$$

$$v^2(r + h) = 2grh \qquad \text{Multiplying by } r + h$$

$$v^2 r + v^2 h = 2grh$$

$$v^2 h = 2grh - v^2 r$$

$$v^2 h = r\left(2gh - v^2\right)$$

$$\frac{v^2 h}{2gh - v^2} = r$$

75.
$$d(n) = 0.75\sqrt{2.8n}$$

$$\frac{4}{3}d(n) = \sqrt{2.8n} \qquad \text{Dividing by 0.75 or 3/4}$$

$$\left[\frac{4}{3}d(n)\right]^2 = \left(\sqrt{2.8n}\right)^2$$

$$\frac{16}{9}[d(n)]^2 = 2.8n$$

$$n = \frac{40}{63}[d(n)]^2$$

77.
$$\frac{x + \sqrt{x + 1}}{x - \sqrt{x + 1}} = \frac{5}{11}$$

$$11\left(x + \sqrt{x + 1}\right) = 5\left(x - \sqrt{x + 1}\right)$$

$$11x + 11\sqrt{x + 1} = 5x - 5\sqrt{x + 1}$$

$$16\sqrt{x + 1} = -6x$$

$$8\sqrt{x + 1} = -3x$$

$$\left(8\sqrt{x + 1}\right)^2 = (-3x)^2$$

$$64(x + 1) = 9x^2$$

$$64x + 64 = 9x^2$$

$$0 = 9x^2 - 64x - 64$$

$$0 = (9x + 8)(x - 8)$$

$$9x + 8 = 0 \qquad or \qquad x - 8 = 0$$
$$9x = -8 \qquad or \qquad x = 8$$
$$x = -\frac{8}{9} \qquad or \qquad x = 8$$

Since $-\frac{8}{9}$ checks but 8 does not, the solution is $-\frac{8}{9}$.

79.
$$\left(z^2 + 17\right)^{3/4} = 27$$

$$\left[\left(z^2 + 17\right)^{3/4}\right]^{4/3} = \left(3^3\right)^{4/3}$$

$$z^2 + 17 = 3^4$$

$$z^2 + 17 = 81$$

$$z^2 - 64 = 0$$

$$(z + 8)(z - 8) = 0$$

$$z = -8 \quad or \quad z = 8$$

Both −8 and 8 check. They are the solutions.

81.
$$\sqrt{8 - b} = b\sqrt{8 - b}$$

$$\left(\sqrt{8 - b}\right)^2 = \left(b\sqrt{8 - b}\right)^2$$

$$(8 - b) = b^2(8 - b)$$

$$0 = b^2(8 - b) - (8 - b)$$

$$0 = (8 - b)(b^2 - 1)$$

$$0 = (8 - b)(b + 1)(b - 1)$$

$$8 - b = 0 \quad or \quad b + 1 = 0 \quad or \quad b - 1 = 0$$
$$8 = b \quad or \quad b = -1 \quad or \quad b = 1$$

Since the numbers 8 and 1 check but −1 docs not, 8 and 1 are the solutions.

83. We find the values of x for which $g(x) = 0$.

$$6x^{1/2} + 6x^{-1/2} - 37 = 0$$

$$6\sqrt{x} + \frac{6}{\sqrt{x}} = 37$$

$$\left(6\sqrt{x} + \frac{6}{\sqrt{x}}\right)^2 = 37^2$$

$$36x + 72 + \frac{36}{x} = 1369$$

$$36x^2 + 72x + 36 = 1369x \qquad \text{Multiplying by } x$$

$$36x^2 - 1297x + 36 = 0$$

$$(36x - 1)(x - 36) = 0$$

$$36x - 1 = 0 \quad or \quad x - 36 = 0$$
$$36x = 1 \quad or \quad x = 36$$
$$x = \frac{1}{36} \quad or \quad x = 36$$

Both numbers check. The x-intercepts are $\left(\frac{1}{36},\ 0\right)$ and $(36, 0)$.

85. *Graphing Calculator Exercise*

Exercise Set 7.7

1. The correct choice is (d) Right.

3. The correct choice is (e) Square roots.

5. The correct choice is (f) 30°-60°-90°.

7. $a = 5, b = 3$
Find c.
$c^2 = a^2 + b^2$ Pythagorean theorem
$c^2 = 5^2 + 3^2$ Substituting
$c^2 = 25 + 9$
$c^2 = 34$
$c = \sqrt{34}$ Exact answer
$c \approx 5.831$ Approximation

9. $a = 9, b = 9$
Observe that the legs have the same length, so this is an isosceles right triangle. Then we know that the length of the hypotenuse is the length of a leg times $\sqrt{2}$, or $9\sqrt{2}$, or approximately 12.728.

11. $b = 15, c = 17$
Find a.
$a^2 + b^2 = c^2$ Pythagorean theorem
$a^2 + 15^2 = 17^2$ Substituting
$a^2 + 225 = 289$
$a^2 = 64$
$a = 8$

13. $a^2 + b^2 = c^2$ Pythagorean theorem
$\left(4\sqrt{3}\right)^2 + b^2 = 8^2$
$16 \cdot 3 + b^2 = 64$
$48 + b^2 = 64$
$b^2 = 16$
$b = 4$
The other leg is 4 m long.

15. $a^2 + b^2 = c^2$ Pythagorean theorem
$1^2 + b^2 = \left(\sqrt{20}\right)^2$ Substituting
$1 + b^2 = 20$
$b^2 = 19$
$b = \sqrt{19}$
$b \approx 4.359$

The length of the other leg is $\sqrt{19}$ in., or about 4.359 in.

17. Observe that the length of the hypotenuse, $\sqrt{2}$, is $\sqrt{2}$ times the length of the given leg, 1 m. Thus, we have an isosceles right triangle and the length of the other leg is also 1 m.

19. We have a right triangle with legs of 150 ft and 200 ft. Let $d = $ the length of the diagonal, in feet. We use the Pythagorean theorem to find d.
$150^2 + 200^2 = d^2$
$22,500 + 40,000 = d^2$
$62,500 = d^2$
$250 = d$
Clare travels 250 ft across the parking lot.

21. We have a right triangle with legs of 800 ft and 60 ft. Let $d = $ the length of the diagonal, in feet. We use the Pythagorean theorem to find d.
$800^2 + 60^2 = d^2$
$640,000 + 3600 = d^2$
$643,600 = d^2$
$d = \sqrt{643,600} = 20\sqrt{1609} \approx 802.247$
The zipline was about 802.247 ft.

23. We make a drawing similar to the one in the text.

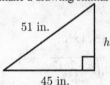

We use the Pythagorean theorem to find h.
$45^2 + h^2 = 51^2$
$2051 + h^2 = 2601$
$h^2 = 576$
$h = 24$
The height of the screen is 24 in.

25. First we will find the diagonal distance, d, in feet, across the room. We make a drawing.

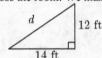

Now we use the Pythagorean theorem.
$12^2 + 14^2 = d^2$
$144 + 196 = d^2$
$340 = d^2$
$d = \sqrt{340} = 2\sqrt{85}$
$d \approx 18.439$
Recall that 4 ft of slack is required on each end. Thus, $\sqrt{340} + 2 \cdot 4$, or $(\sqrt{340} + 8)$ ft, of wire should be purchased. This is about 26.439 ft.

27. The diagonal is the hypotenuse of a right triangle with legs of 70 paces and 40 paces. First we use the Pythagorean theorem to find the length d of the diagonal, in paces.
$70^2 + 40^2 = d^2$
$4900 + 1600 = d^2$
$6500 = d^2$
$d = \sqrt{6500} = 10\sqrt{65}$
$d \approx 80.623$
If Marissa walks along two sides of the quad she takes $70 + 40$, or 110 paces. Then by using the diagonal she saves $(110 - \sqrt{6500})$ paces. This is approximately $110 - 80.623$, or 29.377 paces.

29. Since one acute angle is $45°$, this is an isosceles right triangle with one leg = 5. Then the other leg = 5 also. And the hypotenuse is the length of the a leg times $\sqrt{2}$, or $5\sqrt{2}$.

Exact answer: Leg=5, hypotenuse=$5\sqrt{2}$

Approximation: hypotenuse ≈ 7.071

31. This is a 30-60-90 right triangle with hypotenuse 14. We find the legs:

$2a = 14$, so $a = 7$ and $a\sqrt{3} = 7\sqrt{3}$

Exact answer: shorter leg=7; longer leg=$7\sqrt{3}$

Approximation: longer leg ≈ 12.124

33. This is a 30-60-90 right triangle with one leg = 15. We substitute to find the length of the other leg, a, and the hypotenuse, c.

$$b = a\sqrt{3}$$
$$15 = a\sqrt{3}$$
$$\frac{15}{\sqrt{3}} = a$$
$$\frac{15\sqrt{3}}{3} = a \quad \text{Rationalizing the denominator}$$
$$5\sqrt{3} = a \quad \text{Simplifying}$$
$$c = 2a$$
$$c = 2 \cdot 5\sqrt{3}$$
$$c = 10\sqrt{3}$$

Exact answer: $a = 5\sqrt{3}, c = 10\sqrt{3}$.

Approximations: $a \approx 8.660, c \approx 17.321$

35. This is an isosceles right triangle with hypotenuse 13. The two legs have the same length, a.

$$a\sqrt{2} = 13$$
$$a = \frac{13}{\sqrt{2}} = \frac{13\sqrt{2}}{2}$$

Exact answer: $\frac{13\sqrt{2}}{2}$

Approximation: 9.192

37. This is a 30-60-90 triangle with the shorter leg = 14. We find the longer leg and the hypotenuse.

$a\sqrt{3} = 14\sqrt{3}$, and $2a = 2 \cdot 14 = 28$.

Exact answer: longer leg $= 14\sqrt{3}$, hypotenuse = 28

Approximation: longer leg ≈ 24.249

39. h is the longer leg of a 30-60-90 right triangle with shorter leg $= 5$. Then $h = 5\sqrt{3} \approx 8.660$.

41. We make a drawing.

Triangle ABC is an isosceles right triangle with legs of length 7. Then the hypotenuse$=7\sqrt{2} \approx 9.899$.

43. We make a drawing.

Triangle ABC is an isosceles right triangle with hypotenuse $= 15$. Then $a = \frac{15\sqrt{2}}{2} \approx 10.607$.

45. We will express all distances in feet. Recall that $1 \text{ mi} = 5280 \text{ ft}$.

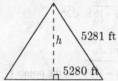

We use the Pythagorean theorem to find h.

$$h^2 + (5280)^2 = (5281)^2$$
$$h^2 + 27,878,400 = 27,888,961$$
$$h^2 = 10,561$$
$$h = \sqrt{10,561}$$
$$h \approx 102.767$$

The height of the bulge is $\sqrt{10,561}$ ft, or about 102.767 ft.

47. We make a drawing.

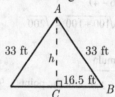

The base of the lodge is an equilateral triangle, so all the angles are 60°. The altitude bisects one angle and one side. Then the triangle ABC is a 30°-60°-90° right triangle with the shorter leg of length $\frac{33}{2}$, or 16.5 ft, and hypotenuse of length 33. Then the height is the length of the shorter leg times $\sqrt{3}$.

Exact answer: $h = \frac{33\sqrt{3}}{2}$ ft

Approximation: $h \approx 28.579$ ft

If the height of triangle ABC is $\frac{33\sqrt{3}}{2}$ and the base is 33 ft, the area is $\frac{1}{2} \cdot 33 \cdot \frac{33\sqrt{3}}{2} = \frac{1089}{4}\sqrt{3}$ ft^2, or about 471.551 ft^2.

49. We make a drawing.

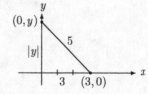

$$|y|^2 + 3^2 = 5^2$$
$$y^2 + 9 = 25$$
$$y^2 = 16$$
$$y = \pm 4$$

The points are $(0, -4)$ and $(0, 4)$.

51. Using the distance formula

$d = \sqrt{(x_2 - x_1)^2 + (y_2 - y_1)^2}$ for the points $(4, 5)$ and $(7, 1)$,

$$d = \sqrt{(7-4)^2 + (1-5)^2}$$
$$= \sqrt{3^2 + (-4)^2} = \sqrt{9+16} = \sqrt{25}$$
$$= 5$$

53. Using the distance formula

$d = \sqrt{(x_2 - x_1)^2 + (y_2 - y_1)^2}$ for the points $(1, -2)$ and $(0, -5)$,

$$d = \sqrt{(1-0)^2 + (-2-(-5))^2}$$
$$= \sqrt{1^2 + 3^2} = \sqrt{1+9} = \sqrt{10}$$
$$\approx 3.162$$

55. Using the distance formula

$d = \sqrt{(x_2 - x_1)^2 + (y_2 - y_1)^2}$ for the points $(6, -6)$ and $(-4, 4)$,

$$d = \sqrt{(6-(-4))^2 + (-6-4)^2}$$
$$= \sqrt{10^2 + (-10)^2} = \sqrt{100+100} = \sqrt{200}$$
$$\approx 14.142$$

57. Using the distance formula

$d = \sqrt{(x_2 - x_1)^2 + (y_2 - y_1)^2}$ for the points $(-9.2, -3.4)$ and $(8.6, -3.4)$,

$$d = \sqrt{(-9.2-8.6)^2 + (-3.4-(-3.4))^2}$$
$$= \sqrt{(-17.8)^2 + 0^2} = \sqrt{316.84}$$
$$\approx 17.8$$

59. Using the distance formula

$d = \sqrt{(x_2 - x_1)^2 + (y_2 - y_1)^2}$ for the points $\left(\frac{5}{6}, -\frac{1}{6}\right)$ and $\left(\frac{1}{2}, \frac{1}{3}\right)$,

$$d = \sqrt{\left(\frac{5}{6} - \frac{1}{2}\right)^2 + \left(-\frac{1}{6} - \frac{1}{3}\right)^2}$$
$$= \sqrt{\left(\frac{2}{6}\right)^2 + \left(-\frac{3}{6}\right)^2} = \sqrt{\frac{4}{36} + \frac{9}{36}} = \sqrt{\frac{13}{36}} = \frac{\sqrt{13}}{6}$$
$$\approx 0.601$$

61. Using the distance formula

$d = \sqrt{(x_2 - x_1)^2 + (y_2 - y_1)^2}$ for the points $(0, 0)$ and $(-\sqrt{6}, \sqrt{6})$,

$$d = \sqrt{(0-(-\sqrt{6}))^2 + (0-\sqrt{6})^2}$$
$$= \sqrt{(\sqrt{6})^2 + (-\sqrt{6})^2} = \sqrt{6+6} = \sqrt{12}$$
$$\approx 3.464$$

63. Using the distance formula

$d = \sqrt{(x_2 - x_1)^2 + (y_2 - y_1)^2}$ for the points $(-2, -40)$ and $(-1, -30)$,

$$d = \sqrt{(-2-(-1))^2 + (-40-(-30))^2}$$
$$= \sqrt{(-1)^2 + (-10)^2} = \sqrt{1+100} = \sqrt{101}$$
$$\approx 10.050$$

65. Using the midpoint formula $\left(\frac{x_1 + x_2}{2}, \frac{y_1 + y_2}{2}\right)$ for the points $(-2, 5)$ and $(8, 3)$,

$\left(\frac{-2+8}{2}, \frac{5+3}{2}\right)$, or $\left(\frac{6}{2}, \frac{8}{2}\right)$, or $(3, 4)$

67. Using the midpoint formula $\left(\frac{x_1 + x_2}{2}, \frac{y_1 + y_2}{2}\right)$ for the points $(2, -1)$ and $(5, 8)$,

$\left(\frac{2+5}{2}, \frac{-1+8}{2}\right)$, or $\left(\frac{7}{2}, \frac{7}{2}\right)$

69. Using the midpoint formula $\left(\frac{x_1 + x_2}{2}, \frac{y_1 + y_2}{2}\right)$ for the points $(-8, -5)$ and $(6, -1)$,

$\left(\frac{-8+6}{2}, \frac{-5+(-1)}{2}\right)$, or $\left(-\frac{2}{2}, \frac{-6}{2}\right)$, or $(-1, -3)$

71. Using the midpoint formula $\left(\frac{x_1 + x_2}{2}, \frac{y_1 + y_2}{2}\right)$ for the points $(-3.4, 8.1)$ and $(4.8, -8.1)$,

$\left(\frac{-3.4+4.8}{2}, \frac{8.1+(-8.1)}{2}\right)$, or $\left(\frac{1.4}{2}, \frac{0}{2}\right)$, or $(0.7, 0)$

73. Using the midpoint formula $\left(\frac{x_1 + x_2}{2}, \frac{y_1 + y_2}{2}\right)$ for the points $\left(\frac{1}{6}, -\frac{3}{4}\right)$ and $\left(-\frac{1}{3}, \frac{5}{6}\right)$,

$\left(\frac{\frac{1}{6} + \left(-\frac{1}{3}\right)}{2}, \frac{-\frac{3}{4} + \frac{5}{6}}{2}\right)$, or $\left(\frac{-\frac{1}{6}}{2}, \frac{\frac{1}{12}}{2}\right)$, or $\left(-\frac{1}{12}, \frac{1}{24}\right)$

75. Using the midpoint formula $\left(\frac{x_1 + x_2}{2}, \frac{y_1 + y_2}{2}\right)$ for the points $(\sqrt{2}, -1)$ and $(\sqrt{3}, 4)$,

$\left(\frac{\sqrt{2}+\sqrt{3}}{2}, \frac{-1+4}{2}\right)$, or $\left(\frac{\sqrt{2}+\sqrt{3}}{2}, \frac{3}{2}\right)$

77. *Writing Exercise.*

79. $y = 2x - 3$

Slope is 2; y-intercept is $(0, -3)$.

81. $8x - 4y = 8$

To find the y-intercept, let $x = 0$ and solve for y.

$$8 \cdot 0 - 4y = 8$$
$$y = -2$$

The y-intercept is $(0, -2)$.

To find the x-intercept, let $y = 0$ and solve for x.

$$8x - 4 \cdot 0 = 8$$
$$x = 1$$

The x-intercept is $(1, 0)$.

Plot these points and draw the line. A third point could be used as a check.

83. $x \geq 1$

Graph the line $x = 1$. Draw the line solid since the inequality symbol is $\geq$. Test the point $(0, 0)$ to determine if it is a solution.

$$x \geq 1$$
$$\frac{}{0 \mid 1}$$
$$\overset{?}{0 \geq 1} \quad \text{FALSE}$$

Since $0 \geq 1$ is false, we shade the half plane that does not contains $(0, 0)$ and obtain the graph.

85. *Writing Exercise.*

87. The length of a side of the hexagon is 72/6, or 12 cm. Then the shaded region is a triangle with base 12 cm. To find the height of the triangle, note that it is the longer leg of a 30°-60°-90° right triangle. Thus its length is the length of the length of the shorter leg times $\sqrt{3}$. The length of the shorter leg is half the length of the base, $\frac{1}{2} \cdot 12$ cm, or 6 cm, so the length of the longer leg is $6\sqrt{3}$ cm. Now we find the area of the triangle.

$$A = \frac{1}{2}bh$$
$$= \frac{1}{2}(12 \text{ cm})(6\sqrt{3} \text{ cm})$$
$$= 36\sqrt{3} \text{ cm}^2$$
$$\approx 62.354 \text{ cm}^2$$

89. We make a drawing.

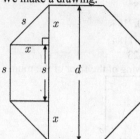

$d = s + 2x$

a. Use the Pythagoran theorem to find x.

$$x^2 + x^2 = s^2$$
$$2x^2 = s^2$$
$$x^2 = \frac{s^2}{2}$$
$$x = \frac{s}{\sqrt{2}} = \frac{s}{\sqrt{2}} \cdot \frac{\sqrt{2}}{\sqrt{2}} = \frac{s\sqrt{2}}{2}$$

Then $d = s + 2x = s + 2\left(\frac{s\sqrt{2}}{2}\right) = s + s\sqrt{2}$.

b. From the drawing we see that an octagon is a composite figure consisting of two trapezoids, one on the right and one on the left, with bases s and d, and height x, and a rectangle with sides s and d. Recall that the area of a trapezoid is $\frac{1}{2}h(b_1 + b_2)$. From part a, we know $d = s + s\sqrt{2}$ and $x = \frac{s\sqrt{2}}{2}$.

$$A = 2 \cdot \frac{1}{2}x[s + d] + sd$$
$$A = \frac{s\sqrt{2}}{2}\left[s + (s + s\sqrt{2})\right] + s(s + s\sqrt{2})$$
$$A = \frac{s\sqrt{2}}{2}(2s + s\sqrt{2}) + s^2 + s^2\sqrt{2}$$
$$A = s^2\sqrt{2} + s^2 + s^2 + s^2\sqrt{2}$$
$$A = 2s^2 + 2s^2\sqrt{2}, \text{ or } 2s^2(1 + \sqrt{2})$$

91. First we find the radius of a circle with an area of 6160 ft^2. This is the length of the hose.

$$A = \pi r^2$$
$$6160 = \pi r^2$$
$$\frac{6160}{\pi} = r^2$$
$$\sqrt{\frac{6160}{\pi}} = r$$
$$44.28 \approx r$$

Now we make a drawing of the room.

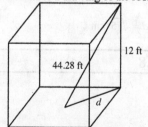

12 ft

44.28 ft

d

We use the Pythagorean theorem to find d.

$$d^2 + 12^2 = 44.28^2$$
$$d^2 + 144 = 1960.7184$$
$$d^2 = 1816.7184$$
$$d \approx 42.623$$

Now we make a drawing of the floor of the room.

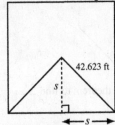

We have an isosceles right triangle with hypotenuse 42.623 ft. We find the length of a side s.

$$s\sqrt{2} = 42.623$$
$$s = \frac{42.623}{\sqrt{2}} \approx 30.14 \text{ ft}$$

Then the length of a side of the room is
$2s = 2(30.14 \text{ ft}) = 60.28$ ft; so the dimensions of the largest square room that meets the given conditions are 60.28 ft by 60.28 ft.

93. We make a drawing.

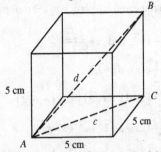

First find the length of a diagonal of the base of the cube. It is the hypotenuse of an isosceles right triangle with legs 5 cm. Then $c = 5\sqrt{2}$ cm.

Triangle ABC is a right triangle with legs of $5\sqrt{2}$ cm and 5 cm and hypotenuse d. Use the Pythagorean theorem to find d, the length of the diagonal that connects two opposite corners of the cube.

$$d^2 = (5\sqrt{2})^2 + 5^2$$
$$d^2 = 25 \cdot 2 + 25$$
$$d^2 = 50 + 25$$
$$d^2 = 75$$
$$d = \sqrt{75}$$

Exact answer: $d = \sqrt{75}$ cm

Exercise Set 7.8

1. False

3. True

5. True

7. False

9. $\sqrt{-100} = \sqrt{-1 \cdot 100} = \sqrt{-1} \cdot \sqrt{100} = i \cdot 10 = 10i$

11. $\sqrt{-5} = \sqrt{-1 \cdot 5} = \sqrt{-1} \cdot \sqrt{5} = i \cdot \sqrt{5}$, or $\sqrt{5}i$

13. $\sqrt{-8} = \sqrt{-1} \cdot \sqrt{4} \cdot \sqrt{2} = i \cdot 2 \cdot \sqrt{2} = 2i\sqrt{2}$, or $2\sqrt{2}i$

15. $-\sqrt{-11} = -\sqrt{-1} \cdot \sqrt{11} = -i \cdot \sqrt{11} = -i\sqrt{11}$, or $-\sqrt{11}i$

17. $-\sqrt{-49} = -\sqrt{-1 \cdot 49} = -\sqrt{-1} \cdot \sqrt{49} = -i \cdot 7 = -7i$

19. $-\sqrt{-300} = -\sqrt{-1} \cdot \sqrt{100} \cdot \sqrt{3} = -i \cdot 10 \cdot \sqrt{3} = -10i\sqrt{3}$, or $-10\sqrt{3}i$

21. $6 - \sqrt{-84} = 6 - \sqrt{-1 \cdot 4 \cdot 21} = 6 - i \cdot 2\sqrt{21} = 6 - 2i\sqrt{21}$, or $6 - 2\sqrt{21}i$

23. $-\sqrt{-76} + \sqrt{-125} = -\sqrt{-1 \cdot 4 \cdot 19} + \sqrt{-1 \cdot 25 \cdot 5}$
$= -i \cdot 2\sqrt{19} + i \cdot 5\sqrt{5} = -2i\sqrt{19} + 5i\sqrt{5} = (-2\sqrt{19} + 5\sqrt{5})i$

25. $\sqrt{-18} - \sqrt{-64} = \sqrt{-1 \cdot 9 \cdot 2} - \sqrt{-1 \cdot 64}$
$= i \cdot 3 \cdot \sqrt{2} - i \cdot 8 = 3i\sqrt{2} - 8i$, or $(3\sqrt{2} - 8)i$

27. $(3 + 4i) + (2 - 7i)$
$= (3 + 2) + (4 - 7)i$ Combining the real and the imaginary parts
$= 5 - 3i$

29. $(9 + 5i) - (2 + 3i) = (9 - 2) + (5 - 3)i$
$= 7 + 2i$

31. $(7 - 4i) - (5 - 3i) = (7 - 5) + [-4 - (-3)]i = 2 - i$

33. $(-5 - i) - (7 + 4i) = (-5 - 7) + (-1 - 4)i = -12 - 5i$

35. $5i \cdot 8i = 40 \cdot i^2 = 40(-1) = -40$

37. $(-4i)(-6i) = 24 \cdot i^2 = 24(-1) = -24$

39. $\sqrt{-36}\sqrt{-9} = \sqrt{-1} \cdot \sqrt{36} \cdot \sqrt{-1} \cdot \sqrt{9} = i \cdot 6 \cdot i \cdot 3$
$= i^2 \cdot 18 = -1 \cdot 18 = -18$

41. $\sqrt{-3}\sqrt{-10} = \sqrt{-1} \cdot \sqrt{3} \cdot \sqrt{-1} \cdot \sqrt{10}$
$= i \cdot \sqrt{3} \cdot i \cdot \sqrt{10}$
$= \sqrt{30} \cdot i^2 = \sqrt{30} \cdot (-1)$
$= -\sqrt{30}$

43. $\sqrt{-6}\sqrt{-21} = \sqrt{-1} \cdot \sqrt{6} \cdot \sqrt{-1} \cdot \sqrt{21}$
$= i \cdot \sqrt{6} \cdot i \cdot \sqrt{21} = i^2\sqrt{126} = -1 \cdot \sqrt{9 \cdot 14}$
$= -3\sqrt{14}$

45. $5i(2 + 6i) = 5i \cdot 2 + 5i \cdot 6i = 10i + 30i^2$
$= 10i - 30 = -30 + 10i$

47. $-7i(3 + 4i) = -7i \cdot 3 - 7i \cdot 4i$
$= -21i - 28i^2$
$= -21i + 28 = 28 - 21i$

49. $(1 + i)(3 + 2i) = 3 + 2i + 3i + 2i^2$
$= 3 + 2i + 3i - 2 = 1 + 5i$

51. $(6-5i)(3+4i) = 18 + 24i - 15i - 20i^2$
$= 18 + 24i - 15i + 20 = 38 + 9i$

53. $(7-2i)(2-6i) = 14 - 42i - 4i + 12i^2$
$= 14 - 42i - 4i - 12 = 2 - 46i$

55. $(3+8i)(3-8i) = 3^2 - (8i)^2$ Difference of squares
$= 9 - 64i^2 = 9 - 64(-1)$
$= 9 + 64 = 73$

57. $(-7+i)(-7-i) = (-7)^2 - i^2$ Difference of squares
$= 49 - i^2 = 49 - (-1)$
$= 49 + 1 = 50$

59. $(4-2i)^2 = 4^2 - 2\cdot 4\cdot 2i + (2i)^2 = 16 - 16i + 4i^2$
$= 16 - 16i - 4 = 12 - 16i$

61. $(2+3i)^2 = 2^2 + 2\cdot 2\cdot 3i + (3i)^2 = 4 + 12i + 9i^2$
$= 4 + 12i - 9 = -5 + 12i$

63. $(-2+3i)^2 = (-2)^2 + 2(-2)(3i) + (3i)^2$
$= 4 - 12i + 9i^2 = 4 - 12i - 9 = -5 - 12i$

65. $\dfrac{10}{3+i} = \dfrac{10}{3+i}\cdot\dfrac{3-i}{3-i}$ Multiplying by 1, using the conjugate
$= \dfrac{30 - 10i}{9 - i^2}$
$= \dfrac{30 - 10i}{9 - (-1)}$
$= \dfrac{30 - 10i}{10}$
$= 3 - i$

67. $\dfrac{2}{3-2i} = \dfrac{2}{3-2i}\cdot\dfrac{3+2i}{3+2i}$ Multiplying by 1, using the conjugate
$= \dfrac{6+4i}{9 - 4i^2}$
$= \dfrac{6+4i}{9 - 4(-1)}$
$= \dfrac{6+4i}{13}$
$= \dfrac{6}{13} + \dfrac{4}{13}i$

69. $\dfrac{2i}{5+3i} = \dfrac{2i}{5+3i}\cdot\dfrac{5-3i}{5-3i} = \dfrac{10i - 6i^2}{25 - 9i^2} = \dfrac{10i+6}{25+9}$
$= \dfrac{10i+6}{34} = \dfrac{6}{34} + \dfrac{10}{34}i = \dfrac{3}{17} + \dfrac{5}{17}i$

71. $\dfrac{5}{6i} = \dfrac{5}{6i}\cdot\dfrac{i}{i} = \dfrac{5i}{6i^2} = \dfrac{5i}{-6} = -\dfrac{5}{6}i$

73. $\dfrac{5-3i}{4i} = \dfrac{5-3i}{4i}\cdot\dfrac{i}{i} = \dfrac{5i - 3i^2}{4i^2} = \dfrac{5i+3}{-4}$
$= -\dfrac{3}{4} - \dfrac{5}{4}i$

75. $\dfrac{7i+14}{7i} = \dfrac{7i}{7i} + \dfrac{14}{7i} = 1 + \dfrac{2}{i} = 1 + \dfrac{2}{i}\cdot\dfrac{i}{i}$
$= 1 + \dfrac{2i}{i^2} = 1 + \dfrac{2i}{-1} = 1 - 2i$

77. $\dfrac{4+5i}{3-7i} = \dfrac{4+5i}{3-7i}\cdot\dfrac{3+7i}{3+7i} = \dfrac{12 + 28i + 15i + 35i^2}{9 - 49i^2}$
$= \dfrac{12 + 28i + 15i - 35}{9 + 49} = \dfrac{-23 + 43i}{58}$
$= -\dfrac{23}{58} + \dfrac{43}{58}i$

79. $\dfrac{2+3i}{2+5i} = \dfrac{2+3i}{2+5i}\cdot\dfrac{2-5i}{2-5i} = \dfrac{4 - 10i + 6i - 15i^2}{4 - 25i^2}$
$= \dfrac{4 - 10i + 6i + 15}{4 + 25} = \dfrac{19 - 4i}{29}$
$= \dfrac{19}{29} - \dfrac{4}{29}i$

81. $\dfrac{3-2i}{4+3i} = \dfrac{3-2i}{4+3i}\cdot\dfrac{4-3i}{4-3i} = \dfrac{12 - 9i - 8i + 6i^2}{16 - 9i^2}$
$= \dfrac{12 - 9i - 8i - 6}{16 + 9} = \dfrac{6 - 17i}{25}$
$= \dfrac{6}{25} - \dfrac{17}{25}i$

83. $i^{32} = (i^2)^{16} = (-1)^{16} = 1$

85. $i^{15} = i^{14}\cdot i = (i^2)^7\cdot i = (-1)^7\cdot i = -i$

87. $i^{42} = (i^2)^{21} = (-1)^{21} = -1$

89. $i^9 = (i^2)^4\cdot i = (-1)^4\cdot i = 1\cdot i = i$

91. $(-i)^6 = (-1\cdot i)^6 = (-1)^6\cdot i^6 = 1\cdot i^6 = (i^2)^3 = (-1)^3 = -1$

93. $(5i)^3 = 5^3\cdot i^3 = 125\cdot i^2\cdot i = 125(-1)(i) = -125i$

95. $i^2 + i^4 = -1 + (i^2)^2 = -1 + (-1)^2 = -1 + 1 = 0$

97. *Writing Exercise.*

99. $x^2 - 100 = (x+10)(x-10)$

101. $2x - 63 + x^2 = x^2 + 2x - 63$
$= (x+9)(x-7)$

103. $w^3 - 4w + 3w^2 - 12$
$= w(w^2 - 4) + 3(w^2 - 4)$
$= (w^2 - 4)(w+3)$
$= (w+2)(w-2)(w+3)$

105. *Writing Exercise.* Yes; every real number a is a complex number $a + bi$ with $b = 0$.

107.

109. $|3+4i| = \sqrt{3^2 + 4^2} = \sqrt{9 + 16} = \sqrt{25} = 5$

111. $|-1+i| = \sqrt{(-1)^2 + 1^2} = \sqrt{1+1} = \sqrt{2}$

113. $g(3i) = \dfrac{(3i)^4 - (3i)^2}{3i - 1} = \dfrac{81i^4 - 9i^2}{-1 + 3i} = \dfrac{81 + 9}{-1 + 3i}$

$= \dfrac{90}{-1 + 3i} = \dfrac{90}{-1 + 3i} \cdot \dfrac{-1 - 3i}{-1 - 3i} = \dfrac{90(-1 - 3i)}{1 - 9i^2}$

$= \dfrac{90(-1 - 3i)}{1 + 9} = \dfrac{90(-1 - 3i)}{10} = \dfrac{9 \cdot 10 \,(-1 - 3i)}{10}$

$= 9(-1 - 3i) = -9 - 27i$

115. First we simplify $g(z)$.

$g(z) = \dfrac{z^4 - z^2}{z - 1} = \dfrac{z^2(z^2 - 1)}{z - 1} = \dfrac{z^2(z + 1)(z - 1)}{z - 1}$

$= \dfrac{z^2(z + 1)(z - 1)}{z - 1} = z^2(z + 1)$

Now we substitute.

$g(5i - 1) = (5i - 1)^2(5i - 1 + 1)$

$= (25i^2 - 10i + 1)(5i)$

$= (-25 - 10i + 1)(5i) = (-24 - 10i)(5i)$

$= -120i - 50i^2 = 50 - 120i$

117. $\dfrac{1}{w - w^2} = \dfrac{1}{\frac{1 - i}{10} - \left(\frac{1 - i}{10}\right)^2}$

$= \dfrac{1}{\frac{1 - i}{10} - \frac{1 - 2i + i^2}{100}} = \dfrac{1}{\frac{10 - 10i - (1 - 2i - 1)}{100}}$

$= \dfrac{1}{\frac{10 - 8i}{100}} = \dfrac{100}{10 - 8i} = \dfrac{50}{5 - 4i}$

$= \dfrac{50}{5 - 4i} \cdot \dfrac{5 + 4i}{5 + 4i} = \dfrac{250 + 200i}{25 + 16}$

$= \dfrac{250 + 200i}{41} = \dfrac{250}{41} + \dfrac{200}{41}i$

119. $(1 - i)^3(1 + i)^3$

$= (1 - i)(1 + i) \cdot (1 - i)(1 + i) \cdot (1 - i)(1 + i)$

$= (1 - i^2)(1 - i^2)(1 - i^2) = (1 + 1)(1 + 1)(1 + 1)$

$= 2 \cdot 2 \cdot 2 = 8$

121. $\dfrac{6}{1 + \frac{3}{i}} = \dfrac{6}{\frac{i + 3}{i}} = \dfrac{6i}{i + 3} = \dfrac{6i}{i + 3} \cdot \dfrac{-i + 3}{-i + 3}$

$= \dfrac{-6i^2 + 18i}{-i^2 + 9} = \dfrac{6 + 18i}{10} = \dfrac{6}{10} + \dfrac{18}{10}i$

$= \dfrac{3}{5} + \dfrac{9}{5}i$

123. $\dfrac{i - i^{38}}{1 + i} = \dfrac{i - (i^2)^{19}}{1 + i} = \dfrac{i - (-1)^{19}}{1 + i} = \dfrac{i - (-1)}{1 + i} = \dfrac{i + 1}{1 + i} = 1$

Chapter 7 Review

1. True

2. False

3. False

4. True

5. True

6. True

7. True

8. False

9. $\sqrt{\dfrac{100}{121}} = \dfrac{\sqrt{100}}{\sqrt{121}} = \dfrac{10}{11}$

10. $-\sqrt{0.36} = -0.6$

11. $f(x) = \sqrt{x + 10}$

$f(15) = \sqrt{15 + 10} = \sqrt{25} = 5$

12. $f(x) = \sqrt{x + 10}$

Since the index is even, the radicand, $x + 10$, must be non-negative. We solve the inequality:

$x + 10 \geq 0$

$x \geq -10$

Domain of $f = \{x \,|\, x \geq -10\}$, or $[-10, \infty)$

13. $\sqrt{64t^2} = \sqrt{(8t)^2} = |8t| = 8|t|$

14. $\sqrt{(c + 7)^2} = |c + 7|$

15. $\sqrt{4x^2 + 4x + 1} = \sqrt{(2x + 1)^2} = |2x + 1|$

16. $\sqrt[5]{-32} = \sqrt[5]{(-2)^5} = -2$

17. $\left(\sqrt[3]{5ab}\right)^4 = (5ab)^{4/3}$

18. $(3a^4)^{1/5} = \sqrt[5]{3a^4}$

19. $\sqrt{x^6 y^{10}} = \left(x^6 y^{10}\right)^{1/2} = x^{6/2} y^{10/2} = x^3 y^5$

20. $\sqrt[6]{(x^2 y)^2} = (x^2 y)^{2/6} = (x^2 y)^{1/3} = \sqrt[3]{x^2 y}$

21. $\left(x^{-2/3}\right)^{3/5} = x^{-\frac{2}{3} \cdot \frac{3}{5}} = x^{-2/5} = \dfrac{1}{x^{2/5}}$

22. $\dfrac{7^{-1/3}}{7^{-1/2}} = 7^{-1/3 + 1/2} = 7^{1/6}$

23. $f(x) = \sqrt{25} \sqrt{(x - 6)^2} = 5|x - 6|$

24. $\sqrt[4]{16x^{20} y^8} = \sqrt[4]{2^4 (x^5)^4 (y^2)^4} = 2x^5 y^2$

25. $\sqrt{250x^3 y^2} = \sqrt{25x^2 y^2 \cdot 10x} = 5xy\sqrt{10x}$

26. $\sqrt{5a}\sqrt{7b} = \sqrt{5a \cdot 7b} = \sqrt{35ab}$

27. $\sqrt[3]{3x^4 b} \sqrt[3]{9xb^2} = \sqrt[3]{3x^4 b \cdot 9xb^2}$

$= \sqrt[3]{27x^3 b^3 \cdot x^2}$

$= 3xb \sqrt[3]{x^2}$

28. $\sqrt[3]{-24x^{10} y^8} \sqrt[3]{18x^7 y^4} = \sqrt[3]{-24x^{10} y^8 \cdot 18x^7 y^4}$

$= \sqrt[3]{-216x^{15} y^{12} \cdot 2x^2}$

$= -6x^5 y^4 \sqrt[3]{2x^2}$

29. $\dfrac{\sqrt[3]{60xy^3}}{\sqrt[3]{10x}} = \sqrt[3]{\dfrac{60xy^3}{10x}} = \sqrt[3]{6y^3} = y\sqrt[3]{6}$

30. $\dfrac{\sqrt{75x}}{2\sqrt{3}} = \dfrac{1}{2}\sqrt{\dfrac{75x}{3}} = \dfrac{1}{2}\sqrt{25x} = \dfrac{5\sqrt{x}}{2}$

31. $\sqrt[4]{\dfrac{48a^{11}}{c^8}} = \dfrac{\sqrt[4]{16a^8 \cdot 3a^3}}{\sqrt[4]{c^8}} = \dfrac{2a^2\sqrt[4]{3a^3}}{c^2}$

32. $5\sqrt[3]{4y} + 2\sqrt[3]{4y} = (5+2)\sqrt[3]{4y} = 7\sqrt[3]{4y}$

33. $2\sqrt{75} - 9\sqrt{3} = 2\sqrt{25 \cdot 3} - 9\sqrt{3} = 2 \cdot 5\sqrt{3} - 9\sqrt{3}$
$\qquad = 10\sqrt{3} - 9\sqrt{3} = (10-9)\sqrt{3} = \sqrt{3}$

34. $\sqrt{50} + 2\sqrt{18} + \sqrt{32} = \sqrt{25 \cdot 2} + 2\sqrt{9 \cdot 2} + \sqrt{16 \cdot 2}$
$\qquad = 5\sqrt{2} + 2 \cdot 3\sqrt{2} + 4\sqrt{2} = 5\sqrt{2} + 6\sqrt{2} + 4\sqrt{2}$
$\qquad = (5+6+4)\sqrt{2} = 15\sqrt{2}$

35. $(3+\sqrt{10})(3-\sqrt{10}) = 3^2 - (\sqrt{10})^2 = 9 - 10 = -1$

36. $(\sqrt{3} - 3\sqrt{8})(\sqrt{5} + 2\sqrt{8})$
$\qquad = \sqrt{3}\sqrt{5} + \sqrt{3} \cdot 2\sqrt{8} - 3\sqrt{8}\sqrt{5} - 3\sqrt{8} \cdot 2\sqrt{8}$
$\qquad = \sqrt{15} + 2\sqrt{3 \cdot 8} - 3\sqrt{8 \cdot 5} - 6\sqrt{64}$
$\qquad = \sqrt{15} + 2\sqrt{4 \cdot 6} - 3\sqrt{4 \cdot 10} - 6 \cdot 8$
$\qquad = \sqrt{15} + 2 \cdot 2\sqrt{6} - 3 \cdot 2\sqrt{10} - 48$
$\qquad = \sqrt{15} + 4\sqrt{6} - 6\sqrt{10} - 48$

37. $\sqrt[4]{x}\sqrt{x} = x^{1/4} \cdot x^{1/2} = x^{1/4+2/4} = x^{3/4} = \sqrt[4]{x^3}$

38. $\dfrac{\sqrt[3]{x^2}}{\sqrt[4]{x}} = \dfrac{x^{2/3}}{x^{1/4}} = x^{2/3-1/4} = x^{5/12} = \sqrt[12]{x^5}$

39. $f(2-\sqrt{a}) = (2-\sqrt{a})^2 = 2^2 - 2 \cdot 2\sqrt{a} + (\sqrt{a})^2$
$\qquad = 4 - 4\sqrt{a} + a$

40. $\sqrt{\dfrac{x}{8y}} = \dfrac{\sqrt{x}}{\sqrt{8y}} = \dfrac{\sqrt{x}}{\sqrt{4 \cdot 2y}} = \dfrac{\sqrt{x}}{2\sqrt{2y}} \cdot \dfrac{\sqrt{2y}}{\sqrt{2y}} = \dfrac{\sqrt{2xy}}{4y}$

41. $\dfrac{4\sqrt{5}}{\sqrt{2}+\sqrt{3}} = \dfrac{4\sqrt{5}}{\sqrt{2}+\sqrt{3}} \cdot \dfrac{\sqrt{2}-\sqrt{3}}{\sqrt{2}-\sqrt{3}} = \dfrac{4\sqrt{10} - 4\sqrt{15}}{2-3}$
$\qquad = -4\sqrt{10} + 4\sqrt{15}$

42. $\dfrac{4\sqrt{5}}{\sqrt{2}+\sqrt{3}} = \dfrac{4\sqrt{5}}{\sqrt{2}+\sqrt{3}} \cdot \dfrac{\sqrt{5}}{\sqrt{5}} = \dfrac{4\sqrt{25}}{\sqrt{10}+\sqrt{15}} = \dfrac{20}{\sqrt{10}+\sqrt{15}}$

43. $\sqrt{y+6} - 2 = 3$
$\qquad \sqrt{y+6} = 5$
$\qquad (\sqrt{y+6})^2 = 5^2$
$\qquad y + 6 = 25$
$\qquad y = 19$

Check: $\dfrac{\sqrt{y+6} - 2 = 3}{\begin{array}{c|c} \sqrt{19+6} - 2 & 3 \\ \sqrt{25} - 2 & \\ 5 - 2 & \end{array}}$
$$\overset{?}{3=3} \quad \text{TRUE}$$

The solution is 19.

44. $(x+1)^{1/3} = -5$
$\qquad \sqrt[3]{x+1} = -5$
$\qquad (\sqrt[3]{x+1})^3 = (-5)^3$
$\qquad x + 1 = -125$
$\qquad x = -126$

Check: $\dfrac{(x+1)^{1/3} = -5}{\begin{array}{c|c} (-126+1)^{1/3} & -5 \\ (-125) & \end{array}}$
$$\overset{?}{-5=-5} \quad \text{TRUE}$$

The solution is -126.

45. $1 + \sqrt{x} = \sqrt{3x-3}$
$\qquad (1+\sqrt{x})^2 = (\sqrt{3x-3})^2$
$\qquad 1 + 2\sqrt{x} + x = 3x - 3$
$\qquad 2\sqrt{x} = 2x - 4$
$\qquad \sqrt{x} = x - 2$
$\qquad (\sqrt{x})^2 = (x-2)^2$
$\qquad x = x^2 - 4x + 4$
$\qquad 0 = x^2 - 5x + 4$
$\qquad 0 = (x-1)(x-4)$
$\qquad x = 1 \ or \ x = 4$

Check:

For $x = 1$: $\dfrac{1+\sqrt{x} = \sqrt{3x-3}}{\begin{array}{c|c} 1+\sqrt{1} & \sqrt{3 \cdot 1 - 3} \\ 1+1 & \sqrt{0} \end{array}}$
$$\overset{?}{2=0} \quad \text{FALSE}$$

For $x = 4$: $\dfrac{1+\sqrt{x} = \sqrt{3x-3}}{\begin{array}{c|c} 1+\sqrt{4} & \sqrt{3 \cdot 4 - 3} \\ 1+2 & \sqrt{9} \end{array}}$
$$\overset{?}{3=3} \quad \text{TRUE}$$

The solution is 4.

46. $f(a) = \sqrt{a+2} + a = 4$
$\qquad \sqrt{a+2} + a = 4$
$\qquad \sqrt{a+2} = 4 - a$
$\qquad (\sqrt{a+2})^2 = (4-a)^2$
$\qquad a + 2 = 16 - 8a + a^2$
$\qquad 0 = a^2 - 9a + 14$
$\qquad 0 = (a-2)(a-7)$
$\qquad a = 2 \ or \ a = 7$

Check:

For $a = 2$: $\sqrt{a+2} + a = 4$

$$\frac{\sqrt{2+2}+2}{2+2}\; \Big|\; 4$$

$$\overset{?}{4=4}\quad \text{TRUE}$$

For $a = 7$: $\sqrt{a+2} + a = 4$

$$\frac{\sqrt{7+2}+7}{3+7}\; \Big|\; 4$$

$$\overset{?}{10=4}\quad \text{FALSE}$$

The solution is 2.

47. Let a represent the side of the square.
Triangle ABC is an isosceles right triangle with hypotenuse 10 cm. Then the length of a side of the square a, is the length of the legs of the triangle. We have

$$a\sqrt{2} = 10$$
$$a = \frac{10}{\sqrt{2}}, \text{ or}$$
$$a = 5\sqrt{2}$$
$$a \approx 7.071$$

The side is $5\sqrt{2}$ cm or about 7.071 cm.

48. Let b represent the base. We use the Pythagorean theorem to find b.

$$b^2 + 2^2 = 6^2$$
$$b^2 + 4 = 36$$
$$b^2 = 32$$
$$b = \sqrt{32} \approx 5.657$$

The base is $\sqrt{32}$ ft or about 5.657 ft.

49. This is a 30°-60°-90° right triangle with hypotenuse 20. Let $a =$ the shorter leg and $b =$ the longer leg.

$2a = 20$, so $a = 10$, and

$b = a\sqrt{3} = 10\sqrt{3} \approx 17.321$.

50. Using the distance formula

$d = \sqrt{(x_2 - x_1)^2 + (y_2 - y_1)^2}$ for the points $(-6, 4)$ and $(-1, 5)$,

$$d = \sqrt{(-1-(-6))^2 + (5-4)^2}$$
$$= \sqrt{5^2 + 1^2} = \sqrt{25+1} = \sqrt{26}$$
$$\approx 5.099$$

51. Using the midpoint formula $\left(\dfrac{x_1 + x_2}{2}, \dfrac{y_1 + y_2}{2}\right)$ for the points $(-7, -2)$ and $(3, -1)$,

$$\left(\frac{-7+3}{2}, \frac{-2+(-1)}{2}\right), \text{ or } \left(\frac{-4}{2}, \frac{-3}{2}\right), \text{ or } \left(-2, -\frac{3}{2}\right)$$

52. $\sqrt{-45} = \sqrt{-1} \cdot \sqrt{9} \cdot \sqrt{5} = 3i\sqrt{5}$ or $3\sqrt{5}\,i$

53. $(-4+3i) + (2-12i) = (-4+2) + (3-12)i = -2 - 9i$

54. $(9-7i) - (3-8i) = (9-3) + (-7+8)i = 6 + i$

55. $(2+5i)(2-5i) = 4 - 25i^2 = 4 + 25 = 29$

56. $i^{34} = \left(i^2\right)^{17} = (-1)^{17} = -1$

57. $(6-3i)(2-i) = 12 - 6i - 6i + 3i^2$
$$= 12 - 6i - 6i - 3 = 9 - 12i$$

58. $\dfrac{7-2i}{3+4i} = \dfrac{7-2i}{3+4i} \cdot \dfrac{3-4i}{3-4i} = \dfrac{21 - 28i - 6i + 8i^2}{9 - 16i^2}$

$$= \frac{21 - 34i - 8}{9 + 16} = \frac{13 - 34i}{25} = \frac{13}{25} - \frac{34}{25}i$$

59. *Writing Exercise.* A complex number $a + bi$ is real when $b = 0$. It is imaginary when $b \neq 0$.

60. *Writing Exercise.* An absolute-value sign must be used to simplify $\sqrt[n]{x^n}$ when n is even, since x may be negative. If x is negative while n is even, the radical expression cannot be simplified to x, since $\sqrt[n]{x^n}$ represents the principal, or nonnegative, root. When n is odd, there is only one root, and it will be positive or negative depending on the sign of x. Thus, there is no absolute-value sign when n is odd.

61. Answers may vary. For example, $\dfrac{2i}{3i}$

62. $\sqrt{11x + \sqrt{6+x}} = 6$

$$\left(\sqrt{11x + \sqrt{6+x}}\right)^2 = (6)^2$$
$$11x + \sqrt{6+x} = 36$$
$$\sqrt{6+x} = 36 - 11x$$
$$\left(\sqrt{6+x}\right)^2 = (36 - 11x)^2$$
$$6 + x = 1296 - 792x + 121x^2$$
$$0 = 121x^2 - 793x + 1290$$
$$0 = (x-3)(121x - 430)$$
$$x = 3 \text{ or } x = \frac{430}{121}$$

Check:

For $x = 3$: $\sqrt{11x + \sqrt{6+x}} = 6$

$$\frac{\sqrt{11\cdot 3 + \sqrt{6+3}}}{\sqrt{33+3}}\; \Big|\; 6$$

$$\overset{?}{6=6}\quad \text{TRUE}$$

For $x = \frac{430}{121}$: $\sqrt{11x + \sqrt{6+x}} = 6$

$$\frac{\sqrt{11\cdot\frac{430}{121} + \sqrt{6+\frac{430}{121}}}}{\sqrt{\frac{430}{11} + \frac{34}{11}}}\; \Big|\; 6$$

$$\sqrt{\frac{464}{11}} \overset{?}{=} 6\quad \text{FALSE}$$

The solution is 3.

63. $\dfrac{2}{1-3i} - \dfrac{3}{4+2i} = \dfrac{2}{1-3i} \cdot \dfrac{1+3i}{1+3i} + \dfrac{-3}{4+2i} \cdot \dfrac{4-2i}{4-2i}$

$\quad = \dfrac{2+6i}{1-9i^2} + \dfrac{-12+6i}{16-4i^2}$

$\quad = \dfrac{2+6i}{1+9} + \dfrac{-12+6i}{16+4}$

$\quad = \dfrac{2+6i}{10} + \dfrac{-12+6i}{20}$

$\quad = \dfrac{1}{5} + \dfrac{6}{10}i + \dfrac{-3}{5} + \dfrac{3}{10}i$

$\quad = -\dfrac{2}{5} + \dfrac{9}{10}i$

64. The isosceles right triangle has hypotenuse 6. Let x represent the leg of the triangle. Using the Pythagorean theorem,

$$x^2 + x^2 = 6^2$$
$$2x^2 = 36$$
$$x^2 = 18$$
$$x = \sqrt{18} \text{ ft}$$

The area of the isosceles right triangle is

$$A = \tfrac{1}{2}x^2 = \tfrac{1}{2}\left(\sqrt{18}\right)^2 = 9 \text{ ft}^2$$

Then in the 30°-60°-90° triangle, a is the shorter leg and we have

$$6 = 2a$$
$$3 = a$$
$$b = a\sqrt{3}$$
$$b = 3\sqrt{3}$$

The area of the 30-60-90 triangle is

$$A = \tfrac{1}{2}\left(3\sqrt{3}\right)(3) = \dfrac{9\sqrt{3}}{2} \text{ ft}^2 \approx 7.794 \text{ ft}^2$$

The area of the isosceles right triangle is larger by about 1.206 ft^2.

65. Using the distance formula

$d = \sqrt{\left(x_2 - x_1\right)^2 + \left(y_2 - y_1\right)^2}$ for the points
(0.456, 1.387) and (4.783, 2.865),

$d = \sqrt{\left(4.783 - 0.456\right)^2 + \left(2.865 - 1.387\right)^2}$

$\quad = \sqrt{\left(4.327\right)^2 + \left(1.478\right)^2} \approx \sqrt{20.907413} \approx 4.5724$

The distance is approximately 4.572 miles.

Chapter 7 Test

1. $\sqrt{50} = \sqrt{25 \cdot 2} = \sqrt{25} \cdot \sqrt{2} = 5\sqrt{2}$

2. $\sqrt[3]{-\dfrac{8}{x^6}} = \dfrac{\sqrt[3]{-8}}{\sqrt[3]{x^6}} = -\dfrac{2}{x^2}$

3. $\sqrt{81a^2} = \sqrt{\left(9a\right)^2} = |9a| = 9|a|$

4. $\sqrt{x^2 - 8x + 16} = \sqrt{\left(x-4\right)^2} = |x-4|$

5. $\sqrt{7xy} = \left(7xy\right)^{1/2}$

6. $\left(4a^3b\right)^{5/6} = \sqrt[6]{\left(4a^3b\right)^5}$

7. $f(x) = \sqrt{2x - 10}$

Since the index is even, the radicand, $2x - 10$, must be non-negative. We solve the inequality:
$$2x - 10 \ge 0$$
$$x \ge 5$$
Domain of $f = \{x | x \ge 5\}$, or $[5, \infty)$

8. $f(x) = x^2$

$f\left(5+\sqrt{2}\right) = \left(5+\sqrt{2}\right)^2 = 5^2 + 2 \cdot 5 \cdot \sqrt{2} + \left(\sqrt{2}\right)^2$

$\quad = 25 + 10\sqrt{2} + 2 = 27 + 10\sqrt{2}$

9. $\sqrt[5]{32x^{16}y^{10}} = \sqrt[5]{2^5 x^{15} y^{10} \cdot x}$

$\quad = \sqrt[5]{2^5 x^{15} y^{10}} \cdot \sqrt[5]{x}$

$\quad = 2x^3 y^2 \sqrt[5]{x}$

10. $\sqrt[3]{4w}\sqrt[3]{4v^2} = \sqrt[3]{4w \cdot 4v^2} = \sqrt[3]{8 \cdot 2v^2 w} = 2\sqrt[3]{2v^2 w}$

11. $\sqrt{\dfrac{100a^4}{9b^6}} = \dfrac{\sqrt{100a^4}}{\sqrt{9b^6}} = \dfrac{10a^2}{3b^3}$

12. $\dfrac{\sqrt[5]{48x^6 y^{10}}}{\sqrt[5]{16x^2 y^9}} = \sqrt[5]{\dfrac{48x^6 y^{10}}{16x^2 y^9}} = \sqrt[5]{3x^4 y}$

13. $\sqrt[4]{x^3}\sqrt{x} = x^{3/4} x^{1/2} = x^{3/4+1/2}$

$\quad = x^{5/4} = x^{1+1/4} = x\sqrt[4]{x}$

14. $\dfrac{\sqrt{y}}{\sqrt[10]{y}} = \dfrac{y^{1/2}}{y^{1/10}} = y^{1/2-1/10} = y^{2/5} = \sqrt[5]{y^2}$

15. $8\sqrt{2} - 2\sqrt{2} = (8-2)\sqrt{2} = 6\sqrt{2}$

16. $\sqrt{50xy} + \sqrt{72xy} - \sqrt{8xy}$

$\quad = \sqrt{25 \cdot 2xy} + \sqrt{36 \cdot 2xy} - \sqrt{4 \cdot 2xy}$

$\quad = 5\sqrt{2xy} + 6\sqrt{2xy} - 2\sqrt{2xy}$

$\quad = 9\sqrt{2xy}$

17. $\left(7+\sqrt{x}\right)\left(2-3\sqrt{x}\right) = 14 - 7 \cdot 3\sqrt{x} + 2\sqrt{x} - \sqrt{x} \cdot 3\sqrt{x}$

$\quad = 14 - 21\sqrt{x} + 2\sqrt{x} - 3\left(\sqrt{x}\right)^2$

$\quad = 14 - 19\sqrt{x} - 3x$

18. $\dfrac{\sqrt[3]{x}}{\sqrt[3]{4y}} = \dfrac{\sqrt[3]{x}}{\sqrt[3]{2^2 y}} \cdot \dfrac{\sqrt[3]{2y^2}}{\sqrt[3]{2y^2}} = \dfrac{\sqrt[3]{2xy^2}}{2y}$

19.
$$6 = \sqrt{x-3} + 5$$
$$1 = \sqrt{x-3}$$
$$1^2 = \left(\sqrt{x-3}\right)^2$$
$$1 = x - 3$$
$$4 = x$$

Check: $6 = \sqrt{x-3} + 5$

$$\begin{array}{c|c} 6 & \sqrt{4-3} + 5 \\ & \sqrt{1} + 5 \\ & \overset{?}{6 = 6} \quad \text{TRUE} \end{array}$$

The solution is 4.

20.
$$x = \sqrt{3x+3} - 1$$
$$x + 1 = \sqrt{3x+3}$$
$$(x+1)^2 = \left(\sqrt{3x+3}\right)^2$$
$$x^2 + 2x + 1 = 3x + 3$$
$$x^2 - x - 2 = 0$$
$$(x+1)(x-2) = 0$$
$$x = -1 \ or \ x = 2$$

Check:

For $x = -1$: $x = \sqrt{3x+3} - 1$

$$\begin{array}{c|c} -1 & \sqrt{3(-1)+3} - 1 \\ & \sqrt{0} - 1 \\ & \overset{?}{-1 = -1} \quad \text{TRUE} \end{array}$$

For $x = 2$: $x = \sqrt{3x+3} - 1$

$$\begin{array}{c|c} 2 & \sqrt{3 \cdot 2 + 3} - 1 \\ & \sqrt{9} - 1 \\ & \overset{?}{2 = 2} \quad \text{TRUE} \end{array}$$

The solutions are -1 and 2.

21.
$$\sqrt{2x} = \sqrt{x+1} + 1$$
$$\sqrt{2x} - 1 = \sqrt{x+1}$$
$$\left(\sqrt{2x} - 1\right)^2 = \left(\sqrt{x+1}\right)^2$$
$$2x - 2\sqrt{2x} + 1 = x + 1$$
$$-2\sqrt{2x} = -x$$
$$\left(-2\sqrt{2x}\right)^2 = (-x)^2$$
$$4(2x) = x^2$$
$$0 = x^2 - 8x$$
$$0 = x(x-8)$$
$$x = 0 \ or \ x = 8$$

Check:

For $x = 0$: $\sqrt{2x} = \sqrt{x+1} + 1$

$$\begin{array}{c|c} \sqrt{2 \cdot 0} & \sqrt{0+1} + 1 \\ & 1 + 1 \\ & \overset{?}{0 = 2} \quad \text{FALSE} \end{array}$$

For $x = 8$: $\sqrt{2x} = \sqrt{x+1} + 1$

$$\begin{array}{c|c} \sqrt{2 \cdot 8} & \sqrt{8+1} + 1 \\ \sqrt{16} & 3 + 1 \\ & \overset{?}{4 = 4} \quad \text{TRUE} \end{array}$$

The solution is 8.

22. We make a drawing.

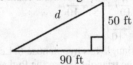

We use the Pythagorean theorem to find d.
$$d^2 = 50^2 + 90^2$$
$$d^2 = 2500 + 8100$$
$$d^2 = 10{,}600$$
$$d = \sqrt{10{,}600} \approx 102.956$$

She jogs $\sqrt{10{,}600}$ ft or about 102.956 ft.

23. This is a 30°-60°-90° right triangle with hypotenuse 10. Let $a =$ the shorter leg and $b =$ the longer leg.
$2a = 10$, so $a = 5$ cm, and
$b = a\sqrt{3} = 5\sqrt{3} \approx 8.660$ cm .

24. Using the distance formula
$$d = \sqrt{(x_2 - x_1)^2 + (y_2 - y_1)^2}$$
for the points (3, 7) and $(-1, 8)$,
$$d = \sqrt{(-1-3)^2 + (8-7)^2}$$
$$= \sqrt{(-4)^2 + (1)^2} = \sqrt{16+1} = \sqrt{17}$$
$$\approx 4.123$$

25. Using the midpoint formula $\left(\dfrac{x_1 + x_2}{2}, \ \dfrac{y_1 + y_2}{2}\right)$ for the points (2, −5) and (1, −7),
$$\left(\frac{2+1}{2}, \ \frac{-5+(-7)}{2}\right), \text{ or } \left(\frac{3}{2}, \frac{-12}{2}\right), \text{ or } \left(\frac{3}{2}, -6\right)$$

26. $\sqrt{-50} = \sqrt{-1} \cdot \sqrt{25} \cdot \sqrt{2} = 5i\sqrt{2}$ or $5\sqrt{2}\, i$

27. $(9+8i) - (-3+6i) = (9+3) + (8-6)i = 12 + 2i$

28. $(4-i)^2 = 4^2 - 2 \cdot 4 \cdot i + (-i)^2 = 16 - 8i + i^2$
$$= 16 - 8i - 1 = 15 - 8i$$

29. $\dfrac{-2+i}{3-5i} = \dfrac{-2+i}{3-5i} \cdot \dfrac{3+5i}{3+5i} = \dfrac{-6-10i+3i+5i^2}{9-25i^2}$
$$= \frac{-6-7i-5}{9+25} = \frac{-11-7i}{34}$$
$$= -\frac{11}{34} - \frac{7}{34}i$$

30. $i^{37} = i^{36} \cdot i = \left(i^2\right)^{18} \cdot i = (-1)^{18} \cdot i = i$

31.

$$\sqrt{2x-2}+\sqrt{7x+4}=\sqrt{13x+10}$$

$$\left(\sqrt{2x-2}+\sqrt{7x+4}\right)^2=\left(\sqrt{13x+10}\right)^2$$

$$2x-2+2\sqrt{2x-2}\sqrt{7x+4}+7x+4=13x+10$$

$$2\sqrt{2x-2}\sqrt{7x+4}=4x+8$$

$$\left(\sqrt{2x-2}\sqrt{7x+4}\right)^2=\left(2x+4\right)^2$$

$$(2x-2)(7x+4)=4x^2+16x+16$$

$$14x^2-6x-8=4x^2+16x+16$$

$$10x^2-22x-24=0$$

$$2\left(5x^2-11x-12\right)=0$$

$$2(5x+4)(x-3)=0$$

$$x=-\frac{4}{5}\ \ or\ \ x=3$$

Check:

For $x=-\frac{4}{5}$:

$$\frac{\sqrt{2x-2}+\sqrt{7x+4}=\sqrt{13x+10}}{\sqrt{2\cdot\left(-\frac{4}{5}\right)+2}+\sqrt{7\left(-\frac{4}{5}\right)+4}\ \Big|\ \sqrt{13\left(-\frac{4}{5}\right)+10}}$$

$$\sqrt{\tfrac{2}{5}}+\sqrt{-\tfrac{8}{5}}\ \Big|\ \sqrt{-\tfrac{2}{5}}$$

Since the values in the check are not real, $-\frac{4}{5}$ is not a solution.

For $x=3$: $\dfrac{\sqrt{2x-2}+\sqrt{7x+4}=\sqrt{13x+10}}{\sqrt{2\cdot3-2}+\sqrt{7\cdot3+4}\ \Big|\ \sqrt{13\cdot3+10}}$

$$\sqrt{4}+\sqrt{25}\ \Big|\ \sqrt{49}$$

$$2+5\ \Big|$$

$$7\overset{?}{=}7\ \ TRUE$$

The solution is 3.

32. $\dfrac{1-4i}{4i(1+4i)^{-1}}=\dfrac{1-4i}{4i}\cdot\dfrac{(1+4i)}{1}$

$$=\frac{1-16i^2}{4i}=\frac{1+16}{4i}$$

$$=\frac{17}{4i}=\frac{17}{4i}\cdot\frac{i}{i}$$

$$=\frac{17i}{4i^2}=-\frac{17}{4}i$$

33. Substitute 180 for $D(h)$, and solve for h.

$$D(h)=1.2\sqrt{h}$$

$$180=1.2\sqrt{h}$$

$$150=\sqrt{h}$$

$$(150)^2=\left(\sqrt{h}\right)^2$$

$$22{,}500=h$$

The pilot must be above 22,500 ft.

Chapter 8

Quadratic Functions and Equations

Exercise Set 8.1

1. The general form of a *quadratic function* is
$$f(x) = ax^2 + bx + c.$$

3. The quadratic equation $ax^2 + bx + c = 0$ is written in *standard form*.

5. If $x^2 = 7$, we know that $x = \sqrt{7}$ or $x = -\sqrt{7}$ because of the principle of *square roots*.

7. $x^2 = 100$
$\quad x = 10 \quad or \quad x = -10 \quad$ Using the principle
$\qquad\qquad\qquad\qquad\qquad\qquad$ of square roots
The solutions are -10 and 10, or ± 10.

9. $p^2 - 50 = 0$
$\quad\quad p^2 = 50 \quad$ Isolating p^2

$\quad p = \sqrt{50} \quad or \quad p = -\sqrt{50} \quad$ Principle of square roots
$\quad p = 5\sqrt{2} \quad or \quad p = -5\sqrt{2}$
The solutions are $5\sqrt{2}$ and $-5\sqrt{2}$ or $\pm 5\sqrt{2}$.

11. $5y^2 = 30$
$\quad\quad y^2 = 6 \quad$ Isolating y^2
$\quad y = \sqrt{6} \ or \ y = -\sqrt{6} \quad$ Principle of square roots
The solutions are $\sqrt{6}$ and $-\sqrt{6}$ or $\pm\sqrt{6}$.

13. $9x^2 - 49 = 0$
$\qquad\quad x^2 = \dfrac{49}{9} \quad$ Isolating x^2

$\quad x = \sqrt{\dfrac{49}{9}} \quad or \quad x = -\sqrt{\dfrac{49}{9}} \quad$ Principle of square roots

$\quad x = \dfrac{7}{3} \quad or \quad x = -\dfrac{7}{3}$

The solutions are $\dfrac{7}{3}$ and $-\dfrac{7}{3}$ or $\pm\dfrac{7}{3}$.

15. $6t^2 - 5 = 0$
$\qquad t^2 = \dfrac{5}{6}$

$\quad t = \sqrt{\dfrac{5}{6}} \quad or \quad t = -\sqrt{\dfrac{5}{6}} \quad$ Principle of square roots

$\quad t = \sqrt{\dfrac{5}{6} \cdot \dfrac{6}{6}} \quad or \quad t = -\sqrt{\dfrac{5}{6} \cdot \dfrac{6}{6}} \quad$ Rationalizing denominators

$\quad t = \dfrac{\sqrt{30}}{6} \quad or \quad t = -\dfrac{\sqrt{30}}{6}$

The solutions are $\sqrt{\dfrac{5}{6}}$ and $-\sqrt{\dfrac{5}{6}}$. This can also be

written as $\pm\sqrt{\dfrac{5}{6}}$ or, if we rationalize the denominator,

$\pm\dfrac{\sqrt{30}}{6}$.

17. $a^2 + 1 = 0$
$\qquad a^2 = -1$

$\quad a = \sqrt{-1} \quad or \quad a = -\sqrt{-1}$
$\quad a = i \qquad or \quad a = -i$
The solutions are i and $-i$ or $\pm i$.

19. $4d^2 + 81 = 0$
$\qquad d^2 = -\dfrac{81}{4}$

$\quad d = \sqrt{-\dfrac{81}{4}} \quad or \quad d = -\sqrt{-\dfrac{81}{4}}$

$\quad d = \dfrac{9}{2}i \qquad or \quad d = -\dfrac{9}{2}i$

The solutions are $\dfrac{9}{2}i$ and $-\dfrac{9}{2}i$ or $\pm\dfrac{9}{2}i$.

21. $(x-3)^2 = 16$
$\quad x - 3 = \sqrt{16} \quad or \quad x - 3 = -\sqrt{16}$
$\quad x - 3 = 4 \qquad or \quad x - 3 = -4$
$\qquad x = 7 \qquad or \qquad x = -1$
The solutions are -1 and 7.

23. $(t+5)^2 = 12$
$\quad t + 5 = \sqrt{12} \qquad or \quad t + 5 = -\sqrt{12}$
$\quad t + 5 = 2\sqrt{3} \qquad or \quad t + 5 = -2\sqrt{3}$
$\qquad t = -5 + 2\sqrt{3} \quad or \qquad t = -5 - 2\sqrt{3}$
The solutions are $-5 + 2\sqrt{3}$ and $-5 - 2\sqrt{3}$, or $-5 \pm 2\sqrt{3}$.

25. $(x+1)^2 = -9$
$\quad x + 1 = \sqrt{-9} \qquad or \quad x + 1 = -\sqrt{-9}$
$\quad x + 1 = 3i \qquad or \quad x + 1 = -3i$
$\qquad x = -1 + 3i \quad or \qquad x = -1 - 3i$
The solutions are $-1 + 3i$ and $-1 - 3i$, or $-1 \pm 3i$.

27. $\left(y + \dfrac{3}{4}\right)^2 = \dfrac{17}{16}$
$\qquad y + \dfrac{3}{4} = \pm\dfrac{\sqrt{17}}{4}$

$\qquad y = -\dfrac{3}{4} \pm \dfrac{\sqrt{17}}{4}$, or $\dfrac{-3 \pm \sqrt{17}}{4}$

The solutions are $-\dfrac{3}{4} \pm \dfrac{\sqrt{17}}{4}$, or $\dfrac{-3 \pm \sqrt{17}}{4}$.

29. $x^2 - 10x + 25 = 64$

$\qquad (x-5)^2 = 64$

$\qquad\qquad x - 5 = \pm 8$

$\qquad\qquad\qquad x = 5 \pm 8$

$\qquad\qquad\qquad x = 13 \ \text{ or } \ x = -3$

The solutions are 13 and −3.

31. $f(x) = x^2$

$\qquad 19 = x^2 \qquad$ Substituting

$\sqrt{19} = x \ \text{ or } \ -\sqrt{19} = x$

The solutions are $\sqrt{19}$ and $-\sqrt{19}$ or $\pm\sqrt{19}$.

33. $\qquad f(x) = 16$

$\quad (x-5)^2 = 16 \qquad$ Substituting

$\quad x - 5 = 4 \ \text{ or } \ x - 5 = -4$

$\qquad\quad x = 9 \ \text{ or } \qquad x = 1$

The solutions are 9 and 1.

35. $\qquad F(t) = 13$

$\quad (t+4)^2 = 13 \qquad$ Substituting

$\quad t + 4 = \sqrt{13} \qquad \text{ or } \quad t + 4 = -\sqrt{13}$

$\qquad\quad t = -4 + \sqrt{13} \ \text{ or } \qquad t = -4 - \sqrt{13}$

The solutions are $-4 + \sqrt{13}$ and $-4 - \sqrt{13}$, or $-4 \pm \sqrt{13}$.

37. $g(x) = x^2 + 14x + 49$

Observe first that $g(0) = 49$. Also observe that when $x = -14$, then

$x^2 + 14x = (-14)^2 - (14)(14) = (14)^2 - (14)^2 = 0$, so

$g(-14) = 49$ as well. Thus, we have $x = 0$ or $x = 14$.

We can also do this problem as follows.

$\qquad\qquad g(x) = 49$

$x^2 + 14x + 49 = 49 \qquad$ Substituting

$\qquad (x+7)^2 = 49$

$x + 7 = 7 \ \text{ or } \ x + 7 = -7$

$\quad x = 0 \ \text{ or } \qquad x = -14$

The solutions are 0 and −14.

39. $x^2 + 16x$

We take half the coefficient of x and square it: Half of 16 is 8, and $8^2 = 64$. We add 64.

$x^2 + 16x + 64 = (x+8)^2$

41. $t^2 - 10t$

We take half the coefficient of t and square it:

Half of −10 is −5, and $(-5)^2 = 25$. We add 25.

$t^2 - 10t + 25 = (t-5)^2$

43. $t^2 - 2t$

We take half the coefficient of t and square it:

$\frac{1}{2}(-2) = -1$, and $(-1)^2 = 1$. We add 1.

$t^2 - 2t + 1 = (t-1)^2$

45. $x^2 + 3x$

We take half the coefficient of t and square it:

$\frac{1}{2}(3) = \frac{3}{2}$, and $\left(\frac{3}{2}\right)^2 = \frac{9}{4}$. We add $\frac{9}{4}$.

$x^2 + 3x + \frac{9}{4} = \left(x + \frac{3}{2}\right)^2$

47. $x^2 + \frac{2}{5}x$

$\frac{1}{2} \cdot \frac{2}{5} = \frac{1}{5}$, and $\left(\frac{1}{5}\right)^2 = \frac{1}{25}$. We add $\frac{1}{25}$.

$x^2 + \frac{2}{5}x + \frac{1}{25} = \left(x + \frac{1}{5}\right)^2$

49. $t^2 - \frac{5}{6}t$

$\frac{1}{2}\left(-\frac{5}{6}\right) = -\frac{5}{12}$, and $\left(-\frac{5}{12}\right)^2 = \frac{25}{144}$. We add $\frac{25}{144}$.

$t^2 - \frac{5}{6}t + \frac{25}{144} = \left(t - \frac{5}{12}\right)^2$

51. $\qquad x^2 + 6x = 7$

$x^2 + 6x + 9 = 7 + 9 \qquad$ Adding 9 to both sides

$\qquad\qquad\qquad\qquad$ to complete the square

$\qquad (x+3)^2 = 16 \qquad$ Factoring

$\qquad\quad x + 3 = \pm 4 \qquad$ Principle of square roots

$\qquad\qquad\quad x = -3 \pm 4$

$x = -3 + 4 \ \text{ or } \ x = -3 - 4$

$x = 1 \qquad\quad \text{ or } \qquad x = -7$

The solutions are 1 and −7.

53. $\qquad t^2 - 10t = -23$

$t^2 - 10t + 25 = -23 + 25 \qquad$ Adding 25 to both sides

$\qquad\qquad\qquad\qquad\qquad$ to complete the square

$\qquad (t-5)^2 = 2 \qquad$ Factoring

$\qquad\quad t - 5 = \pm\sqrt{2} \qquad$ Principle of square roots

$\qquad\qquad\quad t = 5 \pm \sqrt{2}$

The solutions are $5 \pm \sqrt{2}$.

55. $x^2 + 12x + 32 = 0$

$\qquad x^2 + 12x = -32$

$x^2 + 12x + 36 = -32 + 36$

$\qquad (x+6)^2 = 4$

$\qquad\quad x + 6 = \pm 2$

$\qquad\qquad\quad x = -6 \pm 2$

$x = -6 + 2 \ \text{ or } \ x = -6 - 2$

$x = -4 \qquad \text{ or } \qquad x = -8$

The solutions are −8 and −4.

57. $t^2 + 8t - 3 = 0$

$\qquad t^2 + 8t = 3$

$t^2 + 8t + 16 = 3 + 16$

$\qquad (t+4)^2 = 19$

$\qquad\quad t + 4 = \pm\sqrt{19}$

$\qquad\qquad\quad t = -4 \pm \sqrt{19}$

The solutions are $-4 \pm \sqrt{19}$.

59. The value of $f(x)$ must be 0 at any x-intercepts.

$$f(x) = 0$$
$$x^2 + 6x + 7 = 0$$
$$x^2 + 6x = -7$$
$$x^2 + 6x + 9 = -7 + 9$$
$$(x+3)^2 = 2$$
$$x+3 = \pm\sqrt{2}$$
$$x = -3 \pm \sqrt{2}$$

The x-intercepts are $(-3-\sqrt{2},\, 0)$ and $(-3+\sqrt{2},\, 0)$.

61. The value of $g(x)$ must be 0 at any x-intercepts.

$$g(x) = 0$$
$$x^2 + 9x - 25 = 0$$
$$x^2 + 9x = 25$$
$$x^2 + 9x + \frac{81}{4} = 25 + \frac{81}{4}$$
$$\left(x + \frac{9}{2}\right)^2 = \frac{181}{4}$$
$$x + \frac{9}{2} = \pm\frac{\sqrt{181}}{2}$$
$$x = -\frac{9}{2} \pm \frac{\sqrt{181}}{2}$$

The x-intercepts are
$$\left(-\frac{9}{2} - \frac{\sqrt{181}}{2},\, 0\right) \text{ and } \left(-\frac{9}{2} + \frac{\sqrt{181}}{2},\, 0\right).$$

63. The value of $f(x)$ must be 0 at any x-intercepts.

$$f(x) = 0$$
$$x^2 - 10x - 22 = 0$$
$$x^2 - 10x = 22$$
$$x^2 - 10x + 25 = 22 + 25$$
$$(x-5)^2 = 47$$
$$x-5 = \pm\sqrt{47}$$
$$x = 5 \pm \sqrt{47}$$

The x-intercepts are $(5-\sqrt{47},\, 0)$ and $(5+\sqrt{47},\, 0)$.

65.
$$9x^2 + 18x = -8$$
$$x^2 + 2x = -\frac{8}{9} \qquad \text{Dividing both sides by 9}$$
$$x^2 + 2x + 1 = -\frac{8}{9} + 1$$
$$(x+1)^2 = \frac{1}{9}$$
$$x+1 = \pm\frac{1}{3}$$
$$x = -1 \pm \frac{1}{3}$$
$$x = -1 - \frac{1}{3} \quad \text{or} \quad x = -1 + \frac{1}{3}$$
$$x = -\frac{4}{3} \quad \text{or} \quad x = -\frac{2}{3}$$

The solutions are $-\frac{4}{3}$ and $-\frac{2}{3}$.

67.
$$3x^2 - 5x - 2 = 0$$
$$3x^2 - 5x = 2$$
$$x^2 - \frac{5}{3}x = \frac{2}{3} \qquad \text{Dividing both sides by 3}$$
$$x^2 - \frac{5}{3}x + \frac{25}{36} = \frac{2}{3} + \frac{25}{36}$$
$$\left(x - \frac{5}{6}\right)^2 = \frac{49}{36}$$
$$x - \frac{5}{6} = \pm\frac{7}{6}$$
$$x = \frac{5}{6} \pm \frac{7}{6}$$
$$x = \frac{5}{6} - \frac{7}{6} \quad \text{or} \quad x = \frac{5}{6} + \frac{7}{6}$$
$$x = -\frac{1}{3} \quad \text{or} \quad x = 2$$

The solutions are $-\frac{1}{3}$ and 2.

69.
$$5x^2 + 4x - 3 = 0$$
$$5x^2 + 4x = 3$$
$$x^2 + \frac{4}{5}x = \frac{3}{5} \qquad \text{Dividing both sides by 5}$$
$$x^2 + \frac{4}{5}x + \frac{4}{25} = \frac{3}{5} + \frac{4}{25}$$
$$\left(x + \frac{2}{5}\right)^2 = \frac{19}{25}$$
$$x + \frac{2}{5} = \pm\frac{\sqrt{19}}{5}$$
$$x = -\frac{2}{5} \pm \frac{\sqrt{19}}{5}, \text{ or } \frac{-2 \pm \sqrt{19}}{5}$$

The solutions are $-\frac{2}{5} \pm \frac{\sqrt{19}}{5}$, or $\frac{-2 \pm \sqrt{19}}{5}$.

71. The value of $f(x)$ must be 0 at any x-intercepts.

$$f(x) = 0$$
$$4x^2 + 2x - 3 = 0$$
$$4x^2 + 2x = 3$$
$$x^2 + \frac{1}{2}x = \frac{3}{4} \qquad \text{Dividing both sides by 4}$$
$$x^2 + \frac{1}{2}x + \frac{1}{16} = \frac{3}{4} + \frac{1}{16}$$
$$\left(x + \frac{1}{4}\right)^2 = \frac{13}{16}$$
$$x + \frac{1}{4} = \pm\frac{\sqrt{13}}{4}$$
$$x = -\frac{1}{4} \pm \frac{\sqrt{13}}{4}, \text{ or } \frac{-1 \pm \sqrt{13}}{4}$$

The x-intercepts are $\left(-\frac{1}{4} - \frac{\sqrt{13}}{4},\, 0\right)$ and $\left(-\frac{1}{4} + \frac{\sqrt{13}}{4},\, 0\right)$, or $\left(\frac{-1 - \sqrt{13}}{4},\, 0\right)$ and $\left(\frac{-1 + \sqrt{13}}{4},\, 0\right)$.

73. The value of $g(x)$ must be 0 at any x-intercepts.

$$g(x) = 0$$
$$2x^2 - 3x - 1 = 0$$
$$2x^2 - 3x = 1$$
$$x^2 - \frac{3}{2}x = \frac{1}{2} \qquad \text{Dividing both sides by 2}$$
$$x^2 - \frac{3}{2}x + \frac{9}{16} = \frac{1}{2} + \frac{9}{16}$$
$$\left(x - \frac{3}{4}\right)^2 = \frac{17}{16}$$
$$x - \frac{3}{4} = \pm\frac{\sqrt{17}}{4}$$
$$x = \frac{3}{4} \pm \frac{\sqrt{17}}{4}, \text{ or } \frac{3 \pm \sqrt{17}}{4}$$

The x-intercepts are $\left(\frac{3}{4} - \frac{\sqrt{17}}{4}, 0\right)$ and

$\left(\frac{3}{4} + \frac{\sqrt{17}}{4}, 0\right)$, or $\left(\frac{3 - \sqrt{17}}{4}, 0\right)$ and $\left(\frac{3 + \sqrt{17}}{4}, 0\right)$.

75. *Familiarize*. We are already familiar with the compound-interest formula.
Translate. We substitute into the formula.
$$A = P(1 + r)^t$$
$$2205 = 2000(1 + r)^2$$
Carry out. We solve for r.
$$2205 = 2000(1 + r)^2$$
$$\frac{2205}{2000} = (1 + r)^2$$
$$\frac{441}{400} = (1 + r)^2$$
$$\pm\sqrt{\frac{441}{400}} = 1 + r$$
$$\pm\frac{21}{20} = 1 + r$$
$$-\frac{20}{20} \pm \frac{21}{20} = r$$
$$\frac{1}{20} = r \text{ or } -\frac{41}{20} = r$$

Check. Since the interest rate cannot be negative, we need only check $\frac{1}{20}$, or 5%. If \$2000 were invested at 5% interest, compounded annually, then in 2 years it would grow to $\$2000(1.05)^2$, or \$2205. The number 5% checks.
State. The interest rate is 5%.

77. *Familiarize*. We are already familiar with the compound-interest formula.
Translate. We substitute into the formula.
$$A = P(1 + r)^t$$
$$6760 = 6250(1 + r)^2$$
Carry out. We solve for r.
$$\frac{6760}{6250} = (1 + r)^2$$
$$\frac{676}{625} = (1 + r)^2$$
$$\pm\frac{26}{25} = 1 + r$$
$$-\frac{25}{25} \pm \frac{26}{25} = r$$
$$\frac{1}{25} = r \text{ or } -\frac{51}{25} = r$$

Check. Since the interest rate cannot be negative, we need only check $\frac{1}{25}$, or 4%. If \$6250 were invested at 4% interest, compounded annually, then in 2 years it would grow to $\$6250(1.04)^2$, or \$6760. The number 4% checks.
State. The interest rate is 4%.

79. *Familiarize*. We will use the formula $s = 16t^2$.
Translate. We substitute into the formula.
$$s = 16t^2$$
$$3593 = 16t^2$$
Carry out. We solve for t.
$$3593 = 16t^2$$
$$\frac{3593}{16} = t^2$$
$$\sqrt{\frac{3593}{16}} = t \qquad \text{Principle of square roots;}$$
$$\qquad\qquad\qquad \text{rejecting the negative square root}$$
$$15.0 \approx t$$

Check. Since $16(15.0)^2 = 3600 \approx 3593$, our answer checks.
State. It would take an object about 15.0 sec to fall freely from the top.

81. *Familiarize*. We will use the formula $s = 16t^2$.
Translate. We substitute into the formula.
$$s = 16t^2$$
$$890 = 16t^2$$
Carry out. We solve for t.
$$890 = 16t^2$$
$$55.625 = t^2$$
$$\sqrt{55.625} = t \qquad \text{Principle of square roots;}$$
$$\qquad\qquad\qquad \text{rejecting the negative square root}$$
$$7.5 \approx t$$

Check. Since $16(7.5)^2 = 900 \approx 890$, our answer checks.
State. It would take an object about 7.5 sec to fall freely from the bridge to the river.

83. *Writing Exercise*.
1) If the quadratic equation is of the type $x^2 = k$, use the principle of square roots.
2) If the quadratic equation is of the type $ax^2 + bx + c = 0$, $b \neq 0$, use the principle of zero products, if possible.
3) If the quadratic equation is of the type $ax^2 + bx + c = 0$, $b \neq 0$, and factoring is difficult or impossible, solve by completing the square.

85. $3y^2 - 300y = 3y(y^2 - 100)$
$$= 3y(y + 10)(y - 10)$$

87. $6x^2 + 6x + 6 = 6(x^2 + x + 1)$

89. $20x^2 + 7x - 6 = (4x + 3)(5x - 2)$

91. *Writing Exercise.*

93. In order for $x^2 + bx + 81$ to be a square, the following must be true:

$$\left(\frac{b}{2}\right)^2 = 81$$

$$\frac{b^2}{4} = 81$$

$$b^2 = 324$$

$$b = 18 \text{ or } b = -18$$

95. We see that x is a factor of each term, so x is also a factor of $f(x)$. We have

$f(x) = x(2x^4 - 9x^3 - 66x^2 + 45x + 280)$. Since $x^2 - 5$ is a factor of $f(x)$ it is also a factor of

$2x^4 - 9x^3 - 66x^2 + 45x + 280$. We divide to find another factor.

$$
\begin{array}{r}
2x^2 - 9x - 56 \\
x^2 - 5\overline{)2x^4 - 9x^3 - 66x^2 + 45x + 280} \\
\underline{2x^4 \qquad -10x^2} \\
-9x^3 - 56x^2 + 45x \\
\underline{-9x^3 \qquad +45x} \\
-56x^2 \qquad +280 \\
\underline{-56x^2 \qquad +280} \\
0
\end{array}
$$

Then we have $f(x) = x(x^2 - 5)(2x^2 - 9x - 56)$, or

$f(x) = x(x^2 - 5)(2x + 7)(x - 8)$. Now we find the values of a for which $f(a) = 0$.

$$f(a) = 0$$

$$a(a^2 - 5)(2a + 7)(a - 8) = 0$$

$a = 0 \text{ or } a^2 - 5 = 0 \quad \text{or } 2a + 7 = 0 \quad \text{or } a - 8 = 0$

$a = 0 \text{ or } \quad a^2 = 5 \quad \text{or } \quad 2a = -7 \text{ or } \quad a = 8$

$a = 0 \text{ or } \quad a = \pm\sqrt{5} \text{ or } \quad a = -\frac{7}{2} \text{ or } \quad a = 8$

The solutions are 0, $\sqrt{5}$, $-\sqrt{5}$, $-\frac{7}{2}$, and 8.

97. *Familiarize.* It is helpful to list information in a chart and make a drawing. Let r represent the speed of the fishing boat. Then $r - 7$ represents the speed of the barge.

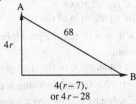

Boat	r	t	d
Fishing	r	4	$4r$
Barge	$r - 7$	4	$4(r - 7)$

Translate. We use the Pythagorean equation:

$$a^2 + b^2 = c^2$$

$$(4r - 28)^2 + (4r)^2 = 68^2$$

Carry out.

$$(4r - 28)^2 + (4r)^2 = 68^2$$

$$16r^2 - 224r + 784 + 16r^2 = 4624$$

$$32r^2 - 224r - 3840 = 0$$

$$r^2 - 7r - 120 = 0$$

$$(r + 8)(r - 15) = 0$$

$$r + 8 = 0 \quad \text{or} \quad r - 15 = 0$$

$$r = -8 \quad \text{or} \qquad r = 15$$

Check. We check only 15 since the speeds of the boats cannot be negative. If the speed of the fishing boat is 15 km/h, then the speed of the barge is $15 - 7$, or 8 km/h, and the distances they travel are $4 \cdot 15$ (or 60) and $4 \cdot 8$ (or 32). $60^2 + 32^2 = 3600 + 1024 = 4624$

$= 68^2$ The values check.

State. The speed of the fishing boat is 15 km/h, and the speed of the barge is 8 km/h.

99. *Graphing Calculator Exercise*

101. *Writing Exercise.* From a reading of the problem we know that we are interested only in positive values of r and it is safe to assume $r \le 1$. We also know that we want to find the value of r for which

$4410 = 4000(1 + r)^2$, so the window must include the y-value 4410. A suitable viewing window might be $[0, 1, 4000, 4500]$, Xscl = 0.1, Yscl = 100.

Connecting the Concepts

1. $x^2 - 3x - 10 = 0$

$(x + 2)(x - 5) = 0$

$x + 2 = 0 \quad \text{or} \quad x - 5 = 0$

$x = -2 \quad \text{or} \qquad x = 5$

2. $x^2 = 121$

$x = \pm 11$

3. $x^2 + 6x = 10$

$x^2 + 6x + 9 = 10 + 9$ Completing the square

$(x + 3)^2 = 19$

$x + 3 = \pm\sqrt{19}$

$x = -3 \pm \sqrt{19}$

4. $x^2 + x - 3 = 0$

$a = 1, b = 1, c = -3$

$$x = \frac{-1 \pm \sqrt{1^2 - 4 \cdot 1 \cdot (-3)}}{2 \cdot 1} = \frac{-1 \pm \sqrt{1 + 12}}{2}$$

$$= \frac{-1 \pm \sqrt{13}}{2} = -\frac{1}{2} \pm \frac{\sqrt{13}}{2}$$

5. $(x + 1)^2 = 2$

$x + 1 = \pm\sqrt{2}$

$x = -1 \pm \sqrt{2}$

6. $x^2 - 10x + 25 = 0$
$(x-5)(x-5) = 0$
$(x-5)^2 = 0$
$x - 5 = 0$
$x = 5$

7. $x^2 - 2x = 6$
$x^2 - 2x + 1 = 6 + 1$
$(x-1)^2 = 7$
$x - 1 = \pm\sqrt{7}$
$x = 1 \pm \sqrt{7}$

8. $4t^2 = 11$
$t^2 = \dfrac{11}{4}$
$t = \pm\sqrt{\dfrac{11}{4}} = \pm\dfrac{\sqrt{11}}{2}$

Exercise Set 8.2

1. True

3. False; see Example 3.

5. False; the quadratic formula yields at most two solutions.

7. $2x^2 + 3x - 5 = 0$
$(2x+5)(x-1) = 0$ Factoring
$2x + 5 = 0$ or $x - 1 = 0$
$x = -\dfrac{5}{2}$ or $x = 1$

The solutions are $-\dfrac{5}{2}$ and 1.

9. $u^2 + 2u - 4 = 0$
$u^2 + 2u = 4$
$u^2 + 2u + 1 = 4 + 1$ Completing the square
$(u+1)^2 = 5$
$u + 1 = \pm\sqrt{5}$ Principle of square roots
$u = -1 \pm \sqrt{5}$
The solutions are $-1 + \sqrt{5}$ and $-1 - \sqrt{5}$.

11. $t^2 + 3 = 6t$
$t^2 - 6t = -3$
$t^2 - 6t + 9 = -3 + 9$
$(t-3)^2 = 6$
$t - 3 = \pm\sqrt{6}$
$t = 3 \pm \sqrt{6}$
The solutions are $3 + \sqrt{6}$ and $3 - \sqrt{6}$.

13. $x^2 = 3x + 5$
$x^2 - 3x - 5 = 0$
$a = 1, \ b = -3, \ c = -5$
$x = \dfrac{-b \pm \sqrt{b^2 - 4ac}}{2a}$
$x = \dfrac{-(-3) \pm \sqrt{(-3)^2 - 4 \cdot 1 \cdot (-5)}}{2 \cdot 1} = \dfrac{3 \pm \sqrt{9 + 20}}{2}$
$x = \dfrac{3 \pm \sqrt{29}}{2} = \dfrac{3}{2} \pm \dfrac{\sqrt{29}}{2}$
The solutions are $\dfrac{3}{2} + \dfrac{\sqrt{29}}{2}$ and $\dfrac{3}{2} - \dfrac{\sqrt{29}}{2}$.

15. $3t(t+2) = 1$
$3t^2 + 6t = 1$
$3t^2 + 6t - 1 = 0$
$a = 3, \ b = 6, \ c = -1$
$t = \dfrac{-b \pm \sqrt{b^2 - 4ac}}{2a}$
$t = \dfrac{-6 \pm \sqrt{6^2 - 4 \cdot 3 \cdot (-1)}}{2 \cdot 3} = \dfrac{-6 \pm \sqrt{36 + 12}}{6}$
$t = \dfrac{-6 \pm \sqrt{48}}{6} = \dfrac{-6 \pm 4\sqrt{3}}{6}$
$t = -\dfrac{6}{6} \pm \dfrac{4\sqrt{3}}{6} = -1 \pm \dfrac{2\sqrt{3}}{3}$
The solutions are $-1 + \dfrac{2\sqrt{3}}{3}$ and $-1 - \dfrac{2\sqrt{3}}{3}$.

17. $\dfrac{1}{x^2} - 3 = \dfrac{8}{x}$,   LCD is x^2
$x^2\left(\dfrac{1}{x^2} - 3\right) = x^2 \cdot \dfrac{8}{x}$
$x^2 \cdot \dfrac{1}{x^2} - x^2 \cdot 3 = 8x$
$1 - 3x^2 = 8x$
$0 = 3x^2 + 8x - 1$
$a = 3, \ b = 8, \ c = -1$
$x = \dfrac{-8 \pm \sqrt{8^2 - 4 \cdot 3 \cdot (-1)}}{2 \cdot 3} = \dfrac{-8 \pm \sqrt{64 + 12}}{6}$
$x = \dfrac{-8 \pm \sqrt{76}}{6} = \dfrac{-8 \pm \sqrt{4 \cdot 19}}{6} = \dfrac{-8 \pm 2\sqrt{19}}{6}$
$x = \dfrac{-4 \pm \sqrt{19}}{3} = -\dfrac{4}{3} \pm \dfrac{\sqrt{19}}{3}$
The solutions are $-\dfrac{4}{3} - \dfrac{\sqrt{19}}{3}$ and $-\dfrac{4}{3} + \dfrac{\sqrt{19}}{3}$.

19. $t^2 + 10 = 6t$
$t^2 - 6t + 10 = 0$
$a = 1, \ b = -6, \ c = 10$
$t = \dfrac{-b \pm \sqrt{b^2 - 4ac}}{2a}$
$t = \dfrac{-(-6) \pm \sqrt{(-6)^2 - 4 \cdot 1 \cdot 10}}{2 \cdot 1} = \dfrac{6 \pm \sqrt{36 - 40}}{6}$
$t = \dfrac{6 \pm \sqrt{-4}}{2} = \dfrac{6 \pm 2i}{2}$
$t = \dfrac{6}{2} \pm \dfrac{2i}{2} = 3 \pm i$
The solutions are $3 + i$ and $3 - i$.

21. $p^2 - p + 1 = 0$
$a = 1,\ b = -1,\ c = 1$

$$p = \frac{-b \pm \sqrt{b^2 - 4ac}}{2a}$$

$$p = \frac{-(-1) \pm \sqrt{(-1)^2 - 4 \cdot 1 \cdot 1}}{2 \cdot 1} = \frac{1 \pm \sqrt{1 - 4}}{2}$$

$$p = \frac{1 \pm \sqrt{-3}}{2} = \frac{1}{2} \pm \frac{\sqrt{3}}{2} i$$

The solutions are $\frac{1}{2} + \frac{\sqrt{3}}{2} i$ and $\frac{1}{2} - \frac{\sqrt{3}}{2} i$.

23. $x^2 + 4x + 6 = 0$

$$x = \frac{-b \pm \sqrt{b^2 - 4ac}}{2a}$$

$$x = \frac{-4 \pm \sqrt{4^2 - 4 \cdot 1 \cdot 6}}{2 \cdot 1} = \frac{-4 \pm \sqrt{16 - 24}}{2}$$

$$x = \frac{-4 \pm \sqrt{-8}}{2} = -\frac{4}{2} \pm \frac{2\sqrt{2}}{2} i$$

$$x = -2 \pm \sqrt{2} i$$

The solutions are $-2 + \sqrt{2}i$ and $-2 - \sqrt{2}i$.

25. $\qquad 12t^2 + 17t = 40$
$12t^2 + 17t - 40 = 0$
$(3t + 8)(4t - 5) = 0$
$3t + 8 = 0 \quad or \quad 4t - 5 = 0$
$\qquad t = -\frac{8}{3} \quad or \qquad t = \frac{5}{4}$

The solutions are $-\frac{8}{3}$ and $\frac{5}{4}$.

27. $\quad 25x^2 - 20x + 4 = 0$
$(5x - 2)(5x - 2) = 0$
$5x - 2 = 0 \quad or \quad 5x - 2 = 0$
$\quad 5x = 2 \quad or \qquad 5x = 2$
$\quad x = \frac{2}{5} \quad or \qquad x = \frac{2}{5}$

The solution is $\frac{2}{5}$.

29. $7x(x + 2) + 5 = 3x(x + 1)$
$7x^2 + 14x + 5 = 3x^2 + 3x$
$4x^2 + 11x + 5 = 0$
$a = 4,\ b = 11,\ c = 5$

$$x = \frac{-11 \pm \sqrt{11^2 - 4 \cdot 4 \cdot 5}}{2 \cdot 4} = \frac{-11 \pm \sqrt{121 - 80}}{8}$$

$$x = \frac{-11 \pm \sqrt{41}}{8} = -\frac{11}{8} \pm \frac{\sqrt{41}}{8}$$

The solutions are $-\frac{11}{8} - \frac{\sqrt{41}}{8}$ and $-\frac{11}{8} + \frac{\sqrt{41}}{8}$.

31. $14(x - 4) - (x + 2) = (x + 2)(x - 4)$
$14x - 56 - x - 2 = x^2 - 2x - 8 \qquad$ Removing
$\qquad\qquad\qquad\qquad\qquad\qquad$ parentheses
$\qquad 13x - 58 = x^2 - 2x - 8$
$\qquad\qquad\quad 0 = x^2 - 15x + 50$
$\qquad\qquad\quad 0 = (x - 10)(x - 5)$

$x - 10 = 0 \quad or \quad x - 5 = 0$
$\quad x = 10 \quad or \qquad x = 5$
The solutions are 10 and 5.

33. $51p = 2p^2 + 72$
$\quad 0 = 2p^2 - 51p + 72$
$\quad 0 = (2p - 3)(p - 24)$
$2p - 3 = 0 \quad or \quad p - 24 = 0$
$\quad p = \frac{3}{2} \quad or \qquad p = 24$

The solutions are $\frac{3}{2}$ and 24.

35. $\qquad x(x - 3) = x - 9$
$\qquad x^2 - 3x = x - 9 \qquad$ Removing parentheses
$\qquad x^2 - 4x = -9$
$x^2 - 4x + 4 = -9 + 4 \qquad$ Completing the square
$\qquad (x - 2)^2 = -5$
$\qquad\quad x - 2 = \pm\sqrt{-5}$
$\qquad\qquad x = 2 \pm \sqrt{5} i$
The solutions are $2 + \sqrt{5}i$ and $2 - \sqrt{5}i$.

37. $\qquad\qquad x^3 - 8 = 0$
$\qquad\qquad x^3 - 2^3 = 0$
$(x - 2)(x^2 + 2x + 4) = 0$
$x - 2 = 0 \quad or \quad x^2 + 2x + 4 = 0$

$x = 2 \quad or \qquad x = \frac{-2 \pm \sqrt{2^2 - 4 \cdot 1 \cdot 4}}{2 \cdot 1}$

$x = 2 \quad or \qquad x = \frac{-2 \pm \sqrt{-12}}{2} = \frac{-2 \pm 2i\sqrt{3}}{2}$

$x = 2 \quad or \qquad x = -\frac{2}{2} \pm \frac{2\sqrt{3}}{2} i$

$x = 2 \quad or \qquad x = -1 \pm \sqrt{3} i$

The solutions are $2,\ -1 + \sqrt{3}i$, and $-1 - \sqrt{3}i$.

39. $\qquad f(x) = 0$
$6x^2 - 7x - 20 = 0$
$(3x + 4)(2x - 5) = 0$
$3x + 4 = 0 \quad or \quad 2x - 5 = 0$
$\quad x = -\frac{4}{3} \quad or \qquad x = \frac{5}{2}$

$f(x) = 0$ for $x = -\frac{4}{3}$ and $x = \frac{5}{2}$.

41. $\qquad\qquad f(x) = 1 \qquad$ Substituting
$\qquad\quad \frac{7}{x} + \frac{7}{x + 4} = 1$

$x(x + 4)\left(\frac{7}{x} + \frac{7}{x + 4}\right) = x(x + 4) \cdot 1$
$\qquad\qquad\qquad\qquad\qquad$ Multiplying by the LCD
$\quad 7(x + 4) + 7x = x^2 + 4x$
$\quad 7x + 28 + 7x = x^2 + 4x$
$\qquad 14x + 28 = x^2 + 4x$
$\qquad\qquad\quad 0 = x^2 - 10x - 28$
$a = 1,\ b = -10,\ c = -28$

$$x = \frac{-(-10) \pm \sqrt{(-10)^2 - 4 \cdot 1 \cdot (-28)}}{2 \cdot 1}$$

$$x = \frac{10 \pm \sqrt{100 + 112}}{2} = \frac{10 \pm \sqrt{212}}{2}$$

$$x = \frac{10 \pm \sqrt{4 \cdot 53}}{2} = \frac{10 \pm 2\sqrt{53}}{2}$$

$$x = 5 \pm \sqrt{53}$$

$f(x) = 1$ for $x = 5 + \sqrt{53}$ and $x = 5 - \sqrt{53}$.

43.
$$F(x) = G(x)$$
$$\frac{3 - x}{4} = \frac{1}{4x}$$
$$4x \cdot \frac{3 - x}{4} = 4x \cdot \frac{1}{4x}$$
$$3x - x^2 = 1$$
$$0 = x^2 - 3x + 1$$
$$x = \frac{-(-3) \pm \sqrt{(-3)^2 - 4 \cdot 1 \cdot 1}}{2 \cdot 1} = \frac{3 \pm \sqrt{5}}{2}$$
$$x = \frac{3}{2} \pm \frac{\sqrt{5}}{2}$$

45. $x^2 + 6x + 4 = 0$
$$x = \frac{-6 \pm \sqrt{6^2 - 4 \cdot 1 \cdot 4}}{2 \cdot 1} = \frac{-6 \pm \sqrt{20}}{2}$$
$$x = \frac{-6 + \sqrt{20}}{2} \approx -0.764$$
$$x = \frac{-6 - \sqrt{20}}{2} \approx -5.236$$

47. $x^2 - 6x + 4 = 0$
$a = 1$, $b = -6$, $c = 4$
$$x = \frac{-(-6) \pm \sqrt{(-6)^2 - 4 \cdot 1 \cdot 4}}{2 \cdot 1} = \frac{6 \pm \sqrt{36 - 16}}{2}$$
$$x = \frac{6 \pm \sqrt{20}}{2}$$
Using a calculator we find that
$$\frac{6 + \sqrt{20}}{2} \approx 5.236 \text{ and } \frac{6 - \sqrt{20}}{2} \approx 0.764.$$
The solutions are approximately 5.236 and 0.764.

49. $2x^2 - 3x - 7 = 0$
$a = 2$, $b = -3$, $c = -7$
$$x = \frac{-(-3) \pm \sqrt{(-3)^2 - 4 \cdot 2 \cdot (-7)}}{2 \cdot 2}$$
$$x = \frac{3 \pm \sqrt{9 + 56}}{4} = \frac{3 \pm \sqrt{65}}{4}$$
Using a calculator we find that
$$\frac{3 + \sqrt{65}}{4} \approx 2.766 \text{ and } \frac{3 - \sqrt{65}}{4} \approx -1.266.$$
The solutions are approximately 2.766 and −1.266.

51. *Writing Exercise.*

53. $(-3x^2 y^6)^0 = 1$

55. $x^{1/4} \cdot x^{2/3} = x^{1/4 + 2/3} = x^{11/12}$

57. $\dfrac{18a^5 bc^{10}}{24a^{-5} bc^3} = \dfrac{3a^{10} c^7}{4}$

59. *Writing Exercise.*

61. $f(x) = \dfrac{x^2}{x - 2} + 1$

To find the x-coordinates of the x-intercepts of the graph of f, we solve $f(x) = 0$.

$$\frac{x^2}{x - 2} + 1 = 0$$
$$x^2 + x - 2 = 0 \quad \text{Multiplying by } x - 2$$
$$(x + 2)(x - 1) = 0$$
$$x = -2 \ \text{ or } \ x = 1$$
The x-intercepts are $(-2, 0)$ and $(1, 0)$.

63.
$$f(x) = g(x)$$
$$\frac{x^2}{x - 2} + 1 = \frac{4x - 2}{x - 2} + \frac{x + 4}{2} \quad \text{Substituting}$$
$$2(x - 2)\left(\frac{x^2}{x - 2} + 1\right) = 2(x - 2)\left(\frac{4x - 2}{x - 2} + \frac{x + 4}{2}\right)$$

Multiplying by the LCD

$$2x^2 + 2(x - 2) = 2(4x - 2) + (x - 2)(x + 4)$$
$$2x^2 + 2x - 4 = 8x - 4 + x^2 + 2x - 8$$
$$2x^2 + 2x - 4 = x^2 + 10x - 12$$
$$x^2 - 8x + 8 = 0$$
$a = 1$, $b = -8$, $c = 8$
$$x = \frac{-(-8) \pm \sqrt{(-8)^2 - 4 \cdot 1 \cdot 8}}{2 \cdot 1} = \frac{8 \pm \sqrt{64 - 32}}{2}$$
$$x = \frac{8 \pm \sqrt{32}}{2} = \frac{8 \pm \sqrt{16 \cdot 2}}{2} = \frac{8 \pm 4\sqrt{2}}{2}$$
$$x = \frac{8}{2} \pm \frac{4\sqrt{2}}{2} = 4 \pm 2\sqrt{2}$$
The solutions are $4 + 2\sqrt{2}$ and $4 - 2\sqrt{2}$.

65. $z^2 + 0.84z - 0.4 = 0$
$a = 1$, $b = 0.84$, $c = -0.4$
$$z = \frac{-0.84 \pm \sqrt{(0.84)^2 - 4 \cdot 1 \cdot (-0.4)}}{2 \cdot 1}$$
$$z = \frac{-0.84 \pm \sqrt{2.3056}}{2}$$
$$z = \frac{-0.84 + \sqrt{2.3056}}{2} \approx 0.339$$
$$z = \frac{-0.84 - \sqrt{2.3056}}{2} \approx -1.179$$
The solutions are approximately 0.339 and −1.179.

67. $\sqrt{2} x^2 + 5x + \sqrt{2} = 0$
$$x = \frac{-5 \pm \sqrt{5^2 - 4 \cdot \sqrt{2} \cdot \sqrt{2}}}{2\sqrt{2}} = \frac{-5 \pm \sqrt{17}}{2\sqrt{2}}, \text{ or}$$
$$x = \frac{-5 \pm \sqrt{17}}{2\sqrt{2}} \cdot \frac{\sqrt{2}}{\sqrt{2}} = \frac{-5\sqrt{2} \pm \sqrt{34}}{4}$$
The solutions are $\dfrac{-5\sqrt{2} \pm \sqrt{34}}{4}$.

69.
$$kx^2 + 3x - k = 0$$
$$k(-2)^2 + 3(-2) - k = 0 \quad \text{Substituting } -2 \text{ for } x$$
$$4k - 6 - k = 0$$
$$3k = 6$$
$$k = 2$$
$$2x^2 + 3x - 2 = 0 \quad \text{Substituting 2 for } k$$
$$(2x - 1)(x + 2) = 0$$
$$2x - 1 = 0 \quad \text{or} \quad x + 2 = 0$$
$$x = \frac{1}{2} \quad \text{or} \qquad x = -2$$

The other solution is $\frac{1}{2}$.

71. *Graphing Calculator Exercise*

Exercise Set 8.3

1. In the quadratic formula, the expression $b^2 - 4ac$ is called the *discriminant*.

3. When $b^2 - 4ac$ is positive, there are *two* solutions.

5. When $b^2 - 4ac$ is a perfect square, the solutions are *rational* numbers.

7. $x^2 - 7x + 5 = 0$
$a = 1, \ b = -7, \ c = 5$
We substitute and compute the discriminant.
$$b^2 - 4ac = (-7)^2 - 4 \cdot 1 \cdot 5$$
$$= 49 - 20$$
$$= 29$$
Since the discriminant is a positive number that is not a perfect square, there are two irrational solutions.

9. $x^2 + 11 = 0$
$a = 1, \ b = 0, \ c = 11$
We substitute and compute the discriminant.
$$b^2 - 4ac = 0^2 - 4 \cdot 1 \cdot 11$$
$$= -44$$
Since the discriminant is negative, there are two imaginary-number solutions.

11. $x^2 - 11 = 0$
$a = 1, \ b = 0, \ c = -11$
We substitute and compute the discriminant.
$$b^2 - 4ac = 0^2 - 4 \cdot 1 \cdot (-11)$$
$$= 44$$
Since the discriminant is a positive number that is not a perfect square, there are two irrational solutions.

13. $4x^2 + 8x - 5 = 0$
$a = 4, \ b = 8, \ c = -5$
We substitute and compute the discriminant.
$$b^2 - 4ac = 8^2 - 4 \cdot 4 \cdot (-5)$$
$$= 64 + 80$$
$$= 144$$
Since the discriminant is a positive number and a perfect square, there are two rational solutions.

15. $x^2 + 4x + 6 = 0$
$a = 1, \ b = 4, \ c = 6$
We substitute and compute the discriminant.
$$b^2 - 4ac = 4^2 - 4 \cdot 1 \cdot 6$$
$$= 16 - 24$$
$$= -8$$
Since the discriminant is negative, there are two imaginary-number solutions.

17. $9t^2 - 48t + 64 = 0$
$a = 9, \ b = -48, \ c = 64$
We substitute and compute the discriminant.
$$b^2 - 4ac = (-48)^2 - 4 \cdot 9 \cdot 64$$
$$= 2304 - 2304$$
$$= 0$$
Since the discriminant is 0, there is just one solution and it is a rational number.

19. $9t^2 + 3t = 0$
Observe that we can factor $9t^2 + 3t$. This tells us that there are two rational solutions. We could also do this problem as follows.
$$b^2 - 4ac = 3^2 - 4 \cdot 9 \cdot 0 = 9$$
Since the discriminant is a positive number and a perfect square, there are two rational solutions.

21. $x^2 + 4x = 8$
$x^2 + 4x - 8 = 0 \quad \text{Standard form}$
$a = 1, \ b = 4, \ c = -8$
We substitute and compute the discriminant.
$$b^2 - 4ac = 4^2 - 4 \cdot 1 \cdot (-8)$$
$$= 16 + 32 = 48$$
Since the discriminant is a positive number that is not a perfect square, there are two irrational solutions.

23. $2a^2 - 3a = -5$
$2a^2 - 3a + 5 = 0 \quad \text{Standard form}$
$a = 2, \ b = -3, \ c = 5$
We substitute and compute the discriminant.
$$b^2 - 4ac = (-3)^2 - 4 \cdot 2 \cdot 5$$
$$= 9 - 40$$
$$= -31$$
Since the discriminant is negative, there are two imaginary-number solutions.

25. $7x^2 = 19x$
$7x^2 - 19x = 0 \quad \text{Standard form}$
$a = 7, \ b = -19, \ c = 0$
We substitute and compute the discriminant.
$$b^2 - 4ac = (-19)^2 - 4 \cdot 7 \cdot 0 = 361$$
Since the discriminant is a positive number and a perfect square, there are two different rational solutions.

27.

$$y^2 + \frac{9}{4} = 4y$$

$$y^2 - 4y + \frac{9}{4} = 0 \qquad \text{Standard form}$$

$$a = 1, \ b = -4, \ c = \frac{9}{4}$$

We substitute and compute the discriminant.

$$b^2 - 4ac = (-4)^2 - 4 \cdot 1 \cdot \frac{9}{4}$$
$$= 16 - 9$$
$$= 7$$

The discriminant is a positive number that is not a perfect square. There are two irrational solutions.

29. The solutions are –5 and 4.

$$x = -5 \quad or \quad x = 4$$
$$x + 5 = 0 \quad or \quad x - 4 = 0$$
$$(x + 5)(x - 4) = 0 \qquad \text{Principle of zero products}$$
$$x^2 + x - 20 = 0 \qquad \text{FOIL}$$

31. The only solution is 3. It must be a repeated solution.

$$x = 3 \quad or \quad x = 3$$
$$x - 3 = 0 \quad or \quad x - 3 = 0$$
$$(x - 3)(x - 3) = 0 \qquad \text{Principle of zero products}$$
$$x^2 - 6x + 9 = 0 \qquad \text{FOIL}$$

33. The solutions are –1 and –3.

$$x = -1 \quad or \quad x = -3$$
$$x + 1 = 0 \quad or \quad x + 3 = 0$$
$$(x + 1)(x + 3) = 0$$
$$x^2 + 4x + 3 = 0$$

35. The solutions are 5 and $\frac{3}{4}$.

$$x = 5 \quad or \quad x = \frac{3}{4}$$

$$x - 5 = 0 \quad or \quad x - \frac{3}{4} = 0$$

$$(x - 5)\left(x - \frac{3}{4}\right) = 0$$

$$x^2 - \frac{3}{4}x - 5x + \frac{15}{4} = 0$$

$$x^2 - \frac{23}{4}x + \frac{15}{4} = 0$$

$$4x^2 - 23x + 15 = 0 \qquad \text{Multiplying by 4}$$

37. The solutions are $-\frac{1}{4}$ and $-\frac{1}{2}$.

$$x = -\frac{1}{4} \quad or \quad x = -\frac{1}{2}$$

$$x + \frac{1}{4} = 0 \quad or \quad x + \frac{1}{2} = 0$$

$$\left(x + \frac{1}{4}\right)\left(x + \frac{1}{2}\right) = 0$$

$$x^2 + \frac{1}{2}x + \frac{1}{4}x + \frac{1}{8} = 0$$

$$x^2 + \frac{3}{4}x + \frac{1}{8} = 0$$

$$8x^2 + 6x + 1 = 0 \qquad \text{Multiplying by 8}$$

39. The solutions are 2.4 and –0.4.

$$x = 2.4 \quad or \quad x = -0.4$$
$$x - 2.4 = 0 \quad or \quad x + 0.4 = 0$$
$$(x - 2.4)(x + 0.4) = 0$$
$$x^2 + 0.4x - 2.4x - 0.96 = 0$$
$$x^2 - 2x - 0.96 = 0$$

41. The solutions are $-\sqrt{3}$ and $\sqrt{3}$.

$$x = -\sqrt{3} \quad or \quad x = \sqrt{3}$$
$$x + \sqrt{3} = 0 \quad or \quad x - \sqrt{3} = 0$$
$$(x + \sqrt{3})(x - \sqrt{3}) = 0$$
$$x^2 - 3 = 0$$

43. The solutions are $2\sqrt{5}$ and $-2\sqrt{5}$.

$$x = 2\sqrt{5} \quad or \quad x = -2\sqrt{5}$$
$$x - 2\sqrt{5} = 0 \quad or \quad x + 2\sqrt{5} = 0$$
$$(x - 2\sqrt{5})(x + 2\sqrt{5}) = 0$$
$$x^2 - (2\sqrt{5})^2 = 0$$
$$x^2 - 4 \cdot 5 = 0$$
$$x^2 - 20 = 0$$

45. The solutions are $4i$ and $-4i$.

$$x = 4i \quad or \quad x = -4i$$
$$x - 4i = 0 \quad or \quad x + 4i = 0$$
$$(x - 4i)(x + 4i) = 0$$
$$x^2 - (4i)^2 = 0$$
$$x^2 + 16 = 0$$

47. The solutions are $2 - 7i$ and $2 + 7i$.

$$x = 2 - 7i \quad or \quad x = 2 + 7i$$
$$x - 2 + 7i = 0 \quad or \quad x - 2 - 7i = 0$$
$$(x - 2) + 7i = 0 \quad or \quad (x - 2) - 7i = 0$$
$$[(x - 2) + 7i][(x - 2) - 7i] = 0$$
$$(x - 2)^2 - (7i)^2 = 0$$
$$x^2 - 4x + 4 - 49i^2 = 0$$
$$x^2 - 4x + 4 + 49 = 0$$
$$x^2 - 4x + 53 = 0$$

49. The solutions are $3 - \sqrt{14}$ and $3 + \sqrt{14}$.

$$x = 3 - \sqrt{14} \quad or \quad x = 3 + \sqrt{14}$$
$$x - 3 + \sqrt{14} = 0 \quad or \quad x - 3 - \sqrt{14} = 0$$
$$(x - 3) + \sqrt{14} = 0 \quad or \quad (x - 3) - \sqrt{14} = 0$$
$$[(x - 3) + \sqrt{14}][(x - 3) - \sqrt{14}] = 0$$
$$(x - 3)^2 - (\sqrt{14})^2 = 0$$
$$x^2 - 6x + 9 - 14 = 0$$
$$x^2 - 6x - 5 = 0$$

51. The solutions are $1 - \frac{\sqrt{21}}{3}$ and $1 + \frac{\sqrt{21}}{3}$.

$$x = 1 - \frac{\sqrt{21}}{3} \quad or \quad x = 1 + \frac{\sqrt{21}}{3}$$

$$x - 1 + \frac{\sqrt{21}}{3} = 0 \quad or \quad x - 1 - \frac{\sqrt{21}}{3} = 0$$

$$(x - 1) + \frac{\sqrt{21}}{3} = 0 \quad or \quad (x - 1) - \frac{\sqrt{21}}{3} = 0$$

$$\left[(x-1)+\frac{\sqrt{21}}{3}\right]\left[(x-1)-\frac{\sqrt{21}}{3}\right]=0$$
$$(x-1)^2-\left(\frac{\sqrt{21}}{3}\right)^2=0$$
$$x^2-2x+1-\frac{21}{9}=0$$
$$x^2-2x+1-\frac{7}{3}=0$$
$$x^2-2x-\frac{4}{3}=0$$
$$3x^2-6x-4=0 \quad \text{Multiplying by 3}$$

53. The solutions are –2, 1, and 5.
$$x=-2 \quad or \quad x=1 \quad or \quad x=5$$
$$x+2=0 \quad or \quad x-1=0 \quad or \quad x-5=0$$
$$(x+2)(x-1)(x-5)=0$$
$$(x^2+x-2)(x-5)=0$$
$$x^3+x^2-2x-5x^2-5x+10=0$$
$$x^3-4x^2-7x+10=0$$

55. The solutions are –1, 0, and 3.
$$x=-1 \quad or \quad x=0 \quad or \quad x=3$$
$$x+1=0 \quad or \quad x=0 \quad or \quad x-3=0$$
$$(x+1)(x)(x-3)=0$$
$$(x^2+x)(x-3)=0$$
$$x^3-3x^2+x^2-3x=0$$
$$x^3-2x^2-3x=0$$

57. *Writing Exercise.*

59. $\sqrt{270a^7b^{12}}=\sqrt{9a^6b^{12}\cdot30a}=3a^3b^6\sqrt{30a}$

61. $\sqrt[3]{x}\sqrt{x}=x^{1/3}\cdot x^{1/2}=x^{1/3+1/2}=x^{5/6}=\sqrt[6]{x^5}$

63. $(2-i)(3+i)=6+2i-3i-i^2=6-i+1=7-i$

65. *Writing Exercise.*

67. The graph includes the points (–3, 0), (0, –3), and (1, 0). Substituting in $y=ax^2+bx+c$, we have three equations.
$$\begin{aligned}0 &= 9a - 3b + c,\\ -3 &= \qquad\qquad c,\\ 0 &= a + b + c\end{aligned}$$
The solution of this system of equations is $a=1$, $b=2$, $c=-3$.

69. a. $kx^2-2x+k=0$; one solution is –3
We first find *k* by substituting –3 for *x*.
$$k(-3)^2-2(-3)+k=0$$
$$9k+6+k=0$$
$$10k=-6$$
$$k=-\frac{6}{10}$$
$$k=-\frac{3}{5}$$

b. Now substitute $-\frac{3}{5}$ for *k* in the original equation.

$$-\frac{3}{5}x^2-2x+\left(-\frac{3}{5}\right)=0$$
$$3x^2+10x+3=0 \quad \text{Multiplying by}-5$$
$$(3x+1)(x+3)=0$$
$$x=-\frac{1}{3} \quad or \quad x=-3$$

The other solution is $-\frac{1}{3}$.

71. a. $x^2-(6+3i)x+k=0$; one solution is 3.
We first find *k* by substituting 3 for *x*.
$$3^2-(6+3i)3+k=0$$
$$9-18-9i+k=0$$
$$-9-9i+k=0$$
$$k=9+9i$$

b. Now we substitute $9+9i$ for *k* in the original equation.
$$x^2-(6+3i)x+(9+9i)=0$$
$$x^2-(6+3i)x+3(3+3i)=0$$
$$[x-(3+3i)][x-3]=0$$
The other solution is $3+3i$.

73. The solutions of $ax^2+bx+c=0$ are
$x=\dfrac{-b\pm\sqrt{b^2-4ac}}{2a}$. When there is just one solution, $b^2-4ac=0$, so $x=\dfrac{-b\pm0}{2a}=-\dfrac{b}{2a}$.

75. We substitute (–3, 0), $\left(\frac{1}{2},0\right)$, and (0,–12) in

$f(x)=ax^2+bx+c$ and get three equations.
$$\begin{aligned}0 &= 9a-3b+c,\\ 0 &= \frac{1}{4}a+\frac{1}{2}b+c,\\ -12 &= c\end{aligned}$$
The solution of this system of equations is $a=8$, $b=20$, $c=-12$.

77. If $-\sqrt{2}$ is one solution then $\sqrt{2}$ is another solution. Then
$$x=-\sqrt{2} \quad or \quad x=\sqrt{2}$$
$$x=\pm\sqrt{2}$$
$$x^2=2 \quad \text{Principle of square roots}$$
$$x^2-2=0$$

79. If $1-\sqrt{5}$ and $3+2i$ are two solutions, then $1+\sqrt{5}$ and $3-2i$ are also solutions. The equation of lowest degree that has these solutions is found as follows.
$$\left[x-(1-\sqrt{5})\right]\left[x-(1+\sqrt{5})\right]\left[x-(3+2i)\right]\left[x-(3-2i)\right]=0$$
$$(x^2-2x-4)(x^2-6x+13)=0$$
$$x^4-8x^3+21x^2-2x-52=0$$

81. *Writing Exercise.*

Exercise Set 8.4

1. c

3. a

5. $A = 4\pi r^2$

$\dfrac{A}{4\pi} = r^2$ Dividing by 4π

$\dfrac{1}{2}\sqrt{\dfrac{A}{\pi}} = r$ Taking the positive square root

7. $A = 2\pi r^2 + 2\pi rh$

$0 = 2\pi r^2 + 2\pi rh - A$ Standard form

$a = 2\pi,\ b = 2\pi h,\ c = -A$

$r = \dfrac{-2\pi h \pm \sqrt{(2\pi h)^2 - 4\cdot 2\pi \cdot(-A)}}{2\cdot 2\pi}$ Using the

                       quadratic formula

$r = \dfrac{-2\pi h \pm \sqrt{4\pi^2 h^2 + 8\pi A}}{4\pi}$

$r = \dfrac{-2\pi h \pm 2\sqrt{\pi^2 h^2 + 2\pi A}}{4\pi}$

$r = \dfrac{-\pi h \pm \sqrt{\pi^2 h^2 + 2\pi A}}{2\pi}$

Since taking the negative square root would result in a negative answer, we take the positive one.

$r = \dfrac{-\pi h + \sqrt{\pi^2 h^2 + 2\pi A}}{2\pi}$

9. $F = \dfrac{Gm_1 m_2}{r^2}$

$Fr^2 = Gm_1 m_2$

$r^2 = \dfrac{Gm_1 m_2}{F}$

$r = \sqrt{\dfrac{Gm_1 m_2}{F}}$

11. $c = \sqrt{gH}$

$c^2 = gH$ Squaring

$\dfrac{c^2}{g} = H$

13. $a^2 + b^2 = c^2$

$b^2 = c^2 - a^2$

$b = \sqrt{c^2 - a^2}$

15. $s = v_0 t + \dfrac{gt^2}{2}$

$0 = \dfrac{gt^2}{2} + v_0 t - s$ Standard form

$a = \dfrac{g}{2},\ b = v_0,\ c = -s$

$t = \dfrac{-v_0 \pm \sqrt{v_0^2 - 4\left(\dfrac{g}{2}\right)(-s)}}{2\left(\dfrac{g}{2}\right)}$

$t = \dfrac{-v_0 \pm \sqrt{v_0^2 + 2gs}}{g}$

Since taking the negative square root would result in a negative answer, we take the positive one.

$t = \dfrac{-v_0 + \sqrt{v_0^2 + 2gs}}{g}$

17. $N = \dfrac{1}{2}\left(n^2 - n\right)$

$N = \dfrac{1}{2}n^2 - \dfrac{1}{2}n$

$0 = \dfrac{1}{2}n^2 - \dfrac{1}{2}n - N$

$a = \dfrac{1}{2},\ b = -\dfrac{1}{2},\ c = -N$

$n = \dfrac{-\left(-\dfrac{1}{2}\right) \pm \sqrt{\left(-\dfrac{1}{2}\right)^2 - 4\cdot\dfrac{1}{2}\cdot(-N)}}{2\left(\dfrac{1}{2}\right)}$

$n = \dfrac{1}{2} \pm \sqrt{\dfrac{1}{4} + 2N}$

$n = \dfrac{1}{2} \pm \sqrt{\dfrac{1 + 8N}{4}}$

$n = \dfrac{1}{2} \pm \dfrac{1}{2}\sqrt{1 + 8N}$

Since taking the negative square root would result in a negative answer, we take the positive one.

$n = \dfrac{1}{2} + \dfrac{1}{2}\sqrt{1 + 8N}$, or $\dfrac{1 + \sqrt{1 + 8N}}{2}$

19. $T = I\sqrt{\dfrac{s}{d}}$

$\dfrac{T}{I} = \sqrt{\dfrac{s}{d}}$ Multiplying by $\dfrac{1}{I}$

$\dfrac{T^2}{I^2} = \dfrac{s}{d}$ Squaring

$dT^2 = I^2 s$ Multiplying by $I^2 d$

$d = \dfrac{I^2 s}{T^2}$ Multiplying by $\dfrac{1}{T^2}$

21. $at^2 + bt + c = 0$

The quadratic formula gives the result.

$t = \dfrac{-b \pm \sqrt{b^2 - 4ac}}{2a}$

23. a. *Familiarize and Translate*. From Example 3, we know

$t = \dfrac{-v_0 + \sqrt{v_0^2 + 19.6s}}{9.8}$.

Carry out. Substituting 500 for s and 0 for v_0, we have

$t = \dfrac{0 + \sqrt{0^2 + 19.6(500)}}{9.8} \approx 10.1$

Check. Substitute 10.1 for t and 0 for v_0 in the original formula. (See Example 3.)

$s = 4.9t^2 + v_0 t = 4.9(10.1)^2 + 0\cdot(10.1)^2 \approx 500$

The answer checks.

State. It takes the bolt about 10.1 sec to reach the ground.

b. **_Familiarize and Translate_**. From Example 3, we know

$$t = \frac{-v_0 + \sqrt{v_0^2 + 19.6s}}{9.8}$$

Carry out. Substitute 500 for s and 30 for v_0.

$$t = \frac{-30 + \sqrt{30^2 + 19.6(500)}}{9.8}$$

$$t \approx 7.49$$

Check. Substitute 30 for v_0 and 7.49 for t in the original formula. (See Example 3.)

$$s = 4.9t^2 + v_0 t = 4.9(7.49)^2 + (30)(7.49) \approx 500$$

The answer checks.

State. It takes the ball about 7.49 sec to reach the ground.

c. **_Familiarize and Translate_**. We will use the formula in Example 3, $s = 4.9t^2 + v_0 t$.

Carry out. Substitute 5 for t and 30 for v_0.

$$s = 4.9(5)^2 + 30(5) = 272.5$$

Check. We can substitute 30 for v_0 and 272.5 for s in the form of the formula we used in part (b).

$$t = \frac{-v_0 + \sqrt{v_0^2 + 19.6s}}{9.8}$$

$$= \frac{-30 + \sqrt{(30)^2 + 19.6(272.5)}}{9.8} = 5$$

The answer checks.

State. The object will fall 272.5 m.

25. **_Familiarize_**. We will use the formula $4.9t^2 = s$.

Translate. Substitute 40 for s.

$$4.9t^2 = 40$$

Carry out. We solve the equation.

$$4.9t^2 = 40$$

$$t^2 = \frac{40}{4.9}$$

$$t = \sqrt{\frac{40}{4.9}}$$

$$t \approx 2.9$$

Check. Substitute 2.9 for t in the formula.

$$s = 4.9(2.9)^2 = 41.209 \approx 40$$

The answer checks.

State. Wyatt will fall for about 2.9 sec before the cord begins to stretch.

27. **_Familiarize_**. We will use the formula $V = 48T^2$.

Translate. Substitute 44 for V.

$$44 = 48T^2$$

Carry out. We solve the equation.

$$44 = 48T^2$$

$$\frac{44}{48} = T^2$$

$$T = \sqrt{\frac{44}{48}}$$

$$T \approx 0.957$$

Check. Substitute 0.957 for T in the formula.

$$V = 48(0.957)^2 \approx 44$$

The answer checks.

State. His hang time is about 0.957 sec.

29. **_Familiarize and Translate_**. We will use the formula in Example 4, $s = 4.9t^2 + v_0 t$.

Carry out. Solve the formula for v_0.

$$s - 4.9t^2 = v_0 t$$

$$\frac{s - 4.9t^2}{t} = v_0$$

Now substitute 51.6 for s and 3 for t.

$$\frac{51.6 - 4.9(3)^2}{3} = v_0$$

$$2.5 = v_0$$

Check. Substitute 3 for t and 2.5 for v_0 in the original formula.

$$s = 4.9(3)^2 + 2.5(3) = 51.6$$

The solution checks.

State. The initial velocity is 2.5 m/sec.

31. **_Familiarize and Translate_**. From Exercise 22 we know

that $r = -1 + \dfrac{-P_2 + \sqrt{P_2^2 + 4AP_1}}{2P_1}$

where A is the total amount in the account after two years, P_1 is the amount of the original deposit, P_2 is deposited at the beginning of the second year, and r is the annual interest rate.

Carry out. Substitute 3200 for P_1, 1800 for P_2, and 5207 for A.

$$r = -1 + \frac{-1800 + \sqrt{(1800)^2 + 4(5207)(3200)}}{2(3200)}$$

Using a calculator we have $r = 0.025$.

Check. Substitute in the original formula in Exercise 22.

$$A = P_1(1 + r)^2 + P_2(1 + r)$$

$$A = 3200(1.025)^2 + 1800(1.025) = 5207$$

The solution checks.

State. The annual interest rate is 0.025 or 2.5%.

33. **_Familiarize_**. We first make a drawing, labeling it with the known and unknown information. We can also organize the information in a table. We let r represent the speed and t the time for the first part of the trip.

r mph t hr $r - 10$ mph $4 - t$ hr

     120 mi   100 mi

Trip	Distance	Speed	Time
1st part	120	r	t
2nd part	100	$r - 10$	$4 - t$

Translate. Using $r = \dfrac{d}{t}$, we get two equations from the table, $r = \dfrac{120}{t}$ and $r - 10 = \dfrac{100}{4 - t}$.

Carry out. We substitute $\dfrac{120}{t}$ for r in the second equation and solve for t.

$$\frac{120}{t} - 10 = \frac{100}{4-t}, \quad \text{LCD is } t(4-t)$$

$$t(4-t)\left(\frac{120}{t} - 10\right) = t(4-t) \cdot \frac{100}{4-t}$$

$$120(4-t) - 10t(4-t) = 100t$$

$$480 - 120t - 40t + 10t^2 = 100t$$

$$10t^2 - 260t + 480 = 0 \quad \text{Standard form}$$

$$t^2 - 26t + 48 = 0 \quad \text{Multiplying by } \frac{1}{10}$$

$$(t-2)(t-24) = 0$$

$$t = 2 \quad \text{or} \quad t = 24$$

Check. Since the time cannot be negative (If $t = 24$, $4-t = -20$.), we check only 2 hr. If $t = 2$, then $4 - t = 2$. The speed of the first part is $\frac{120}{2}$, or 60 mph. The speed of the second part is $\frac{100}{2}$, or 50 mph. The speed of the second part is 10 mph slower than the first part. The value checks.
State. The speed of the first part was 60 mph, and the speed of the second part was 50 mph.

35. Familiarize. We first make a drawing. We also organize the information in a table. We let $r =$ the speed and $t =$ the time of the slower trip.

200 mi	r mph	t hr

200 mi	$r + 10$ mph	$t - 1$ hr

Trip	Distance	Speed	Time
Slower	200	r	t
Faster	200	$r+10$	$t-1$

Translate. Using $t = d/r$, we get two equations from the table:

$$t = \frac{200}{r} \quad \text{and} \quad t - 1 = \frac{200}{r+10}$$

Carry out. We substitute $\frac{200}{r}$ for t in the second equation and solve for r.

$$\frac{200}{r} - 1 = \frac{200}{r+10}, \quad \begin{array}{c}\text{LCD is}\\ r(r+10)\end{array}$$

$$r(r+10)\left(\frac{200}{r} - 1\right) = r(r+10) \cdot \frac{200}{r+10}$$

$$200(r+10) - r(r+10) = 200r$$

$$200r + 2000 - r^2 - 10r = 200r$$

$$0 = r^2 + 10r - 2000$$

$$0 = (r+50)(r-40)$$

$$r = -50 \quad \text{or} \quad r = 40$$

Check. Since negative speed has no meaning in this problem, we check only 40. If $r = 40$, then the time for the slower trip is $\frac{200}{40}$, or 5 hours. If $r = 40$, then $r + 10 = 50$ and the time for the faster trip is $\frac{200}{50}$, or 4 hours. This is 1 hour less time than the slower trip took, so we have an answer to the problem.
State. The speed is 40 mph.

37. Familiarize. We make a drawing and then organize the information in a table. We let $r =$ the speed and $t =$ the time of the Cessna.

600 mi	r mph	t hr

1000 mi	$r + 50$ mph	$t + 1$ hr

Plane	Distance	Speed	Time
Cessna	600	r	t
Beechcraft	1000	$r+50$	$t+1$

Translate. Using $t = d/r$, we get two equations from the table:

$$t = \frac{600}{r} \quad \text{and} \quad t + 1 = \frac{1000}{r+50}$$

Carry out. We substitute $\frac{600}{r}$ for t in the second equation and solve for r.

$$\frac{600}{r} + 1 = \frac{1000}{r+50}, \quad \begin{array}{c}\text{LCD is}\\ r(r+50)\end{array}$$

$$r(r+50)\left(\frac{600}{r} + 1\right) = r(r+50) \cdot \frac{1000}{r+50}$$

$$600(r+50) + r(r+50) = 1000r$$

$$600r + 30{,}000 + r^2 + 50r = 1000r$$

$$r^2 - 350r + 30{,}000 = 0$$

$$(r-150)(r-200) = 0$$

$$r = 150 \quad \text{or} \quad r = 200$$

Check. If $r = 150$, then the Cessna's time is $\frac{600}{150}$, or 4 hr and the Beechcraft's time is $\frac{1000}{150+50}$, or $\frac{1000}{200}$, or 5 hr. If $r = 200$, then the Cessna's time is $\frac{600}{200}$, or 3 hr and the Beechcraft's time is $\frac{1000}{200+50}$, or $\frac{1000}{250}$, or 4 hr. Since the Beechcraft's time is 1 hr longer in each case, both values check. There are two solutions.
State. The speed of the Cessna is 150 mph and the speed of the Beechcraft is 200 mph; or the speed of the Cessna is 200 mph and the speed of the Beechcraft is 250 mph.

39. Familiarize. We make a drawing and then organize the information in a table. We let r represent the speed and t the time of the trip to Hillsboro.

Hillsboro

36 mi	r mph	t hr

36 mi	$r - 3$ mph	$7 - t$ hr

Trip	Distance	Speed	Time
To Hillsboro	36	r	t
Return	36	$r-3$	$7-t$

Translate. Using $t = \frac{d}{r}$, we get two equations from the table,

$$t = \frac{36}{r} \quad \text{and} \quad 7 - t = \frac{36}{r-3}.$$

Carry out. We substitute $\frac{36}{r}$ for t in the second equation and solve for r.

$$7 - \frac{36}{r} = \frac{36}{r-3}, \qquad \text{LCD is } r(r-3)$$

$$r(r-3)\left(7 - \frac{36}{r}\right) = r(r-3) \cdot \frac{36}{r-3}$$

$$7r(r-3) - 36(r-3) = 36r$$

$$7r^2 - 21r - 36r + 108 = 36r$$

$$7r^2 - 93r + 108 = 0$$

$$(7r-9)(r-12) = 0$$

$$r = \frac{9}{7} \quad or \quad r = 12$$

Check. Since negative speed has no meaning in this problem (If $r = \frac{9}{7}$, then $r - 3 = -\frac{12}{7}$.), we check only 12 mph. If $r = 12$, then the time of the trip to Hillsboro is $\frac{36}{12}$, or 3 hr. The speed of the return trip is $12 - 3$, or 9 mph, and the time is $\frac{36}{9}$, or 4 hr. The total time for the round trip is 3 hr + 4 hr, or 7 hr. The value checks.

State. Naoki's speed on the trip to Hillsboro was 12 mph and it was 9 mph on the return trip.

41. *Familiarize*. We make a drawing and organize the information in a table. Let r represent the speed of the boat in still water, and let t represent the time of the trip upriver.

60 mi $r - 3$ mph t hr
—————————————————→ Upriver

Downriver ←————————— 60 mi $r + 3$ mph $9 - t$ hr

Trip	Distance	Speed	Time
Upriver	60	$r-3$	t
Downriver	60	$r+3$	$9-t$

Translate. Using $t = \frac{d}{r}$, we get two equations from the table,

$$t = \frac{60}{r-3} \text{ and } 9 - t = \frac{60}{r+3}.$$

Carry out. We substitute $\frac{60}{r-3}$ for t in the second equation and solve for r.

$$9 - \frac{60}{r-3} = \frac{60}{r+3}$$

$$(r-3)(r+3)\left(9 - \frac{60}{r-3}\right) = (r-3)(r+3) \cdot \frac{60}{r+3}$$

$$9(r-3)(r+3) - 60(r+3) = 60(r-3)$$

$$9r^2 - 81 - 60r - 180 = 60r - 180$$

$$9r^2 - 120r - 81 = 0$$

$$3r^2 - 40r - 27 = 0 \quad \text{Dividing by 3}$$

We use the quadratic formula.

$$r = \frac{-(-40) \pm \sqrt{(-40)^2 - 4 \cdot 3 \cdot (-27)}}{2 \cdot 3}$$

$$r = \frac{40 \pm \sqrt{1924}}{6}$$

$$r \approx 14 \quad or \quad r \approx -0.6$$

Check. Since negative speed has no meaning in this problem, we check only 14 mph. If $r \approx 14$, then the speed upriver is about $14 - 3$, or 11 mph, and the time

is about $\frac{60}{11}$, or 5.5 hr. The speed downriver is about $14 + 3$, or 17 mph, and the time is about $\frac{60}{17}$, or 3.5 hr. The total time of the round trip is $5.5 + 3.5$, or 9 hr. The value checks.

State. The speed of the boat in still water is about 14 mph.

43. *Familiarize*. Let x represent the time it take the spring to fill the pool. Then $x - 8$ represents the time it takes the well to fill the pool. It takes them 3 hr to fill the pool working together, so they can fill $\frac{1}{3}$ of the pool in 1 hr. The spring will fill $\frac{1}{x}$ of the pool in 1 hr, and the well will fill $\frac{1}{x-8}$ of the pool in 1 hr.

Translate. We have an equation.

$$\frac{1}{x} + \frac{1}{x-8} = \frac{1}{3}$$

Carry out. We solve the equation.

We multiply by the LCD, $3x(x-8)$.

$$3x(x-8)\left(\frac{1}{x} + \frac{1}{x-8}\right) = 3x(x-8) \cdot \frac{1}{3}$$

$$3(x-8) + 3x = x(x-8)$$

$$3x - 24 + 3x = x^2 - 8x$$

$$0 = x^2 - 14x + 24$$

$$0 = (x-2)(x-12)$$

Check. Since negative time has no meaning in this problem, 2 is not a solution ($2 - 8 = -6$). We check only 12 hr. This is the time it would take the spring working alone. Then the well would take $12 - 8$, or 4 hr working alone. The well would fill $3\left(\frac{1}{4}\right)$, or $\frac{3}{4}$ of the pool in 3 hr, and the spring would fill $3\left(\frac{1}{12}\right)$, or $\frac{1}{4}$ of the pool in 3 hr. Thus, in 3 hr they would fill $\frac{3}{4} + \frac{1}{4}$ of the pool. This is all of it, so the numbers check.

State. It takes the spring, working alone, 12 hr to fill the pool.

45. We make a drawing and then organize the information in a table. We let r represent Kofi's speed in still water. Then $r - 2$ is the speed upstream and $r + 2$ is the speed downstream. Using $t = \frac{d}{r}$, we let $\frac{1}{r-2}$ represent the time upstream and $\frac{1}{r+2}$ represent the time downstream.

1 mi $r - 2$ mph
—————————————————→ Upstream

Downstream ←————————— 1 mi $r + 2$ mph

Trip	Distance	Speed	Time
Upstream	1	$r-2$	$\frac{1}{r-2}$
Downstream	1	$r+2$	$\frac{1}{r+2}$

Translate. The time for the round trip is 1 hour. We now have an equation.

$$\frac{1}{r-2}+\frac{1}{r+2}=1$$

Carry out. We solve the equation. We multiply by the LCD, $(r-2)(r+2)$.

$$(r-2)(r+2)\left(\frac{1}{r-2}+\frac{1}{r+2}\right)=(r-2)(r+2)\cdot 1$$
$$(r+2)+(r-2)=(r-2)(r+2)$$
$$2r=r^2-4$$
$$0=r^2-2r-4$$

$a=1,\ b=-2,\ c=-4$

$$r=\frac{-(-2)\pm\sqrt{(-2)^2-4\cdot 1(-4)}}{2\cdot 1}$$
$$r=\frac{2\pm\sqrt{4+16}}{2}=\frac{2\pm\sqrt{20}}{2}$$
$$r=\frac{2\pm 2\sqrt{5}}{2}=1\pm\sqrt{5}$$

$1+\sqrt{5}\approx 1+2.236\approx 3.24$
$1-\sqrt{5}\approx 1-2.236\approx -1.24$

Check. Since negative speed has no meaning in this problem, we check only 3.24 mph. If $r\approx 3.24$, then $r-2\approx 1.24$ and $r+2\approx 5.24$. The time it takes to travel upstream is approximately $\frac{1}{1.24}$, or 0.806 hr, and the time it takes to travel downstream is approximately $\frac{1}{5.24}$, or 0.191 hr. The total time is 0.997 which is approximately 1 hour. The value checks.

State. Kofi's speed in still water is approximately 3.24 mph.

47. ***Familiarize***. Let t represent the number of days it takes Katherine to plant new trees, working alone. Then $t+2$ represents the time it takes Julianna to plant new trees, working alone. In 1 day, Katherine does $\frac{1}{t}$ of the job and Julianna does $\frac{1}{t+2}$ of the job.

Translate. Working together, Katherine and Julianna can plant trees in 6 days.

$$6\left(\frac{1}{t}\right)+6\left(\frac{1}{t+2}\right)=1,\ \text{or}\ \frac{6}{t}+\frac{6}{t+2}=1.$$

Carry out. We solve the equation. Multiply on both sides by the LCD, $t(t+2)$.

$$t(t+2)\left(\frac{6}{t}+\frac{6}{t+2}\right)=t(t+2)\cdot 1$$
$$6(t+2)+6t=t(t+2)$$
$$6t+12+6t=t^2+2t$$
$$0=t^2-10t-12$$

$a=1,\ b=-10,\ c=-12$

$$t=\frac{-(-10)\pm\sqrt{(-10)^2-4\cdot 1(-12)}}{2\cdot 1}$$
$$t=\frac{10\pm\sqrt{100+48}}{2}=\frac{10\pm\sqrt{148}}{2}=\frac{10\pm 2\sqrt{37}}{2}$$
$$t=5\pm\sqrt{37}$$

$5+\sqrt{37}\approx 5+6.08\approx 11.08\approx 11$
$5-\sqrt{37}\approx 5-6.08\approx -1.08$

Check. Since negative time has no meaning in this application, we only check 11. If $t=11$, then $t+2=11+2=13$. In 6 days Julianna does $6\cdot\frac{1}{13}$, or $\frac{6}{13}$ of the job. In 6 days the Katherine does $6\cdot\frac{1}{11}$, or $\frac{6}{11}$ of the job. Together they do $\frac{6}{13}+\frac{6}{11}\approx 1$, or 1 entire job. The answer checks.

State. It would take Katherine about 11 days to do the job, working alone.

49. *Writing Exercise.*

51.
$$\frac{1}{2}(x-7)=\frac{1}{3}x+4$$
$$6\cdot\frac{1}{2}(x-7)=6\cdot\left(\frac{1}{3}x+4\right)$$
$$3x-21=2x+24$$
$$x=45$$

53.
$$6|3x+2|=12$$
$$|3x+2|=2$$
$$3x+2=-2\quad or\quad 3x+2=2$$
$$3x=-4\quad or\quad\quad 3x=0$$
$$x=-\frac{4}{3}\quad or\quad\quad\ x=0$$

55.
$$\sqrt{2x-8}=15$$
$$2x-8=225\qquad\text{Squaring}$$
$$2x=233$$
$$x=\frac{233}{2}$$

57. *Writing Exercise.*

59.
$$A=6.5-\frac{20.4t}{t^2+36}$$
$$(t^2+36)A=(t^2+36)\left(6.5-\frac{20.4t}{t^2+36}\right)$$
$$At^2+36A=(t^2+36)(6.5)-(t^2+36)\left(\frac{20.4t}{t^2+36}\right)$$
$$At^2+36A=6.5t^2+234-20.4t$$
$$At^2-6.5t^2+20.4t+36A-234=0$$
$$(A-6.5)t^2+20.4t+(36A-234)=0$$

$a=A-6.5,\ b=20.4,\ c=36A-234$

$$t=\frac{-20.4+\sqrt{(20.4)^2-4(A-6.5)(36A-234)}}{2(A-6.5)}$$
$$t=\frac{-20.4+\sqrt{416.16-144A^2+1872A-6084}}{2(A-6.5)}$$
$$t=\frac{-20.4+\sqrt{-144A^2+1872A-5667.84}}{2(A-6.5)}$$
$$t=\frac{-20.4+\sqrt{144(-A^2+13A-39.36)}}{2(A-6.5)}$$
$$t=\frac{-20.4+12\sqrt{-A^2+13A-39.36}}{2(A-6.5)}$$
$$t=\frac{2(-10.2+6\sqrt{-A^2+13A-39.36})}{2(A-6.5)}$$
$$t=\frac{-10.2+6\sqrt{-A^2+13A-39.36}}{A-6.5}$$

61. *Familiarize.* Let a = the number. Then $a - 1$ is 1 less than a and the reciprocal of that number is $\dfrac{1}{a-1}$.

Also, 1 more than the number is $a + 1$.

Translate.

$$\underbrace{\text{The reciprocal of 1}}_{\frac{1}{(a-1)}} \underset{=}{\underbrace{\text{is}}} \underbrace{\text{1 more than}}_{a+1} \atop \underbrace{\text{less than a number}} \quad \underbrace{\text{the number.}}$$

Carry out. We solve the equation.

$$\frac{1}{a-1} = a+1, \qquad \text{LCD is } a-1$$

$$(a-1)\cdot\frac{1}{a-1} = (a-1)(a+1)$$

$$1 = a^2 - 1$$

$$2 = a^2$$

$$\pm\sqrt{2} = a$$

Check. $\dfrac{1}{\sqrt{2}-1} \approx 2.4142 \approx \sqrt{2}+1$ and

$\dfrac{1}{-\sqrt{2}-1} \approx -0.4142 \approx -\sqrt{2}+1$. The answers check.

State. The numbers are $\sqrt{2}$ and $-\sqrt{2}$, or $\pm\sqrt{2}$.

63.

$$\frac{w}{l} = \frac{l}{w+l}$$

$$l(w+l)\cdot\frac{w}{l} = l(w+l)\cdot\frac{l}{w+l}$$

$$w(w+l) = l^2$$

$$w^2 + lw = l^2$$

$$0 = l^2 - lw - w^2$$

Use the quadratic formula with $a = 1$, $b = -w$, and $c = -w^2$.

$$l = \frac{-(-w)\pm\sqrt{(-w)^2 - 4\cdot 1\cdot(-w^2)}}{2\cdot 1}$$

$$l = \frac{w\pm\sqrt{w^2 + 4w^2}}{2} = \frac{w\pm\sqrt{5w^2}}{2}$$

$$l = \frac{w\pm w\sqrt{5}}{2}$$

Since $\dfrac{w - w\sqrt{5}}{2}$ is negative we use the positive square root: $l = \dfrac{w + w\sqrt{5}}{2}$

65. $mn^4 - r^2 pm^3 - r^2 n^2 + p = 0$

Let $u = n^2$. Substitute and rearrange.

$$mu^2 - r^2 u - r^2 pm^3 + p = 0$$

$$a = m, \ b = -r^2, \ c = -r^2 pm^3 + p$$

$$u = \frac{-(-r^2)\pm\sqrt{(-r^2)^2 - 4\cdot m(-r^2 pm^3 + p)}}{2\cdot m}$$

$$u = \frac{r^2 \pm\sqrt{r^4 + 4m^4 r^2 p - 4mp}}{2m}$$

$$n^2 = \frac{r^2 \pm\sqrt{r^4 + 4m^4 r^2 p - 4mp}}{2m}$$

$$n = \pm\sqrt{\frac{r^2 \pm\sqrt{r^4 + 4m^4 r^2 p - 4mp}}{2m}}$$

67. Let s represent a length of a side of the cube, let S represent the surface area of the cube, and let A represent the surface area of the sphere. Then the diameter of the sphere is s, so the radius r is $s/2$.

From Exercise 5, we know, $A = 4\pi r^2$, so when

$r = s/2$ we have $\qquad A = 4\pi\left(\dfrac{s}{2}\right)^2 = 4\pi\cdot\dfrac{s^2}{4} = \pi s^2$.

From the formula for the surface area of a cube (See Exercise 6.) we know that $S = 6s^2$, so $\dfrac{S}{6} = s^2$ and

then $A = \pi\cdot\dfrac{S}{6}$, or $A(S) = \dfrac{\pi S}{6}$.

Exercise Set 8.5

1. True

3. True

5. Since $\left(\sqrt{p}\right)^2 = p$, use $u = \sqrt{p}$.

7. Since $\left(x^2+3\right)^2 = \left(x^2+3\right)^2$, use $u = x^2 + 3$.

9. Since $\left[(1+t)^2\right]^2 = (1+t)^4$, use $u = (1+t)^2$.

11. $x^4 - 13x^2 + 36 = 0$

Let $u = x^2$ and $u^2 = x^4$.

$u^2 - 13u + 36 = 0$ Substituting u for x^2

$(u-4)(u-9) = 0$

$u - 4 = 0 \quad or \quad u - 9 = 0$

$\quad u = 4 \quad or \qquad u = 9$

Now replace u with x^2 and solve these equations.

$x^2 = 4 \quad or \quad x^2 = 9$

$x = \pm 2 \quad or \quad x = \pm 3$

The numbers 2, –2, 3, and –3 check. They are the solutions.

13. $t^4 - 7t^2 + 12 = 0$

Let $u = t^2$ and $u^2 = t^4$.

$u^2 - 7u + 12 = 0$ Substituting u for t^2

$(u-3)(u-4) = 0$

$u - 3 = 0 \quad or \quad u - 4 = 0$

$\quad u = 3 \quad or \qquad u = 4$

Now replace u with t^2 and solve these equations.

$t^2 = 3 \quad or \quad t^2 = 4$

$t = \pm\sqrt{3} \quad or \quad t = \pm 2$

The numbers $\sqrt{3}, -\sqrt{3}, 2,$ and -2 check. They are the solutions.

15. $4x^4 - 9x^2 + 5 = 0$

Let $u = x^2$ and $u^2 = x^4$.

$4u^2 - 9u + 5 = 0$ Substituting u for x^2

$(4u - 5)(u - 1) = 0$

$4u - 5 = 0$ or $u - 1 = 0$

$u = \dfrac{5}{4}$ or $u = 1$

Now replace u with x^2 and solve these equations.

$x^2 = \dfrac{5}{4}$ or $x^2 = 1$

$x = \pm\sqrt{\dfrac{5}{4}} = \pm\dfrac{\sqrt{5}}{2}$ or $x = \pm 1$

The numbers $\dfrac{\sqrt{5}}{2}, -\dfrac{\sqrt{5}}{2}$, 1, and -1 check. They are the solutions.

17. $(x^2 - 7)^2 - 3(x^2 - 7) + 2 = 0$

Let $u = x^2 - 7$ and $u^2 = (x^2 - 7)^2$.

$u^2 - 3u + 2 = 0$ Substituting

$(u - 1)(u - 2) = 0$

$u = 1$ or $u = 2$

$x^2 - 7 = 1$ or $x^2 - 7 = 2$ Replacing u with $x^2 - 7$

$x^2 = 8$ or $x^2 = 9$

$x = \pm\sqrt{8} = \pm 2\sqrt{2}$ or $x = \pm 3$

The numbers $2\sqrt{2}$, $-2\sqrt{3}$, 3, and -3 check. They are the solutions.

19. $x^4 + 5x^2 - 36 = 0$

Let $u = x^2$ and $u^2 = x^4$.

$u^2 + 5u - 36 = 0$

$(u - 4)(u + 9) = 0$

$u - 4 = 0$ or $u + 9 = 0$

$u = 4$ or $u = -9$

Now replace u with x^2 and solve these equations.

$x^2 = 4$ or $x^2 = -9$

$x = \pm 2$ or $x = \pm\sqrt{-9} = \pm 3i$

The numbers 2, -2, $3i$, and $-3i$ check. They are the solutions.

21. $(n^2 + 6)^2 - 7(n^2 + 6) + 10 = 0$

Let $u = n^2 + 6$ and $u^2 = (n^2 + 6)^2$.

$u^2 - 7u + 10 = 0$

$(u - 2)(u - 5) = 0$

$u - 2 = 0$ or $u - 5 = 0$

$u = 2$ or $u = 5$

Now replace u with $n^2 + 6$ and solve these equations.

$n^2 + 6 = 2$ or $n^2 + 6 = 5$

$n^2 = -4$ or $n^2 = -1$

$n = \pm\sqrt{-4} = \pm 2i$ or $n = \pm\sqrt{-1} = \pm i$

$2i$, $-2i$, i, and $-i$ check. They are the solutions.

23. $w + 4\sqrt{w} - 12 = 0$

Let $u = \sqrt{w}$ and $u^2 = (\sqrt{w})^2 = w$.

$u^2 + 4u - 12 = 0$

$(u + 6)(u - 2) = 0$

$u + 6 = 0$ or $u - 2 = 0$

$u = -6$ or $u = 2$

Now replace u with $\sqrt{w}$ and solve these equations.

$\sqrt{w} = -6$ or $\sqrt{w} = 2$

or $w = 4$

Since the principal square root cannot be negative, only 4 checks as a solution.

25. $r - 2\sqrt{r} - 6 = 0$

Let $u = \sqrt{r}$ and $u^2 = r$.

$u^2 - 2u - 6 = 0$

$u = \dfrac{-(-2) \pm \sqrt{(-2)^2 - 4 \cdot 1 \cdot (-6)}}{2 \cdot 1}$

$u = \dfrac{2 \pm \sqrt{28}}{2} = \dfrac{2 + 2\sqrt{7}}{2}$

$u = 1 \pm \sqrt{7}$

Replace u with $\sqrt{r}$ and solve these equations:

$\sqrt{r} = 1 + \sqrt{7}$ or $\sqrt{r} = 1 - \sqrt{7}$

$(\sqrt{r})^2 = (1 + \sqrt{7})^2$

$r = 1 + 2\sqrt{7} + 7$

$r = 8 + 2\sqrt{7}$

The number $1 - \sqrt{7}$ is not a solution since it is negative.

The number $8 + 2\sqrt{7}$ checks. It is the solution.

27. $(1 + \sqrt{x})^2 + 5(1 + \sqrt{x}) + 6 = 0$

Let $u = 1 + \sqrt{x}$ and $u^2 = (1 + \sqrt{x})^2$.

$u^2 + 5u + 6 = 0$ Substituting

$(u + 3)(u + 2) = 0$

$u = -3$ or $u = -2$

$1 + \sqrt{x} = -3$ or $1 + \sqrt{x} = -2$ Replacing u with $1 + \sqrt{x}$

$\sqrt{x} = -4$ or $\sqrt{x} = -3$

Since the principal square root cannot be negative, this equation has no solution.

29. $x^{-2} - x^{-1} - 6 = 0$

Let $u = x^{-1}$ and $u^2 = x^{-2}$.

$u^2 - u - 6 = 0$ Substituting

$(u - 3)(u + 2) = 0$

$u = 3$ or $u = -2$

Now we replace u with x^{-1} and solve these equations:

$x^{-1} = 3$ or $x^{-1} = -2$

$\dfrac{1}{x} = 3$ or $\dfrac{1}{x} = -2$

$\dfrac{1}{3} = x$ or $-\dfrac{1}{2} = x$

Both $\dfrac{1}{3}$ and $-\dfrac{1}{2}$ check. They are the solutions.

31. $4t^{-2} - 3t^{-1} - 1 = 0$

Let $u = t^{-1}$ and $u^2 = t^{-2}$.

$4u^2 - 3u - 1 = 0$ Substituting

$(4u + 1)(u - 1) = 0$

$4u + 1 = 0$ or $u - 1 = 0$

$u = -\dfrac{1}{4}$ or $u = 1$

Now replace u with t^{-1} and solve these equations.

$t^{-1} = -\dfrac{1}{4}$ or $t^{-1} = 1$

$\dfrac{1}{t} = -\dfrac{1}{4}$ or $\dfrac{1}{t} = 1$

$-4 = t$ or $1 = t$

Both -4 and 1 check. They are the solutions.

33. $t^{2/3} + t^{1/3} - 6 = 0$

Let $u = t^{1/3}$ and $u^2 = t^{2/3}$.

$u^2 + u - 6 = 0$ Substituting

$(u + 3)(u - 2) = 0$

$u = -3$ or $u = 2$

Now we replace u with $t^{1/3}$ and solve these equations:

$t^{1/3} = -3$ or $t^{1/3} = 2$

$t = (-3)^3$ or $t = 2^3$ Raising to the third power

$t = -27$ or $t = 8$

Both -27 and 8 check. They are the solutions.

35. $y^{1/3} - y^{1/6} - 6 = 0$

Let $u = y^{1/6}$ and $u^2 = y^{2/6}$, or $y^{1/3}$.

$u^2 - u - 6 = 0$ Substituting

$(u - 3)(u + 2) = 0$

$u = 3$ or $u = -2$

Now we replace u with $y^{1/6}$ and solve these equations:

$y^{1/6} = 3$ or $y^{1/6} = -2$

$\sqrt[6]{y} = 3$ or $\sqrt[6]{y} = -2$

$y = 3^6$

$y = 729$

The equation $\sqrt[6]{y} = -2$ has no solution since principal sixth roots are never negative. The number 729 checks and is the solution.

37. $t^{1/3} + 2t^{1/6} = 3$

$t^{1/3} + 2t^{1/6} - 3 = 0$

Let $u = t^{1/6}$ and $u^2 = t^{2/6} = t^{1/3}$.

$u^2 + 2u - 3 = 0$ Substituting

$(u + 3)(u - 1) = 0$

$u = -3$ or $u = 1$

$t^{1/6} = -3$ or $t^{1/6} = 1$ Substituting $t^{1/6}$ for u

$t = 1$

Since principal sixth roots are never negative, the

equation $t^{1/6} = -3$ has no solution.
The number 1 checks and is the solution.

39. $(3 - \sqrt{x})^2 - 10(3 - \sqrt{x}) + 23 = 0$

Let $u = 3 - \sqrt{x}$ and $u^2 = (3 - \sqrt{x})^2$.

$u^2 - 10u + 23 = 0$ Substituting

$u = \dfrac{-(-10) \pm \sqrt{(-10)^2 - 4 \cdot 1 \cdot 23}}{2 \cdot 1}$

$u = \dfrac{10 \pm \sqrt{8}}{2} = \dfrac{2 \cdot 5 \pm 2\sqrt{2}}{2}$

$u = 5 \pm \sqrt{2}$

$u = 5 + \sqrt{2}$ or $u = 5 - \sqrt{2}$

Now we replace u with $3 - \sqrt{x}$ and solve these equations:

$3 - \sqrt{x} = 5 + \sqrt{2}$ or $3 - \sqrt{x} = 5 - \sqrt{2}$

$-\sqrt{x} = 2 + \sqrt{2}$ or $-\sqrt{x} = 2 - \sqrt{2}$

$\sqrt{x} = -2 - \sqrt{2}$ or $\sqrt{x} = -2 + \sqrt{2}$

Since both $-2 - \sqrt{2}$ and $-2 + \sqrt{2}$ are negative and principal square roots are never negative, the equation has no solution.

41. $16\left(\dfrac{x-1}{x-8}\right)^2 + 8\left(\dfrac{x-1}{x-8}\right) + 1 = 0$

Let $u = \dfrac{x-1}{x-8}$ and $u^2 = \left(\dfrac{x-1}{x-8}\right)^2$.

$16u^2 + 8u + 1 = 0$ Substituting

$(4u + 1)(4u + 1) = 0$

$u = -\dfrac{1}{4}$

Now we replace u with $\dfrac{x-1}{x-8}$ and solve this equation:

$\dfrac{x-1}{x-8} = -\dfrac{1}{4}$

$4x - 4 = -x + 8$ Multiplying by $4(x - 8)$

$5x = 12$

$x = \dfrac{12}{5}$

The number $\dfrac{12}{5}$ checks and is the solution.

43. The x-intercepts occur where $f(x) = 0$. Thus, we must have $5x + 13\sqrt{x} - 6 = 0$.

Let $u = \sqrt{x}$ and $u^2 = x$.

$5u^2 + 13u - 6 = 0$ Substituting

$(5u - 2)(u + 3) = 0$

$u = \dfrac{2}{5}$ or $u = -3$

Now replace u with $\sqrt{x}$ and solve these equations:

$\sqrt{x} = \dfrac{2}{5}$ or $\sqrt{x} = -3$ has no solution

$x = \dfrac{4}{25}$

The number $\frac{4}{25}$ checks. Thus, the x-intercept is $\left(\frac{4}{25}, 0\right)$.

45. The x-intercepts occur where $f(x) = 0$. Thus, we must have $(x^2 - 3x)^2 - 10(x^2 - 3x) + 24 = 0$.

Let $u = x^2 - 3x$ and $u^2 = (x^2 - 3x)^2$.

$u^2 - 10u + 24 = 0$ Substituting

$(u - 6)(u - 4) = 0$

$u = 6$ *or* $u = 4$

Now replace u with $x^2 - 3x$ and solve these equations:

$x^2 - 3x = 6$ *or* $x^2 - 3x = 4$

$x^2 - 3x - 6 = 0$ *or* $x^2 - 3x - 4 = 0$

$x = \dfrac{-(-3) \pm \sqrt{(-3)^2 - 4(1)(-6)}}{2 \cdot 1}$ *or* $(x-4)(x+1) = 0$

$x = \dfrac{3}{2} \pm \dfrac{\sqrt{33}}{2}$ *or* $x = 4$ *or* $x = -1$

All four numbers check. Thus, the x-intercepts are $\left(\dfrac{3}{2} + \dfrac{\sqrt{33}}{2},\ 0\right)$, $\left(\dfrac{3}{2} - \dfrac{\sqrt{33}}{2},\ 0\right)$, $(4, 0)$, and $(-1, 0)$.

47. The x-intercepts occur where $f(x) = 0$. Thus, we must have $x^{2/5} + x^{1/5} - 6 = 0$.

Let $u = x^{1/5}$ and $u^2 = x^{2/5}$.

$u^2 + u - 6 = 0$ Substituting

$(u + 3)(u - 2) = 0$

$u = -3$ *or* $u = 2$

$x^{1/5} = -3$ *or* $x^{1/5} = 2$ Replacing u with $x^{1/5}$

$x = -243$ $x = 32$ Raising to the fifth power

Both -243 and 32 check. Thus, the x-intercepts are $(-243,\ 0)$ and $(32, 0)$.

49. $f(x) = \left(\dfrac{x^2 + 2}{x}\right)^4 + 7\left(\dfrac{x^2 + 2}{x}\right)^2 + 5$

Observe that, for all real numbers x, each term is positive. Thus, there are no real-number values of x for which $f(x) = 0$ and hence no x-intercepts.

51. *Writing Exercise.*

53. Graph $2x = -5y$.

We find some ordered pairs, plot points, and draw the graph.

x	y
-5	2
0	0
5	-2

55. Graph $2x - 5y = 10$.

We find some ordered pairs, plot points, and draw the graph.

x	y
-5	-4
0	-2
5	0

57. Graph $y - 2 = 3(x - 4)$.

We find some ordered pairs, plot points, and draw the graph.

x	y
0	-10
2	-4
3	-1
4	2
5	5

59. *Writing Exercise.*

61. $3x^4 + 5x^2 - 1 = 0$

Let $u = x^2$ and $u^2 = x^4$.

$3u^2 + 5u - 1 = 0$ Substituting

$u = \dfrac{-5 \pm \sqrt{5^2 - 4 \cdot 3 \cdot (-1)}}{2 \cdot 3}$

$u = \dfrac{-5 \pm \sqrt{37}}{6}$

$x^2 = \dfrac{-5 \pm \sqrt{37}}{6}$ Replacing u with x^2

$x = \pm\sqrt{\dfrac{-5 \pm \sqrt{37}}{6}}$

All four numbers check and are the solutions.

63. $\dfrac{x}{x-1} - 6\sqrt{\dfrac{x}{x-1}} - 40 = 0$

Let $u = \sqrt{\dfrac{x}{x-1}}$ and $u^2 = \dfrac{x}{x-1}$.

$u^2 - 6u - 40 = 0$ Substituting

$(u - 10)(u + 4) = 0$

$u = 10$ *or* $u = -4$

$\sqrt{\dfrac{x}{x-1}} = 10$ *or* $\sqrt{\dfrac{x}{x-1}} = -4$ has no solution

$\dfrac{x}{x-1} = 100$

$x = 100x - 100$ Multiplying by $(x-1)$

$100 = 99x$

$\dfrac{100}{99} = x$

The number $\dfrac{100}{99}$ checks. It is the solution.

65. $a^3 - 26a^{3/2} - 27 = 0$

Let $u = a^{3/2}$.

$u^2 - 26u - 27 = 0$ Substituting

$(u - 27)(u + 1) = 0$

$u = 27$ *or* $u = -1$

Replace u with $a^{3/2}$.

$a^{3/2} = 27$ or $a^{3/2} = -1$ has no solution

$a = 27^{2/3}$

$a = \left(3^3\right)^{2/3}$

$a = 9$

The number 9 checks. It is the solution.

67. $x^6 + 7x^3 - 8 = 0$

Let $u = x^3$.

$u^2 + 7u - 8 = 0$

$(u+8)(u-1) = 0$

$u = -8$ or $u = 1$

$x^3 = -8$ or $x^3 = 1$

$x^3 + 8 = 0$ or $x^3 - 1 = 0$

First solve $x^3 + 8 = 0$.

$x^3 + 8 = 0$

$(x+2)(x^2 - 2x + 4) = 0$

$x + 2 = 0$ or $x^2 - 2x + 4 = 0$

$x = -2$ or $x = 1 \pm \sqrt{3}i$

The solutions of $x^3 - 1 = 0$ are 1 and $-\frac{1}{2} \pm \frac{\sqrt{3}}{2}i$. (See Exercise 66.)

All six numbers check.

69. $-3, -1, 1, 4$

Mid-Chapter Review

1. $x - 7 = \pm\sqrt{5}$

$x = 7 \pm \sqrt{5}$

The solutions are $7 + \sqrt{5}$ and $7 - \sqrt{5}$.

2. $a = 1, b = -2, c = -1$

$x = \dfrac{-(-2) \pm \sqrt{(-2)^2 - 4\cdot1\cdot(-1)}}{2\cdot1}$

$x = \dfrac{2 \pm \sqrt{8}}{2}$

$x = \dfrac{2}{2} \pm \dfrac{2\sqrt{2}}{2}$

The solutions are $1 + \sqrt{2}$ and $1 - \sqrt{2}$.

3. $x^2 + 4x = 21$

$x^2 + 4x - 21 = 0$

$(x+7)(x-3) = 0$ Factoring

$x + 7 = 0$ or $x - 3 = 0$

$x = -7$ or $x = 3$

The solutions are -7 and 3.

4. $t^2 - 196 = 0$

$t^2 = 196$

$t = \pm\sqrt{196} = \pm14$

5. $x^2 = 2x + 5$

$x^2 - 2x + 1 = 5 + 1$

$(x-1)^2 = 6$

$x - 1 = \pm\sqrt{6}$

$x = 1 \pm \sqrt{6}$

The solutions are $1 + \sqrt{6}$ and $1 - \sqrt{6}$.

6. $x^2 = 2x - 5$

$x^2 - 2x + 1 = -5 + 1$

$(x-1)^2 = -4$

$x - 1 = \pm\sqrt{-4}$

$x = 1 \pm 2i$

7. $x^4 = 16$

$x^4 - 16 = 0$

Let $u = x^2$ and $u^2 = x^4$.

$u^2 - 16 = 0$

$(u+4)(u-4) = 0$

$u + 4 = 0$ or $u - 4 = 0$

$u = -4$ or $u = 4$

Replace u with x^2.

$x^2 = 4$ or $x^2 = -4$

$x = \pm2$ or $x = \sqrt{-4} = \pm2i$

The solutions are -2, 2, $-2i$, and $2i$.

8. $(t+3)^2 = 7$

$t + 3 = \pm\sqrt{7}$

$t = -3 \pm \sqrt{7}$

9. $n(n-3) = 2n(n+1)$

$n^2 - 3n = 2n^2 + 2n$

$0 = n^2 + 5n$

$0 = n(n+5)$

$n = 0$ or $n + 5 = 0$

$n = 0$ or $n = -5$

The solutions are -5 and 0.

10. $6y^2 - 7y - 10 = 0$

$(6y+5)(y-2) = 0$

$6y + 5 = 0$ or $y - 2 = 0$

$y = -\dfrac{5}{6}$ or $y = 2$

11. $16c^2 = 7c$

$16c^2 - 7c = 0$

$c(16c - 7) = 0$

$c = 0$ or $16c - 7 = 0$

$c = 0$ or $c = \dfrac{7}{16}$

The solutions are 0 and $\dfrac{7}{16}$.

12. $\quad 3x^2 + 5x = 1$

$3x^2 + 5x - 1 = 0$

$a = 3, b = 5, c = -1$

$x = \dfrac{-5 \pm \sqrt{(5)^2 - 4 \cdot 3 \cdot (-1)}}{2 \cdot 3} = \dfrac{-5 \pm \sqrt{25 + 12}}{6}$

$\quad = \dfrac{-5 \pm \sqrt{37}}{6} = -\dfrac{5}{6} \pm \dfrac{\sqrt{37}}{6}$

13. $\quad (t+4)(t-3) = 18$

$t^2 + t - 12 = 18$

$t^2 + t - 30 = 0$

$(t+6)(t-5) = 0$

$t + 6 = 0 \quad or \quad t - 5 = 0$

$\quad t = -6 \quad or \quad t = 5$

The solutions are -6 and 5.

14. $\left(m^2 + 3\right)^2 - 4\left(m^2 + 3\right) - 5 = 0$

Let $u = m^2 + 3$ and $u^2 = \left(m^2 + 3\right)^2$.

$u^2 - 4u - 5 = 0$

$(u - 5)(u + 1) = 0$

$u - 5 = 0 \quad or \quad u + 1 = 0$

$\quad u = 5 \quad or \quad u = -1$

Replace u with $m^2 + 3$.

$m^2 + 3 = 5 \quad or \quad m^2 + 3 = -1$

$\quad m^2 = 2 \quad or \quad m^2 = -4$

$\quad m = \pm\sqrt{2} \quad or \quad m = \pm\sqrt{-4} = \pm 2i$

15. $x^2 - 8x + 1 = 0$

$a = 1, b = -8, c = 1$

We substitute and compute the discriminant.

$b^2 - 4ac = (-8)^2 - 4 \cdot 1 \cdot 1 = 60$

There are two irrational solutions.

16. $b^2 - 4ac = (-4)^2 - 4 \cdot 3 \cdot (-7) = 100$

There are two rational solutions.

17. $5x^2 - x + 6 = 0$

$a = 5, b = -1, c = 6$

We substitute and compute the discriminant.

$b^2 - 4ac = (-1)^2 - 4 \cdot 5 \cdot 6 = -119$

There are two imaginary-number solutions.

18. $\qquad F = \dfrac{Av^2}{400}$

$400F = Av^2$

$\dfrac{400F}{A} = v^2$

$\sqrt{\dfrac{400F}{A}} = v$

$20\sqrt{\dfrac{F}{A}} = v, \text{ or } v = \dfrac{20\sqrt{FA}}{A}$

19. $\quad D^2 - 2Dd - 2hd = 0$

$D^2 - 2Dd + d^2 = d^2 + 2hd$

$(D - d)^2 = d^2 + 2hd$

$D - d = \sqrt{d^2 + 2hd}$

$D = d + \sqrt{d^2 + 2hd}$

20. Let r represent the speed and t the time of the slower trip.

Trip	Distance	Speed	Time
South	225	r	$\dfrac{225}{r}$
North	225	$r - 30$	$\dfrac{225}{r - 30}$

The time for the round trip is 8 hours. We now have an equation

$\dfrac{225}{r} + \dfrac{225}{r - 30} = 8,$

Solving for r, we get $r = 75$ or $r = 11.25$. Only 75 checks in the original problem. The speed South is 75 mph and the speed North is $75 - 30 = 45$ mph.

Exercise Set 8.6

1. False. The graph of a quadratic function is a parabola.

3. True

5. $f(x) = x^2$

See Example 1 in the text.

7. $f(x) = -2x^2$

We choose some numbers for x and compute $f(x)$ for each one. Then we plot the ordered pairs $(x, f(x))$ and connect them with a smooth curve.

x	$f(x) = -2x^2$
0	0
1	-2
2	-8
-1	-2
-2	-8

9. $g(x) = \frac{1}{3}x^2$

x	$g(x) = \frac{1}{3}x^2$
0	0
1	$\frac{1}{3}$
2	$\frac{4}{3}$
3	3
-1	$\frac{1}{3}$
-2	$\frac{4}{3}$
-3	3

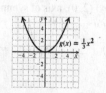

11. $h(x) = -\frac{1}{3}x^2$

Observe that the graph of $h(x) = -\frac{1}{3}x^2$ is the reflection

of the graph of $g(x) = \frac{1}{3}x^2$ across the x-axis. We

graphed $g(x)$ in Exercise 9, so we can use it to graph

$h(x)$. If we did not make this observation we could

find some ordered pairs, plot points, and connect them

with a smooth curve.

x	$h(x) = -\frac{1}{3}x^2$
0	0
1	$-\frac{1}{3}$
2	$-\frac{4}{3}$
3	-3
-1	$-\frac{1}{3}$
-2	$-\frac{4}{3}$
-3	-3

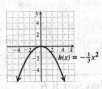

13. $f(x) = \frac{5}{2}x^2$

x	$f(x) = \frac{5}{2}x^2$
0	0
1	$\frac{5}{2}$
2	10
-1	$\frac{5}{2}$
-2	10

15. $g(x) = (x+1)^2 = [x - (-1)]^2$

We know that the graph of $g(x) = (x+1)^2$ looks like

the graph of $f(x) = x^2$ (see Exercise 5) but moved to

the left 1 unit.

Vertex: $(-1, 0)$, axis of symmetry: $x = -1$

17. $f(x) = (x-2)^2$

The graph of $f(x) = (x-2)^2$ looks like the graph of

$f(x) = x^2$ (see Exercise 5) but moved to the right 2

units.

Vertex: $(2, 0)$, axis of symmetry: $x = 2$

19. $g(x) = -(x+1)^2$

The graph of $g(x) = -(x+1)^2$ looks like the graph of

$f(x) = x^2$ (see Exercise 5) but moved to the left 1 unit.

It will also open downward because of the negative

coefficient, -1.

Vertex: $(-1, 0)$, axis of symmetry: $x = -1$

21. $g(x) = -(x-2)^2$

The graph of $g(x) = -(x-2)^2$ looks like the graph of

$f(x) = x^2$ (see Exercise 5) but moved to the right 2

units. It will also open downward because of the

negative coefficient, -1.

Vertex: $(2, 0)$, axis of symmetry: $x = 2$

23. $f(x) = 2(x+1)^2$

The graph of $f(x) = 2(x+1)^2$ looks like the graph of

$h(x) = 2x^2$ (see graph following Example 1) but

moved to the left 1 unit.

Vertex: $(-1, 0)$, axis of symmetry: $x = -1$

25. $g(x) = 3(x-4)^2$

The graph of $g(x) = 3(x-4)^2$ looks like the graph of

$g(x) = 3x^2$ but moved to the right 4 units.

Vertex: $(4, 0)$, axis of symmetry: $x = 4$

27. $h(x) = -\frac{1}{2}(x-4)^2$

The graph of $h(x) = -\frac{1}{2}(x-4)^2$ looks like the graph of

$g(x) = \frac{1}{2}x^2$ (see graph following Example 1) but

moved to the right 4 units. It will also open downward

because of the negative coefficient, $-\frac{1}{2}$.

Vertex: $(4,0)$, axis of symmetry: $x = 4$

$h(x) = -\frac{1}{2}(x-4)^2$

29. $f(x) = \frac{1}{2}(x-1)^2$

The graph of $f(x) = \frac{1}{2}(x-1)^2$ looks like the graph of

$g(x) = \frac{1}{2}x^2$ (see graph following Example 1) but

moved to the right 1 unit.
Vertex: $(1, 0)$, axis of symmetry: $x = 1$

$f(x) = \frac{1}{2}(x-1)^2$

31. $f(x) = -2(x+5)^2 = -2[x-(-5)]^2$

The graph of $f(x) = -2(x+5)^2$ looks like the graph of

$h(x) = 2x^2$ (see the graph following Example 1) but
moved to the left 5 units. It will also open downward
because of the negative coefficient, -2.
Vertex: $(-5, 0)$, axis of symmetry: $x = -5$

$f(x) = -2(x+5)^2$

33. $h(x) = -3\left(x - \frac{1}{2}\right)^2$

The graph of $h(x) = -3\left(x - \frac{1}{2}\right)^2$ looks like the graph of

$f(x) = -3x^2$ (see Exercise 8) but moved to the right

$\frac{1}{2}$ unit.

Vertex: $\left(\frac{1}{2}, 0\right)$, axis of symmetry: $x = \frac{1}{2}$

$h(x) = -3\left(x - \frac{1}{2}\right)^2$

35. $f(x) = (x-5)^2 + 2$

We know that the graph looks like the graph of

$f(x) = x^2$ (see Example 1) but moved to the right 5
units and up 2 units. The vertex is $(5, 2)$, and the axis of

symmetry is $x = 5$. Since the coefficient of $(x-5)^2$ is

positive $(1 > 0)$, there is a minimum function value, 2.

$f(x) = (x-5)^2 + 2$

37. $f(x) = (x+1)^2 - 3$

We know that the graph looks like the graph of

$f(x) = x^2$ (see Example 1) but moved to the left 1 unit
and down 3 units. The vertex is $(-1, -3)$, and the axis of

symmetry is $x = -1$. Since the coefficient of $(x+1)^2$ is

positive $(1 > 0)$, there is a minimum function value, -3.

$f(x) = (x+1)^2 - 3$

39. $g(x) = \frac{1}{2}(x+4)^2 + 1$

We know that the graph looks like the graph of

$f(x) = \frac{1}{2}x^2$ (see graph following Example 1) but

moved to the left 4 units and up 1 unit. The vertex
is $(-4, 1)$, and the axis of symmetry is $x = -4$, and the
minimum function value is 1.

$g(x) = \frac{1}{2}(x+4)^2 + 1$

41. $h(x) = -2(x-1)^2 - 3$

We know that the graph looks like the graph of

$h(x) = 2x^2$ (see graph following Example 1) but
moved to the right 1 unit and down 3 units and turned
upside down. The vertex is $(1, -3)$, and the axis of
symmetry is $x = 1$. The maximum function value is -3.

$h(x) = -2(x-1)^2 - 3$

43. $f(x) = 2(x+3)^2 + 1$

We know that the graph looks like the graph of

$f(x) = 2x^2$ (see graph following Example 1) but
moved to the left 3 units and up 1 unit. The vertex
is $(-3, 1)$, and the axis of symmetry is $x = -3$. The
minimum function value is 1.

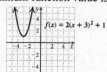

$f(x) = 2(x+3)^2 + 1$

45. $g(x) = -\frac{3}{2}(x-2)^2 + 4$

We know that the graph looks like the graph of

$f(x) = \frac{3}{2}x^2$ (see Exercise 14) but moved to the right 2

units and up 4 units and turned upside down. The
vertex is (2, 4), and the axis of symmetry is $x = 2$, and
the maximum function value is 4.

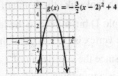

47. $f(x) = 5(x-3)^2 + 9$

The function is of the form $f(x) = a(x-h)^2 + k$ with
$a = 5$, $h = 3$, and $k = 9$. The vertex is (h, k), or $(3, 9)$.
The axis of symmetry is $x = h$, or $x = 3$. Since $a > 0$,
then k, or 9, is the minimum function value.

49. $f(x) = -\frac{3}{7}(x+8)^2 + 2$

The function is of the form $f(x) = a(x-h)^2 + k$ with

$a = -\frac{3}{7}$, $h = -8$, and $k = 2$. The vertex is (h, k), or

$(-8, 2)$. The axis of symmetry is $x = h$, or $x = -8$. Since
$a < 0$, then k, or 2, is the maximum function value.

51. $f(x) = \left(x - \frac{7}{2}\right)^2 - \frac{29}{4}$

The function is of the form $f(x) = a(x-h)^2 + k$ with

$a = 1$, $h = \frac{7}{2}$, and $k = -\frac{29}{4}$. The vertex is (h, k), or

$\left(\frac{7}{2}, -\frac{29}{4}\right)$. The axis of symmetry is $x = h$, or $x = \frac{7}{2}$.

Since $a > 0$, then k, or $-\frac{29}{4}$, is the minimum function

value.

53. $f(x) = -\sqrt{2}(x + 2.25)^2 - \pi$

The function is of the form $f(x) = a(x-h)^2 + k$ with
$a = -\sqrt{2}$, $h = -2.25$, and $k = -\pi$. The vertex is (h, k),
or $(-2.25, -\pi)$. The axis of symmetry is $x = h$, or
$x = -2.25$. Since $a < 0$, then k, or $-\pi$, is the maximum
function value.

55. *Writing Exercise.*

57. $\dfrac{3}{x} + \dfrac{x}{x+2} = \dfrac{3}{x} \cdot \dfrac{x+2}{x+2} + \dfrac{x}{x+2} \cdot \dfrac{x}{x}$

$= \dfrac{3x + 6 + x^2}{x(x+2)}$

$= \dfrac{x^2 + 3x + 6}{x(x+2)}$

59. $\sqrt[3]{8t} - \sqrt[3]{27t} + \sqrt{25t} = 2\sqrt[3]{t} - 3\sqrt[3]{t} + 5\sqrt{t}$
$= -\sqrt[3]{t} + 5\sqrt{t}$

61. $\dfrac{1}{x-1} - \dfrac{x-2}{x+3}$ LCD is $(x-1)(x+3)$

$= \dfrac{1}{x-1} \cdot \dfrac{x+3}{x+3} - \dfrac{x-2}{x+3} \cdot \dfrac{x-1}{x-1}$

$= \dfrac{x+3}{(x-1)(x+3)} - \dfrac{(x-2)(x-1)}{(x-1)(x+3)}$

$= \dfrac{x+3 - x^2 + 3x - 2}{(x-1)(x+3)}$

$= \dfrac{-x^2 + 4x + 1}{(x-1)(x+3)}$

63. *Writing Exercise.*

65. The equation will be of the form $f(x) = \frac{3}{5}(x-h)^2 + k$

with $h = 1$ and $k = 3$:

$f(x) = \frac{3}{5}(x-1)^2 + 3$

67. The equation will be of the form $f(x) = \frac{3}{5}(x-h)^2 + k$

with $h = 4$ and $k = -7$:

$f(x) = \frac{3}{5}(x-4)^2 - 7$

69. The equation will be of the form $f(x) = \frac{3}{5}(x-h)^2 + k$

with $h = -2$ and $k = -5$:

$f(x) = \frac{3}{5}[x-(-2)]^2 + (-5)$, or

$f(x) = \frac{3}{5}(x+2)^2 - 5$

71. Since there is a minimum at (2, 0), the parabola will

have the same shape as $f(x) = 2x^2$. It will be of the

form $f(x) = 2(x-h)^2 + k$ with $h = 2$ and $k = 0$:

$f(x) = 2(x-2)^2$

73. Since there is a maximum at (0, –5), the parabola will

have the same shape as $g(x) = -2x^2$. It will be of the

form $g(x) = -2(x-h)^2 + k$ with $h = 0$ and $k = -5$:

$g(x) = -2(x-0)^2 - 5$, or $g(x) = -2x^2 - 5$.

75. If h is increased, the graph will move to the right.

77. If a is replaced with $-a$, the graph will be reflected
across the x-axis.

79. The maximum value of $g(x)$ is 1 and occurs at the
point $(5, 1)$, so for $F(x)$ we have $h = 5$ and $k = 1$.
$F(x)$ has the same shape as $f(x)$ and has a minimum,
so $a = 3$. Thus, $F(x) = 3(x-5)^2 + 1$.

81. The graph of $y = f(x-1)$ looks like the graph of $y = f(x)$ moved 1 unit to the right.

83. The graph of $y = f(x) + 2$ looks like the graph of $y = f(x)$ moved up 2 units.

85. The graph of $y = f(x+3) - 2$ looks like the graph of $y = f(x)$ moved 3 units to the left and also moved down 2 units.

87. *Graphing Calculator Exercise*

89. *Writing Exercise.* The coefficient of x^2 is negative, so the parabola should open down.

Exercise Set 8.7

1. True; since $a = 3 > 0$, the graph opens upward.

3. True

5. False; the axis of symmetry is $x = \frac{3}{2}$.

7. False; the y-intercept is $(0, 7)$.

9. $\frac{1}{2} \cdot (-8) = -4; \ (-4)^2 = 16$

$$f(x) = x^2 - 8x + 2$$
$$= (x^2 - 8x + 16) - 16 + 2$$
$$= (x - 4)^2 + (-14)$$

11. $\frac{1}{2} \cdot 3 = \frac{3}{2}; \ \left(\frac{3}{2}\right)^2 = \frac{9}{4}$

$$f(x) = x^2 + 3x - 5$$
$$= \left(x^2 + 3x + \frac{9}{4}\right) - \frac{9}{4} - 5$$
$$= \left[x - \left(-\frac{3}{2}\right)\right]^2 + \left(-\frac{29}{4}\right)$$

13. $\frac{1}{2} \cdot 2 = 1; \ 1^2 = 1$

$$f(x) = 3x^2 + 6x - 2$$
$$= 3(x^2 + 2x) - 2$$
$$= 3(x^2 + 2x + 1) + 3(-1) - 2$$
$$= 3[x - (-1)]^2 + (-5)$$

15. $\frac{1}{2} \cdot 4 = 2; \ 2^2 = 4$

$$f(x) = -x^2 - 4x - 7$$
$$= -1(x^2 + 4x) - 7$$
$$= -(x^2 + 4x + 4) + -1(-4) - 7$$
$$= -[x - (-2)]^2 + (-3)$$

17. $\frac{1}{2} \cdot \left(-\frac{5}{2}\right) = -\frac{5}{4}; \ \left(-\frac{5}{4}\right)^2 = \frac{25}{16}$

$$f(x) = 2x^2 - 5x + 10$$
$$= 2\left(x^2 - \frac{5}{2}x\right) + 10$$
$$= 2\left(x^2 - \frac{5}{2}x + \frac{25}{16}\right) + 2\left(-\frac{25}{16}\right) + 10$$
$$= 2\left(x - \frac{5}{4}\right)^2 + \frac{55}{8}$$

19. a. $f(x) = x^2 + 4x + 5$
$$= (x^2 + 4x + 4 - 4) + 5 \quad \text{Adding } 4 - 4$$
$$= (x^2 + 4x + 4) - 4 + 5 \quad \text{Regrouping}$$
$$= (x + 2)^2 + 1$$

The vertex is $(-2, 1)$, the axis of symmetry is $x = -2$, and the graph opens upward since the coefficient 1 is positive. We plot a few points as a check and draw the curve.

b.

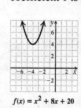

$f(x) = x^2 + 4x + 5$

21. a. $f(x) = x^2 + 8x + 20$
$$= (x^2 + 8x + 16 - 16) + 20 \quad \text{Adding } 16 - 16$$
$$= (x^2 + 8x + 16) - 16 + 20 \quad \text{Regrouping}$$
$$= (x + 4)^2 + 4$$

The vertex is $(-4, 4)$, the axis of symmetry is $x = -4$, and the graph opens upward since the coefficient 1 is positive.

b.

$f(x) = x^2 + 8x + 20$

23. a.
$$h(x) = 2x^2 - 16x + 25$$
$$= 2(x^2 - 8x) + 25 \quad \text{Factoring 2 from the first two terms}$$
$$= 2(x^2 - 8x + 16 - 16) + 25 \quad \text{Adding } 16 - 16 \text{ inside the parentheses}$$
$$= 2(x^2 - 8x + 16) + 2(-16) + 25 \quad \text{Distributing to obtain a trinomial square}$$
$$= 2(x - 4)^2 - 7$$

The vertex is $(4, -7)$, the axis of symmetry is $x = 4$, and the graph opens upward since the coefficient 2 is positive.

b.

$h(x) = 2x^2 - 16x + 25$

25. a.
$$f(x) = -x^2 + 2x + 5$$
$$= -(x^2 - 2x) + 5 \quad \text{Factoring } -1 \text{ from the first two terms}$$
$$= -(x^2 - 2x + 1 - 1) + 5 \quad \text{Adding } 1 - 1 \text{ inside the parentheses}$$
$$= -(x^2 - 2x + 1) - (-1) + 5$$
$$= -(x - 1)^2 + 6$$

The vertex is $(1, 6)$, the axis of symmetry is $x = 1$, and the graph opens downward since the coefficient -1 is negative.

b.

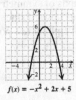

$f(x) = -x^2 + 2x + 5$

27. a.
$$g(x) = x^2 + 3x - 10$$
$$= \left(x^2 + 3x + \frac{9}{4} - \frac{9}{4}\right) - 10$$
$$= \left(x^2 + 3x + \frac{9}{4}\right) - \frac{9}{4} - 10$$
$$= \left(x + \frac{3}{2}\right)^2 - \frac{49}{4}$$

The vertex is $\left(-\frac{3}{2}, -\frac{49}{4}\right)$, the axis of symmetry is $x = -\frac{3}{2}$, and the graph opens upward since the coefficient 1 is positive.

b.

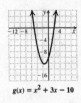

$g(x) = x^2 + 3x - 10$

29. a.
$$h(x) = x^2 + 7x$$
$$= \left(x^2 + 7x + \frac{49}{4}\right) - \frac{49}{4}$$
$$= \left(x + \frac{7}{2}\right)^2 - \frac{49}{4}$$

The vertex is $\left(-\frac{7}{2}, -\frac{49}{4}\right)$, the axis of symmetry is $x = -\frac{7}{2}$, and the graph opens upward since the coefficient 1 is positive.

b.

$h(x) = x^2 + 7x$

31. a.
$$f(x) = -2x^2 - 4x - 6$$
$$= -2(x^2 + 2x) - 6 \quad \text{Factoring}$$
$$= -2(x^2 + 2x + 1 - 1) - 6 \quad \text{Adding } 1 - 1 \text{ inside the parentheses}$$
$$= -2(x^2 + 2x + 1) - 2(-1) - 6$$
$$= -2(x + 1)^2 - 4$$

The vertex is $(-1, -4)$, the axis of symmetry is $x = -1$, and the graph opens downward since the coefficient -2 is negative.

b.

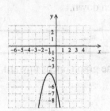

33. a.
$$g(x) = x^2 - 6x + 13$$
$$= (x^2 - 6x + 9 - 9) + 13 \quad \text{Adding } 9 - 9$$
$$= (x^2 - 6x + 9) - 9 + 13 \quad \text{Regrouping}$$
$$= (x - 3)^2 + 4$$

The vertex is $(3, 4)$, the axis of symmetry is $x = 3$, and the graph opens upward since the coefficient 1 is positive. The minimum is 4.

b.

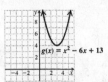

$g(x) = x^2 - 6x + 13$

35. a.
$$g(x) = 2x^2 - 8x + 3$$
$$= 2(x^2 - 4x) + 3 \quad \text{Factoring}$$
$$= 2(x^2 - 4x + 4 - 4) + 3 \quad \text{Adding } 4 - 4$$
$$= 2(x^2 - 4x + 4) + 2(-4) + 3$$
$$= 2(x - 2)^2 - 5$$

The vertex is $(2, -5)$, the axis of symmetry is

$x = 2$, and the graph opens upward since the coefficient 2 is positive. The minimum is -5.

b.

$g(x) = 2x^2 - 8x + 3$

37. a. $f(x) = 3x^2 - 24x + 50$

$= 3(x^2 - 8x) + 50$ Factoring

$= 3(x^2 - 8x + 16 - 16) + 50$ Adding $16 - 16$
                                inside the parentheses

$= 3(x^2 - 8x + 16) - 3 \cdot 16 + 50$

$= 3(x - 4)^2 + 2$

The vertex is $(4, 2)$, the axis of symmetry is $x = 4$, and the graph opens upward since the coefficient 3 is positive. The minimum is 2.

b.

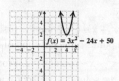

$f(x) = 3x^2 - 24x + 50$

39. a. $f(x) = -3x^2 + 5x - 2$

$= -3\left(x^2 - \frac{5}{3}x\right) - 2$ Factoring

$= -3\left(x^2 - \frac{5}{3}x + \frac{25}{36} - \frac{25}{36}\right) - 2$ Adding $\frac{25}{36} - \frac{25}{36}$
                                inside the parentheses

$= -3\left(x^2 - \frac{5}{3}x + \frac{25}{36}\right) - 3\left(-\frac{25}{36}\right) - 2$

$= -3\left(x - \frac{5}{6}\right)^2 + \frac{1}{12}$

The vertex is $\left(\frac{5}{6}, \frac{1}{12}\right)$, the axis of symmetry is

$x = \frac{5}{6}$, and the graph opens downward since the

coefficient -3 is negative. The maximum is $\frac{1}{12}$.

b.

$f(x) = -3x^2 + 5x - 2$

41. a. $h(x) = \frac{1}{2}x^2 + 4x + \frac{19}{3}$

$= \frac{1}{2}(x^2 + 8x) + \frac{19}{3}$ Factoring

$= \frac{1}{2}(x^2 + 8x + 16 - 16) + \frac{19}{3}$ Adding $16 - 16$
                                inside parentheses

$= \frac{1}{2}(x^2 + 8x + 16) + \frac{1}{2}(-16) + \frac{19}{3}$

$= \frac{1}{2}(x + 4)^2 - \frac{5}{3}$

The vertex is $\left(-4, -\frac{5}{3}\right)$, the axis of symmetry is

$x = -4$, and the graph opens upward since the

coefficient $\frac{1}{2}$ is positive. The minimum is $-\frac{5}{3}$.

b.

$h(x) = \frac{1}{2}x^2 + 4x + \frac{19}{3}$

43. $f(x) = x^2 - 6x + 3$

To find the x-intercepts, solve the equation

$0 = x^2 - 6x + 3$. Use the quadratic formula.

$x = \dfrac{-(-6) \pm \sqrt{(-6)^2 - 4 \cdot 1 \cdot 3}}{2 \cdot 1}$

$x = \dfrac{6 \pm \sqrt{24}}{2} = \dfrac{6 \pm 2\sqrt{6}}{2} = 3 \pm \sqrt{6}$

The x-intercepts are $(3 - \sqrt{6},\ 0)$ and $(3 + \sqrt{6},\ 0)$.

The y-intercept is $(0, f(0))$, or $(0, 3)$.

45. $g(x) = -x^2 + 2x + 3$

To find the x-intercepts, solve the equation

$0 = -x^2 + 2x + 3$. We factor.

$0 = -x^2 + 2x + 3$

$0 = x^2 - 2x - 3$ Multiplying by -1

$0 = (x - 3)(x + 1)$

$x = 3 \text{ or } x = -1$

The x-intercepts are $(-1, 0)$ and $(3, 0)$.

The y-intercept is $(0, g(0))$, or $(0, 3)$.

47. $f(x) = x^2 - 9x$

To find the x-intercepts, solve the equation

$0 = x^2 - 9x$. We factor.

$0 = x^2 - 9x$

$0 = x(x - 9)$

$x = 0 \text{ or } x = 9$

The x-intercepts are $(0, 0)$ and $(9, 0)$.

Since $(0, 0)$ is an x-intercept, we observe that $(0, 0)$ is also the y-intercept.

49. $h(x) = -x^2 + 4x - 4$

To find the x-intercepts, solve the equation

$0 = -x^2 + 4x - 4$. We factor.

$0 = -x^2 + 4x - 4$

$0 = x^2 - 4x + 4$ Multiplying by -1

$0 = (x - 2)(x - 2)$

$x = 2 \text{ or } x = 2$

The x-intercept is $(2, 0)$.

The y-intercept is $(0, h(0))$, or $(0, -4)$.

51. $g(x) = x^2 + x - 5$

To find the x-intercepts, solve the equation

$0 = x^2 + x - 5$. Use the quadratic formula.

$$x = \frac{-1 \pm \sqrt{1^2 - 4 \cdot 1 \cdot (-5)}}{2 \cdot 1}$$

$$x = \frac{-1 \pm \sqrt{21}}{2} = -\frac{1}{2} \pm \frac{\sqrt{21}}{2}$$

The x-intercepts are $\left(-\frac{1}{2} - \frac{\sqrt{21}}{2}, 0\right)$ and

$\left(-\frac{1}{2} + \frac{\sqrt{21}}{2}, 0\right)$.

The y-intercept is $(0, g(0))$, or $(0, -5)$.

53. $f(x) = 2x^2 - 4x + 6$

To find the x-intercepts, solve the equation

$0 = 2x^2 - 4x + 6$. We use the quadratic formula.

$$x = \frac{-(-4) \pm \sqrt{(-4)^2 - 4 \cdot 2 \cdot 6}}{2 \cdot 2}$$

$$x = \frac{4 \pm \sqrt{-32}}{4} = \frac{4 \pm 4i\sqrt{2}}{2} = 2 \pm 2i\sqrt{2}$$

There are no real-number solutions, so there is no x-intercept.

The y-intercept is $(0, f(0))$, or $(0, 6)$.

55. *Writing Exercise*

57. $(x^2 - 7)(x^2 + 3) = x^4 - 4x^2 - 21$

59. $\sqrt[3]{18x^4 y} \cdot \sqrt[3]{6x^2 y} = \sqrt[3]{18x^4 y \cdot 6x^2 y}$

$$= \sqrt[3]{27x^6 \cdot 4y^2}$$

$$= 3x^2 \sqrt[3]{4y^2}$$

61. $\dfrac{4a^2 - b^2}{2ab} \div \dfrac{2a^2 - ab - b^2}{6a^2}$

$$= \frac{4a^2 - b^2}{2ab} \cdot \frac{6a^2}{2a^2 - ab - b^2}$$

$$= \frac{(2a + b)(2a - b)}{2a \cdot b} \cdot \frac{2a \cdot 3a}{(2a + b)(a - b)}$$

$$= \frac{(2a+b)(2a-b)}{2a \cdot b} \cdot \frac{2a \cdot 3a}{(2a+b)(a-b)}$$

$$= \frac{3a(2a - b)}{b(a - b)}$$

63. *Writing Exercise.*

65. a. $f(x) = 2.31x^2 - 3.135x - 5.89$

$= 2.31(x^2 - 1.357142857x) - 5.89$

$= 2.31(x^2 - 1.357142857x$
$\qquad + 0.460459183 - 0.460459183) - 5.89$

$= 2.31(x^2 - 1.357142857x + 0.460459183)$
$\qquad + 2.31(-0.460459183) - 5.89$

$= 2.31(x - 0.678571428)^2 - 6.953660714$

Since the coefficient 2.31 is positive, the function has a minimum value. It is -6.953660714.

b. To find the x-intercepts, solve

$0 = 2.31x^2 - 3.135x - 5.89$.

$$x = \frac{-(-3.135) \pm \sqrt{(-3.135)^2 - 4(2.31)(-5.89)}}{2(2.31)}$$

$$x \approx \frac{3.135 \pm 8.015723611}{4.62}$$

$x \approx -1.056433682 \ \text{ or } \ x \approx 2.413576539$

The x-intercepts are $(-1.056433682, 0)$ and $(2.413576539, 0)$.

The y-intercept is $(0, f(0))$, or $(0, -5.89)$.

67. $f(x) = x^2 - x - 6$

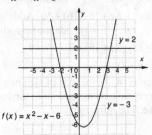

a. The solutions of $x^2 - x - 6 = 2$ are the first coordinates of the points of intersection of the graphs of $f(x) = x^2 - x - 6$ and $y = 2$. From the graph we see that the solutions are approximately -2.4 and 3.4.

b. The solutions of $x^2 - x - 6 = -3$ are the first coordinates of the points of intersection of the graphs of $f(x) = x^2 - x - 6$ and $y = -3$. From the graph we see that the solutions are approximately -1.3 and 2.3.

69. $f(x) = mx^2 - nx + p$

$= m\left(x^2 - \dfrac{n}{m}x\right) + p$

$= m\left(x^2 - \dfrac{n}{m}x + \dfrac{n^2}{4m^2} - \dfrac{n^2}{4m^2}\right) + p$

$= m\left(x - \dfrac{n}{2m}\right)^2 - \dfrac{n^2}{4m} + p$

$= m\left(x - \dfrac{n}{2m}\right)^2 + \dfrac{-n^2 + 4mp}{4m}$, or

$m\left(x - \dfrac{n}{2m}\right)^2 + \dfrac{4mp - n^2}{4m}$

71. The horizontal distance from $(-1, 0)$ to $(3, -5)$ is $|3 - (-1)|$, or 4, so by symmetry the other x-intercept is $(3 + 4, 0)$, or $(7, 0)$. Substituting the three ordered pairs $(-1, 0)$, $(3, -5)$, and $(7, 0)$ in the equation

$f(x) = ax^2 + bx + c$ yields a system of equations:

$0 = a - b + c,$
$-5 = 9a + 3b + c,$
$0 = 49a + 7b + c$

The solution of this system of equations is $\left(\dfrac{5}{16}, -\dfrac{15}{8}, -\dfrac{35}{16}\right)$, so $f(x) = \dfrac{5}{16}x^2 - \dfrac{15}{8}x - \dfrac{35}{16}$.

If we complete the square we find that this function can also be expressed as $f(x) = \dfrac{5}{16}(x - 3)^2 - 5$.

73. $f(x) = |x^2 - 1|$

We plot some points and draw the curve. Note that it will lie entirely on or above the x-axis since absolute value is never negative.

x	$f(x)$
-3	8
-2	3
-1	0
0	1
1	0
2	3
3	8

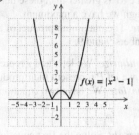

$f(x) = |x^2 - 1|$

75. $f(x) = |2(x-3)^2 - 5|$

We plot some points and draw the curve. Note that it will lie entirely on or above the x-axis since absolute value is never negative.

x	$f(x)$
-1	27
0	13
1	3
2	3
3	5
4	3
5	3
6	13

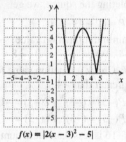

$f(x) = |2(x-3)^2 - 5|$

Exercise Set 8.8

1. True

3. True

5. True

7. *Familiarize and Translate*. We are given the formula
$p(x) = -0.2x^2 + 1.3x + 6.2$.

Carry out. To find the value of x for which $p(x)$ is a maximum, we first find $-\dfrac{b}{2a}$:

$$-\frac{b}{2a} = -\frac{1.3}{2(-0.2)} = 3.25, \text{ or } 3\frac{1}{4}$$

Now we find the maximum value of the function $p(3.25)$:

$$p(3.25) = -0.2(3.25)^2 + 1.3(3.25) + 6.2 = 8.3125$$

The minimum function value of about 8.3 occurs when $x = 3.25$.

Check. We can go over the calculations again. We could also solve the problem again by completing the square. The answer checks.

State. A calf's daily milk consumption is greatest at 3.25 weeks at about 8.3 lb of milk per day.

9. *Familiarize and Translate*. We want to find the value of x for which $C(x) = 0.1x^2 - 0.7x + 2.425$ is a minimum.

Carry out. We complete the square.

$C(x) = 0.1(x^2 - 7x + 12.25) + 2.425 - 1.225$
$C(x) = 0.1(x - 3.5)^2 + 1.2$

The minimum function value of 1.2 occurs when $x = 3.5$.

Check. Check a function value for x less than 3.5 and for x greater than 3.5.

$C(3) = 0.1(3)^2 - 0.7(3) + 2.425 = 1.225$
$C(4) = 0.1(4)^2 - 0.7(4) + 2.425 = 1.225$

Since 1.2 is less than these numbers, it looks as though we have a minimum.

State. The minimum average cost is $1.2 hundred, or $120. To achieve the minimum cost, 3.5 hundred, or 350 dulcimers should be built.

11. *Familiarize*. We make a drawing and label it.

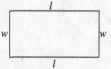

Perimeter: $2l + 2w = 720$ ft

Area: $A = l \cdot w$

Translate. We have a system of equations.

$2l + 2w = 720,$
$A = lw$

Carry out. Solving the first equation for l, we get $l = 360 - w$. Substituting for l in the second equation we get a quadratic function A:

$A = (360 - w)w$
$A = -w^2 + 360w$

Completing the square, we get

$$A = -(w - 180)^2 + 32,400$$

The maximum function value is 32,400. It occurs when w is 180. When $w = 180$, $l = 360 - 180$, or 180.

Check. We check a function value for w less than 180 and for w greater than 180.

$A(179) = -179^2 + 360 \cdot 179 = 32,399$
$A(181) = -181^2 + 360 \cdot 181 = 32,399$

Since 32,400 is greater than these numbers, it looks as though we have a maximum.

State. The maximum area occurs when the dimensions are 180 ft by 180 ft.

13. *Familiarize*. We make a drawing and label it.

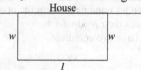

Translate. We have two equations.

$l + 2w = 60,$
$A = lw$

Carry out. Solve the first equation for l.

$l = 60 - 2w$

Substitute for l in the second equation.

$A = (60 - 2w)w$
$A = -2w^2 + 60w$

Completing the square, we get

$A = -2(w - 15)^2 + 450$.

The maximum function value of 450 occurs when $w = 15$. When $w = 15$, $l = 60 - 2 \cdot 15 = 30$.

Check. Check a function value for w less than 15 and for w greater than 15.

$$A(14) = -2 \cdot 14^2 + 60 \cdot 14 = 448$$
$$A(16) = -2 \cdot 16^2 + 60 \cdot 16 = 448$$

Since 450 is greater than these numbers, it looks as though we have a maximum.

State. The maximum area of 450 ft^2 will occur when the dimensions are 15 ft by 30 ft.

15. **Familiarize**. Let x represent the height of the file and y represent the width. We make a drawing.

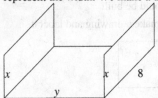

Translate. We have two equations.
$$2x + y = 14$$
$$V = 8xy$$

Carry out. Solve the first equation for y.
$$y = 14 - 2x$$

Substitute for y in the second equation.
$$V = 8x(14 - 2x)$$
$$V = -16x^2 + 112x$$

Completing the square, we get
$$V = -16\left(x - \frac{7}{2}\right)^2 + 196.$$

The maximum function value of 196 occurs when $x = \frac{7}{2}$. When $x = \frac{7}{2}$, $y = 14 - 2 \cdot \frac{7}{2} = 7$.

Check. Check a function value for x less than $\frac{7}{2}$ and for x greater than $\frac{7}{2}$.

$$V(3) = -16 \cdot 3^2 + 112 \cdot 3 = 192$$
$$V(4) = -16 \cdot 4^2 + 112 \cdot 4 = 192$$

Since 196 is greater than these numbers, it looks as though we have a maximum.

State. The file should be $\frac{7}{2}$ in., or 3.5 in. tall.

17. **Familiarize**. We let x and y represent the numbers, and we let P represent their product.

Translate. We have two equations.
$$x + y = 18,$$
$$P = xy$$

Carry out. Solving the first equation for y, we get $y = 18 - x$. Substituting for y in the second equation we get a quadratic function P:
$$P = x(18 - x)$$
$$P = -x^2 + 18x$$

Completing the square, we get
$$P = -(x - 9)^2 + 81.$$

The maximum function value is 81. It occurs when $x = 9$. When $x = 9$, $y = 18 - 9$, or 9.

Check. We can check a function value for x less than 9

and for x greater than 9.
$$P(10) = -10^2 + 18 \cdot 10 = 80$$
$$P(8) = -8^2 + 18 \cdot 8 = 80$$

Since 81 is greater than these numbers, it looks as though we have a maximum.

State. The maximum product of 81 occurs for the numbers 9 and 9.

19. **Familiarize**. We let x and y represent the two numbers, and we let P represent their product.

Translate. We have two equations.
$$x - y = 8,$$
$$P = xy$$

Carry out. Solve the first equation for x.
$$x = 8 + y$$

Substitute for x in the second equation.
$$P = (8 + y)y$$
$$P = y^2 + 8y$$

Completing the square, we get
$$P = (y + 4)^2 - 16.$$

The minimum function value is -16. It occurs when $y = -4$. When $y = -4$, $x = 8 + (-4)$, or 4.

Check. Check a function value for y less than -4 and for y greater than -4.
$$P(-5) = (-5)^2 + 8(-5) = -15$$
$$P(-3) = (-3)^2 + 8(-3) = -15$$

Since -16 is less than these numbers, it looks as though we have a minimum.

State. The minimum product of -16 occurs for the numbers 4 and -4.

21. From the results of Exercises 17 and 18, we might observe that the numbers are -5 and -5 and that the maximum product is 25. We could also solve this problem as follows.

Familiarize. We let x and y represent the two numbers, and we let P represent their product.

Translate. We have two equations.
$$x + y = -10,$$
$$P = xy$$

Carry out. Solve the first equation for y.
$$y = -10 - x$$

Substitute for y in the second equation.
$$P = x(-10 - x)$$
$$P = -x^2 - 10x$$

Completing the square, we get
$$P = -(x + 5)^2 + 25$$

The maximum function value is 25. It occurs when $x = -5$. When $x = -5$, $y = -10 - (-5)$, or -5.

Check. Check a function value for x less than -5 and for x greater than -5.
$$P(-6) = -(-6)^2 - 10(-6) = 24$$
$$P(-4) = -(-4)^2 - 10(-4) = 24$$

Since 25 is greater than these numbers, it looks as though we have a maximum.

State. The maximum product of 25 occurs for the numbers -5 and -5.

23. The data points rise and then fall. The graph appears to represent a quadratic function that opens downward.

Thus a quadratic function $f(x) = ax^2 + bx + c$, $a < 0$, might be used to model the data.

25. The data points rise. The graph does not appear to represent a quadratic function in which the data points would rise and then fall or vice versa. Thus a linear function $f(x) = mx + b$ might be used to model the data.

27. The data points do not represent a linear or quadratic pattern. Thus, it does not appear that the data can be modeled with either a quadratic or a linear function.

29. The data points fall and then rise. The graph appears to represent a quadratic function that opens upward. Thus a quadratic function $f(x) = ax^2 + bx + c$, $a > 0$, might be used to model the data.

31. The data points rise. The graph does not appear to represent a quadratic function in which the data points would rise and then fall or vice versa. Thus a linear function $f(x) = mx + b$ might be used to model the data.

33. The data points fall. The graph does not appear to represent a quadratic function in which the data points would rise and then fall or vice versa. Thus a linear function $f(x) = mx + b$ might be used to model the data.

35. We look for a function of the form

$f(x) = ax^2 + bx + c$. Substituting the data points, we get

$$4 = a(1)^2 + b(1) + c,$$
$$-2 = a(-1)^2 + b(-1) + c,$$
$$13 = a(2)^2 + b(2) + c,$$

or

$$4 = a + b + c,$$
$$-2 = a - b + c,$$
$$13 = 4a + 2b + c.$$

Solving this system, we get
$a = 2$, $b = 3$, and $c = -1$.
Therefore the function we are looking for is

$$f(x) = 2x^2 + 3x - 1.$$

37. We look for a function of the form

$f(x) = ax^2 + bx + c$. Substituting the data points, we get

$$0 = a(2)^2 + b(2) + c,$$
$$3 = a(4)^2 + b(4) + c,$$
$$-5 = a(12)^2 + b(12) + c,$$

or

$$0 = 4a + 2b + c,$$
$$3 = 16a + 4b + c,$$
$$-5 = 144a + 12b + c.$$

Solving this system, we get

$$a = -\frac{1}{4}, b = 3, c = -5.$$

Therefore the function we are looking for is

$$f(x) = -\frac{1}{4}x^2 + 3x - 5.$$

39. a. *Familiarize*. We look for a function of the form

$A(s) = as^2 + bs + c$, where $A(s)$ represents the number of nighttime accidents (for every 200 million km) and s represents the travel speed (in km/h).

Translate. We substitute the given values of s and $A(s)$.

$$400 = a(60)^2 + b(60) + c,$$
$$250 = a(80)^2 + b(80) + c,$$
$$250 = a(100)^2 + b(100) + c,$$

or

$$400 = 3600a + 60b + c,$$
$$250 = 6400a + 80b + c,$$
$$250 = 10,000a + 100b + c.$$

Carry out. Solving the system of equations, we get

$$a = \frac{3}{16}, b = -\frac{135}{4}, c = 1750.$$

Check. Recheck the calculations.
State. The function

$A(s) = \frac{3}{16}s^2 - \frac{135}{4}s + 1750$ fits the data.

b. Find $A(50)$.

$$A(50) = \frac{3}{16}(50)^2 - \frac{135}{4}(50) + 1750 = 531.25$$

About 531 accidents for every 200 million km driven, occur at 50 km/h.

41. *Familiarize*. Think of a coordinate system placed on the drawing in the text with the origin at the point where the arrow is released. Then three points on the arrow's parabolic path are (0, 0), (63, 27), and (126, 0). We look for a function of the form

$h(d) = ad^2 + bd + c$, where $h(d)$ represents the arrow's height and d represents the distance the arrow has traveled horizontally.

Translate. We substitute the values given above for d and $h(d)$.

$$0 = a \cdot 0^2 + b \cdot 0 + c,$$
$$27 = a \cdot 63^2 + b \cdot 63 + c,$$
$$0 = a \cdot 126^2 + b \cdot 126 + c$$

or

$$0 = c,$$
$$27 = 3969a + 63b + c,$$
$$0 = 15,876a + 126b + c$$

Carry out. Solving the system of equations, we get
$a \approx -0.0068$, $b \approx 0.8571$, and $c = 0$.
Check. Recheck the calculations.

State. The function $h(d) = -0.0068d^2 + 0.8571d$ expresses the arrow's height as a function of the distance it has traveled horizontally.

43. *Writing Exercise.*

45. $y = -\frac{1}{3}x + 16$

47. $m = \frac{0-8}{10-4} = -\frac{8}{6} = -\frac{4}{3}$

$$y - 0 = -\frac{4}{3}(x - 10)$$
$$y = -\frac{4}{3}x + \frac{40}{3}$$

49. $2x + y = 3$
$$y = -2x + 3$$

Slope of perpendicular line: $\frac{1}{2}$

$$y = \frac{1}{2}x - 6$$

51. *Writing Exercise.*

53. *Familiarize.* Position the bridge on a coordinate system as shown with the vertex of the parabola at (0, 30).

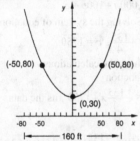

We find a function of the form $y = ax^2 + bx + c$ which represents the parabola containing the points (0, 30), (−50, 80), and (50, 80).

Translate. Substitute for x and y.

$$30 = a \cdot 0^2 + b \cdot 0 + c,$$
$$80 = a(-50)^2 + b(-50) + c,$$
$$80 = a(50)^2 + b(50) + c,$$

or

$$30 = c,$$
$$80 = 2500a - 50b + c,$$
$$80 = 2500a + 50b + c.$$

Carry out. Solving the system of equations, we get
$$a = 0.02, \ b = 0, \ c = 30.$$

The function $y = 0.02x^2 + 30$ represents the parabola.

Because the cable supports are 160 ft apart, the tallest supports are positioned 160/2, or 80 ft, to the left and right of the midpoint. This means that the longest vertical cables occur at $x = -80$ and $x = 80$. For $x = \pm 80$,

$$y = 0.02(\pm 80)^2 + 30$$
$$= 128 + 30$$
$$= 158 \text{ ft}$$

Check. We go over the calculations.
State. The longest vertical cables are 158 ft long.

55. *Familiarize.* Let x represent the number of 25 increases in the admission price. Then $10 + 0.25x$ represents the

admission price, and $80 - x$ represents the corresponding average attendance. Let R represent the total revenue.

Translate. Since the total revenue is the product of the cover charge and the number attending a show, we have the following function for the amount of money the owner makes.
$$R(x) = (10 + 0.25x)(80 - x), \text{ or}$$
$$R(x) = -0.25x^2 + 10x + 800$$

Carry out. Completing the square, we get
$$R(x) = -0.25(x - 20)^2 + 900$$

The maximum function value of 900 occurs when $x = 20$. The owner should charge \$10 + \$0.25(20), or \$15.

Check. We check a function value for x less than 20 and for x greater than 20.
$$R(19) = -0.25(19)^2 + 10 \cdot 19 + 800 = 899.75$$
$$R(21) = -0.25(21)^2 + 10 \cdot 21 + 800 = 899.75$$

Since 900 is greater than these numbers, it looks as though we have a maximum.
State. The owner should charge \$15.

57. *Familiarize.* We add labels to the drawing in the text.

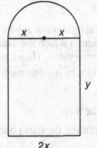

The perimeter of the semicircular portion of the window is $\frac{1}{2} \cdot 2\pi x$, or πx. The perimeter of the rectangular portion is $y + 2x + y$, or $2x + 2y$. The area of the semicircular portion of the window is $\frac{1}{2} \cdot \pi x^2$, or $\frac{\pi}{2}x^2$. The area of the rectangular portion is $2xy$.

Translate. We have two equations, one giving perimeter of the window and the other giving the area.
$$\pi x + 2x + 2y = 24,$$
$$A = \frac{\pi}{2}x^2 + 2xy$$

Carry out. Solve the first equation for y.
$$\pi x + 2x + 2y = 24$$
$$2y = 24 - \pi x - 2x$$
$$y = 12 - \frac{\pi x}{2} - x$$

Substitute for y in the second equation.
$$A = \frac{\pi}{2}x^2 + 2x\left(12 - \frac{\pi x}{2} - x\right)$$
$$A = \frac{\pi}{2}x^2 + 24x - \pi x^2 - 2x^2$$
$$A = -2x^2 - \frac{\pi}{2}x^2 + 24x$$
$$A = -\left(2x + \frac{\pi}{2}\right)x^2 + 24x$$

Completing the square, we get

$$A = -\left(2 + \frac{\pi}{2}\right)\left(x^2 + \frac{24}{-\left(2 + \frac{\pi}{2}\right)}x\right)$$

$$A = -\left(2 + \frac{\pi}{2}\right)\left(x^2 - \frac{48}{4 + \pi}x\right)$$

$$A = -\left(2 + \frac{\pi}{2}\right)\left(x - \frac{24}{4 + \pi}\right)^2 + \left(\frac{24}{4 + \pi}\right)^2$$

The maximum function value occurs when

$x = \dfrac{24}{4 + \pi}$. When $x = \dfrac{24}{4 + \pi}$,

$$y = 12 - \frac{\pi}{2}\left(\frac{24}{4 + \pi}\right) - \frac{24}{4 + \pi}$$

$$= \frac{48 + 12\pi}{4 + \pi} - \frac{12\pi}{4 + \pi} - \frac{24}{4 + \pi} = \frac{24}{4 + \pi}$$

Check. Recheck the calculations.
State. The radius of the circular portion of the window and the height of the rectangular portion should each be

$\dfrac{24}{4 + \pi}$ ft.

59. a. Enter the data and use the quadratic regression operation on a graphing calculator. We get

$m(x) = 0.4344570337x^2 - 25.466889x + 420.6656195$

where x is the number of degrees latitude north.

b. At 46°N, $x = 46$.

$m(46) \approx 169$ cases per 100,000 population

Exercise Set 8.9

1. The solutions of $(x - 3)(x + 2) = 0$ are 3 and -2 and for a test value in $[-2, 3]$, say 0, $(x - 3)(x + 2)$ is negative so the statement is true. (Note that the endpoints must be included in the solution set because the inequality symbol is $\leq$.)

3. The solutions of $(x - 1)(x - 6) = 0$ are 1 and 6. For a value of x less than 1, say 0, $(x - 1)(x - 6)$ is positive; for a value of x greater than 6, say 7, $(x - 1)(x - 6)$ is also positive. Thus, the statement is true. (Note that the endpoints of the intervals are not included because the inequality symbol is $>$.)

5. Since $x + 2 = 0$ when $x = -2$ and $x - 3 = 0$ when $x = 3$, the statement is false.

7. $p(x) \leq 0$ when $-4 \leq x \leq \dfrac{3}{2}$,

$\left[-4, \dfrac{3}{2}\right]$ or $\left\{x \middle| -4 \leq x \leq \dfrac{3}{2}\right\}$

9. $x^4 + 12x > 3x^2 + 4x^2$ is equivalent to

$x^4 - 3x^2 - 4x^2 + 12x > 0$, which is the graph in the text.

$p(x) > 0$ when $(-\infty, -2) \cup (0, 2) \cup (3, \infty)$ or

$\{x \mid x < -2 \text{ or } 0 < x < 2 \text{ or } x > 3\}$

11. $\dfrac{x - 1}{x + 2} < 3$ is equivalent to finding the values of x for which the graph $r(x)$ is less than 3, or below $g(x)$.

$\left(-\infty, -\dfrac{7}{2}\right) \cup (-2, \infty)$ or $\left\{x \middle| x < -\dfrac{7}{2} \text{ or } x > -2\right\}$

13. $(x - 6)(x - 5) < 0$

The solutions of $(x - 6)(x - 5) = 0$ are 5 and 6. They are not solutions of the inequality, but they divide the real number line in a natural way. The product $(x - 6)(x - 5)$ is positive or negative, for values other than 5 and 6, depending on the signs of the factors $x - 6$ and $x - 5$.

$x - 6 > 0$ when $x > 6$ and $x - 6 < 0$ when $x < 6$.
$x - 5 > 0$ when $x > 5$ and $x - 5 < 0$ when $x < 5$
We make a diagram.

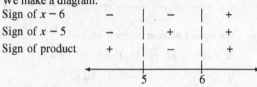

Sign of $x - 6$	$-$		$-$		$+$
Sign of $x - 5$	$-$		$+$		$+$
Sign of product	$+$		$-$		$+$

For the product $(x - 6)(x - 5)$ to be negative, one factor must be positive and the other negative. We see from the diagram that numbers satisfying $5 < x < 6$ are solutions. The solution set of the inequality is $(5, 6)$ or $\{x \mid 5 < x < 6\}$.

15. $(x + 7)(x - 2) \geq 0$

The solutions of $(x + 7)(x - 2) = 0$ are -7 and 2. They divide the number line into three intervals as shown:

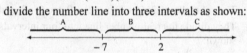

We try test numbers in each interval.
A: Test -8, $f(-8) = (-8 + 7)(-8 - 2) = 10$
B: Test 0, $f(0) = (0 + 7)(0 - 2) = -14$
C: Test 3, $f(3) = (3 + 7)(3 - 2) = 10$
Since $f(-8)$ and $f(3)$ are positive, the function value will be positive for all numbers in the intervals containing -8 and 3. The inequality symbol is $\leq$, so we need to include the endpoints. The solution set is $(-\infty, -7] \cup [2, \infty)$, or $\{x \mid x \leq -7 \text{ or } x \geq 2\}$.

17. $x^2 - x - 2 > 0$
$(x + 1)(x - 2) > 0$ Factoring
The solutions of $(x + 1)(x - 2) = 0$ are -1 and 2. They divide the number line into three intervals as shown:

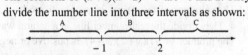

We try test numbers in each interval.
A: Test -2, $f(-2) = (-2 + 1)(-2 - 2) = 4$
B: Test 0, $f(0) = (0 + 1)(0 - 2) = -2$
C: Test 3, $f(3) = (3 + 1)(3 - 2) = 4$
Since $f(-2)$ and $f(3)$ are positive, the function value will be positive for all numbers in the intervals containing -2 and 3. The solution set is $(-\infty, -1) \cup (2, \infty)$, or $\{x \mid x < -1 \text{ or } x > 2\}$.

19. $x^2 + 4x + 4 < 0$

$(x+2)^2 < 0$

Observe that $(x+2)^2 \geq 0$ for all values of x. Thus, the solution set is $\varnothing$.

21. $x^2 - 4x \leq 3$

$x^2 - 4x + 4 \leq 3 + 4$

$(x-2)^2 \leq 7$

$x - 2 \leq \pm\sqrt{7}$

$x \leq 2 \pm \sqrt{7}$

The solutions of $x^2 - 4x - 3 \leq 0$ are $2 \pm \sqrt{7}$. They divide the number line into three intervals as shown:

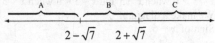

We try test numbers in each interval.

A: Test -1, $f(-1) = (-1)^2 - 4(-1) - 3 = 2$

B: Test 0, $f(0) = 0^2 - 4(0) - 3 = -3$

C: Test 5, $f(5) = 5^2 - 4(5) - 3 = 2$

Since $f(0)$ is negative, the function value will be negative for all numbers in the interval containing 0. The solution set is $\left[2 - \sqrt{7},\ 2 + \sqrt{7}\right]$, or $\left\{x \mid 2 - \sqrt{7} \leq x \leq 2 + \sqrt{7}\right\}$.

23. $3x(x+2)(x-2) < 0$

The solutions of $3x(x+2)(x-2) = 0$ are 0, -2, and 2. They divide the real-number line into four intervals as shown:

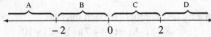

We try test numbers in each interval.

A: Test -3, $f(-3) = 3(-3)(-3+2)(-3-2) = -45$

B: Test -1, $f(-1) = 3(-1)(-1+2)(-1-2) = 9$

C: Test 1, $f(1) = 3(1)(1+2)(1-2) = -9$

D: Test 3, $f(3) = 3(3)(3+2)(3-2) = 45$

Since $f(-3)$ and $f(1)$ are negative, the function value will be negative for all numbers in the intervals containing -3 and 1. The solution set is $(-\infty, -2) \cup (0, 2)$, or $\{x \mid x < -2 \text{ or } 0 < x < 2\}$.

25. $(x-1)(x+2)(x-4) \geq 0$

The solutions of $(x-1)(x+2)(x-4) = 0$ are 1, -2, and 4. They divide the real-number line in a natural way. The product $(x-1)(x+2)(x-4)$ is positive or negative depending on the signs of $x-1$, $x+2$, and $x-4$.

Sign of $x - 1$	$-$	$-$	$+$	$+$
Sign of $x + 2$	$-$	$+$	$+$	$+$
Sign of $x - 4$	$-$	$-$	$-$	$+$
Sign of product	$-$	$+$	$-$	$+$
	-2	1	4	

A product of three numbers is positive when all three factors are positive or when two are negative and one is positive. Since the $\geq$ symbol allows for equality, the endpoints -2, 1, and 4 are solutions. From the chart we

see that the solution set is $[-2, 1] \cup [4, \infty)$, or $\{x \mid -2 \leq x \leq 1 \text{ or } x \geq 4\}$.

27. $f(x) \geq 3$

$7 - x^2 \geq 3$

$-x^2 + 4 \geq 0$

$x^2 - 4 \leq 0$

$(x-2)(x+2) \leq 0$

The solutions of $(x-2)(x+2) = 0$ are 2 and -2. They divide the real-number line as shown below.

Sign of $x - 2$	$-$	$-$	$+$
Sign of $x + 2$	$-$	$+$	$+$
Sign of product	$+$	$-$	$+$
	-2	2	

Because the inequality symbol is $\leq$, we must include the endpoints in the solution set. From the chart, we see that the solution set is $[-2, 2]$, or $\{x \mid -2 \leq x \leq 2\}$.

29. $g(x) > 0$

$(x-2)(x-3)(x+1) > 0$

The solutions of $(x-2)(x-3)(x+1) = 0$ are 2, 3, and -1. They divide the real-number line into four intervals as shown below.

We try test numbers in each interval.

A: Test -2, $f(-2) = (-2-2)(-2-3)(-2+1) = -20$

B: Test 0, $f(0) = (0-2)(0-3)(0+1) = 6$

C: Test $\frac{5}{2}$, $f\left(\frac{5}{2}\right) = \left(\frac{5}{2}-2\right)\left(\frac{5}{2}-3\right)\left(\frac{5}{2}+1\right) = -\frac{7}{8}$

D: Test 4, $f(4) = (4-2)(4-3)(4+1) = 10$

The function value will be positive for all numbers in intervals B and D. The solution set is $(-1, 2) \cup (3, \infty)$, or $\{x \mid -1 < x < 2 \text{ or } x > 3\}$.

31. $F(x) \leq 0$

$x^3 - 7x^2 + 10x \leq 0$

$x(x^2 - 7x + 10) \leq 0$

$x(x-2)(x-5) \leq 0$

The solutions of $x(x-2)(x-5) = 0$ are 0, 2, and 5. They divide the real-number line as shown below.

Sign of x	$-$	$+$	$+$	$+$
Sign of $x - 2$	$-$	$-$	$+$	$+$
Sign of $x - 5$	$-$	$-$	$-$	$+$
Sign of product	$-$	$+$	$-$	$+$
	0	2	5	

Because the inequality symbol is $\leq$ we must include the endpoints in the solution set. From the chart we see that the solution set is $(-\infty, 0] \cup [2, 5]$ or $\{x \mid x \leq 0 \text{ or } 2 \leq x \leq 5\}$.

33. $\dfrac{1}{x-5} < 0$

We write the related equation by changing the $<$ symbol to $=$.

$\dfrac{1}{x-5} = 0$

We solve the related equation.

$(x-5) \cdot \dfrac{1}{x-5} = (x-5) \cdot 0$

$1 = 0$

The related equation has no solution.
Next we find the values that make the denominator 0
by setting the denominator equation to 0 and solving:

$x - 5 = 0$

$x = 5$

We use 5 to divide the number line into two intervals as
shown:

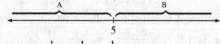

A: Test 0, $\dfrac{1}{0-5} = \dfrac{1}{-5} = -\dfrac{1}{5} < 0$

The number 0 is a solution of the inequality, so the
interval A is part of the solution set.

B: Test 6, $\dfrac{1}{6-5} = 1 \not< 0$

The number 6 is not a solution of the inequality, so the
interval B is part of the solution set.

The solution set is $(-\infty,\ 5)$, or $\{x \mid x < 5\}$.

35. $\dfrac{x+1}{x-3} \ge 0$

Solve the related equation.

$\dfrac{x+1}{x-3} = 0$

$x + 1 = 0$

$x = -1$

Find the values that make the denominator 0.

$x - 3 = 0$

$x = 3$

Use the numbers −1 and 3 to divide the number line
into intervals as shown:

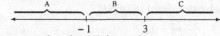

Try test numbers in each interval.

A: Test −2, $\dfrac{-2+1}{-2-3} = \dfrac{-1}{-5} = \dfrac{1}{5} > 0$

The number −2 is a solution of the inequality, so the
interval A is part of the solution set.

B: Test 0, $\dfrac{0+1}{0-3} = \dfrac{1}{-3} = -\dfrac{1}{3} \not> 0$

The number 0 is not a solution of the inequality, so the
interval B is not part of the solution set.

C: Test 4, $\dfrac{4+1}{4-3} = \dfrac{5}{1} = 5 > 0$

The number 4 is a solution of the inequality, so the
interval C is part of the solution set.

The solution set includes intervals A and C. The
number −1 is also included since the inequality symbol
is ≥ and −1 is the solution of the related equation. The

number 3 is not included since $\dfrac{x+1}{x-3}$ is undefined for

$x = 3$. The solution set is $(-\infty, -1] \cup (3,\ \infty)$, or

$\{x \mid x \le -1 \ or \ x > 3\}$.

37. $\dfrac{x+1}{x+6} \ge 1$

Solve the related equation $\dfrac{x+1}{x+6} = 1$

$x + 1 = x + 6$

$1 = 6$

The related equation has no solution.
Find the values that make the denominator 0.

$x + 6 = 0$

$x = -6$

Use the number −6 to divide the number line into two
intervals.

Try test numbers in each interval.

A: Test −7, $\dfrac{-7+1}{-7+6} = \dfrac{-6}{-1} = 6 > 1$.

The number −7 is a solution of the inequality, so the
interval A is part of the solution set.

B: Test 0, $\dfrac{0+1}{0+6} = \dfrac{1}{6} \not> 1$

The number 0 is not a solution of the inequality, so the
interval B is not part of the solution set. The number −6

is not included in the solution set since $\dfrac{x+1}{x+6}$ is

undefined for $x = -6$. The solution set is $(-\infty, -6)$, or
$\{x \mid x < -6\}$.

39. $\dfrac{(x-2)(x+1)}{x-5} \le 0$

Solve the related equation.

$\dfrac{(x-2)(x+1)}{x-5} = 0$

$(x-2)(x+1) = 0$

$x = 2 \ or \ x = -1$

Find the values that make the denominator 0.

$x - 5 = 0$

$x = 5$

Use the numbers 2, −1, and 5 to divide the number line
into intervals as shown:

Try test numbers in each interval.

A: Test −2, $\dfrac{(-2-2)(-2+1)}{-2-5} = \dfrac{-4(-1)}{-7} = -\dfrac{4}{7} \le 0$

Interval A is part of the solution set.

B: Test 0, $\dfrac{(0-2)(0+1)}{0-5} = \dfrac{-2 \cdot 1}{-5} = \dfrac{2}{5} \not\le 0$

Interval B is not part of the solution set.

C: Test 3, $\dfrac{(3-2)(3+1)}{3-5} = \dfrac{1 \cdot 4}{-2} = -2 \le 0$

Interval C is part of the solution set.

D: Test 6, $\dfrac{(6-2)(6+1)}{6-5} = \dfrac{4 \cdot 7}{1} = 28 \not\le 0$

Interval D is not part of the solution set.
The solution set includes intervals A and C. The
numbers −1 and 2 are also included since the inequality
symbol is ≤ and −1 and 2 are the solutions of the related
equation. The number 5 is not included since

$\dfrac{(x-2)(x+1)}{x-5}$ is undefined for $x = 5$. The solution set is

$(-\infty, -1] \cup [2,\ 5)$, or $\{x \mid x \le -1 \ or \ 2 \le x < 5\}$.

41. $\dfrac{x}{x+3} \geq 0$

Solve the related equation.

$$\frac{x}{x+3} = 0$$
$$x = 0$$

Find the values that make the denominator 0.

$$x + 3 = 0$$
$$x = -3$$

Use the numbers 0 and –3 to divide the number line into intervals as shown.

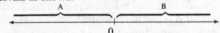

Try test numbers in each interval.

A: Test –4, $\dfrac{-4}{-4+3} = \dfrac{-4}{-1} = 4 \geq 0$

Interval A is part of the solution set.

B: Test –1, $\dfrac{-1}{-1+3} = \dfrac{-1}{2} = -\dfrac{1}{2} \ngeq 0$

Interval B is not part of the solution set.

C: Test 1, $\dfrac{1}{1+3} = \dfrac{1}{4} \geq 0$

The interval C is part of the solution set.
The solution set includes intervals A and C. The number 0 is also included since the inequality symbol is $\geq$ and 0 is the solution of the related equation. The number –3 is not included since $\dfrac{x}{x+3}$ is undefined for $x = -3$. The solution set is $(-\infty, -3) \cup [0, \infty)$, or $\{x \mid x < -3 \text{ or } x \geq 0\}$.

43. $\dfrac{x-5}{x} < 1$

Solve the related equation.

$$\frac{x-5}{x} = 1$$
$$x - 5 = x$$
$$-5 = 0$$

The related equation has no solution.
Find the values that make the denominator 0.

$$x = 0$$

Use the number 0 to divide the number line into two intervals as shown.

Try test numbers in each interval.

A: Test –1, $\dfrac{-1-5}{-1} = \dfrac{-6}{-1} = 6 \nless 1$

Interval A is not part of the solution set.

B: Test 1, $\dfrac{1-5}{1} = \dfrac{-4}{1} = -4 < 1$

Interval B is part of the solution set.
The solution set is $(0, \infty)$ or $\{x \mid x > 0\}$.

45. $\dfrac{x-1}{(x-3)(x+4)} \leq 0$

Solve the related equation.

$$\frac{x-1}{(x-3)(x+4)} = 0$$
$$x - 1 = 0$$
$$x = 1$$

Find the values that make the denominator 0.

$$(x-3)(x+4) = 0$$
$$x = 3 \text{ or } x = -4$$

Use the numbers 1, 3, and –4 to divide the number line into intervals as shown:

Try test numbers in each interval.

A: Test –5, $\dfrac{-5-1}{(-5-3)(-5+4)} = \dfrac{-6}{-8(-1)} = -\dfrac{3}{4} < 0$

Interval A is part of the solution set.

B: Test 0, $\dfrac{0-1}{(0-3)(0+4)} = \dfrac{-1}{-3 \cdot 4} = \dfrac{1}{12} \nleq 0$

Interval B is not part of the solution set.

C: Test 2, $\dfrac{2-1}{(2-3)(2+4)} = \dfrac{1}{-1 \cdot 6} = -\dfrac{1}{6} < 0$

Interval C is part of the solution set.

D: Test 4, $\dfrac{4-1}{(4-3)(4+4)} = \dfrac{3}{1 \cdot 8} = \dfrac{3}{8} \nleq 0$

Interval D is not part of the solution set.
The solution set includes intervals A and C. The number 1 is also included since the inequality symbol is $\leq$ and 1 is the solution of the related equation. The numbers –4 and 3 are not included since $\dfrac{x-1}{(x-3)(x+4)}$ is undefined for $x = -4$ and for $x = 3$.
The solution set is $(-\infty, -4) \cup [1, 3)$, or $\{x \mid x < -4 \text{ or } 1 \leq x < 3\}$.

47. $f(x) \geq 0$

$$\frac{5-2x}{4x+3} \geq 0$$

Solve the related equation.

$$\frac{5-2x}{4x+3} = 0$$
$$5 - 2x = 0$$
$$5 = 2x$$
$$\frac{5}{2} = x$$

Find the values that make the denominator 0.

$$4x + 3 = 0$$
$$4x = -3$$
$$x = -\frac{3}{4}$$

Use the numbers $\dfrac{5}{2}$ and $-\dfrac{3}{4}$ to divide the number line as shown:

Try test numbers in each interval.

A: Test –1, $\dfrac{5-2(-1)}{4(-1)+3} = -7 \ngeq 0$

Interval A is not part of the solution set.

B: Test 0, $\dfrac{5-2 \cdot 0}{4 \cdot 0 + 3} = \dfrac{5}{3} > 0$

Interval B is part of the solution set.

C: Test 3, $\frac{5-2\cdot3}{4\cdot3+3} = -\frac{1}{15} \not\geq 0$

Interval C is not part of the solution set.

The solution set includes interval B. The number $\frac{5}{2}$ is also included since the inequality symbol is $\geq$ and $\frac{5}{2}$ is the solution of the related equation. The number $-\frac{3}{4}$ is not included since $\frac{5-2x}{4x+3}$ is undefined for $x = -\frac{3}{4}$.

The solution set is $\left(-\frac{3}{4}, \frac{5}{2}\right]$, or $\left\{x\middle|-\frac{3}{4} < x \leq \frac{5}{2}\right\}$.

49. $G(x) \leq 1$

$\frac{1}{x-2} \leq 1$

Solve the related equation.

$\frac{1}{x-2} = 1$

$1 = x - 2$

$3 = x$

Find the values of x that make the denominator 0.

$x - 2 = 0$

$x = 2$

Use the numbers 2 and 3 to divide the number line as shown.

A   B   C

            2   3

Try a test number in each interval.

A: Test 0, $\frac{1}{0-2} = -\frac{1}{2} \leq 1$

Interval A is part of the solution set.

B: Test $\frac{5}{2}$, $\frac{1}{\frac{5}{2}-2} = \frac{1}{\frac{1}{2}} = 2 \not\leq 1$

Interval B is not part of the solution set.

C: Test 4, $\frac{1}{4-2} = \frac{1}{2} \leq 1$

Interval C is part of the solution set.

The solution set includes intervals A and C. The number 3 is also included since the inequality symbol is $\leq$ and 3 is the solution of the related equation. The number 2 is not included since $\frac{1}{x-2}$ is undefined for $x = 2$. The solution set is $(-\infty, 2) \cup [3, \infty)$, or $\{x | x < 2 \text{ or } x \geq 3\}$.

51. *Writing Exercise.*

53. *Familiarize.* Let x represent the number of hours spent on educational activities. Then $x + 0.7$ represents the number of hours spent on leisure activities.

Translate.

$x + (x + 0.7) = 7.1$

Carry out. We solve the equation.

$x + (x + 0.7) = 7.1$

$2x + 0.7 = 7.1$

$2x = 6.4$

$x = 3.2$

If $x = 3.2$, then $x + 0.7 = 3.2 + 0.7$, or 3.9.

Check. 3.9 is 0.7 more than 3.2. Also, $3.2 + 3.9 = 7.1$ hours. The answer checks.

State. The student spends 3.2 hr on educational activities.

55. *Familiarize.* Let $n =$ the number of miles. Then the mileage fee is $0.4n$. There is a flat fee of \$70.

Translate. We write an inequality stating that the truck rental costs less than \$90.

$0.4n + 70 < 90$

Carry out.

$0.4n + 70 < 90$

$0.4n < 20$

$n < 50$

Check. We can do a partial check by substituting a value for n less than 50 When $n = 49$, the truck rental costs $0.4(49) + 70$, or \$89.60, so the mileage is less than the budget of \$90. When $n = 51$, the truck rental costs $0.4(51) + 70$, or \$90.40, so the mileage is more expensive than the budget of \$90.

State. The mileage should be no greater than 50 mi.

57. *Writing Exercise.*

59. $x^2 + 2x < 5$

$x^2 + 2x - 5 < 0$

Using the quadratic formula, we find that the solutions of the related equation are $x = -1 \pm \sqrt{6}$. These numbers divide the real-number line into three intervals as shown:

A   B   C

        $-1-\sqrt{6}$ $-1+\sqrt{6}$

We try test numbers in each interval.

A: Test -4, $f(-4) = (-4)^2 + 2(-4) - 5 = 3$

B: Test 0, $f(0) = 0^2 + 2 \cdot 0 - 5 = -5$

C: Test 2, $f(2) = 2^2 + 2 \cdot 2 - 5 = 3$

The function value will be negative for all numbers in interval B. The solution set is $\left(-1-\sqrt{6}, -1+\sqrt{6}\right)$, or $\left\{x\middle|-1-\sqrt{6} < x < -1+\sqrt{6}\right\}$.

61. $x^4 + 3x^2 \leq 0$

$x^2(x^2 + 3) \leq 0$

$x^2 = 0$ for $x = 0$, $x^2 > 0$ for $x \neq 0$, $x^2 + 3 > 0$ for all x

The solution set is $\{0\}$.

63. a. $-3x^2 + 630x - 6000 > 0$

$x^2 - 210x + 2000 < 0$ Multiplying by $-\frac{1}{3}$

$(x - 200)(x - 10) < 0$

The solutions of $f(x) = (x - 200)(x - 10) = 0$ are 200 and 10. They divide the number line as shown:

A   B   C

        10   200

A: Test 0, $f(0) = 0^2 - 210 \cdot 0 + 2000 = 2000$

B: Test 20, $f(20) = 20^2 - 210 \cdot 20 + 2000 = -1800$

C: Test 300, $f(300) = 300^2 - 210 \cdot 300 + 2000$
$$= 29,000$$
The company makes a profit for values of x such that $10 < x < 200$, or for values of x in the interval $(10, 200)$.

b. See part (a). Keep in mind that x must be nonnegative since negative numbers have no meaning in this application.
The company loses money for values of x such that $0 \le x < 10$ or $x > 200$, or for values of x in the interval $[0, 10) \cup (200, \infty)$.

65. We find values of n such that $N \ge 66$ *and* $N \le 300$.
For $N \ge 66$:
$$\frac{n(n-1)}{2} \ge 66$$
$$n(n-1) \ge 132$$
$$n^2 - n - 132 \ge 0$$
$$(n-12)(n+11) \ge 0$$
The solutions of $f(n) = (n-12)(n+11) = 0$ are 12 and -11. They divide the number line as shown:

$$-11 \quad\quad 12$$

However, only positive values of n have meaning in this exercise so we need only consider the intervals shown below:

$$\underset{0}{}\overset{A}{}\quad\underset{12}{}\overset{B}{}$$

A: Test 1, $f(1) = 1^2 - 1 - 132 = -132$

B: Test 20, $f(20) = 20^2 - 20 - 132 = 248$

Thus, $N \ge 66$ for $\{n \mid n \ge 12\}$.

For $N \le 300$:
$$\frac{n(n-1)}{2} \le 300$$
$$n(n-1) \le 600$$
$$n^2 - n - 600 \le 0$$
$$(n-25)(n+24) \le 0$$
The solutions of $f(n) = (n-25)(n+24) = 0$ are 25 and -24. They divide the number line as shown:

$$-24 \quad\quad 25$$

However, only positive values of n have meaning in this exercise so we need only consider the intervals shown below:

$$\underset{0}{}\overset{A}{}\quad\underset{25}{}\overset{B}{}$$

A: Test 1, $f(1) = 1^2 - 1 - 600 = -600$

B: Test 30, $f(30) = 30^2 - 30 - 600 = 270$

Thus, $N \le 300$ (and $n > 0$) for $\{n \mid 0 < n \le 25\}$.

Then $66 \le N \le 300$ for
$\{n \mid n$ is an integer *and* $12 \le n \le 25\}$.

67. From the graph we determine the following:
The solutions of $f(x) = 0$ are -2, 1, and 3.
The solution of $f(x) < 0$ is $(-\infty, -2) \cup (1, 3)$, or $\{x \mid x < -2 \ or \ 1 < x < 3\}$.
The solution of $f(x) > 0$ is $(-2, 1) \cup (3, \infty)$, or $\{x \mid -2 < x < 1 \ or \ x > 3\}$.

69. From the graph we determine the following:
$f(x)$ has no zeros.
The solutions of $f(x) < 0$ are $(-\infty, 0)$, or $\{x \mid x < 0\}$;
The solutions of $f(x) > 0$ are $(0, \infty)$, or $\{x \mid x > 0\}$.

71. From the graph we determine the following:
The solutions of $f(x) = 0$ are -1 and 0.
The solution of $f(x) < 0$ is $(-\infty, -3) \cup (-1, 0)$, or $\{x \mid x < -3 \ or \ -1 < x < 0\}$.
The solution of $f(x) > 0$ is $(-3, -1) \cup (0, 2) \cup (2, \infty)$, or $\{x \mid -3 < x < -1 \ or \ 0 < x < 2 \ or \ x > 2\}$.

73. For $f(x) = \sqrt{x^2 - 4x - 45}$, we find the domain:
$$x^2 - 4x - 45 \ge 0$$
$$(x+5)(x-9) \ge 0$$
The quadratic is nonnegative when $(-\infty, -5] \cup [9, \infty)$, or $\{x \mid x \le -5 \ or \ x \ge 9\}$.

75. For $f(x) = \sqrt{x^2 + 8x}$, we find the domain:
$$x^2 + 8x \ge 0$$
$$x(x+8) \ge 0$$
The quadratic is nonnegative when $(-\infty, -8] \cup [0, \infty)$, or $\{x \mid x \le -8 \ or \ x \ge 0\}$.

77. *Writing Exercise.* Answers may vary.
One example is the rational inequality $\dfrac{a-x}{x-b} \le 0$.

Chapter 8 Review

1. False; a quadratic equation could have one solution.

2. True

3. True

4. True

5. False; the vertex is $(-3, -4)$.

6. True

7. True; since the coefficient of x^2 is -2, the graph opens down and therefore has no minimum.

8. True

9. False; if the quadratic function has two different imaginary-number zeros, the graph will not have any x-intercepts.

10. True

11. $9x^2 - 2 = 0$

$9x^2 = 2$

$x^2 = \frac{2}{9}$

$x = \pm\sqrt{\frac{2}{9}} = \pm\frac{\sqrt{2}}{3}$

The solutions are $-\frac{\sqrt{2}}{3}$ and $\frac{\sqrt{2}}{3}$.

12. $8x^2 + 6x = 0$

$2x(4x + 3) = 0$

$2x = 0 \quad or \quad 4x + 3 = 0$

$x = 0 \quad or \quad x = -\frac{3}{4}$

13. $x^2 - 12x + 36 = 9$

$(x - 6)^2 = 9$

$x - 6 = \pm 3$

$x = 6 \pm 3$

The solutions are 3 and 9.

14. $x^2 - 4x + 8 = 0$

$x^2 - 4x + 4 = 4 - 8$

$(x - 2)^2 = -4$

$x - 2 = \pm 2i$

$x = 2 \pm 2i$

15. $x(3x + 4) = 4x(x - 1) + 15$

$3x^2 + 4x = 4x^2 - 4x + 15$

$0 = x^2 - 8x + 15$

$0 = (x - 3)(x - 5)$

$x - 3 = 0 \quad or \quad x - 5 = 0$

$x = 3 \quad or \quad x = 5$

The solutions are 3 and 5.

16. $x^2 + 9x = 1$

$x^2 + 9x - 1 = 0$

$a = 1, \ b = 9, \ c = -1$

$x = \frac{-9 \pm \sqrt{9^2 - 4 \cdot 1 \cdot (-1)}}{2 \cdot 1} = \frac{-9 \pm \sqrt{85}}{2}$

$x = -\frac{9}{2} \pm \frac{\sqrt{85}}{2}$

17. $x^2 - 5x - 2 = 0$

$a = 1, \ b = -5, \ c = -2$

$x = \frac{-(-5) \pm \sqrt{(-5)^2 - 4 \cdot 1 \cdot (-2)}}{2 \cdot 1} = \frac{5 \pm \sqrt{25 + 8}}{2}$

$x = \frac{5 \pm \sqrt{33}}{2}$

$x \approx -0.372, \ 5.372$

18. $4x^2 - 3x - 1 = 0$

$(4x + 1)(x - 1) = 0$

$4x + 1 = 0 \quad or \quad x - 1 = 0$

$x = -\frac{1}{4} \quad or \quad x = 1$

19. $\frac{1}{2} \cdot (-18) = -9; \quad (-9)^2 = 81$

$x^2 - 18x + 81 = (x - 9)^2$

20. $\frac{1}{2} \cdot \frac{3}{5} = \frac{3}{10}; \quad \left(\frac{3}{10}\right)^2 = \frac{9}{100}$

$x^2 + \frac{3}{5}x + \frac{9}{100} = \left(x + \frac{3}{10}\right)^2$

21. $x^2 - 6x + 1 = 0$

$x^2 - 6x = -1$

$x^2 - 6x + 9 = 9 - 1$

$(x - 3)^2 = 8$

$x - 3 = \pm\sqrt{8}$

$x = 3 \pm \sqrt{8}$

$x = 3 \pm 2\sqrt{2}$

22. $A = P(1 + r)^t$

$2704 = 2500(1 + r)^2$

$\frac{2704}{2500} = (1 + r)^2$

$\pm\frac{26}{25} = 1 + r$

$-\frac{25}{25} \pm \frac{26}{25} = r$

$\frac{1}{25} = r \quad or \quad -\frac{51}{25} = r$

Since the interest rate cannot be negative, it is not a solution. $\frac{1}{25} = 0.04$ The interest rate is 4%.

23. $s = 16t^2$

$443 = 16t^2$

$\frac{443}{16} = t^2$

$\sqrt{\frac{443}{16}} = t \quad$ Principle of square roots; rejecting the negative square root.

$5.3 \approx t$

It will take an object about 5.3 sec to fall.

24. $x^2 + 3x - 6 = 0$

$b^2 - 4ac = 3^2 - 4 \cdot 1 \cdot (-6) = 33$

There are two irrational real numbers.

25. $x^2 + 2x + 5 = 0$

$b^2 - 4ac = 2^2 - 4 \cdot 1 \cdot 5 = -16$

There are two imaginary numbers.

26. The solutions are $3i$ and $-3i$.

$$x = 3i \quad or \quad x = -3i$$
$$x - 3i = 0 \quad or \quad x + 3i = 0$$
$$(x - 3i)(x + 3i) = 0$$
$$x^2 - (3i)^2 = 0$$
$$x^2 + 9 = 0$$

27. The only solution is -5. It must be a repeated solution.

$$x = -5 \quad or \quad x = -5$$
$$x + 5 = 0 \quad or \quad x + 5 = 0$$
$$(x + 5)(x + 5) = 0$$
$$x^2 + 10x + 25 = 0$$

28. Let r represent the plane's speed in still air.

Trip	Distance	Speed	Time
To plant	300	$r - 20$	$\dfrac{300}{r - 20}$
Return	300	$r + 20$	$\dfrac{300}{r + 20}$

Solve the equation $\dfrac{300}{r - 20} + \dfrac{300}{r + 20} = 4$. We get

$r \approx 153$ or $r \approx -2.62$. Only 153 checks in the original problem. The plane's speed in still air is at least 153 mph.

29. *Familiarize*. Let x represent the time it takes Cheri to reply. Then $x + 6$ represents the time it takes Dani to reply. It takes them 4 hr to reply working together, so they can reply to $\dfrac{1}{4}$ of the emails in 1 hr. Cheri will

reply to $\dfrac{1}{x}$ of the emails in 1 hr, and Dani will reply to

$\dfrac{1}{x + 6}$ of the emails in 1 hr.

Translate. We have an equation.

$$\frac{1}{x} + \frac{1}{x + 6} = \frac{1}{4}$$

Carry out. We solve the equation.

We multiply by the LCD, $4x(x + 6)$.

$$4x(x + 6)\left(\frac{1}{x} + \frac{1}{x + 6}\right) = 4x(x + 6) \cdot \frac{1}{4}$$
$$4(x + 6) + 4x = x(x + 6)$$
$$4x + 24 + 4x = x^2 + 6x$$
$$0 = x^2 - 2x - 24$$
$$0 = (x - 6)(x + 4)$$

Check. Since negative time has no meaning in this problem, -4 is not a solution. We check only 6 hr. This is the time it would take Cheri working alone. Then Dani would take $6 + 6$, or 12 hr working alone. Cheri would reply to $4\left(\dfrac{1}{6}\right)$, or $\dfrac{2}{3}$ of the emails in 4 hr, and

Dani would reply to $4\left(\dfrac{1}{12}\right)$, or $\dfrac{1}{3}$ of the emails in 4 hr.

Thus, in 4 hr they would reply to $\dfrac{2}{3} + \dfrac{1}{3}$ of the emails.

This is all of it, so the numbers check.

State. It would take Cheri, working alone, 6 hr to reply to the emails.

30. $x^4 - 13x^2 + 36 = 0$

Let $u = x^2$ and $u^2 = x^4$.

$$u^2 - 13u + 36 = 0$$
$$(u - 4)(u - 9) = 0$$
$$u = 4 \quad or \quad u = 9$$

Now replace u with x^2 and solve these equations.

$$x^2 = 4 \quad or \quad x^2 = 9$$
$$x = \pm 2 \quad or \quad x = \pm 3$$

The intercepts are $(-3, 0)$, $(-2, 0)$, $(2, 0)$, and $(3, 0)$.

31. $15x^{-2} - 2x^{-1} - 1 = 0$

Let $u = x^{-1}$ and $u^2 = x^{-2}$.

$$15u^2 - 2u - 1 = 0$$
$$(5u + 1)(3u - 1) = 0$$
$$5u + 1 = 0 \quad or \quad 3u - 1 = 0$$
$$u = -\frac{1}{5} \quad or \quad u = \frac{1}{3}$$

Replace u with x^{-1}.

$$x^{-1} = -\frac{1}{5} \quad or \quad x^{-1} = \frac{1}{3}$$
$$x = -5 \quad or \quad x = 3$$

The numbers -5 and 3 check. They are the solutions.

32. $\left(x^2 - 4\right)^2 - \left(x^2 - 4\right) - 6 = 0$

Let $u = x^2 - 4$ and $u^2 = \left(x^2 - 4\right)^2$.

$$u^2 - u - 6 = 0$$
$$(u + 2)(u - 3) = 0$$
$$u + 2 = 0 \quad or \quad u - 3 = 0$$
$$u = -2 \quad or \quad u = 3$$

Replace u with $x^2 - 4$.

$$x^2 - 4 = -2 \quad or \quad x^2 - 4 = 3$$
$$x^2 = 2 \quad or \quad x^2 = 7$$
$$x = \pm\sqrt{2} \quad or \quad x = \pm\sqrt{7}$$

The numbers $\sqrt{2}, -\sqrt{2}$, $\sqrt{7}$ and $-\sqrt{7}$ check. They are the solutions.

33. $f(x) = -3(x + 2)^2 + 4$

We know that the graph looks like the graph of

$h(x) = 3x^2$ but moved to the left 2 units and up 4 units and turned upside down. The vertex is $(-2, 4)$, and the axis of symmetry is $x = -2$. The maximum function value is 4.

$f(x) = -3(x + 2)^2 + 4$
Maximum: 4

34. $f(x) = 2x^2 - 12x + 23$

$= 2(x^2 - 6x) + 23$

$= 2(x^2 - 6x + 9) - 18 + 23$

$= 2(x - 3)^2 + 5$

We know that the graph looks like the graph of $h(x) = 2x^2$ but moved to the right 3 units and up 5 units. The vertex is (3, 5), and the axis of symmetry is $x = 3$.

$f(x) = 2x^2 - 12x + 23$

35. $f(x) = x^2 - 9x + 14$

To find the x-intercepts, solve the equation $0 = x^2 - 9x + 14$. We factor.

$0 = x^2 - 9x + 14$

$0 = (x - 2)(x - 7)$

$x = 2 \ or \ x = 7$

The x-intercepts are (2, 0) and (7, 0).

The y-intercept is $(0, f(0))$, or (0, 14).

36. $N = 3\pi\sqrt{\dfrac{1}{p}}$

$N^2 = 9\pi^2 \dfrac{1}{p}$

$p = \dfrac{9\pi^2}{N^2}$

37. $2A + T = 3T^2$

$3T^2 - T - 2A = 0$

$a = 3, \ b = -1, \ c = -2A$

$T = \dfrac{-(-1) \pm \sqrt{(-1)^2 - 4 \cdot 3 \cdot (-2A)}}{2 \cdot 3}$

$T = \dfrac{1 \pm \sqrt{1 + 24A}}{6}$

38. The data points fall. The graph does not appear to represent a quadratic function in which the data points would rise and then fall or vice versa. Thus a linear function $f(x) = mx + b$ might be used to model the data.

39. The data points fall and then rise. The graph appears to represent a quadratic function that opens upward. Thus a quadratic function $f(x) = ax^2 + bx + c, \ a > 0$, might be used to model the data.

40. Since the area is in a corner, only two sides of fencing is needed.

$l + w = 30$

$A = lw$

$l = 30 - w$

$A = (30 - w)w = -w^2 + 30w = -1(w - 15)^2 + 225$

The maximum function value of 225 occurs when

$w = 15$.

When $w = 15$, $l = 30 - 15 = 15$.

Maximum area: 225 ft^2; dimensions: 15 ft by 15 ft

41. a. *Familiarize.* We look for a function of the form $f(x) = ax^2 + bx + c$, where $f(x)$ represents the national debt, in trillions, and x represents the years after 1990.

Translate. We substitute the given values of x and $f(x)$.

$3 = a(0)^2 + b(0) + c,$

$6 = a(10)^2 + b(10) + c,$

$18 = a(25)^2 + b(25) + c,$

or

$3 = 0a + 0b + c,$

$6 = 100a + 10b + c,$

$18 = 625a + 25b + c.$

Carry out. Solving the system of equations, we get $a = \dfrac{1}{50}, \ b = \dfrac{1}{10}, \ c = 3$.

Check. Recheck the calculations.

State. The function $f(x) = \dfrac{1}{50}x^2 + \dfrac{1}{10}x + 3$ fits the data.

b. $2010 - 1990 = 20$

Find $f(20)$.

$f(20) = \dfrac{1}{50}(20)^2 + \dfrac{1}{10}(20) + 3 = 13$

The national debt in 2010 will be about $13,000,000,000,000.

42. $x^3 - 3x > 2x^2$

$x^3 - 2x^2 - 3x > 0$

$x(x + 1)(x - 3) > 0$

The solutions of $x(x + 1)(x - 3)$ are -1, 0, and 3. They divide the real-number line into four intervals as shown:

We try test numbers in each interval.

A Test -2, $f(-2) = -2(-2 + 1)(-2 - 3) = -10$

B: Test $-\dfrac{1}{2}$, $f\left(-\dfrac{1}{2}\right) = -\dfrac{1}{2}\left(-\dfrac{1}{2} + 1\right)\left(-\dfrac{1}{2} - 3\right) = \dfrac{7}{8}$

C: Test 1, $f(1) = 1(1 + 1)(1 - 3) = -4$

D: Test 4, $f(4) = 4(4 + 1)(4 - 3) = 20$

The function value will be positive for all numbers in intervals B and D. The solution set is $(-1, -0) \cup (3, \infty)$, or $\{x \mid -1 < x < 0 \ or \ x > 3\}$.

43. $\dfrac{x - 5}{x + 3} \le 0$

Solve the related equation.

$\dfrac{x - 5}{x + 3} = 0$

$x - 5 = 0$

$x = 5$

Find the values that make the denominator 0.

$$x + 3 = 0$$
$$x = -3$$

Use the numbers 5 and -3 to divide the number line into intervals as shown:

Try test numbers in each interval.

A: Test -4, $\dfrac{-4-5}{-5+3} = \dfrac{-9}{-2} = \dfrac{9}{2} \not< 0$

The number -4 is a not solution of the inequality, so the interval A is not part of the solution set.

B: Test 0, $\dfrac{0-5}{0+3} = \dfrac{-5}{3} = -\dfrac{5}{3} < 0$

The number 0 is a solution of the inequality, so the interval B is part of the solution set.

C: Test 6, $\dfrac{6-5}{6+3} = \dfrac{1}{9} \not< 0$

The number 6 is a not solution of the inequality, so the interval C is not part of the solution set.

The solution set includes interval B. The number 5 is also included since the inequality symbol is $\leq$ and 5 is the solution of the related equation. The number -3 is not included since $\dfrac{x-5}{x+3}$ is undefined for $x = -3$. The solution set is $(-3, 5]$, or $\{x \mid -3 < x \leq 5\}$.

44. *Writing Exercise.* The x-coordinate of the maximum or minimum point lies halfway between the x-coordinates of the x-intercepts.

45. *Writing Exercise.* The first coordinate of each x-intercept of f is a solution of $f(x) = 0$. Suppose the first coordinates of the x-intercepts are a and b. Then $(x - a)$ and $(x - b)$ are factors of $f(x)$. If the graph of a quadratic function has one x-intercept $(a, 0)$, then $(x - a)$ is a repeated factor of $f(x)$.

46. *Writing Exercise.* Completing the square was used to solve quadratic equations and to graph quadratic functions by rewriting the function in the form $f(x) = a(x - h)^2 + k$.

47. Substituting the three ordered pairs $(-3, 0)$, $(5, 0)$, and $(0, -7)$ in the equation $f(x) = ax^2 + bx + c$ yields a system of equations:

$$0 = 9a - 3b + c,$$
$$0 = 25a + 5b + c,$$
$$-7 = 0a + 0b + c$$

The solution of this system of equations is $\left(\dfrac{7}{15}, -\dfrac{14}{5}, -7\right)$, so $f(x) = \dfrac{7}{15}x^2 - \dfrac{14}{5}x - 7$.

48. From Section 8.3, we know the sum of the solutions of $ax^2 + bx + c = 0$ is $-\dfrac{b}{a}$, and the product is $\dfrac{c}{a}$.

$$3x^2 - hx + 4k = 0$$
$$a = 3,\ b = -h,\ c = 4k$$

Substituting

For $-\dfrac{b}{a}$: $-\dfrac{-h}{3} = 20$

$$\dfrac{h}{3} = 20$$
$$h = 60$$

For $\dfrac{c}{a}$: $\dfrac{4k}{3} = 80$

$$4k = 240$$
$$k = 60$$

49. Let x and y represent the positive integers. Since one of the numbers is the square root of the other, we let $y = \sqrt{x}$. To find their average, we find their sum and divide by 2.

$$\dfrac{x + \sqrt{x}}{2} = 171$$
$$x + \sqrt{x} = 342$$
$$x + \sqrt{x} - 342 = 0$$

Let $u = \sqrt{x}$ and $u^2 = x$.

$$u^2 + u - 342 = 0$$
$$(u + 19)(u - 18) = 0$$
$$u = -19 \text{ or } u = 18$$

Substituting: $\sqrt{x} = -19 \text{ or } \sqrt{x} = 18$

We use only $\sqrt{x} = 18$

$$x = 324$$

The numbers are 18 and 324.

Chapter 8 Test

1. $25x^2 - 7 = 0$

$$25x^2 = 7$$
$$x^2 = \dfrac{7}{25}$$
$$x = \pm\sqrt{\dfrac{7}{25}} = \pm\dfrac{\sqrt{7}}{5}$$

The solutions are $-\dfrac{\sqrt{7}}{5}$ and $\dfrac{\sqrt{7}}{5}$.

2. $4x(x - 2) - 3x(x + 1) = -18$

$$4x^2 - 8x - 3x^2 - 3x = -18$$
$$x^2 - 11x + 18 = 0$$
$$(x - 2)(x - 9) = 0$$
$$x - 2 = 0 \quad or \quad x - 9 = 0$$
$$x = 2 \quad or \qquad x = 9$$

3. $x^2 + 2x + 3 = 0$

$a = 1,\ b = 2,\ c = 3$

$$x = \dfrac{-2 \pm \sqrt{2^2 - 4(1)(3)}}{2(1)} = \dfrac{-2 \pm \sqrt{4 - 12}}{2}$$
$$= \dfrac{-2 \pm \sqrt{-8}}{2} = \dfrac{-2 \pm 2i\sqrt{2}}{2} = \dfrac{-2}{2} \pm \dfrac{2i\sqrt{2}}{2}$$
$$= -1 \pm i\sqrt{2} \text{ or } -1 \pm \sqrt{2}i$$

The solutions are $-1 + \sqrt{2}i$ and $-1 - \sqrt{2}i$.

4. $2x+5=x^2$

$5+1=x^2-2x+1$

$6=(x-1)^2$

$\pm\sqrt{6}=x-1$

$1\pm\sqrt{6}=x$

5. $x^{-2}-x^{-1}=\dfrac{3}{4}$

$x^{-2}-x^{-1}-\dfrac{3}{4}=0$

$4x^{-2}-4x^{-1}-3=0$ Clearing fractions

Let $u=x^{-1}$ and $u^2=x^{-2}$.

$4u^2-4u-3=0$

$(2u-3)(2u+1)=0$

$2u-3=0$ or $2u+1=0$

$u=\dfrac{3}{2}$ or $u=-\dfrac{1}{2}$

Now we replace u with x^{-1} and solve these equations:

$x^{-1}=\dfrac{3}{2}$ or $x^{-1}=-\dfrac{1}{2}$

$\dfrac{1}{x}=\dfrac{3}{2}$ or $\dfrac{1}{x}=-\dfrac{1}{2}$

$2=3x$ or $2=-x$

$\dfrac{2}{3}=x$ $-2=x$

The solutions are -2 and $\dfrac{2}{3}$.

6. $x^2+3x=5$

$x^2+3x-5=0$

$a=1,\ b=3,\ c=-5$

$x=\dfrac{-3\pm\sqrt{3^2-4\cdot1\cdot(-5)}}{2\cdot1}=\dfrac{-3\pm\sqrt{29}}{2}$

$x=\dfrac{-3-\sqrt{29}}{2}\approx-4.193$

$x=\dfrac{-3+\sqrt{29}}{2}\approx1.193$

7. Let $f(x)=0$ and solve for x.

$0=12x^2-19x-21$

$0=(4x+3)(3x-7)$

$x=-\dfrac{3}{4}$ or $x=\dfrac{7}{3}$

The solutions are $-\dfrac{3}{4}$ and $\dfrac{7}{3}$.

8. $\dfrac{1}{2}\cdot20=10;\ \ 10^2=100$

$x^2-20x+100=(x-10)^2$

9. $\dfrac{1}{2}\cdot\dfrac{2}{7}=\dfrac{1}{7};\ \ \left(\dfrac{1}{7}\right)^2=\dfrac{1}{49}$

$x^2+\dfrac{2}{7}x+\dfrac{1}{49}=\left(x+\dfrac{1}{7}\right)^2$

10. $x^2+10x+15=0$

$x^2+10x=-15$

$x^2+10x+25=-15+25$

$(x+5)^2=10$

$x+5=\pm\sqrt{10}$

$x=-5\pm\sqrt{10}$

11. $x^2+2x+5=0$

$b^2-4ac=2^2-4(1)(5)=-16$

Two imaginary numbers

12. $x=\sqrt{11}$ or $x=-\sqrt{11}$

$x-\sqrt{11}=0$ or $x+\sqrt{11}=0$

$\left(x-\sqrt{11}\right)\left(x+\sqrt{11}\right)=0$

$x^2-11=0$

13. ***Familiarize***. Let r represent the cruiser's speed in still water. Then $r-4$ is the speed upriver and $r+4$ is the speed downriver. Using $t=\dfrac{d}{r}$, we let $\dfrac{60}{r-4}$ represent the time upriver and $\dfrac{60}{r+4}$ represent the time downriver.

Trip	Distance	Speed	Time
Upriver	60	$r-4$	$\dfrac{60}{r-4}$
Downriver	60	$r+4$	$\dfrac{60}{r+4}$

Translate. We have an equation.

$\dfrac{60}{r-4}+\dfrac{60}{r+4}=8$

Carry out. We solve the equation.

We multiply by the LCD, $(r-4)(r+4)$.

$(r-4)(r+4)\cdot\left(\dfrac{60}{r-4}+\dfrac{60}{r+4}\right)=(r-4)(r+4)\cdot8$

$60(r+4)+60(r-4)=8(r-4)(r+4)$

$60r+240+60r-240=8r^2-128$

$0=8r^2-120r-128$

$0=8(r^2-15r-16)$

$0=8(r-16)(r+1)$

$r=16\ or\ r=-1$

Check. Since negative rate has no meaning in this problem, -1 is not a solution. We check only 16 km/hr. If $r=16$, then the speed upriver is $16-4$, or 12 km/h, and the time is $\dfrac{60}{12}$, or 5 hr. The speed downriver is $16+4$, or 20 km/h, and the time is $\dfrac{60}{20}$, or 3 hr. The total time of the round trip is $5+3$, or 8 hr. The value checks.

State. The speed of the cruiser in still water is 16 km/h.

14. Let x represent the time it takes Dal to assemble the swing set. Then $x+4$ represents the time it takes Kim to do the same job. It takes them 1.5 hr to assemble the swing set working together, so they can complete $\dfrac{2}{3}$ of the job in 1 hr. Dal will assemble $\dfrac{1}{x}$ of the set in

1 hr, and Kim will assemble $\dfrac{1}{x+4}$ of the set in 1 hr.

We solve the equation.

$$\frac{1}{x}+\frac{1}{x+4}=\frac{2}{3}$$

The solutions are –3 and 2. Since negative time has no meaning for this application, Dal can assemble the swing set in 2 hr.

15. $f(x)=x^4-15x^2-16$

To find the x-intercepts, solve the equation

$0=x^4-15x^2-16$.

Let $u=x^2$ and $u^2=x^4$.

$$0=u^2-15u-16$$
$$0=(u-16)(u+1)$$
$$u=16 \ or \ u=-1$$

Replace u with x^2.

$$x^2=16 \quad or \quad x^2=-1 \quad \text{Has no real solutions}$$
$$x=\pm 4$$

The x-intercepts are (–4, 0) and (4, 0).

16. $f(x)=4(x-3)^2+5$

We know that the graph looks like the graph of

$h(x)=4x^2$ but moved to the right 3 units and up 5 units. The vertex is (3, 5), and the axis of symmetry is $x=3$. The minimum function value is 5.

$f(x)=4(x-3)^2+5$
Minimum: 5

17. $f(x)=2x^2+4x-6$
$$=2(x^2+2x)-6$$
$$=2(x^2+2x+1)-6-2$$
$$=2(x+1)^2-8$$

We know that the graph looks like the graph of

$h(x)=2x^2$ but moved to the left 1 unit and down 8 units. The vertex is (–1, –8), and the axis of symmetry is $x=-1$.

$f(x)=2x^2+4x-6$

18. $f(x)=x^2-x-6$

To find the x-intercepts, solve the equation

$0=x^2-x-6$. We factor.

$$0=x^2-x-6$$
$$0=(x+2)(x-3)$$
$$x=-2 \ or \ x=3$$

The x-intercepts are (–2, 0) and (3, 0).
The y-intercept is $(0, f(0))$, or (0, –6).

19.
$$V=\frac{1}{3}\pi\left(R^2+r^2\right)$$
$$\frac{3V}{\pi}=R^2+r^2$$
$$\frac{3V}{\pi}-R^2=r^2$$
$$\sqrt{\frac{3V}{\pi}-R^2}=r$$

We only consider the positive square root as instructed.

20. The data points rise and then fall. The graph appears to represent a quadratic function that opens downward.

21. $C(x)=0.2x^2-1.3x+3.4025$
$C(x)=0.2\left(x^2-6.5x\right)+3.4025$
$C(x)=0.2(x^2-6.5x+10.5625)-0.2(10.5625)+3.4025$
$C(x)=0.2(x-3.25)^2+1.29$

3.25 hundred or 325 cabinets should be built to have a minimum at $1.29 hundred, or $129 per cabinet.

22. Substituting the three ordered pairs (0, 0), (3, 0), and (5, 2) in the equation $f(x)=ax^2+bx+c$ yields a system of equations:

$$0=0a+0b+c,$$
$$0=9a+3b+c,$$
$$2=25a+5b+c$$

The solution of this system of equations is

$\left(\dfrac{1}{5},-\dfrac{3}{5},\,0\right)$, so $f(x)=\dfrac{1}{5}x^2-\dfrac{3}{5}x$.

23.
$$x^2+5x<6$$
$$x^2+5x-6<0$$
$$(x+6)(x-1)<0$$

The solutions of $(x+6)(x-1)=0$ are –6 and 1. They divide the number line into three intervals as shown:

```
            A        B        C
   <----|---------|--------|---->
        -6        1
```

We try test numbers in each interval.

A: Test –7, $f(-7)=(-7+6)(-7-1)=8$

B: Test 0, $f(0)=(0+6)(0-1)=-6$

C: Test 2, $f(2)=(2+6)(2-1)=8$

Since $f(0)$ is negative, the function value will be negative for all numbers in the interval containing 0. Because the symbol is $<$, we do not include the endpoints in the solution. The solution set is $(-6,\ 1)$, or $\{x\,|-6<x<1\}$.

24. $x-\dfrac{1}{x}\geq 0$

Solve the related equation.

$$x-\frac{1}{x}=0$$
$$x^2-1=0$$
$$(x+1)(x-1)=0$$
$$x=-1 \ or \ x=1$$

Find the values of x that make the denominator 0.
$$x=0$$

A: Test -2, $-2 - \dfrac{1}{-2} = -\dfrac{3}{2} \ngeq 0$

B: Test $-\dfrac{1}{2}$, $-\dfrac{1}{2} - \dfrac{1}{-\dfrac{1}{2}} = \dfrac{3}{2} \geq 0$

C: Test $-\dfrac{1}{2}$, $\dfrac{1}{2} - \dfrac{1}{\dfrac{1}{2}} = -\dfrac{3}{2} \ngeq 0$

D: Test 4, $4 - \dfrac{1}{4} = \dfrac{15}{4} \geq 0$

$[-1, 0) \cup [1, \infty)$, or $\{x \,|\, -1 \leq x < 0 \ or \ x \geq 1\}$

25.
$$kx^2 + 3x - k = 0$$
$$k(-2)^2 + 3(-2) - k = 0 \quad \text{Substitute } -2 \text{ for } x$$
$$4k - 6 - k = 0$$
$$3k = 6$$
$$k = 2$$

Then $\quad 2x^2 + 3x - 2 = 0$
$$(x + 2)(2x - 1) = 0$$

The other solution is $\dfrac{1}{2}$.

26. If $-\sqrt{3}$ and $2i$ are solutions, then $\sqrt{3}$ and $-2i$ must also be solutions. A fourth-degree equations can be found as follows:
$$\left[x - (-\sqrt{3})\right]\left[x - \sqrt{3}\right][x - 2i][x + 2i] = 0$$
$$(x^2 - 3)(x^2 + 4) = 0$$
$$x^4 + x^2 - 12 = 0$$

Answers may vary.

27. $x^4 - 4x^2 - 1 = 0$

Let $u = x^2$ and $u^2 = x^4$.
$$u^2 - 4u - 1 = 0$$
$$u^2 - 4u = 1$$
$$u^2 - 4u + 4 = 1 + 4$$
$$(u - 2)^2 = 5$$
$$u - 2 = \pm\sqrt{5}$$
$$u = 2 \pm \sqrt{5}$$

Replace u with x^2.
$$x^2 = 2 + \sqrt{5} \quad or \quad x^2 = 2 - \sqrt{5}$$
$$x = \pm\sqrt{2 + \sqrt{5}} \quad or \quad x = \pm\sqrt{2 - \sqrt{5}}$$

Since $2 - \sqrt{5}$ is negative, we can rewrite it as follows:
$$\pm\sqrt{2 - \sqrt{5}} = \pm\sqrt{\sqrt{5} - 2}\,i$$

The solutions are $\pm\sqrt{\sqrt{5} + 2}$ and $\pm\sqrt{\sqrt{5} - 2}\,i$.

Chapter 9

Exponential Functions and Logarithmic Functions

Exercise Set 9.1

1. True

3. $(g \circ f) = g(f(x)) = x^2 + 3 \neq (x+3)^2$, so the statement is false.

5. False

7. True

9. a. $(f \circ g)(1) = f(g(1)) = f(1-3)$
$= f(-2) = (-2)^2 + 1$
$= 4 + 1 = 5$

b. $(g \circ f)(1) = g(f(1)) = g(1^2 + 1)$
$= g(2) = 2 - 3 = -1$

c. $(f \circ g) = f(g(x)) = f(x-3)$
$= (x-3)^2 + 1 = x^2 - 6x + 9 + 1$
$= x^2 - 6x + 10$

d. $(g \circ f)(x) = g(f(x)) = g(x^2 + 1)$
$= x^2 + 1 - 3 = x^2 - 2$

11. a $(f \circ g)(1) = f(g(1)) = f(2 \cdot 1^2 - 7)$
$= f(-5) = 5(-5) + 1 = -24$

b. $(g \circ f)(1) = g(f(1)) = g(5 \cdot 1 + 1)$
$= g(6) = 2 \cdot 6^2 - 7 = 65$

c. $(f \circ g)(x) = f(g(x)) = f(2x^2 - 7)$
$= 5(2x^2 - 7) + 1 = 10x^2 - 34$

d. $(g \circ f)(x) = g(f(x)) = g(5x + 1)$
$= 2(5x + 1)^2 - 7 = 2(25x^2 + 10x + 1) - 7$
$= 50x^2 + 20x - 5$

13. a. $(f \circ g)(1) = f(g(1)) = f\left(\frac{1}{1^2}\right)$
$= f(1) = 1 + 7 = 8$

b. $(g \circ f)(1) = g(f(1)) = g(1+7) = g(8) = \frac{1}{8^2} = \frac{1}{64}$

c. $(f \circ g)(x) = f(g(x)) = f\left(\frac{1}{x^2}\right) = \frac{1}{x^2} + 7$

d. $(g \circ f)(x) = g(f(x)) = g(x+7) = \frac{1}{(x+7)^2}$

15. a. $(f \circ g)(1) = f(g(1)) = f(1+3)$
$= f(4) = \sqrt{4} = 2$

b. $(g \circ f)(1) = g(f(1)) = g(\sqrt{1})$
$= g(1) = 1 + 3 = 4$

c. $(f \circ g)(x) = f(g(x)) = f(x+3) = \sqrt{x+3}$

d. $(g \circ f)(x) = g(f(x)) = g(\sqrt{x}) = \sqrt{x} + 3$

17. a. $(f \circ g)(1) = f(g(1)) = f\left(\frac{1}{1}\right) = f(1) = \sqrt{4 \cdot 1} = \sqrt{4} = 2$

b. $(g \circ f)(1) = g(f(1)) = g(\sqrt{4 \cdot 1}) = g(\sqrt{4}) = g(2) = \frac{1}{2}$

c. $(f \circ g)(x) = f(g(x)) = f\left(\frac{1}{x}\right) = \sqrt{4 \cdot \frac{1}{x}} = \sqrt{\frac{4}{x}}$

d. $(g \circ f)(x) = g(f(x)) = g(\sqrt{4x}) = \frac{1}{\sqrt{4x}}$

19. a. $(f \circ g)(1) = f(g(1)) = f(\sqrt{1-1})$
$= f(\sqrt{0}) = f(0) = 0^2 + 4 = 4$

b. $(g \circ f)(1) = g(f(1)) = g(1^2 + 4)$
$= g(5) = \sqrt{5-1} = \sqrt{4} = 2$

c. $(f \circ g)(x) = f(g(x)) = f(\sqrt{x-1})$
$= (\sqrt{x-1})^2 + 4 = x - 1 + 4 = x + 3$

d. $(g \circ f)(x) = g(f(x)) = g(x^2 + 4)$
$= \sqrt{x^2 + 4 - 1} = \sqrt{x^2 + 3}$

21. $h(x) = (3x - 5)^4$
This is $3x - 5$ raised to the fourth power, so the two most obvious functions are $f(x) = x^4$ and $g(x) = 3x - 5$.

23. $h(x) = \sqrt{9x + 1}$
We have $9x + 1$ and take the square root of the expression, so the two most obvious functions are $f(x) = \sqrt{x}$ and $g(x) = 9x + 1$.

25. $h(x) = \frac{6}{5x - 2}$
This is 6 divided by $5x - 2$, so two functions that can be used are $f(x) = \frac{6}{x}$ and $g(x) = 5x - 2$.

27. The graph of $f(x) = -x$ is shown below.

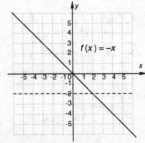

Since there is no horizontal line that crosses the graph more than once, the function is one-to-one.

29. $f(x) = x^2 + 3$

Observe that the graph of this function is a parabola that opens up. Thus, there are many horizontal lines that cross the graph more than once, so the function is not one-to-one. We can also draw the graph as shown below.

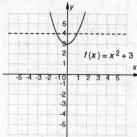

In particular, the line $y = 4$ crosses the graph more than once. The function is not one-to-one.

31. Since there is no horizontal line that crosses the graph more than once, the function is one-to-one.

33. There are many horizontal lines that cross the graph more than once, the function is not one-to-one.

35. a. The function $f(x) = x + 3$ is a linear function that is not constant, so it passes the horizontal-line test. Thus, f is one-to-one.

   b. Replace $f(x)$ by y: $y = x + 3$
   Interchange x and y: $x = y + 3$
   Solve for y: $x - 3 = y$
   Replace y by $f^{-1}(x)$: $f^{-1}(x) = x - 3$

37. a. The function $f(x) = 2x$ is a linear function that is not constant, so it passes the horizontal-line test. Thus, f is one-to-one.

   b. Replace $f(x)$ by y: $y = 2x$
   Interchange x and y: $x = 2y$
   Solve for y: $\dfrac{x}{2} = y$
   Replace y by $f^{-1}(x)$: $f^{-1}(x) = \dfrac{x}{2}$

39. a. The function $g(x) = 3x - 1$ is a linear function that is not constant, so it passes the horizontal-line test. Thus, g is one-to-one.

   b. Replace $g(x)$ by y: $y = 3x - 1$
   Interchange x and y: $x = 3y - 1$
   Solve for y: $x + 1 = 3y$
                              $\dfrac{x + 1}{3} = y$
   Replace y by $g^{-1}(x)$: $g^{-1}(x) = \dfrac{x + 1}{3}$

41. a. The function $f(x) = \frac{1}{2}x + 1$ is a linear function that is not constant, so it passes the horizontal-line test. Thus, f is one-to-one.

   b. Replace $f(x)$ by y: $y = \dfrac{1}{2}x + 1$
   Interchange variables: $x = \dfrac{1}{2}y + 1$
   Solve for y: $x - 1 = \dfrac{1}{2}y$
                              $2x - 2 = y$
   Replace y by $f^{-1}(x)$: $f^{-1}(x) = 2x - 2$

43. a. The graph of $g(x) = x^2 + 5$ is shown below. There are many horizontal lines that cross the graph more than once. For example, the line $y = 8$ crosses the graph more than once. The function is not one-to-one.

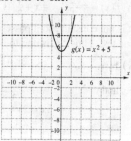

45. a. The function $h(x) = -10 - x$ is a linear function that is not constant, so it passes the horizontal-line test. Thus, h is one-to-one.

   b. Replace $h(x)$ by y: $y = -10 - x$
   Interchange variables: $x = -10 - y$
   Solve for y: $x + 10 = -y$
                              $-x - 10 = y$
   Replace y by $h^{-1}(x)$: $h^{-1}(x) = -x - 10$

47. a. The graph of $f(x) = \dfrac{1}{x}$ is shown below. It passes the horizontal-line test, so the function is one-to-one.

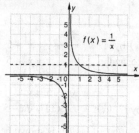

   b. Replace $f(x)$ by y: $y = \dfrac{1}{x}$
   Interchange x and y: $x = \dfrac{1}{y}$
   Solve for y: $xy = 1$
                              $y = \dfrac{1}{x}$
   Replace y by $f^{-1}(x)$: $f^{-1}(x) = \dfrac{1}{x}$

49. a. The graph of $g(x) = 1$ is shown below. The horizontal line $y = 1$ crosses the graph more than once, so the function is not one-to-one.

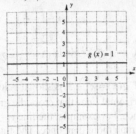

51. a. The function $f(x) = \dfrac{2x+1}{3} = \dfrac{2}{3}x + \dfrac{1}{3}$ is a linear function that is not constant, so it passes the horizontal-line test. Thus, f is one-to-one.

   b. Replace $f(x)$ by y: $y = \dfrac{2x+1}{3}$

       Interchange x and y: $x = \dfrac{2y+1}{3}$

       Solve for y: $3x = 2y+1$

                          $3x - 1 = 2y$

                          $\dfrac{3x-1}{2} = y$

       Replace y by $f^{-1}(x)$: $f^{-1}(x) = \dfrac{3x-1}{2}$

53. a. The graph of $f(x) = x^3 + 5$ is shown below. It passes the horizontal-line test, so the function is one-to-one.

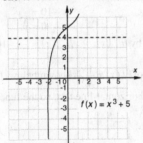

   b. Replace $f(x)$ by y: $y = x^3 + 5$

       Interchange x and y: $x = y^3 + 5$

       Solve for y: $x - 5 = y^3$

                       $\sqrt[3]{x-5} = y$

       Replace y by $f^{-1}(x)$: $f^{-1}(x) = \sqrt[3]{x-5}$

55. a. The graph of $g(x) = (x-2)^3$ is shown below. It passes the horizontal-line test, so the function is one-to-one.

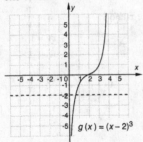

   b. Replace $g(x)$ by y: $y = (x-2)^3$

       Interchange x and y: $x = (y-2)^3$

       Solve for y: $\sqrt[3]{x} = y - 2$

                       $\sqrt[3]{x} + 2 = y$

       Replace y by $g^{-1}(x)$: $g^{-1}(x) = \sqrt[3]{x} + 2$

57. a. The graph of $f(x) = \sqrt{x}$ is shown below. It passes the horizontal-line test, so the function is one-to-one.

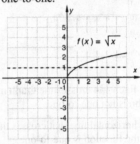

   b. Replace $f(x)$ by y: $y = \sqrt{x}$ (Note that $f(x) \geq 0$)

       Interchange x and y: $x = \sqrt{y}$

       Solve for y: $x^2 = y$

       Replace y by $f^{-1}(x)$: $f^{-1}(x) = x^2; x \geq 0$

59. a. $f(8) = 2(8+12) = 2 \cdot 20 = 40$

      Size 40 in Italy corresponds to size 8 in the U.S.

       $f(10) = 2(10+12) = 2 \cdot 22 = 44$

      Size 44 in Italy corresponds to size 10 in the U.S.

       $f(14) = 2(14+12) = 2 \cdot 26 = 52$

      Size 52 in Italy corresponds to size 14 in the U.S.

       $f(18) = 2(18+12) = 2 \cdot 30 = 60$

      Size 60 in Italy corresponds to size 18 in the U.S.

   b. The function $f(x) = 2(x+12)$ is a linear function that is not constant, so it passes the horizontal-line test and has an inverse that is a function.

      Replace $f(x)$ by y: $y = 2(x+12)$

      Interchange x and y: $x = 2(y+12)$

      Solve for y: $x = 2y+24$

                       $x - 24 = 2y$

                      $\dfrac{x-24}{2} = y$

      Replace y by $f^{-1}(x)$: $f^{-1}(x) = \dfrac{x-24}{2}$ or $\dfrac{x}{2} - 12$

   c. $f^{-1}(40) = \dfrac{40-24}{2} = \dfrac{16}{2} = 8$

      Size 8 in the U.S. corresponds to size 40 in Italy.

       $f^{-1}(44) = \dfrac{44-24}{2} = \dfrac{20}{2} = 10$

      Size 10 in the U.S. corresponds to size 44 in Italy.

       $f^{-1}(52) = \dfrac{52-24}{2} = \dfrac{28}{2} = 14$

      Size 14 in the U.S. corresponds to size 52 in Italy.

       $f^{-1}(60) = \dfrac{60-24}{2} = \dfrac{36}{2} = 18$

      Size 18 in the U.S. corresponds to size 60 in Italy.

61. First graph $f(x) = \frac{2}{3}x + 4$. Then graph the inverse

function by reflecting the graph of $f(x) = \frac{2}{3}x + 4$

across the line $y = x$. The graph of the inverse function can also be found by first finding a formula for the inverse, substituting to find function values, and then plotting points.

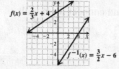

63. Follow the procedure described in Exercise 61 to graph the function and its inverse.

65. Follow the procedure described in Exercise 61 to graph the function and its inverse.

67. Follow the procedure described in Exercise 61 to graph the function and its inverse.

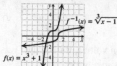

69. Follow the procedure described in Exercise 61 to graph the function and its inverse.

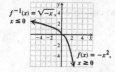

71. We check to see that $\left(f^{-1} \circ f\right)(x) = x$ and

$\left(f \circ f^{-1}\right)(x) = x.$

$(f^{-1} \circ f)(x) = f^{-1}(f(x)) = f^{-1}(\sqrt[3]{x-4})$
$\qquad = \left(\sqrt[3]{x-4}\right)^3 + 4 = x - 4 + 4 = x$

$(f \circ f^{-1})(x) = f(f^{-1}(x)) = f(x^3 + 4)$
$\qquad = \sqrt[3]{x^3 + 4 - 4} = \sqrt[3]{x^3} = x$

73. We check to see that $f^{-1} \circ f(x) = x$ and

$f \circ f^{-1}(x) = x.$

$f^{-1} \circ f(x) = f^{-1}(f(x)) = f^{-1}\left(\frac{1-x}{x}\right) = \dfrac{1}{\dfrac{1-x}{x}+1}$

$\qquad = \dfrac{1}{\dfrac{1-x}{x}+1} \cdot \dfrac{x}{x} = \dfrac{x}{1-x+x} = \dfrac{x}{1} = x$

$f \circ f^{-1}(x) = f(f^{-1}(x)) = f\left(\dfrac{1}{x+1}\right) = \dfrac{1 - \dfrac{1}{x+1}}{\dfrac{1}{x+1}}$

$\qquad = \dfrac{1 - \dfrac{1}{x+1}}{\dfrac{1}{x+1}} \cdot \dfrac{x+1}{x+1} = \dfrac{x+1-1}{1} = \dfrac{x}{1} = x$

75. *Writing Exercise.*

77. $t^{1/5}t^{2/3} = t^{1/5+2/3} = t^{13/15}$

79. $(-3x^{-6}y^4)^{-2} = (-3)^{-2}(x^{-6})^{-2}(y^4)^{-2}$
$\qquad = \dfrac{1}{9}x^{12}y^{-8}$
$\qquad = \dfrac{x^{12}}{9y^8}$

81. $3^3 + 2^2 - (32 \div 4 - 16 \div 8) = 27 + 4 - (8 - 2)$
$\qquad = 31 - 6$
$\qquad = 25$

83. *Writing Exercise.*

85. Reflect the graph of f across the line $y = x$.

87. From Exercise 59(b), we know that a function that converts dress sizes in Italy to those in the United States is $g(x) = \frac{x-24}{2}$. From Exercise 60, we know that a function that converts dress sizes in the United States to those in France is $f(x) = x + 32$. Then a function that converts dress sizes in Italy to those in France is
$h(x) = (f \circ g)(x)$
$h(x) = f\left(\dfrac{x-24}{2}\right)$
$h(x) = \dfrac{x-24}{2} + 32$
$h(x) = \dfrac{x}{2} - 12 + 32$
$h(x) = \dfrac{x}{2} + 20.$

89. *Writing Exercise.*

91. Suppose that $h(x) = (f \circ g)(x)$. First note that for
$I(x) = x$, $(f \circ I)(x) = f(I(x))$ for any function f.

i) $((g^{-1} \circ f^{-1}) \circ h)(x) = ((g^{-1} \circ f^{-1}) \circ (f \circ g))(x)$
$\qquad = ((g^{-1} \circ (f^{-1} \circ f)) \circ g)(x)$
$\qquad = ((g^{-1} \circ I) \circ g)(x)$
$\qquad = (g^{-1} \circ g)(x) = x$

ii) $(h \circ (g^{-1} \circ f^{-1}))(x) = ((f \circ g) \circ (g^{-1} \circ f^{-1}))(x)$
$\qquad = ((f \circ (g \circ g^{-1})) \circ f^{-1})(x)$
$\qquad = ((f \circ I) \circ f^{-1})(x)$
$\qquad = (f \circ f^{-1})(x) = x$

Therefore, $(g^{-1} \circ f^{-1})(x) = h^{-1}(x)$.

93. $(f \circ g)(x) = x$ and $(g \circ f)(x) = x$, so the functions are inverses.

95. $(f \circ g)(x) \neq x$, so the functions are not inverses. (It is also true that $(g \circ f)(x) \neq x$.)

97. (1) C; (2) A; (3) B; (4) D

99. We assume f and g are both linear functions.
Find $f(x)$.
$$m = \frac{6.5 - 6}{7 - 6} = \frac{1}{2}$$
$$y - 6 = \frac{1}{2}(x - 6)$$
$$y = \frac{1}{2}x + 3$$
$$f(x) = \frac{1}{2}x + 3$$

Find $g(x)$.
$$m = \frac{8 - 6}{7 - 6} = 2$$
$$y - 6 = 2(x - 6)$$
$$y = 2x - 6$$
$$g(x) = 2x - 6$$

Determine whether f and g are inverses of each other.
$$f(g(x)) = f(2x - 6) = \frac{1}{2}(2x - 6) + 3 = x - 3 + 3$$
$$= x$$
$$g(f(x)) = g\left(\frac{1}{2}x + 3\right) = 2\left(\frac{1}{2}x + 3\right) - 6 = x + 6 - 6$$
$$= x$$
Yes, f and g are inverses of each other.

101. $(c \circ g)(a)$ makes sense. It represents the cost of sealant for a bamboo floor with area a.

Exercise Set 9.2

1. True

3. True; the graph of $y = f(x - 3)$ is a translation of the graph of $y = f(x)$, 3 units to the right.

5. False; the graph of $y = 3^x$ crosses the y-axis at (0, 1).

7. Graph: $y = f(x) = 3^x$
We compute some function values, thinking of y as $f(x)$, and keep the results in a table.
$$f(0) = 3^0 = 1$$
$$f(1) = 3^1 = 3$$
$$f(2) = 3^2 = 9$$
$$f(-1) = 3^{-1} = \frac{1}{3^1} = \frac{1}{3}$$
$$f(-2) = 3^{-2} = \frac{1}{3^2} = \frac{1}{9}$$

x	y, or $f(x)$
0	1
1	3
2	9
−1	$\frac{1}{3}$
−2	$\frac{1}{9}$

Next we plot these points and connect them with a smooth curve.

$y = f(x) = 3^x$

9. Graph: $y = 6^x$
We compute some function values, thinking of y as $f(x)$, and keep the results in a table.
$$f(0) = 6^0 = 1$$
$$f(1) = 6^1 = 6$$
$$f(2) = 6^2 = 36$$
$$f(-1) = 6^{-1} = \frac{1}{6^1} = \frac{1}{6}$$
$$f(-2) = 6^{-2} = \frac{1}{6^2} = \frac{1}{36}$$

x	y, or $f(x)$
0	1
1	6
2	36
−1	$\frac{1}{6}$
−2	$\frac{1}{36}$

Next we plot these points and connect them with a smooth curve.

$y = 6^x$

11. Graph: $y = 2^x + 1$
We compute some function values, thinking of y as $f(x)$, and keep the results in a table.
$$f(-4) = 2^{-4} + 1 = \frac{1}{2^4} + 1 = \frac{1}{16} + 1 = 1\frac{1}{16}$$
$$f(-2) = 2^{-2} + 1 = \frac{1}{2^2} + 1 = \frac{1}{4} + 1 = 1\frac{1}{4}$$
$$f(0) = 2^0 + 1 = 1 + 1 = 2$$
$$f(1) = 2^1 + 1 = 2 + 1 = 3$$
$$f(2) = 2^2 + 1 = 4 + 1 = 5$$

x	y, or $f(x)$
-4	$1\frac{1}{16}$
-2	$1\frac{1}{4}$
0	2
1	3
2	5

Next we plot these points and connect them with a smooth curve.

13. Graph: $y = 3^x - 2$

We compute some function values, thinking of y as $f(x)$, and keep the results in a table.

$$f(-3) = 3^{-3} - 2 = \frac{1}{3^3} - 2 = \frac{1}{27} - 2 = -\frac{53}{27}$$

$$f(-1) = 3^{-1} - 2 = \frac{1}{3} - 2 = -\frac{5}{3}$$

$$f(0) = 3^0 - 2 = 1 - 2 = -1$$

$$f(1) = 3^1 - 2 = 3 - 2 = 1$$

$$f(2) = 3^2 - 2 = 9 - 2 = 7$$

x	y, or $f(x)$
-3	$-\frac{53}{27}$
-1	$-\frac{5}{3}$
0	-1
1	1
2	7

Next we plot these points and connect them with a smooth curve.

15. Graph: $y = 2^x - 5$

We construct a table of values, thinking of y as $f(x)$. Then we plot the points and connect them with a smooth curve.

$$f(0) = 2^0 - 5 = 1 - 5 = -4$$

$$f(1) = 2^1 - 5 = 2 - 5 = -3$$

$$f(2) = 2^2 - 5 = 4 - 5 = -1$$

$$f(3) = 2^3 - 5 = 8 - 5 = 3$$

$$f(-1) = 2^{-1} - 5 = \frac{1}{2} - 5 = -\frac{9}{2}$$

$$f(-2) = 2^{-2} - 5 = \frac{1}{4} - 5 = -\frac{19}{4}$$

$$f(-4) = 2^{-4} - 5 = \frac{1}{16} - 5 = -\frac{79}{16}$$

x	y, or $f(x)$
0	-4
1	-3
2	-1
3	3
-1	$-\frac{9}{2}$
-2	$-\frac{19}{4}$
-4	$-\frac{79}{16}$

17. Graph: $y = 2^{x-3}$

We construct a table of values, thinking of y as $f(x)$. Then we plot the points and connect them with a smooth curve.

$$f(0) = 2^{0-3} = 2^{-3} = \frac{1}{8}$$

$$f(-1) = 2^{-1-3} = 2^{-4} = \frac{1}{16}$$

$$f(1) = 2^{1-3} = 2^{-2} = \frac{1}{4}$$

$$f(2) = 2^{2-3} = 2^{-1} = \frac{1}{2}$$

$$f(3) = 2^{3-3} = 2^0 = 1$$

$$f(4) = 2^{4-3} = 2^1 = 2$$

$$f(5) = 2^{5-3} = 2^2 = 4$$

x	y, or $f(x)$
0	$\frac{1}{8}$
-1	$\frac{1}{16}$
1	$\frac{1}{4}$
2	$\frac{1}{2}$
3	1
4	2
5	4

19. Graph: $y = 2^{x+1}$

We construct a table of values, thinking of y as $f(x)$. Then we plot the points and connect them with a smooth curve.

$$f(-3) = 2^{-3+1} = 2^{-2} = \frac{1}{4}$$

$$f(-1) = 2^{-1+1} = 2^0 = 1$$

$$f(0) = 2^{0+1} = 2^1 = 2$$

$$f(1) = 2^{1+1} = 2^2 = 4$$

x	y, or $f(x)$
-3	$\frac{1}{4}$
-1	1
0	2
1	4

21. Graph: $y = \left(\frac{1}{4}\right)^x$

We construct a table of values, thinking of y as $f(x)$. Then we plot the points and connect them with a smooth curve.

$f(0) = \left(\frac{1}{4}\right)^0 = 1$

$f(1) = \left(\frac{1}{4}\right)^1 = \frac{1}{4}$

$f(2) = \left(\frac{1}{4}\right)^2 = \frac{1}{16}$

$f(-1) = \left(\frac{1}{4}\right)^{-1} = \frac{1}{\frac{1}{4}} = 4$

$f(-2) = \left(\frac{1}{4}\right)^{-2} = \frac{1}{\frac{1}{16}} = 16$

x	y, or $f(x)$
0	1
1	$\frac{1}{4}$
2	$\frac{1}{16}$
−1	4
−2	16

23. Graph: $y = \left(\frac{1}{3}\right)^x$

We construct a table of values, thinking of y as $f(x)$. Then we plot the points and connect them with a smooth curve.

$f(0) = \left(\frac{1}{3}\right)^0 = 1$

$f(1) = \left(\frac{1}{3}\right)^1 = \frac{1}{3}$

$f(2) = \left(\frac{1}{3}\right)^2 = \frac{1}{9}$

$f(3) = \left(\frac{1}{3}\right)^3 = \frac{1}{27}$

$f(-1) = \left(\frac{1}{3}\right)^{-1} = \frac{1}{\left(\frac{1}{3}\right)^1} = \frac{1}{\frac{1}{3}} = 3$

$f(-2) = \left(\frac{1}{3}\right)^{-2} = \frac{1}{\left(\frac{1}{3}\right)^2} = \frac{1}{\frac{1}{9}} = 9$

$f(-3) = \left(\frac{1}{3}\right)^{-3} = \frac{1}{\left(\frac{1}{3}\right)^3} = \frac{1}{\frac{1}{27}} = 27$

x	y, or $f(x)$
0	1
1	$\frac{1}{3}$
2	$\frac{1}{9}$
3	$\frac{1}{27}$
−1	3
−2	9
−3	27

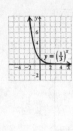

25. Graph: $y = 2^{x+1} - 3$

We construct a table of values, thinking of y as $f(x)$. Then we plot the points and connect them with a smooth curve.

$f(0) = 2^{0+1} - 3 = 2 - 3 = -1$

$f(1) = 2^{1+1} - 3 = 4 - 3 = 1$

$f(2) = 2^{2+1} - 3 = 8 - 3 = 5$

$f(-1) = 2^{-1+1} - 3 = 1 - 3 = -2$

$f(-2) = 2^{-2+1} - 3 = \frac{1}{2} - 3 = -\frac{5}{2}$

$f(-3) = 2^{-3+1} - 3 = \frac{1}{4} - 3 = -\frac{11}{4}$

x	y, or $f(x)$
0	−1
1	1
2	5
−1	−2
−2	$-\frac{5}{2}$
−3	$-\frac{11}{4}$

27. Graph: $x = 6^y$

We can find ordered pairs by choosing values for y and then computing values for x.

For $y = 0$, $x = 6^0 = 1$.

For $y = 1$, $x = 6^1 = 6$.

For $y = -1$, $x = 6^{-1} = \frac{1}{6^1} = \frac{1}{6}$.

For $y = -2$, $x = 6^{-2} = \frac{1}{6^2} = \frac{1}{36}$.

x	y
1	0
6	1
$\frac{1}{6}$	−1
$\frac{1}{36}$	−2

↑　↑ (1) Choose values for y.
(2) Compute values for x.

We plot the points and connect them with a smooth curve.

29. Graph: $x = 3^{-y} = \left(\frac{1}{3}\right)^{y}$

We can find ordered pairs by choosing values for y and then computing values for x. Then we plot these points and connect them with a smooth curve.

For $y = 0$, $x = \left(\frac{1}{3}\right)^{0} = 1$.

For $y = 1$, $x = \left(\frac{1}{3}\right)^{1} = \frac{1}{3}$.

For $y = 2$, $x = \left(\frac{1}{3}\right)^{2} = \frac{1}{9}$.

For $y = -1$, $x = \left(\frac{1}{3}\right)^{-1} = \frac{1}{\frac{1}{3}} = 3$.

For $y = -2$, $x = \left(\frac{1}{3}\right)^{-2} = \frac{1}{\frac{1}{9}} = 9$.

x	y
1	0
$\frac{1}{3}$	1
$\frac{1}{9}$	2
3	−1
9	−2

31. Graph: $x = 4^{y}$

We can find ordered pairs by choosing values for y and then computing values for x. Then we plot these points and connect them with a smooth curve.

For $y = 0$, $x = 4^{0} = 1$.

For $y = 1$, $x = 4^{1} = 4$.

For $y = 2$, $x = 4^{2} = 16$.

For $y = -1$, $x = 4^{-1} = \frac{1}{4}$.

For $y = -2$, $x = 4^{-2} = \frac{1}{16}$.

x	y
1	0
4	1
16	2
$\frac{1}{4}$	−1
$\frac{1}{16}$	−2

33. Graph: $x = \left(\frac{4}{3}\right)^{y}$

We can find ordered pairs by choosing values for y and then computing values for x. Then we plot these points and connect them with a smooth curve.

For $y = 0$, $x = \left(\frac{4}{3}\right)^{0} = 1$.

For $y = 1$, $x = \left(\frac{4}{3}\right)^{1} = \frac{4}{3}$.

For $y = 2$, $x = \left(\frac{4}{3}\right)^{2} = \frac{16}{9}$.

For $y = 3$, $x = \left(\frac{4}{3}\right)^{3} = \frac{64}{27}$.

For $y = -1$, $x = \left(\frac{4}{3}\right)^{-1} = \frac{3}{4}$.

For $y = -2$, $x = \left(\frac{4}{3}\right)^{-2} = \left(\frac{3}{4}\right)^{2} = \frac{9}{16}$.

For $y = -3$, $x = \left(\frac{4}{3}\right)^{-3} = \left(\frac{3}{4}\right)^{3} = \frac{27}{64}$.

x	y
1	0
$\frac{4}{3}$	1
$\frac{16}{9}$	2
$\frac{64}{27}$	3
$\frac{3}{4}$	−1
$\frac{9}{16}$	−2
$\frac{27}{64}$	−3

35. Graph $y = 3^{x}$ (see Exercise 7) and $x = 3^{y}$ (see Exercise 28) using the same set of axes.

37. Graph $y = \left(\frac{1}{2}\right)^{x}$ and $x = \left(\frac{1}{2}\right)^{y}$ (see Exercise 30) using the same set of axes.

39. a. In 0 days (at birth), $t = 0$

$W(0) = 2000(1.0077)^{0} = 2000$ lb

In 30 days, $W(30) = 2000(1.0077)^{30} \approx 2500$ lb

In 60 days, $W(60) = 2000(1.0077)^{60} \approx 3200$ lb

In 90 days, $W(30) = 2000(1.0077)^{90} \approx 4000$ lb

b. Use the function values computed in part (a) and others, if desired, and draw the graph.

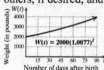

41. a. $P(1) = 21.4(0.914)^{1} \approx 19.6\%$

$P(3) = 21.4(0.914)^{3} \approx 16.3\%$

1 yr = 12 months; $P(12) = 21.4(0.914)^{12} \approx 7.3\%$

b.

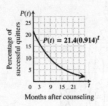

43. a. In 1930, $t = 1930 - 1900 = 30$.

$$P(t) = 150(0.960)^t$$
$$P(30) = 150(0.960)^{30}$$
$$\approx 44.079$$

In 1930, about 44.079 thousand, or 44,079, humpback whales were alive.

In 1960, $t = 1960 - 1900 = 60$.

$$P(t) = 150(0.960)^t$$
$$P(60) = 150(0.960)^{60}$$
$$\approx 12.953$$

In 1960, about 12.953 thousand, or 12,953, humpback whales were alive.

b. Plot the points found in part (a), $(30, 44{,}079)$ and $(60, 12{,}953)$ and additional points as needed and graph the function.

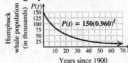

45. a. In 1992, $t = 1992 - 1982 = 10$.

$$P(10) = 5.5(1.047)^{10} \approx 8.706$$

In 1992, about 8.706 thousand, or 8706, humpback whales were alive.

In 2004, $t = 2004 - 1982 = 22$.

$$P(22) = 5.5(1.047)^{22} \approx 15.107$$

In 2004, about 15.107 thousand, or 15,107, humpback whales were alive.

b. Use the function values computed in part (a) and others, if desired, and draw the graph.

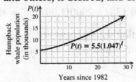

47. a. In 1997, $t = 1997 - 1997 = 0$.

$$M(t) = 50(1.25)^t$$
$$M(0) = 50(1.25)^0 = 50$$

In 1997, about 50 moose were alive.

In 2012, $t = 2012 - 1997 = 15$.

$$M(t) = 50(1.25)^t$$
$$M(15) = 50(1.25)^{15} \approx 1421$$

In 2012, about 1421 moose were alive.

In 2020, $t = 2020 - 1997 = 23$.

$$M(t) = 50(1.25)^t$$
$$M(23) = 50(1.25)^{23} \approx 8470$$

In 2020, about 8470 moose will be alive.

b. Plot the points found in part (a), (0, 50), (15, 1421) and (23, 8470) and additional points as needed and graph the function.

49. *Writing Exercise.*

51. $3x^2 - 48 = 3(x^2 - 16) = 3(x + 4)(x - 4)$

53. $6x^2 + x - 12 = (2x + 3)(3x - 4)$

55. $t^2 - y^2 + 2y - 1$
$$= t^2 - \left(y^2 - 2y + 1\right)$$
$$= t^2 - (y - 1)^2 \qquad \text{Difference of squares}$$
$$= \left[t - (y - 1)\right]\left[t + (y - 1)\right]$$
$$= (t - y + 1)(t + y - 1)$$

57. *Writing Exercise.*

59. Since the bases are the same, the one with the larger exponent is the larger number. Thus $\pi^{2.4}$ is larger.

61. Graph: $f(x) = 2.5^x$

Use a calculator with a power key to construct a table of values. (We will round values of $f(x)$ to the nearest hundredth.) Then plot these points and connect them with a smooth curve.

x	y
0	1
1	2.5
2	6.25
3	15.63
−1	0.4
−2	0.16

63. Graph: $y = 2^x + 2^{-x}$

Construct a table of values, thinking of y as $f(x)$. Then plot these points and connect them with a curve.

$$f(0) = 2^0 + 2^{-0} = 1 + 1 = 2$$

$$f(1) = 2^1 + 2^{-1} = 2 + \frac{1}{2} = 2\frac{1}{2}$$

$$f(2) = 2^2 + 2^{-2} = 4 + \frac{1}{4} = 4\frac{1}{4}$$

$$f(3) = 2^3 + 2^{-3} = 8 + \frac{1}{8} = 8\frac{1}{8}$$

$$f(-1) = 2^{-1} + 2^{-(-1)} = \frac{1}{2} + 2 = 2\frac{1}{2}$$

$$f(-2) = 2^{-2} + 2^{-(-2)} = \frac{1}{4} + 4 = 4\frac{1}{4}$$

$$f(-3) = 2^{-3} + 2^{-(-3)} = \frac{1}{8} + 8 = 8\frac{1}{8}$$

x	y, or $f(x)$
0	2
1	$2\frac{1}{2}$
2	$4\frac{1}{4}$
3	$8\frac{1}{8}$
−1	$2\frac{1}{2}$
−2	$4\frac{1}{4}$
−3	$8\frac{1}{8}$

65. Graph: $y = |2^x - 2|$

We construct a table of values, thinking of y as $f(x)$. Then plot these points and connect them with a curve.

$$f(0) = |2^0 - 2| = |1 - 2| = |-1| = 1$$
$$f(1) = |2^1 - 2| = |2 - 2| = |0| = 0$$
$$f(2) = |2^2 - 2| = |4 - 2| = |2| = 2$$
$$f(3) = |2^3 - 2| = |8 - 2| = |6| = 6$$
$$f(-1) = |2^{-1} - 2| = \left|\frac{1}{2} - 2\right| = \left|-\frac{3}{2}\right| = \frac{3}{2}$$
$$f(-3) = |2^{-3} - 2| = \left|\frac{1}{8} - 2\right| = \left|-\frac{15}{8}\right| = \frac{15}{8}$$
$$f(-5) = |2^{-5} - 2| = \left|\frac{1}{32} - 2\right| = \left|-\frac{63}{32}\right| = \frac{63}{32}$$

x	y, or $f(x)$
0	1
1	0
2	2
3	6
−1	$\frac{3}{2}$
−3	$\frac{15}{8}$
−5	$\frac{63}{32}$

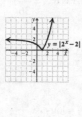

67. Graph: $y = |2^{x^2} - 1|$

We construct a table of values, thinking of y as $f(x)$. Then we plot these points and connect them with a curve.

$$f(0) = |2^{0^2} - 1| = |1 - 1| = 0$$
$$f(1) = |2^{1^2} - 1| = |2 - 1| = 1$$
$$f(2) = |2^{2^2} - 1| = |16 - 1| = 15$$
$$f(-1) = |2^{(-1)^2} - 1| = |2 - 1| = 1$$
$$f(-2) = |2^{(-2)^2} - 1| = |16 - 1| = 15$$

x	y, or $f(x)$
0	0
1	1
2	15
−1	1
−2	15

69.

x	y
0	3
1	1
2	$\frac{1}{3}$
3	$\frac{1}{9}$
−1	9

x	y
3	0
1	1
$\frac{1}{3}$	2
$\frac{1}{9}$	3
9	−1

$y = 3^{-(x-1)}$ $x = 3^{-(y-1)}$

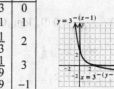

71. Enter the data points (0, 100), (4, 3000) and (8, 14,000) and then use the ExpReg option from the STAT CALC menu of a graphing calculator to find an exponential function that models the data:

$N(t) = 136(1.85)^t$, where $N(t)$ is the number of ruffe, t years after 1984.

In 1990, $t = 1990 - 1984 = 6$

$N(6) \approx 5550$ ruffe.

73. *Writing Exercise.*

75. *Graphing Calculator Exercise*

Exercise Set 9.3

1. The inverse of the function given by $f(x) = 3^x$ is a *logarithmic* function.

3. Logarithm bases are *positive*.

5. $5^2 = 25$, so choice (g) is correct.

7. $5^1 = 5$, so choice (a) is correct.

9. The exponent to which we raise 5 to get 5^x is x, so choice (b) is correct.

11. $5 = 2^x$ is equivalent to $\log_2 5 = x$, so choice (e) is correct.

13. $\log_{10} 1000$ is the exponent to which we raise 10 to get 1000. Since $10^3 = 1000$, $\log_{10} 1000 = 3$.

15. $\log_7 49$ is the exponent to which we raise 7 to get 49. Since $7^2 = 49$, $\log_7 49 = 2$.

17. $\log_5 \frac{1}{25}$ is the exponent to which we raise 5 to get $\frac{1}{25}$. Since $5^{-2} = \frac{1}{25}$, $\log_5 \frac{1}{25} = -2$.

19. Since $8^{-1} = \frac{1}{8}$, $\log_8 \frac{1}{8} = -1$.

21. Since $5^4 = 625$, $\log_5 625 = 4$.

23. Since $7^1 = 7$, $\log_7 7 = 1$.

25. Since $3^0 = 1$, $\log_3 1 = 0$.

27. $\log_6 6^5$ is the exponent to which we raise 6 to get 6^5.

Clearly, this power is 5, so $\log_6 6^5 = 5$.

29. Since $10^{-2} = \frac{1}{100} = 0.01$, $\log_{10} 0.01 = -2$.

31. Since $16^{1/2} = 4$, $\log_{16} 4 = \frac{1}{2}$.

33. Since $9 = 3^2$ and $\left(3^2\right)^{3/2} = 3^3 = 27$, $\log_9 27 = \frac{3}{2}$.

35. Since $1000 = 10^3$ and $\left(10^3\right)^{2/3} = 10^2 = 100$,

$\log_{1000} 100 = \frac{2}{3}$.

37. Since $\log_3 29$ is the power to which we raise 3 to get 29, then 3 raised to this power is 29. That is

$3^{\log_3 29} = 29$.

39. Graph: $y = \log_{10} x$

The equation $y = \log_{10} x$ is equivalent to $10^y = x$. We can find ordered pairs by choosing values for y and computing the corresponding x-values.

For $y = 0$, $x = 10^0 = 1$.

For $y = 1$, $x = 10^1 = 10$.

For $y = 2$, $x = 10^2 = 100$.

For $y = -1$, $x = 10^{-1} = \frac{1}{10}$.

For $y = -2$, $x = 10^{-2} = \frac{1}{100}$.

x, or 10^y	y
1	0
10	1
100	2
$\frac{1}{10}$	-1
$\frac{1}{100}$	-2

↑   ↑ (1) Select y.
(2) Compute x.

We plot the set of ordered pairs and connect the points with a smooth curve.

41. Graph: $y = \log_3 x$

The equation $y = \log_3 x$ is equivalent to $3^y = x$. We can find ordered pairs by choosing values for y and computing the corresponding x-values.

For $y = 0$, $x = 3^0 = 1$.
For $y = 1$, $x = 3^1 = 3$.
For $y = 2$, $x = 3^2 = 9$.
For $y = -1$, $x = 3^{-1} = \frac{1}{3}$.
For $y = -2$, $x = 3^{-2} = \frac{1}{9}$.

x, or 3^y	y
1	0
3	1
9	2
$\frac{1}{3}$	-1
$\frac{1}{9}$	-2

We plot the set of ordered pairs and connect the points with a smooth curve.

43. Graph: $f(x) = \log_6 x$

Think of $f(x)$ as y. Then $y = \log_6 x$ is equivalent to $6^y = x$. We find ordered pairs by choosing values for y and computing the corresponding x-values. Then we plot the points and connect them with a smooth curve.

For $y = 0$, $x = 6^0 = 1$.
For $y = 1$, $x = 6^1 = 6$.
For $y = 2$, $x = 6^2 = 36$.
For $y = -1$, $x = 6^{-1} = \frac{1}{6}$.
For $y = -2$, $x = 6^{-2} = \frac{1}{36}$.

x, or 6^y	y
1	0
6	1
36	2
$\frac{1}{6}$	-1
$\frac{1}{36}$	-2

45. Graph: $f(x) = \log_{2.5} x$

Think of $f(x)$ as y. Then $y = \log_{2.5} x$ is equivalent to $2.5^y = x$. We construct a table of values, plot these points and connect them with a smooth curve.

For $y = 0$, $x = 2.5^0 = 1$.
For $y = 1$, $x = 2.5^1 = 2.5$.
For $y = 2$, $x = 2.5^2 = 6.25$.
For $y = 3$, $x = 2.5^3 = 15.625$.
For $y = -1$, $x = 2.5^{-1} = 0.4$.
For $y = -2$, $x = 2.5^{-2} = 0.16$.

x, or 2.5^y	y
1	0
2.5	1
6.25	2
15.625	3
0.4	-1
0.16	-2

47. Graph $f(x) = 3^x$ (see Exercise Set 9.2, Exercise 7) and $f^{-1}(x) = \log_3 x$ (see Exercise 41 above) on the same

set of axes.

49. $\qquad \downarrow \qquad \downarrow$ The base remains the same.

$x = \log_{10} 8 \Rightarrow 10^x = 8$

$\qquad \uparrow \qquad\qquad \uparrow$ The logarithm is the exponent.

51. $\qquad\qquad \downarrow \quad \downarrow$ The logarithm is the exponent.

$\log_9 9 = 1 \Rightarrow 9^1 = 9$

$\qquad \uparrow \qquad\quad \uparrow$ The base remains the same.

53. $\log_{10} 0.1 = -1$ is equivalent to $10^{-1} = 0.1$.

55. $\log_{10} 7 = 0.845$ is equivalent to $10^{0.845} = 7$.

57. $\log_c m = 8$ is equivalent to $c^8 = m$.

59. $\log_r C = t$ is equivalent to $r^t = C$.

61. $\log_e 0.25 = -1.3863$ is equivalent to $e^{-1.3863} = 0.25$.

63. $\log_r T = -x$ is equivalent to $r^{-x} = T$.

65. $\quad \downarrow \qquad\qquad \downarrow$ The exponent is the logarithm.

$10^2 = 100 \Rightarrow 2 = \log_{10} 100$

$\quad \uparrow \qquad\qquad\quad \uparrow$ The base remains the same.

67. $\quad \downarrow \qquad\qquad \downarrow$ The logarithm is the exponent.

$5^{-3} = \dfrac{1}{125} \Rightarrow -3 = \log_5 \dfrac{1}{125}$

$\quad \uparrow \qquad\qquad\qquad\quad \uparrow$ The base remains the same.

69. $16^{1/4} = 2$ is equivalent to $\dfrac{1}{4} = \log_{16} 2$.

71. $10^{0.4771} = 3$ is equivalent to $0.4771 = \log_{10} 3$.

73. $z^m = 6$ is equivalent to $m = \log_z 6$.

75. $p^t = q$ is equivalent to $t = \log_p q$.

77. $e^3 = 20.0855$ is equivalent to $3 = \log_e 20.0855$.

79. $\log_6 x = 2$

$\quad 6^2 = x$ Converting to an exponential equation

$\quad 36 = x$ Computing 6^2

81. $\log_2 32 = x$

$\quad 2^x = 32$ Converting to an exponential equation

$\quad 2^x = 2^5$

$\quad x = 5$ The exponents must be the same.

83. $\log_x 9 = 1$

$\quad x^1 = 9$ Converting to an exponential equation

$\quad x = 9$ Simplifying x^1

85. $\log_x 7 = \dfrac{1}{2}$

$\quad x^{1/2} = 7$ Converting to an exponential equation

$\quad x = 49$ Squaring both sides

87. $\log_3 x = -2$

$\quad 3^{-2} = x$ Converting to an exponential equation

$\quad \dfrac{1}{9} = x$ Simplifying

89. $\log_{32} x = \dfrac{2}{5}$

$\quad 32^{2/5} = x$ Converting to an exponential equation

$\quad (2^5)^{2/5} = x$

$\quad\quad 4 = x$

91. *Writing Exercise.*

93. $\sqrt{18a^3 b} \sqrt{50ab^7} = \sqrt{18a^3 b \cdot 50ab^7}$

$\qquad\qquad\qquad = \sqrt{900 a^4 b^8}$

$\qquad\qquad\qquad = 30 a^2 b^4$

95. $\sqrt{192x} - \sqrt{75x} = 8\sqrt{3x} - 5\sqrt{3x}$

$\qquad\qquad\qquad = (8-5)\sqrt{3x} = 3\sqrt{3x}$

97. $\dfrac{\sqrt[3]{24xy^8}}{\sqrt[3]{3xy}} = \sqrt[3]{\dfrac{24xy^8}{3xy}} = \sqrt[3]{8y^7} = \sqrt[3]{8y^6} \cdot \sqrt[3]{y} = 2y^2 \sqrt[3]{y}$

99. *Writing Exercise.*

101. Graph: $y = \left(\frac{3}{2}\right)^x$ $\qquad$ Graph: $y = \log_{3/2} x$, or $x = \left(\frac{3}{2}\right)^y$

x	y, or $\left(\frac{3}{2}\right)^x$	x, or $\left(\frac{3}{2}\right)^y$	y
0	1	1	0
1	$\frac{3}{2}$	$\frac{3}{2}$	1
2	$\frac{9}{4}$	$\frac{9}{4}$	2
3	$\frac{27}{8}$	$\frac{27}{8}$	3
-1	$\frac{2}{3}$	$\frac{2}{3}$	-1
-2	$\frac{4}{9}$	$\frac{4}{9}$	-2

103. Graph: $y = \log_3 |x+1|$

x	y
0	0
2	1
8	2
−2	0
−4	1
−10	2

$y = \log_3|x+1|$

105. $\log_4(3x-2) = 2$
$$4^2 = 3x - 2$$
$$16 = 3x - 2$$
$$18 = 3x$$
$$6 = x$$

107. $\log_{10}(x^2 + 21x) = 2$
$$10^2 = x^2 + 21x$$
$$0 = x^2 + 21x - 100$$
$$0 = (x+25)(x-4)$$
$$x = -25 \text{ or } x = 4$$

109. Let $\log_{1/5} 25 = x$. Then
$$\left(\frac{1}{5}\right)^x = 25$$
$$\left(5^{-1}\right)^x = 25$$
$$5^{-x} = 5^2$$
$$-x = 2$$
$$x = -2.$$
Thus, $\log_{1/5} 25 = -2$.

111. $\log_{10}\left(\log_4\left(\log_3 81\right)\right)$
$$= \log_{10}\left(\log_4 4\right) \qquad (\log_3 81 = 4)$$
$$= \log_{10} 1 \qquad\qquad (\log_4 4 = 1)$$
$$= 0$$

113. Let $b = 0$, $x = 1$, and $y = 2$. Then $0^1 = 0^2$, but $1 \neq 2$.
Let $b = 1$, $x = 1$, and $y = 2$. Then $1^1 = 1^2$, but $1 \neq 2$.

Exercise Set 9.4

1. Use the product rule for logarithms.
$\log_7 20 = \log_7(5 \cdot 4) = \log_7 5 + \log_7 4$; choice (e) is correct.

3. Use the quotient rule for logarithms.
$\log_7 \frac{5}{4} = \log_7 5 - \log_7 4$; choice (a) is correct.

5. Use the property of the logarithm of the base to an exponent.
$\log_7 7^4 = 4$; , so choice (c) is correct.

7. $\log_3(81 \cdot 27) = \log_3 81 + \log_3 27$ Using the product rule

9. $\log_4(64 \cdot 16) = \log_4 64 + \log_4 16$ Using the product rule

11. $\log_c rst = \log_c r + \log_c s + \log_c t$ Using the product rule

13. $\log_a 2 + \log_a 10 = \log_a(2 \cdot 10)$ Using the product rule
The result can also be expressed as $\log_a 20$.

15. $\log_c t + \log_c y = \log_c(t \cdot y)$ Using the product rule

17. $\log_a r^8 = 8\log_a r$ Using the power rule

19. $\log_2 y^{1/3} = \frac{1}{3}\log_2 y$ Using the power rule

21. $\log_b C^{-3} = -3\log_b C$ Using the power rule

23. $\log_2 \frac{5}{11} = \log_2 5 - \log_2 11$ Using the quotient rule

25. $\log_b \frac{m}{n} = \log_b m - \log_b n$ Using the quotient rule

27. $\log_a 19 - \log_a 2 = \log_a \frac{19}{2}$ Using the quotient rule

29. $\log_b 36 - \log_b 4 = \log_b \frac{36}{4}$,   Using the quotient rule
or $\log_b 9$

31. $\log_a x - \log_a y = \log_a \frac{x}{y}$ Using the quotient rule

33. $\log_a(xyz)$
$= \log_a x + \log_a y + \log_a z$ Using the product rule

35. $\log_a\left(x^3 z^4\right)$
$= \log_a x^3 + \log_a z^4$ Using the product rule
$= 3\log_a x + 4\log_a z$ Using the power rule

37. $\log_a\left(w^2 x^{-2} y\right)$
$= \log_a w^2 + \log_a x^{-2} + \log_a y$ Using the product rule
$= 2\log_a w - 2\log_a x + \log_a y$ Using the power rule

39. $\log_a \frac{x^5}{y^3 z}$
$= \log_a x^5 - \log_a(y^3 z)$ Using the quotient rule
$= \log_a x^5 - \left(\log_a y^3 + \log_a z\right)$ Using the product rule
$= \log_a x^5 - \log_a y^3 - \log_a z$ Removing the parentheses
$= 5\log_a x - 3\log_a y - \log_a z$ Using the power rule

41. $\log_b \frac{xy^2}{wz^3}$
$= \log_b(xy^2) - \log_b(wz^3)$ Using the quotient rule
$= \log_b x + \log_b y^2 - \left(\log_b w + \log_b z^3\right)$
    Using the product rule
$= \log_b x + \log_b y^2 - \log_b w - \log_b z^3$
    Removing parentheses
$= \log_b x + 2\log_b y - \log_b w - 3\log_b z$
    Using the power rule

43. $\log_a \sqrt{\dfrac{x^7}{y^5 z^8}}$

$= \log_a \left(\dfrac{x^7}{y^5 z^8} \right)^{1/2}$

$= \dfrac{1}{2} \log_a \left(\dfrac{x^7}{y^5 z^8} \right)$ Using the power rule

$= \dfrac{1}{2} \left(\log_a x^7 - \log_a (y^5 z^8) \right)$ Using the quotient rule

$= \dfrac{1}{2} \left[\log_a x^7 - (\log_a y^5 + \log_a z^8) \right]$ Using the
product rule

$= \dfrac{1}{2} \left(\log_a x^7 - \log_a y^5 - \log_a z^8 \right)$ Removing parentheses

$= \dfrac{1}{2} \left(7 \log_a x - 5 \log_a y - 8 \log_a z \right)$ Using the power rule

45. $\log_a \sqrt[3]{\dfrac{x^6 y^3}{a^2 z^7}}$

$= \log_a \left(\dfrac{x^6 y^3}{a^2 z^7} \right)^{1/3}$

$= \dfrac{1}{3} \log_a \left(\dfrac{x^6 y^3}{a^2 z^7} \right)$ Using the power rule

$= \dfrac{1}{3} \left(\log_a (x^6 y^3) - \log_a (a^2 z^7) \right)$ Using the quotient rule

$= \dfrac{1}{3} \left[\log_a x^6 + \log_a y^3 - \left(\log_a a^2 + \log_a z^7 \right) \right]$
   Using the product rule

$= \dfrac{1}{3} \left(\log_a x^6 + \log_a y^3 - \log_a a^2 - \log_a z^7 \right)$
   Removing parentheses

$= \dfrac{1}{3} \left(\log_a x^6 + \log_a y^3 - 2 - \log_a z^7 \right)$

   2 is the number to which
we raise a to get a^2.

$= \dfrac{1}{3} \left(6 \log_a x + 3 \log_a y - 2 - 7 \log_a z \right)$

   Using the power rule

47. $\log_a x + \log_a 3$

$= \log_a 3x$ Using the product rule

49. $8 \log_a x + 3 \log_a z$

$= \log_a x^8 + \log_a z^3$ Using the power rule

$= \log_a \left(x^8 z^3 \right)$ Using the product rule

51. $2 \log_b w - \log_b z - 4 \log_b y$

$= \log_b w^2 - \log_b z - \log_b y^4$ Using the power rule

$= \log_b \dfrac{w^2}{z y^4}$ Using the quotient rule

53. $\log_a x^2 - 2 \log_a \sqrt{x}$

$= \log_a x^2 - \log_a \left(\sqrt{x} \right)^2$ Using the power rule

$= \log_a x^2 - \log_a x$ $\left(\sqrt{x} \right)^2 = x$

$= \log_a \dfrac{x^2}{x}$ Using the quotient rule

$= \log_a x$ Simplifying

55. $\dfrac{1}{2} \log_a x + 5 \log_a y - 2 \log_a x$

$= \log_a x^{1/2} + \log_a y^5 - \log_a x^2$ Using the power rule

$= \log_a x^{1/2} y^5 - \log_a x^2$ Using the product rule

$= \log_a \dfrac{x^{1/2} y^5}{x^2}$ Using the quotient rule

The result can also be expressed as

$\log_a \dfrac{\sqrt{x} y^5}{x^2}$ or as $\log_a \dfrac{y^5}{x^{3/2}}$.

57. $\log_a (x^2 - 9) - \log_a (x + 3)$

$= \log_a \dfrac{x^2 - 9}{x + 3}$ Using the quotient rule

$= \log_a \dfrac{(x+3)(x-3)}{x+3}$

$= \log_a \dfrac{\cancel{(x+3)}(x-3)}{\cancel{x+3}}$ Simplifying

$= \log_a (x - 3)$

59. $\log_b 15 = \log_b (3 \cdot 5)$

$= \log_b 3 + \log_b 5$ Using the product rule

$= 0.792 + 1.161$

$= 1.953$

61. $\log_b \dfrac{3}{5} = \log_b 3 - \log_b 5$ Using the quotient rule

$= 0.792 - 1.161$

$= -0.369$

63. $\log_b \dfrac{1}{5} = \log_b 1 - \log_b 5$ Using the quotient rule

$= 0 - 1.161$ $(\log_b 1 = 0)$

$= -1.161$

65. $\log_b \sqrt{b^3} = \log_b b^{3/2} = \dfrac{3}{2}$ 3/2 is the number to which
we raise b to get $b^{3/2}$.

67. $\log_b 8$

Since 8 cannot be expressed using the numbers 1, 3,
and 5, we cannot find $\log_b 8$ using the given
information.

69. $\log_t t^{10} = 10$ 10 is the exponent to which we raise
t to get t^{10}.

71. $\log_e e^m = m$ m is the exponent to which we raise
e to get e^m.

73. *Writing Exercise.*

75. $|3x + 7| \le 4$

$-4 \le 3x + 7 \le 4$

$-11 \le 3x \le -3$

$-\dfrac{11}{3} \le x \le -1$

The solution is $\left\{ x \left| -\dfrac{11}{3} \le x \le -1 \right. \right\}$ or $\left[-\dfrac{11}{3}, -1 \right]$

77. $x^2 + 4x + 5 = 0$

$x^2 + 4x = -5$

$x^2 + 4x + 4 = -5 + 4$ Completing the square

$(x + 2)^2 = -1$

$x + 2 = \pm\sqrt{-1}$ Principle of square roots

$x = -2 \pm i$

The solutions are $-2 + i$ and $-2 - i$.

79. $x^{1/2} - 6x^{1/4} + 8 = 0$

Let $u = x^{1/4}$.

$u^2 - 6u + 8 = 0$ Substituting

$(u - 4)(u - 2) = 0$

$u = 4$ or $u = 2$

$x^{1/4} = 4$ or $x^{1/4} = 2$

$x = 256$ or $x = 16$ Raising both sides
to the fourth power

Both numbers check. The solutions are 256 and 16.

81. *Writing Exercise.*

83. $\log_a\left(x^8 - y^8\right) - \log_a\left(x^2 + y^2\right)$

$= \log_a \dfrac{x^8 - y^8}{x^2 + y^2}$

$= \log_a \dfrac{\left(x^4 + y^4\right)\left(x^2 + y^2\right)(x + y)(x - y)}{x^2 + y^2}$

$= \log_a\left[\left(x^4 + y^4\right)\left(x^2 - y^2\right)\right]$ Simplifying

$= \log_a\left(x^6 - x^4 y^2 + x^2 y^4 - y^6\right)$

85. $\log_a \sqrt{1 - s^2}$

$= \log_a \left(1 - s^2\right)^{1/2}$

$= \dfrac{1}{2}\log_a\left(1 - s^2\right)$

$= \dfrac{1}{2}\log_a[(1 - s)(1 + s)]$

$= \dfrac{1}{2}\log_a(1 - s) + \dfrac{1}{2}\log_a(1 + s)$

87. $\log_a \dfrac{\sqrt[3]{x^2 z}}{\sqrt[3]{y^2 z^{-2}}}$

$= \log_a \left(\dfrac{x^2 z^3}{y^2}\right)^{1/3}$

$= \dfrac{1}{3}\left(\log_a x^2 z^3 - \log_a y^2\right)$

$= \dfrac{1}{3}\left(2\log_a x + 3\log_a z - 2\log_a y\right)$

$= \dfrac{1}{3}[2 \cdot 2 + 3 \cdot 4 - 2 \cdot 3]$

$= \dfrac{1}{3}(10)$

$= \dfrac{10}{3}$

89. $\log_a x = 2$, so $a^2 = x$.

Let $\log_{1/a} x = n$ and solve for n.

$\log_{1/a} a^2 = n$ Substituting a^2 for x

$\left(\dfrac{1}{a}\right)^n = a^2$

$\left(a^{-1}\right)^n = a^2$

$a^{-n} = a^2$

$-n = 2$

$n = -2$

Thus, $\log_{1/a} x = -2$ when $\log_a x = 2$.

91. $\log_2 80 + \log_2 x = 5$

$\log_2 80x = 5$

$2^5 = 80x$

$\dfrac{32}{80} = x$

$\dfrac{2}{5} = x$

93. True; $\log_a\left(Q + Q^2\right) = \log_a[Q(1 + Q)]$
$= \log_a Q + \log_a(1 + Q) = \log_a Q + \log_a(Q + 1)$

Mid-Chapter Review

1.
$y = 2x - 5$

$x = 2y - 5$

$x + 5 = 2y$

$\dfrac{x + 5}{2} = y$

$f^{-1}(x) = \dfrac{x + 5}{2}$

2. $\log_4 x = 1$

$4^1 = x$

$4 = x$

3. $(f \circ g)(x) = f\left(g(x)\right) = f(x - 5)$
$= (x - 5)^2 + 1$
$= x^2 - 10x + 25 + 1$
$= x^2 - 10x + 26$

4. $h(x) = \sqrt{5x - 3}$

We have $5x - 3$ and take the square root of the
expression, so the two most obvious functions are
$f(x) = \sqrt{x}$ and $g(x) = 5x - 3$.

5.
$y = 6 - x$

$x = 6 - y$

$x - 6 = -y$

$-x + 6 = y$

$g^{-1}(x) = 6 - x$

6. Graph: $f(x) = 2^x + 3$.

7. $\log_4 16 = \log_4 4^2$
$\quad\quad\quad = 2\log_4 4 \quad\quad$ Using the power rule
$\quad\quad\quad = 2 \quad\quad\quad\quad\quad \log_4 4 = 1$

8. $\log_5 \frac{1}{5} = \log_5 5^{-1} = -1\log_5 5 = -1$

9. $\log_{100} 10 = \log_{100} 100^{1/2} = \frac{1}{2}\log_{100} 100 = \frac{1}{2}$

10. $\log_b b = 1$

11. $\log_8 8^{19} = 19\log_8 8 = 19$

12. $\log_t 1 = 0$

13. $\log_x 3 = m$
$\quad\quad x^m = 3$

14. $\log_2 1024 = 10$
$\quad\quad 2^{10} = 1024$

15. $\quad e^t = x$
$\quad \log_e x = t$

16. $\quad 64^{2/3} = 16$
$\quad \log_{64} 16 = \frac{2}{3}$

17. $\log\sqrt{\dfrac{x^2}{yz^3}} = \log\left(\dfrac{x^2}{yz^3}\right)^{1/2} = \frac{1}{2}\log\dfrac{x^2}{yz^3}$
$\quad\quad = \frac{1}{2}\left(\log x^2 - \log y - \log z^3\right)$
$\quad\quad = \frac{1}{2}\left(2\log x - \log y - 3\log z\right)$
$\quad\quad = \log x - \frac{1}{2}\log y - \frac{3}{2}\log z$

18. $\log a - 2\log b - \log c = \log a - \log b^2 - \log c = \log\dfrac{a}{b^2 c}$

19. $\log_x 64 = 3$
$\quad\quad x^3 = 64$
$\quad\quad x^3 = 4^3$
$\quad\quad\; x = 4$

20. $\log_3 x = -1$
$\quad\quad 3^{-1} = x$
$\quad\quad \frac{1}{3} = x$

Exercise Set 9.5

1. True

3. True

5. True; $\log 18 - \log 2 = \log\dfrac{18}{2} = \log 9$

7. True; $\ln 81 = \ln 9^2 = 2\ln 9$

9. True; see Example 7.

11. 0.8451

13. 1.1367

15. Since $10^3 = 1000,\ \log 1000 = 3.$

17. -0.1249

19. 13.0014

21. 50.1187

23. 0.0011

25. 2.1972

27. -5.0832

29. 96.7583

31. 15.0293

33. 0.0305

35. We will use common logarithms for the conversion. Let $a = 10$, $b = 3$, and $M = 28$ and substitute in the change-of-base formula.
$$\log_b M = \frac{\log_a M}{\log_a b}$$
$$\log_3 28 = \frac{\log_{10} 28}{\log_{10} 3}$$
$$\approx \frac{1.447158031}{0.477121254}$$
$$\approx 3.0331$$

37. We will use common logarithms for the conversion. Let $a = 10$, $b = 2$, and $M = 100$ and substitute in the change-of-base formula.
$$\log_2 100 = \frac{\log_{10} 100}{\log_{10} 2} \approx \frac{2}{0.3010} \approx 6.6439$$

39. We will use natural logarithms for the conversion. Let $a = e$, $b = 4$, and $M = 5$ and substitute in the change-of-base formula.
$$\log_4 5 = \frac{\ln 5}{\ln 4} \approx \frac{1.6094}{1.3863} \approx 1.1610$$

41. We will use natural logarithms for the conversion. Let $a = e$, $b = 0.1$, and $M = 2$ and substitute in the change-of-base formula.
$$\log_{0.1} 2 = \frac{\ln 2}{\ln 0.1} \approx \frac{0.6931}{-2.3026} \approx -0.3010$$

43. We will use common logarithms for the conversion. Let $a = 10$, $b = 2$, and $M = 0.1$ and substitute in the change-of-base formula.
$$\log_2 0.1 = \frac{\log_{10} 0.1}{\log_{10} 2} \approx \frac{-1}{0.3010} \approx -3.3219$$

45. We will use natural logarithms for the conversion. Let $a = e$, $b = \pi$, and $M = 10$ and substitute in the change-of-base formula.
$$\log_\pi 10 = \frac{\ln 10}{\ln \pi} \approx \frac{2.3026}{1.1447} \approx 2.0115$$

47. Graph: $f(x) = e^x$

We find some function values with a calculator. We use these values to plot points and draw the graph.

x	e^x
1	2.7
2	7.4
3	20.1
−1	0.4
−2	0.1

The domain is the set of real numbers and the range is $(0, \infty)$.

49. Graph: $f(x) = e^x + 3$

We find some function values, plot points, and draw the graph.

x	$e^x + 3$
0	4
1	5.72
2	10.39
−1	3.37
−2	3.14

The domain is the set of real numbers and the range is $(3, \infty)$.

51. Graph: $f(x) = e^x - 2$

We find some function values, plot points, and draw the graph.

x	$e^x - 2$
0	−1
1	0.72
2	5.4
−1	−1.6
−2	−1.9

The domain is the set of real numbers and the range is $(-2, \infty)$.

53. Graph: $f(x) = 0.5e^x$

We find some function values, plot points, and draw the graph.

x	$0.5e^x$
0	0.5
1	1.36
2	3.69
−1	0.18
−2	0.07

The domain is the set of real numbers and the range is $(0, \infty)$.

55. Graph: $f(x) = 0.5e^{2x}$

We find some function values, plot points, and draw the graph.

x	$0.5e^{2x}$
0	0.5
1	3.69
2	27.30
−1	0.07
−2	0.01

The domain is the set of real numbers and the range is $(0, \infty)$.

57. Graph: $f(x) = e^{x-3}$

We find some function values, plot points, and draw the graph.

x	e^{x-3}
2	0.37
3	1
4	2.72
−2	0.01

The domain is the set of real numbers and the range is $(0, \infty)$.

59. Graph: $f(x) = e^{x+2}$

We find some function values, plot points, and draw the graph.

x	e^{x+2}
−1	2.72
−2	1
−3	0.37
−4	0.14

The domain is the set of real numbers and the range is $(0, \infty)$.

61. Graph: $f(x) = -e^x$

We find some function values, plot points, and draw the graph.

x	$-e^x$
0	−1
1	−2.72
2	−7.39
−1	−0.37
−3	−0.05

The domain is the set of real numbers and the range is $(-\infty, 0)$.

63. Graph: $g(x) = \ln x + 1$

We find some function values, plot points, and draw the graph.

x	$\ln x + 1$
1	1
3	2.10
5	2.61
7	2.95

The domain is $(0, \infty)$ and the range is the set of real numbers.

65. Graph: $g(x) = \ln x - 2$

x	$\ln x - 2$
1	1
3	-0.90
5	-0.39
7	-0.05

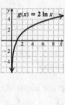

The domain is $(0, \infty)$ and the range is the set of real numbers.

67. Graph: $f(x) = 2 \ln x$

x	$2 \ln x$
0.5	-1.4
1	0
2	1.4
3	2.2
4	2.8
5	3.2
6	3.6

The domain is $(0, \infty)$ and the range is the set of real numbers.

69. Graph: $g(x) = -2 \ln x$

x	$-2 \ln x$
0.5	1.4
1	0
2	-1.4
3	-2.2
4	-2.8
5	-3.2
6	-3.6

The domain is $(0, \infty)$ and the range is the set of real numbers.

71. Graph: $f(x) = \ln(x + 2)$

We find some function values, plot points, and draw the graph.

x	$\ln(x+2)$
1	1.10
3	1.61
5	1.95
-1	0
-2	Undefined

The domain is $(-2, \infty)$ and the range is the set of real numbers.

73. Graph: $g(x) = \ln(x - 1)$

We find some function values, plot points, and draw the graph.

x	$\ln(x-1)$
2	0
3	0.69
4	1.10
6	1.61

The domain is $(1, \infty)$ and the range is the set of real numbers.

75. *Writing Exercise.*

77. $f(-1) = \dfrac{1}{-1+2} = 1$

79. $(g - f)(x) = 5x - 8 - \dfrac{1}{x+2}$

81. For $\dfrac{f(x)}{g(x)} = \dfrac{\dfrac{1}{x+2}}{5x - 8} = \dfrac{1}{(x+2)(5x-8)}$,

$x + 2 = 0 \quad or \quad 5x - 8 = 0$

$x = -2 \quad or \quad x = \dfrac{8}{5}$

The domain is $\{x \mid x$ is a real number *and* $x \neq -2$ *and* $x \neq \frac{8}{5}\}$ or $(-\infty, -2) \cup \left(-2, \frac{8}{5}\right) \cup \left(\frac{8}{5}, \infty\right)$.

83. *Writing Exercise.*

85. We use the change-of-base formula.

$$\log_6 81 = \frac{\log 81}{\log 6}$$

$$= \frac{\log 3^4}{\log(2 \cdot 3)}$$

$$= \frac{4 \log 3}{\log 2 + \log 3}$$

$$\approx \frac{4(0.477)}{0.301 + 0.477}$$

$$\approx 2.452$$

87. We use the change-of-base formula.

$$\log_{12} 36 = \frac{\log 36}{\log 12}$$

$$= \frac{\log(2 \cdot 3)^2}{\log(2^2 \cdot 3)}$$

$$= \frac{2 \log(2 \cdot 3)}{\log 2^2 + \log 3}$$

$$= \frac{2(\log 2 + \log 3)}{2 \log 2 + \log 3}$$

$$\approx \frac{2(0.301 + 0.477)}{2(0.301) + 0.477}$$

$$\approx 1.442$$

89. Use the change-of-base formula with $a = e$ and $b = 10$. We obtain

$$\log M = \frac{\ln M}{\ln 10}.$$

91. $\log(492x) = 5.728$

$10^{5.728} = 492x$

$\dfrac{10^{5.728}}{492} = x$

$1086.5129 \approx x$

93. $\log 692 + \log x = \log 3450$

$\log x = \log 3450 - \log 692$

$\log x = \log \dfrac{3450}{692}$

$x = \dfrac{3450}{692}$

$x \approx 4.9855$

95. a. Domain: $\{x \mid x > 0\}$, or $(0, \infty)$; range:

$\{y \mid y < 0.5135\}$, or $(-\infty, 0.5135)$;

b. $[-1, 5, -10, 5]$;

c.

$y = 3.4 \ln x - 0.25e^x$

97. a. Domain $\{x \mid x > 0\}$, or $(0, \infty)$;

range: $\{y \mid y > -0.2453\}$, or $(-0.2453, \infty)$

b. $[-1, 5, -1, 10]$;

c.

$y = 2x^3 \ln x$

99. *Writing Exercise.*

Connecting the Concepts

1. $\log(2x) = 3$

$10^3 = 2x$

$1000 = 2x$

$500 = x$

2. $2^{x+1} = 2^9$

$x + 1 = 9$

$x = 8$

3. $e^t = 5$

$t = \ln 5 \approx 1.6094$

4. $\log_3(x^2 + 1) = \log_3 26$

$x^2 + 1 = 26$

$x^2 - 25 = 0$

$(x + 5)(x - 5) = 0$

$x = -5 \ \text{or} \ x = 5$

5. $\ln 2x = \ln 8$

$2x = 8$

$x = 4$

6. $3^{5x} = 4$

$\ln 3^{5x} = \ln 4$

$5x \ln 3 = \ln 4$

$x = \dfrac{\ln 4}{5 \ln 3} \approx 0.2524$

7. $\log_2(x + 1) = -1$

$2^{-1} = x + 1$

$\dfrac{1}{2} = x + 1$

$-\dfrac{1}{2} = x$

8. $9^{x+1} = 27^{-x}$

$3^{2(x+1)} = 3^{-3x}$

$2(x + 1) = -3x$

$2x + 2 = -3x$

$5x = -2$

$x = -\dfrac{2}{5}$

Exercise Set 9.6

1. False

3. True

5. If we take the common logarithm on both sides, we see that choice (d) is correct.

7. By the product rule for logarithms,

$\log_5 x + \log_5(x - 2) = \log_5[x(x - 2)] = \log_5(x^2 - 2x)$,

so choice (b) is correct.

9. $3^{2x} = 81$

$3^{2x} = 3^4$

$2x = 4$ Equating the exponents

$x = 2$

11. $4^x = 32$

$2^{2x} = 2^5$

$2x = 5$ Equating the exponents

$x = \dfrac{5}{2}$

13. $2^x = 10$

$\log 2^x = \log 10$ Taking log on both sides

$\log 2^x = 1$ $\log 10 = 1$

$x \log 2 = 1$

$x = \dfrac{1}{\log 2}$

$x \approx 3.322$ Using a calculator

15. $2^{x+5} = 16$

$2^{x+5} = 2^4$

$x + 5 = 4$ Equating the exponents

$x = -1$

17. $8^{x-3} = 19$

$\log 8^{x-3} = \log 19$ Taking log on both sides

$(x - 3)\log 8 = \log 19$

$x - 3 = \dfrac{\log 19}{\log 8}$

$x = \dfrac{\log 19}{\log 8} + 3$

$x \approx 4.416$ Using a calculator

19. $e^t = 50$

$\ln e^t = \ln 50$ Taking ln on both sides

$t = \ln 50$ $\ln e^t = t \ln e = t \cdot 1 = t$

$t \approx 3.912$

21. $e^{-0.02t} = 8$

$\ln e^{-0.02t} = \ln 8$ Taking ln on both sides

$-0.02t = \ln 8$

$t = \dfrac{\ln 8}{-0.02}$

$t \approx -103.972$

23. $4.9^x - 87 = 0$

$4.9^x = 87$

$\log 4.9^x = \log 87$

$x \log 4.9 = \log 87$

$x = \dfrac{\log 87}{\log 4.9}$

$x \approx 2.810$

25. $19 = 2e^{4x}$

$\dfrac{19}{2} = e^{4x}$

$\ln\left(\dfrac{19}{2}\right) = \ln e^{4x}$

$\ln\left(\dfrac{19}{2}\right) = 4x$

$\dfrac{\ln\left(\dfrac{19}{2}\right)}{4} = x$

$0.563 \approx x$

27. $7 + 3e^{-x} = 13$

$3e^{-x} = 6$

$e^{-x} = 2$

$\ln e^{-x} = \ln 2$

$-x = \ln 2$

$x = -\ln 2$

$x \approx -0.693$

29. $\log_3 x = 4$

$x = 3^4$ Writing an equivalent exponential
          equation

$x = 81$

31. $\log_4 x = -2$

$x = 4^{-2}$ Writing an equivalent
                exponential equation

$x = \dfrac{1}{4^2}$, or $\dfrac{1}{16}$

33. $\ln x = 5$

$x = e^5$ Writing an equivalent
             exponential equation

$x \approx 148.413$

35. $\ln 4x = 3$

$4x = e^3$

$x = \dfrac{e^3}{4} \approx 5.021$

37. $\log x = 1.2$ The base is 10.

$x = 10^{1.2}$

$x \approx 15.849$

39. $\ln(2x + 1) = 4$

$2x + 1 = e^4$

$2x = e^4 - 1$

$x = \dfrac{e^4 - 1}{2} \approx 26.799$

41. $\ln x = 1$

$x = e \approx 2.718$

43. $5 \ln x = -15$

$\ln x = -3$

$x = e^{-3} \approx 0.050$

45. $\log_2(8 - 6x) = 5$

$8 - 6x = 2^5$

$8 - 6x = 32$

$-6x = 24$

$x = -4$

The answer checks. The solution is –4.

47. $\log(x - 9) + \log x = 1$ The base is 10.

$\log_{10}[(x-9)(x)] = 1$ Using the product rule

$x(x - 9) = 10^1$

$x^2 - 9x = 10$

$x^2 - 9x - 10 = 0$

$(x + 1)(x - 10) = 0$

$x = -1 \ or \ x = 10$

Check: For -1:

$$\frac{\log(x-9) + \log x = 1}{\log(-1-9) + \log(-1) \ \Big| \ 1}$$

$\log(-10) + \log(-1) \overset{?}{=} 1$ FALSE

For 10:

$$\frac{\log(x-9) + \log x = 1}{\log(10-9) + \log(10) \ \Big| \ 1}$$

$\log 1 + \log 10 \ \Big| $

$0 + 1 \ \Big| $

$1 \overset{?}{=} 1$ TRUE

The number –1 does not check, because negative
numbers do not have logarithms. The solution is 10.

49. $\log x - \log(x + 3) = 1$ The base is 10.

$\log_{10}\dfrac{x}{x+3} = 1$ Using the quotient rule

$\dfrac{x}{x+3} = 10^1$

$x = 10(x + 3)$

$x = 10x + 30$

$-9x = 30$

$x = -\dfrac{10}{3}$

The number $-\dfrac{10}{3}$ does not check. The equation has no

solution.

51. We observe that since $\log(2x + 1) = \log 5$, then

$2x + 1 = 5$

$2x = 4$

$x = 2$

53.
$$\log_4(x+3) = 2 + \log_4(x-5)$$
$$\log_4(x+3) - \log_4(x-5) = 2$$
$$\log_4 \frac{x+3}{x-5} = 2 \quad \text{Using the quotient rule}$$
$$\frac{x+3}{x-5} = 4^2$$
$$\frac{x+3}{x-5} = 16$$
$$x+3 = 16(x-5)$$
$$x+3 = 16x - 80$$
$$83 = 15x$$
$$\frac{83}{15} = x$$

The number $\frac{83}{15}$ checks. It is the solution.

55.
$$\log_7(x+1) + \log_7(x+2) = \log_7 6$$
$$\log_7[(x+1)(x+2)] = \log_7 6 \quad \text{Using the}$$
$$\text{product rule}$$
$$\log_7(x^2 + 3x + 2) = \log_7 6$$
$$x^2 + 3x + 2 = 6 \quad \text{Using the property}$$
$$\text{of logarithmic equality}$$
$$x^2 + 3x - 4 = 0$$
$$(x+4)(x-1) = 0$$
$$x = -4 \quad or \quad x = 1$$

The number 1 checks, but -4 does not.
The solution is 1.

57.
$$\log_5(x+4) + \log_5(x-4) = \log_5 20$$
$$\log_5[(x+4)(x-4)] = \log_5 20 \quad \text{Using the}$$
$$\text{product rule}$$
$$\log_5(x^2 - 16) = \log_5 20$$
$$x^2 - 16 = 20 \quad \text{Using the property}$$
$$\text{of logarithmic equality}$$
$$x^2 = 36$$
$$x = \pm 6$$

The number 6 checks, but -6 does not.
The solution is 6.

59.
$$\ln(x+5) + \ln(x+1) = \ln 12$$
$$\ln[(x+5)(x+1)] = \ln 12$$
$$\ln(x^2 + 6x + 5) = \ln 12$$
$$x^2 + 6x + 5 = 12$$
$$x^2 + 6x - 7 = 0$$
$$(x+7)(x-1) = 0$$
$$x = -7 \quad or \quad x = 1$$

The number -7 does not check, but 1 does.
The solution is 1.

61.
$$\log_2(x-3) + \log_2(x+3) = 4$$
$$\log_2[(x-3)(x+3)] = 4$$
$$(x-3)(x+3) = 2^4$$
$$x^2 - 9 = 16$$
$$x^2 = 25$$
$$x = \pm 5$$

The number 5 checks, but -5 does not.
The solution is 5.

63.
$$\log_{12}(x+5) - \log_{12}(x-4) = \log_{12} 3$$
$$\log_{12} \frac{x+5}{x-4} = \log_{12} 3$$
$$\frac{x+5}{x-4} = 3 \quad \text{Using the property}$$
$$\text{of logarithmic equality}$$
$$x+5 = 3(x-4)$$
$$x+5 = 3x - 12$$
$$17 = 2x$$
$$\frac{17}{2} = x$$

The number $\frac{17}{2}$ checks and is the solution.

65.
$$\log_2(x-2) + \log_2 x = 3$$
$$\log_2[(x-2)(x)] = 3$$
$$x(x-2) = 2^3$$
$$x^2 - 2x = 8$$
$$x^2 - 2x - 8 = 0$$
$$(x-4)(x+2) = 0$$
$$x = 4 \quad or \quad x = -2$$

The number 4 checks, but -2 does not.
The solution is 4.

67. *Writing Exercise.*

69.
$$\frac{a^2-4}{a^2+a} \cdot \frac{a^2-a-2}{a^3} = \frac{(a+2)(a-2)}{a\,\cancel{(a+1)}} \cdot \frac{\cancel{(a+1)}(a-2)}{a^3}$$
$$= \frac{(a+2)(a-2)^2}{a^4}$$

71.
$$\frac{2}{m+1} + \frac{3}{m-5} \quad \text{LCD is } (m+1)(m-5)$$
$$= \frac{2}{m+1} \cdot \frac{m-5}{m-5} + \frac{3}{m-5} \cdot \frac{m+1}{m+1}$$
$$= \frac{2m-10+3m+3}{(m+1)(m-5)}$$
$$= \frac{5m-7}{(m+1)(m-5)}$$

73.
$$\frac{\frac{3}{x} - \frac{2}{xy}}{\frac{2}{x^2} + \frac{1}{xy}} = \frac{\frac{3}{x} - \frac{2}{xy}}{\frac{2}{x^2} + \frac{1}{xy}} \cdot \frac{x^2 y}{x^2 y} = \frac{3xy - 2x}{2y + x} = \frac{x(3y-2)}{2y+x}$$

75. *Writing Exercise.*

77.
$$8^x = 16^{3x+9}$$
$$(2^3)^x = (2^4)^{3x+9}$$
$$2^{3x} = 2^{12x+36}$$
$$3x = 12x + 36$$
$$-36 = 9x$$
$$-4 = x$$

The solution is -4.

79.
$$\log_6(\log_2 x) = 0$$
$$\log_2 x = 6^0$$
$$\log_2 x = 1$$
$$x = 2^1$$
$$x = 2$$

The solution is 2.

81. $\log_5 \sqrt{x^2 - 9} = 1$

$\sqrt{x^2 - 9} = 5^1$

$\left(\sqrt{x^2 - 9}\right)^2 = 5^2$

$x^2 - 9 = 25$

$x^2 = 34$

$x = \pm\sqrt{34}$

The solutions are $\pm\sqrt{34}$.

83. $2^{x^2 + 4x} = \dfrac{1}{8}$

$2^{x^2 + 4x} = \dfrac{1}{2^3}$

$2^{x^2 + 4x} = 2^{-3}$

$x^2 + 4x = -3$

$x^2 + 4x + 3 = 0$

$(x + 3)(x + 1) = 0$

$x = -3 \ \ or \ \ x = -1$

The solutions are -3 and -1.

85. $\log_5 |x| = 4$

$|x| = 5^4$

$|x| = 625$

$x = 625 \ \ or \ \ x = -625$

The solutions are 625 and -625.

87. $\log\sqrt{2x} = \sqrt{\log 2x}$

$\log(2x)^{1/2} = \sqrt{\log 2x}$

$\dfrac{1}{2}\log 2x = \sqrt{\log 2x}$

$\dfrac{1}{4}(\log 2x)^2 = \log 2x$ Squaring both sides

$\dfrac{1}{4}(\log 2x)^2 - \log 2x = 0$

Let $u = \log 2x$.

$\dfrac{1}{4}u^2 - u = 0$

$u\left(\dfrac{1}{4}u - 1\right) = 0$

$u = 0 \qquad or \qquad \dfrac{1}{4}u - 1 = 0$

$u = 0 \qquad or \qquad \dfrac{1}{4}u = 1$

$u = 0 \qquad or \qquad u = 4$

$\log 2x = 0 \quad or \quad \log 2x = 4$ Replacing u
with $\log 2x$

$2x = 10^0 \quad or \quad 2x = 10^4$

$2x = 1 \qquad or \quad 2x = 10,000$

$x = \dfrac{1}{2} \qquad or \qquad x = 5000$

Both numbers check. The solutions are $\dfrac{1}{2}$ and 5000.

89. $3^{x^2} \cdot 3^{4x} = \dfrac{1}{27}$

$3^{x^2 + 4x} = 3^{-3}$

$x^2 + 4x = -3$ The exponents must be equal.

$x^2 + 4x + 3 = 0$

$(x + 1)(x + 3) = 0$

$x = -1 \ \ or \ x = -3$

Both numbers check. The solutions are -1 and -3.

91. $\log x^{\log x} = 25$

$\log x(\log x) = 25$ Using the power rule

$(\log x)^2 = 25$

$\log x = \pm 5$

$x = 10^5 \qquad or \qquad x = 10^{-5}$

$x = 100,000 \quad or \quad x = \dfrac{1}{100,000}$

Both numbers check. The solutions are $100,000$ and

$\dfrac{1}{100,000}$.

93. $\left(81^{x-2}\right)\left(27^{x+1}\right) = 9^{2x-3}$

$\left[\left(3^4\right)^{x-2}\right]\left[\left(3^3\right)^{x+1}\right] = \left(3^2\right)^{2x-3}$

$\left(3^{4x-8}\right)\left(3^{3x+3}\right) = 3^{4x-6}$

$3^{7x-5} = 3^{4x-6}$

$7x - 5 = 4x - 6$

$3x = -1$

$x = -\dfrac{1}{3}$

The solution is $-\dfrac{1}{3}$.

95. $2^y = 16^{x-3} \qquad and \qquad 3^{y+2} = 27^x$

$2^y = \left(2^4\right)^{x-3} \quad and \qquad 3^{y+2} = \left(3^3\right)^x$

$y = 4x - 12 \quad and \qquad y + 2 = 3x$

$12 = 4x - y \quad and \qquad 2 = 3x - y$

Solving this system of equations we get $x = 10$ and $y = 28$. Then $x + y = 10 + 28 = 38$.

97. Find the first coordinate of the point of intersection of $y_1 = \ln x$ and $y_2 = \log x$. The value of x for which the natural logarithm of x is the same as the common logarithm of x is 1.

Exercise Set 9.7

1. b

3. c

5. a. Replace $A(t)$ with 5000 and solve for t.

$A(t) = 77(1.283)^t$

$5000 = 77(1.283)^t$

$\dfrac{5000}{77} = 1.283^t$

$\ln\dfrac{5000}{77} = \ln 1.283^t$

$\ln\dfrac{5000}{77} = t\ln 1.283$

$\dfrac{\ln\dfrac{5000}{77}}{\ln 1.283} = t$

$16.7 \approx t$

The number of known asteroids first reached 5000 was about 17 yr after 1990, or in 2006.

b. $A(0) = 77(1.283)^0 = 77 \cdot 1 = 77$, so to find the doubling time, we replace $A(t)$ with $2(77)$, or

154 and solve for t.

$$154 = 77(1.283)^t$$
$$2 = 1.283^t$$
$$\ln 2 = \ln 1.283^t$$
$$\ln 2 = t \ln 1.283$$
$$\frac{\ln 2}{\ln 1.283} = t$$
$$2.8 \approx t$$

The doubling time is about 2.8 years.

7. a. Replace $A(t)$ with 35,000 and solve for t.

$$A(t) = 29,000(1.03)^t$$
$$35,000 = 29,000(1.03)^t$$
$$1.207 \approx (1.03)^t$$
$$\log 1.207 \approx \log(1.03)^t$$
$$\log 1.207 \approx t \log 1.03$$
$$\frac{\log 1.207}{\log 1.03} \approx t$$
$$6.4 \approx t$$

The amount due will reach \$35,000 after about 6.4 years.

b. Replace $A(t)$ with 2(29,000), or 58,000, and solve for t.

$$58,000 = 29,000(1.03)^t$$
$$2 = (1.03)^t$$
$$\log 2 = \log(1.03)^t$$
$$\log 2 = t \log 1.03$$
$$\frac{\log 2}{\log 1.03} = t$$
$$23.4 \approx t$$

The doubling time is about 23.4 years.

9. a. Replace $f(n)$ with 440 and solve for n.

$$f(n) = 27.5\left(\sqrt[12]{2}\right)^{n-1}$$
$$440 = 27.5\left(\sqrt[12]{2}\right)^{n-1}$$
$$16 = 2^{(n-1)/12}$$
$$\log 16 = \log 2^{(n-1)/12}$$
$$\log 16 = \frac{n-1}{12}\log 2$$
$$\frac{12\log 16}{\log 2} = n-1$$
$$1 + \frac{12\log 16}{\log 2} = n$$
$$49 = n$$

The 49th key has a frequency of 440 Hz.

b. Replace $f(n)$ with 2(27.5), or 55, and solve for n.

$$55 = 27.5\left(\sqrt[12]{2}\right)^{n-1}$$
$$2 = 2^{(n-1)/12}$$
$$\log 2 = \log 2^{(n-1)/12}$$
$$\log 2 = \frac{n-1}{12}\log 2$$
$$12 = n-1$$
$$13 = n$$

From key 1 to key 13 is 12 keys. It takes 12 keys for the frequency to double.

11. a. Substitute 8000 for $A(t)$ and solve for t.

$$A(t) = 2.05(1.8)^t$$
$$8000 = 2.05(1.8)^t$$
$$\frac{8000}{2.05} = (1.8)^t$$
$$\log\frac{8000}{2.05} = \log(1.8)^t$$
$$\log\frac{8000}{2.05} = t\log 1.8$$
$$\frac{\log\frac{8000}{2.05}}{\log 1.8} = t$$
$$14.1 \approx t$$

There was \$8 billion in e-book net sales 14 years after 2002, or in 2016.

b. Replace $V(t)$ with 2(12), or 24, and solve for t.

$$A(t) = 2.05(1.8)^t$$
$$4.1 = 2.05(1.8)^t$$
$$2 = (1.8)^t$$
$$\log 2 = \log(1.8)^t$$
$$\log 2 = t\log 1.8$$
$$\frac{\log 2}{\log 1.8} = t$$
$$1.2 \approx t$$

The doubling time is approximately 1.2 years.

13. $\text{pH} = -\log\left[H^+\right]$

$$= -\log\left(1.3 \times 10^{-5}\right)$$
$$\approx -(-4.886057) \quad \text{Using a calculator}$$
$$\approx 4.9$$

The pH of fresh-brewed coffee is about 4.9.

15. $\text{pH} = -\log[H^+]$

$$7.0 = -\log[H^+]$$
$$-7.0 = \log[H^+]$$
$$10^{-7.0} = [H^+] \quad \text{Converting to an}$$
$$\qquad\qquad\quad \text{exponential equation}$$

The hydrogen ion concentration is 10^{-7} moles per liter.

17. $L = 10 \cdot \log\dfrac{I}{I_0}$

$$= 10 \cdot \log\frac{10}{10^{-12}} = 10 \cdot \log 10^{13}$$
$$= 10 \cdot 13 \cdot \log 10$$
$$= 130$$

The sound level is 130 dB.

19. $$L = 10 \cdot \log\frac{I}{I_0}$$
$$88 = 10 \cdot \log\frac{I}{10^{-12}}$$
$$8.8 = \log\frac{I}{10^{-12}}$$
$$8.8 = \log I - \log 10^{-12}$$
$$8.8 = \log I - (-12)$$
$$-3.2 = \log I$$
$$10^{-3.2} = I$$
$$6.3 \times 10^{-4} \text{ W/m}^2 \approx I$$

The intensity of the sound is 6.3×10^{-4} W/m^2.

21.
$$M = \log\frac{v}{1.34}$$
$$7.5 = \log\frac{v}{1.34}$$
$$10^{7.5} = \frac{v}{1.34}$$
$$1.34\left(10^{7.5}\right) = v$$
$$42,400,000 \approx v$$

Approximately 42.4 million messages per day are sent by that network.

23. a. Substitute 0.025 for k :
$$P(t) = P_0\, e^{0.025t}$$

b. To find the balance after one year, replace P_0 with 5000 and t with 1. We find $P(1)$:
$$P(1) = 5000e^{0.025(1)} = 5000e^{0.025} \approx \$5126.58$$
To find the balance after 2 years, replace P_0 with 5000 and t with 2. We find $P(2)$:
$$P(2) = 5000e^{0.025(2)} = 5000e^{0.05} \approx \$5256.36$$

c. To find the doubling time, replace P_0 with 5000 and $P(t)$ with 10,000 and solve for t.
$$10,000 = 5000e^{0.025t}$$
$$2 = e^{0.025t}$$
$$\ln 2 = \ln e^{0.025t}$$
$$\ln 2 = 0.025t$$
$$\frac{\ln 2}{0.025} = t$$
$$27.7 \approx t$$
The investment will double in about 27.7 years.

25. a. $P(t) = 324e^{0.0073t}$, where $P(t)$ is in millions and t is the number of years after 2016.

b. In 2025, $t = 2025 - 2016 = 9$. Find $P(9)$.
$$P(9) = 324e^{0.0073(9)} = 324e^{0.0657} \approx 346$$
The U.S. population will be about 346 million in 2025.

c. Substitute 400 for $P(t)$ and solve for t.
$$400 = 324e^{0.0073t}$$
$$\frac{400}{324} = e^{0.0073t}$$
$$\ln\frac{400}{324} = \ln e^{0.0073t}$$
$$\ln\frac{400}{324} = 0.0073t$$
$$\frac{\ln\frac{400}{324}}{0.0073} = t$$
$$29 \approx t$$
The U.S. population will reach 400 million about 29 yr after 2016, or in 2045.

27. The exponential growth function is $P(t) = P_0 e^{3.24t}$. We replace $P(t)$ with $2P_0$ and solve for t.
$$P(t) = P_0 e^{0.0324t}$$
$$P2_0 = P_0 e^{0.0324t}$$
$$2 = e^{0.0324t}$$
$$\ln 2 = \ln e^{0.0324t}$$
$$\ln 2 = 0.0324t$$
$$\frac{\ln 2}{0.0324} = t$$
$$21.4 \approx t$$
The doubling time is about 21.4 yr.

29. $Y(x) = 88.5\ln\frac{x}{7.4}$

a. $Y(10) = 88.5\ln\dfrac{10}{7.4} \approx 27$

The world population will reach 10 billion about 27 yr after 2016, or in 2043.

b. $Y(12) = 88.5\ln\dfrac{12}{7.4} \approx 43$

The world population will reach 12 billion about 43 yr after 2016, or in 2059.

c. Plot the points found in parts (a) and (b) and others as necessary and draw the graph.

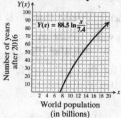

31. $s(t) = 1.1 + 36\ln t$, t is the number of years after 2005.

a. In 2012, $t = 2012 - 2005 = 7$. Find $s(7)$.
$$s(7) = 1.1 + 36\ln 7 \approx 71$$
About 71% of Americans ages 30-49 used social networking sites in 2012.

b. Plot the points found in part (a) and others as necessary and draw the graph.

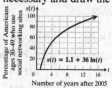

c.
$$95 = 1.1 + 36\ln t$$
$$93.9 = 36\ln t$$
$$\frac{93.9}{36} = \ln t$$
$$e^{93.9/36} = t$$
$$14 \approx t$$
95% of Americans ages 30-49 will use social network sites about 14 yr after 2005, or in 2019.

d. The minimum of the domain is 1. The maximum occurs when 100% is achieved.

$$100 = 1.1 + 36\ln t$$
$$98.9 = 36\ln t$$
$$\frac{98.9}{36} = \ln t$$
$$e^{98.9/36} = t$$
$$15.6 \approx t$$

The domain is [1, 15.6].

33. a. We start with the exponential growth equation. Substituting 12 for P_0, we have

$$P(t) = 12e^{kt}, \text{ where } t \text{ is the number of years after 1967.}$$

To find the exponential growth rate k, observe that the ticket price was \$2670 in 2015, which is 48 years after 1967. We substitute and solve for k.

$$P(t) = 12e^{k \cdot 48}$$
$$2670 = 12e^{48k}$$
$$222.5 = e^{48k}$$
$$\ln 222.5 = \ln e^{48k}$$
$$\ln 222.5 = 48k$$
$$\frac{\ln 222.5}{48} = k$$
$$0.113 \approx k$$

Thus, the exponential growth function is

$$P(t) = 12e^{0.113t}, \text{ where } t \text{ is the number of years after 1967.}$$

b. Substitute 5000 for $P(t)$ and solve for t.

$$5000 = 12e^{0.113t}$$
$$\frac{5000}{12} = e^{0.113t}$$
$$\ln \frac{5000}{12} = \ln e^{0.113t}$$
$$\ln \frac{5000}{12} = 0.113t$$
$$\frac{\ln \frac{5000}{12}}{0.113} = t$$
$$53.4 \approx t$$

The ticket price will reach \$5000 about 53 yr after 1967, or in 2020.

35. a. The function $P(t) = P_0e^{-kt}$, $k > 0$, can be used to model decay. We substitute 11 for $P(t)$ and 1 for t. Then we substitute 2 for $P(t)$ and 6 for t. We have two equations and two unknowns:

$$11 = P_0e^{-k(1)} \text{ and } 2 = P_0e^{-k(6)}$$

Solve each equation for P_0.

$$11 = P_0e^{-k(1)} \text{ and } 2 = P_0e^{-k(6)}$$
$$11e^k = P_0 \qquad 2e^{6k} = P_0$$

Substitute and solve for k.

$$11e^k = 2e^{6k}$$
$$\ln 11e^k = \ln 2e^{6k}$$
$$\ln 11 + \ln e^k = \ln 2 + \ln e^{6k}$$
$$\ln 11 + k = \ln 2 + 6k$$
$$\ln 11 - \ln 2 = 5k$$
$$\frac{\ln 11 - \ln 2}{5} = k$$
$$0.341 \approx k$$

Solve for P_0.

$$P_0 = 11e^{0.341} \approx 15.5$$

Thus, the exponential decay function is

$$P(t) = 15.5e^{-0.341t}, \text{ where } t \text{ is in hours.}$$

b. Substitute 3 for t.

$$P(3) = 15.5e^{-0.341(3)} \approx 5.6$$

After 3 hr, there is about 5.6 mcg/mL concentration.

c. Substitute 4 for $P(t)$ and solve for t.

$$4 = 15.5e^{-0.341t}$$
$$\frac{4}{15.5} = e^{-0.341t}$$
$$\ln \frac{4}{15.5} = \ln e^{-0.341t}$$
$$\ln \frac{4}{15.5} = -0.341t$$
$$\frac{\ln \frac{4}{15.5}}{-0.341} = t$$
$$4 \approx t$$

It will take approximately 4 hr for the dosage to reach 4 mcg/mL.

d. Substitute $\frac{1}{2}(15.5)$, or 7.75 for $P(t)$. Solve for t.

$$7.75 = 15.5e^{-0.341t}$$
$$\frac{1}{2} = e^{-0.341t}$$
$$\ln 0.5 = \ln e^{-0.341t}$$
$$\ln 0.5 = -0.341t$$
$$\frac{\ln 0.5}{-0.341} = t$$
$$2 \approx t$$

The half-life is about 2 hr.

37. We will use the function derived in Example 7:

$$P(t) = P_0e^{-0.00012t}$$

If the seed had lost 21% of its carbon-14 from the initial amount P_0, then $79\%(P_0)$ is the amount present. To find the age t of the seed, we substitute $79\%(P_0)$, or $0.79P_0$ for $P(t)$ in the function above and solve for t.

$$0.79P_0 = P_0e^{-0.00012t}$$
$$0.79 = e^{-0.00012t}$$
$$\ln 0.79 = \ln e^{-0.00012t}$$
$$\ln 0.79 = -0.00012t$$
$$\frac{\ln 0.79}{-0.00012} = t$$
$$1964 \approx t$$

The seed is about 1964 yr old.

39. The function $P(t) = P_0 e^{-kt}$, $k > 0$, can be used to model decay. For iodine-131, $k = 9.6\%$, or 0.096. To find the half-life we substitute 0.096 for k and $\frac{1}{2} P_0$ for $P(t)$, and solve for t.

$$\frac{1}{2} P_0 = P_0 e^{-0.096t}, \text{ or } \frac{1}{2} = e^{-0.096t}$$

$$\ln \frac{1}{2} = \ln e^{-0.096t} = -0.096t$$

$$t = \frac{\ln 0.5}{-0.096} \approx \frac{-0.6931}{-0.096} \approx 7.2 \text{ days}$$

41. a. The function $P(t) = P_0 e^{-kt}$, $k > 0$, can be used to model decay. We substitute $\frac{1}{2} P_0$ for $P(t)$ and 5 for t and solve for the decay rate k.

$$\frac{1}{2} P_0 = P_0 e^{-k \cdot 5}$$

$$\frac{1}{2} = e^{-5k}$$

$$\ln \frac{1}{2} = \ln e^{-5k}$$

$$-\ln 2 = -5k$$

$$\frac{\ln 2}{5} = k$$

$$0.139 \approx k$$

The decay rate is 0.139, or 13.9% per hour.

b. 95% consumed = 5% remains

$$0.05 P_0 = P_0 e^{-0.139t}$$

$$0.05 = e^{-0.139t}$$

$$\ln 0.05 = \ln e^{-0.139t}$$

$$\ln 0.05 = -0.139t$$

$$\frac{\ln 0.05}{-0.139} = t$$

$$21.6 \approx t$$

It will take approximately 21.6 hr for 95% of the caffeine to leave the body.

43. a. We start with the exponential growth equation $V(t) = V_0 e^{kt}$, where t is the number of years after 1989.

Substituting 20.6 for V_0, we have

$$V(t) = 20.6 e^{kt}.$$

To find the exponential growth rate k, observe that the painting sold for $300 million, or $300,000,000 in 2015, or 26 years after 1989. We substitute and solve for k.

$$300 = 20.6 e^{k \cdot 26}$$

$$\frac{300}{20.6} = e^{26k}$$

$$\ln \frac{300}{20.6} = \ln e^{26k}$$

$$\ln \frac{300}{20.6} = 26k$$

$$\frac{\ln \frac{300}{20.6}}{26} = k$$

$$0.103 \approx k$$

The exponential growth function is

$V(t) = 20.6 e^{0.103t}$, where $V(t)$ is in millions of dollars and t is the number of years after 1989.

b. In 2025, $t = 2025 - 1989 = 36$.

$V(36) = 20.6 e^{0.103(36)} \approx \$840$ million

c. To find the doubling time, replace $V(t)$ with 41.2 and solve for t.

$$41.2 = 20.6 e^{0.103t}$$

$$2 = e^{0.103t}$$

$$\ln 2 = \ln e^{0.103t}$$

$$\ln 2 = 0.103t$$

$$\frac{\ln 2}{0.103} = t$$

$$6.7 \approx t$$

The doubling time is about 6.7 years.

d. $1 billion = $1000 million.

$$1000 = 20.6 e^{0.103t}$$

$$\frac{1000}{20.6} = e^{0.103t}$$

$$\ln \frac{1000}{20.6} = \ln e^{0.103t}$$

$$\ln \frac{1000}{20.6} = 0.103t$$

$$\frac{\ln \frac{1000}{20.6}}{0.103} = t$$

$$37.7 \text{ yr} \approx t$$

The painting will be worth $1 billion 37.7 yr after 1989.

45. *Writing Exercise.*

47. $f(x) = 18x + \frac{1}{2}$

49.
$$2x - 3y = 4$$
$$-3y = -2x + 4$$
$$y = \frac{2}{3}x - \frac{4}{3}$$

A line parallel to this line has slope $m = \frac{2}{3}$.

$$y - 7 = \frac{2}{3}(x + 3)$$
$$y - 7 = \frac{2}{3}x + 2$$
$$f(x) = \frac{2}{3}x + 9$$

51. *Writing Exercise.*

53. We will use the exponential growth function $V(t) = V_0 e^{kt}$, where t is the number of years after 2015 and $V(t)$ is in millions of dollars. Substitute 32 for $V(t)$, 0.04 for k, and 8 for t and solve for V_0.

$$V(t) = V_0 e^{kt}$$

$$32 = V_0 e^{0.04(8)}$$

$$\frac{32}{e^{0.32}} = V_0$$

$$23.2 \approx V_0$$

About $23.2 million would need to be invested.

55. a. Substitute 1390 for I and solve for m.

$$m(I) = -(19 + 2.5 \cdot \log I)$$
$$m = -(19 + 2.5 \cdot \log 1390)$$
$$m \approx -26.9$$

The apparent stellar magnitude is about –26.9.

b. Substitute 23 for m and solve for I.

$$m(I) = -(19 + 2.5 \cdot \log I)$$
$$23 = -(19 + 2.5 \cdot \log I)$$
$$-23 = 19 + 2.5 \cdot \log I$$
$$-42 = 2.5 \cdot \log I$$
$$-16.8 = \log I$$
$$10^{-16.8} = I$$
$$1.58 \times 10^{-17} \approx I$$

The intensity is about 1.58×10^{-17} W/m^2.

57. Consider an exponential growth function $P(t) = P_0 e^{kt}$.

Suppose that at time T, $P(T) = 2P_0$. Solve for T:

$$2P_0 = P_0 e^{kT}$$
$$2 = e^{kT}$$
$$\ln 2 = \ln e^{kT}$$
$$\ln 2 = kT$$
$$\frac{\ln 2}{k} = T$$

59. Consider an exponential growth function $P(t) = P_0 e^{kt}$.

Suppose that at time $T = 5.32$, $P(T) = 2P_0$. Solve for k:

$$2P_0 = P_0 e^{kT}$$
$$2 = e^{k5.32}$$
$$\ln 2 = \ln e^{5.32k}$$
$$\ln 2 = 5.32k$$
$$\frac{\ln 2}{5.32} = k$$
$$k \approx 0.1303$$

The growth rate is about 13.03%.

Chapter 9 Review

1. True

2. True

3. True

4. False; the product rule states $\ln(ab) = \ln a + \ln b$.

5. False; log, which has base 10, is not the same as ln, which has base e. In addition, the power rule states

$$\log x^a = a \log x.$$

6. True; this is the quotient rule for logarithms.

7. False; the domain of $f(x) = 3^x$ is all real numbers.

8. False; the domain of $g(x) = \log_2 x$ is $(0, \infty)$.

9. True; if $F(-2) = F(5)$, then the function has the same output for two different inputs, and therefore is not one-to-one.

10. False; the function g is one-to-one if it passes the horizontal-line test.

11. $(f \circ g)(x) = f(g(x)) = f(2x - 3)$
$$= (2x - 3)^2 + 1$$
$$= 4x^2 - 12x + 9 + 1$$
$$= 4x^2 - 12x + 10$$

$(g \circ f)(x) = g(f(x)) = g(x^2 + 1)$
$$= 2(x^2 + 1) - 3$$
$$= 2x^2 + 2 - 3$$
$$= 2x^2 - 1$$

12. $h(x) = \sqrt{3 - x}$

We have $3 - x$ and take the square root of the expression, so the two most obvious functions are $f(x) = \sqrt{x}$ and $g(x) = 3 - x$.

13. $f(x) = 4 - x^2$

The graph of this function is a parabola that opens down. Thus, there are many horizontal lines that cross the graph more than once. In particular, the line $y = -4$ crosses the graph more than once. The function is not one-to-one.

14. Replace $f(x)$ by y: $y = x - 10$
Interchange variables: $x = y - 10$
Solve for y: $x + 10 = y$
Replace y by $f^{-1}(x)$: $f^{-1}(x) = x + 10$

15.
$$y = \frac{3x + 1}{2} \qquad \text{Replace } g(x).$$
$$x = \frac{3y + 1}{2} \qquad \text{Interchange variables.}$$
$$\frac{2x - 1}{3} = y \qquad \text{Solve for } y.$$
$$g^{-1}(x) = \frac{2x - 1}{3} \qquad \text{Replace } y.$$

16.
$$y = 27x^3 \qquad \text{Replace } f(x).$$
$$x = 27y^3 \qquad \text{Interchange variables.}$$
$$\frac{\sqrt[3]{x}}{3} = y \qquad \text{Solve for } y.$$
$$f^{-1}(x) = \frac{\sqrt[3]{x}}{3} \qquad \text{Replace } y.$$

17. Graph $f(x) = 3^x + 1$

We compute some function values, thinking of y as $f(x)$, and keep the results in a table.

$$f(-2) = 3^{-2} + 1 = \frac{1}{9} + 1 = \frac{10}{9}$$
$$f(-1) = 3^{-1} + 1 = \frac{1}{3} + 1 = \frac{4}{3}$$
$$f(0) = 3^0 + 1 = 2$$
$$f(1) = 3^1 + 1 = 4$$
$$f(2) = 3^2 + 1 = 10$$

x	y, or $f(x)$
-2	$\dfrac{10}{9}$
-1	$\dfrac{4}{3}$
0	2
1	4
2	10

$f(x) = 3^x + 1$

18. Graph $x = \left(\dfrac{1}{4}\right)^y$.

$x = \left(\dfrac{1}{4}\right)^y$

19. Graph: $f(x) = \log_5 x$

Think of $f(x)$ as y. Then $y = \log_5 x$ is equivalent to $5^y = x$. We find ordered pairs by choosing values for y and computing the corresponding x-values. Then we plot the points and connect them with a smooth curve.

For $y = 0$, $x = 5^0 = 1$
For $y = 1$, $x = 5^1 = 5$
For $y = 2$, $x = 5^2 = 25$
For $y = -1$, $x = 5^{-1} = \dfrac{1}{5}$
For $y = -2$, $x = 5^{-2} = \dfrac{1}{25}$

x, or 5^y	y
1	0
5	1
25	2
$\dfrac{1}{5}$	-1
$\dfrac{1}{25}$	-2

$y = \log_5 x$

20. $\log_9 81 = \log_9 9^2 = 2$

21. $\log_3 \dfrac{1}{9} = \log_3 3^{-2} = -2\log_3 3 = -2$

22. $\log_2 2^{11} = 11$

23. $\log_{16} 4 = \log_{16} 16^{1/2} = \dfrac{1}{2}\log_{16} 16 = \dfrac{1}{2}$

24. $2^{-3} = \dfrac{1}{8}$ is equivalent to $-3 = \log_2 \dfrac{1}{8}$.

25. $25^{1/2} = 5$ is equivalent to $\dfrac{1}{2} = \log_{25} 5$.

26. $\log_4 16 = x$ is equivalent to $16 = 4^x$.

27. $\log_8 1 = 0$ is equivlant to $1 = 8^0$.

28. $\log_a x^4 y^2 z^3 = \log_a x^4 + \log_a y^2 + \log_a z^3$
$= 4\log_a x + 2\log_a y + 3\log_a z$

29. $\log_a \dfrac{x^5}{yz^2} = \log_a x^5 - \log_a y - \log_a z^2$
$= 5\log_a x - \log_a y - 2\log_a z$

30. $\log \sqrt[4]{\dfrac{z^2}{x^3 y}} = \log\left(\dfrac{z^2}{x^3 y}\right)^{1/4} = \dfrac{1}{4}\log \dfrac{z^2}{x^3 y}$
$= \dfrac{1}{4}\left(\log z^2 - \log x^3 - \log y\right)$
$= \dfrac{1}{4}\left(2\log z - 3\log x - \log y\right)$

31. $\log_a 5 + \log_a 8 = \log_a (5 \cdot 8) = \log_a 40$

32. $\log_a 48 - \log_a 12 = \log_a \dfrac{48}{12} = \log_a 4$

33. $\dfrac{1}{2}\log a - \log b - 2\log c = \log a^{1/2} - \log b - \log c^2$
$= \log \dfrac{a^{1/2}}{bc^2}$

34. $\dfrac{1}{3}\left[\log_a x - 2\log_a y\right] = \dfrac{1}{3}\left[\log_a x - \log_a y^2\right]$
$= \dfrac{1}{3}\left[\log_a \dfrac{x}{y^2}\right] = \log_a \left(\dfrac{x}{y^2}\right)^{1/3}$
$= \log_a \sqrt[3]{\dfrac{x}{y^2}}$

35. $\log_m m = 1$

36. $\log_m 1 = 0$

37. $\log_m m^{17} = 17\log_m m = 17 \cdot 1 = 17$

38. $\log_a 14 = \log_a (2 \cdot 7)$
$= \log_a 2 + \log_a 7$
$= 1.8301 + 5.0999$
$= 6.93$

39. $\log_a \dfrac{2}{7} = \log_a 2 - \log_a 7$
$= 1.8301 - 5.0999$
$= -3.2698$

40. $\log_a 28 = \log_a (2^2 \cdot 7)$
$= \log_a 2^2 + \log_a 7$
$= 2\log_a 2 + \log_a 7$
$= 2(1.8301) + 5.0999$
$= 8.7601$

41. $\log_a 3.5 = \log_a \dfrac{7}{2}$
$= \log_a 7 - \log_a 2$
$= 5.0999 - 1.8301$
$= 3.2698$

42. $\log_a \sqrt{7} = \log_a 7^{1/2} = \dfrac{1}{2}\log_a 7 = \dfrac{1}{2}(5.0999) = 2.54995$

43. $\log_a \frac{1}{4} = \log_a 1 - \log_a 4$
$= 0 - \log_a 2^2$
$= -2\log_a 2$
$= -2(1.8301)$
$= -3.6602$

44. 1.8751

45. 61.5177

46. -1.2040

47. 0.3753

48. $\log_5 50 = \frac{\ln 50}{\ln 5} \approx 2.4307$

49. We will use common logarithms for the conversion. Let $a = 10$, $b = 6$, and $M = 5$ and substitute in the change-of-base formula.
$$\log_6 5 = \frac{\log 5}{\log 6} \approx \frac{0.6890}{0.7782} \approx 0.8982$$

50. Graph $f(x) = e^x - 1$.

The domain is all real numbers and the range is $(-1, \infty)$.

51. Graph $g(x) = 0.6\ln x$.
We find some function values, plot points, and draw the graph.

x	$0.6\ln x$
0.5	-0.42
1	0
2	0.42
3	0.66

The domain is $(0, \infty)$ and the range is all real numbers.

52. $5^x = 125$
$5^x = 5^3$
$x = 3$

53. $3^{2x} = \frac{1}{9}$
$3^{2x} = 3^{-2}$
$2x = -2$
$x = -1$
The solution is –1.

54. $\log_3 x = -4$
$x = 3^{-4} = \frac{1}{81}$

55. $\log_x 16 = 4$
$x^4 = 16$
$x = \pm\sqrt[4]{16} = \pm 2$
Since x cannot be negative, we use only the positive solution. The solution is 2.

56. $\log x = -3$
$x = 10^{-3} = \frac{1}{1000}$

57. $6\ln x = 18$
$\ln x = 3$
$x = e^3 \approx 20.0855$
The solution is e^3 or approximately 20.0855.

58. $4^{2x-5} = 19$
$2x - 5 = \log_4 19$
$2x = \log_4 19 + 5$
$x = \frac{1}{2}(\log_4 19 + 5)$
$x = \frac{1}{2}\left(\frac{\log 19}{\log 4} + 5\right) \approx 3.5620$

59. $2^x = 12$
$x = \log_2 12 = \frac{\log 12}{\log 2} \approx 3.5850$

60. $e^{-0.1t} = 0.03$
$-0.1t = \ln 0.03$
$t = \frac{\ln 0.03}{-0.1} \approx 35.0656$

61. $2\ln x = -6$
$\ln x = -3$
$x = e^{-3} \approx 0.0498$

62. $\log(2x - 5) = 1$
$2x - 5 = 10^1$
$2x = 15$
$x = \frac{15}{2}$

63. $\log_4 x - \log_4(x - 15) = 2$
$\log_4 \frac{x}{x-15} = 2$
$\frac{x}{x-15} = 4^2$
$\frac{x}{x-15} = 16$
$x = 16(x - 15)$
$x = 16x - 240$
$240 = 15x$
$16 = x$

64. $\log_3(x - 4) = 2 - \log_3(x + 4)$
$\log_3(x - 4) + \log_3(x + 4) = 2$
$\log_3(x^2 - 16) = 2$
$x^2 - 16 = 3^2$
$x^2 = 25$
$x = \pm 5$
The value –5 does not check. The solution is 5.

65. $S(t) = 82 - 18\log(t+1)$

a. $S(0) = 82 - 18\log(0+1) = 82 - 18\log 1 = 82$

b. $S(6) = 82 - 18\log(6+1) = 82 - 18\log 7 \approx 66.8$

c. Substitute 54 for $S(t)$ and solve for t.

$$54 = 82 - 18\log(t+1)$$
$$-28 = -18\log(t+1)$$
$$\frac{14}{9} = \log(t+1)$$
$$t+1 = 10^{14/9}$$
$$t = 10^{14/9} - 1 \approx 35$$

The average score will be 54 after about 35 months.

66. a.
$$900 = 1500(0.8)^t$$
$$0.6 = (0.8)^t$$
$$\ln 0.6 = \ln(0.8)^t$$
$$\ln 0.6 = t\ln 0.8$$
$$\frac{\ln 0.6}{\ln 0.8} = t$$
$$2.3\text{ yr} \approx t$$

b. $V(0) = 1500(0.8)^0 = 1500,$ and $0.5(1500) = 750$

$$750 = 1500(0.8)^t$$
$$0.5 = (0.8)^t$$
$$\ln 0.5 = \ln(0.8)^t$$
$$\ln 0.5 = t\ln 0.8$$
$$\frac{\ln 0.5}{\ln 0.8} = t$$
$$3.1\text{ yr} \approx t$$

67. a. We start with the exponential growth equation

$$A(t) = A_0 e^{kt}, \text{ where } t \text{ is the number of}$$
years after 1936.

Substituting \$400 for A_0, we have

$$A(t) = 400e^{kt}.$$

Substitute 77 for t, \$3,000,000 for $A(t)$ and solve for k.

$$A(77) = 400e^{k(77)}$$
$$3,000,000 = 400e^{77k}$$
$$\frac{3,000,000}{400} = e^{77k}$$
$$\ln\frac{3,000,000}{400} = \ln e^{77k}$$
$$\ln\frac{3,000,000}{400} = 77k$$
$$\frac{\ln\frac{3,000,000}{400}}{77} = k$$
$$0.116 \approx k$$

Thus, the exponential growth function is

$$A(t) = 400e^{0.116t}, \text{ where } t \text{ is the number}$$
of years after 1936.

b. In 2020, $t = 2020 - 1936 = 84$

$$A(84) = 400e^{0.116(84)} \approx 6,800,000$$

The value of the nickel in 2020 will be about \$6,800,000.

c. Substitute \$10,000,000 for $A(t)$ and solve for t.

$$10,000,000 = 400e^{0.116t}$$
$$\frac{10,000,000}{400} = e^{0.116t}$$
$$\ln\frac{10,000,000}{400} = \ln e^{0.116t}$$
$$\ln\frac{10,000,000}{400} = 0.116t$$
$$\frac{\ln\frac{10,000,000}{400}}{0.116} = t$$
$$87 \approx t$$

The nickel's value will reach \$10,000,000 about 87 years after 1936, or in 2023.

d. Substitute 2(\$400), or \$800 for $A(t)$ and solve for t.

$$800 = 400e^{0.116t}$$
$$2 = e^{0.116t}$$
$$\ln 2 = \ln e^{0.116t}$$
$$\ln 2 = 0.116t$$
$$\frac{\ln 2}{0.116} = t$$
$$6.0 \approx t$$

The doubling time is about 6.0 years.

68. a. We start with the exponential decay equation

$$C(t) = C_0 e^{-kt}, \text{ where } t \text{ is the number of}$$
years after 1980.

Substituting 22 for C_0, and 6.1% or 0.061 for k, we have the exponential function

$$C(t) = 22e^{-0.061t}, \text{ where } t \text{ is the number}$$
of years after 1980.

b. In 2015, $t = 2015 - 1980 = 35$.

$$C(35) = 22e^{-0.061(35)} \approx \$2.60$$

c.
$$1 = 22e^{-0.061t}$$
$$\frac{1}{22} = e^{-0.061t}$$
$$\ln\frac{1}{22} = \ln e^{-0.061t}$$
$$\ln\frac{1}{22} = -0.061t$$
$$\frac{\ln\frac{1}{22}}{-0.061} = t$$
$$51 \approx t$$

Installation cost per watt will be \$1 about 51 years after 1980, or in 2031.

69. The doubling time of the initial investment P_0 would be $2P_0$ and $t = 6$ years. Substitute this information into the exponential growth formula and solve for k.

$$2P_0 = P_0 e^{k(6)}$$
$$2 = e^{6k}$$
$$\ln 2 = \ln e^{6k}$$
$$\ln 2 = 6k$$
$$\frac{\ln 2}{6} = k$$
$$0.11553 \approx k$$

The rate is 11.553% per year.

70. $2P_0 = 2 \cdot 7600$, $k = 4.2\% = 0.042$, $P_0 = 7600$

Substitute into the exponential growth formula and solve for t.

$$2 \cdot 7600 = 7600e^{0.042t}$$
$$2 = e^{0.042t}$$
$$\ln 2 = \ln e^{0.042t}$$
$$\ln 2 = 0.042t$$
$$\frac{\ln 2}{0.042} = t$$
$$16.5 \approx t$$

$7600 will double in about 16.5 years.

71. We will use the function $P(t) = P_0 e^{-0.00012t}$

If the skull had lost 34% of its carbon-14 from the initial amount P_0, then $66\%(P_0)$ is the amount present. To find the age t of the skull, we substitute $66\%(P_0)$, or $0.66P_0$ for $P(t)$ in the function above and solve for t.

$$0.66P_0 = P_0 e^{-0.00012t}$$
$$0.66 = e^{-0.00012t}$$
$$\ln 0.66 = \ln e^{-0.00012t}$$
$$\ln 0.66 = -0.00012t$$
$$\frac{\ln 0.66}{-0.00012} = t$$
$$3463 \approx t$$

The skull is about 3463 yr old.

72. $\text{pH} = -\log[H^+] = -\log(7.9 \times 10^{-6}) = 5.1$

The pH of coffee is 5.1.

73. The function $P(t) = P_0 e^{-kt}$, $k > 0$, can be used to model decay. We substitute $\frac{1}{2}P_0$ for $P(t)$ and 24,360 for t and solve for the decay rate k.

$$\frac{1}{2}P_0 = P_0 e^{-k \cdot 24,360}$$
$$\frac{1}{2} = e^{-24,360k}$$
$$\ln \frac{1}{2} = \ln e^{-24,360k}$$
$$-\ln 2 = -24,360k$$
$$\frac{\ln 2}{24,360} = k$$
$$0.00002845431776 \approx k$$

So $P(t) = P_0 e^{-0.00002845431776t}$

90% consumed = 10% remains

$$0.10P_0 = P_0 e^{-0.00002845431776t}$$
$$0.10 = e^{-0.00002845431776t}$$
$$\ln 0.10 = \ln e^{-0.00002845431776t}$$
$$\ln 0.10 = -0.00002845431776t$$
$$\frac{\ln 0.10}{-0.00002845431776} = t$$
$$80,922 \approx t$$

A fuel rod of Pu-239 to will lose 90% of its radioactivity in about 80,922 years, or about 80,792 years rounding $k = 0.0000285$.

74. $L = 10 \cdot \log \dfrac{I}{I_0}$

$$= 10 \cdot \log \frac{2.5 \times 10^{-1}}{10^{-12}}$$
$$= 10 \log(2.5 \times 10^{11})$$
$$\approx 114$$

The sound is about 114 dB.

75. *Writing Exercise.* Negative numbers do not have logarithms because logarithm bases are positive, and there is no exponent to which a positive number can be raised to yield a negative number.

76. *Writing Exercise.* If $f(x) = e^x$, then to find the inverse function, we let $y = e^x$ and interchange x and y:

$x = e^y$. If $x = e^y$, then $\log_e x = y$ by the definition of logarithms. Since $\log_e x = \ln x$, we have $y = \ln x$ or $f^{-1}(x) = \ln x$. Thus, $g(x) = \ln x$ is the inverse of $f(x) = e^x$. Another approach is to find $(f \circ g)(x)$ and $(g \circ f)(x)$:

$$(f \circ g)(x) = e^{\ln x} = x, \text{ and}$$
$$(g \circ f)(x) = \ln e^x = x.$$

Thus, g and f are inverse functions.

77. $\ln(\ln x) = 3$
$$\ln x = e^3$$
$$x = e^{e^3}$$

78. $2^{x^2+4x} = \dfrac{1}{8}$ can be written as $2^{x^2+4x} = 2^{-3}$.

So the exponents must be equal.

$$x^2 + 4x = -3$$
$$x^2 + 4x + 3 = 0$$
$$(x+3)(x+1) = 0$$
$$x + 3 = 0 \quad or \quad x + 1 = 0$$
$$x = -3 \quad or \quad x = -1$$

The solutions are -3 and -1.

79. Solve the system:

$$5^{x+y} = 25, \quad (1) \quad or \quad 5^{x+y} = 5^2, \quad (1)$$
$$2^{2x-y} = 64 \quad (2) \qquad \quad 2^{2x-y} = 2^6 \quad (2)$$

From the exponents we have:

$$x + y = 2, \quad (3)$$
$$2x - y = 6 \quad (4)$$

We use elimination.

$$\begin{array}{r} x + y = 2 \\ 2x - y = 6 \\ \hline 3x \quad = 8 \end{array}$$
$$x = \frac{8}{3}$$

From Equation (3) we solve for y.

$$y = 2 - x = 2 - \frac{8}{3} = -\frac{2}{3}$$

The solution is $\left(\frac{8}{3}, -\frac{2}{3}\right)$.

Chapter 9 Test

1. $(f \circ g)(x) = f(g(x)) = f(2x+1)$
$= (2x+1) + (2x+1)^2$
$= 2x + 1 + 4x^2 + 4x + 1$
$= 4x^2 + 6x + 2$
$(g \circ f)(x) = g(f(x)) = g(x + x^2)$
$= 2(x + x^2) + 1$
$= 2x + 2x^2 + 1$
$= 2x^2 + 2x + 1$

2. $h(x) = \dfrac{1}{2x^2 + 1}$

This is 1 divided by $2x^2 + 1$, so two functions that can

be used are $f(x) = \dfrac{1}{x}$ and $g(x) = 2x^2 + 1$.

3. $f(x) = x^2 + 3$

Observe that the graph of this function is a parabola that opens up. Thus, there are many horizontal lines that cross the graph more than once. In particular, the line $y = 4$ crosses the graph more than once. The function is not one-to-one.

4.
$y = 3x + 4$ Replace $f(x)$.
$x = 3y + 4$ Interchange variables.
$\dfrac{x-4}{3} = y$ Solve for y.
$f^{-1}(x) = \dfrac{x-4}{3}$ Replace y.

5.
$y = (x+1)^3$ Replace $g(x)$.
$x = (y+1)^3$ Interchange variables.
$\sqrt[3]{x} - 1 = y$ Solve for y.
$g^{-1}(x) = \sqrt[3]{x} - 1$ Replace y.

6. Graph $f(x) = 2^x - 3$.

7. Graph: $f(x) = \log_7 x$

Think of $f(x)$ as y. Then $y = \log_7 x$ is equivalent to

$7^y = x$. We find ordered pairs by choosing values for y and computing the corresponding x-values. Then we plot the points and connect them with a smooth curve.

For $y = 0$, $x = 7^0 = 1$
For $y = 1$, $x = 7^1 = 7$
For $y = 2$, $x = 7^2 = 49$
For $y = -1$, $x = 7^{-1} = \dfrac{1}{7}$
For $y = -2$, $x = 7^{-2} = \dfrac{1}{49}$

x, or 7^y	y
1	0
7	1
49	2
$\dfrac{1}{7}$	-1
$\dfrac{1}{49}$	-2

8. $\log_5 125 = \log_5 5^3 = 3\log_5 5 = 3$

9. $\log_{100} 10 = \log_{100} 100^{1/2} = \dfrac{1}{2}\log_{100} 100 = \dfrac{1}{2} \cdot 1 = \dfrac{1}{2}$

10. $\log_n n = 1$

11. $\log_c 1 = 0$

12. $5^{-4} = \dfrac{1}{625}$ is equivalent to $-4 = \log_5 \dfrac{1}{625}$.

13. $m = \log_2 \dfrac{1}{2}$ is equivalent to $2^m = \dfrac{1}{2}$.

14. $\log \dfrac{a^3 b^{1/2}}{c^2} = \log a^3 + \log b^{1/2} - \log c^2$
$= 3\log a + \dfrac{1}{2}\log b - 2\log c$

15. $\dfrac{1}{3}\log_a x + 2\log_a z = \log_a x^{1/3} + \log_a z^2$
$= \log_a \sqrt[3]{x} + \log_a z^2$
$= \log_a \left(z^2 \sqrt[3]{x}\right)$

16. $\log_a 14 = \log_a (2 \cdot 7)$
$= \log_a 2 + \log_a 7$
$= 0.301 + 0.845$
$= 1.146$

17. $\log_a 3 = \log_a \dfrac{6}{2}$
$= \log_a 6 - \log_a 2$
$= 0.778 - 0.301$
$= 0.477$

18. $\log_a 16 = \log_a 2^4 = 4\log_a 2 = 4(0.301) = 1.204$

19. 1.3979

20. 0.1585

21. -0.9163

22. 121.5104

23. We will use common logarithms for the conversion. Let $a = 10$, $b = 3$, and $M = 14$ and substitute in the change-of-base formula.

$$\log_3 14 = \frac{\log 14}{\log 3} \approx \frac{1.1461}{0.4771} \approx 2.4022$$

24. Graph $f(x) = e^x + 3$.

The domain is all real numbers and the range is $(3, \infty)$.

25. Graph $g(x) = \ln(x - 4)$.

We find some function values, plot points, and draw the graph.

x	$\ln(x - 4)$
4.5	-0.69
5	0
6	0.69
7	1.10

The domain is $(4, \infty)$ and the range is all real numbers.

26. $2^x = \dfrac{1}{32}$

$2^x = 2^{-5}$

$x = -5$

27. $\log_4 x = \dfrac{1}{2}$

$x = 4^{1/2}$

$x = 2$

28. $\log x = -2$

$x = 10^{-2} = \dfrac{1}{100}$

29. $7^x = 1.2$

$x = \log_7 1.2 = \dfrac{\log 1.2}{\log 7} \approx 0.0937$

30. $\log(x - 3) + \log(x + 1) = \log 5$

$\log(x - 3)(x + 1) = \log 5$

$\log(x^2 - 2x - 3) = \log 5$

$x^2 - 2x - 3 = 5$

$x^2 - 2x - 8 = 0$

$(x - 4)(x + 2) = 0$

$x = 4 \ \ or \ \ x = -2$

The value -2 does not check. The solution is 4.

31. $R = 0.37 \ln P + 0.05$

a. Substitute 384 for P and solve for R.

$R = 0.37 \ln 384 + 0.05 \approx 2.3$

The average walking speed is about 2.3 ft/sec.

b. Substitute 3.0 for R and solve for P.

$3.0 = 0.37 \ln P + 0.05$

$2.95 = 0.37 \ln P$

$\dfrac{2.95}{0.37} = \ln P$

$P = e^{2.95/0.37} \approx 2900$

The population is approximately 2,900,000.

32. a. $P(t) = 186e^{0.026t}$, where $P(t)$ is in millions and t is the number of years after 2016.

b. In 2020, $t = 2020 - 2016 = 4$. Find $P(4)$.

$P(4) = 186e^{0.026(4)} = 186e^{0.104} \approx 206$ million

In 2050, $t = 2050 - 2016 = 34$. Find $P(34)$.

$P(34) = 186e^{0.026(34)} = 186e^{0.884} \approx 450$ million

c. Substitute 500 for $P(t)$ and solve for t.

$500 = 186e^{0.026t}$

$\dfrac{500}{186} = e^{0.026t}$

$\ln \dfrac{500}{186} = \ln e^{0.026t}$

$\ln \dfrac{500}{186} = 0.026t$

$\dfrac{\ln \dfrac{500}{186}}{0.026} = t$

$38 \approx t$

The population will reach 500 million about 38 yr after 2016, or in 2054.

d. $P(0) = 186e^{0.026(0)} = 186$, and $2(186) = 372$

$372 = 186e^{0.026t}$

$2 = e^{0.026t}$

$\ln 2 = 0.026t$

$\dfrac{\ln 2}{0.026} = t$

26.7 yr $\approx t$

33. a. We start with the exponential growth equation $C(t) = C_0 e^{kt}$, where t is the number of years after 2005-2006.

Substituting \$35,106 for C_0, we have

$C(t) = 35,106e^{kt}$.

Substitute 10 for t, \$43,921 for $C(t)$ and solve for k.

$43,921 = 35,106e^{k(10)}$

$\dfrac{43,921}{35,106} = e^{10k}$

$\ln \dfrac{43,921}{35,106} = \ln e^{10k}$

$\ln \dfrac{43,921}{35,106} = 10k$

$\dfrac{\ln \dfrac{43,921}{35,106}}{10} = k$

$0.022 \approx k$

Thus, the exponential growth function is

$C(t) = 35,106e^{0.022t}$, where t is the number of years after 2005-2006.

b. In 2019-2020, $t = 2019 - 2005 = 14$

$C(14) = 35,106e^{0.022(14)} \approx 47,769$

The cost in 2019-2020 will be about \$47,769.

c. Substitute $60,000 for $C(t)$ and solve for t.

$$60,000 = 35,106e^{0.022t}$$

$$\frac{60,000}{35,106} = e^{0.022t}$$

$$\ln\frac{60,000}{35,106} = \ln e^{0.022t}$$

$$\ln\frac{60,000}{35,106} = 0.022t$$

$$\frac{\ln\frac{60,000}{35,106}}{0.022} = t$$

$$24 \approx t$$

Cost for college will be $60,000 about 24 years after 2005-2006, or in 2029-2030.

34. The doubling time of the initial investment P_0 would be $2P_0$ and $t = 16$ years. Substitute this information into the exponential growth formula and solve for k.

$$2P_0 = P_0 e^{k(16)}$$

$$2 = e^{16k}$$

$$\ln 2 = \ln e^{16k}$$

$$\ln 2 = 16k$$

$$\frac{\ln 2}{16} = k$$

$$0.043 \approx k$$

The rate is 4.3% per year.

35. $\text{pH} = -\log[H^+]$

$\quad = -\log(1.0 \times 10^{-7})$

$\quad = 7.0$

The pH of water is 7.0.

36. $\log_5|2x - 7| = 4$

$\quad |2x - 7| = 5^4$

$\quad |2x - 7| = 625$

$2x - 7 = -625 \quad or \quad 2x - 7 = 625$

$\quad\quad x = -309 \quad or \quad\quad\quad x = 316$

37. $\log_a \dfrac{\sqrt[3]{x^2 z}}{\sqrt[3]{y^2 z^{-1}}}$

$= \log_a \sqrt[3]{\dfrac{x^2 z}{y^2 z^{-1}}}$

$= \log_a \sqrt[3]{\dfrac{x^2 z^2}{y^2}}$

$= \log_a \left(\dfrac{x^2 z^2}{y^2}\right)^{1/3}$

$= \dfrac{1}{3}\log_a \dfrac{x^2 z^2}{y^2}$

$= \dfrac{1}{3}\left(\log_a x^2 + \log_a z^2 - \log_a y^2\right)$

$= \dfrac{1}{3}\left(2\log_a x + 2\log_a z - 2\log_a y\right)$

$= \dfrac{1}{3}(2 \cdot 2 + 2 \cdot 4 - 2 \cdot 3)$

$= 2$

Chapter 10

Conic Sections

Exercise Set 10.1

1. Parabolas and circles are examples of *conic sections*.

3. A parabola with a *horizontal* axis of symmetry opens to the right or to the left.

5. In the equation of a circle, the point (h, k) represents the *center* of the circle.

7. $(x-2)^2 + (y+5)^2 = 9$, or $(x-2)^2 + [y-(-5)]^2 = 3^2$, is the equation of a circle with center $(2, -5)$ and radius 3, so choice (f) is correct.

9. $y = (x-2)^2 - 5$ is the equation of a parabola with vertex $(2, -5)$ that opens upward, so choice (c) is correct.

11. $x = (y-2)^2 - 5$ is the equation of a parabola with vertex $(-5, 2)$ that opens to the right, so choice (d) is correct.

13. $y = -x^2$

   This is equivalent to $y = -(x-0)^2 + 0$. The vertex is $(0, 0)$. We choose some x-values on both sides of the vertex and compute the corresponding values of y. The graph opens down, because the coefficient of x^2, -1, is negative.

x	y
0	0
1	-1
2	-4
-1	-1
-2	-4

15. $y = -x^2 + 4x - 5$

   We can find the vertex by computing the first coordinate, $x = -b/2a$, and then substituting to find the second coordinate:

   $$x = -\frac{b}{2a} = -\frac{4}{2(-1)} = 2$$
   $$y = -x^2 + 4x - 5 = -(2)^2 + 4(2) - 5 = -1$$

   The vertex is $(2, -1)$.
   We choose some x-values and compute the corresponding values for y. The graph opens downward because the coefficient of x^2, -1, is negative.

x	y
2	-1
3	-2
4	-5
1	-2
0	-5

17. $x = y^2 - 4y + 2$

   We find the vertex by completing the square.

   $$x = (y^2 - 4y + 4) + 2 - 4$$
   $$x = (y-2)^2 - 2$$

   The vertex is $(-2, 2)$.
   To find ordered pairs, we choose values for y and compute the corresponding values of x. The graph opens to the right, because the coefficient of y^2, 1, is positive.

x	y
7	-1
2	0
-1	1
-2	2
-1	3

19. $x = y^2 + 3$
   $$x = (y-0)^2 + 3$$
   The vertex is $(3, 0)$.
   To find the ordered pairs, we choose y-values and compute the corresponding values for x. The graph opens to the right, because the coefficient of y^2, 1, is positive.

x	y
3	0
4	1
7	2
4	-1
7	-2

21. $x = 2y^2$
   $$x = 2(y-0)^2 + 0$$
   The vertex is $(0, 0)$.
   We choose y-values and compute the corresponding values for x. The graph opens to the right, because the coefficient of y^2, 2, is positive.

x	y
0	0
2	1
2	-1
8	2
8	-2

23. $x = -y^2 - 4y$

   We find the vertex by computing the second coordinate, $y = -b/2a$, and then substituting to find the first coordinate:

$$y = -\frac{b}{2a} = -\frac{-4}{2(-1)} = -2$$

$$x = -y^2 - 4y = -(-2)^2 - 4(-2) = 4$$

The vertex is $(4, -2)$.

We choose y-values and compute the corresponding values for x. The graph opens to the left, because the coefficient of y^2, -1, is negative.

x	y
4	-2
-5	1
0	0
3	-1
3	-3

25. $y = x^2 - 2x + 1$

$y = (x - 1)^2 + 0$

The vertex is $(1, 0)$.

We choose x-values and compute the corresponding values for y. The graph opens upward, because the coefficient of x^2, 1, is positive.

x	y
1	0
0	1
-1	4
2	1
3	4

27. $x = -\frac{1}{2}y^2$

$x = -\frac{1}{2}(y - 0)^2 + 0$

The vertex is $(0, 0)$.

We choose y-values and compute the corresponding values for x. The graph opens to the left, because the coefficient of y^2, $-\frac{1}{2}$, is negative.

x	y
0	0
-2	2
-8	4
-2	-2
-8	-4

29. $x = -y^2 + 2y - 1$

We find the vertex by computing the second coordinate, $y = -b/2a$, and then substituting to find the first coordinate.

$$y = -\frac{b}{2a} = -\frac{2}{2(-1)} = 1$$

$$x = -y^2 + 2y - 1 = -(1)^2 + 2(1) - 1 = 0$$

The vertex is $(0, 1)$.

We choose y-values and compute the corresponding values for x. The graph opens to the left, because the

coefficient of y^2, -1, is negative.

x	y
-4	3
-1	2
0	1
-1	0
-4	-1

31. $x = -2y^2 - 4y + 1$

We find the vertex by completing the square.

$$x = -2(y^2 + 2y) + 1$$

$$x = -2(y^2 + 2y + 1) + 1 + 2$$

$$x = -2(y + 1)^2 + 3$$

The vertex is $(3, -1)$.

We choose y-values and compute the corresponding values for x. The graph opens to the left, because the coefficient of y^2, -2, is negative.

x	y
3	-1
1	-2
-5	-3
1	0
-5	1

33. $(x - h)^2 + (y - k)^2 = r^2$ Standard form

$(x - 0)^2 + (y - 0)^2 = 8^2$ Substituting

$x^2 + y^2 = 64$ Simplifying

35. $(x - h)^2 + (y - k)^2 = r^2$ Standard form

$(x - 7)^2 + (y - 3)^2 = (\sqrt{6})^2$ Substituting

$(x - 7)^2 + (y - 3)^2 = 6$

37. $(x - h)^2 + (y - k)^2 = r^2$

$[x - (-4)]^2 + (y - 3)^2 = (3\sqrt{2})^2$

$(x + 4)^2 + (y - 3)^2 = 18$

39. $(x - h)^2 + (y - k)^2 = r^2$

$[x - (-5)]^2 + [y - (-8)]^2 = (10\sqrt{3})^2$

$(x + 5)^2 + (y + 8)^2 = 300$

41. Since the center is $(0, 0)$, we have

$(x - 0)^2 + (y - 0)^2 = r^2$ or $x^2 + y^2 = r^2$

The circle passes through $(-3, 4)$. We find r^2 by substituting -3 for x and 4 for y.

$$(-3)^2 + 4^2 = r^2$$

$$9 + 16 = r^2$$

$$25 = r^2$$

Then $x^2 + y^2 = 25$ is an equation of the circle.

43. Since the center is (–4, 1), we have

$$\left[x-(-4)\right]^2+(y-1)^2=r^2, \text{ or}$$
$$(x+4)^2+(y-1)^2=r^2.$$

The circle passes through (–2, 5). We find r^2 by substituting –2 for x and 5 for y.

$$(-2+4)^2+(5-1)^2=r^2$$
$$4+16=r^2$$
$$20=r^2$$

Then $(x+4)^2+(y-1)^2=20$ is an equation of the circle.

45. We write standard form.

$$(x-0)^2+(y-0)^2=1^2$$

The center is (0, 0), and the radius is 1.

47. $$(x+1)^2+(y+3)^2=49$$
$$\left[x-(-1)\right]^2+\left[y-(-3)\right]^2=7^2 \quad \text{Standard form}$$

The center is (–1, –3), and the radius is 7.

$(x+1)^2+(y+3)^2=49$

49. $$(x-4)^2+(y+3)^2=10$$
$$(x-4)^2+\left[y-(-3)\right]^2=\left(\sqrt{10}\right)^2$$

The center is (4, –3), and the radius is $\sqrt{10}$.

$(x-4)^2+(y+3)^2=10$

51. $$x^2+y^2=8$$
$$(x-0)^2+(y-0)^2=\left(\sqrt{8}\right)^2 \quad \text{Standard form}$$

The center is (0, 0), and the radius is $\sqrt{8}$, or $2\sqrt{2}$.

53. $$(x-5)^2+y^2=\frac{1}{4}$$
$$(x-5)^2+(y-0)^2=\left(\frac{1}{2}\right)^2 \quad \text{Standard form}$$

The center is (5, 0), and the radius is $\frac{1}{2}$.

55. $$x^2+y^2+8x-6y-15=0$$
$$x^2+8x+y^2-6y=15$$
$$(x^2+8x+16)+(y^2-6y+9)=15+16+9 \quad \text{Com-}$$
$$\text{pleting the square twice}$$
$$(x+4)^2+(y-3)^2=40$$
$$\left[x-(-4)\right]^2+(y-3)^2=\left(\sqrt{40}\right)^2$$
$$\text{Standard form}$$

The center is (–4, 3), and the radius is $\sqrt{40}$, or $2\sqrt{10}$.

$x^2+y^2+8x-6y-15=0$

57. $$x^2+y^2-8x+2y+13=0$$
$$x^2-8x+y^2+2y=-13$$
$$(x^2-8x+16)+(y^2+2y+1)=-13+16+1 \quad \text{Com-}$$
$$\text{pleting the square twice}$$
$$(x-4)^2+(y+1)^2=4$$
$$(x-4)^2+\left[y-(-1)\right]^2=2^2$$
$$\text{Standard form}$$

The center is (4, –1), and the radius is 2.

$x^2+y^2-8x+2y+13=0$

59. $$x^2+y^2+10y-75=0$$
$$x^2+y^2+10y=75$$
$$x^2+(y^2+10y+25)=75+25$$
$$(x-0)^2+(y+5)^2=100$$
$$(x-0)^2+\left[y-(-5)\right]^2=10^2$$

The center is (0, –5), and the radius is 10.

$x^2+y^2+10y-75=0$

61. $$x^2+y^2+7x-3y-10=0$$
$$x^2+7x+y^2-3y=10$$
$$\left(x^2+7x+\frac{49}{4}\right)+\left(y^2-3y+\frac{9}{4}\right)=10+\frac{49}{4}+\frac{9}{4}$$
$$\left(x+\frac{7}{2}\right)^2+\left(y-\frac{3}{2}\right)^2=\frac{98}{4}$$
$$\left[x-\left(-\frac{7}{2}\right)\right]^2+\left(y-\frac{3}{2}\right)^2=\left(\sqrt{\frac{98}{4}}\right)^2$$

The center is $\left(-\frac{7}{2}, \frac{3}{2}\right)$, and the radius is $\sqrt{\frac{98}{4}}$, or

$\frac{\sqrt{98}}{2}$, or $\frac{7\sqrt{2}}{2}$.

$x^2 + y^2 + 7x - 3y - 10 = 0$

63.
$$36x^2 + 36y^2 = 1$$
$$x^2 + y^2 = \frac{1}{36} \quad \text{Multiplying by } \frac{1}{36}$$
$$\text{on both sides}$$
$$(x-0)^2 + (y-0)^2 = \left(\frac{1}{6}\right)^2$$

The center is $(0, 0)$, and the radius is $\frac{1}{6}$.

$36x^2 + 36y^2 = 1$

65. *Writing Exercise.*

67. $\sqrt[4]{48x^7y^{12}} = \sqrt[4]{16x^4y^{12} \cdot 3x^3} = 2xy^3\sqrt[4]{3x^3}$

69. $\dfrac{\sqrt{200x^4w^2}}{\sqrt{2w}} = \sqrt{\dfrac{200x^4w^2}{2w}} = \sqrt{100x^4w} = 10x^2\sqrt{w}$

71. $\sqrt{8} - 2\sqrt{2} + \sqrt{12} = \sqrt{4 \cdot 2} - 2\sqrt{2} + \sqrt{4 \cdot 3}$
$$= 2\sqrt{2} - 2\sqrt{2} + 2\sqrt{3}$$
$$= 2\sqrt{3}$$

73. *Writing Exercise.*

75. We make a drawing of the circle with center $(3, -5)$ and tangent to the *y*-axis.

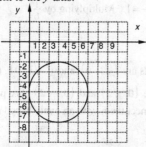

We see that the circle touches the *y*-axis at $(0, -5)$. Hence the radius is the distance between $(0, -5)$ and $(3, -5)$, or $\sqrt{(3-0)^2 + [-5-(-5)]^2}$, or 3. Now we write the equation of the circle.
$$(x-h)^2 + (y-k)^2 = r^2$$
$$(x-3)^2 + [y-(-5)]^2 = 3^2$$
$$(x-3)^2 + (y+5)^2 = 9$$

77. First we use the midpoint formula to find the center:
$\left(\dfrac{7+(-1)}{2}, \dfrac{3+(-3)}{2}\right)$, or $\left(\dfrac{6}{2}, \dfrac{0}{2}\right)$, or $(3, 0)$

The length of the radius is the distance between the center $(3, 0)$ and either endpoint of a diameter. We will use endpoint $(7, 3)$ in the distance formula:
$$r = \sqrt{(7-3)^2 + (3-0)^2} = \sqrt{25} = 5$$

Now we write the equation of the circle:
$$(x-h)^2 + (y-k)^2 = r^2$$
$$(x-3)^2 + (y-0)^2 = 5^2$$
$$(x-3)^2 + y^2 = 25$$

79. For the outer circle, $r^2 = \dfrac{81}{4}$. For the inner circle,

$r^2 = 16$. The area of the red zone is the difference between the areas of the outer and inner circles. Recall that the area A of a circle with radius r is given by the formula $A = \pi r^2$.
$$\pi \cdot \frac{81}{4} - \pi \cdot 16 = \frac{81}{4}\pi - \frac{64}{4}\pi = \frac{17}{4}\pi$$

The area of the red zone is $\dfrac{17}{4}\pi$ m^2, or about

13.4 m^2.

81. Superimposing a coordinate system on the snowboard as in Exercise 80, and observing that $1160/2 = 580$, we know that three points on the circle are $(-580, 0)$, $(0, 23.5)$ and $(580, 0)$. Let $(0, k)$ represent the center of the circle. Use the fact that $(0, k)$ is equidistant from $(-580, 0)$ and $(0, 23.5)$.

$$\sqrt{(-580-0)^2 + (0-k)^2} = \sqrt{(0-0)^2 + (23.5-k)^2}$$
$$\sqrt{336{,}400 + k^2} = \sqrt{552.25 - 47k + k^2}$$
$$336{,}400 + k^2 = 552.25 - 47k + k^2$$
$$335{,}847.75 = -47k$$
$$-7145.7 \approx k$$

Then to find the radius, we find the distance from the center $(0, -7145.7)$ to any one of the three known points on the circle. We use $(0, 23.5)$.

$$r = \sqrt{(0-0)^2 + (-7145.7 - 23.5)^2} \approx 7169 \text{ mm}$$

83. a. When the circle is positioned on a coordinate system as shown in the text, the center lies on the *y*-axis. To find the center, we will find the point on the *y*-axis that is equidistant from $(-4, 0)$ and $(0, 2)$. Let $(0, y)$ be this point.

$$\sqrt{[0-(-4)]^2 + (y-0)^2} = \sqrt{(0-0)^2 + (y-2)^2}$$
$$4^2 + y^2 = 0^2 + (y-2)^2$$
$$\text{Squaring both sides}$$
$$16 + y^2 = y^2 - 4y + 4$$
$$12 = -4y$$
$$-3 = y$$

The center of the circle is $(0, -3)$.

b. We find the radius of the circle.

$$(x-0)^2 + [y-(-3)]^2 = r^2 \quad \text{Standard form}$$
$$x^2 + (y+3)^2 = r^2$$
$$(-4)^2 + (0+3)^2 = r^2 \quad \text{Substituting}$$
$$16 + 9 = r^2 \quad (-4,\ 0) \text{ for } (x,\ y)$$
$$25 = r^2$$
$$5 = r$$

The radius is 5 ft.

85. We write the equation of a circle with center $(0, 30.6)$ and radius 24.3: $x^2 + (y-30.6)^2 = 590.49$

87. Substitute 6 for N.

$$H = \frac{D^2 N}{2.5} = \frac{D^2 \cdot 6}{2.5} = 2.4D^2$$

Find some ordered pairs for $2.5 \le D \le 8$ and draw the graph.

Using the graph, a horse power of 120, on the vertical axis, relates to a diameter of 7 in., on the horizontal axis.

89. *Writing Exercise.*

Exercise Set 10.2

1. Ellipse A

3. Center B

5. True

7. True

9. $\dfrac{x^2}{1} + \dfrac{y^2}{4} = 1$

$$\frac{x^2}{1^2} + \frac{y^2}{2^2} = 1$$

The x-intercepts are $(1, 0)$ and $(-1, 0)$, and the y-intercepts are $(0, 2)$ and $(0, -2)$. We plot these points and connect them with an oval-shaped curve.

11. $\dfrac{x^2}{25} + \dfrac{y^2}{9} = 1$

$$\frac{x^2}{5^2} + \frac{y^2}{3^2} = 1$$

The x-intercepts are $(5, 0)$ and $(-5, 0)$, and the y-intercepts are $(0, 3)$ and $(0, -3)$. We plot these points and connect them with an oval-shaped curve.

$$\frac{x^2}{25} + \frac{y^2}{9} = 1$$

13. $4x^2 + 9y^2 = 36$

$$\frac{1}{36}(4x^2 + 9y^2) = \frac{1}{36}(36) \quad \text{Multiplying by } \frac{1}{36}$$

$$\frac{x^2}{9} + \frac{y^2}{4} = 1$$

$$\frac{x^2}{3^2} + \frac{y^2}{2^2} = 1$$

The x-intercepts are $(-3, 0)$ and $(3, 0)$, and the y-intercepts are $(0, -2)$ and $(0, 2)$. We plot these points and connect them with an oval-shaped curve.

$4x^2 + 9y^2 = 36$

15. $16x^2 + 9y^2 = 144$

$$\frac{x^2}{9} + \frac{y^2}{16} = 1 \quad \text{Multiplying by } \frac{1}{144}$$

$$\frac{x^2}{3^2} + \frac{y^2}{4^2} = 1$$

The x-intercepts are $(3, 0)$ and $(-3, 0)$, and the y-intercepts are $(0, 4)$ and $(0, -4)$. We plot these points and connect them with an oval-shaped curve.

$16x^2 + 9y^2 = 144$

17. $2x^2 + 3y^2 = 6$

$$\frac{x^2}{3} + \frac{y^2}{2} = 1 \quad \text{Multiplying by } \frac{1}{6}$$

$$\frac{x^2}{(\sqrt{3})^2} + \frac{y^2}{(\sqrt{2})^2} = 1$$

The x-intercepts are $(\sqrt{3},\ 0)$ and $(-\sqrt{3},\ 0)$, and the y-intercepts are $(0,\ \sqrt{2})$ and $(0, -\sqrt{2})$. We plot these points and connect them with an oval-shaped curve.

$2x^2 + 3y^2 = 6$

19. $5x^2 + 5y^2 = 125$

Observe that the x^2- and y^2-terms have the same coefficient. We divide both sides of the equation by 5 to obtain $x^2 + y^2 = 25$. This is the equation of a circle with center (0, 0) and radius 5.

$5x^2 + 5y^2 = 125$

21. $3x^2 + 7y^2 - 63 = 0$

$3x^2 + 7y^2 = 63$

$\dfrac{x^2}{21} + \dfrac{y^2}{9} = 1$ Multiplying by $\dfrac{1}{63}$

$\dfrac{x^2}{\left(\sqrt{21}\right)^2} + \dfrac{y^2}{3^2} = 1$

The x-intercepts are $\left(\sqrt{21},\ 0\right)$ and $\left(-\sqrt{21},\ 0\right)$, or about $(4.583, 0)$ and $(-4.583, 0)$. The y-intercepts are $(0, 3)$ and $(0, -3)$. We plot these points and connect them with an oval-shaped curve.

$3x^2 + 7y^2 - 63 = 0$

23. $16x^2 = 16 - y^2$

$16x^2 + y^2 = 16$

$\dfrac{x^2}{1} + \dfrac{y^2}{16} = 1$

The x-intercepts are $(1, 0)$ and $(-1, 0)$, and the y-intercepts are $(0, 4)$ and $(0, -4)$. We plot these points and connect them with an oval-shaped curve.

$16x^2 = 16 - y^2$

25. $16x^2 + 25y^2 = 1$

Note that $16 = \dfrac{1}{\frac{1}{16}}$ and $25 = \dfrac{1}{\frac{1}{25}}$. Thus, we can rewrite the equation:

$\dfrac{x^2}{\frac{1}{16}} + \dfrac{y^2}{\frac{1}{25}} = 1$

$\dfrac{x^2}{\left(\frac{1}{4}\right)^2} + \dfrac{y^2}{\left(\frac{1}{5}\right)^2} = 1$

The x-intercepts are $\left(\frac{1}{4},\ 0\right)$ and $\left(-\frac{1}{4},\ 0\right)$, and the

y-intercepts are $\left(0,\ \frac{1}{5}\right)$ and $\left(0, -\frac{1}{5}\right)$. We plot these

points and connect them with an oval-shaped curve.

$16x^2 + 25y^2 = 1$

27. $\dfrac{(x-3)^2}{9} + \dfrac{(y-2)^2}{25} = 1$

$\dfrac{(x-3)^2}{3^2} + \dfrac{(y-2)^2}{5^2} = 1$

The center of the ellipse is $(3, 2)$. Note that $a = 3$ and $b = 5$. We locate the center and then plot the points $(3+3,\ 2)$ $(3-3,\ 2)$, $(3,\ 2+5)$, and $(3,\ 2-5)$, or $(6, 2)$, $(0, 2)$, $(3, 7)$, and $(3, -3)$. Connect these points with an oval-shaped curve.

$\frac{(x-3)^2}{9} + \frac{(y-2)^2}{25} = 1$

29. $\dfrac{(x+4)^2}{16} + \dfrac{(y-3)^2}{49} = 1$

$\dfrac{(x-(-4))^2}{4^2} + \dfrac{(y-3)^2}{7^2} = 1$

The center of the ellipse is $(-4, 3)$. Note that $a = 4$ and $b = 7$. We locate the center and then plot the points $(-4+4,\ 3)$, $(-4-4,\ 3)$, $(-4,\ 3+7)$, and $(-4,\ 3-7)$, or $(0,\ 3)$, $(-8, 3)$, $(-4, 10)$, and $(-4, -4)$. Connect these points with an oval-shaped curve.

$\frac{(x+4)^2}{16} + \frac{(y-3)^2}{49} = 1$

31. $12(x-1)^2 + 3(y+4)^2 = 48$

$\dfrac{(x-1)^2}{4} + \dfrac{(y+4)^2}{16} = 1$

$\dfrac{(x-1)^2}{2^2} + \dfrac{(y-(-4))^2}{4^2} = 1$

The center of the ellipse is $(1, -4)$. Note that $a = 2$ and $b = 4$. We locate the center and then plot the points $(1+2, -4)$, $(1-2, -4)$, $(1, -4+4)$, and $(1, -4-4)$, or $(3, -4)$, $(-1, -4)$, $(1, 0)$, and $(1, -8)$. Connect these points with an oval-shaped curve.

$12(x-1)^2 + 3(y+4)^2 = 48$

33. $4(x+3)^2 + 4(y+1)^2 - 10 = 90$

$4(x+3)^2 + 4(y+1)^2 = 100$

Observe that the x^2- and y^2-terms have the same coefficient. Dividing both sides by 4, we have

$(x+3)^2 + (y+1)^2 = 25$.

This is the equation of a circle with center $(-3, -1)$ and radius 5.

$4(x+3)^2 + 4(y+1)^2 - 10 = 90$

35. *Writing Exercise.*

37. $x^2 - 5x + 3 = 0$

$x = \dfrac{5 \pm \sqrt{25 - 4 \cdot 1 \cdot 3}}{2 \cdot 1}$ Using the quadratic formula

$= \dfrac{5 \pm \sqrt{13}}{2}$

39. $\dfrac{4}{x+2} + \dfrac{3}{2x-1} = 2$ Note $x \neq -2$ and $x \neq \dfrac{1}{2}$

$4(2x-1) + 3(x+2) = 2(x+2)(2x-1)$

$8x - 4 + 3x + 6 = 4x^2 + 6x - 4$

$4x^2 - 5x - 6 = 0$

$(4x+3)(x-2) = 0$

$4x + 3 = 0 \quad or \quad x - 2 = 0$

$x = -\dfrac{3}{4} \quad or \quad\quad x = 2$

41. $x^2 = 11$

$x = \pm\sqrt{11}$

43. *Writing Exercise.*

45. Plot the given points.

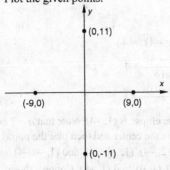

From the location of these points, we see that the ellipse that contains them is centered at the origin with $a = 9$ and $b = 11$. We write the equation of the ellipse:

$\dfrac{x^2}{9^2} + \dfrac{y^2}{11^2} = 1$

$\dfrac{x^2}{81} + \dfrac{y^2}{121} = 1$

47. Plot the given points.

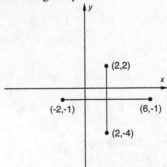

The midpoint of the segment from $(-2, -1)$ to $(6, -1)$ is $\left(\dfrac{-2+6}{2}, \dfrac{-1-1}{2}\right)$, or $(2, -1)$. The midpoint of the segment from $(2, -4)$ to $(2, 2)$ is $\left(\dfrac{2+2}{2}, \dfrac{-4+2}{2}\right)$, or $(2, -1)$. Thus, we can conclude that $(2, -1)$ is the center of the ellipse. The distance from $(-2, -1)$ to $(2, -1)$ is

$\sqrt{[2-(-2)]^2 + [-1-(-1)]^2} = \sqrt{16} = 4$, so $a = 4$. The distance from $(2, 2)$ to $(2, -1)$ is $\sqrt{(2-2)^2 + (-1-2)^2}$

$= \sqrt{9} = 3$, so $b = 3$. We write the equation of the ellipse.

$\dfrac{(x-2)^2}{4^2} + \dfrac{(y-(-1))^2}{3^2} = 1$

$\dfrac{(x-2)^2}{16} + \dfrac{(y+1)^2}{9} = 1$

49. We have a vertical ellipse centered at the origin with $a = 6/2$, or 3, and $b = 10/2$, or 5. Then the equation is $\dfrac{x^2}{3^2} + \dfrac{y^2}{5^2} = 1$, or $\dfrac{x^2}{9} + \dfrac{y^2}{25} = 1$.

51. a. Let $F_1 = (-c, 0)$ and $F_2 = (c, 0)$. Then the sum of the distances from the foci to P is $2a$. By the distance formula,

$\sqrt{(x+c)^2 + y^2} + \sqrt{(x-c)^2 + y^2} = 2a$, or

$\sqrt{(x+c)^2 + y^2} = 2a - \sqrt{(x-c)^2 + y^2}$.

Squaring, we get

$(x+c)^2 + y^2 = 4a^2 - 4a\sqrt{(x-c)^2 + y^2} + (x-c)^2 + y^2$,

or $x^2 + 2cx + c^2 + y^2$

$= 4a^2 - 4a\sqrt{(x-c)^2 + y^2} + x^2 - 2cx + c^2 + y^2$.

Thus

$-4a^2 + 4cx = -4a\sqrt{(x-c)^2 + y^2}$

$a^2 - cx = a\sqrt{(x-c)^2 + y^2}$.

Squaring again, we get

$a^4 - 2a^2cx + c^2x^2 = a^2(x^2 - 2cx + c^2 + y^2)$

$a^4 - 2a^2cx + c^2x^2 = a^2x^2 - 2a^2cx + a^2c^2 + a^2y^2$,

or

$$x^2(a^2 - c^2) + a^2 y^2 = a^2(a^2 - c^2)$$
$$\frac{x^2}{a^2} + \frac{y^2}{a^2 - c^2} = 1.$$

b. When P is at $(0, b)$, it follows that $b^2 = a^2 - c^2$.

Substituting, we have $\frac{x^2}{a^2} + \frac{y^2}{b^2} = 1$.

53. For the given ellipse, $a = 6/2$, or 3, and $b = 2/2$, or 1. The patient's mouth should be at a distance $2c$ from the light source, where the coordinates of the foci of the ellipse are $(-c, 0)$ and $(c, 0)$. From Exercise 51(b), we know $b^2 = a^2 - c^2$. We use this to find c.

$$b^2 = a^2 - c^2$$
$$1^2 = 3^2 - c^2 \quad \text{Substituting}$$
$$c^2 = 8$$
$$c = \sqrt{8}$$

Then $2c = 2\sqrt{8} \approx 5.66$. The patient's mouth should be about 5.66 ft from the light source.

55.
$$x^2 - 4x + 4y^2 + 8y - 8 = 0$$
$$x^2 - 4x + 4y^2 + 8y = 8$$
$$x^2 - 4x + 4(y^2 + 2y) = 8$$
$$(x^2 - 4x + 4 - 4) + 4(y^2 + 2y + 1 - 1) = 8$$
$$(x^2 - 4x + 4) + 4(y^2 + 2y + 1) = 8 + 4 + 4 \cdot 1$$
$$(x - 2)^2 + 4(y + 1)^2 = 16$$
$$\frac{(x - 2)^2}{16} + \frac{(y + 1)^2}{4} = 1$$
$$\frac{(x - 2)^2}{4^2} + \frac{(y - (-1))^2}{2^2} = 1$$

The center of the ellipse is $(2, -1)$. Note that $a = 4$ and $b = 2$. We locate the center and then plot the points $(2 + 4, -1)$, $(2 - 4, -1)$, $(2, -1 + 2)$, $(2, -1 - 2)$, or $(6, -1)$, $(-2, -1)$, $(2, 1)$, and $(2, -3)$. Connect these points with an oval-shaped curve.

$\frac{(x-2)^2}{16} + \frac{(y+1)^2}{4} = 1$

57. *Graphing Calculator Exercise*

Connecting the Concepts

1. $y = 3(x - 4)^2 + 1$ parabola
Vertex: $(4, 1)$
Axis of symmetry: $x = 4$

2. $x = y^2 + 2y + 3$ parabola
$x = y^2 + 2y + 1 + 2$
$x = (y + 1)^2 + 2$
Vertex: $(2, -1)$
Axis of symmetry: $y = -1$

3. $(x - 3)^2 + (y - 2)^2 = 5$ circle
Center: $(3, 2)$

4. $\quad x^2 + 6x + y^2 + 10y = 12$ circle
$x^2 + 6x + 9 + y^2 + 10y + 25 = 12 + 9 + 25$
$\quad (x + 3)^2 + (y + 5)^2 = 46$
Center: $(-3, -5)$

5. $\frac{x^2}{144} + \frac{y^2}{81} = 1$ ellipse
x-intercepts: $(-12, 0)$ and $(12, 0)$
y-intercepts: $(0, 9)$ and $(0, -9)$

6. $\frac{x^2}{9} - \frac{y^2}{121} = 1$ hyperbola
Vertices: $(-3, 0)$ and $(3, 0)$

7. $4y^2 - x^2 = 4$ hyperbola
$\frac{y^2}{1} - \frac{x^2}{4} = 1$
Vertices: $(0, -1)$ and $(0, 1)$

8. $\frac{y^2}{9} - \frac{x^2}{4} = 1$ hyperbola
$a = 2, b = 3$
Asymptotes: $y = \frac{3}{2}x$ and $y = -\frac{3}{2}x$

Exercise Set 10.3

1. Asymptote B

3. Branch A

5. Hyperbola F

7. $\frac{y^2}{16} - \frac{x^2}{16} = 1$

$\frac{y^2}{4^2} - \frac{x^2}{4^2} = 1$

$a = 4$ and $b = 4$, so the asymptotes are $y = \frac{4}{4}x$ and

$y = -\frac{4}{4}x$, or $y = x$ and $y = -x$. We sketch them.

Replacing x with 0 and solving for y, we get $y = \pm 4$, so the intercepts are $(0, 4)$ and $(0, -4)$.
We plot the intercepts and draw smooth curves through them that approach the asymptotes.

$\frac{y^2}{16} - \frac{x^2}{16} = 1$

9. $\dfrac{x^2}{4} - \dfrac{y^2}{25} = 1$

$\dfrac{x^2}{2^2} - \dfrac{y^2}{5^2} = 1$

$a = 2$ and $b = 5$, so the asymptotes are $y = \dfrac{5}{2}x$ and

$y = -\dfrac{5}{2}x$. We sketch them.

Replacing y with 0 and solving for x, we get $x = \pm 2$, so the intercepts are (2, 0) and (−2, 0).
We plot the intercepts and draw smooth curves through them that approach the asymptotes.

$\dfrac{x^2}{4} - \dfrac{y^2}{25} = 1$

11. $\dfrac{y^2}{36} - \dfrac{x^2}{9} = 1$

$\dfrac{y^2}{6^2} - \dfrac{x^2}{3^2} = 1$

$a = 3$ and $b = 6$, so the asymptotes are $y = \dfrac{6}{3}x$ and

$y = -\dfrac{6}{3}x$, or $y = 2x$ and $y = -2x$. We sketch them.

Replacing x with 0 and solving for y, we get $y = \pm 6$, so the intercepts are (0, 6) and (0, −6).
We plot the intercepts and draw smooth curves through them that approach the asymptotes.

$\dfrac{y^2}{36} - \dfrac{x^2}{9} = 1$

13. $y^2 - x^2 = 25$

$\dfrac{y^2}{25} - \dfrac{x^2}{25} = 1$

$\dfrac{y^2}{5^2} - \dfrac{x^2}{5^2} = 1$

$a = 5$ and $b = 5$, so the asymptotes are $y = \dfrac{5}{5}x$ and

$y = -\dfrac{5}{5}x$, or $y = x$ and $y = -x$. We sketch them.

Replacing x with 0 and solving for y, we get $y = \pm 5$, so the intercepts are (0, 5) and (0, −5).
We plot the intercepts and draw smooth curves through them that approach the asymptotes.

$y^2 - x^2 = 25$

15. $25x^2 - 16y^2 = 400$

$\dfrac{x^2}{16} - \dfrac{y^2}{25} = 1$ Multiplying by $\dfrac{1}{400}$

$\dfrac{x^2}{4^2} - \dfrac{y^2}{5^2} = 1$

$a = 4$ and $b = 5$, so the asymptotes are $y = \dfrac{5}{4}x$ and

$y = -\dfrac{5}{4}x$. We sketch them.

Replacing y with 0 and solving for x, we get $x = \pm 4$, so the intercepts are (4, 0) and (−4, 0).
We plot the intercepts and draw smooth curves through them that approach the asymptotes.

$25x^2 - 16y^2 = 400$

17. $xy = -6$

$y = -\dfrac{6}{x}$ Solving for y

We find some solutions, keeping the results in a table.

x	y
$\dfrac{1}{6}$	36
1	−6
6	−1
12	$-\dfrac{1}{2}$
$-\dfrac{1}{6}$	36
−1	6
−6	1
−12	$\dfrac{1}{2}$

Note that we cannot use 0 for x. The x-axis and the y-axis are the asymptotes.

19. $xy = 4$

$y = \dfrac{4}{x}$ Solving for y

We find some solutions, keeping the results in a table.

x	y
$\dfrac{1}{2}$	8
1	4
4	1
8	$\dfrac{1}{2}$
$-\dfrac{1}{2}$	−8
−1	−4
−2	−2
−4	−1

Note that we cannot use 0 for x. The x-axis and the y-axis are the asymptotes.

21. $xy = -2$

$y = -\dfrac{2}{x}$ Solving for y

x	y
$\dfrac{1}{2}$	-4
1	-2
2	-1
4	$-\dfrac{1}{2}$
$-\dfrac{1}{2}$	4
-1	2
-2	1
-4	$\dfrac{1}{2}$

Note that we cannot use 0 for x. The x-axis and the y-axis are the asymptotes.

23. $xy = 1$

$y = \dfrac{1}{x}$ Solving for y

x	y
$\dfrac{1}{4}$	4
$\dfrac{1}{2}$	2
1	1
2	$\dfrac{1}{2}$
4	$\dfrac{1}{4}$
$-\dfrac{1}{4}$	-4
$-\dfrac{1}{2}$	-2
-1	-1
-2	$-\dfrac{1}{2}$
-4	$-\dfrac{1}{4}$

Note that we cannot use 0 for x. The x-axis and the y-axis are the asymptotes.

25. $x^2 + y^2 - 6x + 10y - 40 = 0$

Completing the square twice, we obtain an equivalent equation:
$$\left(x^2 - 6x\right) + \left(y^2 + 10y\right) = 40$$
$$\left(x^2 - 6x + 9\right) + \left(y^2 + 10y + 25\right) = 40 + 9 + 25$$
$$(x-3)^2 + (y+5)^2 = 74$$

The graph is a circle.

27. $9x^2 + 4y^2 - 36 = 0$

$$9x^2 + 4y^2 = 36$$
$$\dfrac{x^2}{4} + \dfrac{y^2}{9} = 1$$

The graph is an ellipse.

29. $4x^2 - 9y^2 - 72 = 0$

$$4x^2 - 9y^2 = 72$$
$$\dfrac{x^2}{18} - \dfrac{y^2}{8} = 1$$

The graph is a hyperbola.

31. $y^2 = 20 - x^2$

$$x^2 + y^2 = 20$$

The graph is a circle.

33. $x - 10 = y^2 - 6y$

$$x - 10 + 9 = y^2 - 6y + 9$$
$$x - 1 = (y-3)^2$$

The graph is a parabola.

35. $x - \dfrac{3}{y} = 0$

$$x = \dfrac{3}{y}$$
$$xy = 3$$

The graph is a hyperbola.

37. $y + 6x = x^2 + 5$

$$y = x^2 - 6x + 5$$

The graph is a parabola

39. $25y^2 = 100 + 4x^2$

$$25y^2 - 4x^2 = 100$$
$$\dfrac{y^2}{4} - \dfrac{x^2}{25} = 1$$

The graph is a hyperbola.

41. $3x^2 + y^2 - x = 2x^2 - 9x + 10y + 40$

$$x^2 + y^2 + 8x - 10y = 40$$

Both variables are squared, so the graph is not a parabola. The plus sign between x^2 and y^2 indicates that we have either a circle or an ellipse. Since the coefficients of x^2 and y^2 are the same, the graph is a circle.

43. $16x^2 + 5y^2 - 12x^2 + 8y^2 - 3x + 4y = 568$

$$4x^2 + 13y^2 - 3x + 4y = 568$$

Both variables are squared, so the graph is not a parabola. The plus sign between x^2 and y^2 indicates that we have either a circle or an ellipse. Since the coefficients of x^2 and y^2 are different, the graph is an ellipse.

45. *Writing Exercise.*

47. $(16 - y^4) = (4 + y^2)(4 - y^2)$
$$= (4 + y^2)(2 + y)(2 - y)$$

49. $10c^3 - 80c^2 + 150c = 10c(c^2 - 8c + 15)$
$$= 10c(c-3)(c-5)$$

51. $8t^4 - 8t = 8t(t^3 - 1) = 8t(t-1)(t^2 + t + 1)$

53. *Writing Exercise.*

55. Since the intercepts are (0, 6) and (0, –6), we know

that the hyperbola is of the form $\dfrac{y^2}{b^2} - \dfrac{x^2}{a^2} = 1$ and that

$b = 6$. The equations of the asymptotes tell us that
$b / a = 3$, so

$$\dfrac{6}{a} = 3$$
$$a = 2.$$

The equation is $\dfrac{y^2}{6^2} - \dfrac{x^2}{2^2} = 1$, or $\dfrac{y^2}{36} - \dfrac{x^2}{4} = 1$.

57. $\dfrac{(x-5)^2}{36} - \dfrac{(y-2)^2}{25} = 1$

$\dfrac{(x-5)^2}{6^2} - \dfrac{(y-2)^2}{5^2} = 1$

$h = 5, k = 2, a = 6, b = 5$
Center: (5, 2)
Vertices: (5 – 6, 2) and (5 + 6, 2), or (–1, 2) and (11, 2)
Asymptotes: $y - 2 = \frac{5}{6}(x - 5)$ and $y - 2 = -\frac{5}{6}(x - 5)$

$\dfrac{(x-5)^2}{36} - \dfrac{(y-2)^2}{25} = 1$

59. $8(y+3)^2 - 2(x-4)^2 = 32$

$\dfrac{(y+3)^2}{4} - \dfrac{(x-4)^2}{16} = 1$

$\dfrac{(y-(-3))^2}{2^2} - \dfrac{(x-4)^2}{4^2} = 1$

$h = 4, k = -3, a = 4, b = 2$
Center: (4, –3)
Vertices: (4, –3 + 2) and (4, – 3 – 2), or (4, –1) and (4,–5)

Asymptotes: $y - (-3) = \frac{2}{4}(x - 4)$ and

$y - (-3) = -\frac{2}{4}(x - 4)$, or $y + 3 = \frac{1}{2}(x - 4)$ and

$y + 3 = -\frac{1}{2}(x - 4)$

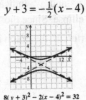

$8(y + 3)^2 - 2(x - 4)^2 = 32$

61. $4x^2 - y^2 + 24x + 4y + 28 = 0$

$4(x^2 + 6x) - (y^2 - 4y) = -28$

$4(x^2 + 6x + 9 - 9) - (y^2 - 4y + 4 - 4) = -28$

$4(x^2 + 6x + 9) - (y^2 - 4y + 4) = -28 + 4 \cdot 9 - 4$

$4(x+3)^2 - (y-2)^2 = 4$

$\dfrac{(x+3)^2}{1} - \dfrac{(y-2)^2}{4} = 1$

$\dfrac{(x-(-3))^2}{1^2} - \dfrac{(y-2)^2}{2^2} = 1$

$h = -3, k = 2, a = 1, b = 2$
Center: (–3, 2)
Vertices: (–3 – 1, 2), and (–3 + 1, 2), or (–4, 2) and (–2, 2)

Asymptotes: $y - 2 = \frac{2}{1}(x - (-3))$ and

$y - 2 = -\frac{2}{1}(x - (-3))$, or $y - 2 = 2(x + 3)$ and

$y - 2 = -2(x + 3)$

$4x^2 - y^2 + 24x + 4y + 28 = 0$

63. *Graphing Calculator Exercise*

Mid-Chapter Review

1. $(x^2 - 4x) + (y^2 + 2y) = 6$
$(x^2 - 4x + 4) + (y^2 + 2y + 1) = 6 + 4 + 1$
$(x - 2)^2 + (y + 1)^2 = 11$

The center of the circle is (2,–1).
The radius is $\sqrt{11}$.

2. a. Is there both an x^2-term and a y^2-term? Yes.

b. Do both the x^2-term and the y^2-term have the same sign? No.

c. The graph of the equation is a hyperbola.

3. $[x - (-4)]^2 + (y - 9)^2 = (2\sqrt{5})^2$
$(x + 4)^2 + (y - 9)^2 = 20$

4. $x^2 - 10x + y^2 + 2y = 10$
$x^2 - 10x + 25 + y^2 + 2y + 1 = 10 + 25 + 1$
$(x - 5)^2 + (y + 1)^2 = 36$

Center: (5, –1); radius: 6

5. $x^2 + y^2 = 36$ is a circle.

$x^2 + y^2 = 36$

6. $y = x^2 - 5$ is a parabola.

$y = x^2 - 5$

7. $\dfrac{x^2}{25} + \dfrac{y^2}{49} = 1$ is an ellipse.

8. $\dfrac{x^2}{25} - \dfrac{y^2}{49} = 1$ is a hyperbola.

9. $x = (y+3)^2 + 2$ is a parabola.

10. $4x^2 + 9y^2 = 36$

$\dfrac{x^2}{9} + \dfrac{y^2}{4} = 1$ is an ellipse.

11. $xy = -4$ is a hyperbola.

12. $(x+2)^2 + (y-3)^2 = 1$ is a circle.

13. $x^2 + y^2 - 8y - 20 = 0$

$x^2 + y^2 - 8y + 16 = 20 + 16$

$x^2 + (y-4)^2 = 36$ is a circle.

14. $x = y^2 + 2y$

$x = y^2 + 2y + 1 - 1$

$x = (y+1)^2 - 1$ is a parabola.

15. $16y^2 - x^2 = 16$

$\dfrac{y^2}{1} - \dfrac{x^2}{16} = 1$ is a hyperbola.

16. $x = \dfrac{9}{y}$

$xy = 9$ is a hyperbola.

Exercise Set 10.4

1. True

3. False

5. True

7. $x^2 + y^2 = 41$,   (1)

$y - x = 1$ (2)

First solve Equation (2) for y.

$y = x + 1$ (3)

Then substitute $x + 1$ for y in Equation (1) and solve for x.

$$x^2 + y^2 = 41$$
$$x^2 + (x+1)^2 = 41$$
$$x^2 + x^2 + 2x + 1 = 41$$
$$2x^2 + 2x - 40 = 0$$
$$x^2 + x - 20 = 0 \quad \text{Multiplying by } \tfrac{1}{2}$$
$$(x+5)(x-4) = 0$$

$x + 5 = 0$ or $x - 4 = 0$ Principle of zero products

$x = -5$ or $x = 4$

Now substitute these numbers in Equation (3) and solve for y.

For $x = -5$, $y = -5 + 1 = -4$

For $x = 4$, $y = 4 + 1 = 5$

The pairs $(-5, -4)$ and $(4, 5)$ check, so they are the solutions.

9. $4x^2 + 9y^2 = 36$,   (1)
      $3y + 2x = 6$ (2)

First solve Equation (2) for y.

$$3y = -2x + 6$$
$$y = -\tfrac{2}{3}x + 2 \quad (3)$$

Then substitute $-\tfrac{2}{3}x + 2$ for y in Equation (1) and solve for x.

$$4x^2 + 9y^2 = 36$$
$$4x^2 + 9\left(-\tfrac{2}{3}x + 2\right)^2 = 36$$
$$4x^2 + 9\left(\tfrac{4}{9}x^2 - \tfrac{8}{3}x + 4\right) = 36$$
$$4x^2 + 4x^2 - 24x + 36 = 36$$
$$8x^2 - 24x = 0$$
$$x^2 - 3x = 0$$
$$x(x - 3) = 0$$
$$x = 0 \ \text{ or } \ x = 3$$

Now substitute these numbers in Equation (3) and solve for y.

For $x = 0$, $y = -\tfrac{2}{3} \cdot 0 + 2 = 2$

For $x = 3$, $y = -\tfrac{2}{3} \cdot 3 + 2 = 0$

The pairs $(0, 2)$ and $(3, 0)$ check, so they are the solutions.

11. $y^2 = x + 3$,   (1)
      $2y = x + 4$ (2)

First solve Equation (2) for x.

    $2y - 4 = x$ (3)

Then substitute $2y - 4$ for x in Equation (1) and solve for y.

$$y^2 = x + 3$$
$$y^2 = (2y - 4) + 3$$
$$y^2 = 2y - 1$$
$$y^2 - 2y + 1 = 0$$
$$(y - 1)(y - 1) = 0$$
$$y - 1 = 0 \quad \text{or} \quad y - 1 = 0$$
$$y = 1 \quad \text{or} \qquad y = 1$$

Now substitute 1 for y in Equation (3) and solve for x.

$$2 \cdot 1 - 4 = x$$
$$-2 = x$$

The pair $(-2, 1)$ checks. It is the solution.

13. $x^2 - xy + 3y^2 = 27$,   (1)
          $x - y = 2$ (2)

First solve Equation (2) for y.

    $x - 2 = y$ (3)

Then substitute $x - 2$ for y in Equation (1) and solve for x.

$$x^2 - xy + 3y^2 = 27$$
$$x^2 - x(x - 2) + 3(x - 2)^2 = 27$$
$$x^2 - x^2 + 2x + 3x^2 - 12x + 12 = 27$$
$$3x^2 - 10x - 15 = 0$$

$$x = \frac{-(-10) \pm \sqrt{(-10)^2 - 4(3)(-15)}}{2 \cdot 3}$$
$$x = \frac{10 \pm \sqrt{100 + 180}}{6} = \frac{10 \pm \sqrt{280}}{6}$$
$$x = \frac{10 \pm 2\sqrt{70}}{6} = \frac{5 \pm \sqrt{70}}{3}$$

Now substitute these numbers in Equation (3) and solve for y.

For $x = \dfrac{5 + \sqrt{70}}{3}$, $y = \dfrac{5 + \sqrt{70}}{3} - 2 = \dfrac{-1 + \sqrt{70}}{3}$

For $x = \dfrac{5 - \sqrt{70}}{3}$, $y = \dfrac{5 - \sqrt{70}}{3} - 2 = \dfrac{-1 - \sqrt{70}}{3}$

The pairs $\left(\dfrac{5 + \sqrt{70}}{3}, \dfrac{-1 + \sqrt{70}}{3}\right)$ and

$\left(\dfrac{5 - \sqrt{70}}{3}, \dfrac{-1 - \sqrt{70}}{3}\right)$ check, so they are the solutions.

15. $x^2 + 4y^2 = 25$,   (1)
       $x + 2y = 7$ (2)

First solve Equation (2) for x.

    $x = -2y + 7$ (3)

Then substitute $-2y + 7$ for x in Equation (1) and solve for y.

$$x^2 + 4y^2 = 25$$
$$(-2y + 7)^2 + 4y^2 = 25$$
$$4y^2 - 28y + 49 + 4y^2 = 25$$
$$8y^2 - 28y + 24 = 0$$
$$2y^2 - 7y + 6 = 0$$
$$(2y - 3)(y - 2) = 0$$
$$y = \tfrac{3}{2} \ \text{ or } \ y = 2$$

Now substitute these numbers in Equation (3) and solve for x.

For $y = \tfrac{3}{2}$, $x = -2 \cdot \tfrac{3}{2} + 7 = 4$

For $y = 2$, $x = -2 \cdot 2 + 7 = 3$

The pairs $\left(4, \tfrac{3}{2}\right)$ and $(3, 2)$ check, so they are the solutions.

17. $x^2 - xy + 3y^2 = 5$,   (1)
         $x - y = 2$ (2)

First solve Equation (2) for y.

    $x - 2 = y$ (3)

Then substitute $x - 2$ for y in Equation (1) and solve for x.

$$x^2 - xy + 3y^2 = 5$$
$$x^2 - x(x - 2) + 3(x - 2)^2 = 5$$
$$x^2 - x^2 + 2x + 3x^2 - 12x + 12 = 5$$
$$3x^2 - 10x + 7 = 0$$
$$(3x - 7)(x - 1) = 0$$
$$x = \tfrac{7}{3} \ \text{ or } \ x = 1$$

Now substitute these numbers in Equation (3) and solve for y.

For $x = \frac{7}{3}$, $y = \frac{7}{3} - 2 = \frac{1}{3}$

For $x = 1$, $y = 1 - 2 = -1$

The pairs $\left(\frac{7}{3}, \frac{1}{3}\right)$ and $(1, -1)$ check, so they are the solutions.

19. $3x + y = 7,$ (1)

$4x^2 + 5y = 24$ (2)

First solve Equation (1) for y.

$y = 7 - 3x$ (3)

Then substitute $7 - 3x$ for y in Equation (2) and solve for x.

$$4x^2 + 5y = 24$$
$$4x^2 + 5(7 - 3x) = 24$$
$$4x^2 + 35 - 15x = 24$$
$$4x^2 - 15x + 11 = 0$$
$$(4x - 11)(x - 1) = 0$$

$$x = \frac{11}{4} \quad \text{or} \quad x = 1$$

Now substitute these numbers into Equation (3) and solve for y.

For $x = \frac{11}{4}$, $y = 7 - 3 \cdot \frac{11}{4} = -\frac{5}{4}$

For $x = 1$, $y = 7 - 3 \cdot 1 = 4$

The pairs $\left(\frac{11}{4}, -\frac{5}{4}\right)$ and $(1, 4)$ check, so they are the solutions.

21. $a + b = 6,$ (1)

$ab = 8$ (2)

First solve Equation (1) for a.

$a = -b + 6$ (3)

Then substitute $-b + 6$ for a in Equation (2) and solve for b.

$(-b + 6)b = 8$

$-b^2 + 6b = 8$

$0 = b^2 - 6b + 8$

$0 = (b - 2)(b - 4)$

$b - 2 = 0 \quad \text{or} \quad b - 4 = 0$

$b = 2 \quad \text{or} \quad b = 4$

Now substitute these numbers in Equation (3) and solve for a.

For $b = 2$, $a = -2 + 6 = 4$

For $b = 4$, $a = -4 + 6 = 2$

The pairs $(4, 2)$ and $(2, 4)$ check, so they are the solutions.

23. $2a + b = 1,$ (1)

$b = 4 - a^2$ (2)

Equation (2) is already solved for b. Substitute $4 - a^2$ for b in Equation (1) and solve for a.

$2a + 4 - a^2 = 1$

$0 = a^2 - 2a - 3$

$0 = (a - 3)(a + 1)$

$a = 3 \quad \text{or} \quad a = -1$

Substitute these numbers in Equation (2) and solve for b.

For $a = 3$, $b = 4 - 3^2 = -5$

For $a = -1$, $b = 4 - (-1)^2 = 3$

The pairs $(3, -5)$ and $(-1, 3)$ check, so they are the solutions.

25. $a^2 + b^2 = 89,$ (1)

$a - b = 3$ (2)

First solve Equation (2) for a.

$a = b + 3$ (3)

Then substitute $b + 3$ for a in Equation (1) and solve for b.

$$(b + 3)^2 + b^2 = 89$$
$$b^2 + 6b + 9 + b^2 = 89$$
$$2b^2 + 6b - 80 = 0$$
$$b^2 + 3b - 40 = 0$$
$$(b + 8)(b - 5) = 0$$

$b = -8 \quad \text{or} \quad b = 5$

Substitute these numbers in Equation (3) and solve for a.

For $b = -8$, $a = -8 + 3 = -5$

For $b = 5$, $a = 5 + 3 = 8$

The pairs $(-5, -8)$ and $(8, 5)$ check, so they are the solutions.

27. $y = x^2,$ (1)

$x = y^2$ (2)

Equation (1) is already solved for y. Substitute x^2 for y in Equation (2) and solve for x.

$x = y^2$

$x = \left(x^2\right)^2$

$x = x^4$

$0 = x^4 - x$

$0 = x(x^3 - 1)$

$0 = x(x - 1)(x^2 + x + 1)$

$x = 0 \quad \text{or} \quad x = 1 \quad \text{or} \quad x = \dfrac{-1 \pm \sqrt{1^2 - 4 \cdot 1 \cdot 1}}{2}$

$x = 0 \quad \text{or} \quad x = 1 \quad \text{or} \quad x = -\dfrac{1}{2} \pm \dfrac{\sqrt{3}}{2} i$

Substitute these numbers in Equation (1) and solve for y.

For $x = 0$, $y = 0^2 = 0$

For $x = 1$, $y = 1^2 = 1$

For $x = -\dfrac{1}{2} + \dfrac{\sqrt{3}}{2} i$, $y = \left(-\dfrac{1}{2} + \dfrac{\sqrt{3}}{2} i\right)^2 = -\dfrac{1}{2} - \dfrac{\sqrt{3}}{2} i$

For $x = -\dfrac{1}{2} - \dfrac{\sqrt{3}}{2} i$, $y = \left(-\dfrac{1}{2} - \dfrac{\sqrt{3}}{2} i\right)^2 = -\dfrac{1}{2} + \dfrac{\sqrt{3}}{2} i$

The pairs $(0, 0)$, $(1, 1)$, $\left(-\dfrac{1}{2} + \dfrac{\sqrt{3}}{2} i, \ -\dfrac{1}{2} - \dfrac{\sqrt{3}}{2} i\right)$,

and $\left(-\dfrac{1}{2} - \dfrac{\sqrt{3}}{2} i, \ -\dfrac{1}{2} + \dfrac{\sqrt{3}}{2} i\right)$ check, so they are the solutions.

29. $x^2 + y^2 = 16$,   (1)
$x^2 - y^2 = 16$ (2)

Here we use the elimination method.

$\begin{array}{l} x^2 + y^2 = 16 \quad (1) \\ \underline{x^2 - y^2 = 16} \quad (2) \\ 2x^2 \quad\quad = 32 \text{ Adding} \\ \quad\quad x^2 = 16 \\ \quad\quad x = \pm 4 \end{array}$

If $x = 4$, $x^2 = 16$, and if $x = -4$, $x^2 = 16$, so substituting 4 or −4 in Equation (2) gives us

$\begin{array}{l} x^2 + y^2 = 16 \\ 16 + y^2 = 16 \\ \quad\quad y^2 = 0 \\ \quad\quad y = 0 \end{array}$

The pairs (4, 0) and (−4, 0) check. They are the solutions.

31. $x^2 + y^2 = 25$,   (1)
$xy = 12$ (2)

First we solve Equation (2) for y.

$\begin{array}{l} xy = 12 \\ y = \dfrac{12}{x} \end{array}$

Then we substitute $\dfrac{12}{x}$ for y in Equation (1) and solve for x.

$$x^2 + y^2 = 25$$
$$x^2 + \left(\frac{12}{x}\right)^2 = 25$$
$$x^2 + \frac{144}{x^2} = 25$$
$$x^4 + 144 = 25x^2 \quad \text{Multiplying by } x^2$$
$$x^4 - 25x^2 + 144 = 0$$
$$u^2 - 25u + 144 = 0 \quad\quad \text{Letting } u = x^2$$
$$(u - 9)(u - 16) = 0$$
$$u = 9 \quad or \quad u = 16$$

We now substitute x^2 for u and solve for x.

$$x^2 = 9 \quad or \quad x^2 = 16$$
$$x = \pm 3 \quad or \quad x = \pm 4$$

Since $y = 12 / x$, if $x = 3$, $y = 4$; if $x = -3$, $y = -4$; if $x = 4$, $y = 3$; and if $x = -4$, $y = -3$. The pairs (3, 4), $(-3, -4)$, (4, 3), and $(-4, -3)$ check. They are the solutions.

33. $x^2 + y^2 = 9$,   (1)
$25x^2 + 16y^2 = 400$ (2)

$\begin{array}{l} -16x^2 - 16y^2 = -144 \quad \text{Multiplying (1) by } -16 \\ \underline{25x^2 + 16y^2 = 400} \\ 9x^2 \quad\quad\quad = 256 \quad \text{Adding} \\ \quad\quad x = \pm\dfrac{16}{3} \end{array}$

$\dfrac{256}{9} + y^2 = 9$ Substituting in (1)

$$y^2 = 9 - \frac{256}{9}$$
$$y^2 = -\frac{175}{9}$$
$$y = \pm\sqrt{-\frac{175}{9}} = \pm\frac{5\sqrt{7}}{3}i$$

The pairs $\left(\dfrac{16}{3}, \dfrac{5\sqrt{7}}{3}i\right)$, $\left(\dfrac{16}{3}, -\dfrac{5\sqrt{7}}{3}i\right)$, $\left(-\dfrac{16}{3}, \dfrac{5\sqrt{7}}{3}i\right)$, and $\left(-\dfrac{16}{3}, -\dfrac{5\sqrt{7}}{3}i\right)$ check. They are the solutions.

35. $x^2 + y^2 = 14$,   (1)
$\begin{array}{l} \underline{x^2 - y^2 = 4} \quad (2) \\ 2x^2 \quad\quad = 18 \quad \text{Adding} \\ \quad\quad x^2 = 9 \\ \quad\quad x = \pm 3 \end{array}$

$9 + y^2 = 14$ Substituting in Eq. (1)
$$y^2 = 5$$
$$y = \pm\sqrt{5}$$

The pairs $(-3, -\sqrt{5})$, $(-3, \sqrt{5})$, $(3, -\sqrt{5})$, and $(3, \sqrt{5})$ check. They are the solutions.

37. $x^2 + y^2 = 10$,   (1)
$xy = 3$ (2)

First we solve Equation (2) for y.

$\begin{array}{l} xy = 3 \\ y = \dfrac{3}{x} \end{array}$

Then we substitute $\dfrac{3}{x}$ for y in Equation (1) and solve for x.

$$x^2 + y^2 = 10$$
$$x^2 + \left(\frac{3}{x}\right)^2 = 10$$
$$x^2 + \frac{9}{x^2} = 10$$
$$x^4 + 9 = 10x^2 \quad \text{Multiplying by } x^2$$
$$x^4 - 10x^2 + 9 = 0$$
$$u^2 - 10u + 9 = 0 \quad\quad \text{Letting } u = x^2$$
$$(u - 1)(u - 9) = 0$$
$$u = 1 \quad or \quad u = 9$$
$$x^2 = 1 \quad or \quad x^2 = 9 \quad \text{Substitute } x^2 \text{ for } u$$
$$x = \pm 1 \quad or \quad x = \pm 3 \quad \text{and solve for } x.$$

$y = 3/x$, so if $x = 1$, $y = 3$; if $x = -1$, $y = -3$; if $x = 3$, $y = 1$; if $x = -3$, $y = -1$. The pairs (1, 3), (−1, −3), (3, 1), and (−3, −1) check. They are the solutions.

39. $x^2 + 4y^2 = 20$,   (1)
$xy = 4$ (2)

First we solve Equation (2) for y.

$$y = \frac{4}{x}$$

Then we substitute $\frac{4}{x}$ for y in Equation (1) and solve for x.

$$x^2 + 4\left(\frac{4}{x}\right)^2 = 20$$
$$x^2 + \frac{64}{x^2} = 20$$
$$x^4 + 64 = 20x^2$$
$$x^4 - 20x^2 + 64 = 0$$
$$u^2 - 20u + 64 = 0 \quad \text{Letting } u = x^2$$
$$(u - 16)(u - 4) = 0$$
$$u = 16 \quad or \quad u = 4$$
$$x^2 = 16 \quad or \quad x^2 = 4$$
$$x = \pm 4 \quad or \quad x = \pm 2$$

$y = 4/x$, so if $x = 4$, $y = 1$; if $x = -4$, $y = -1$; if $x = 2$, $y = 2$; and if $x = -2$, $y = -2$. The pairs $(4, 1)$, $(-4, -1)$, $(2, 2)$, and $(-2, -2)$ check. They are the solutions.

41. $2xy + 3y^2 = 7$, (1)
$3xy - 2y^2 = 4$ (2)

$$\begin{array}{ll} 6xy + 9y^2 = 21 & \text{Multiplying (1) by 3} \\ -6xy + 4y^2 = -8 & \text{Multiplying (2) by } -2 \\ \hline 13y^2 = 13 & \\ y^2 = 1 & \\ y = \pm 1 & \end{array}$$

Substitute for y in Equation (1) and solve for x.
When $y = 1$: $\quad 2 \cdot x \cdot 1 + 3 \cdot 1^2 = 7$
$$2x = 4$$
$$x = 2$$
When $y = -1$: $\quad 2 \cdot x \cdot (-1) + 3(-1)^2 = 7$
$$-2x = 4$$
$$x = -2$$

The pairs $(2, 1)$ and $(-2, -1)$ check. They are the solutions.

43. $4a^2 - 25b^2 = 0$, (1)
$2a^2 - 10b^2 = 3b + 4$ (2)

$$\begin{array}{ll} 4a^2 - 25b^2 = 0 & \\ -4a^2 + 20b^2 = -6b - 8 & \text{Multiplying (2) by } -2 \\ \hline -5b^2 = -6b - 8 & \end{array}$$

$$0 = 5b^2 - 6b - 8$$
$$0 = (5b + 4)(b - 2)$$
$$b = -\frac{4}{5} \quad or \quad b = 2$$

Substitute for b in Equation (1) and solve for a.
When $b = -\frac{4}{5}$: $\quad 4a^2 - 25\left(-\frac{4}{5}\right)^2 = 0$
$$4a^2 = 16$$
$$a^2 = 4$$
$$a = \pm 2$$

When $b = 2$: $\quad 4a^2 - 25(2)^2 = 0$
$$4a^2 = 100$$
$$a^2 = 25$$
$$a = \pm 5$$

The pairs $\left(2, -\frac{4}{5}\right)$, $\left(-2, -\frac{4}{5}\right)$, $(5, 2)$ and $(-5, 2)$ check. They are the solutions.

45. $ab - b^2 = -4$, (1)
$ab - 2b^2 = -6$ (2)

$$\begin{array}{ll} ab - b^2 = -4 & \\ -ab + 2b^2 = 6 & \text{Multiplying (2) by } -1 \\ \hline b^2 = 2 & \end{array}$$
$$b = \pm\sqrt{2}$$

Substitute for b in Equation (1) and solve for a.
When $b = \sqrt{2}$: $\quad a(\sqrt{2}) - (\sqrt{2})^2 = -4$
$$a\sqrt{2} = -2$$
$$a = -\frac{2}{\sqrt{2}} = -\sqrt{2}$$
When $b = -\sqrt{2}$: $\quad a(-\sqrt{2}) - (-\sqrt{2})^2 = -4$
$$-a\sqrt{2} = -2$$
$$a = \frac{-2}{-\sqrt{2}} = \sqrt{2}$$

The pairs $(-\sqrt{2}, \sqrt{2})$ and $(\sqrt{2}, -\sqrt{2})$ check. They are the solutions.

47. *Familiarize*. We first make a drawing. We let l and w represent the length and width, respectively.

Translate. The perimeter is 28 cm.
$$2l + 2w = 28, \text{ or } l + w = 14$$
Using the Pythagorean theorem we have another equation.
$$l^2 + w^2 = 10^2, \text{ or } l^2 + w^2 = 100$$
Carry out. We solve the system:
$$l + w = 14, \quad (1)$$
$$l^2 + w^2 = 100 \quad (2)$$
First solve Equation (1) for w.
$$w = 14 - l \quad (3)$$
Then substitute $14 - l$ for w in Equation (2) and solve for l.
$$l^2 + w^2 = 100$$
$$l^2 + (14 - l)^2 = 100$$
$$l^2 + 196 - 28l + l^2 = 100$$
$$2l^2 - 28l + 96 = 0$$
$$l^2 - 14l + 48 = 0$$
$$(l - 8)(l - 6) = 0$$
$$l = 8 \text{ or } l = 6$$
If $l = 8$, then $w = 14 - 8$, or 6. If $l = 6$, then

$w = 14 - 6$, or 8. Since the length is usually considered to be longer than the width, we have the solution $l = 8$ and $w = 6$, or $(8, 6)$.

Check. If $l = 8$ and $w = 6$, then the perimeter is $2 \cdot 8 + 2 \cdot 6$, or 28. The length of a diagonal is $\sqrt{8^2 + 6^2}$, or $\sqrt{100}$, or 10. The numbers check.

State. The length is 8 cm, and the width is 6 cm.

49. Familiarize. Let l = the length and w = the width of the rectangle.

Translate. The perimeter is 6 in., so we have one equation:

$$2l + 2w = 6, \text{ or } l + w = 3$$

Using the Pythagorean theorem we have another equation.

$$l^2 + w^2 = (\sqrt{5})^2, \text{ or } l^2 + w^2 = 5$$

Carry out. We solve the system of equations:

$$l + w = 3, \quad (1)$$
$$l^2 + w^2 = 5 \quad (2)$$

Solve Equation (1) for l: $l = 3 - w$. Substitute $3 - w$ for l in Equation (2) and solve for w.

$$l^2 + w^2 = 5$$
$$(3 - w)^2 + w^2 = 5$$
$$9 - 6w + w^2 + w^2 = 5$$
$$2w^2 - 6w + 4 = 0$$
$$2(w^2 - 3w + 2) = 0$$
$$2(w - 2)(w - 1) = 0$$
$$w = 2 \text{ or } w = 1$$

For $w = 2$, $l = 3 - 2 = 1$.
For $w = 1$, $l = 3 - 1 = 2$.

Check. The solutions are $(1, 2)$ and $(2, 1)$. We choose the larger number for the length. $2 + 1 = 3$, and $2^2 + 1^2 = 5$. The solution checks.

State. The length is 2 in. and the width is 1 in.

51. Familiarize. We first make a drawing. Let l = the length and w = the width of the cargo area, in feet.

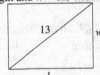

Translate. The cargo area must be 60 ft^2, so we have one equation:

$$lw = 60$$

The Pythagorean equation gives us another equation:

$$l^2 + w^2 = 13^2, \text{ or } l^2 + w^2 = 169$$

Carry out. We solve the system of equations.

$$lw = 60, \quad (1)$$
$$l^2 + w^2 = 169 \quad (2)$$

First solve Equation (1) for w:

$$lw = 60$$
$$w = \frac{60}{l} \quad (3)$$

Then substitute $60 / l$ for w in Equation (2) and solve for l.

$$l^2 + w^2 = 169$$
$$l^2 + \left(\frac{60}{l}\right)^2 = 169$$
$$l^2 + \frac{3600}{l^2} = 169$$
$$l^4 + 3600 = 169l^2$$
$$l^4 - 169l^2 + 3600 = 0$$

Let $u = l^2$ and $u^2 = l^4$ and substitute.

$$u^2 - 169u + 3600 = 0$$
$$(u - 144)(u - 25) = 0$$
$$u = 144 \quad \text{or} \quad u = 25$$
$$l^2 = 144 \quad \text{or} \quad l^2 = 25 \quad \text{Replacing } u \text{ with } l^2$$
$$l = \pm 12 \quad \text{or} \quad l = \pm 5$$

Since the length cannot be negative, we consider only 12 and 5. We substitute in Equation (3) to find w. When $l = 12$, $w = 60/12 = 5$; when $l = 5$, $w = 60/5 = 12$. Since we usually consider length to be longer than width, we check the pair $(12, 5)$.

Check. If the length is 12 ft and the width is 5 ft, then the area is $12 \cdot 5$, or 60 ft^2. Also $12^2 + 5^2 = 144 + 25 = 169 = 13^2$. The answer checks.

State. The length is 12 ft and the width is 5 ft.

53. Familiarize. Let x and y represent the numbers.

Translate. The product of the numbers is 90, so we have

$$xy = 90 \quad (1)$$

The sum of the squares of the numbers is 261, so we have

$$x^2 + y^2 = 261 \quad (2)$$

Carry out. We solve the system of equations.

$$xy = 90, \quad (1)$$
$$x^2 + y^2 = 261 \quad (2)$$

First solve Equation (1) for y:

$$xy = 90$$
$$y = \frac{90}{x} \quad (3)$$

Then substitute $90/x$ for y in Equation (2) and solve for x.

$$x^2 + y^2 = 261$$
$$x^2 + \left(\frac{90}{x}\right)^2 = 261$$
$$x^2 + \frac{8100}{x^2} = 261$$
$$x^4 + 8100 = 261x^2$$
$$x^4 - 261x^2 + 8100 = 0$$

Let $u = x^2$ and $u^2 = x^4$ and substitute.

$$u^2 - 261u + 8100 = 0$$
$$(u - 36)(u - 225) = 0$$
$$u = 36 \quad \text{or} \quad u = 225$$
$$x^2 = 36 \quad \text{or} \quad x^2 = 225 \quad \text{Replacing } u \text{ with } x^2$$
$$x = \pm 6 \quad \text{or} \quad x = \pm 15$$

We use Equation (3) to find y.
When $x = 6$, $y = 90/6 = 15$;

when $x = -6$, $y = 90/(-6) = -15$;
when $x = 15$, $y = 90/15 = 6$;
when $x = -15$, $y = 90/(-15) = -6$. We see that the numbers can be 6 and 15 or –6 and –15.

Check. $6 \cdot 15 = 90$ and $6^2 + 15^2 = 261$; also

$-6(-15) = 90$ and $(-6)^2 + (-15)^2 = 261$. The solutions check.

State. The numbers are 6 and 15 or –6 and –15.

55. **Familiarize**. Let $l =$ the length and $w =$ the width of the screen, in inches.

Translate. The area must be 90 in^2, so we have one equation: $lw = 90$
The Pythagorean equation gives us another equation:

$$l^2 + w^2 = \sqrt{200.25}^2, \text{ or } l^2 + w^2 = 200.25$$

Carry out. We solve the system of equations.

$$lw = 90, \qquad (1)$$
$$l^2 + w^2 = 200.25 \quad (2)$$

First solve Equation (1) for l:

$$l = \frac{90}{w} \quad (3)$$

Then substitute $90/w$ for l in Equation (2) and solve for w.

$$l^2 + w^2 = 200.25$$
$$\left(\frac{90}{w}\right)^2 + w^2 = 200.25$$
$$\frac{8100}{w^2} + w^2 = 200.25$$
$$8100 + w^4 = 200.25w^2$$
$$w^4 - 200.25w^2 + 8100 = 0$$

Let $u = w^2$ and $u^2 = w^4$ and substitute.

$$u^2 - 200.25u + 8100 = 0$$
$$(u - 56.25)(u - 144) = 0$$
$$u = 56.25 \quad or \quad u = 144$$
$$w^2 = 56.25 \quad or \quad w^2 = 144 \quad \text{Replacing } u \text{ with } w^2$$
$$w = \pm 7.5 \quad or \quad w = \pm 12$$

Since the length cannot be negative, we consider only 7.5 and 12. We substitute in Equation (3) to find l. When $w = 7.5$, $l = 90/7.5 = 12$; when $w = 12$, $l = 90/12 = 7.5$. Since we usually consider length to be longer than width, we check the pair (12, 7.5).

Check. If the length is 12 in and the width is 7.5 in, then the area is $12 \cdot 7.5$, or 90 in^2. Also

$12^2 + 7.5^2 = 144 + 56.25 = 200.25 = \sqrt{200.25}^2$. The answer checks.

State. The length is 12 in and the width is 7.5 in.

57. **Familiarize**. Let $l =$ the length and $w =$ the width of the rectangle area, in meters.

Translate. The area must be $\sqrt{3} \text{ m}^2$, so we have one equation: $lw = \sqrt{3}$
The Pythagorean equation gives us another equation:

$$l^2 + w^2 = 2^2, \text{ or } l^2 + w^2 = 4$$

Carry out. We solve the system of equations.

$$lw = \sqrt{3}, \quad (1)$$
$$l^2 + w^2 = 4 \qquad (2)$$

First solve Equation (1) for l:

$$l = \frac{\sqrt{3}}{w} \quad (3)$$

Then substitute $\sqrt{3}/w$ for l in Equation (2) and solve for w.

$$l^2 + w^2 = 4$$
$$\left(\frac{\sqrt{3}}{w}\right)^2 + w^2 = 4$$
$$\frac{3}{w^2} + w^2 = 4$$
$$3 + w^4 = 4w^2$$
$$w^4 - 4w^2 + 3 = 0$$

Let $u = w^2$ and $u^2 = w^4$ and substitute.

$$u^2 - 4u + 3 = 0$$
$$(u - 1)(u - 3) = 0$$
$$u = 1 \quad or \quad u = 3$$
$$w^2 = 1 \quad or \quad w^2 = 3 \qquad \text{Replacing } u \text{ with } w^2$$
$$w = \pm 1 \quad or \quad w = \pm\sqrt{3}$$

Since the length cannot be negative, we consider only 1 and $\sqrt{3}$. We substitute in Equation (3) to find l. When $w = \sqrt{3}$, $l = \sqrt{3}/\sqrt{3} = 1$; when $w = 1$, $l = \sqrt{3}/1 = \sqrt{3}$. Since we usually consider length to be longer than width, we check the pair $\left(\sqrt{3},\ 1\right)$.

Check. If the length is $\sqrt{3}$ m and the width is 1 m, then the area is $\sqrt{3} \cdot 1$, or $\sqrt{3} \text{ m}^2$. Also

$\sqrt{3}^2 + 1^2 = 3 + 1 = 4 = 2^2$. The answer checks.

State. The length is $\sqrt{3}$ m and the width is 1 m.

59. **Writing Exercise**.

61. $(3a^{-4})^2 (2a^{-5})^{-1} = 3^2 (a^{-4})^2 (2)^{-1} (a^{-5})^{-1}$
$$= 9a^{-8} 2^{-1} a^5$$
$$= \frac{9}{2a^3}$$

63. $\log 10{,}000 = \log 10^4 = 4\log 10 = 4$

65. $-10^2 \div 2 \cdot 5 - 3 = -100 \div 2 \cdot 5 - 3$
$$= -50 \cdot 5 - 3$$
$$= -250 - 3$$
$$= -253$$

67. **Writing Exercise**.

69. $p^2 + q^2 = 13,$ (1)

$\dfrac{1}{pq} = -\dfrac{1}{6}$ (2)

Solve Equation (2) for p.

$\dfrac{1}{q} = -\dfrac{p}{6}$

$-\dfrac{6}{q} = p$

Substitute $-6/q$ for p in Equation (1) and solve for q.

$\left(-\dfrac{6}{q}\right)^2 + q^2 = 13$

$\dfrac{36}{q^2} + q^2 = 13$

$36 + q^4 = 13q^2$

$q^4 - 13q^2 + 36 = 0$

$u^2 - 13u + 36 = 0$ Letting $u = q^2$

$(u-9)(u-4) = 0$

$u = 9$ or $u = 4$

$x^2 = 9$ or $x^2 = 4$

$x = \pm 3$ or $x = \pm 2$

Since $p = -6/q$, if $q = 3$, $p = -2$; if $q = -3$, $p = 2$; if $q = 2$, $p = -3$; and if $q = -2$, $p = 3$. The pairs $(-2, 3)$, $(2, -3)$, $(-3, 2)$, and $(3, -2)$ check. They are the solutions.

71. *Familiarize*. Let l = the length of the rectangle, in feet, and let w = the width.

Translate. 100 ft of fencing is used, so we have

$l + w = 100$.   (1)

The area is 2475 ft^2, so we have

$lw = 2475$.   (2)

Carry out. Solving the system of equations, we get $(55, 45)$ and $(45, 55)$. Since length is usually considered to be longer than width, we have $l = 55$ and $w = 45$.

Check. If the length is 55 ft and the width is 45 ft, then $55 + 45$, or 100 ft, of fencing is used. The area is $55 \cdot 45$, or 2475 ft^2. The answer checks.

State. The length of the rectangle is 55 ft, and the width is 45 ft.

73. *Familiarize*. We let x and y represent the length and width of the base of the box, in inches, respectively. Make a drawing.

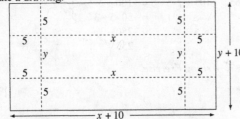

The dimensions of the metal sheet are $x + 10$ and $y + 10$.

Translate. The area of the sheet of metal is 340 in^2, so we have

$(x+10)(y+10) = 340$.   (1)

The volume of the box is 350 in^3, so we have

$x \cdot y \cdot 5 = 350$.   (2)

Carry out. Solving the system of equations, we get $(10, 7)$ and $(7, 10)$. Since length is usually considered to be longer than width, we have $l = 10$ and $w = 7$.

Check. The dimensions of the metal sheet are $10 + 10$, or 20, and $7 + 10$, or 17, so the area is $20 \cdot 17$, or 340 in^2. The volume of the box is $7 \cdot 10 \cdot 5$, or 350 in^3. The answer checks.

State. The dimensions of the box are 10 in. by 7 in. by 5 in.

Chapter 10 Review

1. True

2. False

3. False

4. True

5. True

6. True

7. False

8. True

9. $(x+3)^2 + (y-2)^2 = 16$

$[x-(-3)]^2 + (y-2)^2 = 4^2$ Standard form

The center is $(-3, 2)$ and the radius is 4.

10. $(x-5)^2 + y^2 = 11$

Center: $(5, 0)$; radius: $\sqrt{11}$

11. $x^2 + y^2 - 6x - 2y + 1 = 0$

$x^2 - 6x + y^2 - 2y = -1$

$(x^2 - 6x + 9) + (y^2 - 2y + 1) = -1 + 9 + 1$

$(x-3)^2 + (y-1)^2 = 9$

$(x-3)^2 + (y-1)^2 = 3^2$

The center is $(3, 1)$ and the radius is 3.

12. $x^2 + y^2 + 8x - 6y = 20$

$x^2 + 8x + y^2 - 6y = 20$

$(x^2 + 8x + 16) + (y^2 - 6y + 9) = 20 + 16 + 9$

$(x+4)^2 + (y-3)^2 = 45$

Center: $(-4, 3)$; radius: $\sqrt{45}$ or $3\sqrt{5}$

13. $(x-h)^2 + (y-k)^2 = r^2$

$[x-(-4)]^2 + (y-3)^2 = 4^2$

$(x+4)^2 + (y-3)^2 = 16$

14. $(x-7)^2 + [y-(-2)]^2 = (2\sqrt{5})^2$

$(x-7)^2 + (y+2)^2 = 20$

15. Circle
$$5x^2 + 5y^2 = 80$$
$$x^2 + y^2 = 16$$
Center: (0, 0); radius: 4

$5x^2 + 5y^2 = 80$

16. Ellipse
$$9x^2 + 2y^2 = 18$$
$$\frac{x^2}{2} + \frac{y^2}{9} = 1$$

$9x^2 + 2y^2 = 18$

17. Parabola
$$y = -x^2 + 2x - 3$$
$$x = -\frac{b}{2a} = -\frac{2}{2(-1)} = 1$$
$$y = -x^2 + 2x - 3 = -1^2 + 2 \cdot 1 - 3 = -2$$
The vertex is $(1, -2)$. The graph opens downward
because the coefficient of x^2, -1 is negative.

x	y
-1	-6
0	-3
1	-2
2	-3
3	-6

$y = -x^2 + 2x - 3$

18. Hyperbola
$$\frac{y^2}{9} - \frac{x^2}{4} = 1$$

$\frac{y^2}{9} - \frac{x^2}{4} = 1$

19. Hyperbola
$$xy = 9$$
$$y = \frac{9}{x} \quad \text{Solving for } y$$

x	y
1	9
3	3
9	1
$\frac{1}{2}$	18
-1	-9
-3	-3
-9	-1
$-\frac{1}{2}$	-18

Note that we cannot
use 0 for x. The x-axis
and the y-axis are
the asymptotes.

$xy = 9$

20. Parabola
$$x = y^2 + 2y - 2$$

$x = y^2 + 2y - 2$

21. Ellipse
$$\frac{(x+1)^2}{3} + (y-3)^2 = 1 \quad 3 > 1 \text{ ellipse is horizontal}$$
The center is $(-1, 3)$. Note $a = \sqrt{3}$ and $b = 1$.
Vertices: $(-1, 2)$, $(-1, 4)$, $\left(-1-\sqrt{3},\ 3\right)$, $\left(-1+\sqrt{3},\ 3\right)$

$\frac{(x+1)^2}{3} + (y-3)^2 = 1$

22. Circle
$$x^2 + y^2 + 6x - 8y - 39 = 0$$
$$x^2 + 6x + y^2 - 8y = 39$$
$$\left(x^2 + 6x + 9\right) + \left(y^2 - 8y + 16\right) = 39 + 9 + 16$$
$$(x+3)^2 + (y-4)^2 = 64$$

$x^2 + y^2 + 6x - 8y - 39 = 0$

23. $x^2 - y^2 = 21,$ (1)
     $x + y = 3$ (2)
First we solve Equation (2) for y.
$$y = -x + 3 \quad (3)$$
Then we substitute $-x + 3$ for y in Equation (1) and
solve for x.

$$x^2 - (-x+3)^2 = 21$$
$$x^2 - x^2 + 6x - 9 = 21$$
$$6x = 30$$
$$x = 5$$

Substitute 5 for x in Equation (3).
$$y = -5 + 3 = -2$$
The solution is $(5, -2)$.

24. $x^2 - 2x + 2y^2 = 8$,   (1)
       $2x + y = 6$ (2)

First we solve Equation (2) for y.
$$y = -2x + 6 \quad (3)$$

Then we substitute $-2x + 6$ for y in Equation (1) and solve for x.
$$x^2 - 2x + 2(-2x+6)^2 = 8$$
$$x^2 - 2x + 8x^2 - 48x + 72 = 8$$
$$9x^2 - 50x + 64 = 0$$
$$(x-2)(9x-32) = 0$$
$$x = 2 \ \ or \ \ x = \frac{32}{9}$$

Now we substitute these numbers for x in Equation (3).
For $x = 2$, $y = -2 \cdot 2 + 6 = 2$
For $x = \frac{32}{9}$, $y = -2\left(\frac{32}{9}\right) + 6 = -\frac{10}{9}$

The solutions are $(2, 2)$ and $\left(\frac{32}{9}, -\frac{10}{9}\right)$.

25. $x^2 - y = 5$,   (1)
       $2x - y = 5$ (2)

First we solve Equation (2) for y.
$$y = 2x - 5 \quad (3)$$

Here we multiply Equation (2) by -1 and then add.
$$x^2 - y = 5$$
$$\underline{-2x + y = -5} \quad \text{Multiplying by } -1$$
$$x^2 - 2x = 0 \quad \text{Adding}$$
$$x(x-2) = 0$$
$$x = 0 \text{ or } x = 2$$

Now we substitute these numbers for x in Equation (3).
If $x = 0$, $y = 2 \cdot 0 - 5 = -5$
If $x = 2$, $y = 2 \cdot 2 - 5 = -1$
The solutions are $(0, -5)$ and $(2, -1)$.

26. $x^2 + y^2 = 25$ (1)
$$\underline{\frac{x^2 - y^2 = 7}{2x^2} \quad (2)}$$
$$2x^2 = 32$$
$$x^2 = 16$$
$$x = \pm 4$$

Simultaneously substitute 4 and –4 for x in Equation (1).
$$(\pm 4)^2 + y^2 = 25$$
$$16 + y^2 = 25$$
$$y^2 = 9 \ \text{ or } \ y = \pm 3$$
The solutions are $(4, -3)$, $(4, 3)$, $(-4, -3)$, and $(-4, 3)$.

27. $x^2 - y^2 = 3$,   (1)
       $y = x^2 - 3$ (2)

First we solve Equation (2) for x.
$$y + 3 = x^2$$
$$\pm\sqrt{y+3} = x \quad (3)$$

Adding Equations (1) and (2).
$$x^2 - y^2 = 3 \quad (1)$$
$$\underline{-x^2 + y = -3 \quad (2)}$$
$$-y^2 + y = 0 \quad \text{Adding}$$
$$-y(y-1) = 0$$
$$y = 0 \text{ or } y = 1$$

Now we substitute these numbers for y in Equation (3).
For $y = 0$, $x = \pm\sqrt{0+3} = \pm\sqrt{3}$
For $y = 1$, $x = \pm\sqrt{1+3} = \pm\sqrt{4} = \pm 2$
The solutions are $\left(\sqrt{3}, 0\right)$, $\left(-\sqrt{3}, 0\right)$, $(-2, 1)$ and $(2, 1)$.

28. $x^2 + y^2 = 18$,   (1)
       $2x + y = 3$ (2)

First we solve Equation (2) for y.
$$y = -2x + 3 \quad (3)$$

Then we substitute $-2x + 3$ for y in Equation (1) and solve for x.
$$x^2 + (-2x+3)^2 = 18$$
$$x^2 + 4x^2 - 12x + 9 = 18$$
$$5x^2 - 12x - 9 = 0$$
$$(x-3)(5x+3) = 0$$
$$x = 3 \ \ or \ \ x = -\frac{3}{5}$$

Now we substitute these numbers for x in Equation (3).
For $x = 3$, $y = -2 \cdot 3 + 3 = -3$
For $x = -\frac{3}{5}$, $y = -2\left(-\frac{3}{5}\right) + 3 = \frac{21}{5}$
The solutions are $(3, -3)$ and $\left(-\frac{3}{5}, \frac{21}{5}\right)$.

29. $x^2 + y^2 = 100$,   (1)
       $2x^2 - 3y^2 = -120$ (2)

We use the elimination method.
$$-2x^2 - 2y^2 = -200 \quad \text{Multiplying (1) by } -2$$
$$\underline{2x^2 - 3y^2 = -120 \quad (2)}$$
$$-5y^2 = -320 \quad \text{Adding}$$
$$y^2 = 64$$
$$y = \pm 8$$

Solve Equation (1) for x.
$$x^2 + y^2 = 100$$
$$x^2 = 100 - y^2$$
$$x = \pm\sqrt{100 - y^2} \quad (3)$$

Since x is solved in terms of y^2, we need only substitute once in Equation (3).
For $y^2 = 64$, $x = \pm\sqrt{100 - 64} = \pm\sqrt{36} = \pm 6$
The solutions are $(6, 8)$, $(6, -8)$, $(-6, 8)$ and $(-6, -8)$.

30. $x^2 + 2y^2 = 12$,   (1)
      $xy = 4$ (2)

First we solve Equation (2) for y.

$$y = \frac{4}{x}$$

Then we substitute $\frac{4}{x}$ for y in Equation (1) and solve for x.

$$x^2 + 2\left(\frac{4}{x}\right)^2 = 12$$
$$x^2 + \frac{32}{x^2} = 12$$
$$x^4 + 32 = 12x^2$$
$$x^4 - 12x^2 + 32 = 0$$
$$u^2 - 12u + 32 = 0 \quad \text{Letting } u = x^2$$
$$(u - 8)(u - 4) = 0$$
$$u = 8 \quad or \quad u = 4$$
$$x^2 = 8 \quad or \quad x^2 = 4$$
$$x = \pm 2\sqrt{2} \quad or \quad x = \pm 2$$

$y = 4 / x$, so if $x = 2\sqrt{2}$, $y = \sqrt{2}$; if $x = -2\sqrt{2}$, $y = -\sqrt{2}$; if $x = 2$, $y = 2$; and if $x = -2$, $y = -2$. The pairs $\left(2\sqrt{2},\ \sqrt{2}\right)$, $\left(-2\sqrt{2},\ -\sqrt{2}\right)$, $(2, 2)$, and $(-2,-2)$ check. They are the solutions.

31. Familiarize. Let l = the length and w = the width of the bandstand.
Translate.

   Perimeter: $2l + 2w = 38$, or $l + w = 19$
   Area: $lw = 84$

Carry out. We solve the system:
Solve the first equation for l: $l = 19 - w$.
Substitute $19 - w$ for l in the second equation and solve for w.

$$(19 - w)w = 84$$
$$19w - w^2 = 84$$
$$0 = w^2 - 19w + 84$$
$$0 = (w - 7)(w - 12)$$
$$w = 7 \quad or \quad w = 12$$

If $w = 7$, then $l = 19 - 7$, or 12. If $w = 12$, then $l = 19 - 12$, or 7. Since length is usually considered to be longer than width, we have the solution $l = 12$ and $w = 7$, or $(12, 7)$
Check. If $l = 12$ and $w = 7$, the area is $12 \cdot 7$, or 84. The perimeter is $2 \cdot 12 + 2 \cdot 7$, or 38. The numbers check.
State. The length is 12 m and the width is 7 m.

32. Let l and w represent the length and width, respectively. Solve the system:

$$l^2 + w^2 = 15^2,$$
$$lw = 108$$

The solutions are $(12, 9)$, $(-12, -9)$, $(9, 12)$, and $(-9,-12)$. Only $(12, 9)$ and $(9, 12)$ have meaning in this problem. Choosing the larger number as the length, we have the solution. The length is 12 in., and the width is 9 in.

33. Familiarize. Let x represent the length of a side of the mounting board and y represent the length of a side of the square mirror.
Translate.

   $4x = 4y + 12$, or $x = y + 3$

   $x^2 = y^2 + 39$

Carry out. We solve the system of equations.

   $x = y + 3$ (1)
   $x^2 = y^2 + 39$ (2)

We substitute $y + 3$ for x into Equation (2).

$$(y + 3)^2 = y^2 + 39$$
$$y^2 + 6y + 9 = y^2 + 39$$
$$6y + 9 = 39$$
$$6y = 30$$
$$y = 5$$

Then $x = 5 + 3 = 8$.
Check. If $x = 8$ and $y = 5$, then $4 \cdot 5 + 12 = 32 = 4 \cdot 8$, and $5^2 + 39 = 64 = 8^2$. The numbers check. The perimeter of the mounting board is $4 \cdot 8$, or 32, and the perimeter of the mirror is $4 \cdot 5$, or 20.
State. The perimeter of the mounting board is 32 cm and 20 cm, respectively.

34. Let x and y represent the radii of the two circles. Solve the system:

   $\pi x^2 + \pi y^2 = 130\pi$ Sum of the areas
   $2\pi x - 2\pi y = 16\pi$ Difference in circumferences

The system can be simplified to:

   $x^2 + y^2 = 130$
   $x - y = 8$

The solution $(-11, -3)$ has no meaning for this application, so the only solution is $(3, 11)$.
The radius of one circle is 3 ft, and the other is 11 ft.

35. Writing Exercise.

36. Writing Exercise.

37. $4x^2 - x - 3y^2 = 9$,   (1)
      $-x^2 + x + y^2 = 2$ (2)

We use the elimination method.

$$\begin{array}{rl} 4x^2 - x - 3y^2 = 9 & (1) \\ -3x^2 + 3x + 3y^2 = 6 & \text{Multiplying (2) by 3} \\ \hline x^2 + 2x = 15 & \text{Adding} \end{array}$$

$$x^2 + 2x - 15 = 0$$
$$(x + 5)(x - 3) = 0$$
$$x = -5 \quad or \quad x = 3$$

Solving Equation (2) for y:

$$-x^2 + x + y^2 = 2$$
$$y^2 = x^2 - x + 2$$
$$y = \pm\sqrt{x^2 - x + 2}$$

For $x = -5$, $y = \pm\sqrt{(-5)^2 - (-5) + 2} = \pm\sqrt{32} = \pm 4\sqrt{2}$

For $x = 3$, $y = \pm\sqrt{3^2 - 3 + 2} = \pm\sqrt{8} = \pm 2\sqrt{2}$

The solutions are $(-5, -4\sqrt{2})$, $(-5, 4\sqrt{2})$, $(3, -2\sqrt{2})$,

$(3, 2\sqrt{2})$.

38. The three points are equidistant from the center of the circle, (h, k). Using each of the three points in the equation of a circle, we get three different equations.

For $(-2, -4)$, $\quad [x - (-2)]^2 + [y - (-4)]^2 = r^2$

$$(x + 2)^2 + (y + 4)^2 = r^2$$
$$x^2 + 4x + 4 + y^2 + 8y + 16 = r^2$$
$$x^2 + 4x + y^2 + 8y + 20 = r^2 \quad (1)$$

For $(5, -5)$, $\quad (x - 5)^2 + [y - (-5)]^2 = r^2$

$$(x - 5)^2 + (y + 5)^2 = r^2$$
$$x^2 - 10x + 25 + y^2 + 10y + 25 = r^2$$
$$x^2 - 10x + y^2 + 10y + 50 = r^2 \quad (2)$$

For $(6, 2)$, $\quad (x - 6)^2 + (y - 2)^2 = r^2$

$$x^2 - 12x + 36 + y^2 - 4y + 4 = r^2$$
$$x^2 - 12x + y^2 - 4y + 40 = r^2 \quad (3)$$

Since the radius is equal, we can set Equation (1) equal to Equation (2) and simplify.

$$x^2 + 4x + y^2 + 8y + 20 = x^2 - 10x + y^2 + 10y + 50$$
$$14x - 2y = 30$$
$$7x - y = 15 \quad (4)$$

Next, we set Equation (1) equal to Equation (3).

$$x^2 + 4x + y^2 + 8y + 20 = x^2 - 12x + y^2 - 4y + 40$$
$$16x + 12y = 20$$
$$4x + 3y = 5 \quad (5)$$

We solve the system of Equations (4) and (5) using the elimination method.

$$
\begin{array}{ll}
21x - 3y = 45 & \text{Multiplying (4) by 3} \\
\underline{4x + 3y = 5} & \text{(5)} \\
25x = 50 & \text{Adding} \\
x = 2 & \\
y = -1 &
\end{array}
$$

The center of the circle is $(2, -1)$. Thus the equation is

$$(x - 2)^2 + (y + 1)^2 = r^2.$$

We may choose any of the three points on the circle to determine r^2.

$$(6 - 2)^2 + (2 + 1)^2 = r^2$$
$$4^2 + 3^2 = r^2$$
$$25 = r^2$$

The equation of the circle is $(x - 2)^2 + (y + 1)^2 = 25$.

39. From the x-intercepts, $(-10, 0)$ and $(10, 0)$, we know $a = 10$; from the y-intercepts, $(0, -1)$ and $(0, 1)$, we know
$b = 1$. So, we have the equation:

$$\frac{x^2}{10^2} + \frac{y^2}{1^2} = 1$$

$$\frac{x^2}{100} + \frac{y^2}{1} = 1$$

Chapter 10 Test

1. For circle with center $(3, -4)$ and radius $2\sqrt{3}$,

$$(x - 3)^2 + [y - (-4)]^2 = (2\sqrt{3})^2$$
$$(x - 3)^2 + (y + 4)^2 = 12$$

2. $\quad (x - 4)^2 + (y + 1)^2 = 5$

$$(x - 4)^2 + [y - (-1)]^2 = (\sqrt{5})^2$$

Center: $(4, -1)$; radius: $\sqrt{5}$

3. $\quad x^2 + y^2 + 4x - 6y + 4 = 0$

$$x^2 + 4x + y^2 - 6y = -4$$
$$(x^2 + 4x + 4) + (y^2 - 6y + 9) = -4 + 4 + 9$$
$$(x + 2)^2 + (y - 3)^2 = 3^2$$

The center is $(-2, 3)$ and the radius is 3.

4. Parabola

$$y = x^2 - 4x - 1$$

5. Circle

$$x^2 + y^2 + 2x + 6y + 6 = 0$$
$$x^2 + 2x + y^2 + 6y = -6$$
$$(x^2 + 2x + 1) + (y^2 + 6y + 9) = -6 + 1 + 9$$
$$(x + 1)^2 + (y + 3)^2 = 4$$

The center is $(-1, -3)$ and the radius is 2.

6. Hyperbola

$$\frac{x^2}{16} - \frac{y^2}{9} = 1$$

$\frac{x^2}{16} - \frac{y^2}{9} = 1$

7. Ellipse

$16x^2 + 4y^2 = 64 \quad 4 < 16$ ellipse is vertical

$$\frac{x^2}{4} + \frac{y^2}{16} = 1$$

The center is $(0, 0)$. Note $a = 2$ and $b = 4$.

Vertices: (0, 2), (0, –2), (4, 0), and (–4, 0)

$16x^2 + 4y^2 = 64$

8. Hyperbola

$xy = -5$

$y = \dfrac{-5}{x}$ Solving for y

$xy = -5$

9. Parabola

$x = -y^2 + 4y$

$y = -\dfrac{b}{2a} = -\dfrac{4}{2(-1)} = 2$

$x = -y^2 + 4y = -2^2 + 4 \cdot 2 = 4$

The vertex is (4, 2). The graph opens to the left because the coefficient of y^2, –1 is negative.

x	y
0	0
3	1
4	2
3	3
0	4

$x = -y^2 + 4y$

(4, 2)

10. $x^2 + y^2 = 36$,   (1)

$3x + 4y = 24$ (2)

First solve Equation (2) for y.

$4y = -3x + 24$

$y = -\dfrac{3}{4}x + 6$ (3)

Then substitute $-\dfrac{3}{4}x + 6$ for y in Equation (1) and solve for x.

$x^2 + y^2 = 36$

$x^2 + \left(-\dfrac{3}{4}x + 6\right)^2 = 36$

$x^2 + \dfrac{9}{16}x^2 - 9x + 36 = 36$

$\dfrac{25}{16}x^2 - 9x = 0$

$25x^2 - 144x = 0$

$x(25x - 144) = 0$

$x = 0$ or $x = \dfrac{144}{25}$

Now substitute these numbers in Equation (3) and solve for y.

For $x = 0$, $y = -\dfrac{3}{4} \cdot 0 + 6 = 6$

For $x = \dfrac{144}{25}$, $y = -\dfrac{3}{4} \cdot \dfrac{144}{25} + 6 = \dfrac{42}{25}$

The pairs (0, 6) and $\left(\dfrac{144}{25}, \dfrac{42}{25}\right)$ check, so they are the solutions.

11. $x^2 - y = 3$,   (1)

$2x + y = 5$ (2)

Use the elimination method.

$x^2 - y = 3$ (1)

$\underline{2x + y = 5}$ (2)

$x^2 + 2x = 8$ Adding

$x^2 + 2x - 8 = 0$

$(x + 4)(x - 2) = 0$

$x = -4$ or $x = 2$

Substitute for x in Equation (2).

$y = 5 - 2x$ Solving Eq. (2) for y

For $x = -4$, $y = 5 - 2(-4) = 13$

For $x = 2$, $y = 5 - 2(2) = 1$

The pairs (–4, 13) and (2, 1) check. They are the solutions.

12. $x^2 - 2y^2 = 1$,   (1)

$xy = 6$ (2)

First we solve Equation (2) for y.

$xy = 6$

$y = \dfrac{6}{x}$

Then we substitute $\dfrac{6}{x}$ for y in Equation (1) and solve for x.

$x^2 - 2y^2 = 1$

$x^2 - 2\left(\dfrac{6}{x}\right)^2 = 1$

$x^2 - \dfrac{72}{x^2} = 1$

$x^4 - 72 = x^2$ Multiplying by x^2

$x^4 - x^2 - 72 = 0$

$u^2 - u - 72 = 0$ Letting $u = x^2$

$(u - 9)(u + 8) = 0$

$u = 9$ or $u = -8$

We now substitute x^2 for u and solve for x.

$x^2 = 9$ or $x^2 = -8$

$x = \pm 3$ or $x = \pm 2\sqrt{2}i$

Since $y = 6/x$, if $x = 3$, $y = 2$; if $x = -3$, $y = -2$; if $x = 2\sqrt{2}i$, $y = -\dfrac{3\sqrt{2}}{2}i$; and if $x = -2\sqrt{2}i$, $y = \dfrac{3\sqrt{2}}{2}i$.

The pairs (3, 2), (–3, –2), $\left(-2\sqrt{2}i, \dfrac{3\sqrt{2}}{2}i\right)$ and $\left(2\sqrt{2}i, -\dfrac{3\sqrt{2}}{2}i\right)$ check. They are the solutions.

13. $x^2 + y^2 = 10$,   (1)

$x^2 = y^2 + 2$ (2)

Substitute $y^2 + 2$ for x^2 in Equation (1) and solve for y.

$$x^2 + y^2 = 10$$
$$y^2 + 2 + y^2 = 10$$
$$2y^2 - 8 = 0$$
$$2(y^2 - 4) = 0$$
$$2(y + 2)(y - 2) = 0$$
$$y = -2 \ or \ y = 2$$

Now substitute these numbers in Equation (2) and solve for x.

For $y = -2$, $\quad x^2 = (-2)^2 + 2 = 6$, so $x = \pm\sqrt{6}$

For $y = 2$, $\quad x^2 = 2^2 + 2 = 6$, so $x = \pm\sqrt{6}$

The pairs $(-\sqrt{6}, -2)$, $(-\sqrt{6}, 2)$, $(\sqrt{6}, -2)$, $(\sqrt{6}, 2)$ check, so they are the solutions.

14. Let l = the length and w = the width of the bookmark. We solve the system:
$$lw = 22,$$
$$l^2 + w^2 = (5\sqrt{5})^2$$

The negative solutions $(-2, -11)$ and $(-11, -2)$ are not included since they have no meaning in this application. The remaining solutions are $(11, 2)$ and $(2, 11)$. We choose the larger number to be the length, so the length is 11 and the width is 2.

15. **Familiarize**. We let x = the length of a side of one square, in meters, and y = the length of a side of the other square. Make a drawing.

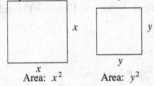

x $\qquad$ y

x

Area: x^2 $\qquad$ Area: y^2

Translate. The sum of the areas is 8 m^2, so we have
$$x^2 + y^2 = 8 . \ (1)$$

The difference of the areas is 2 m^2, so we have
$$x^2 - y^2 = 2 . \ (2)$$

Carry out. We solve the system of equations.

$$
\begin{array}{ll}
x^2 + y^2 = 8 & (1) \\
\underline{x^2 - y^2 = 2} & (2) \\
2x^2 \quad\ = 10 & \text{Adding} \\
x^2 = 5 & \\
x = \pm\sqrt{5} &
\end{array}
$$

Since the length cannot be negative we consider only $\sqrt{5}$. We substitute $\sqrt{5}$ for x in Equation (1) and solve for y.

$$(\sqrt{5})^2 + y^2 = 8$$
$$5 + y^2 = 8$$
$$y^2 = 3$$
$$y = \pm\sqrt{3}$$

Again we consider only the positive number.

Check. If the lengths of the sides of the squares are $\sqrt{5}$ m and $\sqrt{3}$ m, the areas of the squares are $(\sqrt{5})^2$,

or 5 m^2, and $(\sqrt{3})^2$, or 3 ft^2, respectively. Then $5 \text{ m}^2 + 3 \text{ m}^2 = 8 \text{ m}^2$, and $5 \text{ m}^2 - 3 \text{ m}^2 = 2 \text{ m}^2$, so the answer checks.

State. The lengths of the sides of the squares are $\sqrt{5}$ m and $\sqrt{3}$ m.

16. Let l = the length and w = the width of the floor. We solve the system:
$$l^2 + w^2 = (40)^2$$
$$2l + 2w = 112$$

The solutions are $(32, 24)$ and $(24, 32)$. We choose the larger number to be the length, so the length is 32 ft and the width is 24 ft.

17. **Familiarize**. Let p = the principal and r = the interest rate. We recall the formula $I = prt$. Since the time is one year, $t = 1$, and the formula simplifies to $I = pr$.

Translate.

Brett invested p dollars at interest rate r with \$72 in interest.

Erin invested $p + 240$ dollars at $\frac{5}{6}r$ interest rate with \$72 in interest. We have two equations.

$$72 = pr \qquad\qquad (1)$$
$$72 = (p + 240)\frac{5}{6}r \quad (2)$$

Carry out. We solve the system of equations using substitution.

We solve Equation (1) for r: $r = \dfrac{72}{p}$.

We substitute $\dfrac{72}{p}$ for r in Equation (2) and solve for p.

$$72 = (p + 240)\frac{5}{6}\left(\frac{72}{p}\right)$$
$$72 = (p + 240)\frac{60}{p}$$
$$72p = 60p + 14,400$$
$$12p = 14,400$$
$$p = 1200$$

$$r = \frac{72}{1200} = 0.06 \text{ or } 6\%$$

Check. $1200 \cdot 0.06 = 72$ and

$(1200 + 240)\dfrac{5}{6} \cdot 0.06 = 72$, so the numbers check.

State. The principal was \$1200 and the interest rate was 6%.

18. Plot the given points. From the location of these points, we see that the ellipse which contains them is centered at $(6, 3)$. The distance from $(1, 3)$ to $(6, 3)$ is 5, so $a = 5$. The distance from $(6, 6)$ to $(6, 3)$ is 3, so $b = 3$. We write the equation of the ellipse.

$$\frac{(x - 6)^2}{5^2} + \frac{(y - 3)^2}{3^2} = 1$$

$$\frac{(x - 6)^2}{25} + \frac{(y - 3)^2}{9} = 1$$

19. Let x and y represent the numbers. We solve the system.

$$x + y = 36 \quad (1)$$
$$xy = 4 \quad (2)$$

The numbers are $18 - 8\sqrt{5}$ and $18 + 8\sqrt{5}$.
The sum of the reciprocals is

$$\frac{1}{18 + 8\sqrt{5}} + \frac{1}{18 - 8\sqrt{5}} = 9.$$

20. Let the actor be in the center at $(0, 0)$. Using the information, we have $(-4, 0)$ and $(4, 0)$ and $(0, -7)$ and $(0, 7)$. Thus, $a = 4$ and $b = 7$. We write the equation of the ellipse.

$$\frac{x^2}{4^2} + \frac{y^2}{7^2} = 1$$
$$\frac{x^2}{16} + \frac{y^2}{49} = 1$$

Chapter 11

Sequences, Series, and the Binomial Theorem

Exercise Set 11.1

1. B

3. B

5. A

7. f

9. d

11. c

13. $a_n = 5n + 3$
$a_8 = 5 \cdot 8 + 3 = 40 + 3 = 43$

15. $a_n = (3n + 1)(2n - 5)$
$a_9 = (3 \cdot 9 + 1)(2 \cdot 9 - 5) = 28 \cdot 13 = 364$

17. $a_n = (-1)^{n-1}(3.4n - 17.3)$
$a_{12} = (-1)^{12-1}[3.4(12) - 17.3] = -23.5$

19. $a_n = 3n^2(9n - 100)$
$a_{11} = 3 \cdot 11^2(9 \cdot 11 - 100) = 3 \cdot 121(-1) = -363$

21. $a_n = \left(1 + \dfrac{1}{n}\right)^2$
$a_{20} = \left(1 + \dfrac{1}{20}\right)^2 = \left(\dfrac{21}{20}\right)^2 = \dfrac{441}{400}$

23. $a_n = 3n - 1$
$a_1 = 3 \cdot 1 - 1 = 2$
$a_2 = 3 \cdot 2 - 1 = 5$
$a_3 = 3 \cdot 3 - 1 = 8$
$a_4 = 3 \cdot 4 - 1 = 11$
$a_{10} = 3 \cdot 10 - 1 = 29$
$a_{15} = 3 \cdot 15 - 1 = 44$

25. $a_n = n^2 + 2$
$a_1 = 1^2 + 2 = 3$
$a_2 = 2^2 + 2 = 6$
$a_3 = 3^2 + 2 = 11$
$a_4 = 4^2 + 2 = 18$
$a_{10} = 10^2 + 2 = 102$
$a_{15} = 15^2 + 2 = 227$

27. $a_n = \dfrac{n}{n+1}$
$a_1 = \dfrac{1}{1+1} = \dfrac{1}{2}$
$a_2 = \dfrac{2}{2+1} = \dfrac{2}{3}$
$a_3 = \dfrac{3}{3+1} = \dfrac{3}{4}$
$a_4 = \dfrac{4}{4+1} = \dfrac{4}{5}$
$a_{10} = \dfrac{10}{10+1} = \dfrac{10}{11}$
$a_{15} = \dfrac{15}{15+1} = \dfrac{15}{16}$

29. $a_n = \left(-\dfrac{1}{2}\right)^{n-1}$
$a_1 = \left(-\dfrac{1}{2}\right)^{1-1} = 1$
$a_2 = \left(-\dfrac{1}{2}\right)^{2-1} = -\dfrac{1}{2}$
$a_3 = \left(-\dfrac{1}{2}\right)^{3-1} = \dfrac{1}{4}$
$a_4 = \left(-\dfrac{1}{2}\right)^{4-1} = -\dfrac{1}{8}$
$a_{10} = \left(-\dfrac{1}{2}\right)^{10-1} = -\dfrac{1}{512}$
$a_{15} = \left(-\dfrac{1}{2}\right)^{15-1} = \dfrac{1}{16,384}$

31. $a_n = \dfrac{(-1)^n}{n}$
$a_1 = \dfrac{(-1)^1}{1} = -1$
$a_2 = \dfrac{(-1)^2}{2} = \dfrac{1}{2}$
$a_3 = \dfrac{(-1)^3}{3} = -\dfrac{1}{3}$
$a_4 = \dfrac{(-1)^4}{4} = \dfrac{1}{4}$
$a_{10} = \dfrac{(-1)^{10}}{10} = \dfrac{1}{10}$
$a_{15} = \dfrac{(-1)^{15}}{15} = -\dfrac{1}{15}$

33. $a_n = (-1)^n (n^3 - 1)$

$a_1 = (-1)^1 (1^3 - 1) = 0$

$a_2 = (-1)^2 (2^3 - 1) = 7$

$a_3 = (-1)^3 (3^3 - 1) = -26$

$a_4 = (-1)^4 (4^3 - 1) = 63$

$a_{10} = (-1)^{10} (10^3 - 1) = 999$

$a_{15} = (-1)^{15} (15^3 - 1) = -3374$

35. 2, 4, 6, 8, 10,...

These are even integers beginning with 2, so the general term could be $2n$.

37. −1, 1,−1, 1,...

−1 and 1 alternate, beginning with −1, so the general term could be $(-1)^n$.

39. 1,−2, 3,−4,...

These are the first four natural numbers, but with alternating signs, beginning with a positive number. The general term could be $(-1)^{n+1} \cdot n$.

41. 3, 5, 7, 9,...

These are odd integers beginning with 3, so the general term could be $2n + 1$.

43. 0, 3, 8, 15, 24,...

We can see a pattern if we write the sequence $1^2 - 1, 2^2 - 1, 3^2 - 1, 4^2 - 1, 5^2 - 1,...$ The general term could be $n^2 - 1$, or $(n+1)(n-1)$.

45. $\dfrac{1}{2}, \dfrac{2}{3}, \dfrac{3}{4}, \dfrac{4}{5}, \dfrac{5}{6},...$

These are fractions in which the denominator is 1 greater than the numerator. Also, each numerator is 1 greater than the preceding numerator. The general term could be $\dfrac{n}{n+1}$.

47. 0.1, 0.01, 0.001, 0.0001,...

This is negative powers of 10, or positive powers of 0.1, so the general term is 10^{-n}, or $(0.1)^n$.

49. −1, 4, −9, 16,...

This is the squares of the first four natural numbers, but with alternating signs, beginning with a negative number. The general terms could be $(-1)^n \cdot n^2$.

51. −1, 2,−3, 4,−5, 6,...

$S_{10} = -1 + 2 - 3 + 4 - 5 + 6 - 7 + 8 - 9 + 10 = 5$

53. $1, \dfrac{1}{10}, \dfrac{1}{100}, \dfrac{1}{1000},...$

$S_6 = 1 + \dfrac{1}{10} + \dfrac{1}{100} + \dfrac{1}{1000} + \dfrac{1}{10,000} + \dfrac{1}{100,000}$

$= 1.11111$, or $1\dfrac{11,111}{100,000}$

55. $\displaystyle\sum_{k=1}^{5} \dfrac{1}{2k} = \dfrac{1}{2 \cdot 1} + \dfrac{1}{2 \cdot 2} + \dfrac{1}{2 \cdot 3} + \dfrac{1}{2 \cdot 4} + \dfrac{1}{2 \cdot 5}$

$= \dfrac{1}{2} + \dfrac{1}{4} + \dfrac{1}{6} + \dfrac{1}{8} + \dfrac{1}{10}$

$= \dfrac{60}{120} + \dfrac{30}{120} + \dfrac{20}{120} + \dfrac{15}{120} + \dfrac{12}{120}$

$= \dfrac{137}{120}$

57. $\displaystyle\sum_{k=0}^{4} 10^k = 10^0 + 10^1 + 10^2 + 10^3 + 10^4$

$= 1 + 10 + 100 + 1000 + 10,000$

$= 11,111$

59. $\displaystyle\sum_{k=2}^{8} \dfrac{k}{k-1}$

$= \dfrac{2}{2-1} + \dfrac{3}{3-1} + \dfrac{4}{4-1} + \dfrac{5}{5-1} + \dfrac{6}{6-1} + \dfrac{7}{7-1} + \dfrac{8}{8-1}$

$= \dfrac{2}{1} + \dfrac{3}{2} + \dfrac{4}{3} + \dfrac{5}{4} + \dfrac{6}{5} + \dfrac{7}{6} + \dfrac{8}{7}$

$= \dfrac{1343}{140}$

61. $\displaystyle\sum_{k=1}^{8} (-1)^{k+1} 2^k$

$= (-1)^{1+1} 2^1 + (-1)^{2+1} 2^2 + (-1)^{3+1} 2^3 + (-1)^{4+1} 2^4$

$\quad + (-1)^{5+1} 2^5 + (-1)^{6+1} 2^6 + (-1)^{7+1} 2^7 + (-1)^{8+1} 2^8$

$= 2 - 4 + 8 - 16 + 32 - 64 + 128 - 256$

$= -170$

63. $\displaystyle\sum_{k=0}^{5} (k^2 - 2k + 3)$

$= (0^2 - 2 \cdot 0 + 3) + (1^2 - 2 \cdot 1 + 3) + (2^2 - 2 \cdot 2 + 3)$

$\quad + (3^2 - 2 \cdot 3 + 3) + (4^2 - 2 \cdot 4 + 3) + (5^2 - 2 \cdot 5 + 3)$

$= 3 + 2 + 3 + 6 + 11 + 18$

$= 43$

65. $\displaystyle\sum_{k=3}^{5} \dfrac{(-1)^k}{k(k+1)} = \dfrac{(-1)^3}{3(3+1)} + \dfrac{(-1)^4}{4(4+1)} + \dfrac{(-1)^5}{5(5+1)}$

$= \dfrac{-1}{3 \cdot 4} + \dfrac{1}{4 \cdot 5} + \dfrac{-1}{5 \cdot 6}$

$= -\dfrac{1}{12} + \dfrac{1}{20} - \dfrac{1}{30}$

$= -\dfrac{4}{60} = -\dfrac{1}{15}$

67. $\dfrac{2}{3} + \dfrac{3}{4} + \dfrac{4}{5} + \dfrac{5}{6} + \dfrac{6}{7}$

This is a sum of fractions in which the denominator is one greater than the numerator. Also, each numerator is 1 greater than the preceding numerator. Sigma notation is

$$\sum_{k=1}^{5} \dfrac{k+1}{k+2}.$$

69. $1 + 4 + 9 + 16 + 25 + 36$

This is the sum of the squares of the first six natural

numbers. Sigma notation is

$$\sum_{k=1}^{6} k^2.$$

71. $4 - 9 + 16 - 25 + \ldots + (-1)^n n^2$

This is a sum of terms of the form $(-1)^k k^2$, beginning with $k = 2$ and continuing through $k = n$. Sigma notation is

$$\sum_{k=2}^{n} (-1)^k k^2.$$

73. $6 + 12 + 18 + 24 + \ldots$

This is the sum of all the positive multiples of 6. It is an infinite series. Sigma notation is

$$\sum_{k=1}^{\infty} 6k.$$

75. $\dfrac{1}{1 \cdot 2} + \dfrac{1}{2 \cdot 3} + \dfrac{1}{3 \cdot 4} + \dfrac{1}{4 \cdot 5} + \ldots$

This is a sum of fractions in which the numerator is 1 and the denominator is a product of two consecutive integers. The larger integer in each product is the smaller integer in the succeeding product. It is an infinite series. Sigma notation is

$$\sum_{k=1}^{\infty} \dfrac{1}{k(k+1)}.$$

77. *Writing Exercise.*

79. $\dfrac{t^3 + 1}{t + 1} = \dfrac{(t+1)(t^2 - t + 1)}{t + 1} = t^2 - t + 1$

81. $\dfrac{3}{a^2 + a} + \dfrac{4}{2a^2 - 2} = \dfrac{3}{a(a+1)} + \dfrac{2 \cdot 2}{2(a^2 - 1)}$

$= \dfrac{3}{a(a+1)} + \dfrac{2}{(a+1)(a-1)}$

$= \dfrac{3}{a(a+1)} \cdot \dfrac{a-1}{a-1} + \dfrac{2}{(a+1)(a-1)} \cdot \dfrac{a}{a}$

$= \dfrac{3a - 3 + 2a}{a(a+1)(a-1)}$

$= \dfrac{5a - 3}{a(a+1)(a-1)}$

83. $\dfrac{x^2 - 6x + 8}{4x + 12} \cdot \dfrac{x + 3}{x^2 - 4} = \dfrac{(x-2)(x-4)}{4(x+3)} \cdot \dfrac{x+3}{(x+2)(x-2)}$

$= \dfrac{x - 4}{4(x + 2)}$

85. *Writing Exercise.*

87. $a_1 = 1,\ a_{n+1} = 5a_n - 2$

$a_1 = 1$

$a_2 = 5 \cdot 1 - 2 = 3$

$a_3 = 5 \cdot 3 - 2 = 13$

$a_4 = 5 \cdot 13 - 2 = 63$

$a_5 = 5 \cdot 63 - 2 = 313$

$a_6 = 5 \cdot 313 - 2 = 1563$

89. Find each term by multiplying the preceding term by 0.80: \$2500, \$2000, \$1600, \$1280, \$1024, \$819.20, \$655.36, \$524.29, \$419.43, \$335.54.

91. $a_n = (-1)^n$

This sequence is of the form $-1, 1, -1, 1, \ldots$ Each pair of terms adds to 0. S_{100} has 50 such pairs, so $S_{100} = 0$. S_{101} consists of the 50 pairs in S_{100} that add to 0 as well as a_{101}, or -1, so $S_{101} = -1$.

93. $a_n = i^n$

$a_1 = i^1 = i$

$a_2 = i^2 = -1$

$a_3 = i^3 = i^2 \cdot i = -1 \cdot i = -i$

$a_4 = i^4 = \left(i^2\right)^2 = (-1)^2 = 1$

$a_5 = i^5 = \left(i^2\right)^2 \cdot i = (-1)^2 \cdot i = 1 \cdot i = i$

$S_5 = i - 1 - i + 1 + i = i$

95. Enter $y_1 = x^5 - 14x^4 + 6x^3 + 416x^2 - 655x - 1050$. Then scroll through a table of values. We see that $y_1 = 6144$ when $x = 11$, so the 11th term of the sequence is 6144.

Exercise Set 11.2

1. $5 + 7 + 9 + 11$ is an example of an *arithmetic series*.

3. In $5, 7, 9, 11$, the *first term* is 5.

5. $8, 13, 18, 23, \ldots$

$a_1 = 8$

$d = 5$ $\quad (13 - 8 = 5,\ 18 - 13 = 5,\ 23 - 18 = 5)$

7. $7, 3, -1, -5, \ldots$

$a_1 = 7$

$d = -4$ $\quad (3 - 7 = -4,\ -1 - 3 = -4,\ -5 - (-1) = -4)$

9. $\dfrac{3}{2}, \dfrac{9}{4}, 3, \dfrac{15}{4}, \ldots$

$a_1 = \dfrac{3}{2}$

$d = \dfrac{3}{4}$ $\quad \left(\dfrac{9}{4} - \dfrac{3}{2} = \dfrac{3}{4},\ 3 - \dfrac{9}{4} = \dfrac{3}{4}\right)$

11. \$8.16, \$8.46, \$8.76, \$9.06, \ldots

$a_1 = \$8.16$

$d = \$0.30$ $\quad$ (\$8.46 − \$8.16 = \$0.30, \$8.76 − \$8.46 = \$0.30, \$9.06 − \$8.76 = \$0.30)

13. $10, 18, 26, \ldots$

$a_1 = 10,\ d = 8,$ and $n = 19$

$a_n = a_1 + (n - 1)d$

$a_{19} = 10 + (19 - 1)8 = 10 + 18 \cdot 8 = 10 + 144 = 154$

15. $8, 2, -4, \ldots$

$a_1 = 8,\ d = -6,$ and $n = 18$

$a_n = a_1 + (n - 1)d$

$a_{18} = 8 + (18 - 1)(-6) = 8 + 17(-6) = 8 - 102 = -94$

17. $1200, $964.32, $728.64, ...
$a_1 = \$1200$, $d = \$964.32 - \$1200 = -\$235.68$
and $n = 13$
$$a_n = a_1 + (n-1)d$$
$$a_{13} = \$1200 + (13-1)(-\$235.68)$$
$$= \$1200 + 12(-\$235.68) = \$1200 - \$2828.16$$
$$= -\$1628.16$$

19. $a_1 = 10$, $d = 8$
$$a_n = a_1 + (n-1)d$$
Let $a_n = 210$, and solve for n.
$$210 = 10 + (n-1)8$$
$$210 = 10 + 8n - 8$$
$$210 = 2 + 8n$$
$$208 = 8n$$
$$26 = n$$
The 26th term is 210.

21. $a_1 = 8$, $d = -6$
$$a_n = a_1 + (n-1)d$$
$$-328 = 8 + (n-1)(-6)$$
$$-328 = 8 - 6n + 6$$
$$-328 = 14 - 6n$$
$$-342 = -6n$$
$$57 = n$$
The 57th term is -328.

23. $a_n = a_1 + (n-1)d$
$a_{18} = 8 + (18-1)10$ Substituting 18 for n,
                          8 for a_1, and 10 for d
$$= 8 + 17 \cdot 10$$
$$= 8 + 170$$
$$= 178$$

25. $a_n = a_1 + (n-1)d$
$33 = a_1 + (8-1)4$ Substituting 33 for a_8,
                     8 for n, and 4 for d
$$33 = a_1 + 28$$
$$5 = a_1$$

27. $a_n = a_1 + (n-1)d$
$-76 = 5 + (n-1)(-3)$ Substituting -76 for a_n,
                      5 for a_1, and -3 for d
$$-76 = 5 - 3n + 3$$
$$-76 = 8 - 3n$$
$$-84 = -3n$$
$$28 = n$$

29. We know that $a_{17} = -40$ and $a_{28} = -73$. We would have to add d eleven times to get from a_{17} to a_{28}. That is,
$$-40 + 11d = -73$$
$$11d = -33$$
$$d = -3.$$
Since $a_{17} = -40$, we subtract d sixteen times to get a_1.
$$a_1 = -40 - 16(-3) = -40 + 48 = 8$$
We write the first five terms of the sequence:
8, 5, 2, -1, -4.

31. $a_{13} = 13$ and $a_{54} = 54$
Observe that for this to be true, $a_1 = 1$ and $d = 1$.

33. $1 + 5 + 9 + 13 + ...$
Note that $a_1 = 1$, $d = 4$, and $n = 20$. Before using the formula for S_n, we find a_{20}:
$$a_{20} = 1 + (20-1)4 \quad \text{Substituting into}$$
$$\text{the formula for } a_n$$
$$= 1 + 19 \cdot 4$$
$$= 77$$
Then using the formula for S_n,
$$S_{20} = \frac{20}{2}(1 + 77) = 10(78) = 780.$$

35. The sum is $1 + 2 + 3 + ... + 249 + 250$. This is the sum of the arithmetic sequence for which $a_1 = 1$, $a_n = 250$, and $n = 250$. We use the formula for S_n.
$$S_n = \frac{n}{2}(a_1 + a_n)$$
$$S_{250} = \frac{250}{2}(1 + 250) = 125(251) = 31,375$$

37. The sum is $2 + 4 + 6 + ... + 98 + 100$. This is the sum of the arithmetic sequence for which $a_1 = 2$, $a_n = 100$, and $n = 50$. We use the formula for S_n.
$$S_n = \frac{n}{2}(a_1 + a_n)$$
$$S_{50} = \frac{50}{2}(2 + 100) = 25(102) = 2550$$

39. The sum is $6 + 12 + 18 + ... + 96 + 102$. This is the sum of the arithmetic sequence for which $a_1 = 6$, $a_n = 102$, and $n = 17$. We use the formula for S_n.
$$S_n = \frac{n}{2}(a_1 + a_n)$$
$$S_{17} = \frac{17}{2}(6 + 102) = \frac{17}{2}(108) = 918$$

41. Before using the formula for S_n, we find a_{20}:
$$a_{20} = 4 + (20-1)5 \quad \text{Substituting into}$$
$$\text{the formula for } a_n$$
$$= 4 + 19 \cdot 5$$
$$= 99$$
Then using the formula for S_n,
$$S_{20} = \frac{20}{2}(4 + 99) = 10(103) = 1030.$$

43. *Familiarize*. We want to find the fifteenth term and the sum of an arithmetic sequence with $a_1 = 7$, $d = 2$, and $n = 15$. We will first use the formula for a_n to find a_{15}. This result is the number of musicians in the last row. Then we will use the formula for S_n to find S_{15}. This is the total number of musicians.
Translate. Substituting into the formula for a_n, we have
$$a_{15} = 7 + (15-1)2.$$

Carry out. We first find a_{15}.

$$a_{15} = 7 + 14 \cdot 2 = 35$$

Then use the formula for S_n to find S_{15}.

$$S_{15} = \frac{15}{2}(7 + 35) = \frac{15}{2}(42) = 315$$

Check. We can do the calculations again. We can also do the entire addition.

$$7 + 9 + 11 + \ldots + 35.$$

State. There are 35 musicians in the last row, and there are 315 musicians altogether.

45. *Familiarize*. We want to find the sum of the arithmetic sequence $36 + 32 + \ldots + 4$. Note that $a_1 = 36$, and $d = -4$. We will first use the formula for a_n to find n. Then we will use the formula for S_n.

Translate. Substituting into the formula for a_n, we have

$$4 = 36 + (n - 1)(-4).$$

Carry out. We solve for n.

$$4 = 36 + (n - 1)(-4)$$
$$4 = 36 - 4n + 4$$
$$4 = 40 - 4n$$
$$-36 = -4n$$
$$9 = n$$

Now we find S_9.

$$S_9 = \frac{9}{2}(36 + 4) = \frac{9}{2}(40) = 180$$

Check. We can do the calculations again. We can also do the entire addition.

$$36 + 32 + \ldots + 4.$$

State. There are 180 stones in the pyramid.

47. *Familiarize*. We want to find the sum of the arithmetic sequence with $a_1 = 10$¢, $d = 10$¢, and $n = 31$. First we will find a_{31} and then we will find S_{31}.

Translate. Substituting in the formula for a_n, we have

$$a_{31} = 10 + (31 - 1)(10).$$

Carry out. First we find a_{31}.

$$a_{31} = 10 + 30 \cdot 10 = 10 + 300 = 310$$

Then we use the formula for S_n to find S_{31}.

$$S_{31} = \frac{31}{2}(10 + 310) = \frac{31}{2}(320) = 4960$$

Check. We can do the calculations again.
State. The amount saved is 4960¢, or $49.60.

49. *Writing Exercise*.
$$1 + 2 + 3 + \ldots + 100$$
$$= (1 + 100) + (2 + 99) + (3 + 98) + \ldots + (50 + 51)$$
$$= \underbrace{101 + 101 + 101 + \ldots + 101}_{50 \text{ addends of } 101}$$
$$= 50 \cdot 101$$
$$= 5050$$

51. Using the slope-intercept form, where $m = \frac{1}{3}$ and $b = 10$, we have $y = \frac{1}{3}x + 10$.

53. Rewrite the equation.
$$2x + y = 8$$
$$y = -2x + 8$$
The slope of the parallel line is –2.
Use point-slope form.
$$y - y_1 = m(x - x_1)$$
$$y - 0 = -2(x - 5)$$
$$y = -2x + 10$$

55. A circle with center $(0, 0)$ and radius 4 is $x^2 + y^2 = 16$.

57. *Writing Exercise*.

59. Find the sequence with $a_1 = 150$. When $n = 21$, $a_{21} = 135$, find d. Note that we could use any of the other points from the table.
$$a_n = a_1 + (n - 1)d$$
$$135 = 150 + (21 - 1)d$$
$$-15 = 20d$$
$$-0.75 = d$$

Substitute to find the general term of the sequence.
$$a_n = a_1 + (n - 1)d$$
$$a_n = 150 + (n - 1)(-0.75)$$
$$a_n = 150 - 0.75n + 0.75$$
$$a_n = -0.75n + 150.75$$

61. Let d = the common difference. Since p, m, and q form an arithmetic sequence, $m = p + d$, and $q = p + 2d$. Then
$$\frac{p + q}{2} = \frac{p + (p + 2d)}{2} = p + d = m.$$

63. Each integer from 501 through 750 is 500 more than the corresponding integer from 1 through 250. There are 250 integers from 501 through 750, so their sum is the sum of the integers from 1 to 250 plus $250 \cdot 500$. From Exercise 39, we know that the sum of the integers from 1 through 250 is 31,375. Thus, we have
$$31,375 + 250 \cdot 500, \text{ or } 156,375.$$

Connecting the Concepts

1. $d = 112 - 115 = -3$, or
$d = 109 - 112 = -3$, or
$d = 106 - 109 = -3$

2. $\frac{1}{3}$, $-\frac{1}{6}$, $\frac{1}{12}$, $-\frac{1}{24}$, ...

$$r = \frac{-\frac{1}{6}}{\frac{1}{3}} = -\frac{1}{6} \cdot \frac{3}{1} = -\frac{1}{2}$$

3. 10, 15, 20, 25, ...
Note that $a_1 = 10$, $d = 5$, and $n = 21$.
$$a_{21} = 10 + (21 - 1)5 = 10 + 100 = 110$$

4. 5, 10, 20, 40, ...

$a_1 = 5$, and $r = \dfrac{10}{5} = 2$

$a_n = a_1 r^{n-1}$

$a_8 = 5(2)^{8-1} = 5(2)^7 = 5(128) = 640$

5. $2 + 12 + 22 + 32 + ...$

$a_1 = 1$, $d = 10$, and $n = 30$.

$a_{30} = 2 + (30-1)10 = 292$

$S_{30} = \dfrac{30}{2}(2 + 292) = 4410$

6. $\$100 + \$100(1.03) + \$100(1.03)^2 + ...$

$a_1 = \$100$, $n = 10$, and $r = \dfrac{\$100(1.03)}{\$100} = 1.03$

$S_n = \dfrac{a_1\left(1 - r^n\right)}{1 - r}$

$S_{10} = \dfrac{\$100\left[1 - (1.03)^{10}\right]}{1 - 1.03} \approx \dfrac{\$100(1 - 1.3439164)}{-0.03}$

$\approx \$1146.39$

7. $0.9 + 0.09 + 0.009 + ...$

$|r| = \left|\dfrac{0.09}{0.9}\right| = |0.1| = 0.1 < 1$, so the series has a limit.

$S_\infty = \dfrac{0.9}{1 - 0.1} = \dfrac{0.9}{0.9} = 1$

Thus, $0.9 + 0.09 + 0.009 + ... = 1$.

8. $0.9 + 9 + 90 + ...$

$|r| = \left|\dfrac{9}{0.9}\right| = |10| = 10 \not< 1$, so the series does not have a limit.

Exercise Set 11.3

1. The list 16, 8, 4, 2, 1, ... is an *infinite geometric sequence*.

3. For $16 + 8 + 4 + 2 + 1 + \cdots$, the common *ratio* is *less* than 1, so the limit *does* exist.

5. $\dfrac{a_{n+1}}{a_n} = 2$, so this is a geometric sequence.

7. $\dfrac{a_{n+1}}{a_n} = 5$, so this is a geometric series.

9. $\dfrac{a_{n+1}}{a_n} = -\dfrac{1}{2}$, so this is a geometric series.

11. 10, 20, 40, 80, ...

$\dfrac{20}{10} = 2$, $\dfrac{40}{20} = 2$, $\dfrac{80}{40} = 2$

$r = 2$

13. 6, −0.6, 0.06, −0.006, ...

$-\dfrac{0.6}{6} = -0.1$, $\dfrac{0.06}{-0.6} = -0.1$, $\dfrac{-0.006}{0.06} = -0.1$

$r = -0.1$

15. $\dfrac{1}{2}$, $-\dfrac{1}{4}$, $\dfrac{1}{8}$, $-\dfrac{1}{16}$, ...

$\dfrac{-\frac{1}{4}}{\frac{1}{2}} = -\dfrac{1}{4} \cdot \dfrac{2}{1} = -\dfrac{2}{4} = -\dfrac{1}{2}$, $\dfrac{\frac{1}{8}}{-\frac{1}{4}} = \dfrac{1}{8} \cdot \left(-\dfrac{4}{1}\right) = -\dfrac{4}{8} = -\dfrac{1}{2}$,

$\dfrac{-\frac{1}{16}}{\frac{1}{8}} = -\dfrac{1}{16} \cdot \dfrac{8}{1} = -\dfrac{8}{16} = -\dfrac{1}{2}$

$r = -\dfrac{1}{2}$

17. 75, 15, 3, $\dfrac{3}{5}$, ...

$\dfrac{15}{75} = \dfrac{1}{5}$, $\dfrac{3}{15} = \dfrac{1}{5}$, $\dfrac{\frac{3}{5}}{3} = \dfrac{3}{5} \cdot \dfrac{1}{3} = \dfrac{1}{5}$

$r = \dfrac{1}{5}$

19. $\dfrac{1}{m}$, $\dfrac{6}{m^2}$, $\dfrac{36}{m^3}$, $\dfrac{216}{m^4}$, ...

$\dfrac{\frac{6}{m^2}}{\frac{1}{m}} = \dfrac{6}{m^2} \cdot \dfrac{m}{1} = \dfrac{6}{m}$, $\dfrac{\frac{36}{m^3}}{\frac{6}{m^2}} = \dfrac{36}{m^3} \cdot \dfrac{m^2}{6} = \dfrac{6}{m}$

$\dfrac{\frac{216}{m^4}}{\frac{36}{m^3}} = \dfrac{216}{m^4} \cdot \dfrac{m^3}{36} = \dfrac{6}{m}$

$r = \dfrac{6}{m}$

21. 2, 6, 18, ...

$a_1 = 2$, $n = 7$, and $r = \dfrac{6}{2} = 3$

We use the formula $a_n = a_1 r^{n-1}$.

$a_7 = 2 \cdot 3^{7-1} = 2 \cdot 3^6 = 2 \cdot 729 = 1458$

23. $\sqrt{3}$, 3, $3\sqrt{3}$, ...

$a_1 = \sqrt{3}$, $n = 10$, and $r = \dfrac{3\sqrt{3}}{3} = \sqrt{3}$

$a_n = a_1 r^{n-1}$

$a_{10} = \sqrt{3}\left(\sqrt{3}\right)^{10-1} = \sqrt{3}\left(\sqrt{3}\right)^9 = \left(\sqrt{3}\right)^{10} = 243$

25. $-\dfrac{8}{243}$, $\dfrac{8}{81}$, $-\dfrac{8}{27}$, ...

$a_1 = -\dfrac{8}{243}$, $n = 14$, and $r = \dfrac{\frac{8}{81}}{-\frac{8}{243}} = \dfrac{8}{81}\left(-\dfrac{243}{8}\right) = -3$

$a_n = a_1 r^{n-1}$

$a_{14} = -\dfrac{8}{243}(-3)^{14-1} = -\dfrac{8}{243}(-3)^{13}$

$= -\dfrac{8}{243}(-1,594,323) = 52,488$

27. $1000, $1040, $1081.60, ...

$a_1 = \$1000$, $n = 10$, and $r = \dfrac{1040}{1000} = 1.04$

$a_n = a_1 r^{n-1}$

$a_{10} = \$1000(1.04)^{10-1} \approx \$1000(1.423311812)$
$\qquad \approx \$1423.31$

29. $1, 5, 25, 125,...$

$a_1 = 1$, and $r = \dfrac{5}{1} = 5$

$a_n = a_1 r^{n-1}$

$a_n = 1 \cdot 5^{n-1} = 5^{n-1}$

31. $1, -1, 1, -1,...$

$a_1 = 1$, and $r = \dfrac{-1}{1} = -1$

$a_n = a_1 r^{n-1}$

$a_n = 1(-1)^{n-1} = (-1)^{n-1}$, or $(-1)^{n+1}$

33. $\dfrac{1}{x}, \dfrac{1}{x^2}, \dfrac{1}{x^3},...$

$a_1 = \dfrac{1}{x}$, and $r = \dfrac{\frac{1}{x^2}}{\frac{1}{x}} = \dfrac{1}{x^2} \cdot \dfrac{x}{1} = \dfrac{1}{x}$

$a_n = a_1 r^{n-1}$

$a_n = \dfrac{1}{x}\left(\dfrac{1}{x}\right)^{n-1} = \dfrac{1}{x} \cdot \dfrac{1}{x^{n-1}} = \dfrac{1}{x^{1+n-1}} = \dfrac{1}{x^n}$, or x^{-n}

35. $6 + 12 + 24 + ...$

$a_1 = 6$, $n = 9$, and $r = \dfrac{12}{6} = 2$

$S_n = \dfrac{a_1(1 - r^n)}{1 - r}$

$S_9 = \dfrac{6(1 - 2^9)}{1 - 2} = \dfrac{6(1 - 512)}{-1} = \dfrac{6(-511)}{-1} = 3066$

37. $\dfrac{1}{18} - \dfrac{1}{6} + \dfrac{1}{2} - ...$

$a_1 = \dfrac{1}{18}$, $n = 7$, and $r = \dfrac{-\frac{1}{6}}{\frac{1}{18}} = -\dfrac{1}{6} \cdot \dfrac{18}{1} = -3$

$S_n = \dfrac{a_1(1 - r^n)}{1 - r}$

$S_7 = \dfrac{\frac{1}{18}\left[1 - (-3)^7\right]}{1 - (-3)} = \dfrac{\frac{1}{18}(1 + 2187)}{4} = \dfrac{\frac{1}{18}(2188)}{4}$
$\quad = \dfrac{1}{18}(2188)\left(\dfrac{1}{4}\right) = \dfrac{547}{18}$

39. $1 + x + x^2 + x^3 + ...$

$a_1 = 1$, $n = 8$, and $r = \dfrac{x}{1}$, or x

$S_n = \dfrac{a_1(1 - r^n)}{1 - r}$

$S_8 = \dfrac{1(1 - x^8)}{1 - x} = \dfrac{(1 + x^4)(1 - x^4)}{1 - x}$
$\quad = \dfrac{(1 + x^4)(1 + x^2)(1 - x^2)}{1 - x}$
$\quad = \dfrac{(1 + x^4)(1 + x^2)(1 + x)(1 - x)}{1 - x}$
$\quad = (1 + x^4)(1 + x^2)(1 + x)$

41. $\$200 + \$200(1.06) + \$200(1.06)^2 + ...$

$a_1 = \$200$, $n = 16$, and $r = \dfrac{\$200(1.06)}{\$200} = 1.06$

$S_n = \dfrac{a_1(1 - r^n)}{1 - r}$

$S_{16} = \dfrac{\$200\left[1 - (1.06)^{16}\right]}{1 - 1.06} \approx \dfrac{\$200(1 - 2.540351685)}{-0.06}$
$\qquad \approx \$5134.51$

43. $18 + 6 + 2 + ...$

$|r| = \left|\dfrac{6}{18}\right| = \left|\dfrac{1}{3}\right| = \dfrac{1}{3}$, and since $|r| < 1$, the series does have a limit.

$S_\infty = \dfrac{a_1}{1 - r} = \dfrac{18}{1 - \frac{1}{3}} = \dfrac{18}{\frac{2}{3}} = 18 \cdot \dfrac{3}{2} = 27$

45. $7 + 3 + \dfrac{9}{7} + ...$

$|r| = \left|\dfrac{3}{7}\right| = \dfrac{3}{7}$, and since $|r| < 1$, the series does have a limit.

$S_\infty = \dfrac{a_1}{1 - r} = \dfrac{7}{1 - \frac{3}{7}} = \dfrac{7}{\frac{4}{7}} = 7 \cdot \dfrac{7}{4} = \dfrac{49}{4}$

47. $3 + 15 + 75 + ...$

$|r| = \left|\dfrac{15}{3}\right| = |5| = 5$, and since $|r| \not< 1$, the series does not have a limit.

49. $4 - 6 + 9 - \dfrac{27}{2} + ...$

$|r| = \left|\dfrac{-6}{4}\right| = \left|-\dfrac{3}{2}\right| = \dfrac{3}{2}$, and since $|r| \not< 1$, the series does not have a limit.

51. $0.43 + 0.0043 + 0.000043 + ...$

$|r| = \left|\dfrac{0.0043}{0.43}\right| = |0.01| = 0.01$, and since $|r| < 1$, the series does have a limit.

$S_\infty = \dfrac{a_1}{1 - r} = \dfrac{0.43}{1 - 0.01} = \dfrac{0.43}{0.99} = \dfrac{43}{99}$

53. $\$500(1.02)^{-1} + \$500(1.02)^{-2} + \$500(1.02)^{-3} + \ldots$

$|r| = \left|\dfrac{\$500(1.02)^{-2}}{\$500(1.02)^{-1}}\right| = \left|(1.02)^{-1}\right| = (1.02)^{-1}$, or $\dfrac{1}{1.02}$,

and since $|r| < 1$, the series does have a limit.

$S_\infty = \dfrac{a_1}{1-r} = \dfrac{\$500(1.02)^{-1}}{1-\left(\dfrac{1}{1.02}\right)} = \dfrac{\dfrac{\$500}{1.02}}{\dfrac{0.02}{1.02}} = \dfrac{\$500}{1.02} \cdot \dfrac{1.02}{0.02}$

$\quad = \$25,000$

55. $0.5555\ldots = 0.5 + 0.05 + 0.005 + 0.0005 + \ldots$

This is an infinite geometric series with $a_1 = 0.5$.

$|r| = \left|\dfrac{0.05}{0.5}\right| = |0.1| = 0.1 < 1$, so the series has a limit.

$S_\infty = \dfrac{a_1}{1-r} = \dfrac{0.5}{1-0.1} = \dfrac{0.5}{0.9} = \dfrac{5}{9}$

Fractional notation for $0.5555\ldots$ is $\dfrac{5}{9}$.

57. $3.4646\ldots = 3 + 0.4646\ldots$

$0.464646\ldots = 0.46 + 0.0046 + 0.000046 + \ldots$

$|r| = \left|\dfrac{0.0046}{0.46}\right| = |0.01| = 0.01 < 1$, so the series has a limit.

$S_\infty = \dfrac{a_1}{1-r} = \dfrac{0.46}{1-0.01} = \dfrac{0.46}{0.99} = \dfrac{46}{99}$

Fractional notation for $0.4646\ldots$ is $\dfrac{46}{99}$.

Fractional notation for $3.4646\ldots$ is $3 + \dfrac{46}{99} = \dfrac{343}{99}$.

59. $0.15151515\ldots = 0.15 + 0.0015 + 0.000015 + \ldots$

This is an infinite geometric series with $a_1 = 0.15$.

$|r| = \left|\dfrac{0.0015}{0.15}\right| = |0.01| = 0.01 < 1$, so the series has a limit.

$S_\infty = \dfrac{a_1}{1-r} = \dfrac{0.15}{1-0.01} = \dfrac{0.15}{0.99} = \dfrac{15}{99} = \dfrac{5}{33}$

Fractional notation for $0.15151515\ldots$ is $\dfrac{5}{33}$.

61. *Familiarize*. The rebound distances form a geometric sequence:

$$\dfrac{1}{4} \times 20, \ \left(\dfrac{1}{4}\right)^2 \times 20, \ \left(\dfrac{1}{4}\right)^3 \times 20, \ldots,$$

or $5, \ \dfrac{1}{4} \times 5, \ \left(\dfrac{1}{4}\right)^2 \times 5, \ldots$

The height of the 6th rebound is the 6th term of the sequence.

Translate. We will use the formula $a_n = a_1 r^{n-1}$, with $a_1 = 5$, $r = \dfrac{1}{4}$, and $n = 6$:

$$a_6 = 5\left(\dfrac{1}{4}\right)^{6-1}$$

Carry out. We calculate to obtain $a_6 = \dfrac{5}{1024}$.

Check. We can do the calculation again.

State. It rebounds $\dfrac{5}{1024}$ ft the 6th time.

63. *Familiarize*. In one year, the population will be $100,000 + 0.03(100,000)$, or $(1.03)100,000$. In two years, the population will be $(1.03)100,000 + 0.03(1.03)100,000$, or $(1.03)^2 100,000$. Thus, the populations form a geometric sequence:

$100,000, \ (1.03)100,000, \ (1.03)^2 100,000, \ldots$

The population in 15 years will be the 16th term of the sequence.

Translate. We will use the formula $a_n = a_1 r^{n-1}$ with $a_1 = 100,000$, $r = 1.03$, and $n = 16$:

$$a_{16} = 100,000(1.03)^{16-1}$$

Carry out. We calculate to obtain $a_{16} \approx 155,797$.

Check. We can do the calculation again.

State. In 15 years the population will be about $155,797$.

65. *Familiarize*. At the end of each minute the population is 96% of the previous population. We have a geometric sequence:

$5000, \ 5000(0.96), \ 5000(0.96)^2, \ldots$

The number of fruit flies remaining alive after 15 minutes is given by the 16th term of the sequence.

Translate. We use the formula $a_n = a_1 r^{n-1}$ with $a_1 = 5000$, $r = 0.96$, and $n = 16$:

$$a_{16} = 5000(0.96)^{16-1}$$

Carry out. We calculate to obtain $a_{16} \approx 2710$.

Check. We can do the calculation again.

State. About 2710 flies will be alive after 15 min.

67. *Familiarize*. The lengths of the falls form a geometric sequence:

$$556, \ 556\left(\dfrac{3}{4}\right), \ 556\left(\dfrac{3}{4}\right)^2, \ 556\left(\dfrac{3}{4}\right)^3, \ldots$$

The total length of the first 6 falls is the sum of the first six terms of this sequence. The heights of the rebounds also form a geometric sequence:

$$556\left(\dfrac{3}{4}\right), \ 556\left(\dfrac{3}{4}\right)^2, \ 556\left(\dfrac{3}{4}\right)^3, \ldots \text{ or}$$

$$417, \ 417\left(\dfrac{3}{4}\right), \ 417\left(\dfrac{3}{4}\right)^2, \ldots$$

When the ball hits the ground for the 6th time, it will have rebounded 5 times. Thus the total length of the rebounds is the sum of the first five terms of this sequence.

Translate. We use the formula $S_n = \dfrac{a_1(1-r^n)}{1-r}$ twice, once with $a_1 = 556$, $r = \dfrac{3}{4}$, and $n = 6$ and a second time with $a_1 = 417$, $r = \frac{3}{4}$, and $n = 5$.

$D = $ Length of falls + length of rebounds

$$= \frac{556\left[1 - \left(\frac{3}{4}\right)^6\right]}{1 - \frac{3}{4}} + \frac{417\left[1 - \left(\frac{3}{4}\right)^5\right]}{1 - \frac{3}{4}}$$

Carry out. We use a calculator to obtain $D \approx 3100.35$.
Check. We can do the calculations again.
State. The ball will have traveled about 3100.35 ft.

69. *Familiarize*. The heights of the stack form a geometric sequence:

$$0.02, \ 0.02(2), \ 0.02(2)^2, \ ...$$

The height of the stack after it is doubled 10 times is given by the 11th term of this sequence.
Translate. We have a geometric sequence with $a_1 = 0.02$, $r = 2$, and $n = 11$. We use the formula

$$a_n = a_1 r^{n-1}.$$

Carry out. We substitute and calculate.

$$a_{11} = 0.02\left(2^{11-1}\right)$$
$$a_{11} = 0.02(1024) = 20.48$$

Check. We can do the calculations again.
State. The final stack will be 20.48 in. high.

71. *Writing Exercise*.

73. $|x - 3| = 11$

$$x - 3 = -11 \quad or \quad x - 3 = 11$$
$$x = -8 \quad or \quad x = 14$$

The solution set is $\{-8, 14\}$.

75. $|3x - 7| \geq 1$

$$3x - 7 \leq -1 \quad or \quad 1 \leq 3x - 7$$
$$3x \leq 6 \quad or \quad 8 \leq 3x$$
$$x \leq 2 \quad or \quad \frac{8}{3} \leq x$$

The solution set is $\left\{x \,\middle|\, x \leq 2 \ or \ x \geq \frac{8}{3}\right\}$, or

$$\left(-\infty, -2\right] \cup \left[\frac{8}{3}, \infty\right).$$

77. $x^2 - 5x - 14 < 0$
$(x + 2)(x - 7) < 0$

The solutions of $(x + 2)(x - 7) = 0$ are -2 and 7. They divide the number line into three intervals as shown:

We try test numbers in each interval.
A: Test -3, $f(-3) = (-3 + 2)(-3 - 7) = 10$
B: Test 0, $f(0) = (0 + 2)(0 - 7) = -14$
C: Test 8, $f(8) = (8 + 2)(8 - 7) = 10$
Since $f(-3)$ and $f(8)$ are positive, the function value will be negative for all numbers in the interval containing -2 and 7. The inequality symbol is $<$, so we do not include the endpoints. The solution set is
$(-2, 7)$, or $\{x \mid -2 < x < 7\}$.

79. *Writing Exercise*.

81. $\sum\limits_{k=1}^{\infty} 6(0.9)^k = 6(0.9) + 6(0.9)^2 + 6(0.9)^3 + ...$

$$|r| = \left|\frac{6(0.9)^2}{6(0.9)}\right| = |0.9| = 0.9 < 1, \text{ so the series has a limit.}$$

$$S_\infty = \frac{a_1}{1 - r} = \frac{6(0.9)}{1 - 0.9} = \frac{5.4}{0.1} = 54$$

83. $x^2 - x^3 + x^4 - x^5 + ...$

This is a geometric series with $a_1 = x^2$ and $r = -x$.

$$S_n = \frac{a_1\left(1 - r^n\right)}{1 - r} = \frac{x^2\left[1 - (-x)^n\right]}{1 - (-x)} = \frac{x^2\left[1 - (-x)^n\right]}{1 + x}$$

85. The length of a side of the first square is 16 cm. The length of a side of the next square is the length of the hypotenuse of a right triangle with legs 8 cm and 8 cm, or $8\sqrt{2}$ cm. The length of a side of the next square is the length of the hypotenuse of a right triangle with legs $4\sqrt{2}$ cm and $4\sqrt{2}$ cm, or 8 cm. The areas of the squares form a sequence:

$$(16)^2, \ \left(8\sqrt{2}\right)^2, \ (8)^2, \ ..., \text{ or}$$
$$256, \ 128, \ 64, \ ...$$

This is a geometric series with $a_1 = 256$ and $r = \frac{1}{2}$.

We find the sum of the infinite geometric series
$256 + 128 + 64 + ...$

$$S_\infty = \frac{a_1}{1 - r} = \frac{256}{1 - \frac{1}{2}} = \frac{256}{\frac{1}{2}} = 512 \text{ cm}^2$$

87. *Writing Exercise*.

Mid-Chapter Review

1. $a_n = a_1 + (n - 1)d$
   $n = 14$, $a_1 = -6$, $d = 5$
   $a_{14} = -6 + (14 - 1)5$
   $a_{14} = 59$

2. $a_n = a_1 r^{n-1}$
   $n = 7$, $a_1 = \frac{1}{9}$, $r = -3$
   $a_7 = \frac{1}{9} \cdot (-3)^{7-1}$
   $a_7 = 81$

3. $a_n = n^2 - 5n$
   $a_{20} = 20^2 - 5 \cdot 20 = 300$

4. $\frac{1}{2}, \ \frac{1}{3}, \ \frac{1}{4}, \ \frac{1}{5}, \ ...$

   The general term is $\frac{1}{n+1}$.

5. 1, 2, 3, 4, …

Note that $a_1 = 1$, $d = 1$, and $n = 12$. Before using the formula to find S_{12}, we find a_{12}.

$$a_{12} = 1 + (12 - 1)1 = 12$$

Then using the formula for S_n,

$$S_{12} = \frac{12}{2}(1 + 12) = 6(13) = 78.$$

6. $\displaystyle\sum_{k=2}^{5} k^2 = 2^2 + 3^2 + 4^2 + 5^2 = 4 + 9 + 16 + 25 = 54$

7. $1 - 2 + 3 - 4 + 5 - 6 = \displaystyle\sum_{k=1}^{6} (-1)^{k+1} \cdot k$

8. $a_n = a_1 + (n-1)d$

$22 = 10 + (n-1)0.2$

$22 = 10 + 0.2n - 0.2$

$12.2 = 0.2n$

$61 = n$

22 is the 61st term.

9. $a_n = a_1 + (n-1)d$

$a_{25} = 9 + (25 - 1)(-2) = 9 + (24)(-2) = -39$

10. $a_n = a_1 r^{n-1}$

$n = 12$, $a_1 = 1000$, $r = \dfrac{1}{10}$

$a_{12} = 1000 \cdot \left(\dfrac{1}{10}\right)^{12-1}$

$a_{12} = \dfrac{1}{10^8}$

11. 2, −2, 2, −2, …

$a_1 = 2$, and $r = \dfrac{-2}{2} = -1$

$a_n = 2(-1)^{n+1}$

12. $|r| = \left|\dfrac{-20}{100}\right| = \dfrac{1}{5} < 1$, so the series has a limit.

$$S_\infty = \frac{100}{1 - \left(-\dfrac{1}{5}\right)} = \frac{100}{\dfrac{6}{5}} = \frac{250}{3} = 83\frac{1}{3}$$

13. $\$1 + \$2 + \$3 + \$4 + \dots$

This is an arithmetic sequence $a_1 = 1$, $d = 1$, and $n = 30$. Before using the formula to find S_{30}, we find a_{30}.

$$a_{30} = 1 + (30 - 1)1 = 30$$

Then using the formula for S_n,

$$S_{30} = \frac{30}{2}(1 + 30) = 15(31) = 465.$$

She earns $465.

14. $\$1 + \$2 + \$4 + \$8 + \dots$

This is a geometric sequence. $a_1 = 1$, $n = 30$, $r = \dfrac{2}{1} = 2$

Then using the formula for S_n,

$$S_n = \frac{a_1(1 - r^n)}{1 - r}$$

$$S_{30} = \frac{1(1 - 2^{30})}{1 - 2} = \frac{1 - 1,073,741,824}{-1} = 1,073,741,823$$

He earns $1,073,741,823.

Exercise Set 11.4

1. The expression $(x + y)^2$ is a *binomial* squared.

3. The *first* number in every row of Pascal's triangle is 1.

5. 8! is an example of *factorial* notation.

7. The last term in the expansion of $(x + 2)^5$ is 2^5, or 32.

9. In the expansion of $(a + b)^9$, the sum of the exponents in each term is 9.

11. $4! = 4 \cdot 3 \cdot 2 \cdot 1 = 24$

13. $10! = 10 \cdot 9 \cdot 8 \cdot 7 \cdot 6 \cdot 5 \cdot 4 \cdot 3 \cdot 2 \cdot 1 = 3,628,800$

15. $\dfrac{10!}{8!} = \dfrac{10 \cdot 9 \cdot 8!}{8!} = 10 \cdot 9 = 90$

17. $\dfrac{9!}{4!5!} = \dfrac{9 \cdot 8 \cdot 7 \cdot 6 \cdot 5!}{4!5!} = \dfrac{9 \cdot 8 \cdot 7 \cdot 6}{4 \cdot 3 \cdot 2 \cdot 1} = 3 \cdot 7 \cdot 6 = 126$

19. $\dbinom{10}{4} = \dfrac{10!}{6!4!} = \dfrac{10 \cdot 9 \cdot 8 \cdot 7 \cdot 6!}{6!4!} = \dfrac{10 \cdot 9 \cdot 8 \cdot 7}{4 \cdot 3 \cdot 2 \cdot 1} = 10 \cdot 3 \cdot 7 = 210$

21. $\dbinom{9}{9} = \dfrac{9!}{0!9!} = \dfrac{9!}{1 \cdot 9!} = \dfrac{9!}{9!} = 1$

23. $\dbinom{30}{2} = \dfrac{30!}{28!2!} = \dfrac{30 \cdot 29 \cdot 28!}{28!2!} = \dfrac{30 \cdot 29}{2 \cdot 1} = 15 \cdot 29 = 435$

25. $\dbinom{40}{38} = \dfrac{40!}{2!38!} = \dfrac{40 \cdot 39 \cdot 38!}{2!38!} = \dfrac{40 \cdot 39}{2 \cdot 1} = 20 \cdot 39 = 780$

27. Expand $(a - b)^4$.

We have $a = a$, $b = -b$, and $n = 4$.

Form 1: We use the fifth row of Pascal's triangle:

  1   4   6   4   1

$(a - b)^4 = 1 \cdot a^4 + 4a^3(-b) + 6a^2(-b)^2 + 4a(-b)^3 + 1 \cdot (-b)^4$

$= a^4 - 4a^3b + 6a^2b^2 - 4ab^3 + b^4$

Form 2:

$(a - b)^4 = \dbinom{4}{0}a^4 + \dbinom{4}{1}a^3(-b) + \dbinom{4}{2}a^2(-b)^2$

$\qquad + \dbinom{4}{3}a(-b)^3 + \dbinom{4}{4}(-b)^4$

$= \dfrac{4!}{4!0!}a^4 + \dfrac{4!}{3!1!}a^3(-b) + \dfrac{4!}{2!2!}a^2(-b)^2$

$\qquad + \dfrac{4!}{1!3!}a(-b)^3 + \dfrac{4!}{0!4!}(-b)^4$

$= a^4 - 4a^3b + 6a^2b^2 - 4ab^3 + b^4$

29. Expand $(p+w)^7$.

We have $a = p$, $b = w$, and $n = 7$.

Form 1: We use the 8th row of Pascal's triangle:

$$1 \quad 7 \quad 21 \quad 35 \quad 35 \quad 21 \quad 7 \quad 1$$

$$(p+w)^7 = p^7 + 7p^6w^1 + 21p^5w^2 + 35p^4w^3$$
$$+ 35p^3w^4 + 21p^2w^5 + 7pw^6 + w^7$$

Form 2:

$$(p+q)^7$$
$$= \binom{7}{0}p^7 + \binom{7}{1}p^6w^1 + \binom{7}{2}p^5w^2 + \binom{7}{3}p^4w^3$$
$$+ \binom{7}{4}p^3w^4 + \binom{7}{5}p^2w^5 + \binom{7}{6}pw^6 + \binom{7}{7}w^7$$
$$= \frac{7!}{7!0!}p^7 + \frac{7!}{6!1!}p^6w^1 + \frac{7!}{5!2!}p^5w^2 + \frac{7!}{4!3!}p^4w^3$$
$$+ \frac{7!}{3!4!}p^3w^4 + \frac{7!}{2!5!}p^2w^5 + \frac{7!}{1!6!}pw^6 + \frac{7!}{0!7!}w^7$$
$$= p^7 + 7p^6w^1 + 21p^5w^2 + 35p^4w^3$$
$$+ 35p^3w^4 + 21p^2w^5 + 7pw^6 + w^7$$

31. Expand $(3c-d)^7$.

We have $a = 3c$, $b = -d$, and $n = 7$.

Form 1: We use the 8th row of Pascal's triangle:

$$1 \quad 7 \quad 21 \quad 35 \quad 35 \quad 21 \quad 7 \quad 1$$

$$(3c-d)^7$$
$$= (3c)^7 + 7(3c)^6(-d)^1 + 21(3c)^5(-d)^2 + 35(3c)^4(-d)^3$$
$$+ 35(3c)^3(-d)^4 + 21(3c)^2(-d)^5 + 7(3c)(-d)^6 + (-d)^7$$
$$= 2187c^7 - 5103c^6d + 5103c^5d^2 - 2835c^4d^3$$
$$+ 945c^3d^4 - 189c^2d^5 + 21cd^6 - d^7$$

Form 2:

$$(3c-d)^7$$
$$= \binom{7}{0}(3c)^7 + \binom{7}{1}(3c)^6(-d)^1 + \binom{7}{2}(3c)^5(-d)^2$$
$$+ \binom{7}{3}(3c)^4(-d)^3 + \binom{7}{4}(3c)^3(-d)^4 + \binom{7}{5}(3c)^2(-d)^5$$
$$+ \binom{7}{6}(3c)(-d)^6 + \binom{7}{7}(-d)^7$$
$$= \frac{7!}{7!0!}(3c)^7 + \frac{7!}{6!1!}(3c)^6(-d)^1 + \frac{7!}{5!2!}(3c)^5(-d)^2$$
$$+ \frac{7!}{4!3!}(3c)^4(-d)^3 + \frac{7!}{3!4!}(3c)^3(-d)^4 + \frac{7!}{2!5!}(3c)^2(-d)^5$$
$$+ \frac{7!}{1!6!}(3c)(-d)^6 + \frac{7!}{0!7!}(-d)^7$$
$$= 2187c^7 - 5103c^6d + 5103c^5d^2 - 2835c^4d^3$$
$$+ 945c^3d^4 - 189c^2d^5 + 21cd^6 - d^7$$

33. Expand $(t^{-2}+2)^6$.

We have $a = t^{-2}$, $b = 2$, and $n = 6$.

Form 1: We use the 7th row of Pascal's triangle:

$$1 \quad 6 \quad 15 \quad 20 \quad 15 \quad 6 \quad 1$$

$$(t^{-2}+2)^6$$
$$= 1 \cdot (t^{-2})^6 + 6(t^{-2})^5(2)^1 + 15(t^{-2})^4(2)^2 + 20(t^{-2})^3(2)^3$$
$$+ 15(t^{-2})^2(2)^4 + 6(t^{-2})^1(2)^5 + 1 \cdot (2)^6$$
$$= t^{-12} + 12t^{-10} + 60t^{-8} + 160t^{-6} + 240t^{-4} + 192t^{-2} + 64$$

Form 2:

$$(t^{-2}+2)^6$$
$$= \binom{6}{0}(t^{-2})^6 + \binom{6}{1}(t^{-2})^5(2)^1 + \binom{6}{2}(t^{-2})^4(2)^2$$
$$+ \binom{6}{3}(t^{-2})^3(2)^3 + \binom{6}{4}(t^{-2})^2(2)^4$$
$$+ \binom{6}{5}(t^{-2})^1(2)^5 + \binom{6}{6}(2)^6$$
$$= \frac{6!}{6!0!}(t^{-2})^6 + \frac{6!}{5!1!}(t^{-2})^5(2)^1 + \frac{6!}{4!2!}(t^{-2})^4(2)^2$$
$$+ \frac{6!}{3!3!}(t^{-2})^3(2)^3 + \frac{6!}{2!4!}(t^{-2})^2(2)^4$$
$$+ \frac{6!}{1!5!}(t^{-2})^1(2)^5 + \frac{6!}{0!6!}(2)^6$$
$$= t^{-12} + 12t^{-10} + 60t^{-8} + 160t^{-6} + 240t^{-4} + 192t^{-2} + 64$$

35. Expand $\left(3s + \dfrac{1}{t}\right)^9$.

We have $a = 3s$, $b = \dfrac{1}{t}$, and $n = 9$.

Form 1: We use the tenth row of Pascal's triangle:

$$1 \quad 9 \quad 36 \quad 84 \quad 126 \quad 126 \quad 84 \quad 36 \quad 9 \quad 1$$

$$\left(3s + \frac{1}{t}\right)^9$$
$$= 1 \cdot (3s)^9 + 9(3s)^8\left(\frac{1}{t}\right)^1 + 36(3s)^7\left(\frac{1}{t}\right)^2 + 84(3s)^6\left(\frac{1}{t}\right)^3$$
$$+ 126(3s)^5\left(\frac{1}{t}\right)^4 + 126(3s)^4\left(\frac{1}{t}\right)^5 + 84(3s)^3\left(\frac{1}{t}\right)^6$$
$$+ 36(3s)^2\left(\frac{1}{t}\right)^7 + 9(3s)^1\left(\frac{1}{t}\right)^8 + 1 \cdot \left(\frac{1}{t}\right)^9$$
$$= 19{,}683s^9 + \frac{59{,}049s^8}{t} + \frac{78{,}732s^7}{t^2} + \frac{61{,}236s^6}{t^3}$$
$$+ \frac{30{,}618s^5}{t^4} + \frac{10{,}206s^4}{t^5} + \frac{2268s^3}{t^6} + \frac{324s^2}{t^7} + \frac{27s}{t^8} + \frac{1}{t^9}$$

Form 2:

$$\left(3s + \frac{1}{t}\right)^9$$
$$= \binom{9}{0}(3s)^9 + \binom{9}{1}(3s)^8\left(\frac{1}{t}\right)^1 + \binom{9}{2}(3s)^7\left(\frac{1}{t}\right)^2$$
$$+ \binom{9}{3}(3s)^6\left(\frac{1}{t}\right)^3 + \binom{9}{4}(3s)^5\left(\frac{1}{t}\right)^4 + \binom{9}{5}(3s)^4\left(\frac{1}{t}\right)^5$$
$$+ \binom{9}{6}(3s)^3\left(\frac{1}{t}\right)^6 + \binom{9}{7}(3s)^2\left(\frac{1}{t}\right)^7 + \binom{9}{8}(3s)^1\left(\frac{1}{t}\right)^8$$
$$+ \binom{9}{9}\left(\frac{1}{t}\right)^9$$
$$= \frac{9!}{9!0!}(3s)^9 + \frac{9!}{8!1!}(3s)^8\left(\frac{1}{t}\right)^1 + \frac{9!}{7!2!}(3s)^7\left(\frac{1}{t}\right)^2$$
$$+ \frac{9!}{6!3!}(3s)^6\left(\frac{1}{t}\right)^3 + \frac{9!}{5!4!}(3s)^5\left(\frac{1}{t}\right)^4 + \frac{9!}{4!5!}(3s)^4\left(\frac{1}{t}\right)^5$$
$$+ \frac{9!}{3!6!}(3s)^3\left(\frac{1}{t}\right)^6 + \frac{9!}{2!7!}(3s)^2\left(\frac{1}{t}\right)^7 + \frac{9!}{1!8!}(3s)^1\left(\frac{1}{t}\right)^8$$
$$+ \frac{9!}{0!9!}\left(\frac{1}{t}\right)^9$$

$$= 19{,}683s^9 + \frac{59{,}049s^8}{t} + \frac{78{,}732s^7}{t^2}$$
$$+ \frac{61{,}236s^6}{t^3} + \frac{30{,}618s^5}{t^4} + \frac{10{,}206s^4}{t^5}$$
$$+ \frac{2268s^3}{t^6} + \frac{324s^2}{t^7} + \frac{27s}{t^8} + \frac{1}{t^9}$$

37. Expand $\left(x^3 - 2y\right)^5$.

We have $a = x^3$, $b = -2y$, and $n = 5$.

Form 1: We use the 6th row of Pascal's triangle:

$$1 \quad 5 \quad 10 \quad 10 \quad 5 \quad 1$$

$$\left(x^3 - 2y\right)^5$$
$$= 1 \cdot \left(x^3\right)^5 + 5\left(x^3\right)^4(-2y)^1 + 10\left(x^3\right)^3(-2y)^2$$
$$+ 10\left(x^3\right)^2(-2y)^3 + 5\left(x^3\right)^1(-2y)^4 + 1 \cdot (-2y)^5$$
$$= x^{15} - 10x^{12}y + 40x^9y^2 - 80x^6y^3 + 80x^3y^4 - 32y^5$$

Form 2:

$$\left(x^3 - 2y\right)^5$$
$$= \binom{5}{0}\left(x^3\right)^5 + \binom{5}{1}\left(x^3\right)^4(-2y) + \binom{5}{2}\left(x^3\right)^3(-2y)^2$$
$$+ \binom{5}{3}\left(x^3\right)^2(-2y)^3 + \binom{5}{4}\left(x^3\right)(-2y)^4 + \binom{5}{5}(-2y)^5$$
$$= \frac{5!}{5!0!}\left(x^3\right)^5 + \frac{5!}{4!1!}\left(x^3\right)^4(-2y) + \frac{5!}{3!2!}\left(x^3\right)^3(-2y)^2$$
$$+ \frac{5!}{2!3!}\left(x^3\right)^2(-2y)^3 + \frac{5!}{1!4!}\left(x^3\right)(-2y)^4 + \frac{5!}{0!5!}(-2y)^5$$
$$= x^{15} - 10x^{12}y + 40x^9y^2 - 80x^6y^3 + 80x^3y^4 - 32y^5$$

39. Expand $\left(\sqrt{5} + t\right)^6$.

We have $a = \sqrt{5}$, $b = t$, and $n = 6$.

Form 1: We use the 7th row of Pascal's triangle:

$$1 \quad 6 \quad 15 \quad 20 \quad 15 \quad 6 \quad 1$$

$$\left(\sqrt{5} + t\right)^6$$
$$= 1 \cdot \left(\sqrt{5}\right)^6 + 6\left(\sqrt{5}\right)^5 t^1 + 15\left(\sqrt{5}\right)^4 t^2 + 20\left(\sqrt{5}\right)^3 t^3$$
$$+ 15\left(\sqrt{5}\right)^2 t^4 + 6\left(\sqrt{5}\right)^1 t^5 + 1 \cdot t^6$$
$$= 125 + 150\sqrt{5}t + 375t^2 + 100\sqrt{5}t^3 + 75t^4 + 6\sqrt{5}t^5 + t^6$$

Form 2:

$$\left(\sqrt{5} + t\right)^6$$
$$= \binom{6}{0}\left(\sqrt{5}\right)^6 + \binom{6}{1}\left(\sqrt{5}\right)^5 t^1 + \binom{6}{2}\left(\sqrt{5}\right)^4 t^2 + \binom{6}{3}\left(\sqrt{5}\right)^3 t^3$$
$$+ \binom{6}{4}\left(\sqrt{5}\right)^2 t^4 + \binom{6}{5}\left(\sqrt{5}\right)^1 t^5 + \binom{6}{6}t^6$$
$$= \frac{6!}{6!0!}\left(\sqrt{5}\right)^6 + \frac{6!}{5!1!}\left(\sqrt{5}\right)^5 t^1 + \frac{6!}{4!2!}\left(\sqrt{5}\right)^4 t^2 + \frac{6!}{3!3!}\left(\sqrt{5}\right)^3 t^3$$
$$+ \frac{6!}{2!4!}\left(\sqrt{5}\right)^2 t^4 + \frac{6!}{1!5!}\left(\sqrt{5}\right)^1 t^5 + \frac{6!}{0!6!}t^6$$
$$= 125 + 150\sqrt{5}t + 375t^2 + 100\sqrt{5}t^3 + 75t^4 + 6\sqrt{5}t^5 + t^6$$

41. Expand $\left(\frac{1}{\sqrt{x}} - \sqrt{x}\right)^6$.

We have $a = \frac{1}{\sqrt{x}}$, $b = -\sqrt{x}$, and $n = 6$.

Form 1: We use the 7th row of Pascal's triangle:

$$1 \quad 6 \quad 15 \quad 20 \quad 15 \quad 6 \quad 1$$

$$\left(\frac{1}{\sqrt{x}} - \sqrt{x}\right)^6$$
$$= 1 \cdot \left(\frac{1}{\sqrt{x}}\right)^6 + 6\left(\frac{1}{\sqrt{x}}\right)^5\left(-\sqrt{x}\right)^1 + 15\left(\frac{1}{\sqrt{x}}\right)^4\left(-\sqrt{x}\right)^2$$
$$+ 20\left(\frac{1}{\sqrt{x}}\right)^3\left(-\sqrt{x}\right)^3 + 15\left(\frac{1}{\sqrt{x}}\right)^2\left(-\sqrt{x}\right)^4$$
$$+ 6\left(\frac{1}{\sqrt{x}}\right)^1\left(-\sqrt{x}\right)^5 + 1 \cdot \left(-\sqrt{x}\right)^6$$
$$= x^{-3} - 6x^{-2} + 15x^{-1} - 20 + 15x - 6x^2 + x^3$$

Form 2:

$$\left(\frac{1}{\sqrt{x}} - \sqrt{x}\right)^6$$
$$= \binom{6}{0}\left(\frac{1}{\sqrt{x}}\right)^6 + \binom{6}{1}\left(\frac{1}{\sqrt{x}}\right)^5\left(-\sqrt{x}\right)^1 + \binom{6}{2}\left(\frac{1}{\sqrt{x}}\right)^4\left(-\sqrt{x}\right)^2$$
$$+ \binom{6}{3}\left(\frac{1}{\sqrt{x}}\right)^3\left(-\sqrt{x}\right)^3 + \binom{6}{4}\left(\frac{1}{\sqrt{x}}\right)^2\left(-\sqrt{x}\right)^4$$
$$+ \binom{6}{5}\left(\frac{1}{\sqrt{x}}\right)^1\left(-\sqrt{x}\right)^5 + \binom{6}{6}\left(-\sqrt{x}\right)^6$$
$$= \frac{6!}{6!0!}\left(\frac{1}{\sqrt{x}}\right)^6 + \frac{6!}{5!1!}\left(\frac{1}{\sqrt{x}}\right)^5\left(-\sqrt{x}\right)^1 + \frac{6!}{4!2!}\left(\frac{1}{\sqrt{x}}\right)^4\left(-\sqrt{x}\right)^2$$
$$+ \frac{6!}{3!3!}\left(\frac{1}{\sqrt{x}}\right)^3\left(-\sqrt{x}\right)^3 + \frac{6!}{2!4!}\left(\frac{1}{\sqrt{x}}\right)^2\left(-\sqrt{x}\right)^4$$
$$+ \frac{6!}{1!5!}\left(\frac{1}{\sqrt{x}}\right)^1\left(-\sqrt{x}\right)^5 + \frac{6!}{0!6!}\left(-\sqrt{x}\right)^6$$
$$= x^{-3} - 6x^{-2} + 15x^{-1} - 20 + 15x - 6x^2 + x^3$$

43. Find the 3rd term of $(a + b)^6$.

First we note that $3 = 2 + 1$, $a = a$, $b = b$, and $n = 6$.

Then the 3rd term of the expansion of $(a + b)^6$ is

$$\binom{6}{2}a^{6-2}b^2, \text{ or } \frac{6!}{4!2!}a^4b^2, \text{ or } 15a^4b^2.$$

45. Find the 12th term of $(a - 3)^{14}$.

First we note that $12 = 11 + 1$, $a = a$, $b = -3$, and $n = 14$.

Then the 12th term of the expansion of $(a - 3)^{14}$ is

$$\binom{14}{11}a^{14-11} \cdot (-3)^{11} = \frac{14!}{3!11!}a^3(-177{,}147)$$
$$= 364a^3(-177{,}147)$$
$$= -64{,}481{,}508a^3$$

47. Find the 5th term of $\left(2x^3 + \sqrt{y}\right)^8$.

First we note that $5 = 4 + 1$, $a = 2x^3$, $b = \sqrt{y}$, and $n = 8$.

Then the 5th term of the expansion of $\left(2x^3 + \sqrt{y}\right)^8$ is

$$\binom{8}{4}\left(2x^3\right)^{8-4}\left(\sqrt{y}\right)^4 = \frac{8!}{4!4!}\left(2x^3\right)^4\left(\sqrt{y}\right)^4$$
$$= 70\left(16x^{12}\right)\left(y^2\right)$$
$$= 1120x^{12}y^2$$

49. The expansion of $\left(2u+3v^2\right)^{10}$ has 11 terms so the 6th term is the middle term. Note that $6=5+1$, $a=2u$, $b=3v^2$, and $n=10$. Then the 6th term of the expansion of $\left(2u+3v^2\right)^{10}$ is

$$\binom{10}{5}(2u)^{10-5}\left(3v^2\right)^5 = \frac{10!}{5!5!}(2u)^5\left(3v^2\right)^5$$
$$= 252\left(32u^5\right)\left(243v^{10}\right)$$
$$= 1,959,552u^5v^{10}$$

51. The 9th term of $(x-y)^8$ is the last term, y^8.

53. *Writing Exercise.*

55. Graph $y=x^2-5$.
This is a parabola with vertex $(0,-5)$.

57. Graph $y \geq x-5$.
Use a solid line to form the line $y=x-5$. Since the test point $(0,0)$ is a solution, shade this side of the line.

59. Graph $f(x)=\log_5 x$.

61. *Writing Exercise.*

63. Consider the set of 5 elements $\{a, b, c, d, e\}$. List all the subsets of size 3:
$\{a, b, c\}$, $\{a, b, d\}$, $\{a, b, e\}$, $\{a, c, d\}$, $\{a, c, e\}$,
$\{a, d, e\}$, $\{b, c, d\}$, $\{b, c, e\}$, $\{b, d, e\}$, $\{c, d, e\}$.

There are exactly 10 subsets of size 3 and $\binom{5}{3}=10$, so

there are exactly $\binom{5}{3}$ ways of forming a subset of size 3 from a set of 5 elements.

65. Find the sixth term of $(0.15+0.85)^8$.

$$\binom{8}{5}(0.15)^{8-5}(0.85)^5 = \frac{8!}{3!5!}(0.15)^3(0.85)^5 \approx 0.084$$

67. Find and add the 7th through 9th terms of $(0.15+0.85)^9$.

$$\binom{8}{6}(0.15)^2(0.85)^6+\binom{8}{7}(0.15)(0.85)^7+\binom{8}{8}(0.85)^8$$
$$\approx 0.89$$

69. $\dbinom{n}{n-r} = \dfrac{n!}{[n-(n-r)]!(n-r)!} = \dfrac{n!}{r!(n-r)!} = \dbinom{n}{r}$

71. The expansion of $\left(x^2-6y^{3/2}\right)^6$ has 7 terms, so the 4th term is the middle term.

$$\binom{6}{3}\left(x^2\right)^3\left(-6y^{3/2}\right)^3 = \frac{6!}{3!3!}\left(x^6\right)\left(-216y^{9/2}\right)$$
$$= -4320x^6y^{9/2}$$

73. The $(r+1)$st term of $\left(\sqrt[3]{x}-\dfrac{1}{\sqrt{x}}\right)^7$ is

$\binom{7}{r}\left(\sqrt[3]{x}\right)^{7-r}\left(-\dfrac{1}{\sqrt{x}}\right)^r$. The term containing $\dfrac{1}{x^{1/6}}$ is the term in which the sum of the exponents is $-1/6$. That is,

$$\left(\frac{1}{3}\right)(7-r)+\left(-\frac{1}{2}\right)(r) = -\frac{1}{6}$$
$$\frac{7}{3}-\frac{r}{3}-\frac{r}{2} = -\frac{1}{6}$$
$$-\frac{5r}{6} = -\frac{15}{6}$$
$$r = 3$$

Find the $(3+1)$st, or 4th term.

$$\binom{7}{3}\left(\sqrt[3]{x}\right)^4\left(-\frac{1}{\sqrt{x}}\right)^3 = \frac{7!}{4!3!}\left(x^{4/3}\right)\left(-x^{-3/2}\right) = -35x^{-1/6}$$

or $-\dfrac{35}{x^{1/6}}$.

75. The degree of $\left(x^3+3\right)^4$ is the degree of $\left(x^3\right)^4 = x^{12}$, or 12.

Chapter 11 Review

1. False; the next term of the arithmetic sequence 10, 15, 20, ... is $20+5$, or 25.

2. True; the next term of the geometric sequence 2, 6, 18, 54, ... is $54 \cdot 3$, or 162.

3. True.

4. False; for $a_n = 3n-1$, $a_{17} = 3(17)-1 = 50$.

5. False; a geometric sequence has a common *ratio*.

6. True; $|r| = \left|\dfrac{-5}{10}\right| = \left|-\dfrac{1}{2}\right| = \dfrac{1}{2} < 1$, so the series has a limit.

7. False; $n! = n \cdot (n-1) \cdot (n-2) \cdot ... \cdot 3 \cdot 2 \cdot 1$.

8. False; $(x+y)^{17}$ has 18 terms.

9. $a_n = 10n-9$
$a_1 = 10 \cdot 1 - 9 = 1$
$a_2 = 10 \cdot 2 - 9 = 11$
$a_3 = 10 \cdot 3 - 9 = 21$
$a_4 = 10 \cdot 4 - 9 = 31$
$a_8 = 10 \cdot 8 - 9 = 71$
$a_{12} = 10 \cdot 12 - 9 = 111$

10. $a_n = \dfrac{n-1}{n^2+1}$

$a_1 = \dfrac{1-1}{1^2+1} = 0$

$a_2 = \dfrac{2-1}{2^2+1} = \dfrac{1}{5}$

$a_3 = \dfrac{3-1}{3^2+1} = \dfrac{2}{10} = \dfrac{1}{5}$

$a_4 = \dfrac{4-1}{4^2+1} = \dfrac{3}{17}$

$a_8 = \dfrac{8-1}{8^2+1} = \dfrac{7}{65}$

$a_{12} = \dfrac{12-1}{12^2+1} = \dfrac{11}{145}$

11. $-5,\ -10,\ -15,\ -20,\dots$

These are negative multiples of 5 beginning with -5, so the general term could be $-5n$.

12. $-1,\ 3,\ -5,\ 7,\ -9,\dots$

These are odd integers beginning with 1, but with alternating signs, beginning with a negative number.

The general terms could be $(-1)^n(2n-1)$.

13. $\displaystyle\sum_{k=1}^{5}(-2)^k = (-2)^1 + (-2)^2 + (-2)^3 + (-2)^4 + (-2)^5$

$= -2 + 4 + (-8) + 16 + (-32) = -22$

14. $\displaystyle\sum_{k=2}^{7}(1-2k) = (1-2\cdot2) + (1-2\cdot3) + (1-2\cdot4) + (1-2\cdot5)$
$+ (1-2\cdot6) + (1-2\cdot7)$

$= -3 + (-5) + (-7) + (-9) + (-11) + (-13)$
$= -48$

15. $7 + 14 + 21 + 28 + 35 + 42$

This is the sum of the first six positive multiples of 7. It is an finite series. Sigma notation is

$$\sum_{k=1}^{6} 7k.$$

16. $\dfrac{-1}{2} + \dfrac{1}{4} + \dfrac{-1}{8} + \dfrac{1}{16} + \dfrac{-1}{32}$

This is a sum of fractions in which the numerator is 1, but with alternating signs, and the denominator is the first five powers of two. Sigma notation is

$$\sum_{k=1}^{5}\dfrac{(-1)^k}{2^k}\ \text{ or }\ \sum_{k=1}^{5}\dfrac{1}{(-2)^k}.$$

17. $-3, -7, -11, \dots$

$a_1 = -3,\ d = -4,\ \text{and } n = 14$

$a_n = a_1 + (n-1)d$

$a_{14} = -3 + (14-1)(-4) = -3 + 13(-4) = -55$

18. $a_n = a_1 + (n-1)d$

$14 = 11 + (16-1)d$ Substituting 14 for a_{16},
16 for n, and 11 for a_1

$14 = 11 + 15d$

$3 = 15d$

$\dfrac{1}{5} = d$

19. We know that $a_8 = 20$ and $a_{24} = 100$. We would have to add d sixteen times to get from a_8 to a_{24}. That is,

$20 + 16d = 100$

$16d = 80$

$d = 5.$

Since $a_8 = 20,$ we subtract d seven times to get a_1.

$a_1 = 20 - 7(5) = 20 - 35 = -15$

20. Before using the formula for S_n, we find a_{17}:

$a_{17} = -8 + (17-1)(-3)$ Substituting into
the formula for a_n

$= -8 + 16(-3)$

$= -56$

Then using the formula for S_n,

$S_{17} = \dfrac{17}{2}(-8 + -56) = \dfrac{17}{2}(-64) = -544.$

21. The sum is $5 + 10 + 15 + \dots + 495 + 500$. This is the sum of the arithmetic sequence for which $a_1 = 5$, $a_n = 500$, and $n = 100$. We use the formula for S_n.

$S_n = \dfrac{n}{2}(a_1 + a_n)$

$S_{100} = \dfrac{100}{2}(5 + 500) = 50(505) = 25{,}250$

22. $2,\ 2\sqrt{2},\ 4,\dots$

$a_1 = 2,\ n = 20,\ \text{and } r = \dfrac{2\sqrt{2}}{2} = \sqrt{2}$

$a_n = a_1 r^{n-1}$

$a_{20} = 2\left(\sqrt{2}\right)^{20-1} = 2\left(\sqrt{2}\right)^{19} = 2\left(512\sqrt{2}\right) = 1024\sqrt{2}$

23. $r = \dfrac{30}{40} = \dfrac{3}{4}$

24. $-2,\ 2,\ -2,\dots$

$a_1 = -2,\ \text{and } r = \dfrac{2}{-2} = -1$

$a_n = a_1 r^{n-1}$

$a_n = -2(-1)^{n-1} = 2(-1)^n$

25. $3, \frac{3}{4}x, \frac{3}{16}x^2, \ldots$

$a_1 = 3$, and $r = \dfrac{\frac{3}{4}x}{3} = \dfrac{3x}{4} \cdot \dfrac{1}{3} = \dfrac{x}{4}$

$a_n = a_1 r^{n-1}$

$a_n = 3\left(\dfrac{x}{4}\right)^{n-1}$

26. $3 + 15 + 75 + \ldots$

$a_1 = 3$, $n = 6$, and $r = \dfrac{15}{3} = 5$

$S_n = \dfrac{a_1\left(1 - r^n\right)}{1 - r}$

$S_6 = \dfrac{3\left(1 - 5^6\right)}{1 - 5} = \dfrac{3(1 - 15{,}625)}{-4} = \dfrac{3(-15{,}624)}{-4} = 11{,}718$

27. $3x - 6x + 12x - \ldots$

$a_1 = 3x$, $n = 12$, and $r = \dfrac{-6x}{3x} = -2$

$S_n = \dfrac{a_1\left(1 - r^n\right)}{1 - r}$

$S_{12} = \dfrac{3x\left(1 - (-2)^{12}\right)}{1 - (-2)} = \dfrac{3x(1 - 4096)}{3} = x(-4095)$

$\phantom{S_{12}} = -4095x$

28. $6 + 3 + 1.5 + 0.75 + \ldots$

$|r| = \left|\dfrac{3}{6}\right| = \left|\dfrac{1}{2}\right| = \dfrac{1}{2} < 1$, the series has a limit.

$S_\infty = \dfrac{a_1}{1 - r} = \dfrac{6}{1 - \frac{1}{2}} = \dfrac{6}{\frac{1}{2}} = 6 \cdot \dfrac{2}{1} = 12$

29. $7 - 4 + \dfrac{16}{7} - \ldots$

$|r| = \left|\dfrac{-4}{7}\right| = \dfrac{4}{7} < 1$, the series has a limit.

$S_\infty = \dfrac{a_1}{1 - r} = \dfrac{7}{1 - \left(-\frac{4}{7}\right)} = \dfrac{7}{\frac{11}{7}} = 7 \cdot \dfrac{7}{11} = \dfrac{49}{11}$

30. $-\dfrac{1}{2} + \dfrac{1}{2} + \left(-\dfrac{1}{2}\right) + \dfrac{1}{2} + \ldots$

$|r| = \left|\dfrac{\frac{1}{2}}{-\frac{1}{2}}\right| = |-1| = 1$, and since $|r| \not< 1$, the series does

not have a limit.

31. $0.04 + 0.08 + 0.16 + 0.32 + \ldots$

$|r| = \left|\dfrac{0.08}{0.04}\right| = |2| = 2$, and since $|r| \not< 1$, the series does

not have a limit.

32. $\$2000 + \$1900 + \$1805 + \$1714.75 \ldots$

$|r| = \left|\dfrac{\$1900}{\$2000}\right| = \left|\dfrac{19}{20}\right| = \dfrac{19}{20} < 1$, the series has a limit.

$S_\infty = \dfrac{a_1}{1 - r} = \dfrac{\$2000}{1 - \frac{19}{20}} = \dfrac{\$2000}{\frac{1}{20}} = \$2000 \cdot \dfrac{20}{1} = \$40{,}000$

33. $0.5555\ldots = 0.5 + 0.05 + 0.005 + 0.0005 + \ldots$

This is an infinite geometric series with $a_1 = 0.5$.

$|r| = \left|\dfrac{0.05}{0.5}\right| = |0.1| = 0.1 < 1$, so the series has a limit.

$S_\infty = \dfrac{a_1}{1 - r} = \dfrac{0.5}{1 - 0.1} = \dfrac{0.5}{0.9} = \dfrac{5}{9}$

Fractional notation for $0.5555\ldots$ is $\dfrac{5}{9}$.

34. $1.454545\ldots = 1 + 0.45 + 0.0045 + 0.000045 + \ldots$

$|r| = \left|\dfrac{0.0045}{0.45}\right| = |0.01| = 0.01 < 1$, so the series has a

limit.

$S_\infty = \dfrac{a_1}{1 - r} = \dfrac{0.45}{1 - 0.01} = \dfrac{0.45}{0.99} = \dfrac{45}{99} = \dfrac{5}{11}$

Fractional notation for $1.4545\ldots$ is $1 + \dfrac{5}{11} = \dfrac{16}{11}$.

35. *Familiarize*. A $\$0.40$ raise every 3 months for 8 years, is 32 raises. We want to find the 33th term of an arithmetic sequence with $a_1 = \$11.50$, $d = \$0.40$, and $n = 33$. We will use the formula for a_n to find a_{33}. This result is the hourly wage after 8 years.

Translate. Substituting into the formula for a_n, we have

$a_{33} = \$11.50 + (33 - 1)(\$0.40)$.

Carry out. We find a_{33}.

$a_{33} = \$11.50 + 32(\$0.40) = \$24.30$

Check. We can do the calculations again.

State. After 8 years, Tyrone earns $\$24.30$ an hour.

36. We go from 42 poles in the bottom layer to one pole in the top layer, where each layer has 1 pole less than the layer below it, so there must be 42 layers.

Thus, we want to find the sum of the first 42 terms of an arithmetic sequence with $a_1 = 42$ and $a_{42} = 1$.

$S_{42} = \dfrac{42}{2}(42 + 1) = 21(43) = 903$ poles

37. *Familiarize*. At the end of each year, the interest at 4% will be added to the previous year's amount.

We have a geometric sequence:

$\$12{,}000$, $\$12{,}000(1.04)$, $\$12{,}000(1.04)^2, \ldots$

We are looking for the amount after 7 years.

Translate. We use the formula $a_n = a_1 r^{n-1}$ with

$a_1 = \$12{,}000$, $r = 1.04$, and $n = 8$:

$a_8 = \$12{,}000(1.04)^{8-1}$

Carry out. We calculate to obtain $a_8 \approx \$15,791.18$.

Check. We can do the calculation again.

State. After 7 years, the amount of the loan will be about $15,791.18.

38. $S_\infty = \dfrac{12}{1-\frac{1}{3}} = \dfrac{12}{\frac{2}{3}} = \dfrac{12}{1} \cdot \dfrac{3}{2} = 18$

The total distance is 18 m, the fall distance is 12 m, so the rebound distance is $18 - 12$, or 6 m.

The total rebound distance is 6 m.

39. $7! = 7 \cdot 6 \cdot 5 \cdot 4 \cdot 3 \cdot 2 \cdot 1 = 5040$

40. $\dbinom{10}{3} = \dfrac{10!}{7!3!} = \dfrac{10 \cdot 9 \cdot 8 \cdot 7!}{7!3!} = \dfrac{10 \cdot 9 \cdot 8}{3 \cdot 2 \cdot 1} = 10 \cdot 3 \cdot 4 = 120$

41. $\dbinom{20}{2} a^{20-2} b^2 = 190 a^{18} b^2$

42. Expand $(x - 2y)^4$.

We have $a = x$, $b = -2y$, and $n = 4$.

Form 1: We use the fifth row of Pascal's triangle:

$\quad$ 1 $\quad$ 4 $\quad$ 6 $\quad$ 4 $\quad$ 1

$(x-2y)^4$

$= 1 \cdot x^4 + 4x^3(-2y) + 6x^2(-2y)^2 + 4x(-2y)^3 + 1 \cdot (-2y)^4$

$= x^4 - 8x^3y + 24x^2y^2 - 32xy^3 + 16y^4$

Form 2:

$(x-2y)^4 = \dbinom{4}{0}x^4 + \dbinom{4}{1}x^3(-2y) + \dbinom{4}{2}x^2(-2y)^2$

$\qquad + \dbinom{4}{3}x(-2y)^3 + \dbinom{4}{4}(-2y)^4$

$= \dfrac{4!}{4!0!}x^4 + \dfrac{4!}{3!1!}x^3(-2y) + \dfrac{4!}{2!2!}x^2(-2y)^2$

$\qquad + \dfrac{4!}{1!3!}x(-2y)^3 + \dfrac{4!}{0!4!}(-2y)^4$

$= x^4 - 8x^3y + 24x^2y^2 - 32xy^3 + 16y^4$

43. *Writing Exercise.* For a geometric sequence with $|r| < 1$, as n gets larger, the absolute value of the terms gets smaller, since $|r^n|$ gets smaller.

44. *Writing Exercise.* The first form of the binomial theorem draws the coefficients from Pascal's triangle; the second form uses factorial notation. The second form avoids the need to compute all preceding rows of Pascal's triangle, and is generally easier to use when only one term of an expression is needed. When several terms of an expansion are needed and n is not large (say, $n \le 8$), it is often easier to use Pascal's triangle.

45. $1 - x + x^2 - x^3 + \dots$

$a_1 = 1$, $n = n$, and $r = \dfrac{-x}{1} = -x$

$S_n = \dfrac{1(1-(-x)^n)}{1-(-x)} = \dfrac{1-(-x)^n}{1+x}$

46. Expand $(x^{-3} + x^3)^5$.

We have $a = x^{-3}$, $b = x^3$, and $n = 5$.

Form 1: We use the 6th row of Pascal's triangle:

$\quad$ 1 $\quad$ 5 $\quad$ 10 $\quad$ 10 $\quad$ 5 $\quad$ 1

$(x^{-3}+x^3)^5 = 1 \cdot (x^{-3})^5 + 5(x^{-3})^4(x^3)^1 + 10(x^{-3})^3(x^3)^2$

$\qquad + 10(x^{-3})^2(x^3)^3 + 5(x^{-3})^1(x^3)^4 + 1 \cdot (x^3)^5$

$= x^{-15} + 5x^{-9} + 10x^{-3} + 10x^3 + 5x^9 + x^{15}$

Form 2:

$(x^{-3}+x^3)^5 = \dbinom{5}{0}(x^{-3})^5 + \dbinom{5}{1}(x^{-3})^4(x^3)$

$\qquad + \dbinom{5}{2}(x^{-3})^3(x^3)^2 + \dbinom{5}{3}(x^{-3})^2(x^3)^3$

$\qquad + \dbinom{5}{4}(x^{-3})(x^3)^4 + \dbinom{5}{5}(x^3)^5$

$= \dfrac{5!}{5!0!}(x^{-3})^5 + \dfrac{5!}{4!1!}(x^{-3})^4(x^3)$

$\qquad + \dfrac{5!}{3!2!}(x^{-3})^3(x^3)^2 + \dfrac{5!}{2!3!}(x^{-3})^2(x^3)^3$

$\qquad + \dfrac{5!}{1!4!}(x^{-3})(x^3)^4 + \dfrac{5!}{0!5!}(x^3)^5$

$= x^{-15} + 5x^{-9} + 10x^{-3} + 10x^3 + 5x^9 + x^{15}$

Chapter 11 Test

1. $a_n = \dfrac{1}{n^2 + 1}$

$a_1 = \dfrac{1}{1^2 + 1} = \dfrac{1}{2}$

$a_2 = \dfrac{1}{2^2 + 1} = \dfrac{1}{5}$

$a_3 = \dfrac{1}{3^2 + 1} = \dfrac{1}{10}$

$a_4 = \dfrac{1}{4^2 + 1} = \dfrac{1}{17}$

$a_5 = \dfrac{1}{5^2 + 1} = \dfrac{1}{26}$

$a_{12} = \dfrac{1}{12^2 + 1} = \dfrac{1}{145}$

2. $\dfrac{4}{3}, \dfrac{4}{9}, \dfrac{4}{27}, \dots$

These are positive fractions in which the numerator is 4 and the denominator is a power of 3. The general terms could be $4\left(\dfrac{1}{3}\right)^n$.

3. $\displaystyle\sum_{k=2}^{5} (1 - 2^k) = (1 - 2^2) + (1 - 2^3) + (1 - 2^4) + (1 - 2^5)$

$\qquad = -3 + (-7) + (-15) + (-31) = -56$

4. $1 + (-8) + 27 + (-64) + 125$

This is the sum of the cubes of the first five natural numbers, but with alternating signs. Sigma notation is

$\displaystyle\sum_{k=1}^{5} (-1)^{k+1} k^3.$

5. $\frac{1}{2}$, 1, $\frac{3}{2}$, 2, ...

$$a_1 = \frac{1}{2}, \; d = \frac{1}{2}, \; \text{and } n = 13$$

$$a_n = a_1 + (n-1)d$$

$$a_{13} = \frac{1}{2} + (13-1)\left(\frac{1}{2}\right) = \frac{1}{2} + 12\left(\frac{1}{2}\right) = \frac{13}{2}$$

6. We know that $a_5 = 16$ and $a_{10} = -3$. We would have to add d five times to get from a_5 to a_{10}. That is,

$$16 + 5d = -3$$
$$5d = -19$$
$$d = -3.8.$$

Since $a_5 = 16$, we subtract d four times to get a_1.

$$a_1 = 16 - 4(-3.8) = 16 + 15.2 = 31.2$$

7. The sum is $24 + 36 + 48 + ... + 228 + 240$. This is the sum of the arithmetic sequence for which $a_1 = 24$, $a_n = 240$, and $n = 19$. We use the formula for S_n.

$$S_n = \frac{n}{2}(a_1 + a_n)$$

$$S_{19} = \frac{19}{2}(24 + 240) = \frac{19}{2}(264) = 2508$$

8. -3, 6, -12,...

$$a_1 = -3, \; n = 10, \; \text{and } r = \frac{6}{-3} = -2$$

$$a_n = a_1 r^{n-1}$$

$$a_{10} = -3(-2)^{10-1} = -3(-2)^9 = -3(-512) = 1536$$

9. $r = \frac{15}{22\frac{1}{2}} = \frac{15}{1} \cdot \frac{2}{45} = \frac{2}{3}$

10. 3, 9, 27,...

$$a_1 = 3, \; \text{and } r = \frac{9}{3} = 3$$

$$a_n = a_1 r^{n-1}$$

$$a_n = 3(3)^{n-1} = 3^n$$

11. $11 + 22 + 44 + ...$

$$a_1 = 11, \; n = 9, \; \text{and } r = \frac{22}{11} = 2$$

$$S_n = \frac{a_1(1 - r^n)}{1 - r}$$

$$S_9 = \frac{11(1 - 2^9)}{1 - 2} = \frac{11(1 - 512)}{-1} = \frac{11(-511)}{-1} = 5621$$

12. $0.5 + 0.25 + 0.125 + ...$

$|r| = \left|\dfrac{0.25}{0.5}\right| = |0.5| = 0.5 < 1$, the series has a limit.

$$S_\infty = \frac{a_1}{1 - r} = \frac{0.5}{1 - 0.5} = \frac{0.5}{0.5} = 1$$

13. $0.5 + 1 + 2 + 4 + ...$

$|r| = \left|\dfrac{1}{0.5}\right| = |2| = 2$, and since $|r| \not< 1$, the series does not have a limit.

14. $\$1000 + \$80 + \$6.40 + ...$

$|r| = \left|\dfrac{\$80}{\$1000}\right| = |0.08| = 0.08 < 1$, the series has a limit.

$$S_\infty = \frac{a_1}{1 - r} = \frac{\$1000}{1 - 0.08} = \frac{\$1000}{0.92} = \frac{\$25,000}{23} \approx \$1086.96$$

15. $0.85858585... = 0.85 + 0.0085 + 0.000085 + ...$

This is an infinite geometric series with $a_1 = 0.85$.

$|r| = \left|\dfrac{0.0085}{0.85}\right| = |0.01| = 0.01 < 1$, so the series has a limit.

$$S_\infty = \frac{a_1}{1 - r} = \frac{0.85}{1 - 0.01} = \frac{0.85}{0.99} = \frac{85}{99}$$

Fractional notation for $0.85858585...$ is $\dfrac{85}{99}$.

16. *Familiarize.* We want to find the seventeenth term of an arithmetic sequence with $a_1 = 31$, $d = 2$, and $n = 17$. We will use the formula for a_n to find a_{17}. This result is the number of seats in the 17th row. *Translate.* Substituting into the formula for a_n, we have

$$a_{17} = 31 + (17 - 1)2.$$

Carry out. We find a_{17}.

$$a_{17} = 31 + 16 \cdot 2 = 63$$

Check. We can do the calculations again.
State. There are 63 seats in the 17th row.

17. The sum is $\$100 + \$200 + \$300 + ... + \$1800$. This is the sum of the arithmetic sequence for which $a_1 = \$100$, $a_n = \$1800$, and $n = 18$. We use the formula for S_n:

$$S_n = \frac{n}{2}(a_1 + a_n)$$

$$S_{18} = \frac{18}{2}(\$100 + \$1800) = 9(\$1900) = \$17,100$$

Her uncle gave her a total of $\$17,100$.

18. *Familiarize.* At the end of each week, the price will be 95% of the previous week's price.
We have a geometric sequence:

$$\$10,000, \; \$10,000(0.95), \; \$10,000(0.95)^2, ...$$

We are looking for the price after 10 weeks.

Translate. We use the formula $a_n = a_1 r^{n-1}$ with $a_1 = \$10,000$, $r = 0.95$, and $n = 11$:

$$a_{10} = \$10,000(0.95)^{11-1}$$

Carry out. We calculate to obtain $a_{11} \approx \$5987.37$.

Check. We can do the calculation again.
State. After 10 weeks, the price of the boat will be about $\$5987.37$.

19. $S_\infty = \dfrac{18}{1-\dfrac{2}{3}} = \dfrac{18}{\dfrac{1}{3}} = \dfrac{18}{1} \cdot \dfrac{3}{1} = 54$

The total distance is 54 m, the fall distance is 18 m, so the rebound distance is $54 - 18$, or 36 m.
The total rebound distance is 36 m.

20. $\dbinom{12}{9} = \dfrac{12!}{3!9!} = \dfrac{12 \cdot 11 \cdot 10 \cdot 9!}{3!9!} = \dfrac{12 \cdot 11 \cdot 10}{3 \cdot 2 \cdot 1} = 4 \cdot 11 \cdot 5 = 220$

21. Expand $(x-3y)^5$.

We have $a = x$, $b = -3y$, and $n = 5$.

Form 1: We use the 6th row of Pascal's triangle:

$\quad$ 1 $\quad$ 5 $\quad$ 10 $\quad$ 10 $\quad$ 5 $\quad$ 1

$(x-3y)^5$
$= 1 \cdot x^5 + 5x^4(-3y)^1 + 10x^3(-3y)^2$
$\quad + 10x^2(-3y)^3 + 5x^1(-3y)^4 + 1 \cdot (-3y)^5$
$= x^5 - 15x^4y + 90x^3y^2 - 270x^2y^3 + 405xy^4 - 243y^5$

Form 2:

$(x-3y)^5$
$= \dbinom{5}{0}x^5 + \dbinom{5}{1}x^4(-3y) + \dbinom{5}{2}x^3(-3y)^2$
$\quad + \dbinom{5}{3}x^2(-3y)^3 + \dbinom{5}{4}x(-3y)^4 + \dbinom{5}{5}(-3y)^5$
$= \dfrac{5!}{5!0!}x^5 + \dfrac{5!}{4!1!}x^4(-3y) + \dfrac{5!}{3!2!}x^3(9y^2)$
$\quad + \dfrac{5!}{2!3!}x^2(-27y^3) + \dfrac{5!}{1!4!}x(81y^4) + \dfrac{5!}{0!5!}(-243y^5)$
$= x^5 - 15x^4y + 90x^3y^2 - 270x^2y^3 + 405xy^4 - 243y^5$

22. $\dbinom{12}{3}a^{12-3}x^3 = 220a^9x^3$

23. The sum is $2 + 4 + 6 + \ldots + 2n$. This is the sum of the arithmetic sequence for which $a_1 = 2$, $a_n = 2n$, and $n = n$. We use the formula for S_n.

$$S_n = \frac{n}{2}(a_1 + a_n)$$
$$S_n = \frac{n}{2}(2 + 2n) = \frac{n}{2}[2(1+n)] = n(n+1)$$

24. $1 + \dfrac{1}{x} + \dfrac{1}{x^2} + \dfrac{1}{x^3} + \ldots$

$a_1 = 1$, $n = n$, and $r = \dfrac{\dfrac{1}{x}}{1} = \dfrac{1}{x}$

$$S_n = \frac{1\left(1 - \left(\dfrac{1}{x}\right)^n\right)}{1 - \dfrac{1}{x}} = \frac{1 - \left(\dfrac{1}{x}\right)^n}{1 - \dfrac{1}{x}}, \text{ or } \frac{x^n - 1}{x^{n-1}(x-1)}$$